Thrust Fault-Related Folding

AAPG Memoir 94

Edited by
K. McClay, J. H. Shaw, and J. Suppe

ISBN13: 978-0-89181-376-7

AAPG Editor: Stephen E. Laubach
AAPG Geoscience Director: James B. Blankenship

On the cover: The background image on the cover is a seismic section through the frontal thrust fault-related fold system of the Cascade accretionary prism, offshore western Canada. The large photo shows folded Lower Eocene platformal carbonate units that form the Mediano detachment anticline, Ainsa basin, Spanish Pyrenees. The Cotiella thrust sheet and the axial zone of the Pyrenees form the background. The small photo shows a detachment fold formed by Miocene siltstones and mudstones, Quilitak anticline, frontal fold system, southern Tianshan mountains, western China.

This publication is available from:

The AAPG Bookstore
P.O. Box 979
Tulsa, OK U.S.A. 74101-0979
Phone: 1-918-584-2555 or 1-800-364-AAPG (U.S.A. only)
E-mail: bookstore@aapg.org
www.aapg.org

Canadian Society of Petroleum Geologists
600, 640 – 8th Avenue S.W.
Calgary, Alberta T2P 1G7
Canada
Phone: 1-403-264-5610
Fax: 1-403-264-5898
E-mail: reception@cspg.org
www.cspg.org

Geological Society Publishing House
Unit 7, Brassmill Enterprise Centre
Brassmill Lane, Bath BA13JN
United Kingdom
Phone: +44-1225-445046
Fax: +44-1225-442836
E-mail: sales@geoloc.org.uk
www.geolsoc.org.uk

Affiliated East-West Press Private Ltd.
G-1/16 Ansari Road, Darya Gaaj
New Delhi 110-002
India
Phone: +91-11-23279113
Fax: +91-11-23260538
E-mail: affiliate@vsnl.com

The American Association of Petroleum Geologists Books Refereeing Procedures

The Association makes every effort to ensure that the scientific and production quality of its books matches that of its journals. Since 1937, all book proposals have been refereed by specialist reviewers as well as by the Association's Publications Committee. If the referees identify weaknesses in the proposal, these must be addressed before the proposal is accepted.

Once the book is accepted, the Association Book Editors ensure that the volume editors follow strict guidelines on refereeing and quality control. We insist that individual papers can only be accepted after satisfactory review by two independent referees. The questions on the review forms are similar to those for the AAPG Bulletin. The referees' forms and comments must be available to the Association's Book Editors on request.

Although many of the books result from meetings, the editors are expected to commission papers that were not presented at the meeting to ensure that the book provides a balanced coverage of the subject. Being accepted for presentation at the meeting does not guarantee inclusion in the book.

More information about submitting a proposal and producing a book for The American Association of Petroleum Geologists can be found on its web site: www.aapg.org.

Table of Contents

Foreword

This volume arises from papers and posters presented at the International Conference on *Fault-Related Folding in Foreland Basins* held in Beijing in 2005 and organized by Vice President Chengzhao Jia of the Research Institute of Petroleum Exploration and Development (RIPED) of PetroChina and Professor John Suppe of Princeton University and the Tarim Oilfield Company of PetroChina, and researchers from Nanjing and Zhejiang Universities in China. The conference was organized to bring together a significant fraction of the worldwide experts on the study and analysis of thrust-related fold systems and to provide an introduction to the petroleum-bearing fold belts of western China to this international community.

More than 200 attendees saw presentations on a wide range of fold and thrust belts from subaerial fold belts in western China, the Andes, the Alps, the Rocky Mountains of North America, the Zagros of Iran, Taiwan, and elsewhere. Folds developed in deep-water settings such as the South Atlantic passive margins as well as the Niger delta, northwest Borneo, and the Gulf of Mexico were also discussed. The three-day meeting in Beijing was followed by a seven-day field trip organized by PetroChina to the actively growing frontal fold belts of the northern Tarim, Turfan, and southern Junggar basins in westernmost China.

Manuscripts on these topics were submitted after the conference and key papers are presented in this volume. The topics in this volume build upon previously published key AAPG volumes such as Memoir 82—*Thrust Tectonics and Hydrocarbon Systems* (2004) and AAPG Seismic Atlas Studies in Geology 53—*Seismic Interpretation of Contractional Fault-Related Folds* (2005).

Acknowledgments

The conference organizers are thanked for their generous support. Nanjing and Zhejiang Universities and, in particular, PetroChina through RIPED and also the Tarim Oilfield Company of PetroChina are thanked for supporting the conference and the field trip.

Papers submitted for this Memoir were kindly refereed by the following people, many of whom reviewed several manuscripts and gave generously of their time. The editors and AAPG are extremely grateful for their help.

Jurgen Adam
Richard Allmendinger
Alejandro Amilibia
Jean-Philippe Avouac
Clem Chase
Peter Cobbold
Ernesto Cristallini
Ian Davison
Leonardo Duerto
Fred Dula
Thomas Flottmann
Stuart Hardy
Robert Hooper
Jake Hossack
Martin Insley
Rene Manceda
Steve Matthews
Van Mount
Josep Munoz
Josep Poblet
Victor Ramos
Lisandro Rojas
Greg Schoenborn
John Shaw
Fabrizio Storti
John Suppe
Bob Thompson
Jose de Vera
Jaume Verges
Thomas Zapata

About the Editors

Ken McClay has a B.Sc. (honors) degree and a D.Sc. degree from Adelaide University, and a M.Sc. degree and Ph.D. degree from Imperial College London. He is a professor of structural geology and the Director of the Fault Dynamics Research Group at Royal Holloway University of London. His research has focused upon the dynamics of inverted, extensional, and strike-slip terrains and fault systems in fold-thrust belts. This involves the integration of field studies, seismic interpretation, remote sensing, and scaled physical modeling to develop quantitative 4-D models for fault systems in sedimentary basins. McClay established the analog modeling laboratories at Royal Holoway University of London. He and his research students have carried out numerous collaborative research projects with the international petroleum industry. He has run numerous short courses for the petroleum industry and has published widely in international journals.

John H. Shaw is the Harry C. Dudley Professor of Structural & Economic Geology and Chair of the Department of Earth & Planetary Sciences at Harvard University. Prior to joining the faculty at Harvard, Shaw received his Ph.D. from Princeton University working under the direction of Professor Suppe, and subsequently worked for Texaco's Exploration & Production Technology Department in Houston, Texas. During his time in the industry, Shaw participated in a number of exploration and development projects in Nigeria, Angola, Canada, Italy, Indonesia, Malaysia, China, Colombia, Venezuela, Argentina, and the United States. Shaw's research interests are centered on investigating the nature of faulting and fault-related folding in the crust, with applications to petroleum trap and reservoir characterization and regional earthquake hazards assessment. His research generally involves analysis of 2- and 3-D seismic reflection, well, remote sensing, and surface geologic data, and he employs a variety of kinematic and mechanical modeling approaches to investigating structures. His current efforts include studies of the structure and petroleum systems of the deep-water Niger Delta and basins in western China, and the development of new mechanical approaches to performing 3-D structural restorations. These restoration methods are being applied in Saudi Arabia, Qatar, China, and other areas to investigate patterns of natural fractures in petroleum fields that can aid in reservoir characterization and development. Shaw regularly leads an AAPG field course in the Canadian Rockies, and offers a short course on seismic interpretation methods based on an AAPG Seismic Atlas (Studies in Geology 53), which is co-authored by Professors Connors and Suppe.

John Suppe is a structural geologist with wide experience in deformed petroleum basins, including Gulf of Mexico, Niger Delta, California, Philippines, Venezuela, Taiwan, and Tarim, Junggar, and Sichuan basins in China. He and his collaborators are the originators of many of the concepts of fault-related folding applied in compressive and extensional environments, as well as key concepts of interpreting growth strata for folding histories. The application of these structural concepts to practical seismic interpretation is introduced in an AAPG seismic atlas (Studies in Geology 53), edited by John Shaw, Chris Connors, and Suppe. He has also contributed key advances in the understanding mechanics of thrust belts and state of the stress in the crust. Two of his contributions have been awarded the Best Publication Award in Structural Geology & Tectonics of the Geological Society of America. Suppe was named "Highly Cited Researcher" by Science Citation Index. Suppe was born and raised in Los Angeles, attended University of California Riverside studying under K. J. Hsü, received a Ph.D. from Yale University in 1969 studying under John Rodgers and was a post-doc at UCLA with Gary Ernst. He was Professor of Geology at Princeton University from 1971-2007, was named Blair Professor of Geology in 1988, and served as Department Chair. He now is Distinguished Chair Research Professor at National Taiwan University where he has an international research group. He has been visiting professor at Caltech twice, University of Barcelona, Ludwigs Maximillian University in Munich, National Taiwan University twice, and is an honorary professor of Nanjing University. He was a Guest Investigator of the NASA Magellan Mission to Venus. He was elected member of the U.S. National Academy of Sciences in 1995. He has received the Career Contribution Award in Structural Geology & Tectonics from the Geological Society of America, the Research Prize of the Alexander von Humboldt Foundation, the Wilber Cross Medal from Yale University, and is a Guggenheim Fellow.

1

McClay, Ken, 2011, Introduction to thrust fault-related folding, *in* K. McClay, J. H. Shaw, and J. Suppe, eds., Thrust fault-related folding: AAPG Memoir 94, p. 1 – 19.

Introduction to Thrust Fault-related Folding

Ken McClay
Fault Dynamics Research Group, Department of Earth Sciences, Royal Holloway University of London, Egham, Surrey, United Kingdom

ABSTRACT

The 16 chapters presented in this memoir cover some of the recent advances made in the descriptions and analysis of thrust-related fold systems. The chapters include kinematic and geometric analyses of fault-propagation folding, trishear folding, detachment folding, wedge-thrust fold systems, and basement-involved thrust systems. Examples are given from the Zagros fold belt of Iran; the sub-Andean fold belt of western Argentina; the frontal fold belt of the Tianshan, northern margin of the Tarim Basin in western China; from the fold and thrust belts of the Spanish Pyrenees, Taiwan, Wyoming, and southern California; as well as the deep-water fold belts of offshore Brazil and the Niger Delta. In particular, several chapters focus on new analyses of detachment folding as well as improved kinematic and geometric models of thrust-related folding. Improved seismic imaging combined with theoretical, numerical, and analog modeling, plus the detailed field studies as presented in this volume, indicates the range of challenges and the strategies that can be brought to bear in integrating the well-established geometric and kinematic models of thrust fault-related folding with mechanical models to account for the natural complexities of real-world structures. I hope that the readers of this memoir will find these new ideas and concepts relevant for the exploration and exploitation of hydrocarbon systems in fold and thrust belts worldwide.

INTRODUCTION

Thrust fault-related folds form hydrocarbon traps in many fold and thrust belts, from subaerial fold belts such as the Zagros of Iran and Iraq (e.g., Beydoun et al., 1992; Sherkati et al., 2006; Cooper, 2007), and the sub-Andean fold and thrust belt of South America (Figure 1a) (Echavarria et al., 2003; also see Tankard et al., 1995), to deep-water fold belts such as offshore northwest Borneo (e.g., James, 1984; Sandal, 1996) (Figure 1b) and offshore west Africa in the Niger Delta (e.g., Figure 1c) (Ajakaiye and Bally, 2002; Bilotti and Shaw, 2005; Corredor et al., 2005a, b).

Understanding and recognizing the two-dimensional (2-D) and three-dimensional (3-D) geometries of thrust fault-related fold structures and how hanging-wall folds develop are vital for the exploration and exploitation of hydrocarbons in both subaerial and submarine fold belts. In particular, the geometric and kinematic models of thrust-related folding have been instrumental in guiding the interpretation of structural styles of thrust belts in the subsurface (e.g., Suppe, 1983, 1985; Suppe et al., 1992, 2004; Shaw et al., 2005) and for the construction of balanced cross sections (Guzofski et al., 2009). Many of these models have been incorporated into the algorithms used in computer programs for cross section

DOI:10.1306/13251330M9450

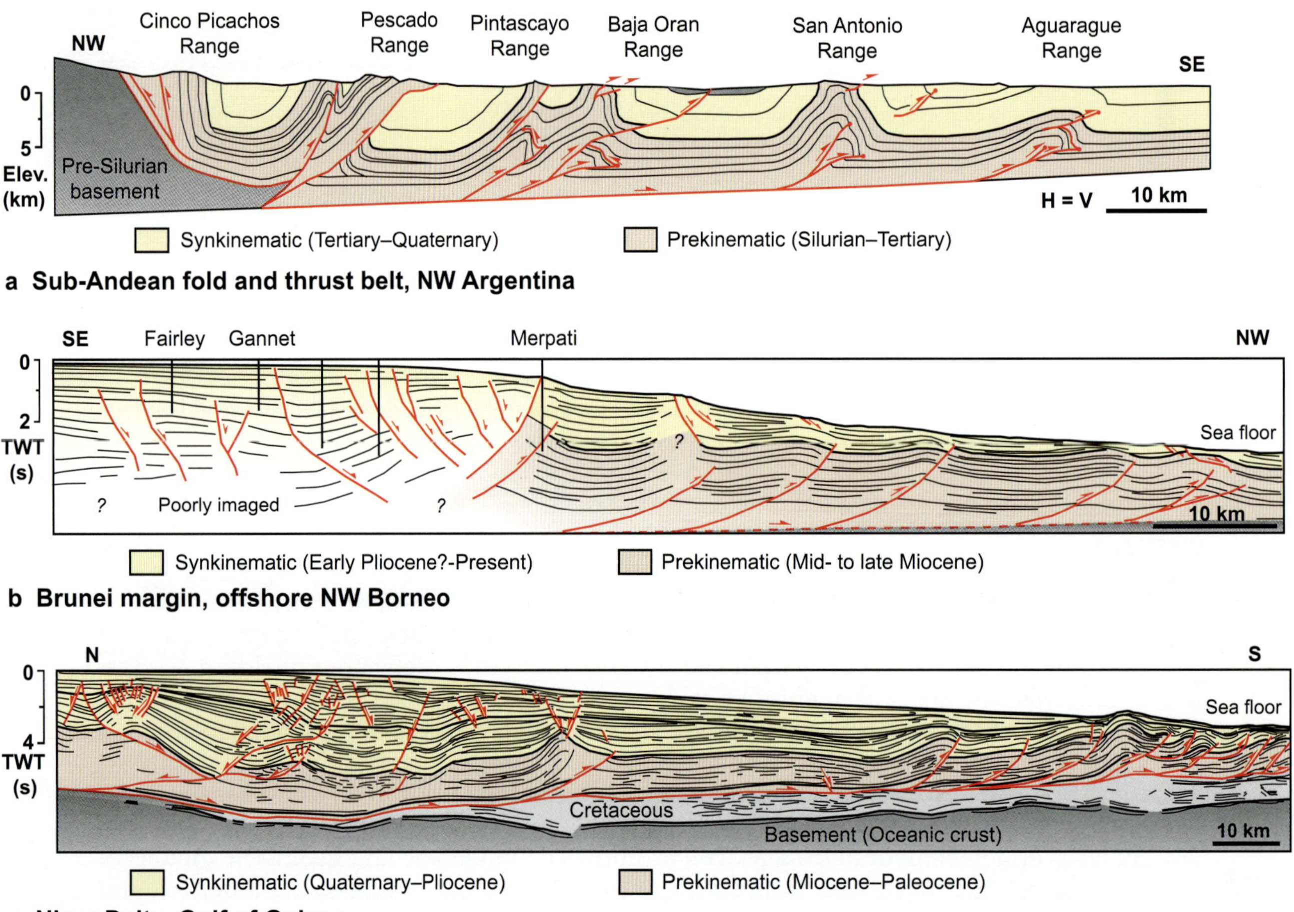

Figure 1. Fold-thrust structures in three hydrocarbon-bearing fold and thrust belts. TWT (s) = two-way traveltime in seconds; H = V; horizontal = vertical scales; Elev. = elevation. (a) The sub-Andean fold and thrust belt of northwestern Argentina. Significant hydrocarbon accumulations have been found in the complex fault-related fold systems in the frontal sectors of the belt (modified from Echavarria et al., 2003). (b) Schematic section through the near-shore to deep-water offshore Brunei margin, northwest Borneo. Thrust-related folds are developed in the slope and toe section of the Barram delta complex due to regional contraction (modified from Sandal, 1996). (c) The eastern lobe of the Niger Delta, Gulf of Guinea, showing gravity-driven contraction and the development of thrust fault-related folds at the delta toe region (modified from Ajakaiye and Bally, 2002).

construction, for retrodeforming and balancing cross sections, as well as for predictive forward modeling of strain and fracture distributions within hydrocarbon reservoirs in thrust-related fold structures.

Since the benchmark articles of Suppe (1983, 1985) and coworkers (Suppe and Medwedeff, 1990) on predictive geometric models of thrust-related, fault-bend folds and fault-propagation folds, considerable research and many publications have focused on evaluating, refining, and applying these models to thrust terranes in both two and three dimensions (e.g., Suppe, 1983; Suppe and Medwedeff, 1990; Erslev, 1991, 1993; Jordan and Noack, 1992; Suppe et al., 1992, 2004; Narr and Suppe, 1994; Epard and Groshong, 1995; Homza and Wallace, 1995; Poblet and McClay, 1996; Erslev and Mayborn, 1997; Hardy and Ford, 1997; Medwedeff and Suppe, 1997; Poblet et al., 1997; Storti and Poblet, 1997; Allmendinger, 1998; Zehnder and Allmendinger, 2000; Atkinson and Wallace, 2003; Mitra, 2003; Gonzalez-Mieres and Suppe, 2006; Hardy and Connors, 2006).

The inclusion of synkinematic growth strata into these kinematic models of thrust-related folding (e.g., the benchmark article of Suppe et al., 1992) was a major step forward and focused the attention on the possibilities of extracting quantitative information from thrust-related fold systems both in subaerial and submarine environments (e.g., Hardy et al., 1996, Poblet et al., 1997; Storti and Poblet, 1997; Suppe et al., 1997, 2004; Allmendinger, 1998; Shaw et al., 2005; Gonzalez-Mieres and Suppe, 2006). The different kinematic models for these thrust-related folds give different predictions for the rates and distributions of instantaneous uplift of

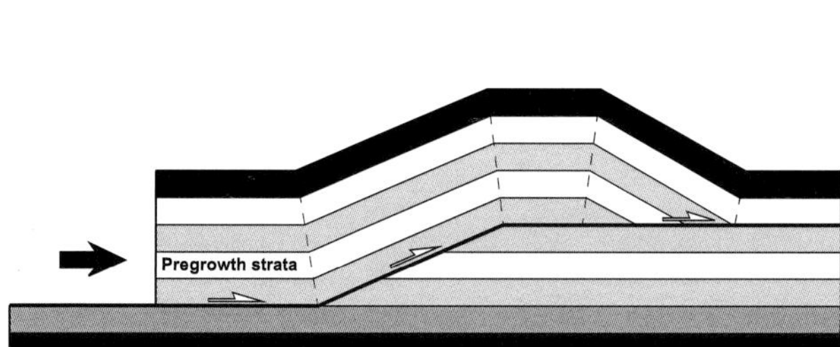

a) Fault-bend fold

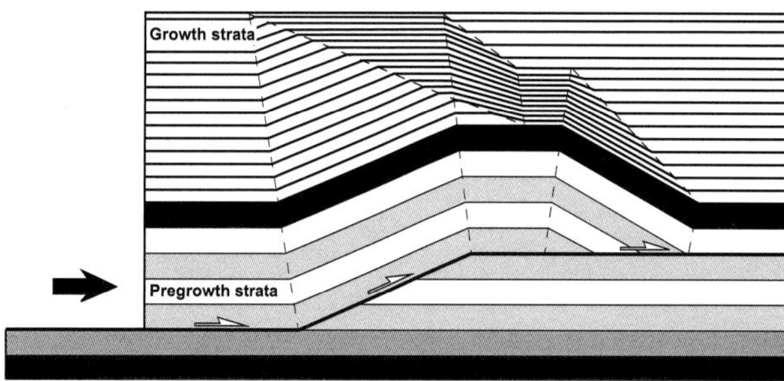

b) Fault-bend fold with growth strata

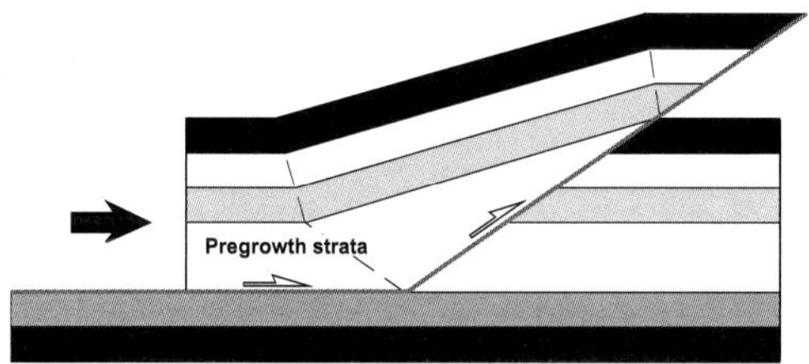

c) Pure shear fault-bend fold

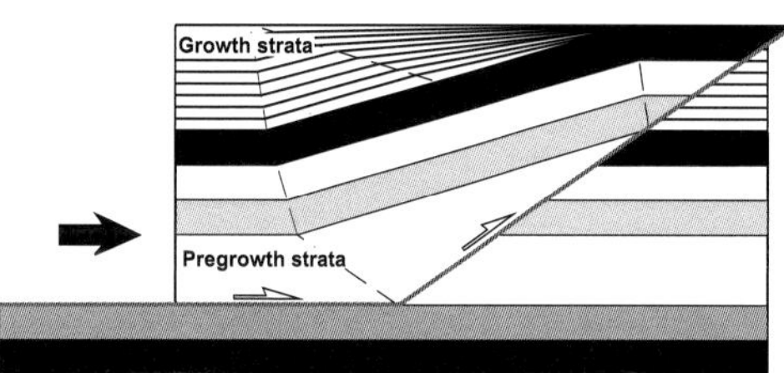

d) Pure shear fault-bend fold with growth strata

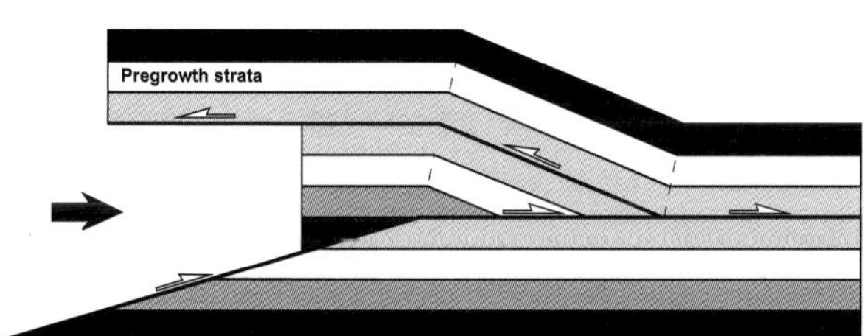

e) Wedge-thrust fold

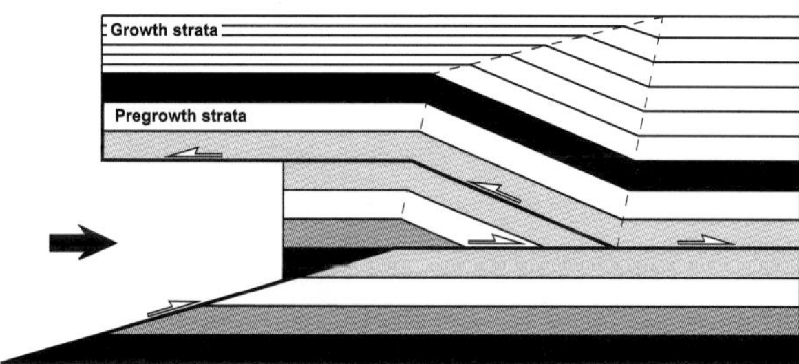

f) Wedge-thrust fold with growth strata

Figure 2. Thrust-related fault-bend folds with and without synkinematic growth strata. (a) Kink-band fault-bend fold (modified from Suppe, 1983). (b) Kink-band fault-bend fold with growth strata (modified from Suppe et al., 1992). (c) Kink-band, pure shear fault-bend fold (modified from Suppe et al., 2004). (d) Kink-band, pure shear fault-bend fold with growth strata (modified from Suppe et al., 2004). (e) Wedge-thrust fold system (modified from Medwedeff, 1989; Shaw et al., 2005). (f) Wedge-thrust fold system with growth strata (modified from Medwedeff, 1989; Shaw et al., 2005).

the fold crest relative to total fold uplift. The resultant growth stratal patterns in general will be different for each of these various kinematic models and therefore may be used to infer the style of fault-related folding and also to quantify fault slip and fold uplift rates in detail (e.g., Hardy et al., 1996; Rivero and Shaw, 2011; Yue et al., 2011).

In 2005, the AAPG benchmark publication on seismic interpretation in contractional settings by Shaw et al. (2005) captured many of the recent advances in applying these geometric models to guide seismic interpretation and analysis of contraction-related fold structures. This current memoir presents further advances in the understanding of thrust fault-related folds, particularly applying other techniques such as numerical and physical modeling as well as documenting different fold styles in a variety of contractional regimes, from delta toe-thrust belts to subaerial fold-belt systems. New theoretical as well as practical analyses of detachment fold systems are presented. New techniques are presented for the interpretation, analysis, and understanding of detachment folds that are common in deep-water fold belts as well as in subaerial fold belts.

This memoir includes chapters that analyze fault-propagation folding, detachment folding, wedge-thrust systems, basement-involved thrust systems, basin-inversion-related thrusting and folding as well as deep-water fold belts. Examples are given from the Zagros fold belt of Iran; the Spanish Pyrenees; the southern

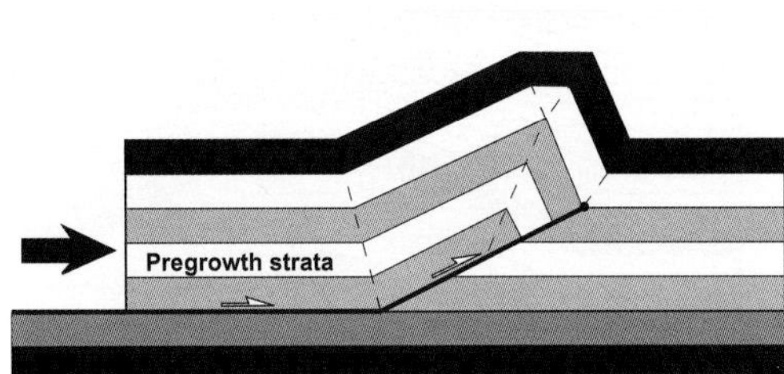

a) Fault-propagation fold

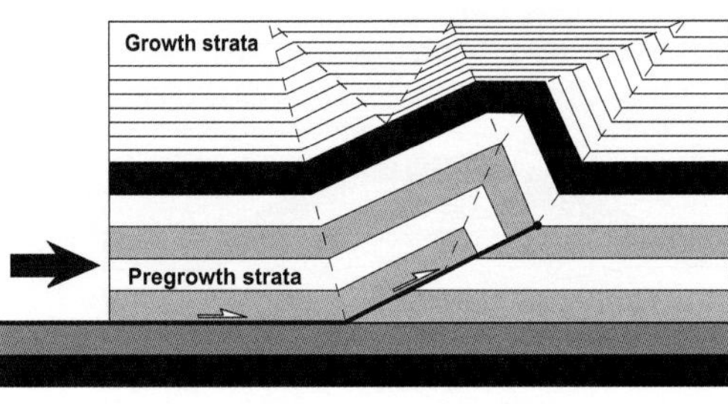

b) Fault-propagation fold with growth strata

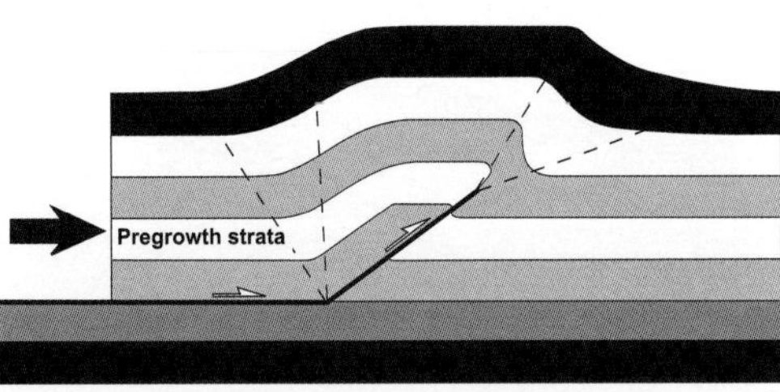

c) Trishear fault-propagation fold

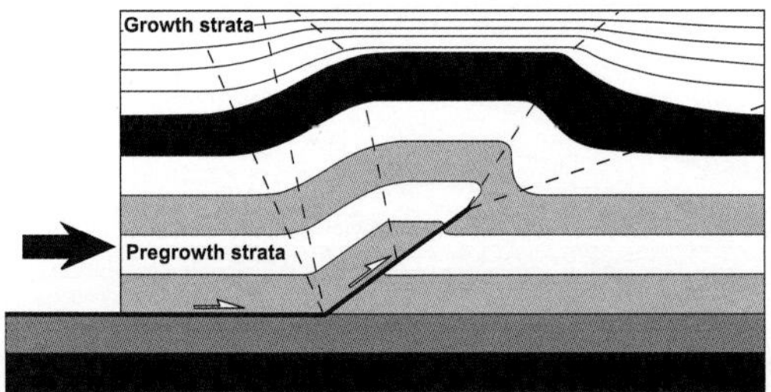

d) Trishear fault-propagation fold with growth strata

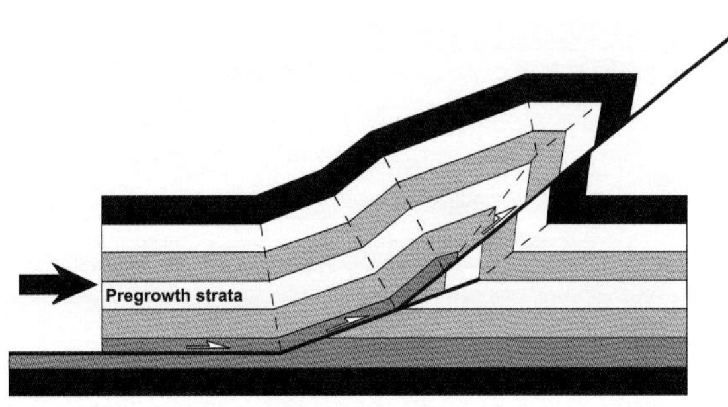

e) High-angle forelimb break-thrust fault-propagation fold

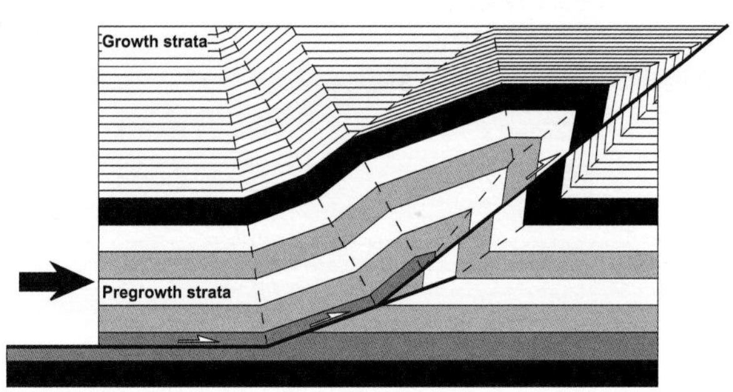

f) High-angle forelimb break-thrust fault-propagation fold with growth strata

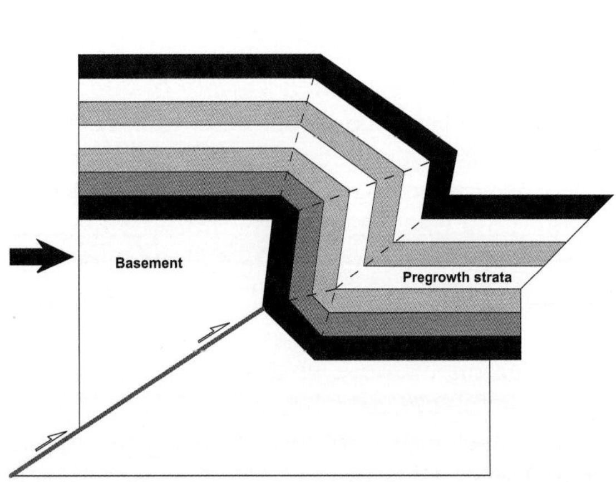

g) Basement-involved fault-propagation fold

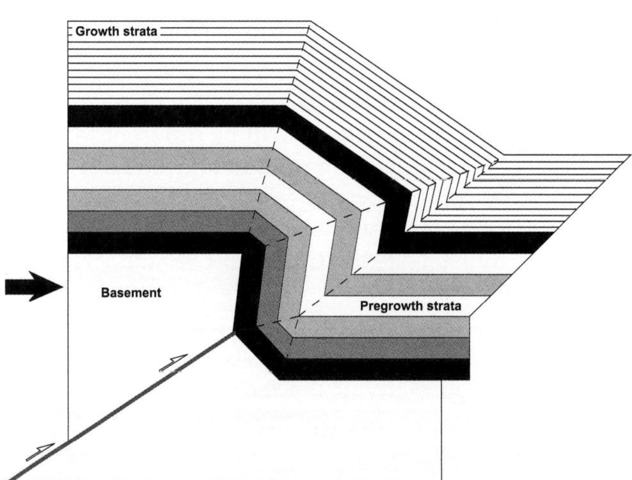

h) Basement-involved fault-propagation fold with growth strata

Tianshan fold and thrust belt of western China; and fold and thrust belts in Taiwan, Argentina, Wyoming, offshore California as well as offshore Brazil and the offshore Niger delta.

THRUST-RELATED FOLD SYSTEMS

In this part of the introductory chapter, the principal types of kink-band, thrust-related fold systems are briefly reviewed and their key geometric characteristics are highlighted. Figure 2 summarizes the finite deformed state geometries of many of the kinematic models for fault-bend fold-type structures that have been published in the past two decades, whereas Figure 3 shows examples of fault-propagation fold systems. Figure 4 summarizes some of the key elements of detachment fold systems. Examples of these thrust-related fold geometries are given in Figures 5–10.

In these kink-band-style models, the dominant deformation mechanism is assumed to be bed-parallel slip between the layers (e.g., Suppe, 1983, 1985; Suppe and Medwedeff, 1990; Suppe et al., 1992, 1997; Shaw et al., 2005). In general, the layer thickness is constant except in some detachment fold models where the ductile flow of material into and out of the fold core may be required (cf. Poblet and McClay, 1996).

Many of these geometric fold models are self-similar in that, as the fold amplifies, the limb dips remain constant and the limb lengths change by kink-band migration (Suppe, 1983; Suppe et al., 1992). In these self-similar models, fold amplification does not change the fundamental geometric architecture of the fold through time. In contrast, geometric models of folds where the limb dip changes as the fold amplifies are termed "non-self-similar" whereby the fold geometry changes through time mostly by progressive or episodic limb rotation (Hardy and Poblet, 1994; Poblet and McClay, 1996; Poblet et al., 2004). These progressive changes in limb dip result in fanning growth strata on both the front limbs and backlimbs as the fold amplifies (e.g., Figure 4b, c) (Poblet et al., 1997).

Seismic examples of thrust fault-related folds are shown in Figure 5, and outcrop examples are shown in Figure 6. Comparative natural examples of growth stratal geometries are shown in Figure 7. Scaled physical modeling (e.g., Wu and McClay, 2011) and numerical modeling (Hardy and Allmendinger, 2011) have been extensively used to simulate thrust-related fold systems, and typical examples are shown in Figures 8 and 9.

Fault-bend Folds

Fault-bend folds are generated above steps in the underlying thrust fault-producing folds in the hanging wall, whereas the footwall section remains undeformed (Figure 2). Suppe (1983) published a benchmark article on the geometric analysis of kink-band-style fault-bend folds demonstrating the key relationships between ramp geometries and the overlying hanging-wall anticline and syncline folds (e.g., Figure 2a). In these kink-band models of fault-bend folding, the thrust fault has propagated across the entire section, and subsequent displacement of the hanging wall over this stepped fault trajectory produces kink-band folds with constant limb dips. Amplification of the fold occurs through kink-band migration as the hanging wall is translated over the stepped footwall thrust surface (Figure 2a). Deformation within the fold is accomplished by flexural slip between the layered units. Fold axial surfaces bisect the interlimb angles between different dip domains of the fold limbs.

In 1992, Suppe et al. added growth sequences to these fault-bend fold models (Figure 2b) and demonstrated how quantitative analyses of these synkinematic units could be used to determine fold uplift and shortening rates. Suppe et al. (2004) further modified the theory of kink-band fault-bend folding to include folds where the thrust fault trajectory cuts across the layers with resultant internal shearing and named these "shear fault-bend folds" (Figure 2c). This kinematic model of shear fault-bend folding accounts for the geometries of fault-bend folds where the hanging-wall strata dip at lower angles than the underlying thrust ramp as well as for fanning growth strata on the back limbs of such folds (e.g., Figures 2d, 5a).

Shaw et al. (2005) discussed the various geometries of kink-band folds produced by wedge thrusts where an allochthonous section is thrust into an undeformed section with an overlying back thrust (Figure 2d, e). Wedge thrusts are commonly developed at the frontal sections and terminations of fold and thrust belts in highly anisotropic strata such as in the deep-water turbidites at the front of the Mahakam delta (Figure 5b).

Figure 3. Fault-propagation folds with and without synkinematic growth stratal architectures. (a) Fault-propagation fold; (b) fault-propagation fold with growth strata; (c) fore- and backlimb trishear fault-propagation fold; (d) trishear fault-propagation fold with growth strata; (e) high-angle forelimb break-thrust fault-propagation fold; (f) high-angle forelimb break-thrust fault-propagation fold with growth strata; (g) basement-involved fault-propagation fold; (h) basement-involved fault-propagation fold with growth strata (modified from Suppe et al., 1992, 2004; Narr and Suppe, 1994; Poblet et al., 1997; Shaw et al., 2005).

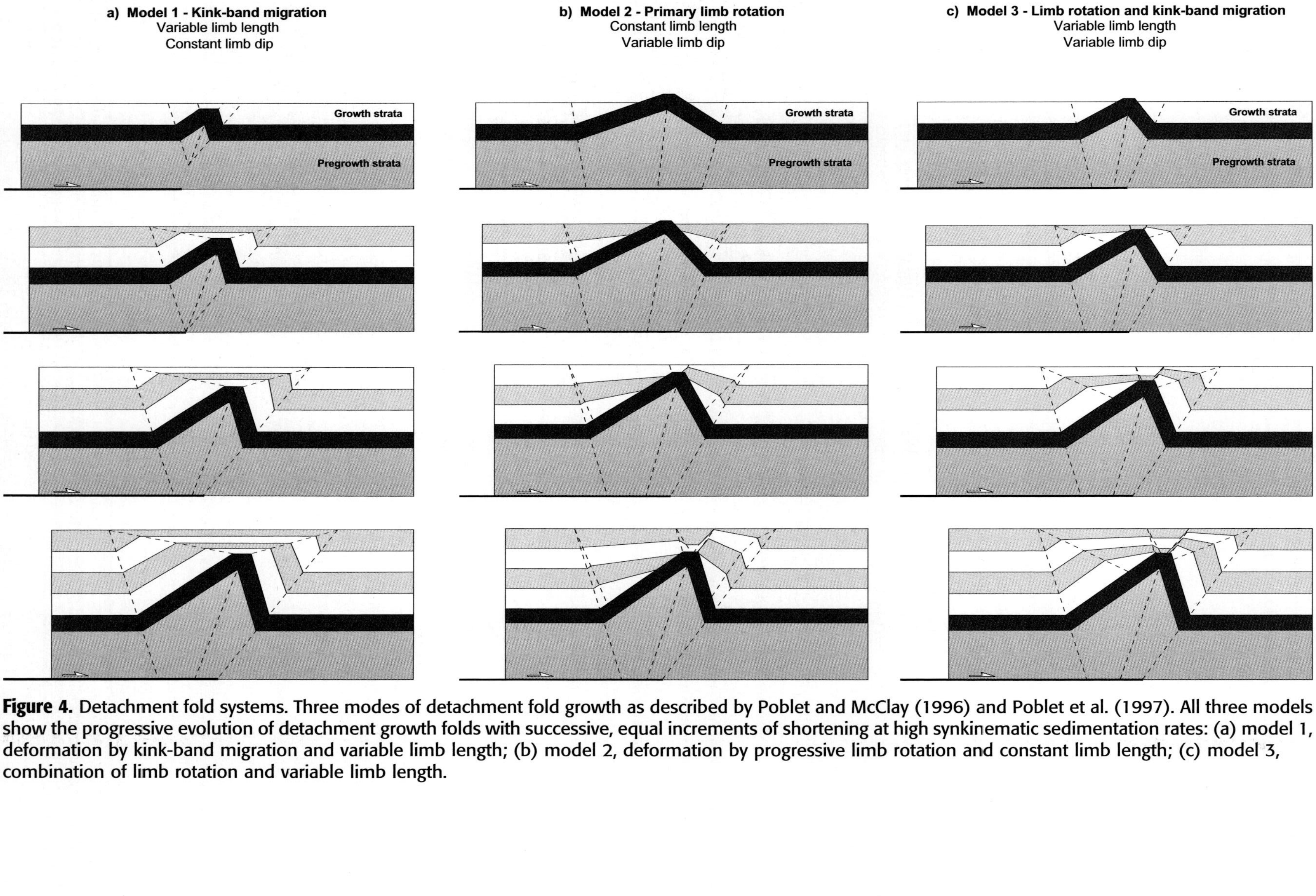

Figure 4. Detachment fold systems. Three modes of detachment fold growth as described by Poblet and McClay (1996) and Poblet et al. (1997). All three models show the progressive evolution of detachment growth folds with successive, equal increments of shortening at high synkinematic sedimentation rates: (a) model 1, deformation by kink-band migration and variable limb length; (b) model 2, deformation by progressive limb rotation and constant limb length; (c) model 3, combination of limb rotation and variable limb length.

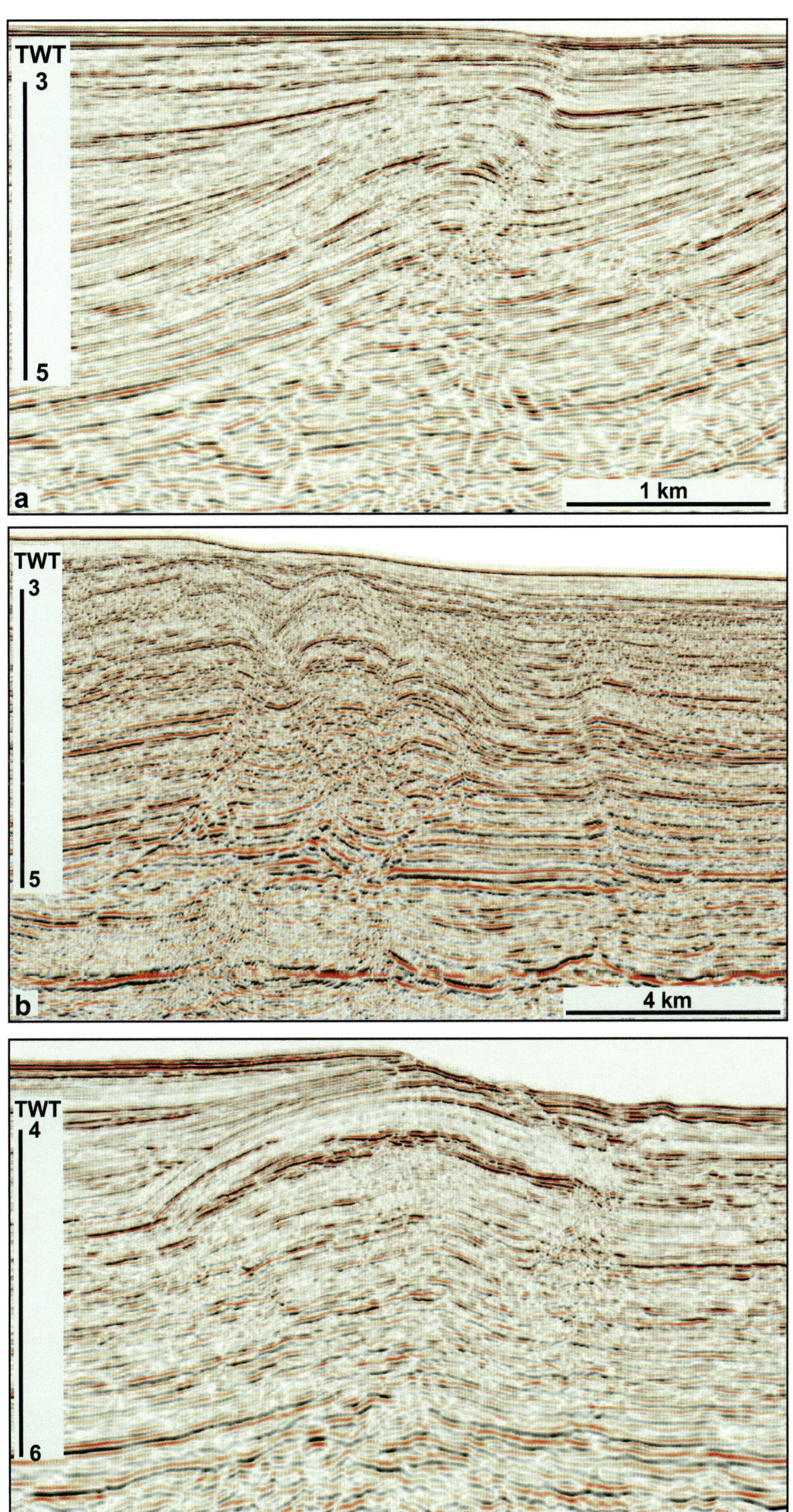

Figure 5. Seismic examples of thrust-related fold systems. Vertical scales are two-way traveltime (TWT) in seconds. (a) Fault-propagation fold, deep-water Niger Delta (seismic data courtesy of CGGVeritas). (b) Fault-propagation folds and wedge thrust folds, Mahakam fold belt, Makassar Straits, Indonesia (seismic data courtesy of TGS Nopec). (c) Detachment fold, deep-water Niger Delta (seismic data courtesy of CGGVeritas).

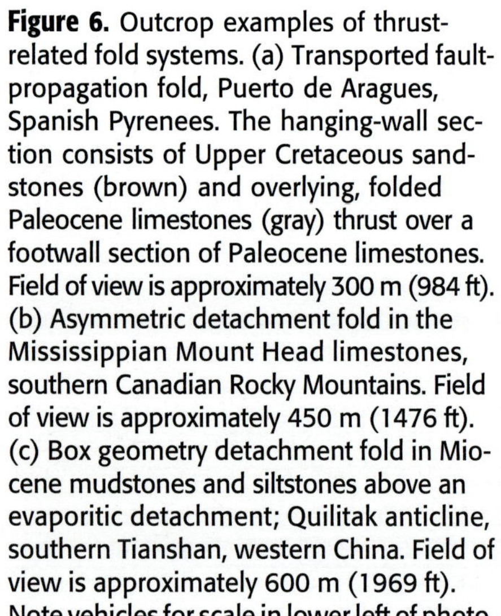
Figure 6. Outcrop examples of thrust-related fold systems. (a) Transported fault-propagation fold, Puerto de Aragues, Spanish Pyrenees. The hanging-wall section consists of Upper Cretaceous sandstones (brown) and overlying, folded Paleocene limestones (gray) thrust over a footwall section of Paleocene limestones. Field of view is approximately 300 m (984 ft). (b) Asymmetric detachment fold in the Mississippian Mount Head limestones, southern Canadian Rocky Mountains. Field of view is approximately 450 m (1476 ft). (c) Box geometry detachment fold in Miocene mudstones and siltstones above an evaporitic detachment; Quilitak anticline, southern Tianshan, western China. Field of view is approximately 600 m (1969 ft). Note vehicles for scale in lower left of photo.

Fault-propagation Folds

Figure 3 summarizes some of the key, kink-band-style geometric models of thrust fault-propagation folding. Suppe (1985) and Suppe and Medwedeff (1990) formalized the kink-band geometric model for thrust fault-propagation folding where a hanging-wall anticline-syncline fold pair forms at the leading edge of the thrust fault. This fold pair is tied to the fault tip line, which is located on the ramp and climbs up stratigraphic section in the direction of tectonic transport with progressive fault displacement (Figure 3a). This fault-propagation fold model develops by kink-band migration such that the limb dips form instantaneously as the fold starts to develop and do not change as the fold amplifies (i.e., constant limb dip). Fault-propagation folds are very common in fold and thrust belts particularly where steep to overturned front limbs occur (e.g., Figures 5a, 7c, d). In the simplest kink-band model for a fault-propagation fold with a high rate of synkinematic sedimentation, characteristic growth triangles with inclined axial surfaces are developed (Figure 3b). These are significantly different from the growth geometries formed by simple fault-bend folds (Figure 2b) and may possibly be used to differentiate between these two models for thrust-fault-related folding.

Erslev (1991) introduced the concept of "trishear" fault-propagation folding whereby the front limb of a fault-propagation fold developed by progressive limb rotation with differential shear between two inclined axial surfaces (Figure 3c). In this model, the dips of the anticline front limb increased downward toward the fold tip line. The addition of growth strata and the development of models with triangular shear in the backlimb synclinal hinge region resulted in fault-propagation fold models where fanning growth strata were developed on both the front limbs and backlimbs (e.g., Figure 3d). Fault-propagation folds with fanning growth strata on both the front limb and backlimb are common in deep-water fold belts where syntectonic sedimentation rates are high and the structures are preserved (e.g., Figure 7c, d).

Transported (or translated) fault-propagation folds (Jamison, 1987; Shaw et al., 2005) are commonly developed where the anticline front limb becomes faulted and the previously formed fold is carried forward on the new fault (Figure 3e, f). This is termed a forelimb break thrust and the resultant fold geometry a hanging-wall transported fault-propagation fold (e.g., Figure 6a). In these transported fault-propagation folds, a small footwall syncline commonly remains underneath the break thrust (Figure 3e, f).

Basement-involved fault-propagation folds are formed where steep thrust or reverse faults in the basement propagate upward causing folding in the overlying strata (e.g., Figure 3g, h) (Narr and Suppe, 1994). These are commonly formed by inversion of preexisting extensional fault systems and generate complex basement uplifts such as those found in the Laramide uplifts of Wyoming or those in the Sierras Pampenas in western Argentina (Erslev, 1993).

Detachment Folds

It has been 100 yr since Buxtorf and others described the classic detachment folds formed in the Jura above the Triassic evaporites (cf. Buxtorf, 1907, 1916). Since then, various studies have described and discussed detachment folds in a variety of subaerial fold belts (Laubscher, 1965, 2008; Jamison, 1987), but it is only relatively recently that they have become more widely recognized in both subaerial fold belts and also in deep-water fold belts (e.g., Poblet and McClay, 1996; Poblet et al., 1997; Mitra, 2002, 2003; Shaw et al., 2005; Gonzalez-Mieres and Suppe, 2006).

Figure 4 summarizes the three principal detachment fold models discussed by Poblet and McClay (1996) together with the addition of high sedimentation rate and growth stratigraphy (Poblet et al., 1997). These detachment fold models are characterized by a competent upper layer, or lid, above a ductile decollement unit. It is generally assumed that during shortening, the ductile decollement unit initially flows into and subsequently out of the fold core as the fold amplifies and the section shortens.

These three simple models of detachment folds produce identical final fold geometries for the prekinematic competent layer but with different synkinematic growth stratal architectures (Figure 4) (Poblet et al., 1997). Model 1 (Figure 4a) has constant limb dip and amplifies by kink-band migration as shortening progresses. Model 2 (Figure 4b) has constant limb lengths and amplifies by progressive limb rotation, whereas model 3 involves both limb lengthening and limb rotation during fold amplification (Figure 4c).

The synkinematic stratal architectures associated with the three different kinematic models produce significantly different patterns that may be used to distinguish between the models. Models 2 and 3 show characteristic fanning growth stratal patterns associated with progressive limb rotation (Figure 4b, c). Note that these models may generate local synclines in the growth strata directly above the crest of the detachment anticline (Figure 4b, c).

Detailed studies of many fold and thrust belts, however, show that detachment fold systems may be much more complex than the simple models shown in Figure 4. Hybrid structures (cf. Marrett and Bentham, 1997; McClay, 2004) and transported thrust-related folds (cf. Jamison, 1987; Mitra, 2003; Shaw et al., 2005) have been

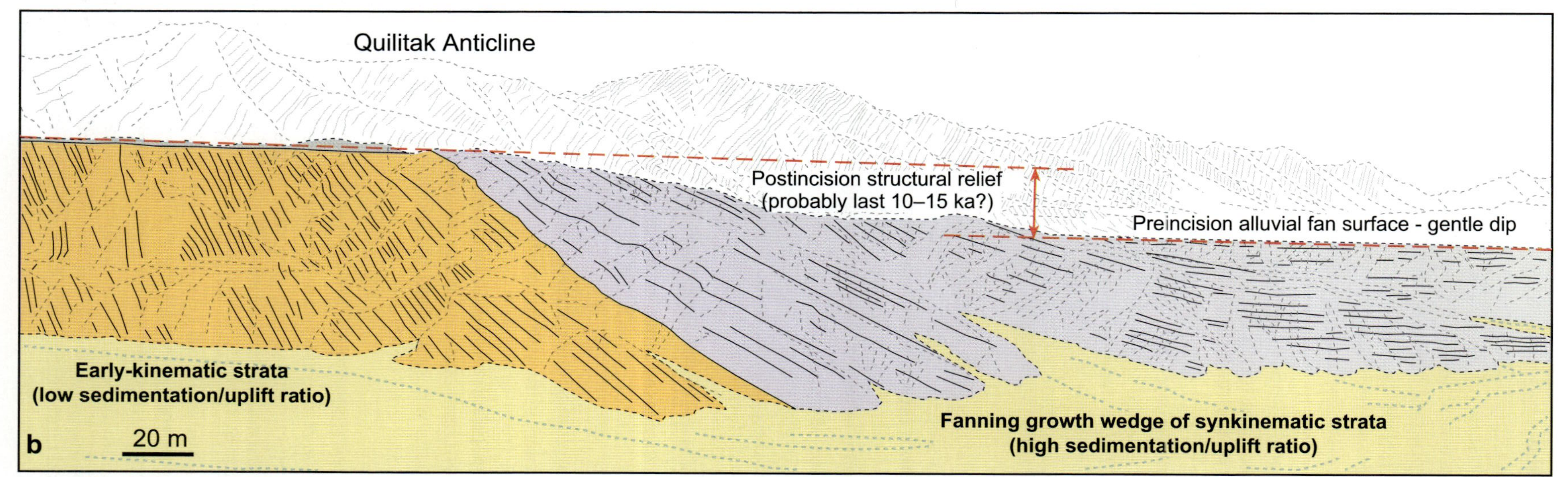

Figure 7. Examples of growth strata in fold and thrust belts. Vertical scales (c, d) are two-way traveltime (TWT) in seconds. (a) Alluvial fan growth wedges at the front of the Quilitak fold, Kuqa fold belt, frontal Tianshan ranges, western China. Light-colored early-kinematic gravels and sands (deposited at lower sedimentation rate relative to uplift rate) are overlain by successively younger, gray alluvial fan gravels and sands (deposited at higher sedimentation rate relative to uplift rate). The growth strata show fanning geometries with dips decreasing upward to the gently dipping preincision alluvial surface (after He et al., 2005; Hubert-Ferrari et al., 2007). (b) Interpretation of the growth wedge on the frontal limb of the Quilitak anticline showing light-colored early-kinematic gravels and sands overlain by an upwardly fanning wedge of Quaternary gray gravels and sands. The folding has been modeled by Hubert-Ferrari et al. (2007) as produced by progressive kink-band migration through an approximately 100-m (328-ft)-wide hinge zone.

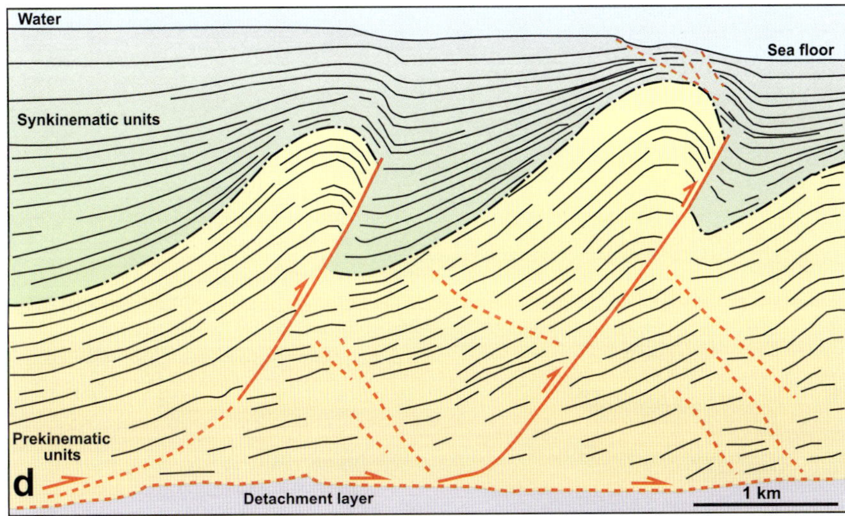

Figure 7. (cont.). (c) Seismic section through growth folds, frontal section of the toe-thrust fold belt, southwestern Niger Delta (seismic data courtesy of CGGVeritas). (d) Interpretation of the seismic section in panel c showing the synkinematic growth wedges on the front limbs and backlimbs of the fault-propagation folds.

recognized (Figure 10a). Mitra (2003) developed geometric and kinematic models for faulted symmetric and asymmetric detachment folds. Gonzalez-Mieres and Suppe (2006) analyzed low-amplitude detachment folds that grew by bed shortening and thickening, with almost no flexure, similar to the pure-shear detachment fold model of Groshong and Epard (1994).

Faulted detachment folds are characteristic in subaerial fold belts developed above salt detachments such as in the Zagros (see Mitra, 2003; McQuarrie, 2004; Sherkati et al., 2006; Vergés et al., 2011) and in deep-water fold and thrust belts where fold limbs are cut by low displacement thrusts (Figure 5c) (Shaw et al., 2005). Figure 10a shows a boxlike detachment fold where one limb has been cut by a thrust such that an asymmetric transported detachment fold has been developed. Note also the smaller back thrust that has displaced the backlimb producing a complex roof thrust system similar to the systems originally described by Buxtorf (1916) in the Jura and analyzed by Suppe (2011).

Outcrop examples of detachment folds in interbedded shales and sandstones commonly show complex imbrication and folding in their cores (Figure 10b, c). Numerical models of detachment folds in strata formed by alternating competent and incompetent units (such as turbiditic sandstones and shales) develop outer-arc extensional fault systems around the anticlinal hinges and smaller scale folds and out-of-syncline thrusts in the synclinal cores (Figure 9a). Similar structural styles are also formed in analog models of detachment folds (Figure 8c). These complexities in the cores of detachment folds are commonly poorly imaged on seismic sections such as those in the Niger Delta (Figure 5c).

In this memoir, descriptions and detailed analyses of various types of thrust fault-related folds are presented, and these chapters are briefly introduced below.

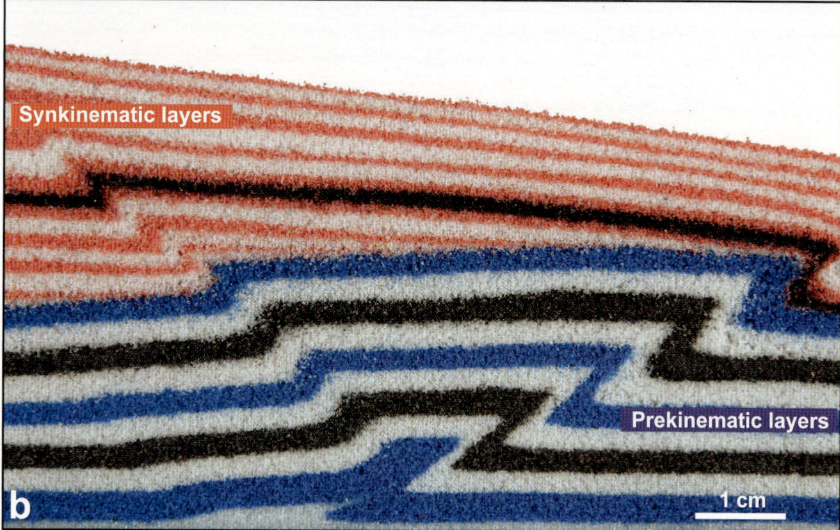

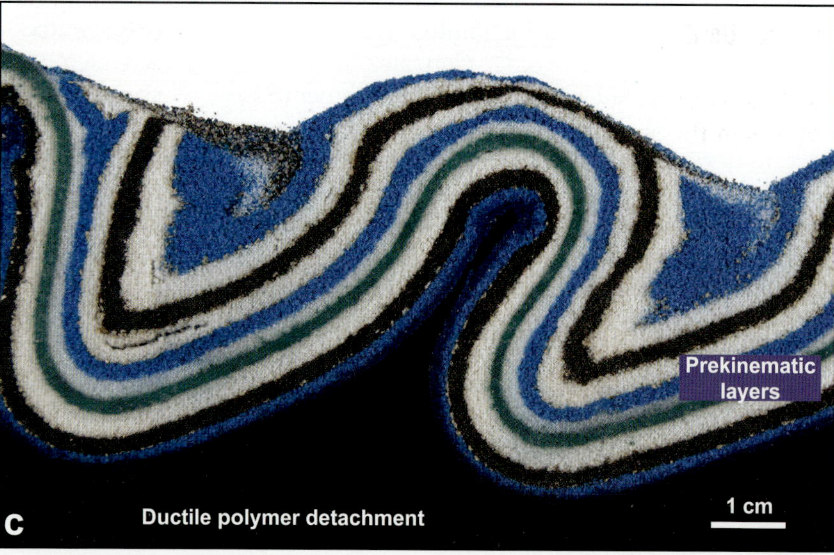

Figure 8. Scaled analog model examples of thrust-related fold systems. (a) Fault-propagation fold with rotated front limb and with the thrust ramp cutting across the hanging-wall strata. (b) Hybrid faulted detachment and fault-propagation fold with synkinematic growth strata onlapping the top limb of the hanging-wall box fold anticline. (c) Detachment folds showing lift-off geometries above a basal ductile polymer detachment layer. Ductile polymer has been squeezed out from the core of the anticline. Note also the small out-of-syncline thrusts in the core of the right syncline.

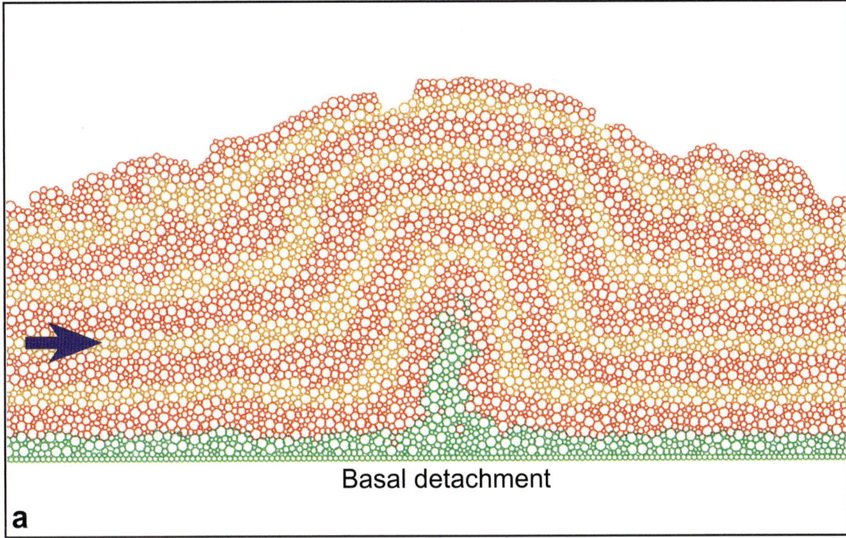

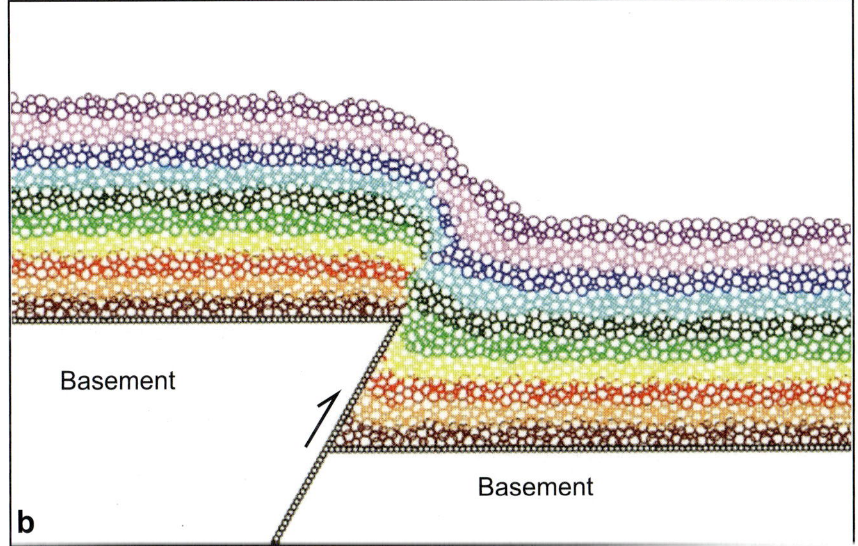

Figure 9. Distinct element numerical models (numerical sandbox) by Hardy and co-workers show how the mechanical stratigraphy and basement involvement control the fold geometries in the upper layers. Models are courtesy of Stuart Hardy (see also Hardy and Allmendinger, 2011). (a) Distinct element model of a detachment fold; (b) distinct element model of a basement-involved fault-propagation fold.

THIS VOLUME

The following sections introduce the 15 subsequent chapters in this memoir.

Thrust Fault-related Fold Models: Detachment Folds, Fault-propagation Folds, Shear Fault-bend Folds, and Wedge-thrust Systems

Chapter 2 by Suppe (2011) summarizes a new 2-D analysis of detachment folds, quantifying fold growth and evolution. The 2-D models of detachment folding analyzed in this chapter include classic "closed systems" where all of the strain is accommodated within the fold and "open systems" where out-of-the-plane flow of a weak ductile detachment layer accommodated the shortening or where a "roof" detachment or thrust occurs and accommodated shortening away from the principal fold. An early stage of pure shear folding is recognized before significant fold amplification and uplift develop. Examples from the classic cross sections of the Grechenberg railway tunnel in the Jura mountains drawn by Buxtorf (1916) are used to illustrate this new approach to understand detachment folds in thrust terranes.

Chapter 3 by Gonzales-Mieres and Suppe (2011) follows on from Chapter 2 by presenting new strategies for extracting detailed shortening histories of detachment folds based on area-of-relief measurements in growth strata, including the complexities of multiple detachments and flow of salt. These authors analyze active detachment folds in accretionary prisms such as the Nankai and Cascadia prisms as well as on the frontal fold system in the Kuqa fold belt, southern margin of the

Figure 10. Detachment folds: examples of complex and hybrid structures. (a) Detachment fold with the front limb thrust in shale and limestone of the Pennsylvanian Calico Bluff Formation, Yukon River, Alaska. The cliff is approximately 250 m (820 ft) high (photo by Kevin Pogue, courtesy of Keck Geology Consortium). (b) Inclined boxlike detachment fold with inward-dipping limbs and internal folding accommodating space problems in the core of the anticline, Midterhuken peninsula, Spitsbergen (Maher et al., 1986). The units are Triassic shales and sandstones with the detachment surface sitting on top of the lower Bravaisberget formation sandstones, with intense chevron folding in black shales and thin siltstones above the detachment. The upper more competent "lid" is the sandy Upper Triassic De Geerdalen Formation. The height from the base of the photo to the uppermost peak is approximately 780 m (2559 ft) (photo courtesy of Alvar Braathen). (c) Detachment fold with thrusting and duplication in the core, North Cornwall. Note the complex accommodation structures around the hinge and core above the detachment surface. Units are Namurian age thick-bedded sandstones and interbedded shales. Field of view is approximately 5 m (16 ft).

Tianshan range, western China, and on the Agbami detachment fold, offshore Niger Delta. Episodic fold growth was documented in these examples with fault slip rates lower than the regional shortening rates.

Chapter 4 by Vergés et al. (2011) presents an elegant detailed description and analysis of the large Kabir Kuh fold structure in the southwest Zagros of Iran. This chapter focuses on the influences of multiple detachments and mechanical stratigraphy on the evolution of the fold system. This detailed analysis highlights the internal complexity of fold structures in what has previously been considered to be rather simple concentric-style folds in the simply folded belt of the southern Zagros (cf. Beydoun et al., 1992; McQuarrie, 2004; Sherkati et al., 2006).

Hardy and Allmendinger (2011) in Chapter 5 review the application of "trishear" kinematic models (cf. Erslev, 1991, 1993) to the study of fault-propagation folds with a focus on contractional tectonic settings. "Trishear" deformation has been used to develop models of progressive rotation of the front limbs of fault-propagation folds (Hardy and Ford, 1997; Allmendinger, 1998; Hardy and McClay, 1999) as well as forming fanning growth stratal wedges in these folds. The addition of backlimb trishear deformation in these kinematic models by Cristallini and Allmendinger (2002) has allowed the trishear model to be widely applied to fault-propagation growth folds where fanning growth strata are found on both the front limbs and backlimbs of a fault-propagation fold.

In Chapter 6, Tavani and Storti (2011) discuss the tip-line strain effects of the so-called "double-edge fault-propagation folding" model, a geometric variant of the model of faulting developing first within the competent layer and then propagating both up to the surface as well as down to the underlying detachment. This model may offer an alternative explanation to the "shear fault-bend fold" model (Suppe et al., 2004; Shaw et al., 2005) to explain fault-fold systems where the backlimb hanging-wall panel is not parallel to the fault at depth.

Alonso et al. (2011) in Chapter 7 describe a detailed analysis of folding mechanisms in a fault-propagation fold using key unconformities within the growth strata. This analysis shows how internal deformation may be distributed within an evolving fault-propagation fold with synkinematic growth strata. The results of this type of analysis are particularly important where growth strata are involved and where nonparallel layers of the synkinematic sequence are folded during uplift and amplification. In some circumstances, this may produce synclines in the growth sequence directly on top of anticlines in the pregrowth sequence (cf. Alonso, 1989).

Active Thrust-related Folding and Faulting

Chapters 8 and 9 both focus on active folding developed above thrust ramps. In both of these chapters, active thrust deformation is analyzed to validate the geometric models if the hanging wall folds.

Yue et al. (2011, Chapter 8) use 2-D seismic profiles and well-constrained surface geometries in pregrowth strata and deformed flights of fluvial terraces to develop plausible kinematic models through the active fold belt in western Taiwan. The authors describe two adjacent active thrust ramps in western Taiwan involving the same stratigraphic sequence and detachment; one is a large slip ramp (~14 km [8.7 mi]) with classic fault-bend fold geometries without limb rotation and the other is a low-slip ramp (1.7 km [1 mi]) with a hanging-wall limb rotation characteristic of shear fault-bend folding, suggesting a possible evolution in folding mechanism with increasing slip. The authors use the analysis of terrace folding, uplift, and tilting, together with the coseismic displacements in the 1999 Chi-Chi earthquake to validate these end-member kinematic models for thrust evolution.

Chapter 9 by Rivero and Shaw (2011) examines the thrust-related fold systems developed in the California borderlands, offshore southwest California, which has been considered a classic transpressional province. The authors relate the geometries of folds seen in seismic data to active earthquake faulting and active thrust systems associated with shortening and basin inversion in this complex contractional and strike-slip region. They propose that structural wedging, shear fault-bend folding, and fault-propagation folding are the most appropriate models to describe the evolution of these fold systems.

Basement-involved Thrust Systems

Wang et al. (2011, Chapter 10) integrated surface and subsurface data to produce a series of balanced geoseismic sections across the Kuqa fold belt, southern Tianshan ranges (on the northern edge of the Tarim Basin), western China. Here, the foreland basin deformation is linked to steep basement-involved, thrust uplifts of the main Tianshan ranges. In the foreland basin, both shear fault-bend folds (Suppe et al., 2004) and detachment fold systems were recognized. The frontal fold, the Yaken anticline, is active, deforming alluvial terraces and is interpreted to be an early stage in the formation of a detachment fold. Retrodeformed sections show how this complex, basement-involved fold and thrust belt developed and how active surface structures may be linked to thrust faulting at depth. This contribution is a significant new description of this important hydrocarbon-bearing fold and thrust belt.

Chapter 11 by Kraemer et al. (2011) analyzes the structural evolution of the Andean fold and thrust belt between 35°S and 36°S in western Argentina. Here, the Miocene–Pliocene deformation involves both basement thrusts as well as salt detachments in the cover strata. Syncontractional depocenters were bounded by active structures. Using a critical Coulomb wedge model, the authors explain the reactivation and uplift of hinterland structures during supercritical wedge times and enhanced deposition of synorogenic clastics during subcritical wedge times when reduced hinterland uplift was observed. This evolution is similar to that described in the analog models of Wu and McClay (2011, Chapter 14).

Mount et al. (2011) (Chapter 12) present a detailed analysis of wedge thrust systems involving basement in the Hanna Basin, Wyoming, western United States. Using long-offset seismic data, they describe two forms of wedge models where displacement on a deeper fault is transferred up section onto an oppositely dipping shallow fault. In the Hanna Basin example shown in this chapter, both basement-detached and basement-involved wedge structures are developed. Such indentor wedge structures are commonly seen at uplifted basin margins and at the termination of fold-thrust belts.

Quantitative Three-dimensional Analysis

Mencos et al. (2011) in Chapter 13 show how detailed field mapping combined with full 3-D construction of structural surfaces was essential in developing accurate 3-D geological and geometric models of growth-thrust-related detachment folds in the central Spanish Pyrenees. The authors incorporate surface geological data, digital elevation models, and orthophotos as well as seismic and well data to build a constrained 3-D geological model of a complex inversion-related anticline in the Bóixols thrust system of the central Spanish Pyrenees.

Scaled Analog Modeling of Thrust-related Fold Systems

Scaled physical laboratory models have been used extensively to illuminate the diverse issues in the kinematic evolution of fold and thrust belts. In Chapter 14, Wu and McClay (2011) describe the results of a detailed scaled analog modeling program designed to investigate the dynamic feedback and interactions of surface processes (synkinematic erosion and synkinematic sedimentation) on model fold and thrust belts as well as on individual folds within them. The analog models of fold and thrust belts form critically tapered Coulomb wedges (e.g., Davis et al., 1983). Within these wedges, synchronous thrusting on two or more active faults was focused toward the wedge front. In contrast, syntectonic erosion reduced the wedge taper, thereby promoting renewed thrusting at the rear of the wedge. Syntectonic sedimentation promoted focused deformation at the front of the wedge and changed the geometries of the thrust fault-related folds (e.g., Figure 7). These kinematic models show the complex slip history of fault-related folds in thrust belts and the episodic and cyclic nature of thrusting in these systems.

Deep-water Fold and Thrust Belts

In recent years, considerable exploration efforts have focused on deep-water fold belts where hanging-wall folds above either shale or salt detachments have been the main targets. Chapters 15 and 16 examine fold belt development offshore northeastern Brazil and in the offshore Nile Delta, respectively.

Chapter 15 by Zalán (2011) describes the 2-D and 3-D geometries of deep-water fold belts offshore northeastern Brazil. Here, the detachment is above "overpressured" shale systems. Complex imbricate thrust systems with hanging-wall fault-propagation folds are described and visualized in 3-D.

Chapter 16 by Krueger and Grant (2011) presents a detailed study of the fold systems in the offshore Niger Delta. These authors demonstrate that fault-propagation folds in the Niger Delta, once initiated, grew rapidly along strike reaching near-ultimate length early in the structural history and then subsequently amplified with the last stages of deformation focused in the central sections of the folds. This pattern of fold growth is attributed to the focusing of elevated pore-fluid pressures into the central section of the thrust-related fold system. The study has significant implications for fold growth mechanisms as well as for the timings of fold amplification and hence trapping of hydrocarbons in the toe-thrust systems of the Niger Delta.

SUMMARY

The chapters presented in this memoir cover some of the advances made in the study of thrust-related fold systems made over the past decade. Improved seismic imaging combined with theoretical, numerical, and analog modeling as well as detailed field studies has illuminated the challenges involved in integrating the well-established geometric and kinematic models of fault-related folding with mechanical models as well as being able to account for structures found in real-world examples of hydrocarbon-bearing structures. I hope that the readers of this memoir will find these new ideas and concepts relevant for the exploration and exploitation of hydrocarbon systems in fold and thrust belts worldwide.

ACKNOWLEDGMENTS

Ken McClay gratefully acknowledges the generous assistance of Jonny Wu, Jose de Vera, Hannah Rogers, and Nicola Scaselli with drafting and compiling the seismic images shown in this review. Alvar Braathen is thanked for use of the image in Figure 10b and Robert Burger for permission to use the photograph in Figure 10a. Stuart Hardy is thanked for the images of distinct element models of fault-related fold systems. Critical comments by John Suppe, John Shaw, Jose de Vera, and Stuart Hardy were greatly appreciated.

REFERENCES CITED

Ajakaiye, D. E., and A. W. Bally, 2002, Some structural styles on reflection profiles from offshore Niger Delta: AAPG Continuing Education Course Note 41, Course manual and atlas of structural styles from the Niger Delta, 107 p.

Allmendinger, R. W., 1998, Inverse and forward numerical modeling of trishear fault propagation folds: Tectonics, v. 17, no. 4, p. 640–656, doi:10.1029/98TC01907.

Alonso, J. L., 1989, Fold reactivation involving angular unconformable sequences: Theoretical analysis and natural examples from the Cantabrian zone (northwest Spain): Tectonophysics, v. 170, p. 57–77, doi:10.1016/0040-1951(89)90103-0.

Alonso, J. L., F. Colombo, and O. Riba, 2011, Folding mechanisms in a fault-propagation fold inferred from the analysis of unconformity angles: The Sant Llorenç growth structure (Pyrenees, Spain), *in* K. McClay, J. H. Shaw, and J. Suppe, eds., Thrust fault-related folding: AAPG Memoir 94, p. 137–151.

Atkinson, P. K., and W. K. Wallace, 2003, Competent unit thickness variation in detachment folds in the northeastern Brooks Range, Alaska: Geometric analysis and a conceptual model: Journal of Structural Geology, v. 25, no. 10, p. 1751–1771, doi:10.1016/S0191-8141(03)00003-8.

Beydoun, Z. R., M. W. Hughes Clarke, and R. Stoneley, 1992, Petroleum in the Zagros Basin: A Late Tertiary foreland basin overprinted onto the outer edge of a vast hydrocarbon-rich Paleozoic–Mesozoic passive-margin shelf, *in* R. W. Macqueen and D. A. Leckie, eds., Foreland basins and fold belts: AAPG Memoir 55, p. 309–339.

Bilotti, F., and J. H. Shaw, 2005, Deep-water Niger Delta fold and thrust belt modeled as a critical-taper wedge: The influence of elevated basal fluid pressure on structural styles: AAPG Bulletin, v. 89, no. 11, p. 1475–1491, doi:10.1306/06130505002.

Buxtorf, A., 1907, Zur Tektonik des Kettenjura: Beiricht der Versammlung des Oberrheinischen Geologischen Veireins, v. 40, p. 29–38.

Buxtorf, A., 1916, Prognosen und Befunde beim Hauenstein-basis–und Grechenberbtunnel und die Bedeutung der letzeren fur die Geologie des Jurabirges: Verhandlungen der naturforschenden Gesellschaft Basel, v. 27, p. 184–254.

Cooper, M. A., 2007, Structural style and hydrocarbon prospectivity in fold and thrust belts: A global review, *in* A. C. Ries, R. W. H. Butler, and R. H. Graham, eds., Deformation of the continental crust: The legacy of Mike Coward: Geological Society (London) Special Publication 272, p. 44–472.

Corredor, F., J. Shaw, and F. Bilotti, 2005a, Structural styles in the deep-water fold and thrust belts of the Niger Delta: AAPG Bulletin, v. 89, no. 6, p. 753–780, doi:10.1306/02170504074.

Corredor, F., J. Shaw, and J. Suppe, 2005b, Shear fault-bend folding, deep water Niger Delta, *in* J. Shaw, C. Connors, and J. Suppe, eds., Seismic interpretation of contractional fault-related folds: AAPG Studies in Geology 53, p. 87–92.

Cristallini, E. O., and R. W. Allmendinger, 2002, Backlimb trishear: A kinematic model for curved folds developed over angular fault bends: Journal of Structural Geology, v. 24, no. 2, p. 289–295.

Davis, D., J. Suppe, and F. A. Dahlen, 1983, Mechanics of fold-and-thrust-belts and accretionary wedges: Journal of Geophysical Research, v. 88, no. B2, p. 1153–1172, doi:10.1029/JB088iB02p01153.

Echavarria, L., R. Hernandez, R. Allmendinger, and J. Reynolds, 2003, Subandean thrust and fold belt of northwestern Argentina: Geometry and timing of the Andean evolution: AAPG Bulletin, v. 87, no. 6, p. 965–985.

Epard, J. L., and R. H. Groshong, 1995, Kinematic model of detachment folding including limb rotation, fixed hinges and layer-parallel strain: Tectonophysics, v. 247, no. 1–4, p. 85–103, doi:10.1016/0040-1951(94)00266-C.

Erslev, E. A., 1991, Trishear fault-propagation folding: Geology, v. 19, no. 6, p. 617–620, doi:10.1130/0091-7613(1991)019<0617:TFPF>2.3.CO;2.

Erslev, E. A., 1993, Thrusts, back-thrusts, and detachment of Laramide foreland arches, *in* C. J. Schmidt, R. Chase, and E. A. Erslev, eds., Laramide basement deformation in the Rocky Mountain foreland of the western United States: Geological Society of America Special Paper, v. 280, p. 339–358.

Erslev, E. A., and K. R. Mayborn, 1997, Multiple geometries and modes of fault-propagation folding in the Canadian thrust belt: Journal of Structural Geology, v. 19, no. 3–4, p. 321–335, doi:10.1016/S0191-8141(97)83027-1.

Gonzalez-Mieres, R., and J. Suppe, 2006, Relief and shortening in detachment folds: Journal of Structural Geology, v. 28, no. 10, p. 1785–1807, doi:10.1016/j.jsg.2006.07.001.

Gonzalez-Mieres, R., and J. Suppe, 2011, Shortening histories in active detachment folds based on area-of-relief methods, *in* K. McClay, J. H. Shaw, and J. Suppe, eds., Thrust fault-related folding: AAPG Memoir 94, p. 39–67.

Groshong Jr., R. H., and J. L. Epard, 1994, Role of strain in area-constant detachment folding: Journal of Structural Geology, v. 16, p. 613–618, doi:10.1016/0191-8141(94)90113-9.

Guzofski, C. A., J. P. Mueller, J. H. Shaw, P. Muron, D. A. Medwedeff, F. Bilotti, and C. Rivero, 2009, Insights into the mechanisms of fault-related folding provided by volumetric structural restorations using spatially varying mechanical constraints: AAPG Bulletin, v. 93, no. 4, p. 479–502.

Hardy, S., and R. W. Allmendinger, 2011, Trishear: A review of kinematics, mechanics, and applications, *in* K. McClay, J. H. Shaw, and J. Suppe, eds., Thrust fault-related folding: AAPG Memoir 94, p. 95–119.

Hardy, S., and C. D. Connors, 2006, A velocity description of shear fault-bend folding: Journal of Structural Geology, v. 28, no. 3, p. 536–543, doi:10.1016/j.jsg.2005.12.015.

Hardy, S., and M. Ford, 1997, Numerical modeling of trishear fault propagation folding: Tectonics, v. 16, no. 5, p. 841–854, doi:10.1029/97TC01171.

Hardy, S., and K. McClay, 1999, Kinematic modelling of extensional forced folding: Journal of Structural Geology, v. 21, p. 695–702.

Hardy, S., and J. Poblet, 1994, Geometric and numerical model of progressive limb rotation in detachment folds: Geology, v. 22, no. 4, p. 371–374.

Hardy, S., J. Poblet, K. R. McClay, and D. Waltham, 1996, Mathematical modeling of growth strata associated with fault-related fold structures: Geological Society (London) Special Publication 99, p. 265–282.

He, D., J. Suppe, G. Yang, S. Guan, S. Huang, X. Shi, X. Wang, and C. Zhang, 2005, Guidebook for fieldtrip in south and north Tianshan foreland basin, Xinjiang Uygur Autonomous Region, China: International Conference on Theory and Application of Fault-Related Folding in Foreland Basins, Beijing, China, 78 p.

Homza, T. X., and W. K. Wallace, 1995, Geometric and kinematic models for detachment folds with fixed and variable detachment depths: Journal of Structural Geology, v. 17, no. 4, p. 575–588, doi:10.1016/0191-8141(94)00077-D.

Hubert-Ferrari, A., J. Suppe, R. Gonzalez-Mieres, and X. Wang, 2007, Mechanisms of active folding of the landscape (southern Tian Shan, China): Journal of Geophysical Research, v. 112, B03S09, p. 1–39, doi:10.1029/2006JB004362.

James, D. M. D., 1984, The geology and hydrocarbon resources of Negara Brunei Darussalam: Muzium Brunei, Bandar Seri Begawan, Brunei, p. 164.

Jamison, W. R., 1987, Geometric analysis of fold development in overthrust terranes: Journal of Structural Geology, v. 9, no. 2, p. 207–219.

Jordan, P., and T. Noack, 1992, Hanging wall geometry of overthrusts emanating from ductile decollements, *in* K. McClay, ed., Thrust tectonics: London, United Kingdom, Chapman & Hall, p. 311–318.

Kraemer, P., J. Silvestro, F. Achilli, and W. Brinkworth, 2011, Kinematics of a hybrid thick-thin-skinned fold and thrust belt recorded in Neogene syntectonic top-wedge basins, southern central Andes between 35° and 36°S, Malargüe, Argentina, *in* K. McClay, J. H. Shaw, and J. Suppe, eds., Thrust fault-related folding: AAPG Memoir 94, p. 245–270.

Krueger, S. W., and N. T. Grant, 2011, The growth history of toe thrusts of the Niger delta and the role of pore pressure, *in* K. McClay, J. H. Shaw, and J. Suppe, eds., Thrust fault-related folding: AAPG Memoir 94, p. 357–390.

Laubscher, H., 1965, Ein kinematisches modell der jurafaltung: Eclogae Geologicae Helvetiae, v. 58, p. 231–318.

Laubscher, H., 2008, The Grechenberg conundrum in the Swiss Jura: A case for the centenary of the thin-skin decollement nappe model (Buxtorf 1907): Swiss Journal of Geosciences, v. 101, p. 41–60, doi:10.1007/s00015-008-1248-2.

Maher, H. D., C. Craddock, and K. Maher, 1986, Kinematics of Tertiary structures in Upper Paleozoic and Mesozoic strata on Midterhuken, west Spitsbergen: Geological Society of America Bulletin, v. 97, p. 1411–1421.

Marrett, R., and P. A. Bentham, 1997, Geometric analysis of hybrid fault-propagation/detachment folds: Journal of Structural Geology, v. 19, p. 243–248, doi:10.1016/S0191-8141(96)00092-2.

McClay, K. R., 2004, Thrust tectonics and hydrocarbon systems: Introduction, *in* K. R. McClay, ed., Thrust tectonics and hydrocarbon systems: AAPG Memoir 82, p. ix–xx.

McQuarrie, N., 2004, Crustal-scale geometry of Zagros fold-thrust belt, Iran: Journal of Structural Geology, v. 26, no. 3, p. 519–535, doi:10.1016/j.jsg.2003.08.009.

Medwedeff, D., 1989, Growth fault-bend folding at southeast Lost Hills, San Joaquin Valley, California: AAPG Bulletin, v. 73, no. 1, p. 54–67.

Medwedeff, D. A., and J. Suppe, 1997, Multibend fault-bend folding: Journal of Structural Geology, v. 19, no. 3–4, p. 279–292, doi:10.1016/S0191-8141(97)83026-X.

Mencos, J., J. A. Muñoz, and S. Hardy, 2011, Three-dimensional geometry and forward numerical modeling of the Sant Corneli anticline (southern Pyrenees, Spain), *in* K. McClay, J. H. Shaw, and J. Suppe, eds., Thrust fault-related folding: AAPG Memoir 94, p. 283–300.

Mitra, S., 2002, Structrural models of faulted detachment folds: AAPG Bulletin, v. 86, no. 9, p. 1673–1694.

Mitra, S., 2003, A unified kinematic model for the evolution of detachment folds: Journal of Structural Geology, v. 25, no. 10, p. 1659–1673, doi:10.1016/S0191-8141(02)00198-0.

Mount, V. S., K. W. Martindale, T. W. Griffith, and J. O. D. Byrd, 2011, Basement-involved contractional wedge structural styles: Examples from the Hanna Basin, Wyoming, *in* K. McClay, J. H. Shaw, and J. Suppe, eds., Thrust fault-related folding: AAPG Memoir 94, p. 271–281.

Narr, W., and J. Suppe, 1994, Kinematics of basement-involved compressive structures: American Journal of Science, v. 294, no. 7, p. 802–860.

Poblet, J., and K. McClay, 1996, Geometry and kinematics of single-layer detachment folds: AAPG Bulletin, v. 80, no. 7, p. 1085–1109.

Poblet, J., K. McClay, F. Storti, and J. A. Munoz, 1997, Geometries of syntectonic sediments associated with single-layer detachment folds: Journal of Structural Geology, v. 19, no. 3–4, p. 369–381, doi:10.1016/S0191-8141(96)00113-7.

Poblet, J., M. Bulnes, K. McClay, and S. Hardy, 2004, Plots of crestal structural relief and fold area versus shortening—A graphical technique to unravel the kinematics of thrust-related folds, *in* K. R. McClay, ed., Thrust tectonics and hydrocarbon systems: AAPG Memoir 82, p. 372–399.

Rivero, C., and J. H. Shaw, 2011, Active folding and blind-thrust faulting induced by basin inversion processes, inner California borderlands, *in* K. McClay, J. H. Shaw, and J. H. Suppe, eds., Thrust fault-related folding: AAPG Memoir 94, p. 187–214.

Sandal, S. T., 1996, The geology and hydrocarbon resources of Negara Brunei Darussalam 1996 revision: Bandar Seri Begawan, Brunei Darussalam, Syabas, 243 p.

Shaw, J. H., C. D. Connors, and J. Suppe, 2005, Seismic interpretation of contractional fault-related folds: AAPG Studies in Geology 53, 156 p.

Sherkati, S., J. Letouzey, and D. Frizon de Lamotte, 2006, Central Zagros fold-thrust belt (Iran): New insights from seismic data, field observation, and sandbox modeling: Tectonics, v. 25, p. TC4007, doi:10.1029/2004TC001766.

Storti, F., and J. Poblet, 1997, Growth stratal architectures associated to decollement folds and fault-propagation folds. Inferences on fold kinematics: Tectonophysics, v. 282, no. 1–4, p. 353–373, doi:10.1016/S0040-1951(97)00230-8.

Suppe, J., 1983, Geometry and kinematics of fault-bend folding: American Journal of Science, v. 283, no. 7, p. 684–721.

Suppe, J., 1985, Principles of structural geology: New Jersey, Prentice Hall, 537 p.

Suppe, J., 2011, Mass balance and thrusting in detachment folds, *in* K. McClay, J. H. Shaw, and J. Suppe, eds., Thrust fault-related folding: AAPG Memoir 94, p. 21–37.

Suppe, J., and D. A. Medwedeff, 1990, Geometry and kinematics of fault-propagation folding: Eclogae Geologicae Helvetiae, v. 83, no. 3, p. 409–454.

Suppe, J., G. T. Chou, and S. C. Hook, 1992, Rates of folding and faulting determined from growth strata, *in* K. McClay, ed., Thrust tectonics: London, United Kingdom, Chapman & Hall, p. 105–121.

Suppe, J., F. Sabat, J. A. Muños, J. Poblet, E. Roca, and J. Verges, 1997, Bed-by-bed fold growth by kink-band migration: Sant Llorenç de Morunys, eastern Pyrenees: Journal of Structural Geology, v. 19, no. 3–4, p. 443–461, doi:10.1016/S0191-8141(96)00103-4.

Suppe, J., C. D. Connors, and Y. K. Zhang, 2004, Shear fault-bend folding, *in* K. McClay, ed., Thrust tectonics and hydrocarbon systems: AAPG Memoir 82, p. 303–323.

Tankard, A. J., R. Suarez-Soruco, and H. J. Welsink, 1995, Petroleum basins of South America: AAPG Memoir 62, 792 p.

Tavani, S., and F. Storti, 2011, Layer-parallel shortening temples associated with double-edge fault-propagation folding, *in* K. McClay, J. H. Shaw, and J. Suppe, eds., Thrust fault-related folding: AAPG Memoir 94, p. 121–135.

Vergés, J., M. G. H. Goodarzi, H. Emami, R. Karpuz, J. Efstathiou, and P. Gillespie, 2011, Multiple detachment folding in Pusht-e Kuh arc, Zagros: Role of mechanical stratigraphy, *in* K. McClay, J. H. Shaw, and J. Suppe, eds., Thrust fault-related folding: AAPG Memoir 94, p. 69–94.

Wang, X., J. Suppe, S. Guan, A. Hubert-Ferrari, R. Gonzalez-Mieres, and C. Jia, 2011, Cenozoic structure and tectonic evolution of the Kuqa fold belt, southern Tianshan, China, *in* K. McClay, J. H. Shaw, and J. Suppe, eds., Thrust fault-related folding: AAPG Memoir 94, p. 215–243.

Wu, J. E., and K. R. McClay, 2011, Two-dimensional analog modeling of fold and thrust belts: Dynamic interactions with syncontractional sedimentation and erosion, *in* K. McClay, J. H. Shaw, and J. Suppe, eds., Thrust fault-related folding: AAPG Memoir 94, p. 301–333.

Yue, L.-F., J. Suppe, and J.-H. Hung, 2011, Two contrasting kinematic styles of active folding above thrust ramps, western Taiwan, *in* K. McClay, J. H. Shaw, and J. Suppe, eds., Thrust fault-related folding: AAPG Memoir 94, p. 153–186.

Zalán, P. V., 2011, Fault-related folding in the deep waters of the equatorial margin of Brazil, *in* K. McClay, J. H. Shaw, and J. Suppe, eds., Thrust fault-related folding: AAPG Memoir 94, p. 335–355.

Zehnder, A. T., and R. W. Allmendinger, 2000, Velocity field for the trishear model: Journal of Structural Geology, v. 22, no. 8, p. 1009–1014, doi:10.1016/S0191-8141(00)00037-7.

2

Suppe, John, 2011, Mass balance and thrusting in detachment folds, *in* K. McClay, J. H. Shaw, and J. Suppe, eds., Thrust fault-related folding: AAPG Memoir 94, p. 21 – 37.

Mass Balance and Thrusting in Detachment Folds

John Suppe[1]
Department of Geosciences, National Taiwan University, Taipei, Taiwan

ABSTRACT

Detachment folds seem conceptually the simplest of all fault-related folds, yet they continue to offer substantial and rich challenges that are symptomatic of fault-related folding in general. The standard picture of detachment folds goes back at least to August Buxtorf's widely reproduced 1916 regional cross sections of the Swiss Jura in which he interpreted the Jura box folds to be regionally detached from the basement along a weak layer of Triassic evaporites. Most theoretical and conceptual models of detachment folding have followed Buxtorf's model with a twofold mechanical stratigraphy composed of a competent flexural lid conserving bed length overlying a weak basal detachment layer that conserves only volume. However, this standard model of detachment folding has a fundamental and classic problem from a balancing perspective, which is outlined in this chapter. Furthermore, the classic Jura box folds drawn by Buxtorf were in fact not at all constrained by subsurface data, except along the Grenchenberg railroad tunnel where a surprising and much more complex structure was encountered, involving a strongly folded thrust fault that was intersected three times in the 8.5-km (5.3-mi)-long tunnel. Hans Laubscher argued that such complexly folded thrust faults are in fact typical of many Jura box folds. Analogous structures have been observed in Canada, Mexico, and Spain. I argue here on the basis of mass-balance considerations that detachment folds with folded thrusts are one of several theoretically expected modes of detachment folding. More broadly, the classical models of detachment folding assume a closed-system behavior in which all shortening is locally consumed in the fold. These classical models include (1) local flow of a weak basal layer, for example, salt (Wiltschko and Chapple mechanism) and (2) pure-shear detachment folding (Groshong and Epard mechanism). Here we discuss several additional models of detachment folding, including open-system behavior with (3) far-field flow of the weak basal layer, (4) roof detachments above the basal layer, and (5) roof ramps above the basal layer such as the folded thrust fault of the Grenchenberg anticline. In this final mechanism, the flexural lid or other more competent stratigraphic intervals shorten by a combination of fault slip and flexure, whereas the less competent intervals, including the basal layer, shorten by flow. All of these mechanisms are end members capable of existing in combination within actual folds; several examples combine two or three of these mechanisms.

[1]*Also at*: Department of Geosciences, Princeton University, Princeton, New Jersey, U.S.A.

DOI:10.1306/13251331M94389

INTRODUCTION

Many of our current conceptual models of detachment folding can be traced back at least to a classic regional cross section of August Buxtorf (1916) through the Swiss Jura fold belt, shown in part in Figure 1. This very influential and widely reproduced section became the standard image of Jura box folding, containing Buxtorf's interpretation of anticlines such as Vorburgkette, Vellerat, and Moutier as simple not-quite-isoclinal straight-limbed anticlines in competent Jurassic carbonates, detached from their substrate along the evaporites of the Triassic Muschelkalk. These folds would now be called detachment folds (Jamison, 1987), and much recent geometric and mechanical modeling of detachment folding has adopted a twofold Jura-like mechanical stratigraphy very similar to Buxtorf's interpretation, which is composed of a competent flexural lid overlying a basal incompetent interval that shows strong thickness variations and detaches the fold from the substrate. For example, Figure 2 from Poblet and McClay (1996) shows a typical model arrangement with an overlying competent layer that conserves bed length and a basal detachment layer that conserves only area.

Interestingly, Buxtorf's interpretation of most of the anticlines in this cross section was based entirely on surface geology and in reality was not strongly constrained by data. The detached nature of the Jura fold belt is now very well documented by seismic imaging (Sommaruga, 1997), but the subsurface structure of individual steep-limbed Jura folds is still not generally well constrained by data. Even worldwide, it is only in exceptional circumstances that we have relatively complete images of high-amplitude detachment folds, which come from mountainside exposures (e.g., Canadian Rockies, Figure 3), down-plunge projection of refolded anticlines such as in the Pyrenees (Poblet and Hardy, 1995) and Mexico (Rico, 1999), and coal-mine mapping. Furthermore, a few exceptional seismic images of box folds similar to Buxtorf's interpretation have been published recently (e.g., Camerlo and Benson, 2006).

Nevertheless, long before the development of advanced seismic imaging, Buxtorf's 1916 section (Figure 1) was taken seriously because it did incorporate substantial subsurface data from the excavation of an 8565-m (28,100-ft)-long railroad tunnel through the Jura Mountains. Buxtorf's regional structural synthesis and regional cross sections of the Jura came at the end of a 70-page report that focused almost entirely on the surprising details of the subsurface geology encountered during excavation of the Grenchenberg railroad tunnel through the Graitery and Grenchenberg anticlines (Figure 1). Buxtorf's report contains a fascinating detailed documentation of the changing structural interpretations as the excavations progressed. The geology of the tunnel through Graitery anticline was very close to what had been predicted from extrapolation of surface data, but the Grenchenberg anticline was radically different. An unexpected major folded thrust fault was encountered three times in excavation of the tunnel and found to correspond at the land surface to what before the excavation had been considered to be a minor thrust or landslide based on interpretation of the surface geology in the core of the syncline between Graitery and Grenchenberg anticlines (Figure 1). The Grenchenberg anticline was central to Buxtorf's argument that the Jura was regionally detached because the Triassic Muschelkalk evaporitic detachment was encountered in the tunnel excavation in the hanging wall of the folded thrust.

Buxtorf considered the existence of the folded thrust within the Grenchenberg anticline to be essentially fortuitous. He published a detailed model of the progressive development of the anticline in which he showed an earlier fault-bend fold ramp anticline being folded into a later detachment fold to produce a composite Grenchenberg anticline. A central argument of the present chapter is that such folded thrust ramps are predicted from a mass-balance consideration and are in many cases an integral part of the detachment folding process, forming simultaneously with the detachment fold in more competent stratigraphic intervals.

CLASSIC PROBLEM OF DETACHMENT FOLDING

Models of simple Buxtorf-like detachment folds such as that of Poblet and McClay (1996) shown in Figure 2 might be called toy models in the sense that they attempt to reduce detachment folding to its essential behavior, especially from a mass-balance perspective. Such models do not seek to include all the complexities of actual detachment folds such as the mountainside example in the Canadian Rockies (Figure 3) or even necessarily predict the actual shapes of such folds. They are mostly concerned with the mass balance and bed-length balance in such systems. Surprisingly, the behavior of toy models with a Jura-like mechanical stratigraphy is not a trivial problem. This nontrivial nature can be immediately sensed from the existence of a very large number of important articles that have been published on the problem (including Laubscher, 1962, 1965; Wiltschko and Chapple, 1977; Jamison, 1987; Mitra and Namson, 1989; Groshong and Epard, 1994; Hardy and Poblet, 1994; Homza and Wallace, 1995; Poblet and Hardy, 1995; Poblet and McClay, 1996; Mitra, 2002, 2003; Atkinson and Wallace, 2003, etc.).

Without reviewing this extensive literature, in essence, a fundamental incompatibility is observed in

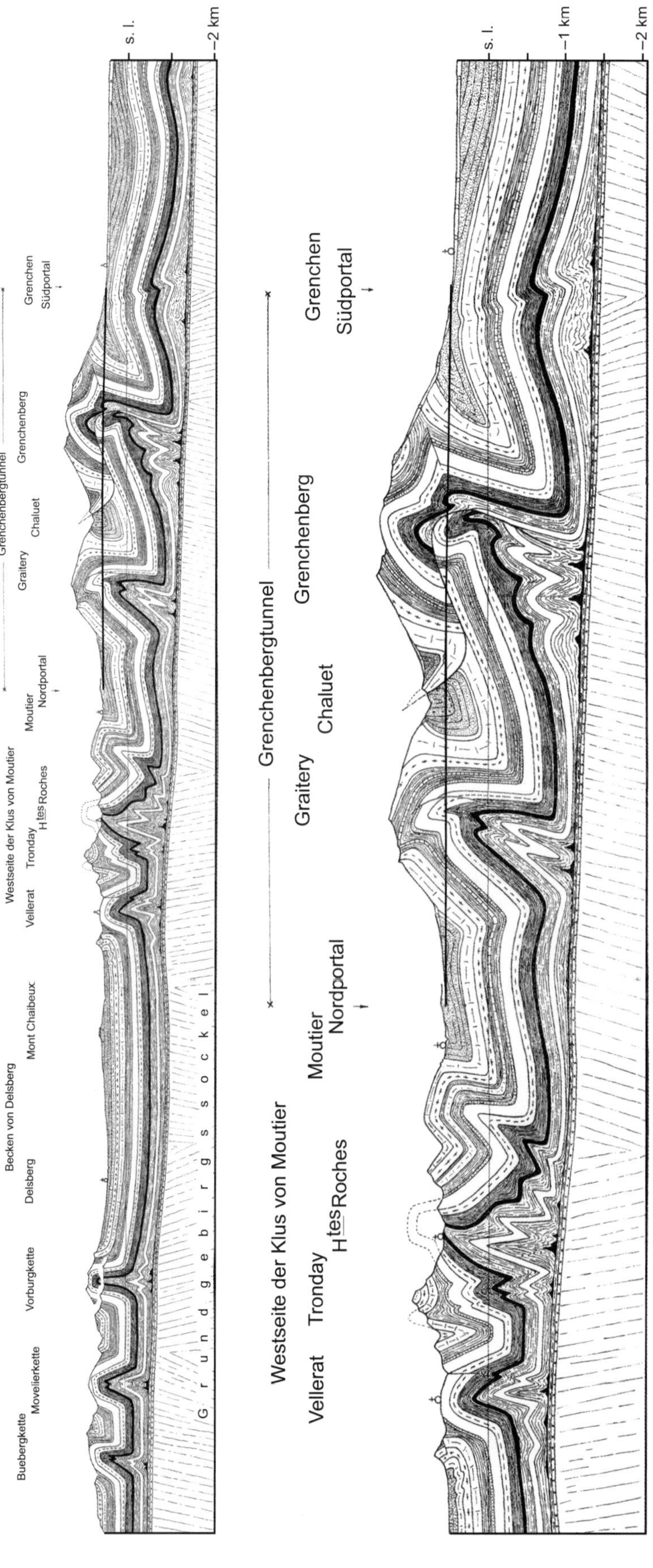

Figure 1. Buxtorf's (1916) widely reproduced cross section of the Swiss Jura along the alignment of the Grenchenberg railroad tunnel.

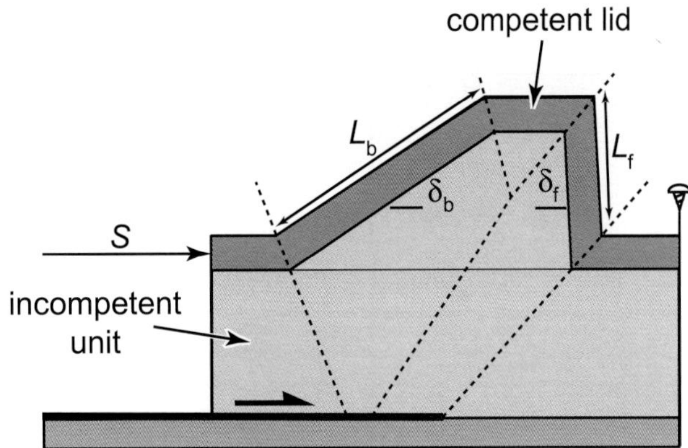

Figure 2. Simple toy model of a detachment fold in a Jura-like stratigraphy composed of (1) an overlying competent flexural lid that conserves bed length and layer thickness and (2) a weak basal detachment layer of finite thickness that conserves only area (modified from Poblet and McClay, 1996).

deforming toy models of detachment folds composed of the two-fold Jura-like mechanical stratigraphy, which was first clearly enunciated by Wiltschko and Chapple (1977). This incompatibility is still an issue today. In essence, it is challenging to deform such a model in a geologically reasonable way such that it conserves area in the incompetent basal detachment layer and simultaneously conserves bed length and layer thickness in the competent flexural lid. Making geologically reasonable Buxtorf-like models is not easy.

The essence of the problem of detachment folding in a Jura-like mechanical stratigraphy can be more easily shown if we adopt a specific toy model. Here we adopt a symmetric angular fold model composed of a flexural lid conserving bed length and layer thickness and a basal detachment layer of finite thickness, h_b, as shown in Figure 4. Wiltschko and Chapple (1977) presented a very similar model using a sinusoidal fold shape. The essential behavior of such models for our purpose is insensitive to the details of the fold shape. Wiltschko and Chapple pointed out that, at the early stages of deformation, the area of the fold core, A_c, grows very rapidly relative to the area displaced in the shortening of the basal layer, A_b; therefore, many fold cores at the early stages of growth appear to contain a positive excess area, $\Delta A_e = A_c - A_b$. In contrast, at a late nearly isoclinal stage similar to Buxtorf's interpretations, the area of the fold core has become very much less than the area of shortening of the basal layer. In general, if you draw a cross section using the Buxtorf model assuming a Jura-like mechanical stratigraphy, it will probably suffer from this problem of incompatibility of deformation between the flexural lid and the basal detachment layer, such that $A_c \neq A_b$.

Let us look at this problem a little more closely in the case of our specific toy model (Figure 4). The area of a competent fold core of limb dip δ is proportional to the square of the limb length, L^2, and is independent of the basal layer thickness, h_b

$$A_c = L^2 \sin\delta \cos\delta \tag{1}$$

In contrast, the area of shortening of the basal detachment layer is proportional to limb length, L, times the basal layer thickness, h_b.

$$A_b = \overline{S} h_b = 2h_b L(1 - \cos\delta) \tag{2}$$

where $\overline{S}$ is the mean shortening of the interval h_b, which normally has been assumed to be equal to the flexural shortening, $S_f = 2L(1 - \cos\delta)$, at the top of the basal layer. These two equations are in general incompatible because the two areas grow at different rates with increasing limb dip.

Initially, the area of the fold core, A_c, increases rapidly because equation 1 is approximately $L^2 \sin\delta$ at low limb dips. As the limb dip increases, the area reaches maximum at 45° after which it decreases going to zero at the 90° limb dip of an isoclinal box fold (Figure 4). In contrast, the area of shortening, A_b, of the basal detachment initially increases very slowly (Figure 4) because $L(1 - \cos\delta)$ in equation 2 grows slowly at low limb dips. Therefore, in the early stages of fold growth, a predicted positive excess area to the fold core exists, as pointed out by Wiltschko and Chapple (1977)

$$\Delta \mathrm{A_e} = \mathrm{A_c} - \mathrm{A_b} \tag{3}$$

and at high-limb dips, a negative excess area is predicted. In general, for a given limb length and layer thickness, only one limb dip exists for which there is no excess area, such as points f_1 and f_2 in Figure 4.

We might be tempted to conclude that a Jura-like fold model requires a continuously changing ratio of basal layer thickness to limb length, h_b/L, during fold growth, which is obtained by equating equations 1 and 2.

$$\frac{h_b}{L} = \frac{\sin\delta\cos\delta}{2(1-\cos\delta)} \tag{4}$$

A graph of this equation is given in Figure 5, which shows that folds would have to form initially with very short limbs relative to the basal layer thickness, which seems mechanically implausible. Furthermore, many low-amplitude detachment folds have h_b/L ratios that are several orders of magnitude smaller than the predictions of this theory, such as the natural folds shown in Figure 5. Therefore, the strict Jura-like model of

Figure 3. Mountainside photo of an asymmetric detachment fold outlined in snow, Halfway River, British Columbia, Canadian Rockies (photo courtesy of Robert Thompson). The view is approximately 3 km wide.

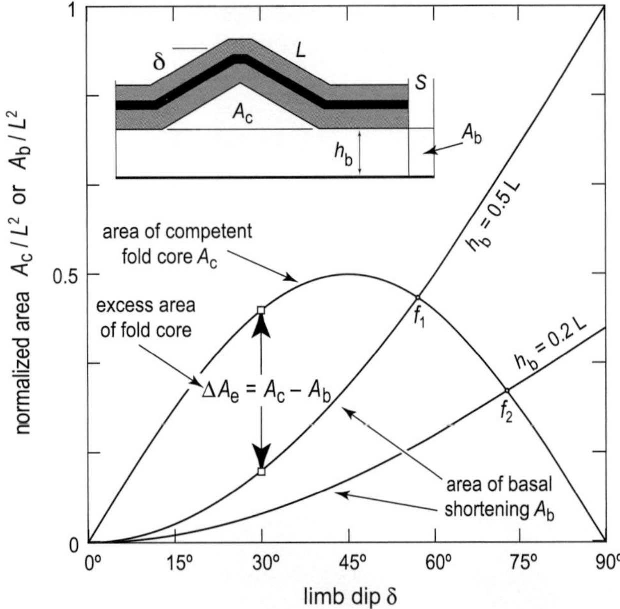

Figure 4. Compatibility problem of detachment folding in a Jura-like stratigraphy composed of a flexural lid that conserves bed length and a deformable basal detachment layer that conserves only area. Generally there is an excess area to the fold core, $\Delta A_e \neq 0$.

detachment folding seems generally unsatisfactory. There have been two geologically successful classic solutions to this problem, which are summarized below.

CLASSIC SOLUTIONS TO THE CLASSIC PROBLEM

Local-flow Solution

The first classic solution to the problem of excess area in the cores of detachment folds is due to Wiltschko and Chapple (1977) who first clearly pointed out the existence of this problem. Their article "Flow of Weak Rocks in Appalachian Plateau Folds" focused on a set of broad low-amplitude detachment folds that are developed above a basal layer of Silurian salt (shown as data in Figure 5). They proposed that this basal detachment layer had flowed extensively relative to the flexural lid, moving salt from the synclines into the anticlinal cores. They showed evidence for this flow in the large thickness variations in the Silurian salt in excess of probable primary stratigraphic thickness variations. Here we call this the local-flow solution, which has been addressed in a variety of theoretical models (e.g., Wiltschko and Chapple, 1977; Homza and Wallace, 1995, 1997).

From a mass-balance perspective, local flow produces a component of negative area of structural relief, A_s, associated with the withdrawal of material of the detachment layer from beneath the synclines, which contributes to the area of the anticlinal core, A_c (Figure 6a). Equation 3 then becomes

$$\Delta A_e = (A_c - A_s) - A_b \tag{5}$$

which for conservation of area in two dimensions is equal to zero. In some environments of salt tectonics, three-dimensional flow is an important contribution to $(A_c - A_s)$.

The area of the fold core, A_c, in folds that form by the local-flow mechanism will show an apparent excess area, ΔA_e, of magnitude A_s if the area of shortening, A_s, is computed using the initial stratigraphic thickness of the basal layer instead of the present thickness below the withdrawal synclines. However, the apparent excess area is not a unique signature of the local-flow mechanism because in a later section, we show that several other mechanisms produce an apparent excess area to the fold core at the top of the basal layer.

Pure-shear Solution

The second classical solution is that of pure shear, which relaxes the original assumption that bed length and

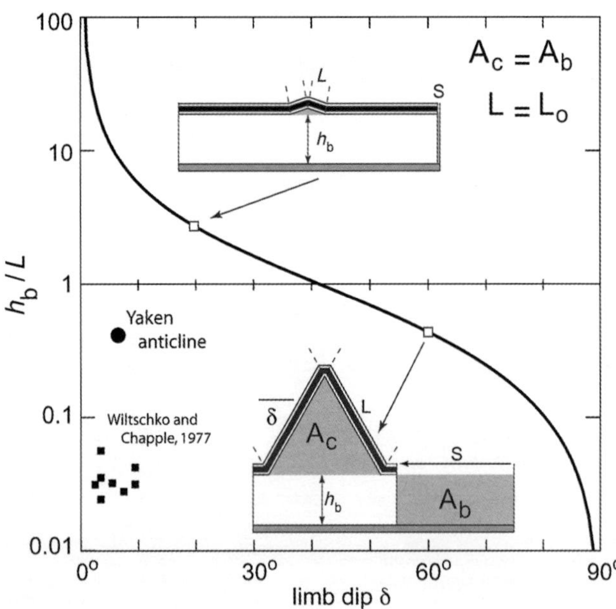

Figure 5. Principal constant-volume closed-system solution for detachment folding in a Jura-like mechanical stratigraphy predicts fold shapes lying along the curve (equation 4). Many natural detachment folds do not agree with this solution, including the fold of the Appalachian Plateau studied by Wiltschko and Chapple (1977) and the Yaken anticline in the southern Tianshan (Hubert-Ferrari et al., 2005).

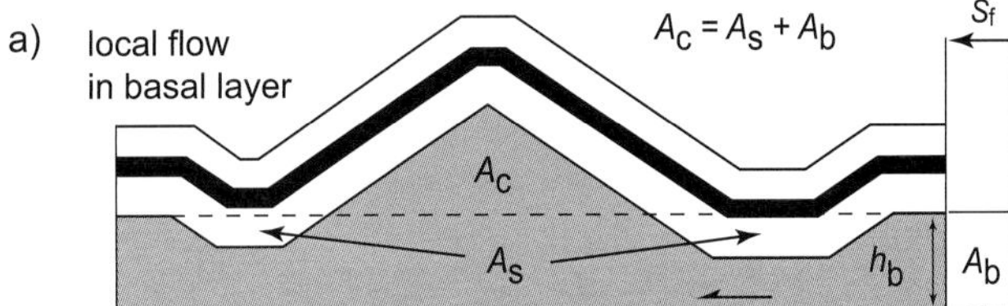

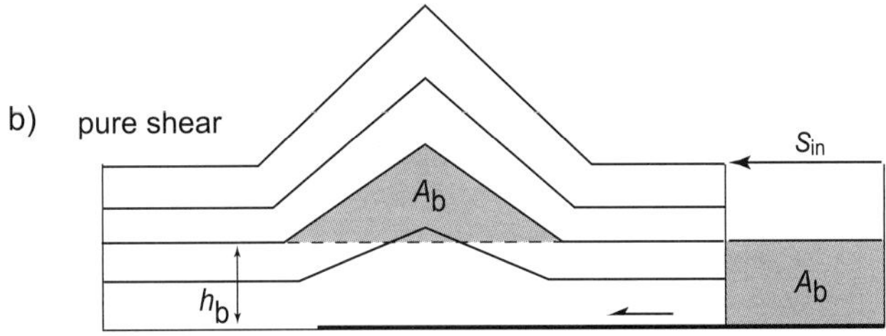

Figure 6. Local-flow and pure-shear solutions to the problem of mass balance in detachment folding.

layer thickness must be conserved. This solution was proposed by Groshong and Epard (1994) and Epard and Groshong (1995) who demonstrated that strain, especially at low amplitude, has a large impact on the problem of detachment folding. No weak basal layer is required for the pure-shear solution, although models that include both pure shear and local flow of a basal layer have been developed successfully (Atkinson and Wallace, 2003). Significant pure shear is likely even in the Appalachian Plateau examples studied by Wiltschko and Chapple (1977) based on the widespread observation of layer-parallel pure-shear shortening in this region (Engelder and Engelder, 1977; Geiser and Engelder, 1983). In some structures, layer-parallel horizontal compaction is quantitatively important in addition to constant-volume pure shear (Gonzalez-Mieres and Suppe, 2006).

The essence of the pure-shear solution (Figure 6b) is that, in addition to the mean flexural component of shortening, $\overline{S}_f$, which is $2L(1 - \cos\delta)$ in equation 2, a mean layer-parallel strain component, $\overline{S}_\varepsilon$, is observed at every stratigraphic height, h_b, above the detachment that is sufficient to cause the area of shortening to be equal to the area of the fold core, $A_c = A_b$. Therefore, the area of shortening of a layer of height h_b above the detachment becomes the sum of a mean flexural component, $\overline{S}_f = 2L(1 - \cos\delta)$, and a mean strain component, $\overline{S}_\varepsilon = 2L\left[\left(1/\overline{\mathbf{S}}_\varepsilon\right) - 1\right]$.

$$A_b = 2\overline{S}_f h_b + 2\overline{S}_\varepsilon h_b = 2h_b L\left[\left(1/\overline{\mathbf{S}}_\varepsilon\right) - \cos\delta\right] \quad (6)$$

where the mean layer-parallel horizontal strain (stretch) for the layer is $\overline{\mathbf{S}}_\varepsilon = L/L_o$ and L_o is the undeformed limb length. By equating with equation 1, we find the mean strain (stretch) required for the conservation of area:

$$\overline{\mathbf{S}}_\varepsilon = \frac{1}{(L/2h_b)\sin\delta\cos\delta + \cos\delta} \quad (7)$$

A key prediction of the pure-shear theory that follows from equation 7 is that, at low-limb dips (~10°), the mean flexural or bed-length shortening, $\overline{S}_f$, should be one to two orders of magnitude smaller than the total shortening, $\overline{S} = \overline{S}_f + \overline{S}_\varepsilon$.

Several detachment folds have been shown to agree well with the predictions of the pure-shear mechanism (Groshong and Epard, 1994; Epard and Groshong, 1995; Hubert-Ferrari et al., 2005; Gonzalez-Mieres and Suppe, 2006). In particular, these pure-shear folds at low amplitude show a flexural or bed-length shortening that is one to two orders of magnitude less than the total shortening (Gonzalez-Mieres and Suppe, 2006). In essence, if pure shear dominates the shortening, there is little flexural deflection of the layers beyond what is required by the change in magnitude of pure shear from the crest to the flanks of the fold.

An example of a fold that agrees well with the predictions of these classic solutions is the actively growing 100-km-long Yaken anticline in the southern Tianshan thrust belt of western China, which is shown in Figure 7 (Hubert-Ferrari et al., 2005, 2007; Gonzalez-Mieres and Suppe, 2006, 2011). By studying the area of the fold core and the bed-length shortening as a function of height above the detachment, it has been shown that Yaken has a total shortening, $\overline{S} = \overline{S}_f + \overline{S}_\varepsilon$, of 1200 m (3937 ft) that is consumed almost entirely by layer-parallel simple shear. The component of bed-length shortening, $\overline{S}_f$, is only about 50 m (164 ft) (Gonzalez-Mieres and Suppe, 2006). In addition, the Yaken anticline has a basal evaporitic detachment layer that has an apparent excess area of about 0.8 km^2 (0.3 mi^2), which is ascribed to the flow of the basal detachment layer (Hubert-Ferrari et al., 2005). This excess area is about 25% of the area of the fold core at a height of 2 km (1.2 mi) above the detachment, so it is quantitatively important. Therefore, the Yaken anticline combines the two classic mechanisms of detachment folding, basal flow and pure shear.

THREE NONCLASSICAL SOLUTIONS

The classic formulations and solutions to the problem of mass balance in detachment folds have all been closed-system solutions, such as Figures 2 and 6. These solutions consider that all the shortening entering the confines of the fold is consumed locally in fold growth

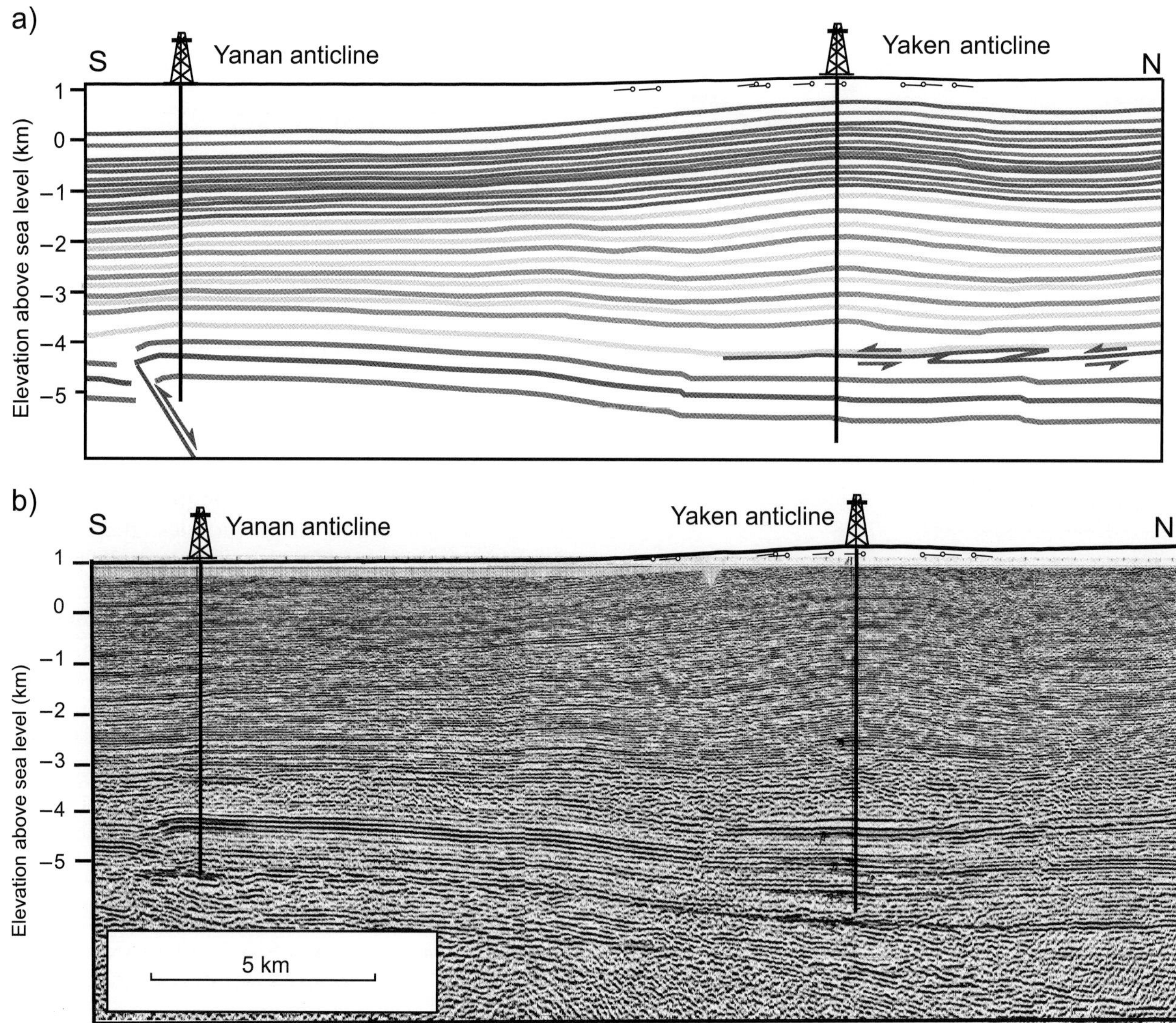

Figure 7. Yaken detachment fold (modified from Hubert-Ferrari et al., 2005).

and that all the excess area of the fold core is locally derived, as in the local-flow solution (Figure 6a). Here we consider several additional geologically likely solutions in which the apparent excess area, ΔA_e, of the fold core is associated with an open-system mass balance. Any of these solutions can be combined with the classical mechanisms of pure shear and local flow in actual structures.

Figure 8 shows the mass balance of three open-system solutions. If we were to analyze the folds assuming closed-system behavior, all would have the same apparent excess area, ΔA_e, based on a mismatch between the local flexural shortening, S_f, of the flexural lid and shortening of the basal layer, $\overline{S}_b$. They also have the same flexural shortening, $S_f = 2h_b L(1 - \cos\delta)$, within the lid at and above the top of the basal layer of thickness h_b, which leads to the same apparent area of shortening of the basal layer that is consumed in the fold, $A_b = \overline{S}_b h_b$, where $\overline{S}_b = S_f$. Finally, the area of the fold core, A_c, at the top of a basal layer is identical in each solution and equal to the sum of the apparent basal shortening and the apparent excess area, $A_c = A_b + \Delta A_e$.

In addition, for these open-system nonclassical solutions, we need to consider the displacement or shortening, S_{in}, that enters the fold above the basal layer and the shortening that exits the fold to be taken up elsewhere, S_{out} (Figure 8). The difference between S_{in} and S_{out} is the shortening consumed locally in flexural folding at and above the top of the basal layer, S_f. The three solutions are as follows, each of which allows for different but compatible shortening of the weak basal layer and the competent flexural lid.

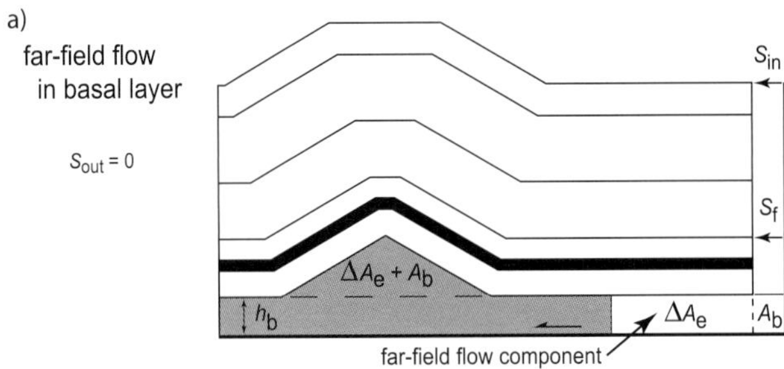

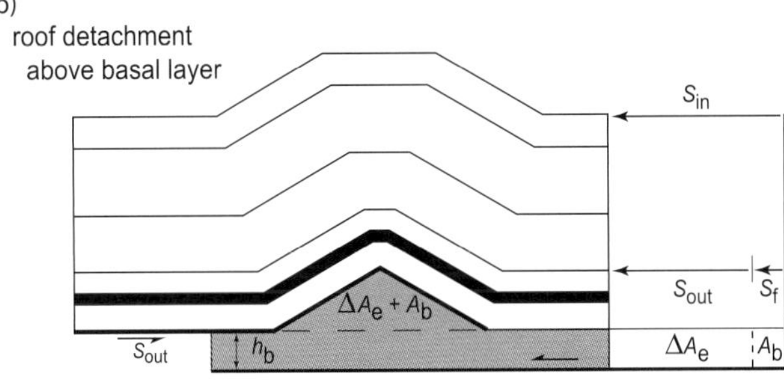

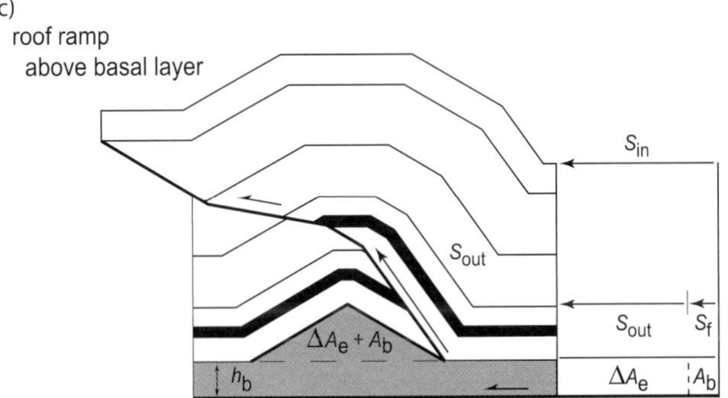

Figure 8. Mass balance in three types of open-system detachment folds.

Far-field Flow Solution

In the solution of far-field flow (Figure 8a), an injection of material in the basal layer from outside the two-dimensional system in the amount of the excess area, ΔA_e, is observed. Therefore, the effective area of shortening of the basal layer is $A_b + \Delta A_e$ and equal to the area of the fold core at the top of the basal layer, A_c. The shortening of the flexural lid, S_f, is equal to the input shortening, S_{in}, because no output shortening of the flexural lid exists, $S_{out} = 0$. In contrast with the classic local-flow solution discussed above, no withdrawal of material from the adjacent synclines is observed (compare Figures 6a, 8a). However, in actual structures, there could be a complete continuum of solutions between the end-member local-flow and far-field flow solutions.

The Agbami anticline in offshore Niger Delta (Figure 9a) has been interpreted as dominated by far-field flow in the basal detachment layer of the over-pressured Akata shale (Bilotti et al., 2005; Gonzalez-Mieres and Suppe, 2006). The shortening of the strata above the top of the Akata Formation is 2.4 km (1.5 mi), of which 1.5 km (0.9 mi) is flexural and 0.93 km (0.58 mi) is pure-shear bedding-parallel shortening (Gonzalez-Mieres and Suppe, 2006). This measured shortening within the fold would supply only 2.4–6 km^2 (0.9–2.3 mi^2) of the observed 10-km^2 (3.9-mi^2) area of the fold core at the top of the Akata Formation. (The uncertainty in the area of shortening comes from the unknown thickness of the basal layer caused by lack of clear imaging within the Akata Formation; the thickness of the basal layer is only constrained to be 1–2.5 km [0.6–1.5 mi], the upper bound given by the depth to the top oceanic crust, which is well imaged.) Therefore, an excess area of the fold core of 4 to 7.6 km^2 (1.5 to 2.9 mi^2) exists. The classic solution of local flow of shale in the basal layer can be ruled out because the well-imaged syncline to the west of Agbami does not show the required relief in thickness of the Akata Formation or relief in thickness within the growth strata. In contrast, the far-field flow solution is kinematically possible for Agbami (Figure 9a). However, such far-field flow of shale, which is sometimes contemplated by analogy with flow of salt, is unfortunately not a well-tested hypothesis in the literature despite that shale diapirs have been interpreted for other structures of the Niger Delta (e.g., Hooper et al., 2002). Other open-system solutions are possible as discussed below.

Roof-detachment Solution

In the roof-detachment solution (Figure 8b), the total shortening of the basal layer is S_{in}, and the area of shortening of the basal layer is $S_{in}h_b = \Delta A_e + A_b$. In contrast, only a small fraction of the shortening, S_f/S_{in}, is consumed in the overlying flexural sequence, with most of the input shortening exiting the system along a detachment at the roof of the basal layer.

$$S_{out} = \frac{\Delta A_e}{h_b} = S_{in} - S_f \tag{8}$$

The roof-detachment solution functions like a duplex with a roof and floor detachment at the top and bottom of the basal layer. Note that the excess area divided by the height of the layer, $\Delta A_e/h_b$, is the amount of slip exiting the structure.

The Agbami anticline can be interpreted using the roof-detachment solution (Figure 9b). The seismic image alone does not allow us to distinguish between the far-field flow and roof-detachment solutions; more regional data are required. Well-developed detachments near the top and base of the Akata Formation are imaged regionally (Corredor et al., 2005a, b) in agreement

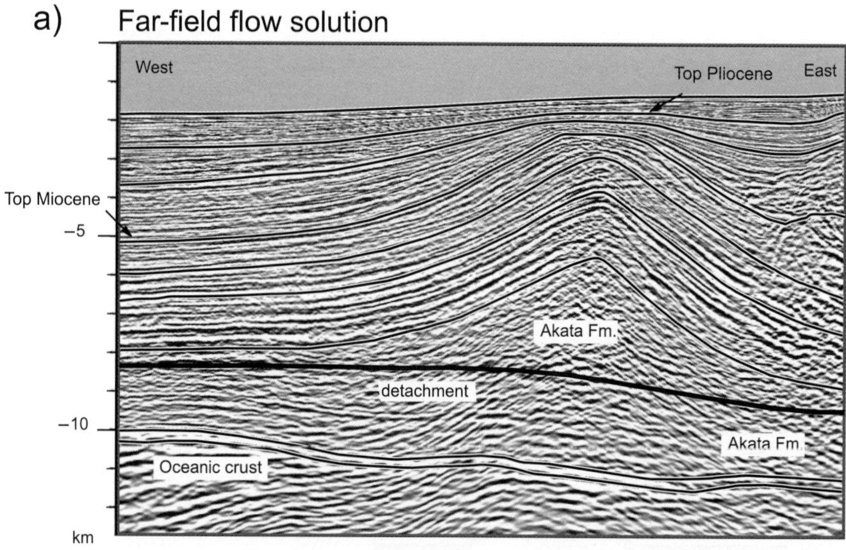

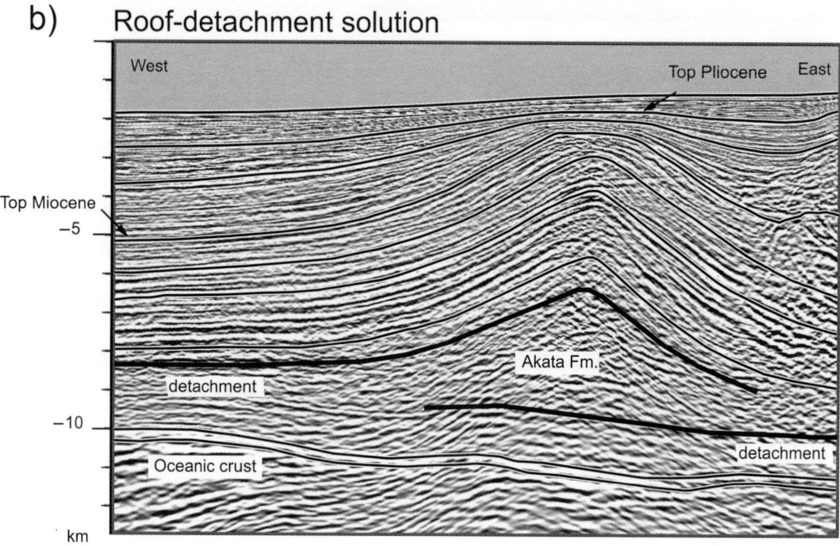

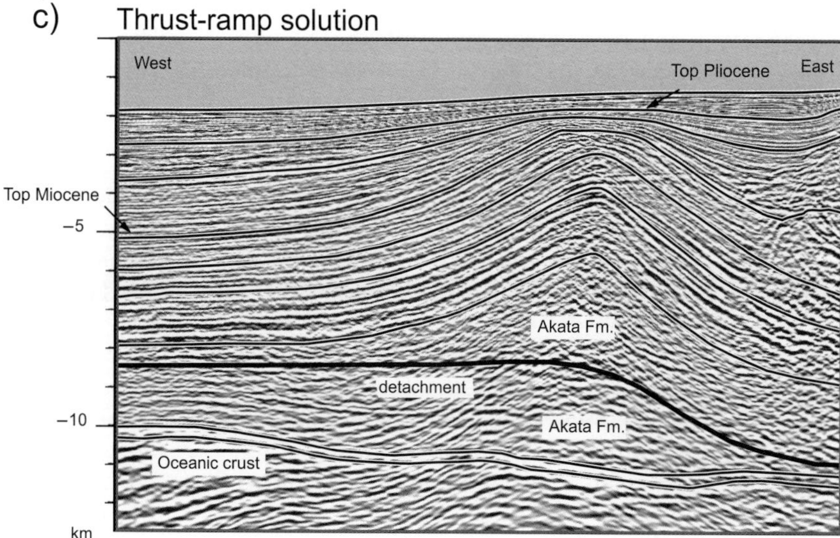

Figure 9. Three open-system solutions to the mass balance of the Agbami anticline, deep-water Niger Delta (seismic data from Bilotti et al., 2005). Fm. = Formation.

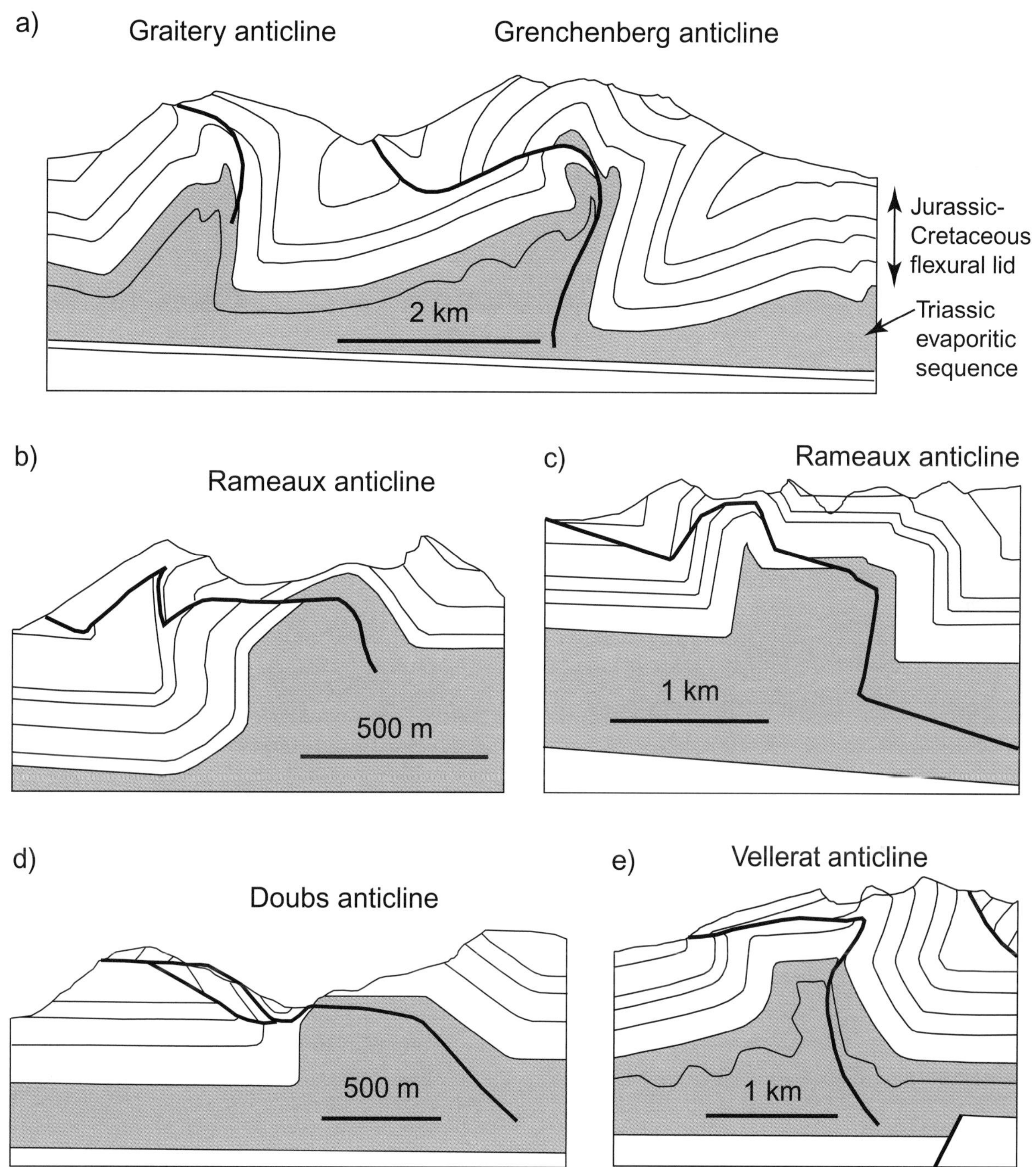

Figure 10. Folded thrust ramps in the competent roof sequence of the Jura fold belt. (a) Grenchenberg anticline of Buxtorf (1916). (b, c) Raimeux anticline (modified from Laubscher, 1977a). (d) Doubs anticline (modified from M. Suter in Laubscher, 1977b). (e) Vellerat anticline (modified from Heckendorn, 1974). Note that with the exception of the Grenchenberg anticline, these folded thrust ramps were observed only in outcrop, although some benefit from fold plunges. The fault interpretations are shown as published but could be modified in light of the roof-ramp model (Figures 8b, 14) to root into a detachment near the top of the weak basal layer (see also Figure 11).

with the roof-detachment solution. The amount of slip that is predicted to exit Agbami along the roof detachment is 2.4 km (1.5 mi) from equation 8. The shortening of this magnitude associated with the near top of the Akata detachment is observed about 10 km (6.2 mi) to the west of the Agbami anticline (cf. seismic line in

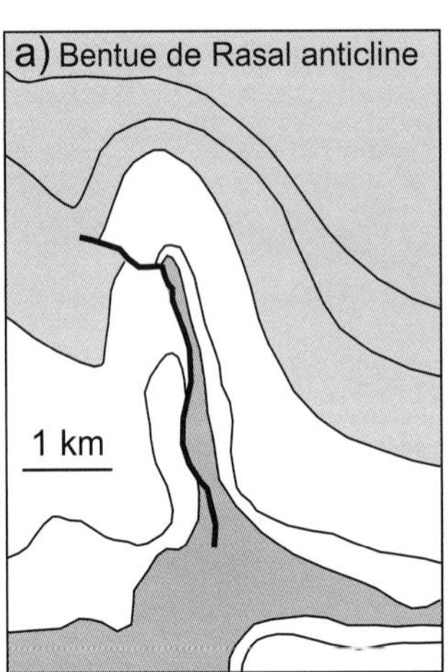

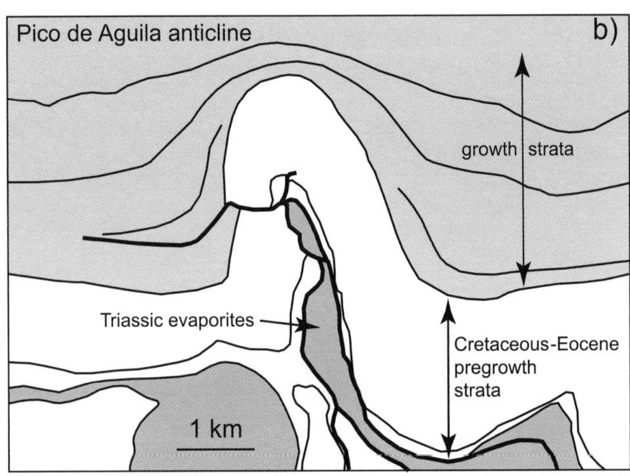

Figure 11. Roof ramps in the Bentue de Rasal and Pico de Aguila detachment folds Jaca basin of the Spanish Pyrenees. (a) Map of the north-plunging Bentue de Rasal detachment fold (compiled from Anastasio, 1987; Millán et al., 1994). (c) Map of the north-plunging Pico de Aguila detachment fold (compiled from Puigdefàbregas, 1975; Anastasio, 1987).

Shaw et al., 2005, p. 56), which suggests that the roof-detachment solution is probably correct.

The details of the internal shortening of the Akata Formation are not well imaged in the present seismic line of the Agbami anticline. As emphasized by Groshong and Epard (1994), the deformation in the cores of detachment folds can occur in a variety of forms and scales, including those that are not normally thought of in association with detachment folding, such as discrete fault-ramp imbrication and duplex structures. For example, the core of Agbami could be the effect of an approximately 4.8-km (3-mi) shortening in a 2-km (1.2-mi)-thick basal layer accomplished as a somewhat flattened simple or complex fault-bend fold at the Akata level with the roof strata deforming by a combination of flexure (1.5-km [0.9-mi] shortening) and layer-parallel pure shear (0.93-km [0.58-mi] shortening) (Figure 9c).

Roof-ramp Solution

The roof-ramp solution is very similar to the roof-detachment solution, but all shortening is consumed locally. In both cases, an output displacement, S_{out}, on a roof detachment exists. This displacement must eventually break to the surface in some structure in the roof sequence that consumes slip. In the roof-detachment solution, this roof structure is separated from the detachment fold in question. For example, it is consumed in roof structures about 10 km (6.2 mi) to the west in the case of Agbami. In contrast, in the roof-ramp solution, the roof structures are in the immediate vicinity of the detachment fold. In Figure 8c, the roof ramp directly overlies the detachment fold and is folded by it.

Note that a roof-ramp structure is not a preexisting structure folded by a later detachment fold, as was the interpretation of Buxtorf (1916) for the folded thrust in the Grenchenberg anticline. The amount of slip on the roof ramp, S_{out}, is directly determined at every step in the deformation by the ratio of excess area of the fold core to the thickness of the basal layer, as given by equation 8.

Many detachment folds are qualitatively similar to the roof-ramp model, although most examples are not imaged well enough to make a quantitative test through equation 8. The cross section of the Grenchenberg anticline of Buxtorf (1916) is qualitatively similar to the predictions of the roof-ramp solution (Figure 10a). Laubscher (1977a, b) argued that many Jura box folds contain complexly folded thrust faults similar to the Grenchenberg anticline (Figure 10b–d). The Vallerat anticline, which was one of the box folds interpreted by Buxtorf (Figure 1), contains a folded thrust fault (Figure 10e). All of these Jura structures are plausibly roof-ramp structures.

A region of classic box folds exists on the southern flank of the Jaca basin in the External Sierra of the Spanish Pyrenees where the north-plunging anticlines in map view appear as distorted cross sections by virtue of later refolding (Puigdefàbregas, 1975; Anastasio, 1987, 1992; Poblet and Hardy, 1995; Anastasio and Holl, 2001). The stratigraphy is similar to the Jura with a basal detachment layer dominated by Triassic evaporates overlain by more competent Mesozoic carbonates and lower Tertiary strata that record the growth of these structures. Several of these structures, including the Bentue de Rasal and Pico de Aguila anticlines, show folded thrust faults in their cores that step up from the Triassic evaporates to the top of the competent sequence, as shown in Figure 11 (Puigdefàbregas, 1975; Anastasio, 1987; Millán et al., 1994; Anastasio and Holl, 2001). These folded thrusts are plausibly roof-ramp structures.

A mountainside exposure of a thrust ramp emanating from a detachment horizon at the crest of a box fold in the Canadian Rockies is shown in Figure 12. The underlying shale-rich strata show apparent thickness changes, whereas the strata overlying the detachment are more competent carbonates. Therefore, a roof-ramp

Figure 12. Roof ramp in box fold in Canadian Rockies, Monkman Pass, British Columbia (photo courtesy of Robert Thompson). The view is approximately 1 km wide.

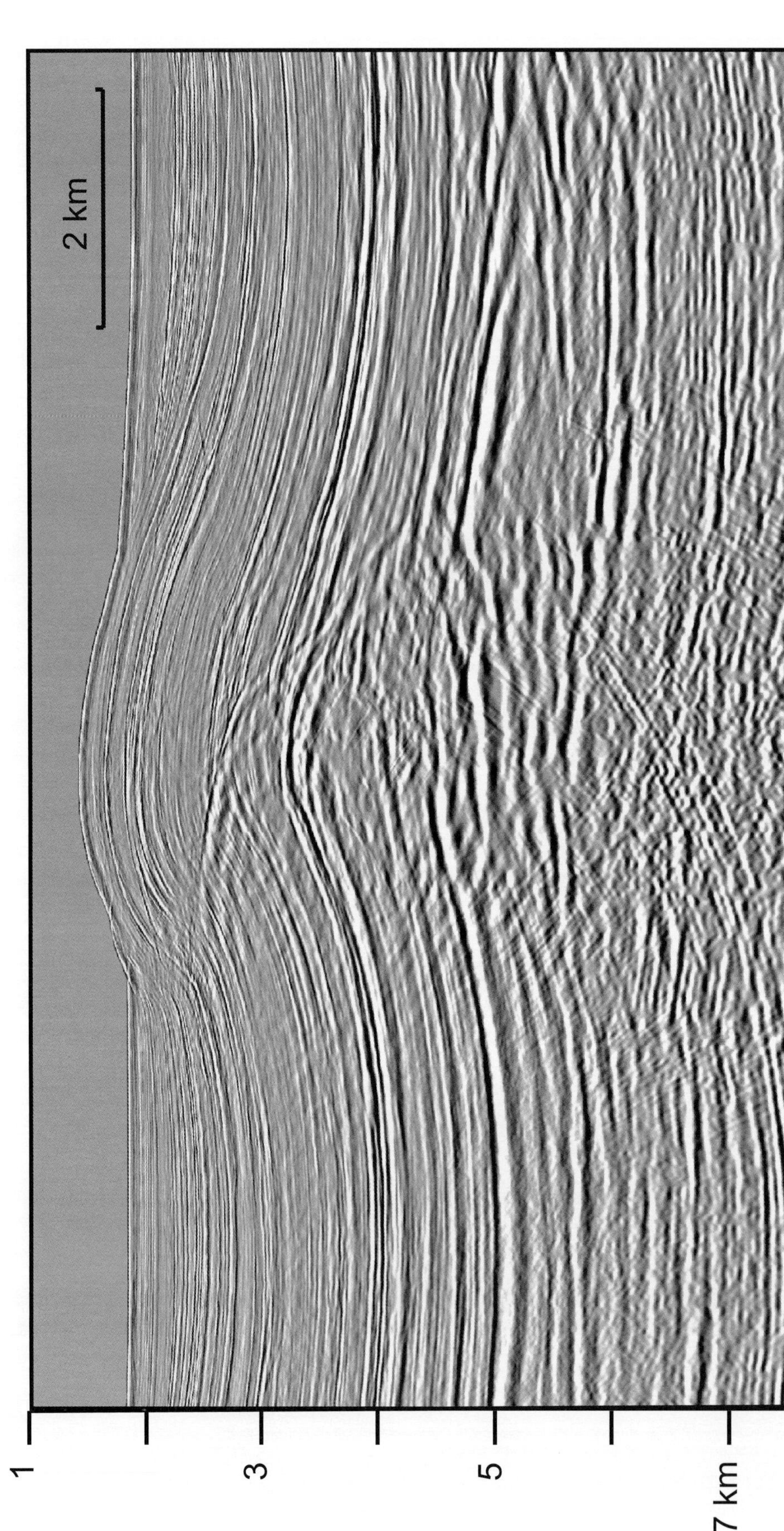

Figure 13. Roof ramp at 2.5- to 4.2-km (1.5- to 2.6-mi) depth in the Mexican Ridges fold belt, deep-water Gulf of Mexico (Román-Ramos et al., 2001).

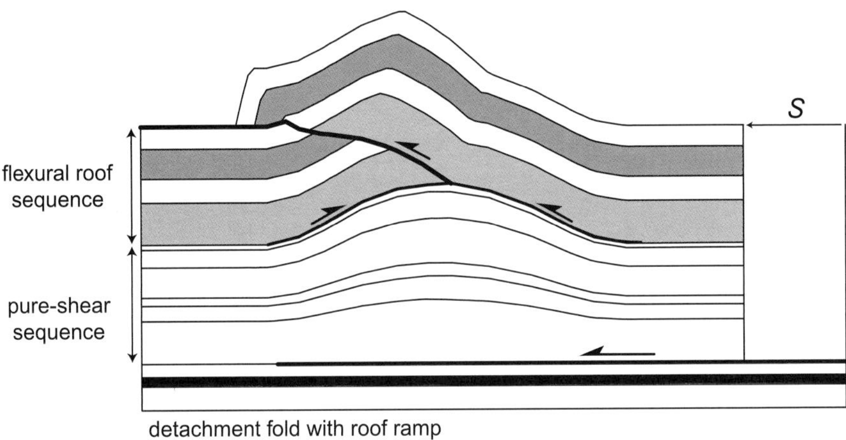

Figure 14. Model of a detachment fold in which the basal stratigraphy has deformed in pure shear and the upper stratigraphy by flexure, plus fault slip on a thrust ramp. The flexural shortening plus fault slip in the flexural lid equals the pure-shear shortening of the basal layer, as in Figure 8c.

interpretation seems plausible. Note that roof ramps at a variety of scales, controlled by the scales of mechanical stratigraphy, seem plausible (cf. Mitra, 2002).

A final example is a seismic image of a folded roof ramp from the southern Mexican Ridges fold belt offshore of Poza Rica, Veracruz, Mexico (Figure 13). The prominent folded thrust emanates from the upper of two regionally mapped detachments that bound the upper Eocene–upper Oligocene shale section in the Mexican Ridges fold belt (Román-Ramos et al., 2001, their figure 3). The structure is interpreted to be analogous to the simple balanced model shown in Figure 14 of a pure-shear detachment fold overlain by a competent sequence that deforms by a combination of flexure and fault-bend folding. In essence, more and less competent parts of the stratigraphic sequence may deform by different proportions of pure shear and flexure, with the difference being taken up by fault slip.

Although none of these examples of folded thrusts in detachment folds have been studied quantitatively, they indicated the widespread nature of the phenomenon. This observation, combined with the mass-balance considerations of detachment folding (Figures 6, 8), suggests that roof ramps should be a widespread process in detachment folding.

DISCUSSION

In this chapter, we have shown on the basis of mass-balance considerations and natural examples five plausible modes of detachment folding: (1) local flow of the weak basal layer, for example, salt (Wiltschko and Chapple, 1977); (2) pure-shear detachment folding (Groshong and Epard, 1994); (3) far-field flow of the weak basal layer; (4) roof detachments above the basal layer; and (5) roof ramps above the basal layer. These five modes have the property of functioning as end-member solutions. Therefore, natural folds can form by linear combinations of these mechanisms. In our examples, both Yaken and Agbami anticlines have been shown quantitatively to simultaneously involve two or three of these mechanisms.

In closing, I want to emphasize that these mass-balance considerations say relatively little about the actual shapes of detachment folds or the mechanics of their formation. These issues await future developments, most likely involving a combination of quantitative analysis of well-imaged structures and mechanical modeling.

ACKNOWLEDGMENTS

I am grateful to Bob Thompson of the Geological Survey of Canada and Josep Poblet of the University of Oviedo for their thoughtful and careful reviews. In addition, I thank Bob Thompson and Mario Aranda for providing images of detachment folds.

REFERENCES CITED

Anastasio, D. J., 1987, Thrusting, halotectonics and sedimentation in the External Sierra, southern Pyrenees, Spain: Ph.D. thesis, Johns Hopkins University, Baltimore, Maryland, 181 p.

Anastasio, D. J., 1992, Structural evolution of the External Sierra, southern Pyrenees, Spain, *in* S. Mitra, G. W. Fisher, O. M. Phillips, S. M. Stanley, and D. F. Strobel, eds., Structural geology of fold and thrust belts: Baltimore, Maryland, Johns Hopkins University Press, 254 p.

Anastasio, D. J., and J. E. Holl, 2001, Transverse fold evolution in the External Sierra, southern Pyrenees, Spain: Journal of Structural Geology, v. 23, p. 379–392, doi:10.1016/S0191-8141(00)00102-4.

Atkinson, P. K., and W. K. Wallace, 2003, Competent unit

thickness variation in detachment folds of the northeastern Brooks Range, Alaska, geometric analysis and conceptual model: Journal of Structural Geology, v. 25, p. 1751–1771, doi:10.1016/S0191-8141(03)00003-8.

Bilotti, F., J. Shaw, R. Cupich, and R. Lakings, 2005, Detachment fold, Niger Delta, *in* J. Shaw, C. Connors, and J. Suppe, eds., Seismic interpretation of contractional fault-related folds: AAPG Studies in Geology 53, p. 156.

Buxtorf, A., 1916, Prognosen und befunde beim Hauenstembasis und Grenchenbergtunnel und die Bedeutung der letzen für die Geologie des Juragebirges: Verhandlungen der Naturforschenden Gesellschaft Basel, v. 27, p. 184–254, http://www.museums-gesellschaft.ch/tripoli/tr_geologie.html (accessed January 2, 2010).

Camerlo, R. H., and E. F. Benson, 2006, Geometric and seismic interpretation of the Perdido fold belt: Northwestern deep water Gulf of Mexico: AAPG Bulletin, v. 90, p. 363–386, doi:10.1306/10120505003.

Corredor, F., J. H. Shaw, and F. Bilotti, 2005a, Structural styles in the deep-water fold and thrust belts of the Niger Delta: AAPG Bulletin, v. 89, p. 753–780, doi:10.1306/02170504074.

Corredor, F., J. H. Shaw, and J. Suppe, 2005b, Shear fault-bend folding, deep-water Niger Delta, *in* J. H. Shaw, C. Connors, and J. Suppe, eds., Seismic interpretation of contractional fault-related folds: AAPG Studies in Geology 53, p. 87–92.

Engelder, T., and R. Engelder, 1977, Fossil distortion and decollement tectonics of the Appalachian Plateau: Geology, v. 5, p. 457–460, doi:10.1130/0091-7613(1977)5<457:FDADTO>2.0.CO;2.

Epard, J.-L., and R. Groshong, 1995, Kinematic model of detachment folding including limb rotation, fixed hinges and layer-parallel strain: Tectonophysics, v. 247, p. 85–103, doi:10.1016/0040-1951(94)00266-C.

Geiser, P., and T. Engelder, 1983, The distribution of layer parallel shortening in the Appalachian foreland of New York and Pennsylvania: Evidence for two non-coaxial phases of deformation of the Alleghany orogeny, *in* R. D. Hatcher Jr., H. Williams, and I. Zeitz, eds., Contributions to the tectonics and geophysics of mountain chains: Geological Society of America Memoir 158, p. 161–175.

Gonzalez-Mieres, R., and J. Suppe, 2006, Relief and shortening in detachment folds: Journal of Structural Geology, v. 28, p. 1785–1807, doi:10.1016/j.jsg.2006.07.001.

Gonzalez-Mieres, R., and J. Suppe, 2011, Shortening histories in active detachment folds based on area-of-relief methods, *in* K. R. McClay, J. H. Shaw, and J. Suppe, eds., Thrust fault-related folding: AAPG Memoir 94, p. 39–67.

Groshong Jr., R. H., and J. L. Epard, 1994, Role of strain in area-constant detachment folding: Journal of Structural Geology, v. 16, p. 613–618, doi:10.1016/0191-8141(94)90113-9.

Hardy, S., and J. Poblet, 1994, Geometric and numerical model of progressive limb rotation in detachment folds: Geology, v. 22, p. 371–374, doi:10.1130/0091-7613(1994)022<0371:GANMOP>2.3.CO;2.

Heckendorn, W., 1974, Zur Tecktonik der Vellerat-Antiklinale (Berner Jura): Beiträge zur Geologishen Karte der Schweiz, v. NF147, 49 p.

Homza, T. X., and W. K. Wallace, 1995, Geometric and kinematic models for detachment folds with fixed and variable detachment depths: Journal of Structural Geology, v. 17, p. 575–588, doi:10.1016/0191-8141(94)00077-D.

Homza, T. X., and W. K. Wallace, 1997, Detachment folds with fixed hinges and variable detachment depth, northeast Brooks Range, Alaska: Journal of Structural Geology, v. 19, p. 337–354, doi:10.1016/S0191-8141(96)00118-6.

Hooper, R. J., R. J. Fitzsimmons, N. Grant, and B. C. Vendeville, 2002, The role of deformation in controlling depositional patterns in the south-central Niger Delta, west Africa: Journal of Structural Geology, v. 24, p. 847–859, doi:10.1016/S0191-8141(01)00122-5.

Hubert-Ferrari, A., J. Suppe, X. Wang, and C. Jia, 2005, The Yakeng detachment fold, south Tianshan, China, *in* J. Shaw, C. Connors, and J. Suppe, eds., Seismic interpretation of contractional fault-related folds: AAPG Studies in Geology 53, 156 p.

Hubert-Ferrari, A., J. Suppe, R. Gonzalez-Mieres, and X. Wang, 2007, Mechanisms of active folding of the landscape (southern Tianshan, China): Journal of Geophysical Research, v. 112, no. B03S09, 39 p., doi:10.1029/2006JB004362.

Jamison, W. R., 1987, Geometric analysis of fold development in overthrust terranes: Journal of Structural Geology, v. 9, p. 207–219, doi:10.1016/0191-8141(87)90026-5.

Laubscher, H. P., 1962, Die Zweiphasenhypothese der Jurafaltung: Ecologae Geologicae Helvetiae, v. 55, p. 1–22.

Laubscher, H. P., 1965, Ein Kinematisches Modell der Jurafaltung: Ecologae Geologicae Helvetiae, v. 58, p. 231–318.

Laubscher, H. P., 1977a, An intriguing example of a folded thrust in the Jura: Ecologae Geologicae Helvetiae, v. 70, p. 97–104.

Laubscher, H. P., 1977b, Fold development in the Jura: Tectonophysics, v. 37, p. 337–362, doi:10.1016/0040-1951(77)90056-7.

Millán, H., M. Aurell, and A. Meléndez, 1994, Synchronous detachment folds and coeval sedimentation in the Prepyrenean External Sierras (Spain): A case study for a tectonic origin of sequences and system tracts: Sedimentology, v. 41, p. 1001–1024, doi:10.1111/j.1365-3091.1994.tb01437.x.

Mitra, S., 2002, Structural models of faulted detachment folds: AAPG Bulletin, v. 61, p. 653–670.

Mitra, S., 2003, A unified kinematic model for the evolution of detachment folds: Journal of Structural Geology, v. 86, p. 1673–1694.

Mitra, S., and J. Namson, 1989, Equal-area balancing: American Journal of Science, v. 289, p. 563–599.

Poblet, J., and S. Hardy, 1995, Reverse modeling of detachment folds, application to the Pico del Aguila anticline in the south central Pyrenees (Spain): Journal of Structural Geology, v. 17, p. 1707–1724, doi:10.1016/0191-8141(95)00059-M.

Poblet, J., and K. McClay, 1996, Geometry and kinematics of single-layer detachment folds: AAPG Bulletin, v. 80, p. 1085–1109.

Puigdefàbregas, C., 1975, La sedimentación molásica en la cuenca de Jaca: Pirineos, v. 104, p. 1–188.

Rico, L., 1999, Geometric and kinematic evolution of a complete detachment fold in a natural cross-section: M.S. thesis, University of Texas, Austin, 294 p.

Román-Ramos, J. R., L. E. Salomón-Mora, M. Aranda-García, C. Rosas-Lara, and T. Cárdenas-Domínguez, 2001, Structural styles and gasiferous potential at the western slope of the Gulf of México: 4th Associación Mexicana de Geólogos Petroleros/AAPG International Hedberg Meeting, Veracruz, Mexico, 7 p.

Shaw, J. H., C. Connors, and J. Suppe, 2005, Part I: Structural interpretation methods, *in* J. H. Shaw, C. Connors, and J. Suppe, eds., Seismic interpretation of contractional fault-related folds: AAPG Studies in Geology 53, p. 1–58.

Sommaruga, A., 1997, Geology of the central Jura and the Molasse Basin: New insight into an evaporite-based foreland fold and thrust belt: Mémoire de la Société Neuchételoise des Sciences Naturelles 12, 176 p.

Wiltschko, D. V., and W. M. Chapple, 1977, Flow of weak rocks in Appalachian Plateau folds: AAPG Bulletin, v. 61, p. 653–670.

3

Gonzalez-Mieres, Ramon, and John Suppe, 2011, Shortening histories in active detachment folds based on area-of-relief methods, *in* K. McClay, J. H. Shaw, and J. Suppe, eds., Thrust fault-related folding: AAPG Memoir 94, p. 39–67.

Shortening Histories in Active Detachment Folds Based on Area-of-relief Methods

Ramon Gonzalez-Mieres[1]

Department of Geosciences, Princeton University, Princeton, New Jersey, U.S.A.

John Suppe[2]

Department of Geosciences, National Taiwan University, Taipei, Taiwan

ABSTRACT

The thickness variations and shapes of strata deposited over actively growing folds provide a convolved record of the interaction of deformation and sedimentation. Here we show how key elements of the history of shortening can be extracted using area-of-relief measurements on many stratigraphic horizons in well-imaged structures. In pregrowth strata, the shortening is equal to the vertical gradient in the area of structural relief, $S = \mathrm{d}A/\mathrm{d}z$. In growth strata, this relationship must be modified because the observed gradient in the area of relief now includes the effects of both structural thickening and stratigraphic thinning caused by deposition over the growing structure, $S = (\mathrm{d}A/\mathrm{d}z)_{\mathrm{obs}} - (\mathrm{d}A/\mathrm{d}z)_{\mathrm{strat}}$. This stratigraphic thinning cannot be measured directly from thickness variations because of later structural thickening but can be determined through consideration of the ratio of the mean shortening rate to the sedimentation rate, $\mathrm{d}\bar{S}/\mathrm{d}H$, which is a key parameter. We apply these concepts to a set of actively growing detachment folds: Nankai Trough, Japan; Cascadia, offshore Oregon; Yaken anticline, western China; and Agbami anticline, Niger Delta. These examples show a considerable diversity and complexity, including the effects of excess area in weak basal detachment layers, multiple detachment levels, and large differences in $\mathrm{d}\bar{S}/\mathrm{d}H$. All show not only constant shortening as a function of height in pregrowth strata and approximately constant shortening rates, $\mathrm{d}\bar{S}/\mathrm{d}H$, over substantial stratigraphic thicknesses, but also abrupt and large increases in shortening rate by factors of 5–10 and significant hiatuses in fold growth. The typical shortening rates in these frontal structures are in the range of 0.1–3 mm/yr (0.003–0.12 in./yr) and represent only 1–10% of the regional plate tectonic rates of their larger tectonic settings (2–6 cm/yr [0.8–2.4 in./yr]), indicating that shortening is not currently concentrated in these frontal zones of these mountain belts and accretionary wedges.

[1]*Present address*: Saudi Arabian Chevron, Chevron Global Upstream and Gas, Houston, Texas, U.S.A.
[2]*Also at*: Department of Geosciences, Princeton University, Princeton, New Jersey, U.S.A.

DOI:10.1306/13251332M943428

INTRODUCTION

Our primary source of information on the deformation histories of individual structures viewed at seismic resolution comes from the shapes and thickness variations of sedimentary layers deposited over the structures during deformation. These growth strata display distinctive geometric patterns that reflect both the rates of deformation relative to sedimentation and the specific underlying fold amplification mechanisms, such as limb rotation, kink-band migration, or more complex mechanisms (e.g., Riba, 1976; Suppe et al., 1992). The strategies that people have taken in analyzing growth strata can be roughly divided into two groups, although significant overlap exists. (1) To a considerable degree, growth strata have been analyzed qualitatively or semiquantitatively by comparing images of structures with simple kinematic models that seek to capture the essence of specific deformation mechanisms such as fault-bend folding, shear fault-bend folding, fault-propagation folding, trishear fault-propagation folding, and detachment folding (Suppe et al., 1992; Rowan et al., 1993; Poblet and Hardy, 1995; Hardy and Ford, 1997; Poblet et al., 1997; Rowan, 1997; Storti and Poblet, 1997; Allmendinger, 1998). This approach has been especially fruitful in demonstrating the existence of diverse kinematic mechanisms of fold growth and has in some cases provided quantitative data on timing and deformation rates (e.g., Shaw and Suppe, 1994). (2) An alternative approach has been to directly extract the history of growth from analysis of changes in elevation and layer thickness across a structure using well data or seismic images, independent of models of folding mechanisms. Such techniques are widely used in practical analysis, such as the classic expansion indices or vertical-separation diagrams across growth normal faults and in fault-seal analysis (Bischke, 1994; Bischke et al., 1999; Masaferro et al., 1999; Tearpock and Bischke, 2003; Poblet et al., 2004) or progressive bed-by-bed restoration (e.g., Nunns, 1991; Suppe et al., 1997; Novoa et al., 2000).

In this chapter, we adopt this second strategy of direct data analysis mostly independent of models of folding mechanisms. Our approach is to apply recent advances in shortening analysis based on measurements of the area of structural relief on many stratigraphic horizons in well-imaged structures. Here we show how these techniques can be applied to growth strata, with examples of detachment folds. We obtain high-resolution histories of detachment fold growth that not only show constant rates of shortening relative to sedimentation, but also show large changes in shortening rate as well as hiatuses in fold growth.

The modern analysis of the area of structural relief began with Epard and Groshong (1993), showing that it is possible to determine shortening with considerable accuracy by measuring the area of structural relief across a structure on many well-imaged stratigraphic horizons. The mean shortening, $\overline{S}$, is simply the change in area of structural relief, ΔA, over a stratigraphic interval divided by the difference in height, ΔH, within pregrowth strata, assuming conservation of area

$$\overline{S} = \Delta A/\Delta H \tag{1}$$

Typically, the mean shortening has been determined by linear regression of data from many horizons (e.g., Figure 1). These techniques have been applied to a variety of compressional and extensional structures (Epard and Groshong, 1993, 1995; Groshong, 1994, 1996; Pashin et al., 1995; Adam et al., 2004; Higuera-Díaz et al., 2005; Hubert-Ferrari et al., 2005; Gonzalez-Mieres and Suppe, 2006).

Note that equation 1 concerns only area balancing in two dimensions over a stratigraphic interval, ΔH. None of the traditional assumptions concerning conservation of bed length or layer thickness is made in this chapter. Nevertheless, any changes in bed length can be determined from line-length measurement in combination with area-of-relief data using the techniques developed by Gonzalez-Mieres and Suppe (2006), who applied these techniques to the pregrowth strata of the four examples of this chapter. The pregrowth strata of our first three examples show a flexural or bed-length shortening that is only a few percent of the total shortening, indicating that the shortening is almost entirely heterogeneous pure shear. In contrast, our final example shows layer-parallel stretching. More complex structures have been studied; the Pakuashan anticline analyzed by Yue et al. (2011) shows conservation of bed length at the lower stratigraphic levels and an approximately 50% mixture of flexural and pure-shear shortening at the upper stratigraphic levels. Our models conserve area and illustrate general properties of equation 1 applied to area balance in growth strata, independent of assumptions of conservation of bed length.

We limit our consideration in the following discussion specifically to detachment folding because it leads to somewhat simpler notation and because our examples are detachment folds. For this reason, we explicitly speak of the mean shortening, $\overline{S}$, as being the area of structural relief, A, of a horizon divided by its height, H, above the detachment horizon.

$$\overline{S} = A/H \tag{2}$$

Nevertheless, all the basic concepts that we present are more general; for example, they have been applied with appropriate modification to ramp anticlines (e.g.,

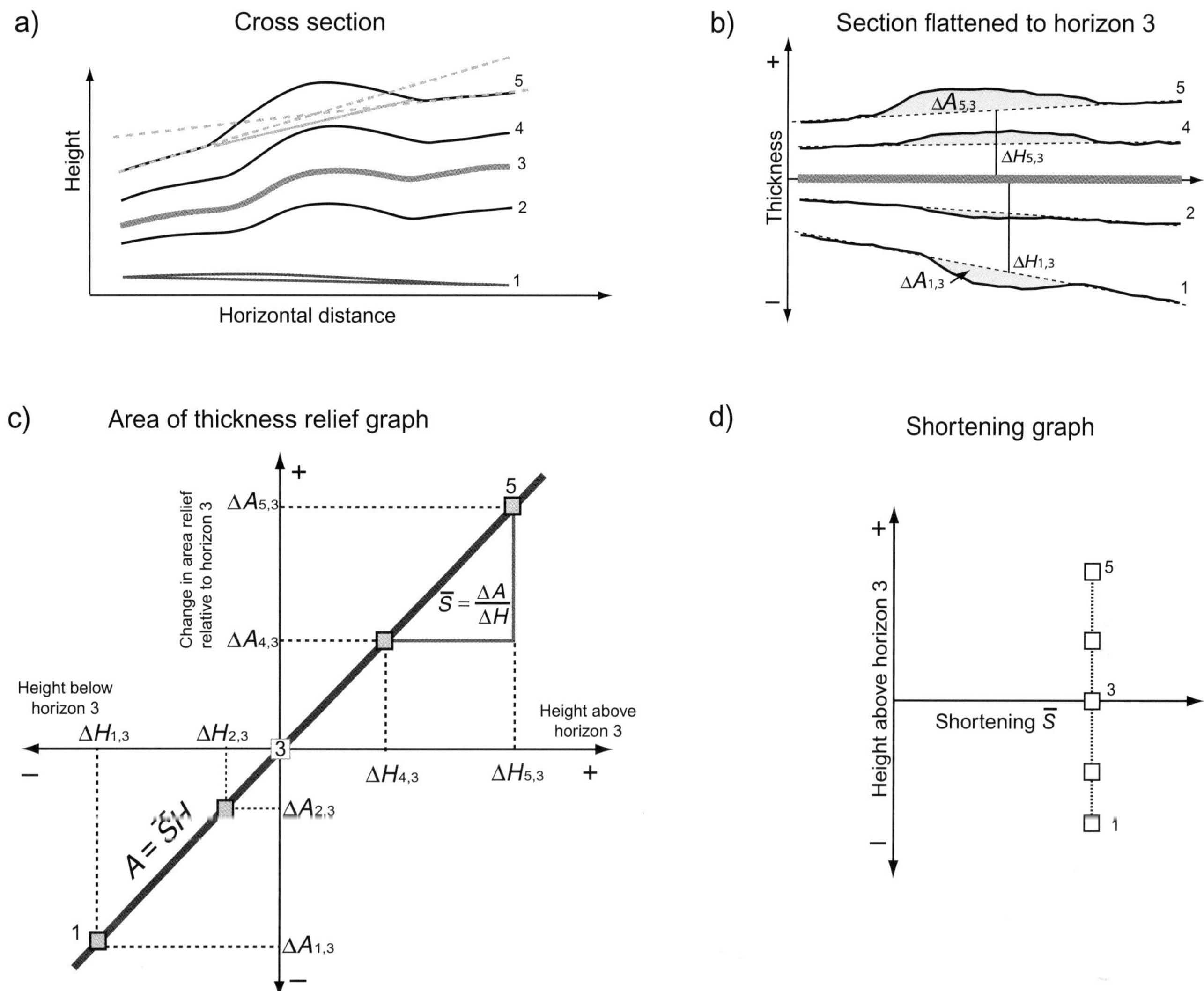

Figure 1. Thickness-relief method for measuring the area of relief and shortening. (a) The choice of the regional stratigraphic gradient of a horizon is required for measuring the area of relief but is commonly ambiguous in depth cross sections in contrast with simple models. The ambiguity exists because the present shape of a horizon is a convolution of primary stratigraphic geometry, the structure of interest, and other deformation below the structure. (b) As suggested by Hubert-Ferrari et al. (2005), these ambiguities in many cases disappear if the structure is flattened to a suitable stratigraphic horizon, in this case horizon 3. This flattening is done by vertical shear such that vertical thickness and area are conserved. In the flattened domain, the regional stratigraphic gradients can be relatively easily identified in many cases, such that the differences in the area of structural relief, ΔA, and height, ΔH, between nearby horizons can be measured unambiguously. (c, d) The mean shortening as a function of height is determined from the vertical gradient in the area of relief, $\overline{S} = \Delta A/\Delta H$, typically by linear regression. Modified from Gonzalez-Mieres and Suppe (2006).

Yue et al., 2011). The mean shortening can be directly calculated from equation 2 for each of the data points, $\langle A_i, H_i \rangle$, for both pregrowth and growth strata of simple detachment folds; however, not all detachment folds are simple, including most of our examples. In these more complex cases, we must adopt other strategies to estimate $\overline{S}$.

In practice, the area of structural relief is measured by subtracting the regional stratigraphic gradient from the present deformed shape of the horizon. However, the choice of the regional stratigraphic gradient is subjective (Figure 1a) and can have significant uncertainty because of primary stratigraphic thickness variations, warping of the basement, and interference from adjacent folds. For this reason, Hubert-Ferrari et al. (2005) introduced a technique of measuring the area of relief in what they called the thickness domain, which is simply the structure flattened to a well-behaved stratigraphic horizon while conserving area, which is accomplished by vertical simple shear, conserving vertical thickness (Figure 1b). Unambiguously identifying the regional stratigraphic gradients in the thickness domain is typically much easier. Typically, measurements of the area of relief are best made using different flattening horizons for different depositional sequences within the structure. The resulting measurements are changes in area of relief relative to an adjacent flattening horizon, which are then combined to form a comprehensive data set on a common basis giving area of relief as a function of height, using methods described by Gonzalez-Mieres and Suppe (2006). These so-called thickness-relief methods (Figure 1b–d) typically provide significantly more robust measurements than those obtained in the depth domain (Figure 1a), particularly in low-relief or stratigraphically complex structures.

If the data are sufficiently good, it is possible to determine not only the mean shortening, $\overline{S}$, discussed above, but also the shortening as a function of height, $S(z)$, which is the vertical gradient in the area of relief within pregrowth strata.

$$S(z) = (\mathrm{d}A/\mathrm{d}z)_{t\,=\,\mathrm{const}} \tag{3}$$

This equation is written as the vertical gradient in the area of relief at a fixed time (t) to emphasize that it holds only for pregrowth strata, in which time since inception of deformation is constant for all layers. Growth strata are more complex because time since inception is varying vertically (see equation 27).

Accurate determination of the shortening, $S(z)$, in pregrowth strata from equation 3 is somewhat more difficult in contrast with mean shortening, $\overline{S}(z)$, because our primary measurement is the area of structural relief, $A(z)$, which is the integral of the shortening.

$$A(z) = \overline{S}(z)H = \int_0^H S(z)\mathrm{d}z \tag{4}$$

Furthermore, we cannot reliably measure $S(z)$ from the deformed bed lengths because bed length and layer thickness are typically not conserved in these structures; for example, in our first three detachment folds, the bed-length or flexural shortening is only a few percent of the total shortening because they have been deformed by heterogeneous pure shear (Gonzalez-Mieres and Suppe, 2006).

Keep in mind the distinction between the shortening, $S(z)$, and the mean shortening, $\overline{S}(z)$, because the extent to which vertical gradients in shortening exist in structures is still an open question (e.g., Epard and Groshong, 1993; Suppe et al., 2004). In cases in which shortening is constant as a function of height, the mean shortening and the shortening become equal, $\overline{S} = S$. Methods for determining $S(z)$ from equation 3 are given by Gonzalez-Mieres and Suppe (2006), who applied them to the pregrowth strata of our examples. The distinction between mean shortening, $\overline{S}(z)$, and shortening, $S(z)$, is illustrated with a forward-kinematic model of a detachment fold (Figure 2a, b). The model is made slightly complex by including a basal zone of simple shear, which results in $\overline{S}(z) < S(z)$ at every stratigraphic level (Figure 2b). We also show $S(z)$ and $\overline{S}(z)$ for several intermediate stages in the deformation history of this model (deformation since t_1, t_2, and t_3 in Figure 2b).

APPLICATION TO GROWTH STRATA

We now consider how to determine shortening from a set of measurements of the area of structural relief as a function of height in growth strata, $\langle A_i, H_i \rangle$. The determination of present mean shortening for individual data points, $\overline{S}(z) = A(z)/H$, is the same for growth and pregrowth strata using equation 2. However, the other relationships between shortening and area of relief (equations 1, 3) no longer hold within growth strata because an additional process contributes to the change in area of relief across a growth interval. Here we present a brief explanation of the issues; a more complete development is given in the Appendix.

Let us consider the change in area of relief, $\Delta A_{n,n+1}$, across a layer or stratigraphic interval within a growth sequence bounded by mapped horizons n and $n + 1$, as shown in Figure 3. Here our simplified illustration, which assumes neither paleobathymetry nor compaction, shows that the change in relief between adjacent horizons is no longer entirely structural. Part of this relief is primary stratigraphic thickness variation caused by growth. During the time of deposition, Δt, of this layer, horizon n deforms by an amount $\Delta A_{n,\Delta t}$ controlled by processes below that horizon (Figure 3a). In contrast, horizon $n + 1$ at the top of the layer has not yet undergone any deformation. This reduces the observed change in area of relief across the layer, $\Delta A_{n,n+1}$, relative to the actual area of shortening of the layer, $S_{n+1}\Delta H_{n,n+1}$, that has occurred since the deposition of its upper horizon. We now have the observed change in area of relief across the growth layer (Figure 3) as the shortening component minus the stratigraphic component, still assuming that area is conserved for this illustration.

$$\Delta A_{n,n+1} = S_{n+1}\Delta H_{n,n+1} - \Delta A_{n,\Delta t} \tag{5}$$

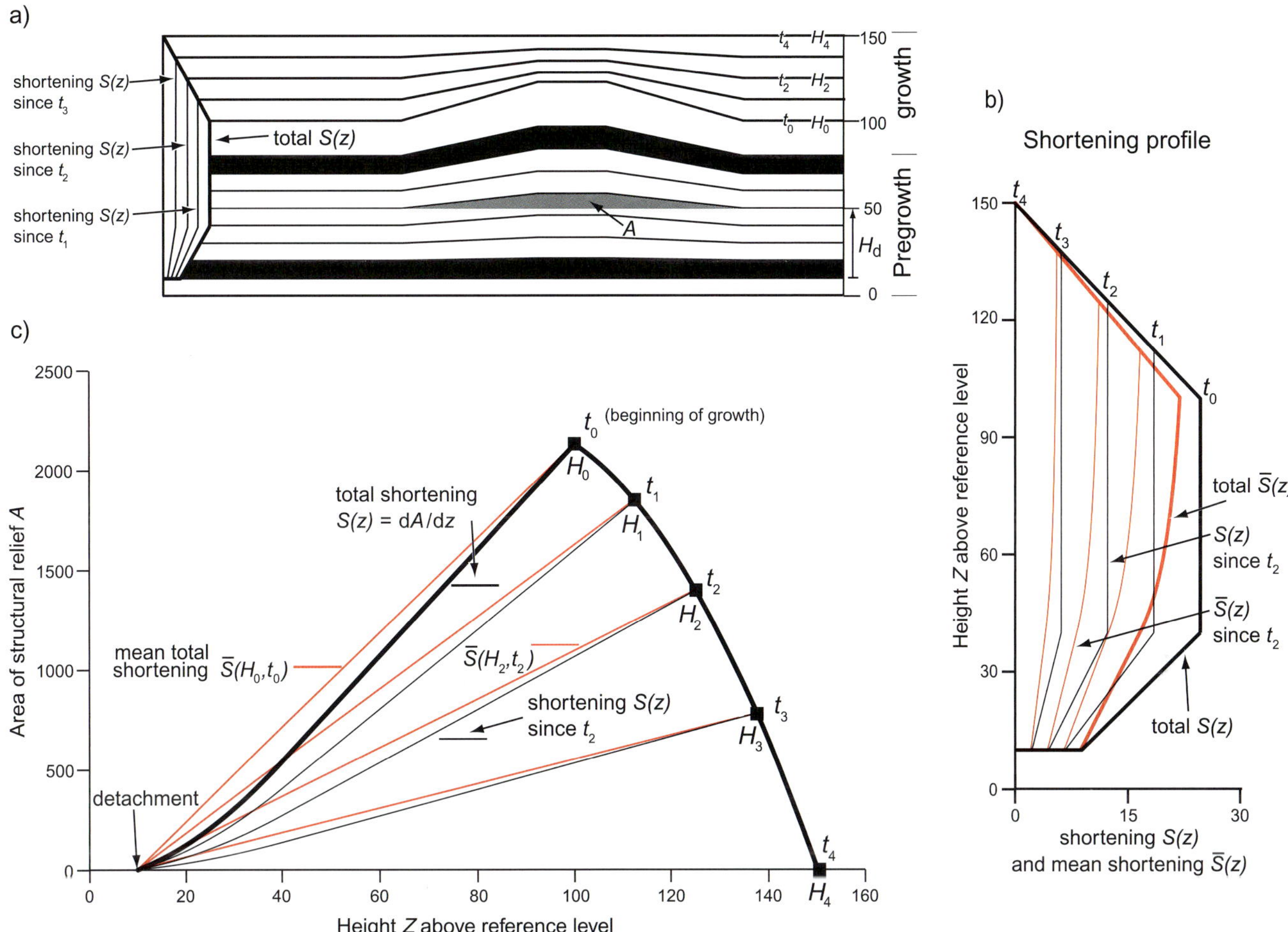

Figure 2. (a) Detachment fold model with a basal zone of simple shear above a detachment fault. Within the growth section, a linear upward decrease in shortening, $S(z)$, as a function of height, z, is observed. The height above the detachment is H_d with an associated area of relief, A. (b) Graph of shortening as a function of height. Note that shortening, $S(z)$ (black lines), and mean shortening, $\bar{S}(z)$ (red lines), are not equal because of the basal zone of simple shear. The heavy black and red lines give the present shortening (since t_0), which can be determined from area-of-relief data. The light black and red lines give the shortening since earlier times (t_1, t_2, t_3), which normally cannot be determined from relief data. (c) Graph of the area of structural relief, A, as a function of height above the reference level (bold line). The mean shortening is $\bar{S}(z) = A/H_d$, shown by the slopes of the red lines. In contrast, the shortening within the pregrowth strata is $S(z) = dA/dz$, which is everywhere greater than the mean shortening because of the simple shear. The area-of-relief profiles at three stages in the growth (since t_1, t_2, t_3) are given by the thin black lines, which are not observable with normal area-of-relief data.

These two components of the change in area of relief are shown in the shortening profile of the figure in the Appendix of this paper, which corresponds to the fold model shown in Figure 2.

We should to mention that these two components of the vertical change in area of relief cannot in general be directly observed. (1) The primary stratigraphic thickness variation cannot be measured directly using the present thickness variations because of later thickness changes due to deformation, which are important in many detachment folds. For example, the bed-length shortening in our examples is more than an order of magnitude smaller than the actual shortening; these folds conserve neither bed length nor layer thickness as shown by Gonzalez-Mieres and Suppe (2006). (2) The shortening, S_{n+1}, cannot be determined directly from the vertical gradient in the area of relief (equation 3) because that would require data on the net change in area of relief as a function of height at fixed time since deposition, which is not observable for growth strata, as shown in the Appendix of this chapter. Therefore, we take a different approach.

$\Delta A_{n,n+1} = -\Delta A_{n,\Delta t}$

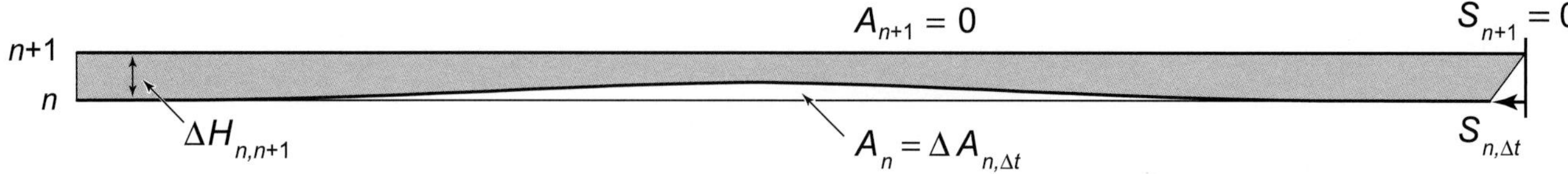

$\Delta A_{n,n+1} = S_{n+1}\Delta H_{n,n+1} - \Delta A_{n,\Delta t}$

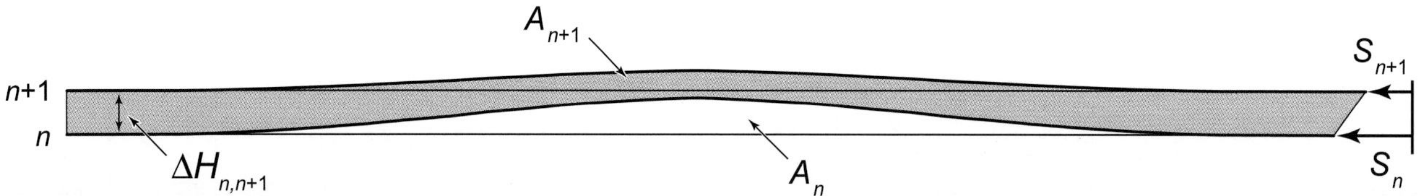

Figure 3. The change in the area of relief across a growth layer is the combination of stratigraphic and structural components. (a) During the time of deposition, Δt, of a growth layer, horizon n at its base has already accumulated an area of relief, $\Delta A_{n,\Delta t}$, when horizon $n+1$ at its top is just deposited and still undeformed. At this stage, the change in area of relief across the layer, $\Delta A_{n,n+1} = -\Delta A_{n,\Delta t}$, is entirely stratigraphic and decreasing upward. (b) After some shortening, S_{n+1}, the change in area of relief is composed of the combination of the structural and stratigraphic components, $\Delta A_{n,n+1} = S_{n+1}\Delta H_{n,n+1} - \Delta A_{n,\Delta t}$.

In the specific case of a simple detachment fold, which is a fold without excess area and with a single detachment level, we know that the stratigraphic component of the change in relief across a growth bed (Figure 3) is simply the change in mean shortening of the underlying structure during the deposition of the bed times the depth to detachment, $\Delta A_{n,\Delta t} = H_n \Delta \bar{S}_{n,\Delta t}$. Therefore, equation 5 can be recast for simple detachment folds as the vertical gradient in the observed area of relief within growth strata

$$(\mathrm{d}A/\mathrm{d}H)_{\mathrm{growth}} = S(z) - H\frac{\mathrm{d}\bar{S}}{\mathrm{d}H} \tag{6}$$

as outlined in the Appendix. The key new quantity in this equation is $\mathrm{d}\bar{S}/\mathrm{d}H$, which can be thought of as the instantaneous ratio of the mean shortening rate, $\mathrm{d}\bar{S}/\mathrm{d}t$, to the sedimentation rate, $\mathrm{d}H/\mathrm{d}t$, during growth (both are positive). One strategy that allows us to estimate the shortening $S(z)$ from equation 6 involves fitting models to data to find the best-fitting $\mathrm{d}\bar{S}/\mathrm{d}H$ ratios.

The importance of the ratio of shortening to sedimentation, $\mathrm{d}\bar{S}/\mathrm{d}H$, for understanding area-of-relief data, $\langle H_i, A_i\rangle$, in growth strata can be illustrated with simple models of detachment folds. Figure 4 shows a set of models with identical shortening within pregrowth strata but with different rates of upward decrease in mean shortening within the growth strata. At high rates of upward decrease in shortening (Figure 4a, high $H(\mathrm{d}\bar{S}/\mathrm{d}H)$), an upward decrease in the area of relief is observed ($\mathrm{d}A/\mathrm{d}H < 0$), as might be generally expected for growth strata because

$$\mathrm{d}A/\mathrm{d}H = S - H\frac{\mathrm{d}\bar{S}}{\mathrm{d}H} < 0$$

Thus, a maximum in area of relief at the pregrowth-growth boundary exists. In contrast, at high sedimentation rates relative to deformation rate (Figure 4b, c, low $H(\mathrm{d}\bar{S}/\mathrm{d}H)$), an initial upward increase in the area of relief ($\mathrm{d}A/\mathrm{d}H > 0$) exists such that

$$\mathrm{d}A/\mathrm{d}H = S - H\frac{\mathrm{d}\bar{S}}{\mathrm{d}H} > 0$$

A maximum in area of relief is reached at $\mathrm{d}A/\mathrm{d}H = 0$ where the competing structural and stratigraphic contributions to the relief are equal in magnitude, $S = H(\mathrm{d}\bar{S}/\mathrm{d}H)$, after which the area of relief decreases upward, $\mathrm{d}A/\mathrm{d}H < 0$. Several of our examples (Yaken and Agbami anticlines) show this phenomenon of a maximum in area of relief within the growth strata (cf. Figure 4b, c) instead of at the pregrowth-growth boundary (Figure 4a, b).

In the following section, we apply these concepts to four actively growing detachment folds: Nankai Trough, Japan; Cascadia, offshore Oregon; Yaken anticline,

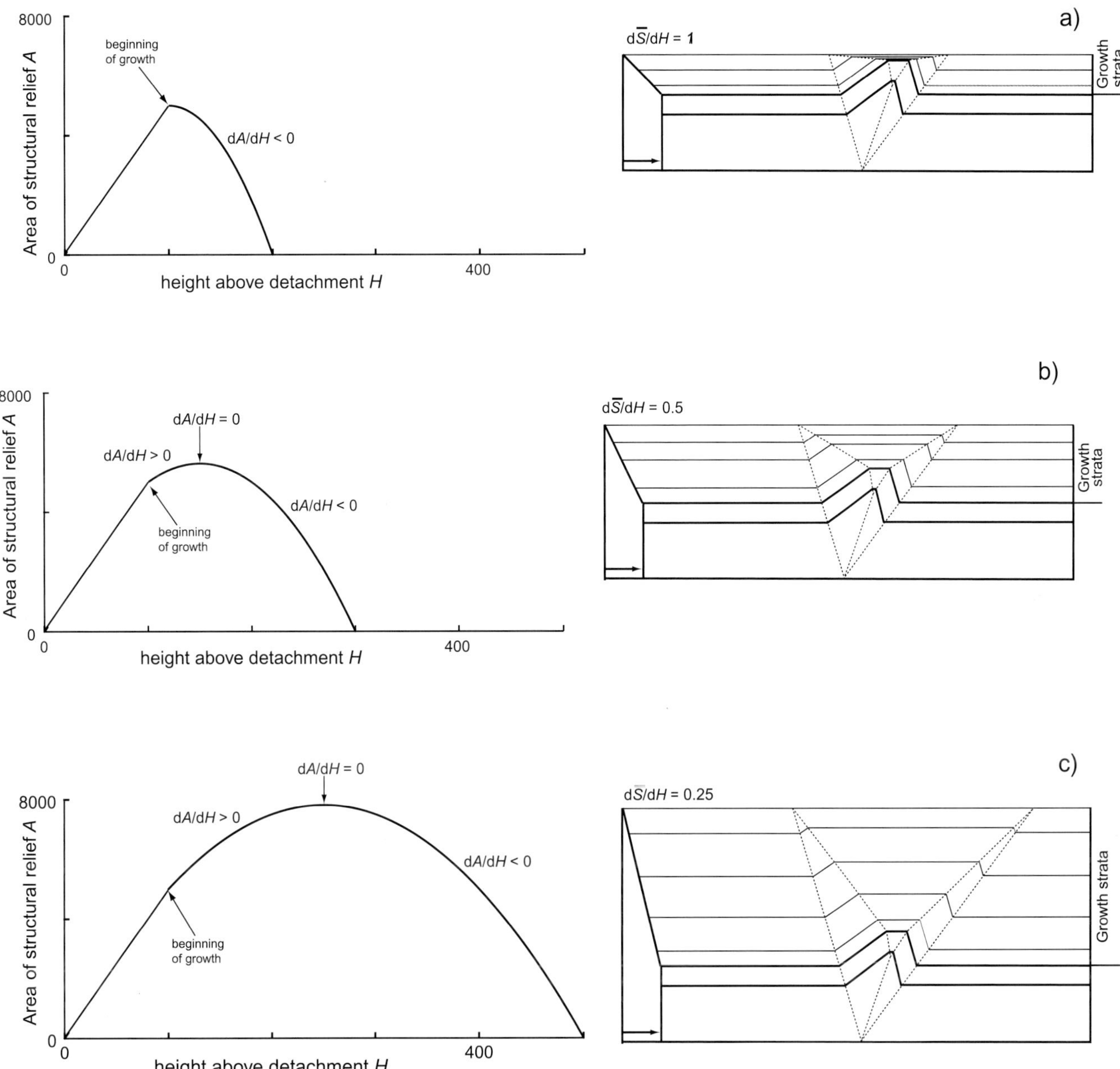

Figure 4. Effect of sedimentation on the rate of change of area of relief, dA/dH, in growth strata and on the shapes of area-of-relief graphs. Three detachment fold models are shown, all of which have the same total shortening, $\overline{S}$, in pregrowth strata and the same constant shortening rate, $d\overline{S}/dt$, but different constant sedimentation rates, dH/dt, which gives different rates of upward decrease in shortening, $d\overline{S}/dH$. (a) At low sedimentation rates, the area of relief decreases rapidly upward within the growth section, $dA/dH < 0$, and the maximum area of relief marks the beginning of growth. (c, d) At sufficiently high sedimentation rates, the area of relief initially increases upward within the growth section, $dA/dH > 0$, until a maximum in the area of relief is reached when $dA/dH = 0$, which is followed by an upward decrease in relief, $dA/dH < 0$. These models are qualitatively similar to our examples. This illustration is for a detachment fold model that conserves bed length (modified from Poblet et al., 1997); however, the behavior is quite general because it is simply a reflection of area balance with increasing thickness (see equation 6). Similar behaviors are shown by our examples, which do not conserve bed length.

western China; and Agbami anticline, Niger Delta. In a previous publication, we analyzed the pregrowth shortening of these structures in considerable detail (Gonzalez-Mieres and Suppe, 2006). We found these examples to be a bit simpler than our foregoing discussion in the sense that all four show pregrowth shortening, $S(z)$, and mean shortening, $\overline{S}(z)$, that are equal and constant as a function of height; that is, no layer-parallel simple shear

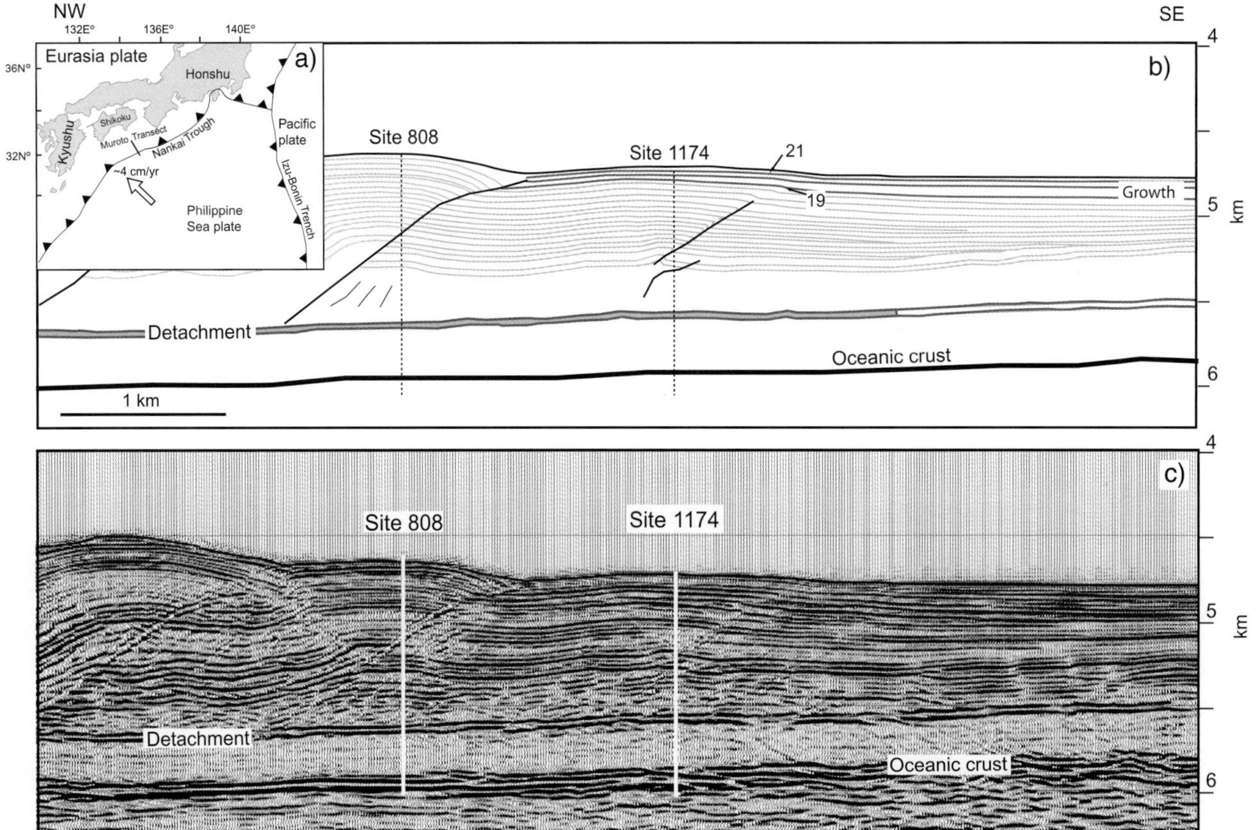

Figure 5. Active detachment fold lying seaward of the frontal thrust ramp of the Nankai Trough accretionary wedge, Japan. (a) Location map of the seismic line along the Nankai Muroto transect. (b) Interpretation of the seismic profile and location of Ocean Drilling Project borehole sites 808 and 1174. The growth sequence is highlighted and starts at horizon 19. (c) Depth-migrated seismic profile 62-8 Muroto transect (horizontal scale = vertical scale) (modified from Moore et al., 1991).

of the sort is shown in our model (Figure 2). Nevertheless, considerable variation in deformational behavior between these examples, including several complexities that go beyond what we have discussed so far, is observed, such as excess area within a basal detachment layer of finite thickness, horizontal compaction, multiple detachment levels, changes in rate of deformation, and hiatuses in deformation.

The data used in the following examples consist of two-dimensional (2-D) seismic images in depth, with borehole constraints in two cases. We assume that the images are undistorted and correctly interpreted. Any errors in the images or horizon mapping will in general map into the shortening estimates in a complex way. Full analysis of errors and error propagation is beyond the scope of this already-long chapter; the stated formal errors are discussed by Gonzalez-Mieres and Suppe (2006). The goal of this chapter is to present and illustrate with examples the basic structural principles and strategies of determining shortening histories from area-of-relief data.

NATURAL EXAMPLES AND RESULTS

Nankai Trough, Japan

The Nankai Trough is the locus of convergence between the Philippine Sea plate and the Eurasia plate offshore southwest Japan (Le Pichon et al., 1987; Shipboard Scientific Party, 2001a) (Figure 5a). A large accretionary wedge has developed since the middle to late Miocene, which is composed of a deformed sequence of trench turbidites and underlying hemipelagic sediments, originally deposited seaward of the trench in the Shikoku basin (Taira et al., 1992; Kamata and Kodama, 1994; Moore et al., 2001). The rate of convergence has been estimated at 2–4 cm/yr (0.8–1.6 in./yr) for the last 6 Ma (Seno, 1977; Karig and Angevine, 1986); however, Global Positioning System measurement and recent stratigraphic studies suggest a higher rate of convergence of 5–6 cm/yr (2–2.4 in./yr) for the last 3 Ma (Miyazaki and Heki, 2001; Shipboard Scientific Party, 2001a).

The active detachment fold (Figure 5) that is our focus is at the very toe of the Nankai accretionary prism in the protothrust zone ahead of the frontal-most thrust ramp (Moore et al., 1991, 2001). This is an area of distributed deformation with the principal visible structure being a single low-amplitude, 2–3-km (1.2–1.9-mi)-wide detachment fold with a slightly faulted core that runs for more than 100 km (62 mi) along the landward edge of the trench (Moore et al., 2001; Shipboard Scientific Party, 2001a). The depth-migrated 2-D seismic section that we have studied is from Moore et al. (1990) and lies along the extensively studied Cape Muroto transect of the Nankai Trough, which is a site of several extensively logged and cored Ocean Drilling Project (ODP) holes (Moore et al., 1990, 1991; Shipboard Scientific Party, 2001a; Gulick et al., 2004).

The ODP hole 1174 lies near the crest of the detachment fold (Figure 5) along the seismic line and penetrates the entire stratigraphic section into the top of the oceanic crust, providing excellent depth calibration and stratigraphic identification of seismic reflectors. The shape of the fold is well constrained seismically in the upper zone of relatively continuous reflectors that represents the trench-fill turbidite sequence, which includes the growth sequence of concern to us here. The trench-fill sequence in the hole extends upward to 4 m (13 ft) below the sea bottom where it is overlain by hemipelagic sediments that record the full emergence of the fold above the zone of trench sedimentation (Shipboard Scientific Party, 2001b). An additional 18 m (59 ft) of transitional facies may record the beginnings of emergence. At present, the fold crest has a relief of about 60 m (197 ft) above the regional bathymetric gradient of the trench.

The detachment (Figure 5) is located just above a pair of strong flat reflectors lying above a semitransparent zone of hemipelagic sediments that overlie the oceanic crust (Moore et al., 1990, 2001). This same seismically determined detachment level and associated strong reflectors extend far under the accretionary wedge and are seen to be the base of numerous seismically imaged thrust ramps (Moore et al., 1990; Gulick et al., 2004). The precise location of the detachment has been confirmed by drilling at sites ODP-808 and ODP-1174. Hole 1174 shows a detachment interval of 33 m (108 ft), between 807 and 840 m (2648 and 2759 ft) from the top of the well, giving a very strong independent constraint on the depth of the detachment for our analysis of this fold (Shipboard Scientific Party, 2001b).

We have mapped 21 seismic horizons and determined their areas of relief using the thickness-relief methods. The area-of-relief graph is shown in Figure 6. Our previous analysis of the pregrowth strata shows a vertically nearly homogeneous shortening, $\overline{S} = S$, of 77 ± 6 m (253 ± 20 ft) and a bed-length shortening that is more than an order of magnitude smaller (~2–3 m [6–10 ft]), which indicates that the bed length is not conserved such that the deformation is dominated by pure shear instead of flexure shortening (Gonzalez-Mieres and Suppe, 2006). The beginning of growth at H = ~750 m (2461 ft) above the detachment (ΔH_{growth} = ~100 m [328 ft]) is marked by an abrupt change to the upward-decreasing area of structural relief, which is qualitatively similar to our model at a high rate of shortening relative to sedimentation shown in Figure 4a.

The growth strata display (Figure 6b) an initial linear decrease in shortening with height ($d\overline{S}/dH = 0.5$) followed by an acceleration in shortening rate associated with the bathymetric emergence of the anticline, ($d\overline{S}/dH = \sim 6$). Approximately 45% of the shortening of the Nankai detachment fold has occurred since emergence, which represents only the last approximately 5–10% of growth sedimentation as recorded by off-structure sediment thicknesses to the south. This postemergent deformation has produced a current relief of the crest of the fold of about 60 m (197 ft) above the regional, which is large relative to the total growth sedimentation (ΔH_{growth} = ~100 m [328 ft]).

The approximately linear trend to the pre-emergent shortening data (Figure 6b) suggests that both sedimentation and deformation rates were roughly constant since the beginning of growth, which was at approximately 0.16 Ma based on the interpolation of a magnetostratigraphic calibration at 0.78 Ma from ODP-1174 (Shipboard Scientific Party, 2001b). The approximately 40 m (131 ft) of pre-emergence shortening in about 150 ka yields an estimated average shortening rate caused by folding of approximately 0.27 mm/yr (0.01 in./yr), which is roughly two orders of magnitude smaller than the plate boundary shortening rates (5–6 cm/yr [2–2.4 in./yr]) discussed above, indicating that deformation in the Nankai margin is not concentrated in the frontal structure. The postemergent shortening rate of about 3.5–4 mm/yr (0.14–0.16 in./yr) is also a small fraction of the plate rate. Other sources of shortening near the toe of the wedge include horizontal compaction (Gonzalez-Mieres and Suppe, 2006) and slip on the frontal thrust ramps (Suppe et al., 2004), but the sum of this frontal deformation is about 1–1.5 cm/yr (0.39–0.59 in./yr) and therefore is still a modest fraction of the plate-tectonic shortening rate.

Cascadia Accretionary Front, Oregon

Our second example is the active frontal detachment fold at the toe of the Cascadia accretionary wedge, offshore central Oregon (Figure 7), which is associated

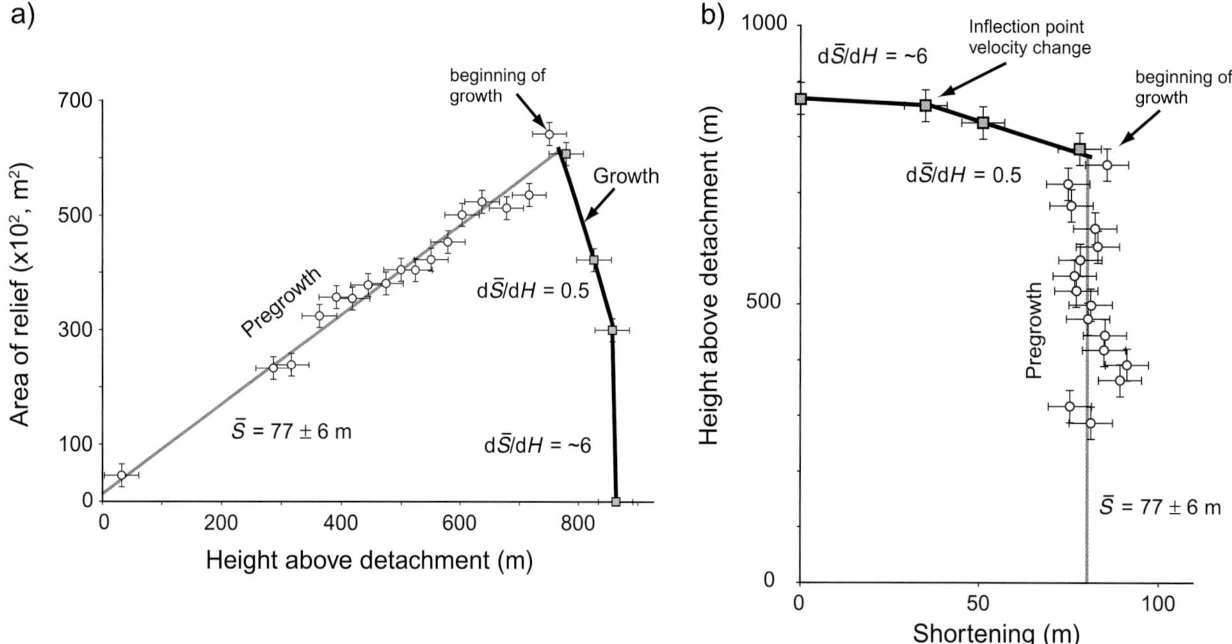

Figure 6. Area-of-relief data and computed shortening for the Nankai Trough detachment fold. (a) Area relief as a function of height above detachment level, which is about 360 m (1181 ft) above the top of the oceanic crust in the Ocean Drilling Project (ODP) 1174 borehole. The gray line represents the linear regression of pregrowth data, giving a mean shortening of $\bar{S} = 77 \pm 6$ m (253 ± 20 ft) and a zero intercept at the detachment indicating a simple detachment fold with vertically homogeneous shortening and no excess area in the basal interval. The growth data are qualitatively similar to the model of Figure 4a, with a maximum in relief at the beginning of growth and a rapid upward decrease in relief. (b) Shortening graph. The data points are computed from $\bar{S} = A/H$. The pregrowth mean shortening was computed in panel a. The early pre-emergent growth sequence shows a relatively constant ratio of shortening rate to sedimentation rate of $d\bar{S}/dH = 0.5$, whereas the postemergent rate is about an order of magnitude larger, $d\bar{S}/dH = \sim 6$.

with the oblique subduction of the Juan de Fuca plate below the North American plate at a rate of approximately 4 cm/yr (1.6 in./yr) along a zone extending from Vancouver Island to northernmost California (Kulm et al., 1984; DeMets et al., 1990; MacKay et al., 1992; Westbrook et al., 1994). The stratigraphic section entering the deformation front is substantially thicker (3–4 km [1.9–2.5 mi]) than the Nankai example (~1 km [0.6 mi]) and is composed of turbidites in which a well-defined fold and thrust belt has developed with substantial along-strike variation, including regions of both seaward and landward vergence (Silver, 1972; Kulm et al., 1984; MacKay et al., 1992; MacKay, 1995; Adam et al., 2004).

The study site is similar to Nankai in that it shows a protothrust zone ahead of a seaward-vergent frontal thrust ramp (Figure 7), but the decollement level is not as immediately obvious as the Japanese case (Figure 5). The protothrust zone is dominated by a low-amplitude detachment fold just seaward of the frontal thrust ramp, but additional more distributed deformation has been documented as indicated by several incipient thrust ramps with barely resolvable slip (Figure 7), which is studied in detail by Cochrane et al. (1994) and Moore et al. (1995), and landward-increasing seismic velocities indicating horizontal compaction (see Gonzalez-Mieres and Suppe, 2006). These incipient thrusts terminate downward at or above the same detachment level as the main frontal thrust ramp, which Moore et al. (1995) traced along strike in other seismic lines. Our analysis of the area of structural relief in pregrowth strata identified this same detachment level, which is about 1300 m (4265 ft) above the top of the oceanic crust, as well as a deeper detachment level close to the base of the stratigraphic section (Gonzalez-Mieres and Suppe, 2006).

We mapped 35 seismic horizons on the depth-seismic line OR-9 of the ODP (Westbrook et al., 1994; Carson et al., 1995). The areas of relief of these horizons were measured using the thickness-relief methods and are shown in Figure 8. The pregrowth strata were analyzed previously (Gonzalez-Mieres and Suppe, 2006). The most obvious feature of the pregrowth data is that a rather abrupt change in slope of the area-height graph

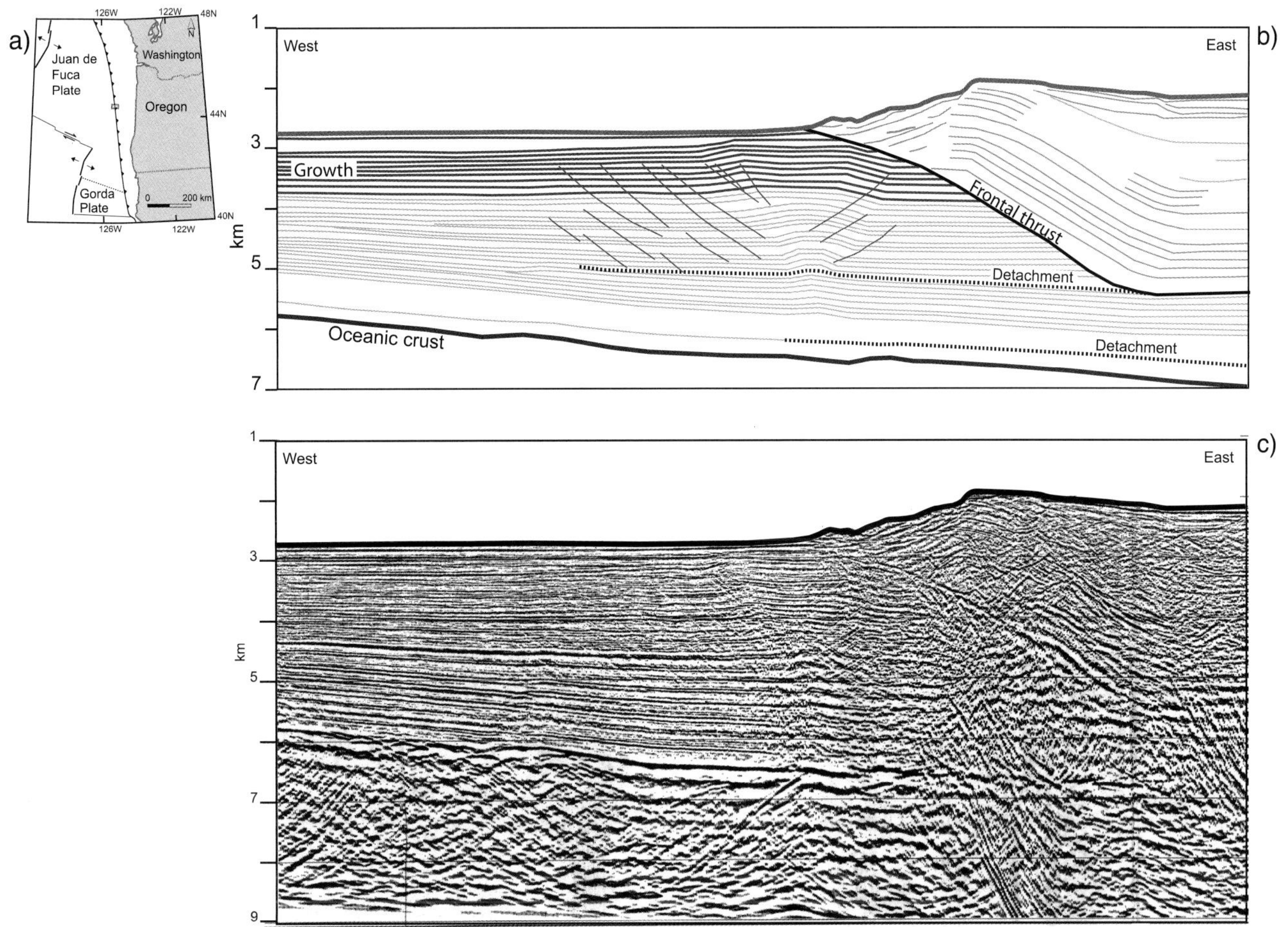

Figure 7. Active detachment fold lying seaward of the frontal thrust ramp of the Cascadia accretionary complex, offshore Oregon. (a) Location map of the seismic line. (b) Interpretation of the seismic profile with growth sequence highlighted. Thrust faults of negligible offset are shown. (c) Depth seismic profile OR-9 (horizontal scale = vertical scale). Modified from Westbrook et al., 1994; Gonzalez-Mieres and Suppe (2006).

at about 1080 m (3543 ft) above the origin is observed, which is about 20 m (66 ft) below the horizon of the detachment mapped seismically by Cochrane et al. (1994) and Moore et al. (1995). Linear regression of the data above this detachment gives a shortening of 256 ± 20 m (840 ± 66 ft). Shortening below the detachment is 84 ± 8 m (275 ± 26 ft). A lower detachment depth is estimated at about 280 m (919 ft) above the top of oceanic crust based on the zero intercept of the linear regression of the area-height data, which defines the origin of our graphs. Gonzalez-Mieres and Suppe (2006) showed that the shortening in the pregrowth section is vertically homogeneous, $S(z) = \overline{S}$, within each fault block; therefore, the fault slip consumed in folding is 84 m (275 ft) on the lower detachment and 172 m (564 ft) (256–84 m [840–275 ft]) on the upper detachment.

With two detachments, the problem of deconvolving the growth history from the area-of-relief data is substantially more complex than in the single-detachment case of Nankai (Appendix). The area of relief, $A(z)$, at every growth horizon is now the sum of the contributions of the two structural levels

$$A(z) = A_L(z) + A_U(z) \tag{7}$$

which leads to ambiguity. For example, we can no longer directly compute the mean shortening for growth strata from their area of relief using equation 2 because a single height above the detachment no longer exists. Furthermore, the vertical gradient in the area of relief in growth strata is now the sum of the effects of each fault block (see equations 30, 31). However, the analysis for Cascadia can be simplified for two reasons. First, because the shortening is vertically homogeneous, $S(z) = \overline{S}$, we can very simply express the total area of relief at any elevation as the sum of the fractional contributions of

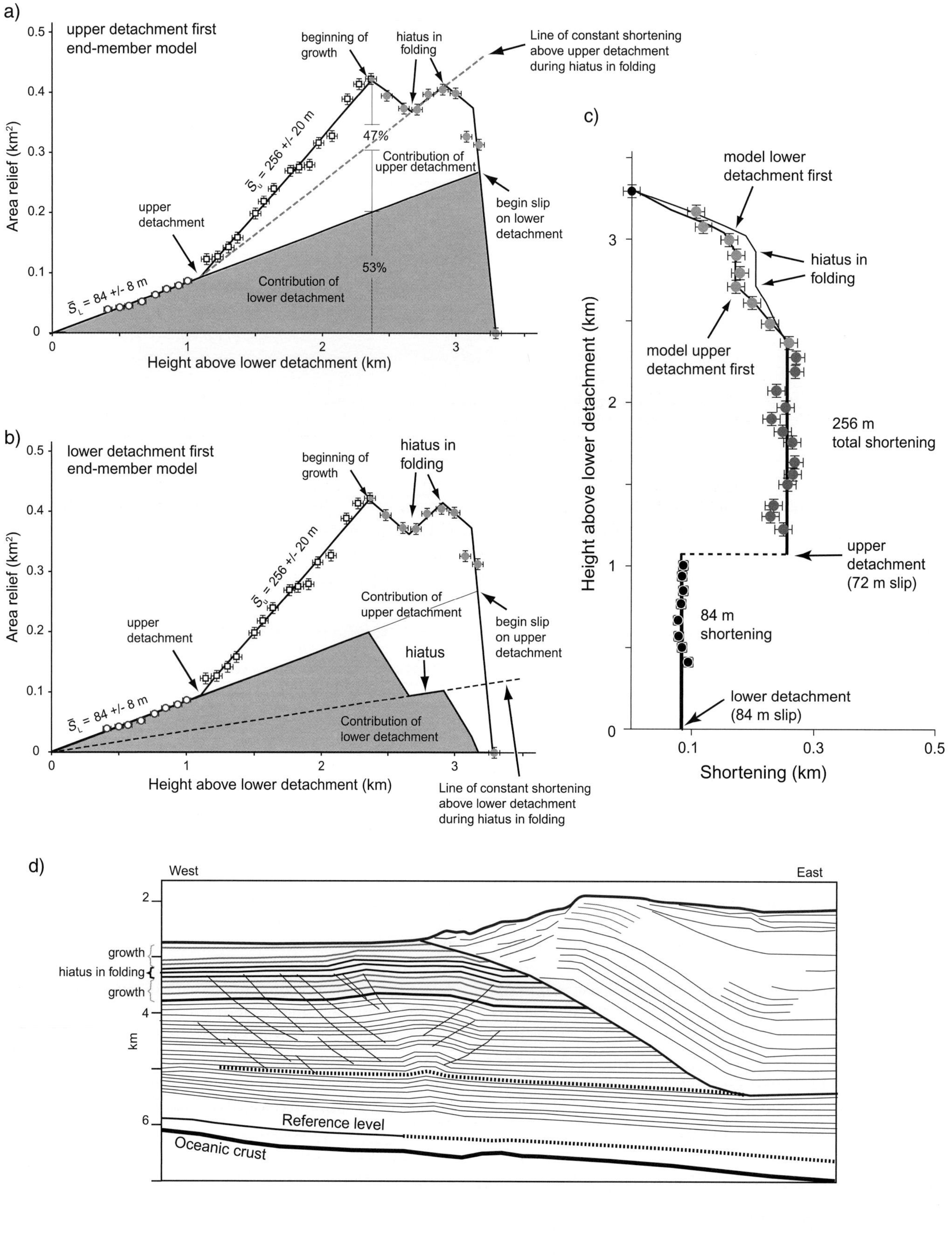

a)

upper detachment first
end-member model

Area relief (km²)

0.5
0.4
0.3
0.2
0.1
0

beginning of growth

hiatus in folding

Line of constant shortening above upper detachment during hiatus in folding

47%

Contribution of upper detachment

begin slip on lower detachment

$\bar{S}_U$ = 256 +/- 20 m

upper detachment

53%

Contribution of lower detachment

$\bar{S}_L$ = 84 +/- 8 m

0 1 2 3

Height above lower detachment (km)

b)

lower detachment first
end-member model

Area relief (km²)

0.5
0.4
0.3
0.2
0.1
0

beginning of growth

hiatus in folding

$\bar{S}_U$ = 256 +/- 20 m

Contribution of upper detachment

upper detachment

begin slip on upper detachment

hiatus

$\bar{S}_L$ = 84 +/- 8 m

Contribution of lower detachment

0 1 2 3

Height above lower detachment (km)

Line of constant shortening above lower detachment during hiatus in folding

c)

model lower detachment first

hiatus in folding

model upper detachment first

256 m total shortening

upper detachment (72 m slip)

84 m shortening

lower detachment (84 m slip)

Height above lower detachment (km)

0
1
2
3

0.1 0.3 0.5

Shortening (km)

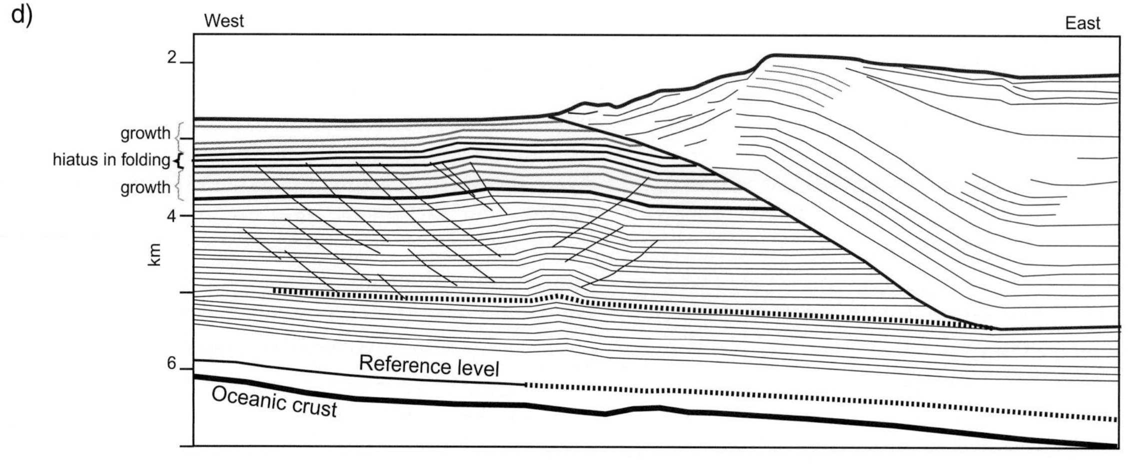

slip consumed in shortening above each detachment (equation 28)

$$A(z) = f_L(z)\overline{S}_L(H_L + H_U) + f_U(z)\overline{S}_U H_U \quad (8)$$

where $\overline{S}_L = 84$ m (275 ft) and $\overline{S}_U = 172$ m (564 ft) are the total shortening consumed in folding for each detachment and $f_L(z)$ and $f_U(z)$ are the fractions of the total shortening contributing to the observed relief of the horizon in question, with the fractions ranging from 1 in pregrowth strata to 0 at the end of growth. We can search for models, $\langle f_L(z), f_U(z)\rangle$, of the history of these fractions that satisfy the data, $\langle A, z = H_L + H_U\rangle$, which we have done. Note that because the models are required to satisfy equation 8 to some arbitrary degree of closeness, they cannot be distinguished on the basis of relief data in contrast with pregrowth data. The shortening-of-growth data shown in Figure 8c were computed using the model and therefore only serve to illustrate where the model is constrained by data. Second, the details of the data for Cascadia, discussed below, allow us to focus on two end-member models in which either the lower or the upper detachment slips first. All other possible models will lie between these two end members. Other geologic constraints will allow us to further limit substantially the range of possible models.

Upper-first End Member

We begin by noting that (1) more than half of the drop in the area of relief occurs in the upper few hundred meters of growth strata and (2) the lower detachment makes a very large contribution to the total area of relief, although the shortening is small (84 m [275 ft]), because of the large height of the growth horizons above the lower detachment. For example, the lower detachment contributes 53% of the total relief at the horizon of the beginning of growth and potentially contributes as much as 85% to the relief of the uppermost growth horizon (Figure 8a). These observations imply that a very simple end-member model is possible in which slip on the lower detachment does not begin until just before the deposition of the uppermost growth horizon (upper-first end member). For this end member, the area-of-relief data for growth strata provide a complete record of the slip on the upper detachment once we subtract the effect of later folding above the lower detachment, as shown in Figure 8a.

Lower-first End Member

Conversely, the other possible end-member model involves slip on the lower detachment until just before the deposition of the uppermost growth horizon (lower-first end member, Figure 8b). In this case, the area-of-relief data for all but the uppermost growth strata record the slip on the lower detachment once we subtract the area of relief produced by 172 m (564 ft) of shortening above the upper detachment, as shown in Figure 8b. Slip models intermediate to these end members are possible (Figure 8c). A key constraint for limiting the range of possible fault slip models comes from the observation of a hiatus in fold growth.

Hiatus in Fold Growth

The area-of-relief data show a hiatus or an extreme reduction in fold growth over an approximately 300-m

Figure 8. Area-of-relief data and computed shortening for the Cascadia detachment fold. (a, b) Area relief as a function of height above a reference level that lies approximately 200 m (656 ft) above the top of the oceanic crust. Area relief data for pregrowth strata show two quasi-linear trends, indicating constant shortening as a function of depth above two detachments (detailed analysis by Gonzalez-Mieres and Suppe, 2006). Linear regression of the lower trend shows a shortening of $\overline{S} = 84 \pm 8$ m (275 ± 26 ft) with a zero intercept defining the reference level. Linear regression of the upper trend gives a shortening of $\overline{S} = 256 \pm 20$ m (840 ± 66 ft). The intersection of the two trends is close to the height of the detachment of the frontal thrust and of the protothrust zone mapped regionally by Moore et al. (1995). Within the growth section, a hiatus in deformation is indicated by data points that lie along lines of constant mean shortening, similar to the pregrowth data points. End-member models of the histories of the contributions of the two detachments are shown: (a) The end-member model with the upper detachment moving entirely before the lower, with the contributed area of relief from the lower detachment shown in gray. (b) The end-member model with the lower detachment moving entirely before the upper, with the contributed area of relief from the lower detachment shown in gray (computed by subtracting the area of relief produced by later shortening of 172 m (564 ft) above the upper detachment). (c) Shortening graph. The pregrowth sequence shows a vertically homogeneous shortening of 84 m (275 ft) for the lower sequence and 256 m (840 ft) for the upper sequence, with their difference (172 m [564 ft]) being the slip on the upper detachment consumed in folding. Two end-member models for the history of growth bound the range of all possible growth models; however, other geologic observations based on the upward and downward terminations of minor faults shown in panel d favor the upper detachment first model (see full discussion in the text). The computation of the shortening data in the growth section is not independent of the model but does indicate where the model is constrained by area-of-relief data (see the text).

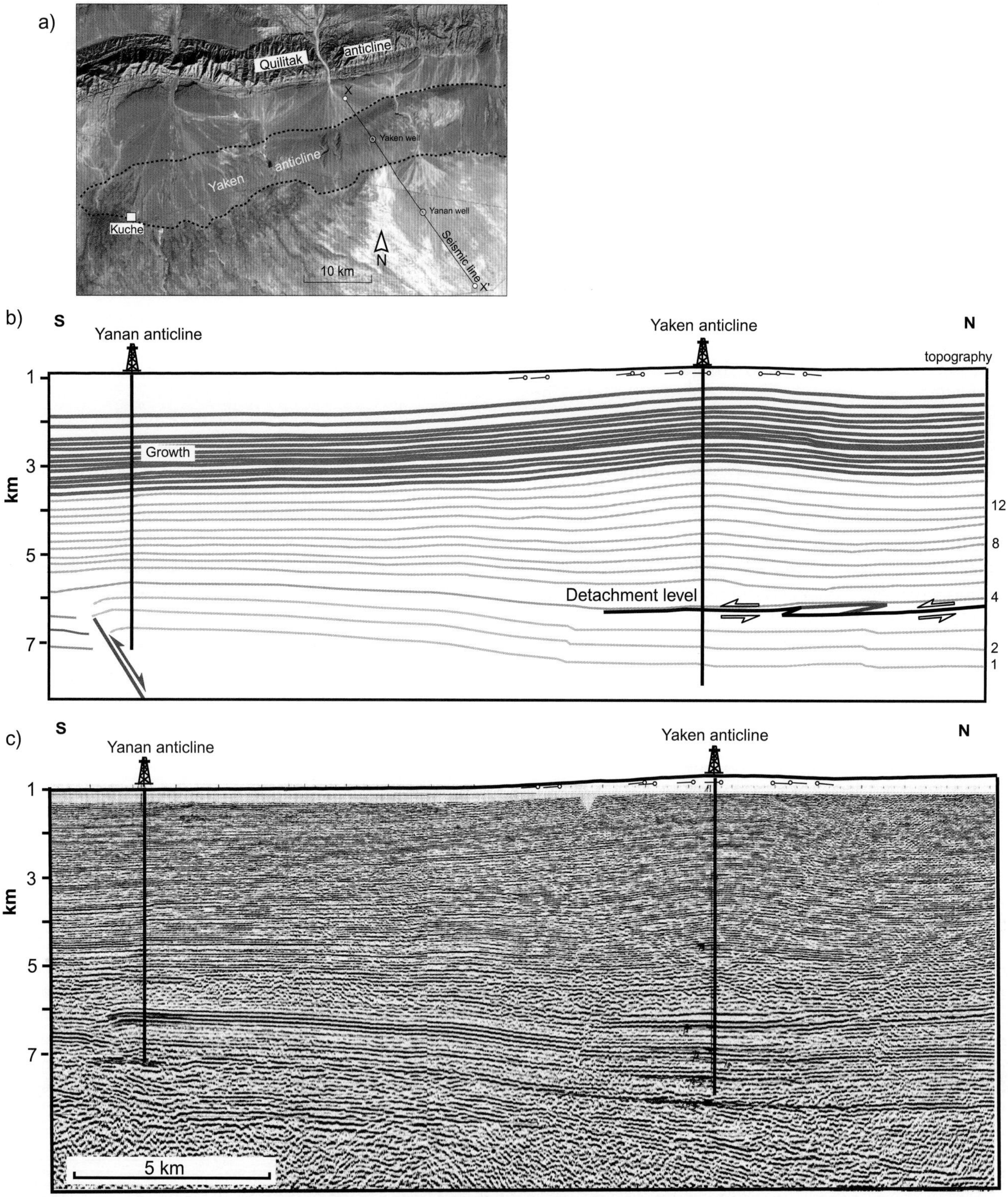

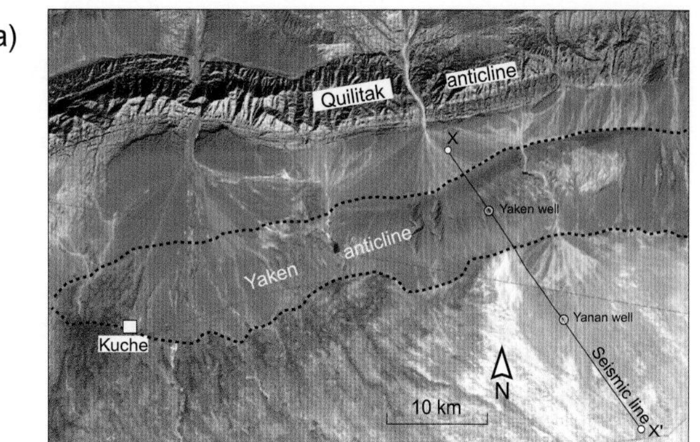

Figure 9. Active Yaken detachment fold at the front of the southern Tianshan thrust belt, near Kuche, western China. (a) Landsat image of the Yaken anticline, showing the location of the seismic line and wells. (b) Interpretation of the seismic profile and location of wells. The growth section is shown with darker lines. (c) Depth seismic profile (horizontal scale = vertical scale). From Hubert-Ferrari et al. (2005).

(984-ft)-thick stratigraphic interval (2.6–2.9 km [1.6–1.8 mi] above the lower detachment), which is indicated by the fact that the data points lie approximately along lines radiating for each detachment (Figure 8a, b), which are lines of constant mean shortening for nongrowth strata. Additionally, the thickness of this stratigraphic interval is essentially constant across the fold showing no sign of stratigraphic growth variation. The shortening since this hiatus is 90 m (295 ft) computed for the upper-first end member and 34 m (111 ft) for the lower-first end member (Figure 8c). (Alternatively, this interval could be viewed as a growth interval with an increase in sedimentation rate by a factor of about 100, yielding a value of $d\overline{S}/dH$ approaching zero. Such a rate, which is estimated to be approximately 100 mm/yr [3.9 in/yr] for 30 ka, is probably excessive.)

The cross section in Figure 8d shows that the stratigraphic interval of the hiatus in fold growth is a zone in which the thrust faults of small displacement terminate upward within the protothrust zone. Therefore, the hiatus in fold growth was the time of broader deformation, including propagation of the seismically imaged faults of barely resolvable slip, which has been shown regionally to be confined above the upper detachment (Cochrane et al., 1994; Moore et al., 1995). Therefore, the hiatus in folding was a time of slip on the upper detachment. This conclusion excludes the lower-first end-member model, which does not have the upper detachment activating until 200–300 m (656–984 ft) above the hiatus in fold growth. There exists a more limited range of intermediate models that are consistent with the upper detachment being active at the time of the hiatus in folding, some of which require oscillations in activity between the two detachments. We favor models close to the upper-first end-member solution involving late activation of the lower detachment (Figure 8a) because such models, particularly those involving simultaneous late slip on both detachments, provide a straightforward mechanism for the observed very rapid upward decrease in the area of the relief during the last few hundred meters of deposition.

The regional average sedimentation rate for the Oregon Cascadia area is about 1 mm/yr (0.04 in./yr) (Kulm et al., 1984; Moore et al., 1995; Underwood and Hoke, 2000). Using this nominal rate and the upper-first end-member model, we estimate a beginning of growth at about 0.93 Ma and a hiatus in fold growth lasting about 0.3 Ma from about 0.66 to 0.36 Ma. The late acceleration in fold growth rate began about 0.16 Ma. The early shortening has a ratio of shortening to sedimentation rate, $d\overline{S}/dH$, of 0.27, which would be associated with the upper detachment, which yields shortening rates for the upper detachment of about 0.27 mm/yr (0.01 in./yr). This value requires about 0.63 Ma to accomplish the 172 m (564 ft) of shortening on the upper detachment, which is essentially all of the time since the beginning of fold growth at about 0.9 Ma, except for the approximately 0.3 Ma of the hiatus in folding. In contrast, the 84 m (275 ft) of shortening on the lower detachment that has accumulated since activation at about 0.16 Ma indicates a lower detachment rate of about 0.5–0.6 mm/yr (0.020–0.024 in./yr). The factor of two contrast in shortening rates between the upper and lower detachments, as well as the 50% greater depth to detachment, accounts for the observed rapid increase in rate of fold growth in the last approximately 150 m (492 ft) of sedimentation, even if the upper detachment were not currently active. As with Nankai, the growth of the Cascadia frontal detachment fold at the present rate of approximately 1 mm/yr (0.04 in./yr) accounts for only a tiny fraction of the present plate convergence of about 4 cm/yr (1.6 in./yr) and indicates a very low propagation rate for the toe of the accretionary wedge.

Yaken Anticline, Southern Tianshan, China

The active Yaken detachment fold (Figure 9) lies at the toe of the southern Tianshan fold and thrust belt and extends for about 100 km (62 mi) along strike east of the city of Kuche at the northern edge of the Tarim Basin in western China (Hubert-Ferrari et al., 2005, 2007). The Tianshan Mountains form a locus of active compression north of Tibet induced by the collision of India with Eurasia and compose one of the largest and most active intracontinental mountain ranges with the highest peak reaching more than 7000 m (22,966 ft) (Molnar and Tapponier, 1977). Geodetic studies indicate that up to 20-mm/yr (0.79-in./yr) shortening is presently accommodated across the Tianshan at the longitude of Kuche (Allen et al., 1991; Reigber et al., 2001). In this region, the southern thrust belt lies within the Kuche subbasin of the Tarim Basin, which is marked by an 8–10-km (5–6.2-mi)-thick Triassic to Holocene continental sequence (Yin et al., 1998; Burchfiel et al., 1999; Wang et al., 2011). The deformation is mechanically dominated by coal detachments in the Jurassic and evaporitic detachments in the middle Tertiary, especially the Jidike formation (Hubert-Ferrari et al., 2005, 2007; Wang et al., 2011).

The Yaken anticline has quite recently emerged topographically as shown by folded alluvial surfaces that still preserve abundant evidence of relict throughgoing drainage networks marked by wind gaps at the topographic crest of the 50–150-m (164–492-ft)-high, 5–7-km (3.1–4.3-mi)-wide broad ridge (Hubert-Ferrari et al., 2005, 2007). Along the few drainages that have incised deeply enough to form water gaps, the bedding dips in outcrop reach a maximum of 4–6°. The Yaken

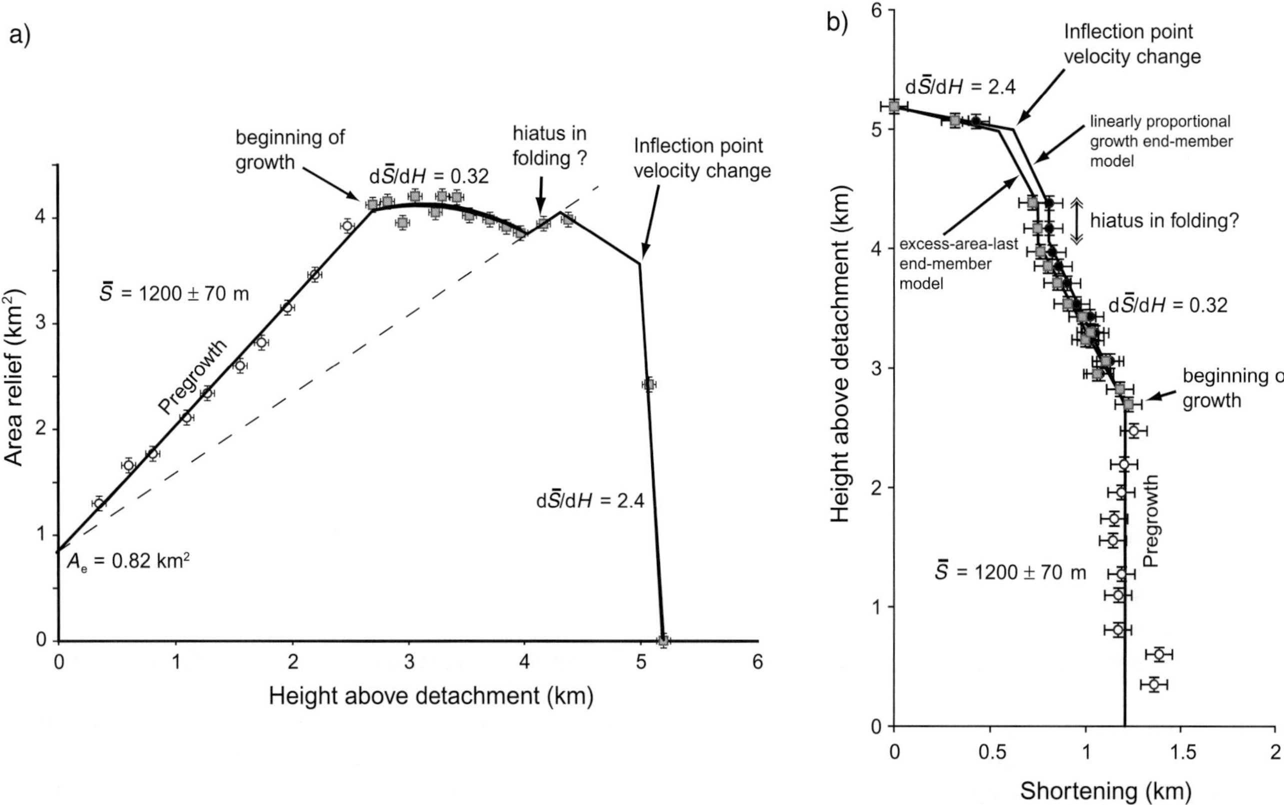

Figure 10. Area-of-relief data and computed shortening for the Yaken anticline. (a) Area relief as a function of height above the detachment level in the evaporitic Jidike formation (data from Hubert-Ferrari et al., 2005). The linear regression of pregrowth data with a mean shortening of 1200 ± 70 m (3937 ± 230 ft) and a nonzero intercept at the detachment indicates a significant excess area in the fold core of $A_e = 0.82$ km^2 (0.32 mi^2). The growth data are qualitatively similar to the model of Figure 4b, with a maximum in relief after the beginning of growth, indicating a high sedimentation rate relative to the shortening rate. A probable hiatus in folding is indicated by data lying along lines of constant $\bar{S}$, less well constrained than the Cascadia hiatus (Figure 8). (b) Shortening graph. The data points are computed from $\bar{S} = (A - A_e)/H$. The early pre-emergent growth sequence shows a relatively constant ratio of shortening rate to sedimentation rate of $\sim d\bar{S}/dH = 0.32$ prior to the probable hiatus in folding, whereas the postemergent rate is about an order of magnitude larger, $d\bar{S}/dH = \sim 2.4$, which is associated with topographic emergence. See the text for discussion of growth models.

anticline in seismic images is seen to be a broad gentle fold with 5–6° limb dips, formed above a detachment in the Jidike formation at a depth of 5–6 km (3.1–3.7 mi), which has been penetrated by several exploratory boreholes (Figure 9). The Jidike detachment level is confirmed on more regional seismic lines that show that slip feeds into this level from a deep thrust ramp under the Quilitak anticline to the north (see Hubert-Ferrari et al., 2007). To the south of Yaken is the basement-involved Yanan structure, which also gently folds upper Tertiary strata but without surface expression.

The thickness-relief method was first proposed and applied by Hubert-Ferrari et al. (2005) as part of an analysis of the Yaken anticline in which they determined the area of relief for 27 horizons (Figure 9). The pregrowth data show a linear relationship between the area of relief and the height, indicating a mean shortening, $\bar{S}$, of 1200 ± 70 m (3937 ± 230 ft) obtained by linear regression (Figure 10a). However, the regression line does not intersect close to the origin of the graph and shows an excess area, A_e, of the fold core at the detachment level of 0.82 km^2 (0.32 mi^2). Hubert-Ferrari et al. (2005) showed that substantial thickness variation within the evaporitic strata of the Jidike formation exists and therefore interpreted this excess area, A_e, to be the result of net flow of the basal evaporite into the core of the fold. This excess area represents a constant area that is added to all overlying pregrowth horizons as described by

$$A(z) = \bar{S}(z)H + A_e \tag{9}$$

In addition, Gonzalez-Mieres and Suppe (2006) showed that shortening is essentially constant as a function of height in the pregrowth strata such that no layer-parallel simple shear and $S(z) = \overline{S}(z)$ (Figures 9, 10) exist. Hubert-Ferrari et al. (2005, 2007) presented an analysis of the growth history of Yaken, which we here extend by making use of the theory presented in the Appendix.

Because the growth strata of Yaken record both the shortening and the development of the excess area, we once again are dealing with the ambiguity of a two-component system of the sort discussed above for Cascadia (see equations 7, 26). In the case of Yaken, we know the total shortening, $\overline{S}(z) = S(z)$, and the total excess area, A_e; therefore, equation 26 becomes (see equation 29)

$$A(z) = f_{\overline{S}}(z)\overline{S}H + f_e(z)A_e \tag{10}$$

where $f_{\overline{S}}(z)$ is the fraction of the total shortening and $f_e(z)$ is the fraction of the excess area that has accumulated since the deposition of the horizon in question. We search for models, $\langle f_{\overline{S}}(z), f_e(z)\rangle$, of the history of these fractions that satisfy the data, $\langle A,z\rangle$.

Three principal end-member models that satisfy the data for Yaken are as follows (Figure 10b): (1) An end-member model is possible in which the excess area developed late, after, or synchronous with topographic emergence such that equation 9 still holds. (2) The other possible end-member model has the excess area developing before any shortening (not shown in Figure 10b). This end member is considered unlikely because it requires the transition between the growth in excess area and onset of shortening to lie within the interval between the beginning of growth and a hiatus in folding (discussed below), which is an interval in which the area-of-relief data are relatively homogeneous. In the other models, this interval is well explained by a nearly constant shortening rate for at least 3.5 Ma, as discussed below. (3) Finally, it is possible to construct structurally reasonable models of detachment folds in which the shortening and growth of excess area are coupled, especially such that $f_{\overline{S}}(z) = f_e(z)$ (see Suppe, 2011; also discussion of the Agbami anticline below). The mean shortening predicted by this model (Figure 10b) is only 10–20% larger than the model of the late excess area because the excess area of Yaken is a modest fraction of the total area of relief at the beginning of growth.

Note that because the models are required to satisfy equation 10 to some arbitrary degree of closeness, they cannot be distinguished based on area-of-relief data in contrast with pregrowth data. The shortening-of-growth data shown in Figure 10b are computed using the models and therefore only serve to illustrate where the models are constrained by data. However, other data may be helpful in distinguishing the models, such as the details of the fold shape. Hubert-Ferrari et al. (2005) argued for model 1 based on the fact that the spatial distribution of the excess area is somewhat offset relative to the spatial distribution of the area of relief that developed prior to topographic emergence. In contrast, the relief of the emergent topographic surface is not offset. Therefore, a significant fraction of the excess area appears to be late. The spatial distribution of growth is in agreement with models 1 and 3, which predict 60–100% late excess area but not in agreement with model 2.

The early growth history of Yaken (Figure 10a) is marked by a maximum in the area of structural relief that does not coincide with the beginning of growth, which is similar to the detachment fold model of Figure 4b. This behavior is characteristic of low shortening rates relative to sedimentation rate. The data are reasonably well fit to a constant shortening to sedimentation ratio of $d\overline{S}/dH = 0.32$ for the first 1.5–2 km (0.9–1.2 mi) of growth sedimentation. In contrast, about 60% of the area of structural relief has developed since topographic emergence, which accounts for only approximately 100 m (328 ft) of basin sedimentation as shown by Hubert-Ferrari et al. (2007). Since topographic emergence, the ratio of shortening to sedimentation has increased by nearly an order of magnitude, ($d\overline{S}/dH = \sim 2.4$).

The details of the transition between the slow early growth of Yaken and the late emergence are not well resolved because of limited near-surface seismic imaging. However, the available seismic imaging shows no onlap or truncation of reflectors within the growth sequence at a seismic resolution. Therefore, no evidence of earlier topographic emergence or acceleration of growth exists. However, limited evidence in the area-of-relief data for a possible hiatus in the uppermost seismic growth horizons is observed, similar to the better constrained hiatus in Cascadia, but resolved by fewer data (Figure 10). The stratigraphic length of the possible hiatus in deformation would be about 200 m (656 ft), which would represent approximately 0.5 Ma given sedimentation rates discussed as follows.

Nearby stratigraphic and magnetostratigraphic studies indicate a nearly constant regional sedimentation rate of 0.43 mm/yr (0.02 in./yr) over the last 10 Ma, apparently controlled regionally by rising base level within the internally drained Tarim Basin (Charreau, 2005; Hubert-Ferrari et al., 2007). This sedimentation rate and seismic stratigraphic correlation indicates that Yaken began deformation at 5.5 Ma with a nearly constant shortening rate of 0.16 mm/yr (0.006 in./yr) until at least 2 Ma. Topographic emergence began about 0.16–0.21 Ma (Hubert-Ferrari et al., 2007). There has been an acceleration in shortening rate by approximately an order of magnitude with a change in $d\overline{S}/dH$ from 0.32 to 2.4. Therefore, during most of its history (5.5 to 0.3–0.2 Ma), Yaken anticline had no topographic expression and a long-term shortening rate, $d\overline{S}/dt$, of about 0.16 mm/yr. The shortening rate since topographic emergence is an order of magnitude larger by approximately 1.7–2.2 mm/yr (0.07–0.09 in./yr). Both shortening rates are a small fraction (1–10%) of the geodetically derived shortening rate across the entire Tianshan discussed above. Furthermore, the front of the Kuche thrust belt has been stationary since the initiation of the Yaken anticline at about 5.5 Ma, showing no forward propagation (Hubert-Ferrari et al., 2007).

Agbami Anticline, Niger Delta

The Agbami anticline is part of the deep-water passive-margin fold and thrust belt of the Niger Delta (Figure 11) and contains one of the larger petroleum discoveries of the last decade (Grimes et al., 2004). This part of the delta lies at 1500-m (4921-ft) water depth and has a total stratigraphic thickness of 9–10 km (5.6–6.2 mi) overlying oceanic crust. The section is composed of an upper well-imaged Miocene to Holocene turbiditic section (0–17 Ma) overlying a 2–3-km (1.2–1.9-mi)-thick seismically poorly imaged basal section dominated by prodelta shales of the Akata Formation (Damuth, 1994; Cohen and McClay, 1996; Bilotti et al., 2005; Corredor et al., 2005a, b). The Akata Formation is thought to be highly overpressured and forms the main detachments associated with thrust ramps in the Niger Delta, dominating its mechanics (Bilotti and Shaw, 2005; Corredor et al., 2005a). Seismic images show a detachment near the base of Akata to be important in the inner thrust belts and a detachment just below the top of Akata to be the most important detachment regionally, including in thrust ramp structures immediately adjacent to Agbami (Corredor et al., 2005a, b). In some structures, the Akata Formation has been interpreted as highly mobile or diapiric (Cohen and McClay, 1996; Hooper et al., 2002).

The Agbami anticline is approximately 10 km (6.2 mi) wide with substantially higher amplitude and limb dips than our previous examples (Figure 11). A strong approximately 4–5° landward-dipping reflector at 10–11-km (6.2–6.8-mi) depth is determined to be the top of the oceanic basement as shown by magnetic depth-to-basement calculations (Bilotti et al., 2005; Corredor et al., 2005b). Reflector geometry within the growth strata shows limb rotation, combined with limb lengthening. The deep structure has been interpreted by Bilotti et al. (2005) as a detachment fold, but the exact nature and level of the detachment are not clearly seen because of problems of multiples in the seismic imaging (Grimes et al., 2004). Nevertheless, faint reflectors of regional dip are apparently visible on the west flank as much as 2 km (1.2 mi) shallower than the top of the oceanic crust and subparallel to it, suggesting the existence of a detachment at that level, which is the regional upper Akata detachment (Figure 11). Our interpreted detachment is about 1 km (0.6 mi) shallower than that of Bilotti et al. (2005). Other more complex detachment fold interpretations are possible for the Agbami anticline, as discussed by Gonzalez-Mieres and Suppe (2006) and Suppe (2011). These different possible deep interpretations are mostly concerned with the origin of the very large excess area of the core of the Agbami anticline.

Thirty-five reflectors were mapped between the top of the oceanic crust and the sea bottom (Figure 11). The pregrowth strata above the Akata Formation give a total shortening, $\overline{S}$, based on a linear regression of 814 ± 110 m (2671 ± 361 ft) with a very large excess area, A_e, of 9.3 km^2 (3.6 mi^2), which accounts for more than 80% of the total area of relief within the pregrowth strata. This estimated excess area is relatively insensitive to the interpreted depth of detachment because of the small shortening; it would be 8.5 km^2 (3.3 mi^2) for the approximately 1-km (0.6-mi) deeper detachment interpreted by Bilotti et al. (2005). Gonzalez-Mieres and Suppe (2006) showed that the shortening in the pregrowth strata is approximately constant as a function of height, $S(z) = \overline{S}(z)$; therefore, no overall component of layer-parallel simple shear exists. There has been modest layer-parallel stretching (~0.6 km [0.4 mi]) in the lid sequence above the excess area of the Akata Formation, as shown by the fact that nominal bed-length shortening exceeds the shortening (Gonzalez-Mieres and Suppe, 2006).

The analysis of the history of growth is essentially the same as the analysis for Yaken because of the two components, shortening and excess area, each contributing to the area of relief in potentially different proportions at each growth horizon (equation 10). Therefore, based on area-of-relief data alone, we are once again reduced to constraining the range of possible histories of these two components instead of determining a unique history. The uncertainties in shortening history are large for Agbami because of its extreme excess area relative to the total area of relief, in contrast with Yaken in which the excess area is a small fraction. The end-member models are the same as Yaken: (1) the shortening-first end member and (2) the excess-area-first end member. Note that the permissible solution space for shortening (Figure 12b) is larger than for excess area (Figure 12c), simply because excess area dominates the area-of-relief data. Note again that because the fractional models are required to satisfy equation 10 to some arbitrary degree of closeness, these models cannot be distinguished based on area-of-relief data in contrast with pregrowth data. The shortening and excess area of growth data shown in Figure 12b and c are computed using the models and therefore only serve to illustrate where the models are constrained by data. Other information is required to further constrain this history, especially more insight into the structural mechanism of the excess area.

One intermediate solution deserves special mention because it involves a well-defined coupling between the shortening and the excess area, which is the linearly proportional growth model (Figure 12b, c), similar to model 3 of Yaken. This is a duplexlike model involving a pair of simultaneously active detachments, a roof detachment near the top of the Akata Formation, and a floor detachment near its base, as discussed by Gonzalez-Mieres and Suppe (2006) and Suppe (2011). Under this solution, the excess area is equal to the slip that exits the structure on the roof detachment, $(\overline{S}_{\text{Akata}} - S_{\text{lid}})$, times the height, ΔH, between the roof and floor detachments

$$A_e = (\overline{S}_{\text{Akata}} - \overline{S}_{\text{lid}})\Delta H$$

where $\overline{S}_{\text{Akata}}$ is the mean total shortening of the Akata Formation between the two detachments and $\overline{S}_{\text{lid}}$ is the shortening consumed in the anticline above the upper detachment, which is assumed in the simplicity of this model to be in fixed proportion. Given a probable ΔH of 2–2.5 km (1.2–1.5 mi), an $\overline{S}_{\text{lid}}$ of 0.8 km (0.5 mi), and an A_e of 9.3 km^2 (3.4 mi^2), we estimate a basal shortening of the Akata Formation of 4.5–5.5 km (2.8–3.4 mi) with a 3.7–4.7-km (2.3–2.9-mi) shortening exiting the structure to the southwest on the upper detachment, where slip on a detachment at this level is observed (Corredor et al., 2005b; Gonzalez-Mieres and Suppe, 2006; Suppe, 2011). Therefore, under this plausible roof-detachment structural solution, we obtained a well-constrained history of both shortening and excess area (Figure 12b, c). The other principal possible structural solution for the observed excess area is a large-scale flow of a mobile Akata Formation from the more interior parts of the Niger Delta thrust belt into the core of the Agbami anticline (Gonzalez-Mieres and Suppe, 2006; Suppe, 2011). This solution is not easily evaluated with the present data because it does not present a well-defined relationship between shortening and growth in excess area. Further understanding of the origin of the excess area requires better imaging of the deep structure.

Each of the growth models has an associated history of shortening to sedimentation, $d\overline{S}/dH$, which is most rapid for the end-member models (0.5–0.7) and slower for the linearly proportional model (~0.25). Given published stratigraphic ages for Agbami from Bilotti et al. (2005), we calculate an average sedimentation rate dH/dt for the growth strata of 1.4 mm/yr (0.05 in./yr). This yields long-term shortening rates, $d\overline{S}/dt$, of about 0.7–1 mm/yr (0.03–0.04 in./yr) for the end-member models. A shortening rate of about 0.4 mm/yr (0.02 in./yr) for lid sequence is given under the linearly proportional roof-thrust growth model and about 2.2–2.8 mm/yr (0.09–0.11 in./yr) for the Akata Formation, with the difference exiting on the roof thrust toward the southwest.

DISCUSSION

The shapes and thickness variations of strata deposited over growing structures are generally recognized to provide constraints on their histories of deformation. However, the exact nature and extent of these constraints are a topic that requires further research. Here we have taken the limited step of exploring how we may use the vertical variation in the area of relief, which is a subset of the available information, to constrain the shortening histories of detachment folds. In particular, we have not made use of data on the actual layer shapes and how they vary vertically in this analysis, which may further constrain the problem significantly in at least some instances (e.g., see the analysis of growth of the Yaken anticline based on a comparison of layer shapes by Hubert-Ferrari et al., 2007).

In pregrowth strata, extracting the total shortening as a function of height from area-of-relief data on many horizons is possible even in the face of such complexities as layer-parallel simple shear, multiple detachments, and excess area in the fold core. This is possible because in pregrowth strata, we are not concerned with how the histories of these different processes have contributed to the present observed area of relief (Gonzalez-Mieres and Suppe, 2006). In contrast, it is substantially more difficult to extract the total shortening of growth horizons for the following three reasons.

1) The vertical gradient in the area of relief in growth strata is no longer entirely deformational; it also includes a stratigraphic component, reflecting changes in primary stratigraphic thickness that result from deposition over the growing structure (Figures 3, 13; equations 5, 6, 24). A key element in deconvolving these two components of the vertical gradient in relief is consideration of $d\overline{S}/dH$, which can be thought of as the ratio of the rate of change in mean shortening, $d\overline{S}/dt$, to the sedimentation rate, dH/dt. Our examples show that $d\overline{S}/dH$ is relatively constant over substantial stratigraphic intervals, which allows the shortening history to be determined by finding the best-fitting $d\overline{S}/dH$ value for a stratigraphic interval. A further indication of relatively constant $d\overline{S}/dH$ is the observation that the main

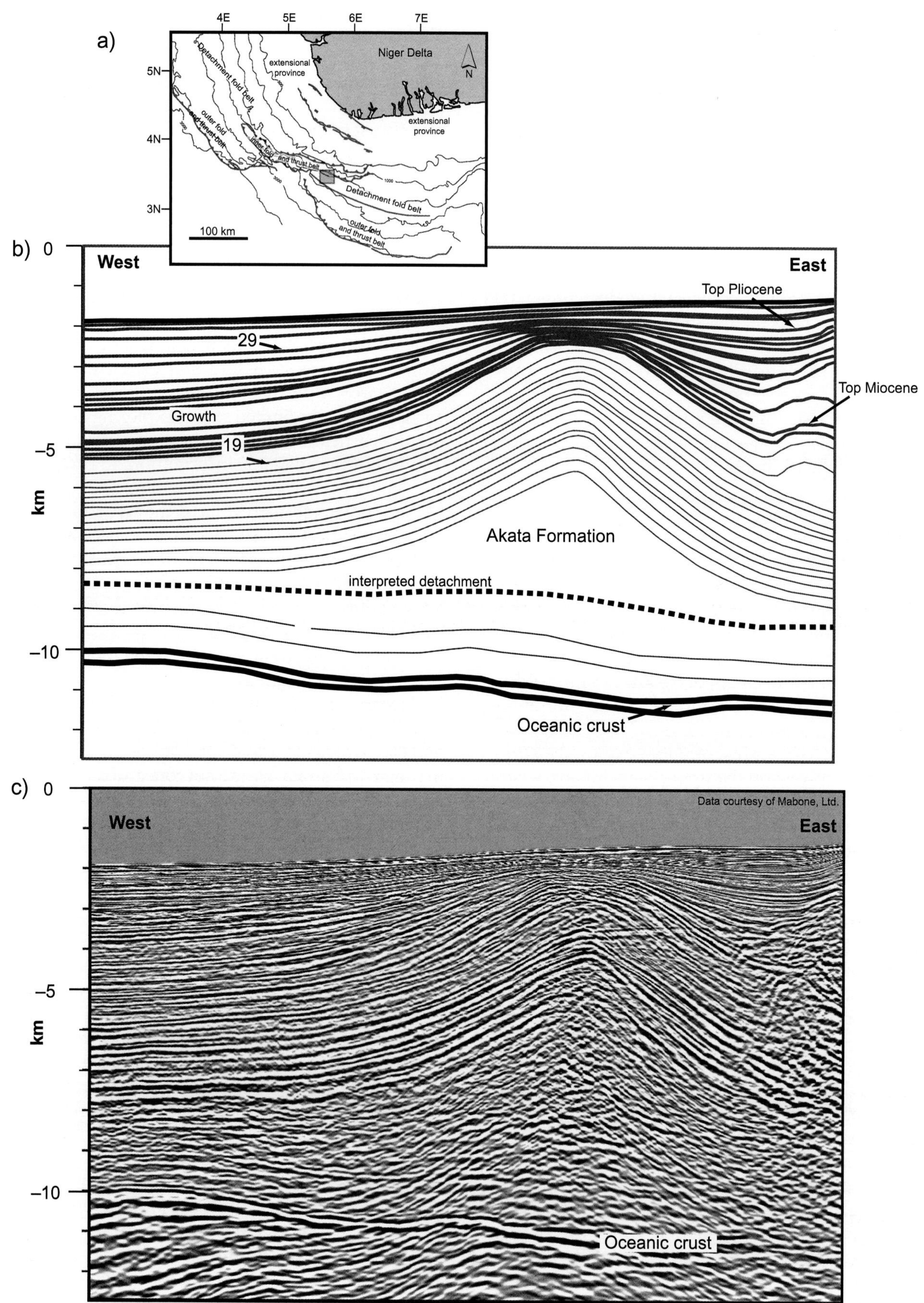

a)
4E
5E
6E
7E
5N
4N
3N
Niger Delta
N
extensional province
extensional province
Detachment fold belt
outer fold and thrust belt
inner fold and thrust belt
Detachment fold belt
outer fold and thrust belt
100 km
b) 0
West
East
Top Pliocene
29
Top Miocene
Growth
19
−5
km
Akata Formation
interpreted detachment
−10
Oceanic crust
c) 0
Data courtesy of Mabone, Ltd.
West
East
−5
km
−10
Oceanic crust

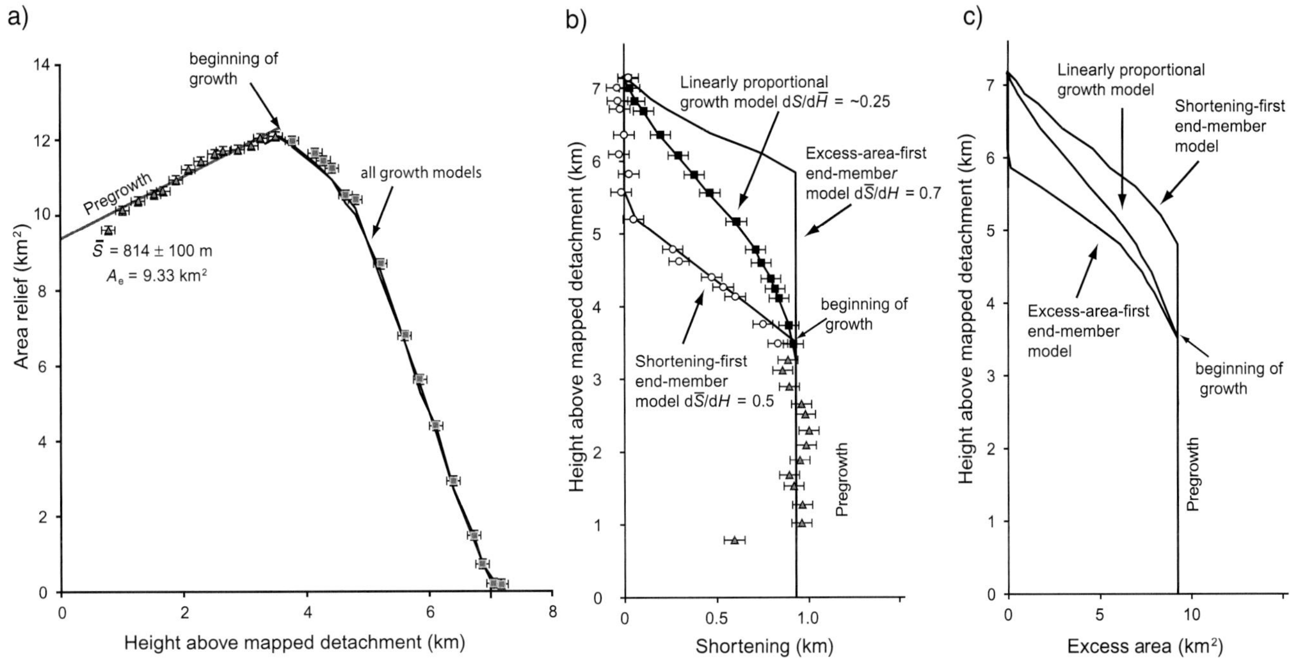

Figure 12. Area-of-relief data and computed shortening and excess area for the Agbami anticline. (a) Area relief as a function of height above the interpreted detachment level (Figure 11). The linear regression of the pregrowth data indicates a shortening of the lid sequence of Agbami of 814 ± 100 m (2671 ± 328 ft) and a nonzero intercept, indicating a very large excess area in the fold core, $A_e = 9.33$ km^2 (3.6 mi^2). (b, c) Shortening and excess area graphs of the Agbami anticline, showing the range of possible growth histories that satisfy the area-of-relief data in panel a. See the text for discussion of growth models.

features of the area-of-relief graph in our examples are analogous to the main features of the simple constant $d\overline{S}/dH$ models of detachment folds shown in Figure 4.

2) If the area of relief is the net effect of several different independent processes, such as slip on multiple detachment levels and excess area, then the observed history of area of relief is the sum of the histories of these different processes, which cannot be directly determined from area-of-relief data alone. However, the range of possible histories of these contributions can be determined (equations 8, 10, 29, 30), as we have shown for Cascadia, Yaken, and Agbami. In the cases of Cascadia and Yaken, the range of solutions is further constrained by other geologic data, and the total range of possible solutions is relatively small (Figures 8, 10). In contrast, the range of shortening histories that are possible for Agbami is large because shortening contributes only 10–15% of the total area of relief in the lid sequence with 85–90% being excess area. However, a regionally plausible structural model that links excess area to shortening provides a well-defined model-based history of shortening for Agbami (Figure 12b).

3) If a component of layer-parallel simple shear exists, then the shortening is vertically heterogeneous such that $\overline{S}(z) \neq S(z)$ and a variety of possible histories of this shear, $S(z,t)$, will satisfy the observed areas of relief of the growth horizons. For example, the history of vertically heterogeneous shortening, $S(z,t)$, given for the model in Figure 2 is only one of the set of possible histories that satisfy the areas of relief, $A(z) = \overline{S}(z)H$, of the growth horizons. A variety of histories of layer-parallel simple shear can be found that produce the same observed areas of relief because the mean shortening of a growth horizon is the integral of the shortening above the

Figure 11. Agbami anticline, offshore Niger Delta (reprinted from Shaw et al., 2005). (a) Location of the seismic profile and the structural framework. (b) Interpretation of the seismic profile. The interpreted detachment level is not well constrained because of difficulties in imaging the Akata Formation. A deeper detachment level was proposed by Bilotti et al. (2005), and other interpretations are possible; see discussion in the text and Suppe (2011). The darker black horizons represent the growth strata. Ages are from Bilotti et al. (2005). (c) Depth seismic profile (horizontal scale = vertical scale) (Bilotti et al., 2005).

detachment (equation 4). We did not have to confront this problem in this study because analysis of the pregrowth strata showed that shortening is approximately constant as a function of height, $\overline{S}(z) \approx S(z)$, with negligible layer-parallel simple shear (Gonzalez-Mieres and Suppe, 2006).

CONCLUSIONS FOR SHORTENING HISTORIES

We have determined well-constrained shortening histories and rates based on area-of-relief data for three of our four examples. The one exception is Agbami, which is not a surprise because shortening is only a second-order contributor (10–15%) to its total area of relief. The remaining structures (Nankai, Cascadia, and Yaken) show shortening histories that have much in common. Furthermore, they show several structural features in common, including their locations at the toes of active thrust belts, their very low limb dips, and the dominance of heterogeneous layer-parallel pure shear with little flexure and no layer-parallel simple shear. However, they are substantially different in the duration of deformation (0.16 to 5.5 Ma), total shortening (77 to 1200 m [253 to 3937 ft]), total stratigraphic thickness (1 to 5.5 km [0.6 to 3.4 mi]), existence of excess area in the basal layer, and nature and number of detachments.

Each of these structures shows episodes in which the rate of shortening relative to sedimentation, $d\overline{S}/dH$, is close to constant, especially in the early stages of their deformation. This suggests that both shortening rates and sedimentation rates are close to constant, which is best corroborated in Yaken where independent evidence of a constant long-term sedimentation rate is observed. The early episodes of constant $d\overline{S}/dH$ have lasted about 3.5 Ma in Yaken, about 0.3 Ma in Cascadia, and about 0.15 Ma in Nankai. In addition, evidence of hiatuses in fold growth that lasted about 0.3 Ma in Cascadia and about 0.5 Ma in Yaken is observed. In the case of Cascadia, the hiatus in fold growth is a time at which the detachment was nevertheless active and deformation spread out more broadly within the protothrust zone at the front of the accretionary wedge. All of these structures were entirely buried by sedimentation during these episodes of constant $d\overline{S}/dH$ as well as during the hiatuses.

All of these structures show a very large acceleration in shortening rate late in their histories. This acceleration in Nankai and Yaken led to their emergence above the regional depositional gradient of the basin. If current shortening rates continue for Cascadia, it will emerge in the near future. This acceleration dominates the area-of-relief data; in terms of the total pregrowth area of relief, 60% has developed in the most recent 5% of time for Yaken, about 50% has developed in the last approximately 20% of time for Cascadia, and 45% has developed in the last approximately 5% of time for Nankai. The acceleration in shortening rate is about an order of magnitude for Yaken (0.16 mm/yr [0.006 in./yr] to ~1.7–2.2 mm/yr [0.07–0.09 in./yr]) and Nankai (~0.27 mm/yr [0.01 in./yr] to ~3.5–4 mm/yr [0.14–0.16 in./yr]). The acceleration of Cascadia appears to be dominated by activation of the deeper detachment level.

Note that the shortening and shortening rates for these structures are based on the assumption of conservation of area; therefore, they represent only the constant-area components of the deformation. In addition, a significant component of horizontal compaction in these frontal structures is observed. Horizontal gradients in seismic velocity show that the horizontal compaction component is larger than the constant-volume component of the total shortening in Nankai and Cascadia (Gonzales-Mieres and Suppe, 2006), and significant horizontal gradients in seismic velocity exist for Yaken as well (A. Hubert-Ferrari, 2002, personal communication). The horizontal compaction appears to be associated with the propagation of the detachment ahead of the frontal detachment zone within the protothrust zone (Cochrane et al., 1994; Moore et al., 1995; Bangs and Gulick, 2005). In the case of Cascadia, the hiatus in shortening of the detachment fold was a time of distributed deformation within the broader protothrust zone, as discussed above. Therefore, some but not all of the variations in deformation rate that we observe could perhaps reflect detachment propagation and compaction episodes.

These frontal structures show constant-area shortening rates that represent a small fraction of the plate tectonic shortening rates of their larger structural settings. The Nankai detachment fold shortening rate (~0.27 mm/yr [0.01 in./yr] to ~3.5–4 mm/yr [0.14–0.16 in./yr]) is 0.5–7% of the 5–6-cm/yr (2–2.4-in./yr) plate rate. The Cascadia shortening rate (~0.27 mm/yr [0.01 in./yr] upper detachment, ~0.5–0.6 mm/yr [0.020–0.024 in./yr] lower detachment, and ~0.8–0.9 mm/yr [0.031–0.035 in./yr] combined) is about 0.7–2% of the 4-cm/yr (1.6-in./yr) plate rate. The Yaken shortening rate (0.16 mm/yr [0.006 in./yr] to ~1.7–2.2 mm/yr [0.07–0.09 in./yr]) is 0.8–10% of the approximately 2-cm/yr (0.8-in./yr) geodetically constrained shortening rate of the Tianshan. Therefore, the shortening of these accretionary wedges and mountain belts is not currently concentrated in their frontal zones. This is in contrast with the observation that the Holocene shortening of the central Nepal Himalayas is almost entirely concentrated in the frontal thrust ramp (Lavé and Avouac, 2000). We also note that the deformation fronts are slowly propagating. In the case of

Cascadia, the deformation front has been fixed for the last 0.9 Ma, and in Yaken, the front has not propagated in the last 5.5 Ma since Yaken initiated. These observations are difficult to explain with steady-state models of mountain belts and accretionary wedges.

ACKNOWLEDGMENTS

We thank Aurélia Hubert-Ferrari, Wang Xin, Frank Bilotti, Mathieu Daëron, and a number of participants at the International Conference on Fault-Related Folding in Beijing for stimulating advice and encouragement on this research. We thank Greg F. Moore of the University of Hawaii for providing the seismic image of the Nankai Trough. Suppe is grateful to the Tectonic Observatory at Caltech, the Ludwig Maximilians University in Munich, and the Alexander von Humboldt Foundation for support during the writing of this manuscript. Finally, we would like to extend our gratitude to Josep Poblet and John Shaw for their insightful reviews of this manuscript, which helped improve the final publication.

APPENDIX

RELATIONSHIPS BETWEEN AREA OF RELIEF AND SHORTENING

Here we present key relationships between shortening and observed areas of relief as a function of height. The relationships are different for pregrowth and growth strata and are presented separately. To simplify this presentation, much of this theory is for simple detachment folds in which the area of structural relief, A, is the mean shortening times the depth to detachment, $A = \overline{S}H$, assuming conservation of area. Nevertheless, the principal differences between pregrowth and growth strata are quite general and are not confined to detachment folds. These relationships can be modified to deal with other structural situations as we do in the final section for detachment folds that develop excess area in their basal stratigraphy (e.g., Yaken and Agbami) or folds with multiple detachments (e.g., Cascadia). Furthermore, we make no assumptions concerning conservation of bed length or layer thickness (see Gonzalez-Mieres and Suppe, 2006, for a full discussion of the meaning of the so-called bed-length shortening).

We consider the case in which shortening, $S(z)$, may vary as a function of stratigraphic height to preserve the distinction between shortening and mean shortening; that is, we allow the possibility that $S(z) \neq \overline{S}(z)$. The geometry of the problem is shown in Figure 13, corresponding to the forward model of a detachment fold shown in Figure 2. In the present deformed state, we can observe a set of areas of structural relief as a function of height, $\langle H_i, A_i \rangle$, from which we can obtain the mean shortening, $\overline{S}(z)$, and perhaps even the shortening, $S(z)$. However, determining the shortening, $S(z)$, is more challenging because the measured area of relief of a horizon is the net effect of all the shortening below that horizon plus any other contributions such as flow within a weak basal layer.

AREA OF RELIEF AND SHORTENING IN PREGROWTH STRATA

Within pregrowth strata (Figure 13a), a change in area of structural relief, $\Delta A_{n,n+1}$, between adjacent stratigraphic horizons n and $n+1$ is observed

$$\Delta A_{n,n+1} = A_{n+1} - A_n \qquad (11)$$

which is a result of the shortening, S, of the layer

$$\Delta A_{n,n+1} = S\Delta H \qquad (12)$$

where $\Delta H = H_{n+1} - H_n$ is its thickness. More generally, the shortening, $S(z)$, as a function of height in pregrowth strata is the vertical gradient in the area of structural relief

$$S(z) = (\mathrm{d}A/\mathrm{d}z)_{\text{pregrowth}} \qquad (13)$$

The total area of structural relief for pregrowth horizons in a simple detachment fold without excess area in the basal layer is

$$A_n = \overline{S}_n H_n \qquad (14)$$

where H_n is the primary height of the horizon above the detachment level and $\overline{S}_n$ is the mean shortening over that interval. The total area of structural relief of horizon $n+1$ is then

$$A_{n+1} = \overline{S}_n H_n + S\Delta H \qquad (15)$$

More generally, the area of structural relief as a function of height, $A(z)$, is the shortening integrated over the height above the detachment.

$$A(z) = \overline{S}(z)H = \int_{z=0}^{H} S(z)\mathrm{d}z \qquad (16)$$

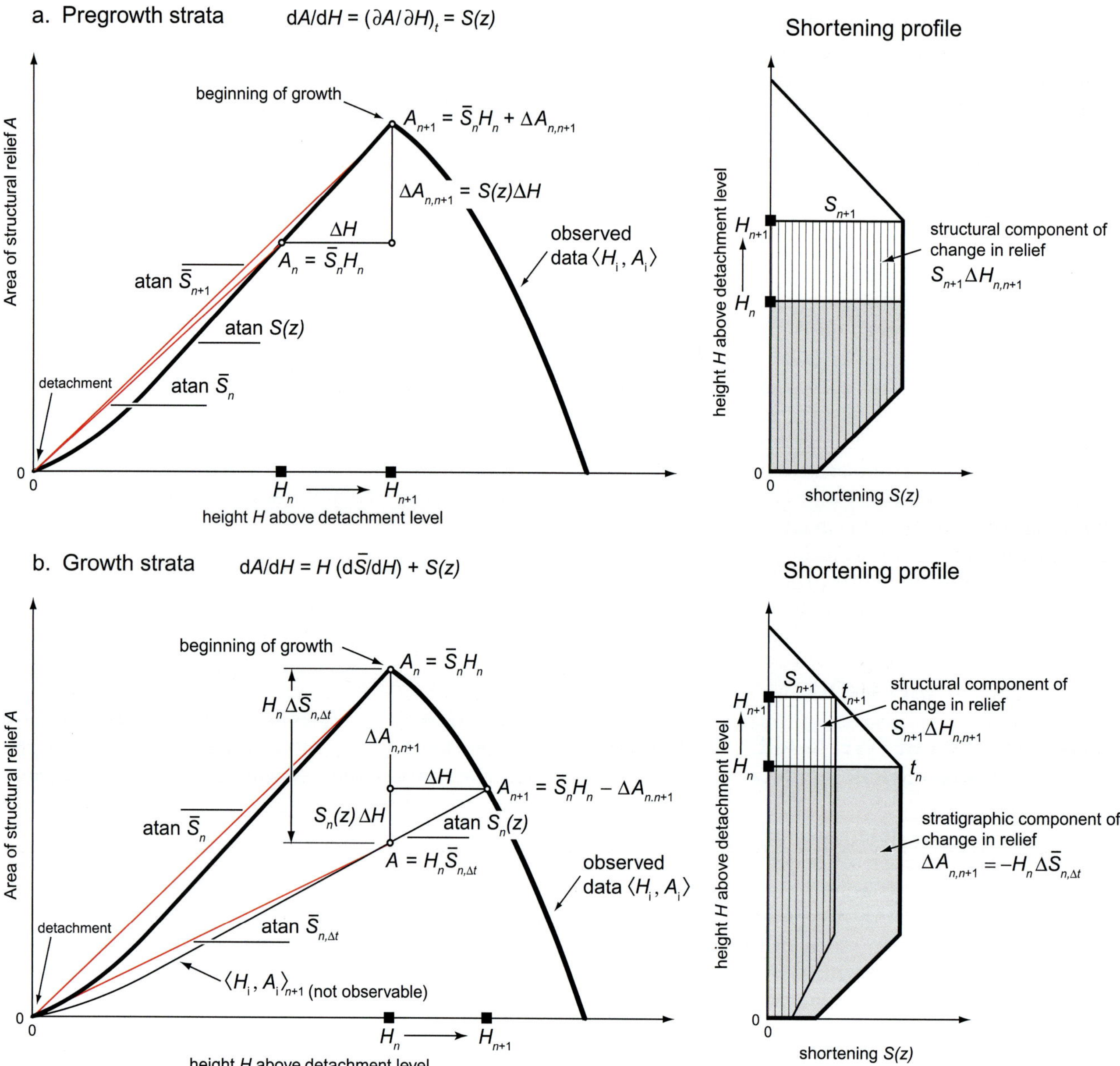

Figure 13. Components of area of relief and shortening for the detachment fold model of Figure 2 used to illustrate the Appendix equations. The principal concern is identifying the origins of the change in area of relief over a stratigraphic interval from H_n to H_{n+1}. (a) In pregrowth strata, a structural increase in the area of relief, $S_{n+1}\Delta H_{n,n+1}$, associated with the increase in height, $\Delta H_{n,n+1}$, is observed. (b) In growth strata, the upward change in area of relief can be either positive or negative (see Figure 4) because two components to the upward change in relief of opposite sign now exist. The structural component is the same as the pregrowth case, which is the area of relief that is generated within the interval, $\Delta H_{n,n+1}$, by the shortening, S_{n+1}, that has occurred since its deposition. The stratigraphic component, $-H_n\Delta\bar{S}_{n,\Delta t}$, is caused by the decrease in mean shortening below horizon H_n during the time of deposition, Δt, of the stratigraphic interval. This stratigraphic component represents a stratigraphic thinning of the growth layer over the structure as shown in Figure 3a. atan = arc tangent.

AREA OF RELIEF AND SHORTENING IN GROWTH STRATA

Within growth strata (Figure 13b), the relationships between shortening and vertical gradient in the area of structural relief given above (equations 12, 13) no longer hold because an additional process exists that contributes to the observed change in the area of relief, $\Delta A_{n,n+1}$, across a growth layer bounded by horizons n and $n+1$. In particular, not all the measured change in area of relief is structural; part of the relief is primary stratigraphic thickness variation due to growth. During the time of deposition, Δt, of this layer, horizon n at its base deforms by an amount, $\Delta A_{n,\Delta t}$, controlled by shortening and other processes below that horizon (Figure 3). This represents a stratigraphic component to the change in relief, $\Delta A_{\text{stratigraphic}}$, that reduces the observed change in area of relief across the layer, relative to the pregrowth case considered in equation 12; we now have

$$\Delta A_{\text{observed}} = \Delta A_{\text{structural}} - \Delta A_{\text{stratigraphic}} \tag{17}$$

$$\Delta A_{n,n+1} = S_{n+1}\Delta H_{n,n+1} - \Delta A_{n,\Delta t} \tag{18}$$

These structural and stratigraphic components of the observed change in relief are shown in the shortening profile of Figure 13b.

In the case of a simple detachment fold, we can specify this stratigraphic component, $\Delta A_{n,\Delta t}$, as

$$\Delta A_{n,\Delta t} = H_n \Delta \overline{S}_{n,\Delta t} \tag{19}$$

where H_n is the height above the detachment and $\Delta \overline{S}_{n,\Delta t}$ is the increment of mean shortening between the detachment and the base of our layer during the time, Δt, of deposition of the layer. Finally, as a result of later deformation, the present observed area of relief of the two horizons will be

$$A_n = \overline{S}_n H_n \tag{20}$$

and from equations 7, 8, and 9

$$A_{n+1} = A_n + S_{n+1}\Delta H_{n,n+1} - H_n \Delta \overline{S}_{n,\Delta t} \tag{21}$$

The change in relief is then

$$\Delta A_{n,n+1} = S_{n+1}\Delta H_{n,n+1} - H_n \Delta \overline{S}_{n,\Delta t} \tag{22}$$

Dividing by the thickness of the layer, we obtain

$$\frac{\Delta A_{n+1}}{\Delta H_{n,n+1}} = S_{n+1} - H_n \frac{\Delta \overline{S}_{n,\Delta t}}{\Delta H_{n,n+1}} \tag{23}$$

More generally, the vertical gradient in the observed area of relief in growth strata is

$$\frac{\mathrm{d}A}{\mathrm{d}z} = S - H\frac{\mathrm{d}\overline{S}}{\mathrm{d}H} \tag{24}$$

where $\mathrm{d}\overline{S}/\mathrm{d}H$ can be interpreted as the ratio of the mean shortening rate, $\mathrm{d}\overline{S}/\mathrm{d}t$, to the sedimentation rate, $\mathrm{d}H/\mathrm{d}t$, both of which may vary during growth. If the shortening is vertically homogeneous, which may be the case for our examples, then $\overline{S} = S$ and equation 24 becomes

$$\frac{\mathrm{d}A}{\mathrm{d}z} = S - H\frac{\mathrm{d}S}{\mathrm{d}H} = \overline{S} - H\frac{\mathrm{d}\overline{S}}{\mathrm{d}H} \tag{25}$$

MULTIPLE DETACHMENTS AND EXCESS AREA

We can write analogous equations to describe the vertical gradients in the area of relief in more complex structural situations, for example, (1) folds containing multiple detachment levels, as in Cascadia, or (2) folds containing excess area of relief in a basal layer, as in Yaken and Agbami. What is common to these complexities is the existence of multiple possible contributions to the observed area of relief of any growth horizon

$$A(z) = A_1(z) + A_2(z) + \cdots \tag{26}$$

Typically, we may know the total contribution, $\langle A_{1\text{tot}}(z), A_{2\text{tot}}(z), \ldots \rangle$, of these processes from analysis of pregrowth strata, but we do not know how they are distributed in time. For example, we may know the total shortening on each detachment and the total excess area, but we do not know when they occurred within the period of growth. We confront this ambiguity by developing end-member models to constrain the range of possible histories of growth that satisfy the data. Each model can be specified as a set of fractions, $\langle f_1(z), f_2(z), \ldots \rangle$, of the total contribution of each process for each growth horizon, such that the observed area of relief is satisfied

$$A(z)_{\text{observed}} = f_1(z)A_{1\text{tot}}(z) + f_2(z)A_{2\text{tot}}(z) + \cdots \tag{27}$$

where the fractions monotonically decrease from 1 in pregrowth strata reaching 0 when the contribution of the component is exhausted.

In the case of two detachments, such as Cascadia, for which we know the total shortening, $\overline{S}(z) = S(z)$, equation 27 becomes

$$A(z) = f_L(z)\overline{S}_L(H_L + H_U) + f_U(z)\overline{S}_U H_U \tag{28}$$

where $\overline{S}_L$ and $\overline{S}_U$ are the total shortenings consumed in folding for the lower and upper detachments, H_L is the thickness of the lower fault block, and H_U is the height of the horizon in question above the upper detachment such that $z = H_L + H_U$. We can then search for models $\langle f_L(z), f_U(z) \rangle$ of the history of these fractions that satisfy the data $\langle A, z \rangle$.

The case of excess area in the basal layer, such as Yaken and Agbami, is similar. We know the total shortening, $\overline{S}(z) = S(z)$, and the total excess area, A_e; therefore, equation 27 becomes

$$A(z) = f_{\overline{S}}(z)\overline{S}H + f_e(z)A_e \quad (29)$$

where $z = H$ is the height of the horizon above the detachment. We can then search for models $\langle f_{\overline{S}}(z), f_e(z)\rangle$ of the history of these fractions that satisfy the data $\langle A,z\rangle$.

In the remainder of this Appendix, we develop expressions for the vertical gradients in the area of relief for these more complex structural situations. As always, the incremental change in area of relief of a growth layer (Figure 3), bounded by horizons n and $n + 1$, is given by equations 17 and 18

$$\Delta A_{n,n+1} = S_{n+1}\Delta H_{n,n+1} - \Delta A_{n,\Delta t}$$

where horizon n at its base deforms by an amount $\Delta A_{n,\Delta t}$ during the time of deposition, Δt, of this layer, controlled by processes below that horizon. These processes may include such complexities as multiple detachment levels and excess area of relief.

In the case of two detachments, $\Delta A_{n,\Delta t}$ is expressed as the sum of the contributions of the upper and lower fault blocks

$$\Delta A_{n,\Delta t} = \Delta A_{L,\Delta t} + \Delta A_{U,\Delta t} = H_L\Delta\overline{S}_{L,\Delta t} + H_U\Delta\overline{S}_{U,\Delta t} \quad (30)$$

where $\Delta A_{L,\Delta t} = H_L\Delta\overline{S}_{L,\Delta t}$ is the contribution of incremental shortening, $\Delta\overline{S}_{L,\Delta t}$, of the lower fault block of thickness, H_L, and $\Delta A_{U,\Delta t} = H_U\Delta\overline{S}_{U,\Delta t}$ is the contribution of incremental shortening, $\Delta\overline{S}_{U,\Delta t}$, of the upper fault block of thickness, H_U. In this case, equation 24 for the vertical gradient in the area of relief in growth strata becomes

$$\frac{dA}{dz} = S(z) - H_L\frac{d\overline{S}_L}{dH} - H_U\frac{d\overline{S}_U}{dH} \quad (31)$$

where $d\overline{S}_L/dH$ and $d\overline{S}_U/dH$ can be considered the ratios of mean shortening rates in the lower and upper fault blocks to the sedimentation rate.

The case of excess area in the basal layer is similar to the case of multiple detachments. The incremental deformation, $\Delta A_{n,\Delta t}$, of horizon n during the deposition of the layer is expressed as the sum of the incremental contribution of excess area, $\Delta A_{e,\Delta t}$, and the shortening, $H_n\Delta\overline{S}_{n,\Delta t}$.

$$\Delta A_{n,\Delta t} = \Delta A_{e,\Delta t} + H_n\Delta\overline{S}_{n,\Delta t} \quad (32)$$

The vertical gradient in the area of relief in growth strata becomes

$$\frac{dA}{dz} = S(z) - \frac{dA_e}{dH} - H\frac{d\overline{S}}{dH} \quad (33)$$

where dA_e/dH can be considered the ratio of the rate of change in excess area to the sedimentation rate.

REFERENCES CITED

Adam, J., D. Klaeschen, N. Kukowski, and E. Flueh, 2004, Upward delamination of Cascadia Basin sediment infill with landward frontal accretion thrusting caused by rapid glacial age material flux: Tectonics, v. 23, p. TC3009, doi:10.1029/2002TC001475.

Allen, M. B., B. F. Windley, C. Zhang, Z. Zhao, and G. R. Wang, 1991, Basin evolution within and adjacent to the Tien Shan range, NW China: Journal of the Geological Society, v. 148, p. 369–378, doi:10.1144/gsjgs.148.2.0369.

Allmendinger, R. W., 1998, Inverse and forward numerical modeling of trishear fault-propagation folds: Tectonics, v. 17, no. 4, p. 640–656, doi:10.1029/98TC01907.

Bangs, N. L. B., and S. P. S. Gulick, 2005, Physical properties along the developing decollement in the Nankai trough: Inferences from 3-D seismic reflection data inversion and leg 190 and 196 drilling data, *in* H. Mikada, G. F. Moore, A. Taira, K. Becker, J. C. Moore, and A. Klaus, eds., Proceedings of the Ocean Drilling Program, Scientific Results, 190/196: http://www-odp.tamu.edu/publications/190196SR/354/354.htm (accessed November 16, 2005).

Bilotti, F., and J. Shaw, 2005, Deep-water Niger Delta fold and thrust belt modeled as a critical-taper wedge: The influence of elevated basal fluid pressure on structural styles: AAPG Bulletin, v. 89, p. 1475–1491, doi:10.1306/06130505002.

Bilotti, F., J. Shaw, R. Cupich, and R. Lakings, 2005, Detachment fold, Niger Delta, *in* J. H. Shaw, C. Connors, and J. Suppe, eds., Seismic interpretation of contractional fault-related folds—An AAPG seismic atlas: AAPG Studies in Geology 53, p. 103–104.

Bischke R. E., 1994, Interpreting sedimentary growth structures from well log and seismic data (with examples): AAPG Bulletin, v. 78, p. 873–892.

Bischke, R. E., W. Finley, and D. J. Tearpock, 1999, Growth analysis (Δd/d): Case histories of the resolution of correlation problems as encountered while mapping around salt: Transactions of the Gulf Coast Association of Geological Societies, v. 49, p. 102–110.

Burchfiel, B. C., E. T. Brown, Q. Deng, X. Feng, J. Li, P. Molnar, J. Shi, Z. Wu, and H. You, 1999, Crustal shortening on the margins of the Tien Shan, Xinjiang, China: International Geology Review, v. 41, p. 665–700.

Carson, B., G. K. Westbrook, R. J. Musgrave, and E. Suess, eds., 1995, Proceedings of the Ocean Drilling Program, Scientific Results, 146 (Pt. 1): College Station, Texas, Ocean Drilling Program, v. 146, 426 p.

Charreau, J., 2005, Evolution tectonique du Tianshan au Cénozoïque liée à la collision Inde-Asie Apports de la magnétostratigraphie et de la géochronologie isotopique U-Th/He: Ph.D. thesis, l'Université de Orléans, Orléans, 279 p.

Cochrane, G. R., J. C. Moore, M. E. McClay, and G. F. Moore, 1994, Velocity and inferred porosity model of the Oregon accretionary prism from multichannel seismic reflection data: Implication on sediment dewatering and overpressure: Journal of Geophysical Research, v. 99, p. 7033–7043, doi:10.1029/93JB03206.

Cohen, H. A., and K. McClay, 1996, Sedimentation and shale tectonics of the northwestern Niger Delta front: Marine and Petroleum Geology, v. 13, p. 313–328, doi:10.1016/0264-8172(95)00067-4.

Corredor, F., J. Shaw, and F. Bilotti, 2005a, Structural styles in the deep-water fold and thrust belts of the Niger Delta: AAPG Bulletin, v. 89, p. 753–780, doi:10.1306/02170504074.

Corredor, F., J. Shaw, and J. Suppe, 2005b, Shear, fault-bend folding, deep water Niger Delta, *in* J. H. Shaw, C. Connors, and J. Suppe, eds., Seismic interpretation of contractional fault-related folds—An AAPG seismic atlas: AAPG Studies in Geology 53, p. 87–92.

Damuth, J. E., 1994, Neogene gravity tectonics and depositional processes on the deep Niger Delta continental margin: Marine and Petroleum Geology, v. 11, p. 320–346, doi:10.1016/0264-8172(94)90053-1.

DeMets, C., R. G. Gordon, D. F. Argus, and S. Stein, 1990, Current plate motion: Geophysical Journal International, v. 101, p. 425–478, doi:10.1111/j.1365-246X.1990.tb06579.x.

Epard, J. L., and J. R. H. Groshong, 1993, Excess area and depth to detachment: AAPG Bulletin, v. 77, p. 1291–1302.

Epard, J. L., and J. R. H. Groshong, 1995, Kinematic model of detachment folding including limb rotation, fixed hinges and layer-parallel strain: Tectonophysics, v. 247, p. 85–103, doi:10.1016/0040-1951(94)00266-C.

Gonzalez-Mieres, R., and J. Suppe, 2006, Relief and shortening in detachment folds: Journal of Structural Geology, v. 28, p. 1785–1807, doi:10.1016/j.jsg.2006.07.001.

Grimes, D., E. Gringer, and J. Spokes, 2004, Agbami field Nigeria-addressing challenges and uncertainty: Houston Geological Society Bulletin, v. 46, p. 19–21.

Groshong, J. R. H., 1994, Area balance, depth to detachment, and strain in extension: Tectonics, v. 13, p. 1488–1497, doi:10.1029/94TC02020.

Groshong, J. R. H., 1996, Construction and validation of extensional cross sections using lost area and strain, with application to the Rhine Graben, *in* P. G. Buchanan and D. A. Nieuwland, eds., Modern developments in structural interpretation, validation and modeling: Geological Society (London) Special Publication 99, p. 79–87.

Gulick, S. S., N. L. Bangs, T. H. Shipley, Y. Nakamura, G. F. Moore, and S. Kuramoto, 2004, Three-dimensional architecture of the Nankai accretionary prism's imbricate thrust zone off Cape Muroto, Japan: Prism reconstruction via en echelon thrust propagation: Journal of Geophysical Research, v. 109, p. B02105.1–B02117, doi:10.1029/2003JB002654.

Hardy, S., and M. Ford, 1997, Numerical modeling of trishear fault propagation folding: Tectonics, v. 16, no. 5, p. 841–854, doi:10.1029/97TC01171.

Higuera-Díaz, I. C., M. P. Fischer, and M. S. Wilkerson, 2005, Geometry and kinematics of the Nuncios detachment fold complex: Implications for lithotectonics in northeastern Mexico: Tectonics, v. 24, p. TC4010, doi:10.1029/2003TC001615.

Hooper, R. J., R. J. Fitzsimmons, N. Grant, and B. C. Vendeville, 2002, The role of deformation in controlling depositional patterns in the south-central Niger Delta, west Africa: Journal of Structural Geology, v. 24, p. 847–859.

Hubert-Ferrari, A., J. Suppe, X. Wang, and C. Jia, 2005, The Yaken detachment fold, south Tianshan, China, *in* J. H. Shaw, C. Connors, and J. Suppe, eds., Seismic interpretation of contractional fault-related folds—An AAPG seismic atlas: AAPG Studies in Geology 53, p. 110–113.

Hubert-Ferrari, A., J. Suppe, R. Gonzalez-Mieres, and X. Wang, 2007, Mechanisms of active folding of the landscape (southern Tianshan, China): Journal of Geophysical Research, v. 122, 39 p., B03S09, doi:10.1029/2006JB004362.

Kamata, H., and K. Kodama, 1994, Tectonics of an arc-arc junction: An example from Kyushu Island at the junction of the southwest Japan arc and the Ryukyu arc: Tectonophysics, v. 233, p. 69–81, doi:10.1016/0040-1951(94)90220-8.

Karig, D. E., and C. L. Angevine, 1986, Geologic constraints on subduction rates in the Nankai trough, *in* H. Kagami et al., eds., Initial reports of the Deep Sea Drilling Project: Washington, D.C., U.S. Printing Office, v. 87, p. 789–796.

Kulm, L. D., P. W. Louhere, and J. S. Peper, eds., 1984, Western North American continental margin and adjacent ocean floor off Oregon and Washington: Woods Hole, Marine Science International, 29 p.

Lavé, J., and J.-P. Avouac, 2000, Active folding of fluvial terraces across the Himalayas of central Nepal: Journal of Geophysical Research, v. 105, p. 5735–5770, doi:10.1029/1999JB900292.

Le Pichon, X., T. Iiyama, H. Chamley, J. Charvet, M. Faure, H. Fujimoto, T. Furuta, Y. Ida, H. Kagami, and S. Lallemant, 1987, Nankai trough and the fossil Shikoku Ridge: Results of Box 6 Kaiko survey: Earth and Planetary Science Letters, v. 83, p. 186–198.

MacKay, M. E., 1995, Structural variation and landward vergence at the toe of the Oregon accretionary prism: Tectonics, v. 14, p. 1309–1320, doi:10.1029/95TC02320.

MacKay, M. E., G. F. Moore, G. R. Cochrane, C. J. Moore, and L. D. Kulm, 1992, Landward vergence and oblique structural trends in the Oregon margin accretionary prism: Implications and effect on fluid flow: Earth and Planetary Science Letters, v. 109, p. 477–491.

Masaferro, J., J. Poblet, M. Bulnes, G. P. Eberli, T. Dixon, and K. R. McClay, 1999, Paleogene–Neogene/present day growth folding in the Bahamian foreland of the Cuban fold and thrust belt: Journal of the Geological Society, v. 156, p. 617–631.

Miyazaki, S. I., and K. Heki, 2001, Crustal velocity field of southwest Japan: Subduction and arc-arc collision: Journal of Geophysical Research, v. 106, p. 4305–4326.

Molnar, P., and P. Tapponier, 1977, Relation of the tectonics of eastern China to the India-Eurasia collision: Application of slip-line field theory to large-scale continental tectonics: Geology, v. 5, p. 212–216, doi:10.1130/0091-7613(1977)5 <212:ROTTOE>2.0.CO;2.

Moore, G. F., T. H. Shipley, P. L. Stoffa, D. E. Karig, A. Taira, S. Kuramoto, H. Tokuyama, and K. Suyehiro, 1990, Structure of the Nankai trough accretionary zone from multichannel seismic reflection data: Journal of Geophysical Research, v. 95, p. 8753–8765.

Moore, G. F., D. Karig, T. H. Shipley, A. Taira, P. L. Stoffa, and W. T. Wood, 1991, Structural framework of the ODP leg 131 area, Nankai trough, *in* A. Taira et al., eds., Proceedings of the Ocean Drilling Program, Initial Reports, 131: College Station, Texas, Ocean Drilling Program, p. 15–20.

Moore, G. F., et al., 2001, Data report: Structural setting of the leg 190 Muroto transect, *in* G. F. Moore et al., eds., Proceedings of the Ocean Drilling Program, Initial Reports, 190, p. 1–14: http://www-odp.tamu.edu/publications/190_IR/VOLUME/CHAPTERS/IR190_02.PDF (accessed September 5, 2005).

Moore, J. C., K. Moran, M. E. MacKay, and H. Tobin, 1995, Frontal thrust, Oregon accretionary prism: geometry, physical properties, and fluid pressure, *in* B. Carson, G. K. Westbrook, R. J. Musgrave, and E. Suess, eds., Proceedings of the Ocean Drilling Project, Scientific Results, 146 (Pt. 1): College Station, Texas, Ocean Drilling Program, p. 359–366.

Novoa, E., J. Suppe, and J. H. Shaw, 2000, Inclined-shear restoration of growth folds: AAPG Bulletin, v. 84, p. 787–804.

Nunns, A. G., 1991, Structural restoration of seismic and geologic sections in extensional regimes: AAPG Bulletin, v. 75, p. 278–297.

Pashin, J. C., J. R. H. Groshong, and S. Wang, 1995, Thin-skinned structures influence gas production in Alabama coalbed methane fields: Intergas '95, International Unconventional Gas Symposium, p. 39–52.

Poblet, J., and S. Hardy, 1995, Reverse modeling of detachment folds; application to the Pico del Aguila anticline in the south central Pyrenees (Spain): Journal of Structural Geology, v. 17, p. 1707–1724, doi:10.1016/0191-8141(95)00059-M.

Poblet, J., K. McClay, F. Storti, and J. A. Muñoz, 1997, Geometries of syntectonic sediments associated with single-layer detachment folds: Journal of Structural Geology, v. 19, p. 369–381, doi:10.1016/S0191-8141(96)00113-7.

Poblet, J., M. Bulnes, K. R. McClay, and S. Hardy, 2004, Plots of crestal structural relief and fold area versus shortening—A graphical technique to unravel the kinematics of thrust-related folds, *in* K. R. McClay, ed., Thrust tectonics and hydrocarbon systems: AAPG Memoir 82, p. 372–399.

Reigber, C., G. W. Michel, R. Galas, D. Angermann, J. Klotz, J. Y. Chen, A. Papschev, R. Arslanov, V. E. Tzurkov, and M. C. Ishanov, 2001, New space geodetic constraints on the distribution of deformation in central Asia: Earth and Planetary Science Letters, v. 191, p. 157.

Riba, O., 1976, Syntectonic unconformities of the Alto Cardener, Spanish Pyrenees: A genetic interpretation: Sedimentary Geology, v. 15, p. 213–233, doi:10.1016/0037-0738(76)90017-8.

Rowan, M. G., 1997, Three-dimensional geometry and evolution of a segmented detachment fold, Mississippi Fan fold belt, Gulf of Mexico: Journal of Structural Geology, v. 19, p. 463–480, doi:10.1016/S0191-8141(96)00098-3.

Rowan, M. G., R. Kligfield, and P. Weimer, 1993, Processes and rates of deformation: Preliminary results from the Mississippi Fan fold belt, deep Gulf of Mexico: Gulf Coast Section SEPM Foundation 14th Annual Research Conference, Programs with Papers, p. 209–218.

Seno, T., 1977, The instantaneous rotation vector of the Philippine sea plate relative to the Eurasian plate: Tectonophysics, v. 42, p. 209–226, doi:10.1016/0040-1951(77)90168-8.

Shaw, J. H., and J. Suppe, 1994, Active faulting and growth folding in the eastern Santa Barbara Channel, California: Geological Society of America Bulletin, v. 106, p. 607–626, doi:10.1130/0016-7606(1994)106<0607:AFAGFI>2.3.CO;2.

Shaw, J. H., C. Connors, and J. Suppe, 2005, Seismic interpretation of contractional fault-related folds—An AAPG seismic atlas: AAPG Studies in Geology 53, p. 103.

Shipboard Scientific Party, 2001a, Leg 190 summary, *in* G. F. Moore et al., eds., Proceedings of the Ocean Drilling Program, Initial Reports, 190, p. 1–87: http://www-odp.tamu.edu/publications/190_IR/VOLUME/CHAPTERS/IR190_01.PDF (accessed September 28, 2005).

Shipboard Scientific Party, 2001b, Site 1174, *in* G.F. Moore et al., eds., Proceedings of the Ocean Drilling Program, Initial Reports, 190, p. 1–149: http://www-odp.tamu.edu/publications/190_IR/VOLUME/CHAPTERS/IR190_05.PDF (accessed September 28, 2005).

Silver, E. A., 1972, Pleistocene tectonic accretion of the continental slope off Washington*1: Marine Geology, v. 13, p. 239–249, doi:10.1016/0025-3227(72)90053-9.

Storti, F., and J. Poblet, 1997, Growth strata architectures associated to decollement folds and fault-propagation folds. Inferences on fold kinematics: Tectonophysics, v. 282, p. 353–373, doi:10.1016/S0040-1951(97)00230-8.

Suppe, J., 2011, Mass balance and thrusting in detachment folds, *in* K. McClay, J. H. Shaw, and J. Suppe, eds., Thrust fault-related folding: AAPG Memoir 94, p. 21–37.

Suppe, J., G. T. Chou, and S. C. Hook, 1992, Rates of folding and faulting determined from growth strata, *in* K. R. McClay, ed., Thrust tectonics: London, United Kingdom, Chapman & Hall, p. 105–121.

Suppe, J., F. Sabat, J. Anton Munoz, J. Poblet, E. Roca, and J. Verges, 1997, Bed-by-bed fold growth by kink-band migration: Sant Llorenç de Morunys, eastern Pyrenees: Journal of Structural Geology, v. 19, p. 443–461, doi:10.1016/S0191-8141(96)00103-4.

Suppe, J., C. D. Connors, and Y. Zhang, 2004, Shear fault-bend folding, *in* K. R. McClay, ed., Thrust tectonics and hydrocarbon systems: AAPG Memoir 82, p. 303–323.

Taira, A., I. Hill, J. Firth, U. Berner, W. Bruckmann, T. Byrne, T. Chabernaud, A. Fisher, J.-P. Foucher, and T. Gamo, 1992, Sediment deformation and hydrogeology of the Nankai trough accretionary prism: Synthesis of shipboard results of ODP leg 131: Earth and Planetary Science Letters, v. 109, p. 431–450.

Tearpock and Bischke, 2003, Applied subsurface geological

mapping with structural methods: Upper Saddle River, Prentice Hall, 822 p.

Underwood, M., and K. D. Hoke, 2000, Composition and provenance of turbidite sand and hemipelagic mud in northwestern Cascadia Basin, *in* A. T. Fisher, E. E. Davis, and C. Escutia, eds., Proceeding of the Ocean Drilling Program, Scientific Results: College Station, Texas, Ocean Drilling Program, v. 168, p. 51–65.

Wang, X., J. Suppe, S. Guan, A. Hubert-Ferrari, R. Gonzalez-Mieres, and C. Jia, 2011, Cenozoic structure and tectonic evolution of the Kuqa fold belt, southern Tianshan, China, *in* K. McClay, J. H. Shaw, and J. Suppe, eds., Thrust fault-related folding: AAPG Memoir 94, p. 215–243.

Westbrook, G. K., et al., 1994, Leg 146 introduction; Cascadia margin, *in* G. K. Westbrook et al., eds., Proceeding of the Ocean Drilling Program, Initial Report: College Station, Texas, Ocean Drilling Program, v. 146, p. 5–14.

Yin, A., S. Nie, P. Craig, T. M. Harrison, F. J. Ryerson, Q. Xianglin, and Y. Geng, 1998, Late Cenozoic tectonic evolution of the southern Chinese Tianshan: Tectonics, v. 17, p. 1–27, doi:10.1029/97TC03140.

Yue, L.-F., J. Suppe, and J.-H. Hung, 2011, Two contrasting kinematic styles of active folding above thrust ramps, western Taiwan, *in* K. McClay, J. Shaw, and J. Suppe, eds., Thrust fault-related folding: AAPG Memoir 94, p. 153–186.

4

Vergés, J., M. G. H. Goodarzi, H. Emami, R. Karpuz, J. Efstathiou, and P. Gillespie, 2011, Multiple detachment folding in Pusht-e Kuh arc, Zagros: Role of mechanical stratigraphy, *in* K. McClay, J. H. Shaw, and J. Suppe, eds., Thrust fault-related folding: AAPG Memoir 94, p. 69–94.

Multiple Detachment Folding in Pusht-e Kuh Arc, Zagros: Role of Mechanical Stratigraphy

J. Vergés, M. G. H. Goodarzi, and H. Emami
Group of Dynamics of the Lithosphere (GDL), Institute of Earth Sciences Jaume Almera, Consejo Superior de Investigaciones Cientificas (CSIC), Barcelona, Spain

R. Karpuz
Österr Mineralöl Verwaltung Exploration & Production, Vienna, Austria

J. Efstathiou
Statoil, Jebel Ali Free Zone (Jafza), Dubai, U.A.E.

P. Gillespie
Statoil, Tectonics and Structural Geology, Forushagen, Stavanger, Norway

ABSTRACT

Field data in combination with interpretation of old seismic lines across the Pusht-e Kuh arc and northwest Dezful embayment in the Zagros fold belt allow the investigation of the geometry of folding at different structural levels. Folds in the Pusht-e Kuh arc are exposed along the upper part of the 7-km (4.34-mi)-thick Competent Group level, which is folded between the main detachment at or near the base of the cover sequence (lower Mobile Group) and the Gachsaran evaporites (upper Mobile Group). Intermediate detachment levels (Triassic Dashtak, middle Cretaceous Garau and Kazhdumi, Paleocene Amiran, and middle Miocene Kalhur formations) within the Competent Group control the geometry of anticlines at surface as well as their variations with depth.

The Kabir Kuh anticline is the largest and highest anticline in the Pusht-e Kuh arc and has been selected to construct a geometrical model to explore the variations of folding style with depth. Results from this anticline constituted a backbone for both a regional study to link the mechanical stratigraphy to the structure and to build up a conceptual model for folding that may apply to the Pusht-e Kuh arc as well as to the Dezful embayment tectonic domains. The Kabir Kuh anticline in its central part displays box fold geometry, characterized by a wide and rounded crestal domain, which is slightly tilted to the southwest. This geometry developed above an intermediate detachment at the level of 1.3-km (0.81-mi)-thick Triassic Dashtak

DOI:10.1306/13251333M942899

evaporites within the Competent Group. Below this detachment, the geometry of the fold changes to more acute with a narrower crestal region. Although conjectural, we propose low-angle thrusting forming a tectonic wedge to accommodate shortening at the deepest part of the anticline (Paleozoic sequence). Subsidiary thrusts, related to fold tightening, may reactivate axial surfaces as well as local detachment levels to propitiate displacements of the crest of the anticline above its forelimb as inferred along the southeastern segment of the Kabir Kuh anticline.

A conceptual model of folding characterized by changes in fold geometry at depth is proposed. A significant outcome of this model is that anticlines in the Passive Group may be displaced, sometimes few kilometers, from anticlines in the Competent Group. In the same way, the internal mechanically weak layers of the Competent Group may form intermediate detachments that can produce a significant change in fold style with depth and displace the upper part of the structure shifting again the position of the anticline crests.

The uplifted Pusht-e Kuh arc is an excellent natural laboratory to investigate potential relationships among mechanical stratigraphy, tectonic structure, time of oil generation, and time of trap development. Fold and thrust geometries investigated in this study are directly applicable to petroleum exploration of recently awarded exploration areas within the Pusht-e Kuh arc and may apply to the rest of the Zagros fold belt.

INTRODUCTION

Because of its great oil and gas potential, the nicely outcropping system of whaleback anticlines along the Zagros fold belt has been the focus of attention for many structural geologists in the last decades (Figure 1). Almost 15% of the total oil reserves of the Middle East hydrocarbon province are concentrated in this large folded region along Iran and Iraqi Kurdistan, which contains the largest oil reserves among any fold and thrust belt in the world (about 50% of total worldwide reserves). Most of these reserves are located in the Dezful embayment with the 45 most important oil fields containing about 8% of the global oil reserves (e.g., Bordenave and Hegre, 2005). However, new exploration is aimed at smaller and generally more complex traps that require a good understanding of the geometries of folds and the mechanisms of folding and thrusting operating at depth.

Despite the stunning shapes of the Zagros folds at the level of the resistant limestones that constitute the two main fractured oil reservoirs (Cenomanian–Turonian Sarvak Formation and Oligocene–early Miocene Asmari Formation) (e.g., Hull and Warman, 1970; McQuillan, 1974), the continuation of these anticlines at depth is very poorly known. This is because of several factors: (1) the large 10–12-km (6.2–7.4-mi)-thick sedimentary cover of the Arabian passive margin, (2) the existence of multiple weak layers in the cover that can be potential detachment levels, (3) the poor quality of existing seismic surveys across the Pusht-e Kuh arc, and (4) the low overall structural relief across the Pusht-e Kuh arc that makes it difficult to detect changes in the tectonic style with depth.

The lack of reliable data at depth allows different structural interpretations of folding. The fold mechanism in the Zagros fold belt has been variably interpreted as detachment folding, fault-related folding, basement fault reactivation, or a combination of these mechanisms (e.g., Sattarzadeh et al., 2000; Blanc et al., 2003; McQuarrie, 2004; Molinaro et al., 2004; Sherkati and Letouzey, 2004). The maps published in the 1960s and 1970s by the then Iranian Oil Operating Companies show cross sections with detachment anticlines, which are sometimes affected by minor thrust faults. These cross sections, however, are limited to the older outcropping sedimentary unit in the section and thus are not showing the deeper structure. In these cross sections, the Miocene evaporites of the Gachsaran Formation were considered to be an intermediate detachment separating levels with two different structural styles (e.g., Dunnington, 1968), and little significance was given to other intermediate detachment levels in the thick Mesozoic and Cenozoic sedimentary cover. Nevertheless, interpretations of folding in the Zagros fold belt have been rapidly evolving in the last decade, and several research groups have been actively working on the topic. Sattarzadeh et al. (2000) indicated the coexistence of different types of folds, including a large group of pure buckle folds that formed linked anticlines as proposed by Price and Cosgrove (1990), and forced folds associated to the inversion of inherited basement normal faults. These thrusts possibly cut through the cover sequence, flattening in the Gachsaran detachment to produce fault-bend folds (Sattarzadeh et al., 2000, their figure 7). Blanc et al. (2003) interpreted the folds of both the Dezful embayment and the Pusht-e Kuh arc to have formed by an interconnected thrust system in which the large

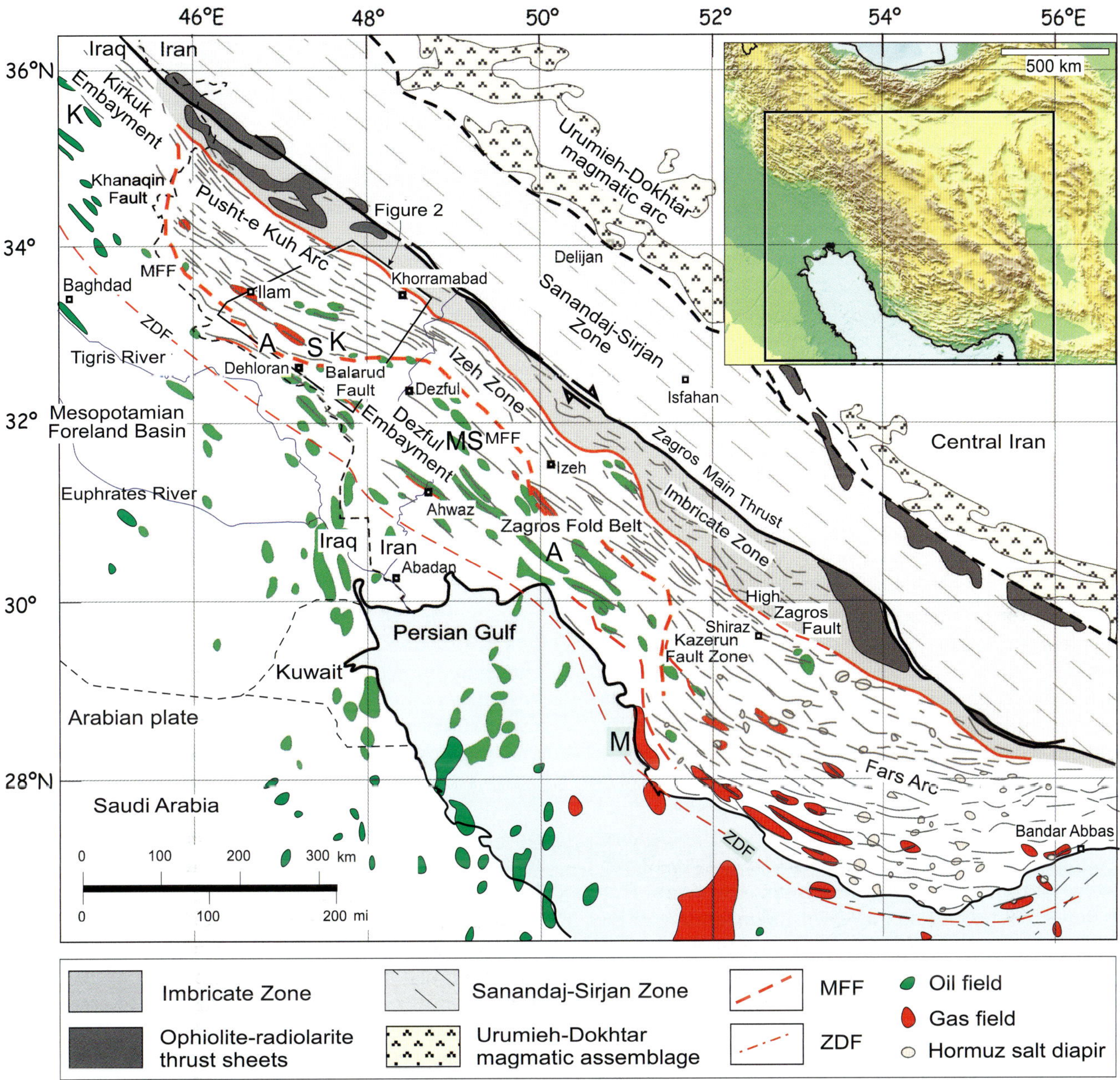

Figure 1. Tectonic map of the Zagros Mountains showing the traces of anticlines in the Zagros fold belt, Mesopotamian foreland basin, and Persian Gulf. The imbricate zone, including ophiolites and volcanic arc outcrops, the Sanandaj-Sirjan zone, and the continental Iran block are also shown. Oil and gas fields display the northwest–southeast Zagros trend, and the roughly north–south Arabian trend, in the Abadan plains and Persian Gulf. The two domains are separated by the Zagros deformational front (ZDF) (modified from Alavi, 2004). MFF = mountain front flexure; A = Anaran anticline; S = Samand anticline; K = Kabir Kuh anticline; M = Mand; MS = Masjid-i Soleyman oil field.

Anaran, Samand, and Kabir Kuh anticlines evolved as thrust ramp anticlines directly above basement uplifts (Figure 1). McQuarrie (2004) presented three regional balanced and restored sections across the Fars arc, Dezful embayment, and Pusht-e Kuh arc (Figure 1). This author suggested, based on the low amounts of shortening of the Zagros fold belt (~15%), the reduced taper, and the comparable structural elevation in the Fars and Pusht-e Kuh arcs, that folds in these two regions are similar and thus both detached above the Hormuz salt. Molinaro et al. (2005) presented three mechanisms that produced folding in the southeastern termination of the Fars arc: multiple detachments, thrust ramps cutting through basement and cover rocks, and basement uplifts. A similar interpretation that combines different internal detachments with high-angle thrusting and basement uplifts has been provided for the Izeh zone and northeast Dezful embayment tectonic domains (Sherkati and Letouzey, 2004) (Figure 1).

Despite the difficulty in drawing an unequivocal cross section at depth, several studies converge on a multidetachment system of folds (e.g., Molinaro et al., 2004; Sherkati and Letouzey, 2004; Sepehr et al., 2006). The current understanding of these folding mechanisms may change rapidly after the accumulation of new data as confirmed by the recent study of Sherkati et al. (2005), who modified their previous interpretations and exclusively used a multiple detachment model for the eastern and central Zagros above a basal ductile level at the Hormuz salt level, with intermediate detachment levels in the Dashtak, Kazhdumi, and Pabdeh-Gurpi levels as well as in the Gachsaran evaporites. Oveisi et al. (2007) concluded that the Mand anticline (M in Figure 1), along the northwest front of the Fars arc, grew as a detachment fold decoupled from basement structures. This conclusion was supported by the growth geometries of the fluviomarine Pleistocene folded terraces that wrap the aniticline.

In this chapter, field information, well data, and available seismic lines from the Pusht-e Kuh arc are used to build up a comprehensive model for the style of multidetachment folding, which is strongly controlled by the position at depth and the areal extent of weak mechanical layers in the stratigraphy. For this purpose, a cross section across the Kabir Kuh anticline, the largest structure of the study area that constituted a long-lived paleogeographic boundary separating two domains with different fold characteristics, was built. The Kabir Kuh section provides clues for a better understanding of the deeper structure of folds in the Pusht-e Kuh arc. The potential extent of the Hormuz evaporites beneath the Pusht-e Kuh arc and the implications of this fold development for hydrocarbon entrapment are also discussed.

ZAGROS FOLD BELT: THE PUSHT-E KUH ARC

The long-lived Late Cretaceous to Pliocene Zagros orogenic belt consists of four northwest-southeast–trending tectonic units separated by large thrust faults (e.g., Stöcklin, 1968; Falcon, 1974; Haynes and McQuillan, 1974; Berberian and King, 1981; Alavi, 1994) (Figure 1). These tectonic units are from the inner to outer part of the belt: the Sanandaj-Sirjan zone, the imbricate zone, the Zagros fold belt, and the Mesopotamian foreland basin along the Tigris River plain that continues along the Persian Gulf (Figure 1). Stöcklin (1968), Berberian and King (1981), and Agard et al. (2005), among others, considered the Zagros main thrust separating the Sanandaj-Sirjan zone from the imbricate zone as the Zagros suture, whereas Alavi (1994) located this suture farther to the northeast between the Sanandaj-Sirjan and Urumieh Dokhtar tectonic domains (Figure 1).

The Sanandaj-Sirjan zone comprises deformed and metamorphosed Paleozoic to Mesozoic rocks intruded by Late Cretaceous to Paleocene plutons (Stöcklin, 1968; Alavi, 1994; Mohajjel and Fergusson, 2000). The Sanandaj-Sirjan zone thrusted southwestward over the imbricate zone above the Zagros main thrust. The imbricate zone, or high Zagros, consists of imbricated tectonic slices involving radiolarite-ophiolite complexes, Mesozoic and Cenozoic sedimentary and volcanic rocks, and thrust sheets from the Sanandaj-Sirjan zone (e.g., Agard et al., 2005). The southwestern boundary of the imbricate zone runs along the high Zagros fault (Berberian, 1995).

The Zagros fold belt (also named simply folded belt) is characterized by a folded 12–14-km (7.4–8.7-mi)-thick sedimentary cover deposited on the northeastern continental border of the Arabian plate (e.g., Falcon, 1974; Colman-Sadd, 1978). Its maximum width is about 280 km (174 mi) in the center of the Fars arc and 230 km (143 mi) in the Pusht-e Kuh arc, incorporating about 17–18 and 12–14 anticlines, respectively. These two belts of anticlines are limited by a major geoflexure (Falcon, 1961), variously named the main front fault (Berberian, 1995), the mountain front flexure (MFF) (McQuarrie, 2004), and the Zagros frontal fault (Sepehr and Cosgrove, 2004). This structural and topographic front, which we will define as MFF, has an irregular geometry that defines tectonic salients or arcs and reentrants or embayments: respectively, from southeast to northwest, the Fars arc (Fars stratigraphic province), the Dezful embayment (Khuzestan stratigraphic province), the Pusht-e Kuh arc (Lurestan stratigraphic province), and the Kirkuk embayment (Kurdistan in Iraq) (Figure 1). The MFF bounding the Pusht-e Kuh arc has an east-west–trending segment along the Balarud fault, a frontal segment along the Anaran anticline, and a north–south segment along the Khanaqin fault. The Balarud fault, the Anaran anticline on top of the MFF, and the Khanaqin fault separate the Pusht-e Kuh arc from the Dezful embayment, the foreland, and the Kirkuk embayment, respectively (Figure 1). Within the Pusht-e Kuh arc, a large number of folds of variable size (12–14 anticlines) show a regular northwest–southeast structural grain very similar to the other tectonic domains of the Zagros fold belt. These anticlines plunge down into the prolific oil provinces of Dezful and Kirkuk reentrants over the MFF boundaries.

The Mesopotamian foreland basin and its southeastern continuation along the Persian Gulf extend in front of the Zagros fold belt. The most proximal part of the foreland basin is characterized by northwest-southeast–trending anticlines buried beneath the large alluvial plains of Iraq and to a lesser extent Iran (e.g., Dunnington, 1968; Mohammed, 2006). Although it is not straightforward to trace the front of the deformation within the foreland basin, a tentative trace of the Zagros deformational front, slightly modified from Alavi (2004), is shown (ZDF in Figure 1). This front separates the northwest-southeast–oriented Zagros folds from the roughly north-south–trending Late Cretaceous inverted structures that are dominantly located in the northern Persian Gulf and below the Abadan plains, facing the Dezful embayment tectonic domain (e.g., Abdollahie Fard et al., 2006).

The anticlines of the Pusht-e Kuh arc are mostly exposed at the level of resistant limestones corresponding to the Asmari Formation or to the upper part of the Bangestan Group. The synclines are commonly tight, and only few of them contain Agha Jari foreland deposits that entirely cover the Dezful embayment (Figure 2). According to the present level of exposed rocks in the cores of the anticlines, five domains can be distinguished across the southeastern region of the Pusht-e Kuh arc, which from southwest to northeast are formed by (1) the Anaran and Samand anticlines cored by top Bangestan Group strata; (2) the Kabir Kuh anticline cored by different levels of Bangestan Group strata; (3) a set of anticlines, which expose units from Gurpi-Amiran to Asmari in their cores; (4) a group of smaller anticlines developed at the Asmari level; and (5) the Khorramabad double anticline (Figure 2).

No traces of major thrust faults at surface in the southeastern region of the Pusht-e Kuh arc are observed, and the altitude of the anticlines and synclines remain fairly regular across the entire region. These characteristics suggest that variations in fold styles across these five selected domains are directly related to variation in the mechanical stratigraphy across the Pusht-e Kuh arc as discussed below.

Amounts of shortening calculated from different balanced and restored cross sections show a maximum of 20% (Blanc et al., 2003) with average values of 15% in both the Pusht-e Kuh arc and Izeh zone reaching minimum values of about 6% in the Dezful embayment (McQuarrie, 2004; Sherkati et al., 2005; Carruba et al., 2006; Sepehr et al., 2006).

MECHANICAL STRATIGRAPHY OF THE PUSHT-E KUH ARC (LURESTAN PROVINCE)

The stratigraphy of the Lurestan Province consists of a 10- to 12-km (6.21- to 7.45-mi)-thick succession that encompasses the Paleozoic and Mesozoic Arabian passive margin deposits followed by the sediments corresponding to the long-lived Cenozoic Zagros orogenic phase (Figure 3). This thick pile of sediments was probably deposited on top of the Proterozoic–Early Cambrian Hormuz evaporites, although this assertion is not directly proved in the Lurestan Province. Most of the stratigraphy described in this section is based on the work of James and Wynd (1965) and Colman-Sadd (1978).

The Paleozoic sequence is best documented in the Izeh zone where it forms the hanging wall of the high Zagros fault (O'B Perry and Setudehnia, 1967). The base of the hanging-wall section is formed by 1160 m (3806 ft) of Cambrian rocks followed by 925 m (3035 ft) of Carboniferous and Permian deposits. These thicknesses agree with results from deep exploration wells in the study area. The Kabir Kuh and the Samand wells penetrated 783 and 1057 m (2569 and 3468 ft) of Permian rocks, respectively (KK and S in Figure 2). The Permian deposits consist of shallow-water carbonates that are cyclically interbedded with evaporites (Beydoun et al., 1992). Older rocks have only been intersected in the Kabir Kuh well, in which about 200 m (656 ft) of Ordovician rocks was drilled. The seismic lines across the southeastern Pusht-e Kuh arc as well as across the northwest frontal regions of the Dezful embayment show relatively deep subparallel folded horizons that are interpreted as corresponding to lower Paleozoic strata deposited within the wide, stable, and long-lived northern passive margin of the Paleo-Tethys border of Gondwana (e.g., Beydoun et al., 1992).

Along the northeastern boundary of the Arabian plate, in the Lurestan Province, the Mesozoic sequence is about 3 km (1.863 mi) thick and encompasses the Triassic Dashtak Formation, the Khami Group (Jurassic–Aptian), and the Bangestan Group (Albian–middle Campanian) (James and Wynd, 1965) (Figure 3). During the Triassic, Jurassic, and Lower Cretaceous, the region was dominated by large carbonate platforms (Khami Group) with associated shallow basins filled with marls, shales, and argillaceous carbonates interbedded with episodic plugs of evaporites (e.g., Triassic Dashtak and Upper Jurassic Gotnia Formation). These anoxic basins provided several large areas of potential source rocks through this long sedimentary sequence (e.g., Murris, 1980; Stoneley, 1990). The Bangestan Group in Lurestan Province represents the continuation of the carbonate platform deposition embracing the Kazhdumi, Sarvak, Surgah, and Ilam formations. The thick rudist limestones of the Sarvak Formation correspond to the lower reservoir in the Zagros oil and gas province. Northeast of the Kabir Kuh anticline, most of this group changes laterally to more basinal dark carbonaceous shales of the Garau Formation with a maximum thickness of 1000 m (3281 ft) (James and Wynd, 1965). Shallow-marine sedimentation continued with the middle Campanian to Paleocene

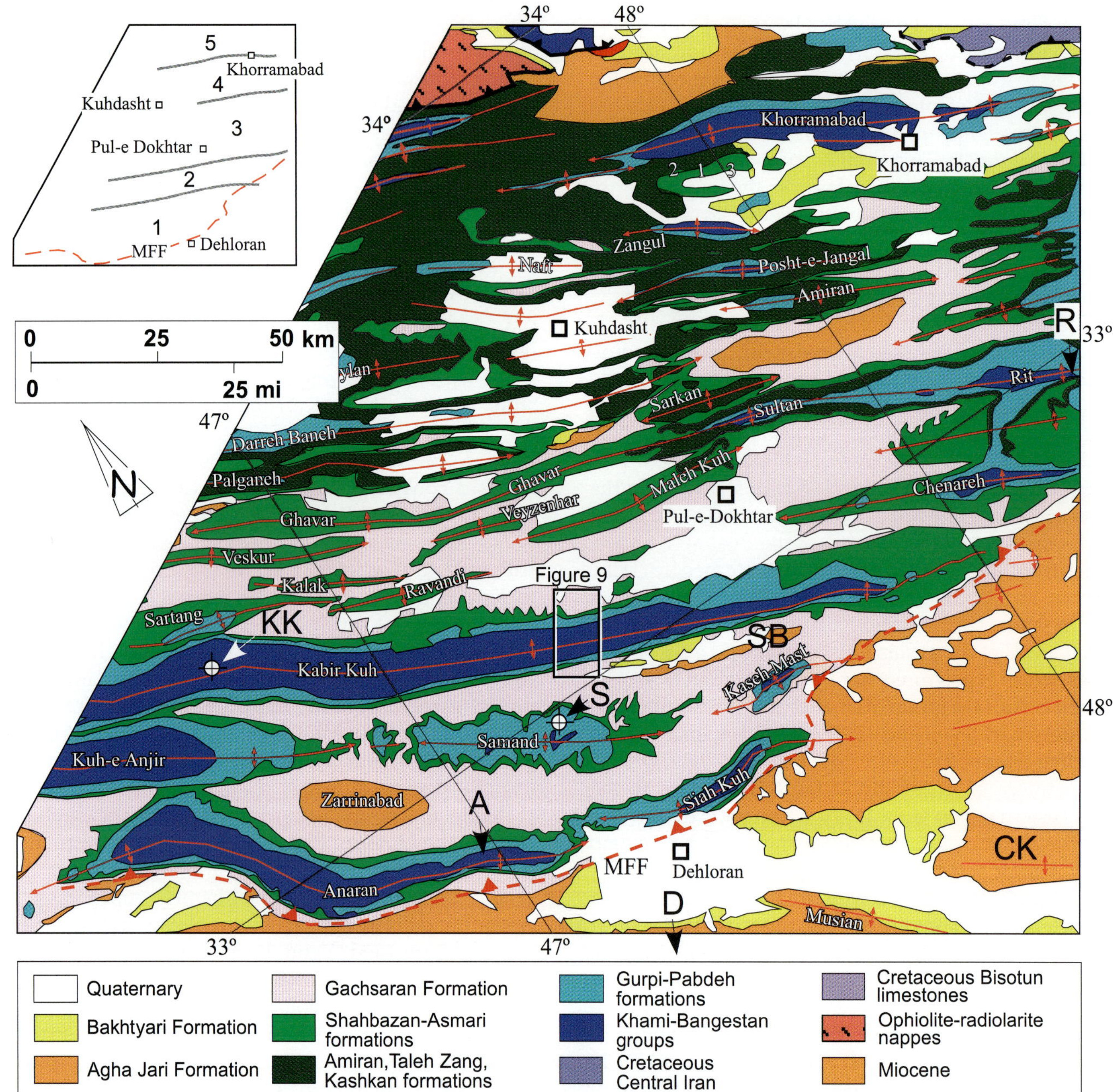

Figure 2. Map of the Pusht-e Kuh arc (Lurestan Province) with traces of anticlines. Additional locations correspond to the Anaran anticline (A), Kabir Kuh oil well (KK), Samand oil well (S), Chachmeh Kush thrust (CK), Dehluran anticline (D), Sarab Bagh syncline (SB), and Rit anticline (R). See also the location map of Figure 9. The inset shows the same region but separated in five domains with different fold characteristics. MFF = mountain front flexure.

Gurpi Formation, including two extensive fossiliferous carbonate members.

During the Paleocene, to the north of the Kabir Kuh anticline, a thick shallowing-upward siliciclastic succession was deposited, which includes the Amiran, Taleh Zang, and Kashkan formations (Homke et al., 2009) (Figure 3). The Shahbazan and Asmari formations, the latter corresponding to the upper main reservoir in the Zagros oil province, are Oligocene to middle Miocene in age to the north of the Kabir Kuh anticline (Homke et al., 2009). Above the middle Miocene Asmari Formation, the 1-km (0.62-mi)-thick and contorted evaporites of the Gachsaran Formation correspond to the lower Fars Group (e.g., Kashfi, 1980; Bahroudi and Koyi,

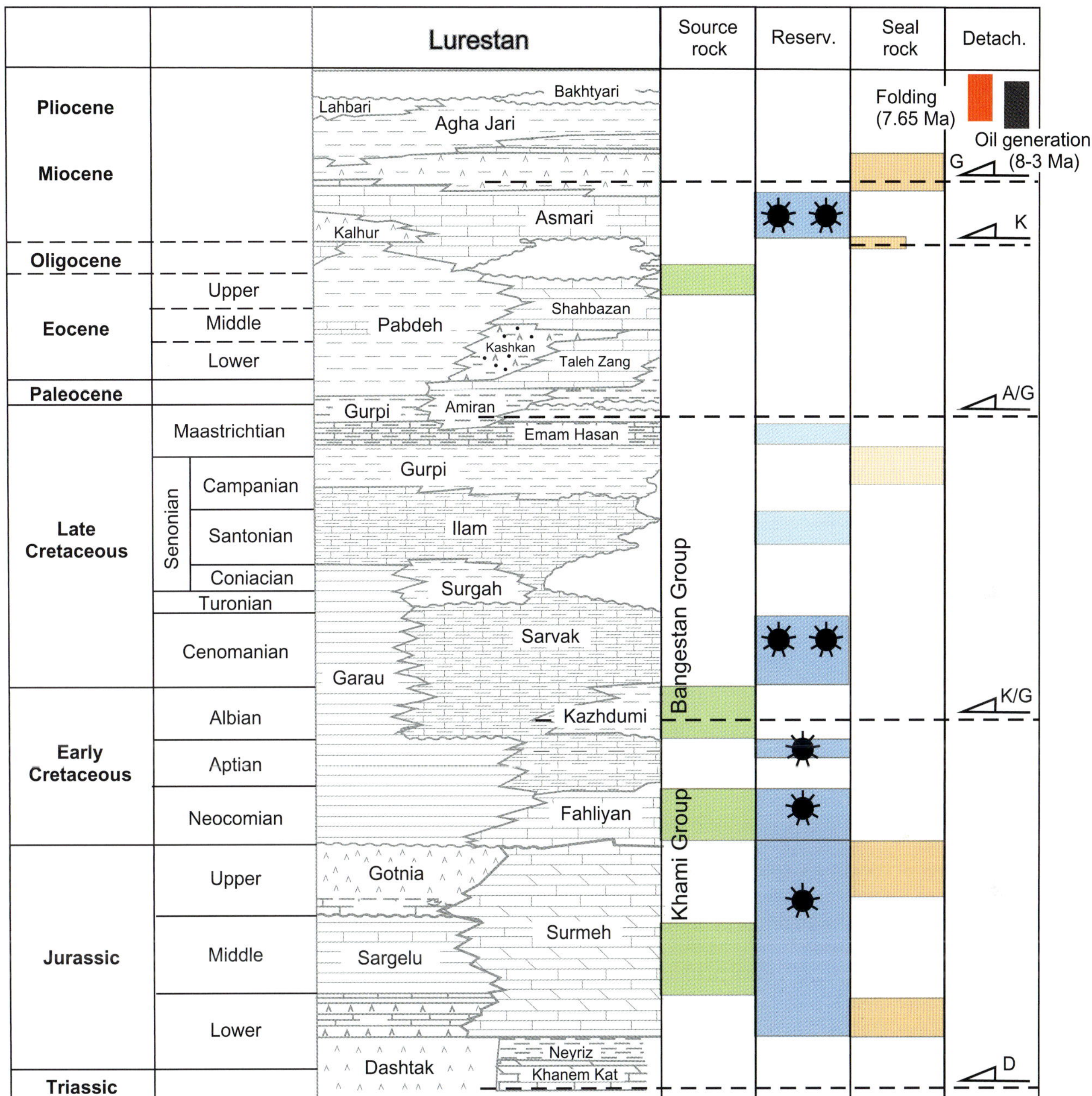

Figure 3. Stratigraphic panel of the Pusht-e Kuh arc (Lurestan Province) based on James and Wynd (1965) combined with the Pusht-e Kuh arc and Dezful embayment hydrocarbon systems (Bordenave and Hegre, 2005), and the tectonic detachment levels. The timing of deformation along the front of the Pusht-e Kuh arc is based on Homke et al. (2004), whereas the timing of oil generation and migration from Mesozoic source rocks in the Dezful embayment is based on Bordenave and Hegre (2005). Detachments are located at Dashtak level (D), Kazhdumi/Garau levels (K/G), Amiran/Gurpi levels (A/G), Kalhur level (K), and Gachsaran level (G). Reserv. = Reservoir; Detach. = Detachment.

2004). These evaporites form an effective seal above the Asmari reservoir. The age of the top Gachsaran evaporites near the front of the Pusht-e Kuh arc is constrained by magnetostratigraphy to be approximately 12.5 Ma, corresponding to the uppermost middle Miocene (Homke et al., 2004). Along the front of the Pusht-e Kuh arc, the approximately 1650-m (5413-ft)-thick lower Agha Jari deposits encompass the interval from 12.8–12.3 Ma to 5.5 Ma. Above these, the fine-grained Lahbari Member (~825-m [2707-ft]-thick unit) has an age from

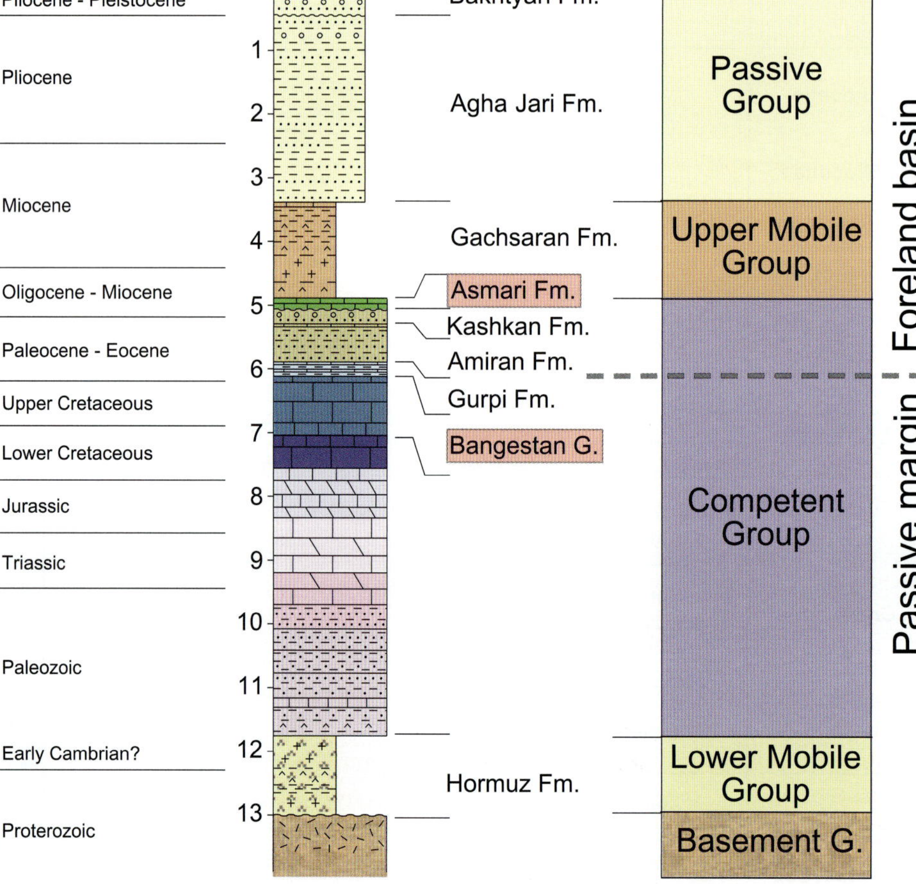

Figure 4. Simplified stratigraphic column of the Lurestan Province actually forming the Pusht-e Kuh arc combined with the division in major structural units with similar mechanical behavior based on O'Brien (1950) and Dunnington (1968). The carbonates of the upper part of the Bangestan Group and the Asmari Formation represent the two main reservoirs in the region. Fm. = Formation; G. = Group.

5.5 to 3 Ma. The age of the conglomeratic Bakhtyari Formation is estimated to be from 3 Ma to at least 2.5 Ma, although its younger extrapolated age is about 1.5 Ma (Homke et al., 2004) (Figure 3).

The mechanical behavior of the 10–12-km (6.2–7.4-mi)-thick sedimentary pile in response to folding was first discussed by O'Brien (1950) and by Dunnington (1968) and has been the focus of several recent articles (e.g., Sattarzadeh et al., 2000; Molinaro et al., 2004; Sherkati et al., 2005; Sepehr et al., 2006, among others). O'Brien (1950) and Dunnington (1968) used regional stratigraphy and divided the succession of the Zagros fold belt in five major structural units having a similar mechanical behavior that from top to bottom are (Figure 4) (1) Passive Group, (2) upper Mobile Group, (3) Competent Group, (4) lower Mobile Group, and (5) Basement Group. The Passive Group consists of 3–4-km (1.86–2.48-mi)-thick foreland clastic deposits corresponding to Agha Jari and Bakhtyari formations. The upper Mobile Group is mostly formed by the thick Gachsaran evaporites. The Competent Group is the thickest tectonic unit with a thickness of almost 6 km (3.73 mi) and consists of limestones, marls, and evaporites from the Oligocene–Miocene Asmari Formation down to the Triassic incorporating the entire Bangestan and Khami groups, as well as the Permian and Paleozoic deposits. The lower Mobile Group is formed by Late Proterozoic–Early Cambrian Hormuz evaporites. At the bottom of this tectonic succession lies the Precambrian metamorphic basement, which forms the Basement Group.

STRUCTURE OF PASSIVE GROUP

Initial observations of the Passive Group were based on field and well data from oil fields mostly located in the Kirkuk embayment (Dunnington, 1968), whereas recently published two-dimensional and three-dimensional seismic data from existing oil fields in the Dezful embayment better highlight the structure of this group at depth (Sherkati et al., 2005; Alaei and Pajchel, 2006; Carruba et al., 2006; Sepehr et al., 2006). These data show disharmonic folding above and below the Gachsaran evaporitic detachment. The highly reflective Asmari limestones at the top of the Competent Group form rounded anticlines that pushed up the evaporitic layers of the

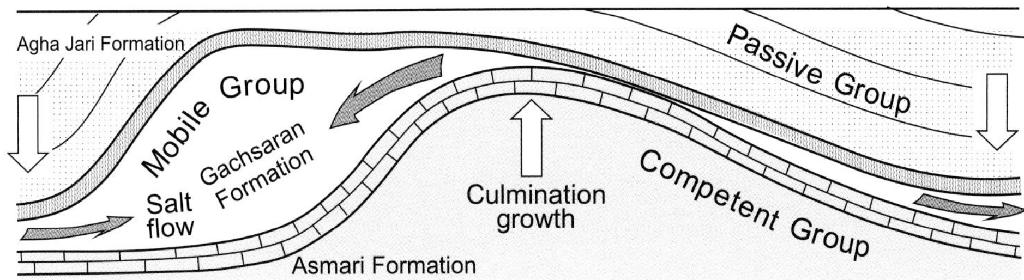

Figure 5. Development of disharmonic folding across the upper Mobile Group (Gachsaran Formation) related to the lateral migration of evaporites by the growth of anticlines in the Competent Group (slightly modified from O'Brien, 1957).

Gachsaran Formation against the base of the more resistant Passive Group producing lateral flow of the evaporites and overthickening in the adjacent synclines (Figure 5). This is very well imaged below the Dehluran plains, where a growth syncline developed above the Dehluran anticline (Figure 6A, and D in Figure 2). In some cases, the growth of an anticline in the Competent Group produced total migration of the evaporites away from the anticline crest and thus caused contact between the top of the Competent Group and the base of the Passive Group (O'Brien, 1950; Dunnington, 1968) (Figure 5). The synchronous growth of folds in the Competent Group and salt flow in the upper Mobile Group influenced the geometry of folding and thrusting in the Passive Group and made it difficult to describe the processes in the different units separately. The partially coeval sedimentation of the Agha Jari deposits in the growth synclines of the Passive Group caused loading of the synclines and triggered further salt migration away from synclines as depicted by vertical white arrows pointing down in Figure 5. Shortening in the Passive Group was accomplished by folding and thrusting with a significant decoupling from the Competent Group. Some of the thrust faults reach maximum displacements of about 5 km (3.1 mi) as in the case of the Chachmeh Kush thrust in the northwestern edge of the Dezful embayment (Figure 6B, and CK in Figure 2).

STRUCTURE OF THE UPPER MOBILE GROUP (GACHSARAN EVAPORITES)

The evaporites of the Gachsaran Formation constitute the cap rock of the Asmari limestone reservoir. This unit, which is from a few hundred meters to up to 2 km (1.24 mi) thick (Edgell, 1996; Bahroudi and Koyi, 2004), forms the upper Mobile Group that separates the Competent Group from the Passive Group. This formation mostly crops out in the Dezful embayment in Iran and in the Kirkuk embayment in Iraq, whereas in the Pusht-e Kuh arc, the formation is only preserved in sparse large synclines (Figure 2). From well data, the Gachsaran Formation can be divided in seven members showing an alternation of salts and anhydrites with thin limestones at the base and gray shales at the top (James and Wynd, 1965). During deformation, evaporites flowed to form large accumulations that can partially govern the geometry of the overlying Passive Group, as reported in the Agha Jari oil field (Ion et al., 1951) (A in Figure 1), in the Masjid-i Soleyman oil field (Edgell, 1996; O'Brien, 1957) (MS in Figure 1), in the Kirkuk anticline (Dunnington, 1968) (K in Figure 1), and in seismic examples from the Dezful embayment (Sherkati and Letouzey, 2004, among others).

The internal structure of the Gachsaran detachment level is complex, but field observations as well as seismic line interpretations show that both the lowermost and uppermost parts of the formation (up to 100 m [328 ft]) stay conformable with the underlying beds of the Asmari Formation and with the overlying beds of the Agha Jari Formation, respectively. These two boundaries between undeformed and deformed Gachsaran evaporites correspond to basal and roof thrusts separating a thick zone of intense folding and thrusting producing significant tectonic thickening of the unit.

STRUCTURE OF COMPETENT GROUP: FOLDING BY MULTIPLE DETACHMENT LEVELS

In the disharmonic fold model based on O'Brien (1950) and Dunnington (1968), the Competent Group is only represented by its top, near the contact between the Asmari and Gachsaran formations (Figure 5). The deep geometries of folds within the 6-km (3.73-mi)-thick Competent Group vary according to the changes in the stratigraphy at both regional and local scale and with the present level of erosion. In the Pusht-e Kuh arc, the Passive Group is only present along scarce synclines, whereas the upper part of the Competent Group is nicely exposed at the level of the more resistant limestones corresponding to the Asmari Formation and to the top of the Bangestan Group (Ilam and Sarvak formations) as described in the next section.

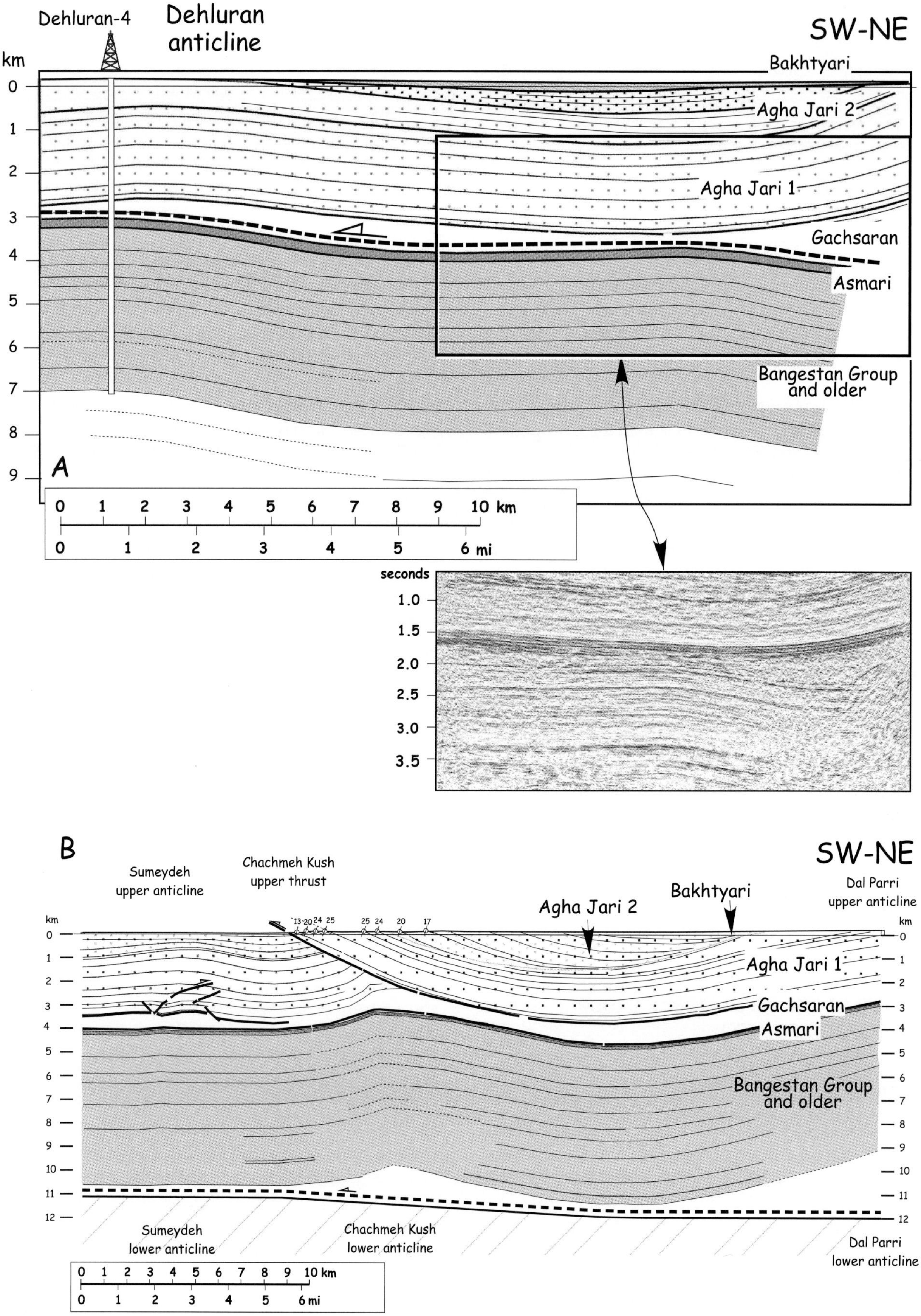

Dehluran-4
Dehluran anticline
SW-NE
km
Bakhtyari
Agha Jari 2
Agha Jari 1
Gachsaran
Asmari
Bangestan Group and older
A
0 1 2 3 4 5 6 7 8 9 10 km
0 1 2 3 4 5 6 mi
seconds
1.0
1.5
2.0
2.5
3.0
3.5

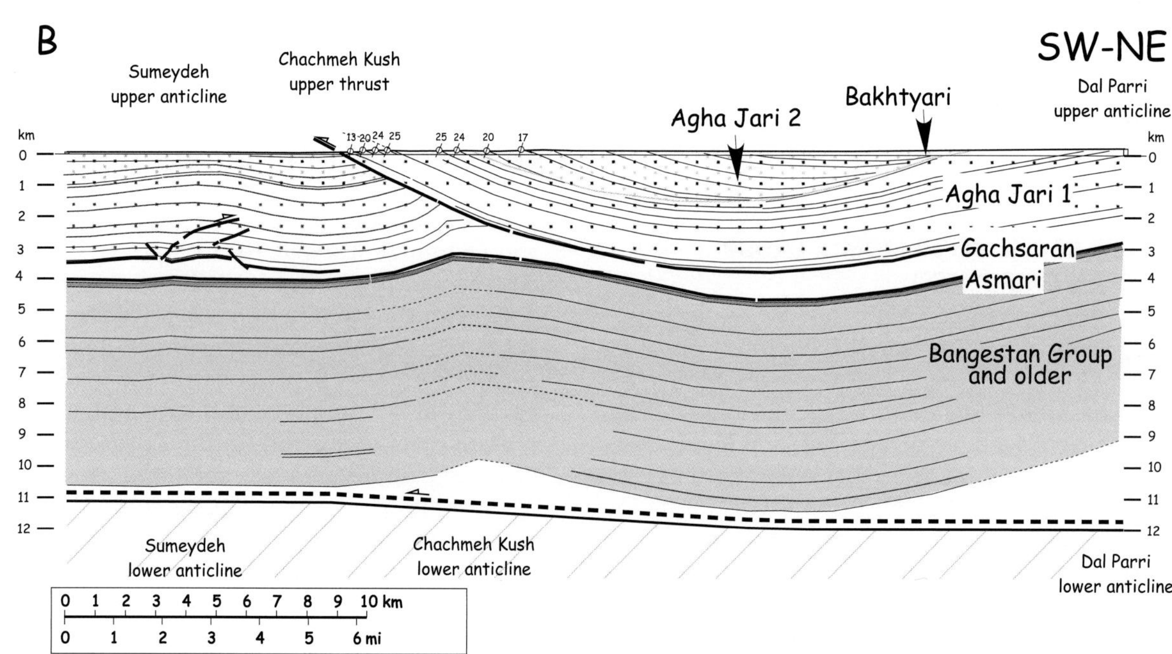

Fold Characteristics in the Southeastern Pusht-e Kuh Arc

At the surface, the folds of the southeastern Pusht-e Kuh arc are expressed as a series of typical whaleback anticlines that exhibit variable lengths, widths, and profile shapes. Twenty-nine anticlines have been analyzed in the southeastern Pusht-e Kuh arc. These folds show a large variability of fold shape, size, and length regardless of their present level of outcrop at Asmari or top of Bangestan limestone ranging from chevron folds, with straight limbs and curvature concentrated at the hinge, to rounded or semielliptical shapes, and interlimb angles from gentle to tight (Figure 7) (Srivastava and Lisle, 2004). All analyzed anticlines show zones of rather low curvature, separated by multiple hinge zones (axial surfaces), and similar lengths for limbs and crestal domains (aspect ratio smaller than 1.2; Srivastava and Lisle, 2004) (Figure 7A). The northwestern segment of the Chenareh anticline is the perfect example of one of these fairly symmetric, broad anticlines in which the low-curvature dip domains are separated by narrow hinge zones (Figure 7B). Most of the anticlines in the southeastern region of the Pusht-e Kuh arc are separated by tight synclines as in between the Sultan and Sarkan anticlines (Figure 7C), or even isoclinal geometries as in between the Amiran and Pusht-e Jangal anticlines, to the northeast of the Sarkan anticline. Thrusting is minor and is considered as a secondary structural feature, accommodating folding. Folds outcropping in the southeastern region of the Pusht-e Kuh arc would correspond to the upper part of the model for concentric folding published by Dahlstrom (1969) (Figure 7D).

Twenty-three anticlines belonging to the southeastern region of the Pusht-e Kuh arc have been analyzed for amplitude and wavelength (Figure 8). Forty-one measurements across selected anticlines are plotted to show amplitude vs. 1/4 wavelength relationship. The plot shows two large well-defined populations of anticlines grouped according to their exposure level (boxes for the Asmari Formation and circles for the top of the Bangestan Group in Figure 8). Although some overlapping exists, the anticlines exposed at the level of Asmari limestones show, in general, smaller fold amplitude and wavelength than the ones exposed at the level of the top Bangestan Group. Within the cluster of anticlines cropping out at the level of limestones of the top of the Bangestan Group, the Kabir Kuh and the Khorramabad anticlines show the maximum amplitude (the Khorramabad anticline corresponds to a double anticline). The Samand and Anaran anticlines show the maximum wavelengths, but they have smaller amplitude than the Kabir Kuh and Khorramabad anticlines. These results fairly coincide with the map division in five domains (Figures 2, 8). The frontal Anaran and Samand anticlines are grouped as domain 1, the Kabir Kuh is selected as domain 2, the anticlines cropping out to the northeast of the Kabir Kuh at either Amiran or top Bangestan levels are identified as domain 3, the wide region with Asmari folding is domain 4, and the Khorramabad double anticline is domain 5 (Figure 1). The change in amplitude and wavelength of these anticlines across the southeastern region of the Pusht-e Kuh arc is related to their structural position within the Competent Group as well as to the occurrence and location of intermediate detachment levels at depth as described hereinafter.

Kabir Kuh Anticline

The Kabir Kuh anticline is the largest fold of the Pusht-e Kuh arc with a length of about 200 km (124 mi) and a maximum elevation of up to 2700 m (8858 ft). The Kabir Kuh anticline constitutes the present drainage divide of the Pusht-e Kuh arc constraining the Seymareh River to the northeast of it (Vergés, 2007) (Figures 2, 9). The Kabir Kuh anticline varies along its length from closed to tight but always shows a box folding profile. Some of the segments of the anticline display two culminations separated by a gentle syncline along the crestal domain. The anticline shows along-strike changes in vergence, with segments verging either to the northeast or to the southwest interspersed with segments of no vergence. The anticline shows no trace of thrust faults at surface except along the southeastern termination of its southwestern flank where both field data and an old seismic line (number 6905) suggest the existence of a southwest-directed thrust fault that places middle Miocene Asmari limestones on top of late Miocene Agha Jari strata with a thrust displacement that can be estimated between 160 and 1000 m (525 and 3281 ft) due to the poor quality of seismic reflectors in the footwall of the thrust. In an attempt to understand the structure of the Kabir Kuh anticline, a balanced cross section was constructed through the central part of the anticline along the secondary road joining the villages of Saleheldin in its

Figure 6. (A) Line drawing and segment of a seismic line across the Dehluran anticline (D in Figure 2). To the northeast of the Dehluran anticline, an anticline in the Competent Group developed beneath the large Dehluran growth syncline. (B) Line drawing across the Chachmeh Kush anticline showing folded reflectors down to 10–11 km (6.2–6.8 mi) at the lower part of the Competent Group, which have been attributed to Paleozoic strata. The Agha Jari 2 deposits, infilling the synclines in both line drawings, show growth geometries that are time equivalent to the Agha Jari growth units in the Changuleh syncline (Homke et al., 2004).

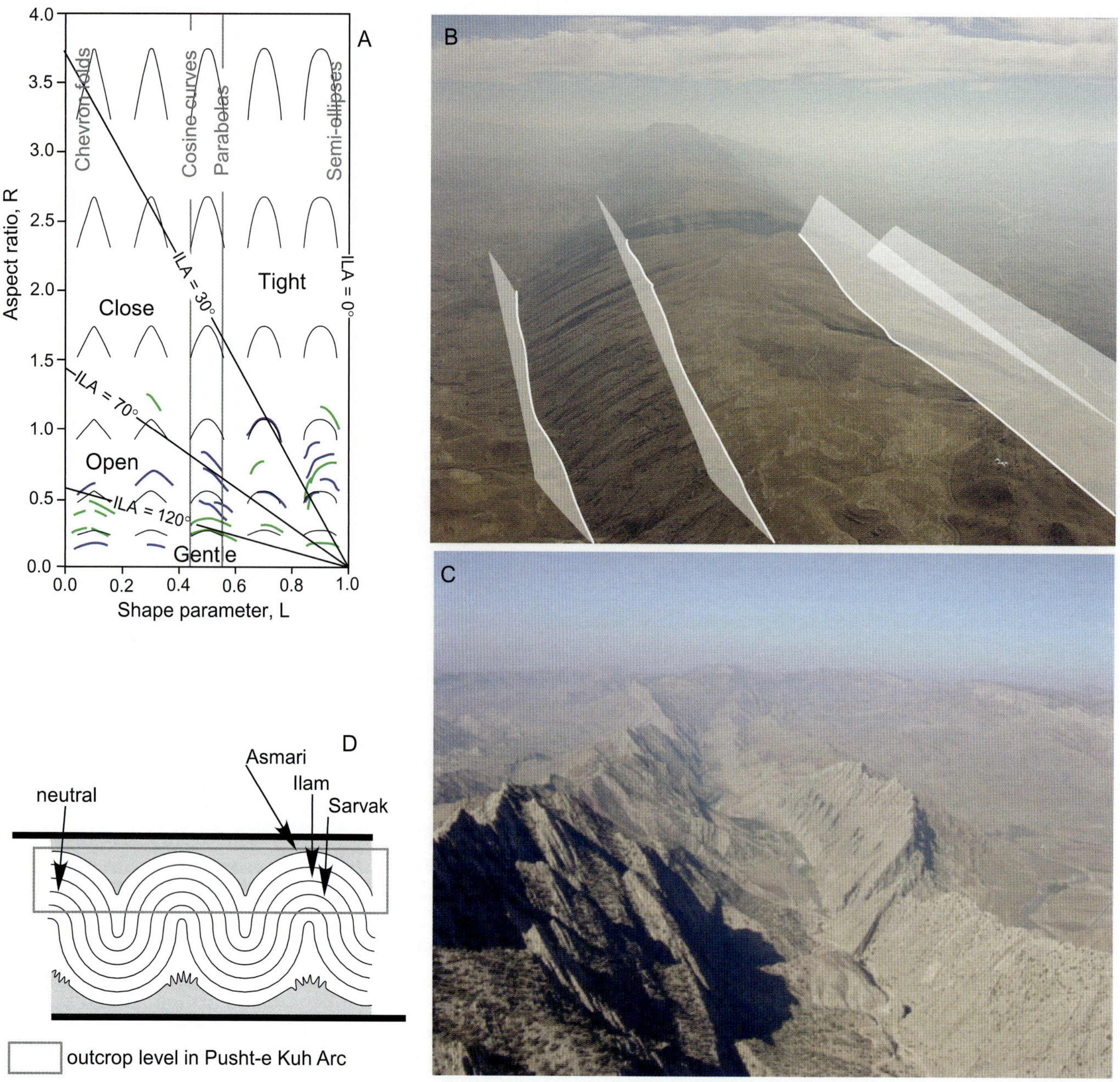

Figure 7. (A) Plot of 29 anticlines of the southeastern region of the Pusht-e Kuh arc relating the fold shape parameter (L) vs. aspect ratio (R) (modified from Srivastava and Lisle, 2004) (ILA = interlimb angle). (B) Aerial view of the northwestern segment of the Chenareh anticline showing a fairly symmetric box fold shape with a wide rounded crestal domain separated by pairs of axial surfaces. The area depicted is 2.62 km (1.63 mi). (C) Aerial view of the tight syncline limiting the Sultan and Sarkan anticlines. The area depicted is 2.78 km (1.68 mi). (D) Position of the structural level cropping out in the southeastern region of the Pusht-e Kuh arc within the conceptual model for concentric folding (Dahlstrom, 1969). See the locations of the Chenareh, Sultan, and Sarkan anticlines in the map of Figure 2.

southwestern flank and Darreh Shahr in its northeastern flank (Figure 9B). The resultant cross section was also constrained by the Kabir Kuh and Samand exploratory oil wells, projected far from the plane of the section (KK and S in Figure 2). No seismic lines are available across the high Kabir Kuh anticline. For this reason, the cross section must be considered as a first approximation waiting for new seismic reflection surveys across the area. The section is constructed using the kink method in part because of the existence of well-recognized dip domains separated by axial surfaces as is described below.

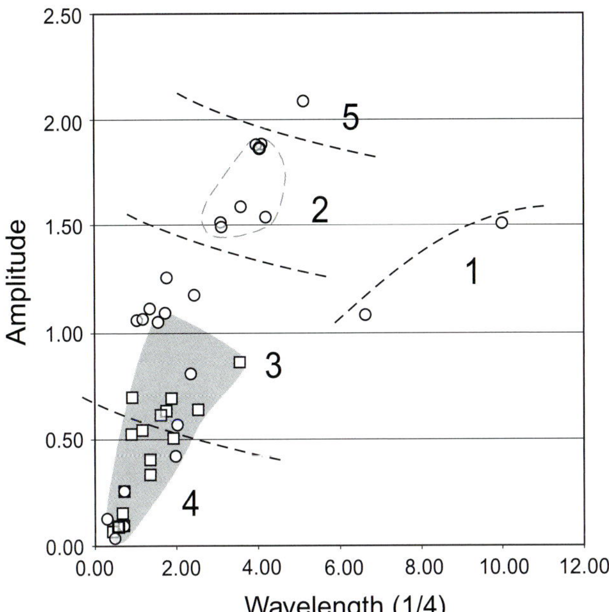

Figure 8. Plot of amplitude vs. wavelength for 23 anticlines of the southeastern region of the Pusht-e Kuh arc (41 measurements; circles and squares show fold characteristics at the top Bangestan and Asmari limestone levels, respectively). Distribution of folds belonging to the five domains identified in the geological map confirms that anticlines show similar amplitude and wavelength within selected domains. The shaded area indicates the zone of overlapping of folding characteristics at the top of the Bangestan and Asmari levels.

THREE-STEP CROSS SECTION CONSTRUCTION

In the first step of cross section construction, field data (this study) and geological maps for the area (MacLeod, 1970) as well as stratigraphic logs from James and Wynd (1965) were used. This segment of the Kabir Kuh anticline is asymmetric with a shorter and steeper forelimb and a longer and gentler backlimb. The anticline shows a rounded geometry that can be separated in dip domains across individual axial surfaces in both the forelimb and the backlimb (Figures 9, 10A). No traces of thrust faults cutting the anticline in this transect are observed. The forelimb shows four dip domains that from top to base correspond to dips of 22° (Gurpi-Pabdeh contact), 50°, 70°, and subvertical (Asmari limestone level) (Figure 10A). The buried length of this near vertical domain is not known, but the geometry of the syncline to the southwest constrains it to be approximately 1000 m (3281 ft) (Figure 11). The backlimb of the anticline is complex with a short and steeper domain with dips of 60° in the highest part of the anticline at the Ilam Formation level that connects to a longer and gentler dip domain with a fairly constant dip of 17° between top Ilam and top Pabdeh formations (Figure 10B). A second steeper dip domain of 34° at the level of the Asmari Formation shapes the backlimb to the northeast (Figure 10B). The crest of the anticline is defined as a near 3-km (1.86-mi)-wide smoothly rounded surface gently dipping toward

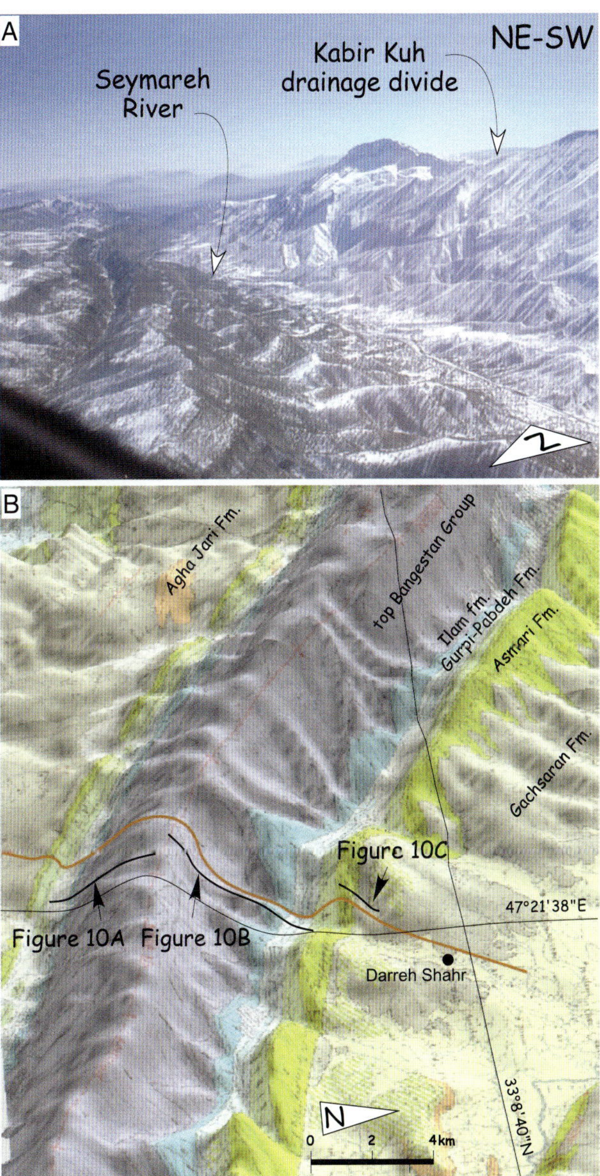

Figure 9. (A) Helicopter view of the southeastern segment of the Kabir Kuh anticline (from the northwest). The 200-km (124-mi)-long Kabir Kuh anticline forms the southwestern drainage divide of the Pusht-e Kuh arc constraining the Seymareh River valley that flows toward the Dezful embayment. (B) Three-dimensional view of the Kabir Kuh geological map (modified from MacLeod, 1970), showing the position of the cross section together with the location of field views of Figure 10. Fm. = Formation.

Figure 10. Field and helicopter views of the flanks of the Kabir Kuh anticline along the cross section of Figure 11A. (A) Different dip domains defining the southwestern forelimb. (B) Northeastern backlimb of the Kabir Kuh anticline at the Gurpi-Pabdeh level showing a general dip of about 17° to the northeast and a short but steep domain separated by two nearly parallel axial surfaces defining the box folding shape. (C) Northeastern flank at the level of the Asmari-Gachsaran contact showing a flap structure with overturned Asmari limestones on top of northeast-dipping Gachsaran evaporites. The scale is in Figure 9.

the southwest at the top Bangestan Group (top of Sarvak to Ilam formations). In the cross section, this wide crestal domain has been separated in multiple dip domains according to dip data (Figure 11A). The southwestern tilt of the crest of the Kabir Kuh anticline in this section locates its highest elevation along the northeastern axial surface that separates the crest from the backlimb. The short and steep dip domain in the upper part of the backlimb together with the broad and gently rounded crestal domain defines the anticline as a box fold at surface (Figure 11A). The forelimb and the backlimb beneath adjacent synclines are located at similar depths constraining the structure of the anticline at depth. Cross section construction emphasizes the slight decrease in thickness of the Gurpi and Pabdeh formations toward the northeast. At depth, the anticline geometry has been constructed using a fairly constant thickness for the Bangestan Group and underlying Garau Formation. The projection at depth of the box folding geometry of the anticline, with a 3-km (1.86-mi)-wide crestal domain and steep domains in both flanks, leads to the prediction of a detachment level at about 2900 m (9514 ft) below the top of the Bangestan Group. The stratigraphic position at this depth roughly corresponds

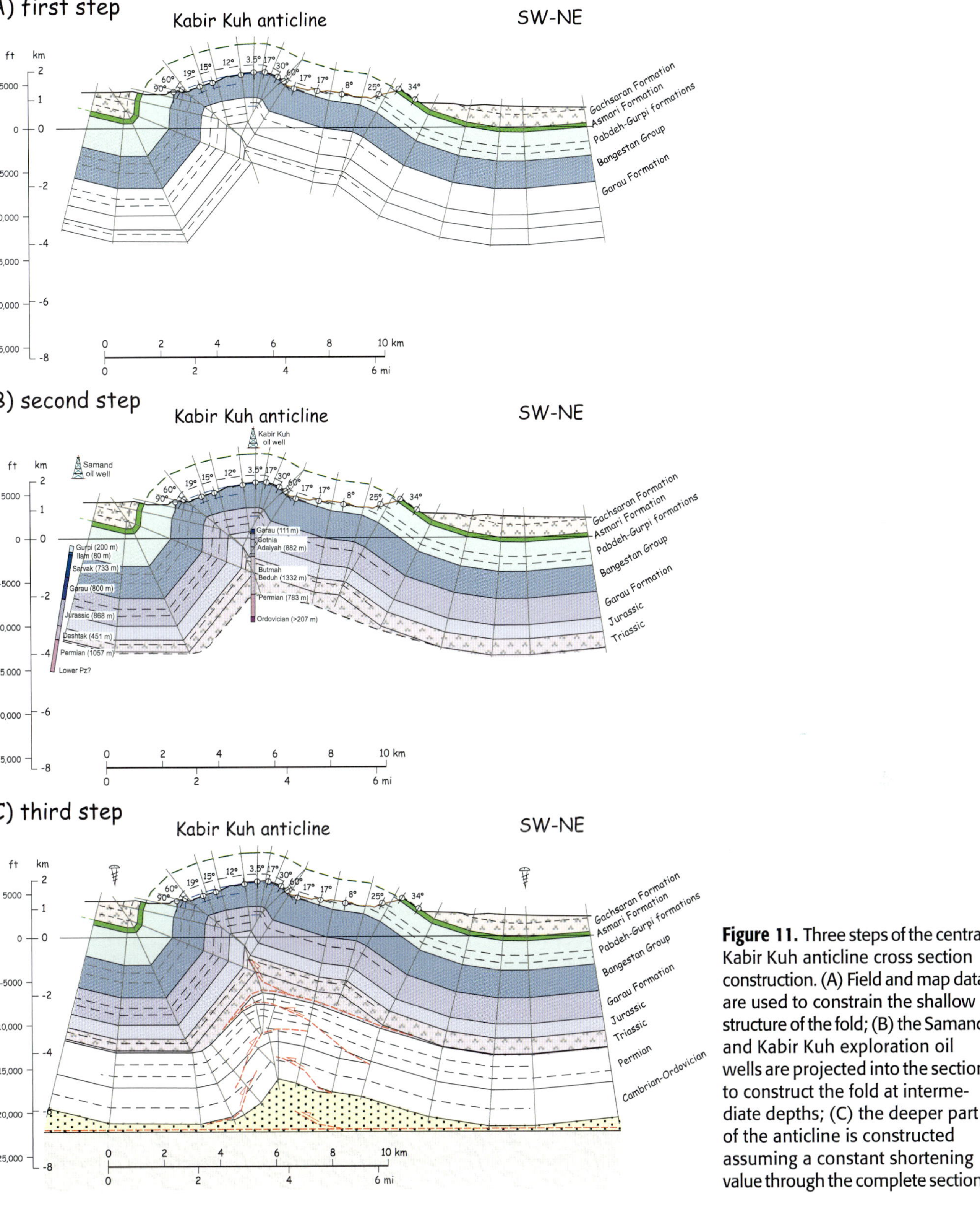

Figure 11. Three steps of the central Kabir Kuh anticline cross section construction. (A) Field and map data are used to constrain the shallow structure of the fold; (B) the Samand and Kabir Kuh exploration oil wells are projected into the section to construct the fold at intermediate depths; (C) the deeper part of the anticline is constructed assuming a constant shortening value through the complete section.

to the Triassic–Jurassic boundary using constant thickness for the Bangestan Group and underlying units across the anticline (Figure 11A).

At surficial levels, the backlimb of the Kabir Kuh anticline is covered by strongly overturned Asmari limestones placed on top of normally dipping Gachsaran evaporites, which form a typical flap structure (e.g., Harrison and Falcon, 1934; Sherkati et al., 2005) (Figure 10C). These slumped Asmari folds are typical of many of the Pusht-e Kuh arc anticlines and use steeply dipping evaporitic (Kalhur Member) or marly units (Pabdeh Formation) to detach and slide down into the adjacent syncline. These gravitational collapses of the flanks of the anticlines represent a very efficient mechanism for anticline denudation, which is occasionally facilitated by the ruptures along the extensional crestal faults. The most catastrophic and gigantic gravitational collapse in the area originated by the complete failure of a 15-km (9.3-mi)-long segment of the northeastern flank of the Kabir Kuh anticline. The collapse produced the massive Seymareh rock avalanche (>160 km^3 [38.4 mi^3]) that filled up the syncline between the Kabir Kuh and the Chenareh anticlines (Harrison and Falcon, 1937). The southeastern edge of this avalanche surpassed the top of the Chenareh anticline (Figure 7B).

In the second step of cross section construction, stratigraphic data from two old exploration oil wells were included: the Kabir Kuh and the Samand wells (KK and S in Figure 2). The Kabir Kuh oil well is projected in our section from 67 km (41.6 mi) to the northwest. This well is located in the northeastern culmination to match the contact between the Garau and the Gotnia formations at depth, which are the outcropping levels at the well rig position (Figure 11B). The Samand oil well is projected in the syncline adjacent to the Kabir Kuh forelimb from about 10 km (6.21 mi). The well has been located by matching the contact between the Ilam and Gurpi formations. According to the Kabir Kuh well information, the inferred detachment at about 2900 m (9514 ft) below the top of the Bangestan Group would correspond to an approximately 1300-m (4265-ft)-thick accumulation of Triassic evaporites of the Dashtak Formation (e.g., Ziegler, 2001). These thick evaporites are distributed along an elongated region, which clearly coincides with the Kabir Kuh anticline at surface. Dashtak Formation rapidly thins toward the southwest as evidenced by its thickness of about 450 m (1476 ft) in the Samand oil well (Figure 11B). In the cross section, the Dashtak evaporites constitute an intermediate detachment level, which affects the Kabir Kuh anticline. This intermediate detachment level increased the amplitude of the Kabir Kuh anticline by internally decoupling the Competent Group at the Dashtak level. Below the evaporites of the Dashtak Formation, the Kabir Kuh oil well shows a thick succession, including the Permian succession (700–1000 m [2296–3281 ft]) and early Paleozoic rocks with a thickness of 207 m (679 ft). These units have been constructed below the Dashtak Formation, and the section has been completed using about 1-km (0.6-mi)-thick early Paleozoic strata, which fits with the estimated thickness observed in seismic lines from the northwest of the Dezful embayment. The dip projection method, applied below the Dashtak detachment, produced a fold with an open to gentle shape and interlimb angle of about 115°. Using this simple approach, the amount of shortening is higher above the Dashtak detachment than below it, being approximately 2280 m (7480 ft) at the top of the Bangestan Group level and only 930 m (3051 ft) at the top of the Permian strata (Figure 11B) The calculated shortening for the top of the Bangestan Group is about 13.5%. If we assume that this shortening must be approximately constant through the entire section across the Kabir Kuh anticline, the structure below the Dashtak detachment must be modified to increase its shortening.

In the third step, the deeper part of the Kabir Kuh anticline was modified (Figure 11C). As stated previously, seismic lines across the Chachmeh Kush structure show numerous gently folded reflectors, corresponding to depths between 8 and 10.5 km (4.97 and 6.52 mi), attributed to early Paleozoic strata (Figure 6B). Seismic data across the Ahwaz anticline also show fairly continuous folded reflectors located down to at least 1 s beneath the Kazhdumi Formation (Sherkati et al., 2005, their figure 7), which can be roughly ascribed to the top of the Permian succession. The observation of deep folding together with the absence of evidence for major thrusting across the Pusht-e Kuh arc suggests that the main folding mechanism is detachment folding above a basal detachment located near to or at the base of the cover succession. Among the ways to increase shortening in gently folded strata at the core of the anticline, the occurrence of a tectonic wedge, formed by a set of opposed low-angle thrust ramps, is considered the most probable. Development of a tectonic wedge produces shortening in the fold core while preserving the gentle anticline geometry and giving similar shortening that was determined for the units located above the Dashtak intermediate detachment (Figure 11C). This hypothetical structure in the tectonic wedge or fishtail has been described in the deepest stiff units outcropping in the Zagros fold belt (Sherkati and Letouzey, 2004, their figure 14) and is typical in most detached fold belts (e.g., Harrison, 1995; Letouzey et al., 1995; Sans and Vergés, 1995). Low-angle ramps in both hanging wall and footwall produce smooth anticlines with low structural relief that is difficult to image at seismic scales but can form the "rabbit ear" structures that are typical of the region (e.g., Sherkati and Letouzey, 2004). This resultant smooth topography would also match with the apparent continuity of reflectors at depth in existing seismic lines. An alternative interpretation would be the occurrence of fore thrusts and back thrusts that shorten and lift the deeper part of the anticline. Below

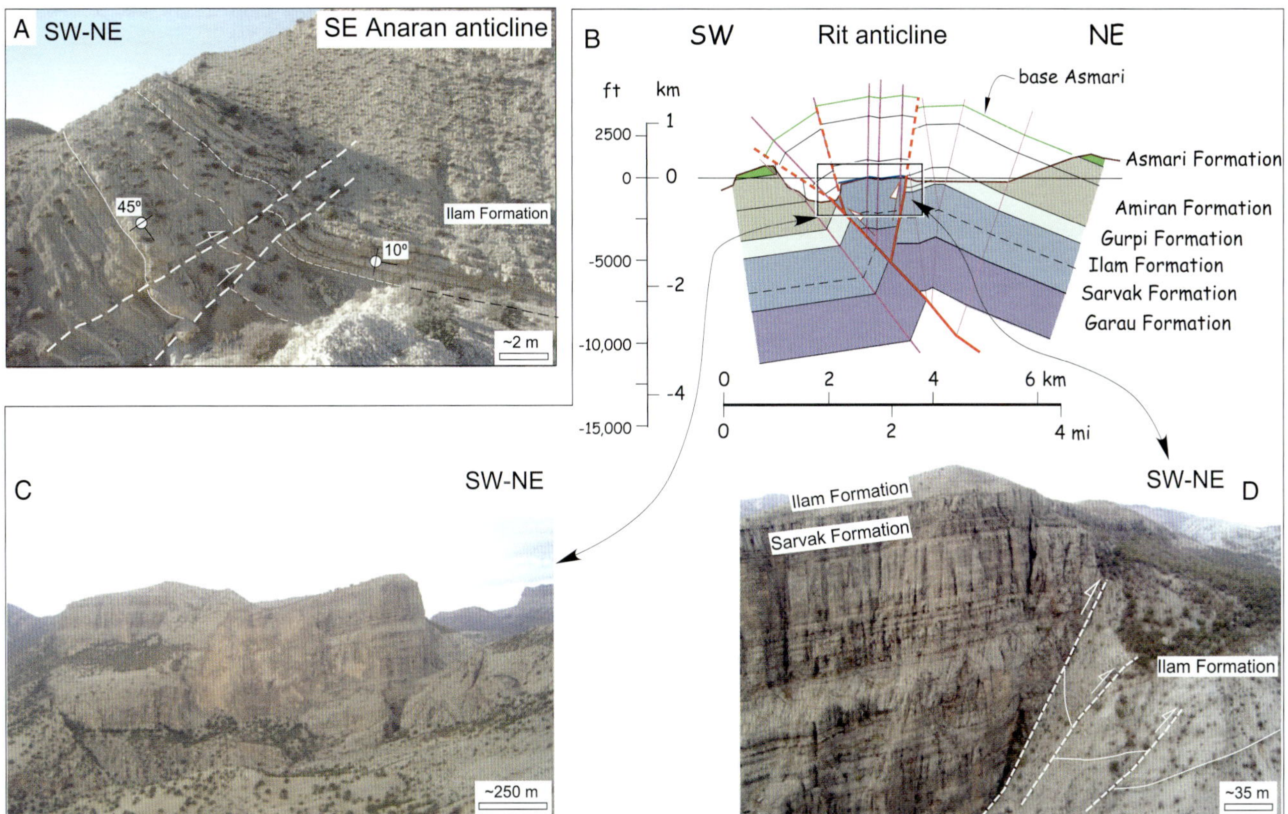

Figure 12. (A) Field view of the northeastern limb of the southeast Anaran anticline showing an abrupt change in dip marked by an axial surface, which corresponds to a thrust with small displacement (A in Figure 2). (B) Cross section across the Rit anticline showing the pop-up structure that uplifts the rigid limestones of the Sarvak Formation above the marls of the Gurpi Formation in the crestal domain of the anticline (R in Figure 2). (C) View of the crestal domain of the Rit anticline from the southeast. (D) View of the backthrust system with steep dips superposing highly fractured limestones of the Sarvak Formation on top of Ilam Formation marls.

this hypothetical cover tectonic wedge structure, the Hormuz evaporites or equivalent may infill the core of the Kabir Kuh anticline reaching, in our section, a maximum potential amplitude of 1.7 km (1.06 mi) at the base of the cover sequence (Figure 11C). This interpretation, which summarizes our present understanding on Pusht-e Kuh arc fold geometries and mechanisms of folding, remains to be tested by good-quality seismic data.

ROLE OF FOLD HINGES (AXIAL SURFACES) IN THRUST DEVELOPMENT

In the Zagros fold belt, the crestal domains of numerous anticlines are bounded by two nearly axial traces defined by two opposed axial surfaces. Lateral changes in folding style from a curved box fold shape to a more chevron fold geometry can cause the termination of one of these two hinges along the strike of the crestal domain, as commonly observed in folds of both the Fars and Pusht-e Kuh arcs. Tilting of the crestal domain can also elevate one of the two hinges, giving the impression that fold is sharper with only one hinge and no crestal domain as shown in the studied section across both the Kabir Kuh anticline and the southeast Anaran anticline. In these two cases, the anticline might be regarded as a northeastern-directed anticline with only one hinge by incorrectly considering the crestal domain as the forelimb of the anticline.

The recognition of the axial surfaces is important because observations in the region indicate that they have an important function in localizing minor thrusting as exemplified along the northeastern flank of the southeast Anaran anticline or along the crestal domain of the Rit anticline (Figure 12). The backlimb of the southeast Anaran anticline shows an abrupt change in dip from more than 45° to only about 15° at the level of the Gurpi-Pabdeh marls (Figure 12A). This change occurs across an approximately 1-m (3-ft) narrow band corresponding to the axial surface. This zone exhibits two parallel thrust

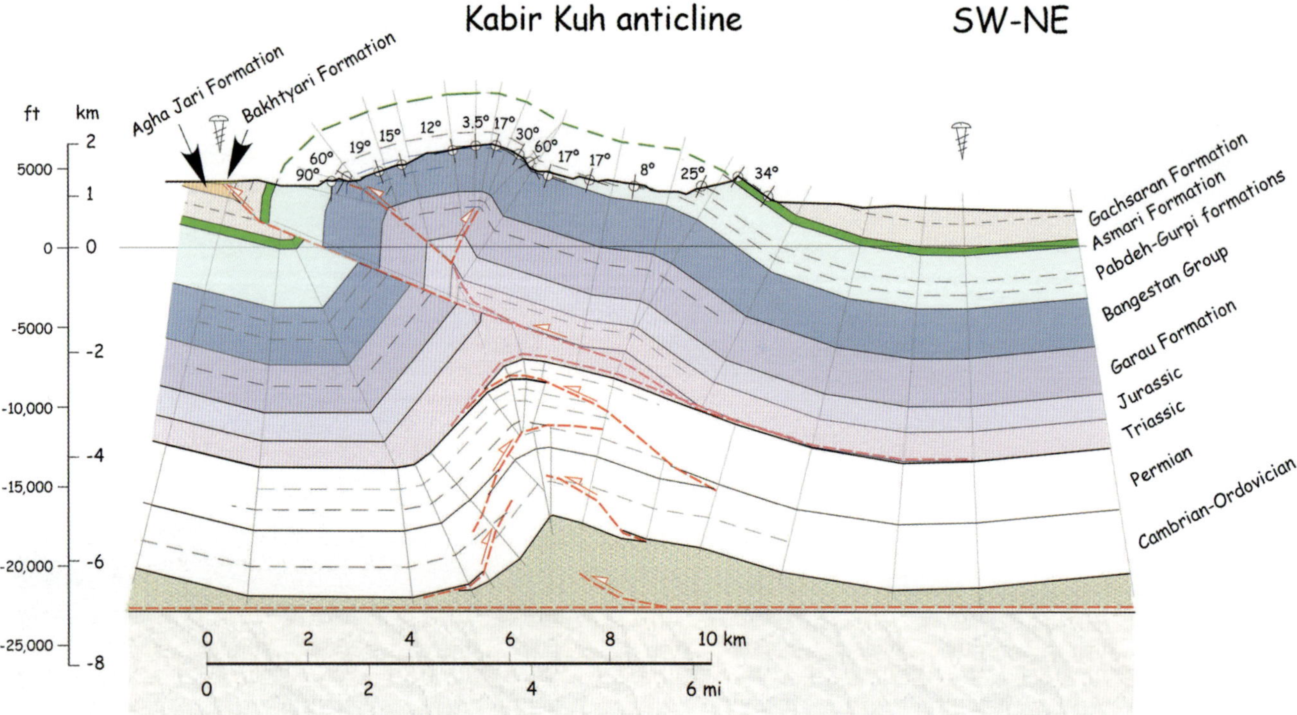

Figure 13. Conceptual cross section constructed combining different examples to show potential thrusting developed by fold tightening. A major thrust displacing the crest of the anticline above its forelimb might develop along an intermediate detachment (Dashtak evaporites in this case) producing an out-of-the-syncline thrust at shallow levels. Thrusting may shift the crest of the anticlines.

planes with a maximum displacement of about 2 m (6 ft), which decreases upward. The Rit anticline shows a more spectacular structure affecting its entire crestal domain constituted by limestones of the Sarvak and Ilam formations (Figure 12B–D, and R in Figure 2). This entire crestal domain has been uplifted above the Gurpi and Amiran formations by a pop-up structure (Figure 12C). The backthrust zone is formed by at least three thrust splays that progressively decrease in dip away from the upper and major thrust. This major thrust that bounds the limestones of the Sarvak Formation dips 80° with a thrust displacement of about 300 m (984 ft). The Gurpi marls in the footwall show a 30-m (98-ft)-thick band with near vertical bedding sandwiched between two major and adjacent back thrusts giving the impression that these two thrusts developed along the axial surfaces bounding a preexisting short limb (dip domain) with near vertical attitude (Figure 12D). Described examples suggest that previously developed axial surfaces had a significant role in focusing thrust planes that dislocated the anticline structure during its late evolution.

Using these observations as well as structural models from other detached fold belts, an interpretation for the southeastern segment of the Kabir Kuh anticline in contact with an approximately 28-km (17.4-mi)-long but narrow syncline (Sarab Bagh syncline; SB in Figure 2) filled up with red beds corresponding to the Agha Jari and discordant Bakhtyari formations is proposed. These Agha Jari red beds are only cropping out along the southwestern limb of the syncline, whereas the northeastern limb is formed by evaporites of the Gachsaran Formation. The contact between these two formations is thus inferred to correspond to a thrust directed to the southwest although the trace of the thrust fault is not mapped and would be, in any case, partially buried beneath the unconformable conglomerates of the Bakhtyari Formation (Setudehnia and O'B Perry, 1967). A potential fold model to account for the existence of a thrust is presented in Figure 13, combining the results from the section in Figure 11C but with the upper part of the Competent Group decoupled and displaced to the southwest above a thrust cutting across the lower part of the steep forelimb dip domain (e.g., Mitra, 2002; Wallace and Homza, 2004) connected to the Dashtak detachment. Additional forelimb axial surfaces may be displaced during thrusting as described for the northeastern limb of the Rit anticline (Figure 12B). In the presented section, thrust propagation may produce out-of-syncline thrusting as has been already documented in Zagros folds (Colman-Sadd, 1978; Beydoun et al., 1992) (Figure 13).

We infer thrusting to initiate as a consequence of tightening of the anticline during increasing shortening. Thrusting can produce displacements of the anticline on top of its forelimb, uplift of the crestal domain by two opposed thrusts as a pop-up as described for the Rit anticline, or both together. This thrust development is interpreted as produced during the final stages of detachment fold growth (Sans and Vergés, 1995).

OTHER INTERMEDIATE DETACHMENTS IN THE COMPETENT GROUP

Additional detachment levels of more restricted extent can be found in the Competent Group of the southeastern region of the Pusht-e Kuh arc. These include the middle Cretaceous Garau Formation, the Paleocene Amiran Formation, and the Miocene Kalhur Member. These intermediate detachment levels have been observed in the field and by seismic interpretation. The middle Cretaceous Garau detachment level, composed of carbonaceous shales, is observed in the Khorramabad anticline where it is expressed as a highly deformed core (Figure 2). The projection of cross sections to depth also supports the existence of a detachment level in correspondence of the Garau Formation, especially to the northeast of the Kabir Kuh anticline.

The Amiran Formation filled up a flexural basin formed to the northeast of the Kabir Kuh anticline (e.g., Alavi, 2004; Homke et al., 2009). The Amiran Formation comprises an approximately 850-m (2789-ft)-thick shallowing-upward siliciclastic sequence that has been dated to the Paleocene in the Amiran anticline, which is located in the center of the Lurestan Province (Homke et al., 2009) (Figure 2). Although not directly observed in the field, the Amiran Formation contains thick, dark mudstone-siltstone deposits that may play as a detachment level. Field data together with seismic interpretation provide significant evidence of the geometry of the Amiran detachment in the Pusht-e Jangal anticline, a few kilometers to the northwest of Kuhdasht city (Figure 2). In the seismic profile, the reflectors corresponding to the Bangestan Group display a 10-km (6.2-mi)-wide antiform that is gently folded in two relatively low-amplitude smaller culminations. When structural dips from the Asmari limestones are combined with the seismic interpretation at the level of the Bangestan Group, a clear folding disharmony emerges from the resultant cross section, superposing a syncline on the Asmari Formation on top of the northeastern anticline culmination at the Bangestan Group level (Figure 14A). The disharmonic folding, however, vanishes toward the southwest of the antiform where Asmari and the top of Bangestan limestones are roughly concordant, implying a change in the detachment level properties (Figure 14A).

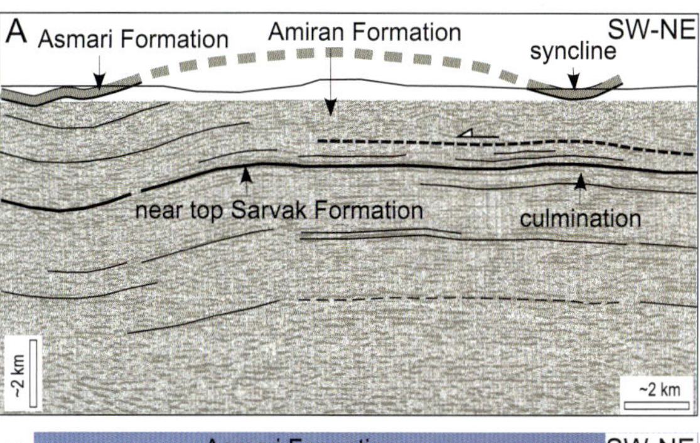

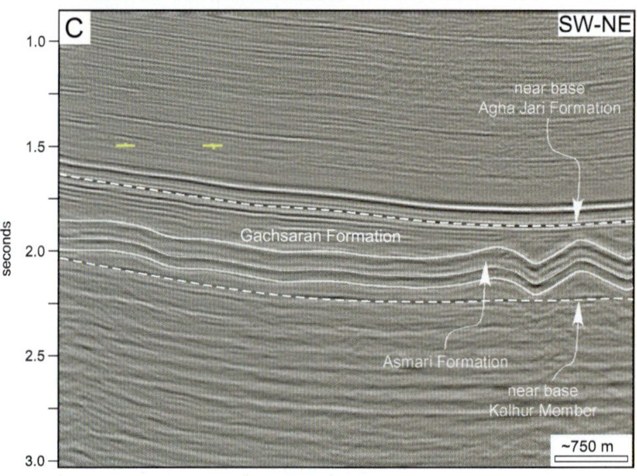

Figure 14. (A) Combination of field cross section and seismic line interpretation across the Pusht-e Jangal anticline near the town of Kuhdasht (see position of town and anticline in Figure 2) showing a significant detachment at the Gurpi-Amiran level. This detachment superposes a set of relatively narrow anticlines on top of a much wider structure at the level of the Sarvak Formation. (B) Asmari limestones tectonically duplicated by low-angle back thrusts in the southwestern flank of the Anaran anticline along its northwestern termination. (C) Prestack time migration seismic line of the northeastern limb of the Changuleh anticline showing the detached folds of the Asmari limestones sandwiched between the Kalhur and the Gachsaran detachment levels.

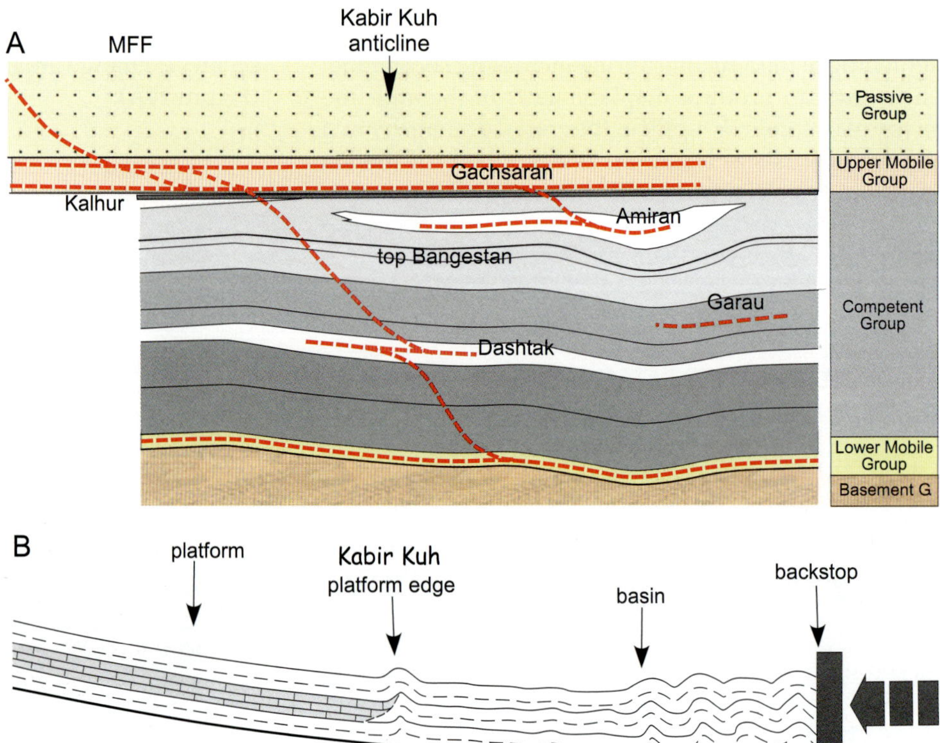

Figure 15. (A) Detachment levels in the Pusht-e Kuh arc related to the Lurestan Province mechanical stratigraphy. (B) Sandbox model, redrawn from Dixon (2004), showing the localization of deformation at the basin-platform boundary (roughly following the Kabir Kuh anticline) above a low-strength basal detachment after approximately 16% of shortening (similar to Pusht-e Kuh arc values). MFF = mountain front flexure; G. = Group.

Whereas the Amiran Formation is restricted to the region northeast of the Kabir Kuh anticline, the Oligocene–Miocene Kalhur evaporites are restricted to the region southwest of it. The Kalhur evaporites constitute a weak level above which the Asmari limestones detach and deform by folding and thrusting as shown by both field examples (Figure 14B) and seismic data (Figure 14C) along the southwestern limb of the Anaran anticline.

CONTROL OF REGIONAL STRATIGRAPHY ON THE FOLDING OF THE PUSHT-E KUH ARC

The regional stratigraphy exerts a first-order control on folding evolution in the Pusht-e Kuh arc (Lurestan Province) as recently summarized for the Fars arc and Dezful embayment (Molinaro et al., 2004; Sherkati et al., 2005; Sepehr et al., 2006, among others). Locally, intermediate detachment levels may also contribute to shape the final geometry of the folds resulting in a large variability of anticline shapes and sizes. In this section, we summarize the interplay between detachment levels and stratigraphy by constructing a panel across the southeastern region of the Pusht-e Kuh arc to determine the position and extent of the detachments (Figure 15A). The idea of this panel is to facilitate the prediction of either folding style, by knowing the position and extent of detachment levels at depth, or the position and extent of a potential detachment by grouping anticlines with similar characteristics at surface as described previously (Figures 2, 8).

In the southeastern region of the Pusht-e Kuh arc, several intermediate detachment levels at Dashtak, Garau, Amiran, and Kalhur levels can be distinguished that complicate the geometry of anticlines by increasing the fold amplitude and tightness (box folds with near vertical flanks) or by developing fore- and back-thrusting, and smaller folds. The deeper detachment levels and the shallower Kalhur detachment are located to the northeast and southwest of the Kabir Kuh anticline, respectively. This asymmetric distribution of detachment levels is consistent with the role of long-term boundary that the 150-km (93-mi)-long Kabir Kuh anticline played from the Mesozoic to present. During the Mesozoic, the anticline represented the boundary between platform and basin facies, whereas during the Cenozoic, it constituted the southwestern limit of the Amiran Basin (Alavi, 2004; Homke et al., 2009).

The change in folding style across the Kabir Kuh anticline is consistent with the lateral passage from platform to basinal facies within the Sarvak Formation (most of the Bangestan Group). These fold variations across sedimentary facies boundaries are comparable with results from analog models (Dixon, 2004) (Figure 15B). Dixon's configuration model MO19 matches the large-scale original heterogeneous depositional arrangement of the study area with weaker and multilayered basin facies close to the backstop (northeast segment) and stronger and more massive carbonate platform facies toward the foreland (southwestern segment). In this model, compression results in a fast propagation of deformation to the platform-basin boundary producing along this limit an early anticline that would correspond to the Kabir Kuh anticline (Figure 15B).

The deeper intermediate detachment levels at the Dashtak and Garau formations are located within the basin to the northeast of the Kabir Kuh anticline. These detachments are deduced by cross section construction and scarce field observations. The most spectacular tight and upright folds (lift-off folds; Mitra, 2003) of the region crop out on the eastern side of the Pusht-e Kuh arc close to the Dezful embayment and Izeh zone. These anticlines show isoclinal folding deforming limestones of the Sarvak Formation detached at the level of the Kazhdumi Formation that thickens toward the Dezful embayment. These isoclinal folds that have been named bullet anticlines by Sherkati et al. (2005) and Sepehr et al. (2006) show local accommodation low-angle flat-ramp-flat thrusts cutting the anticline flanks.

Although preliminary because of the lack of a more consistent data set at depth, the different folding domains can be attributed to different detachment levels at depth. The Anaran and Samand anticlines from domain 1 are detached at the base of the cover succession, as is the Kabir Kuh anticline. However, Kabir Kuh corresponds to domain 2 because of the existence of a thick Dashtak detachment that increases the anticline amplitude. To the northeast, the level of detachment is located at the Garau Formation within domain 3, whereas it climbs up to the Amiran Formation in domain 4. The Khorramabad anticline (domain 5) is again detached at the Garau detachment level as observed in the field. Along the southeastern border of the Pusht-e Kuh arc, the isoclinal folds in the upper part of the Bangestan Group seem to be detached above thick Kazhdumi shales.

The correspondence between singular detachment levels at depth and variations in folding style at surface needs to be tested by a more accurate map distribution of these detachment levels at depth as well as sorting out their degree of overlapping that may be an important factor in folding evolution, subsequent thrusting, and interconnectivity of thrust faults (Casciello et al., 2009).

DISCUSSION

The first part of the discussion examines how different the folds are in the Pusht-e Kuh arc and Dezful embayment with respect to the Fars arc where Hormuz salt is present, whereas in the second part, the characterization of the entire structure of the Pusht-e Kuh arc and its implications for hydrocarbon accumulation are discussed.

Potential Presence of Weak Detachment Beneath the Pusht-e Kuh Arc

A general consensus among structural geologists working in the Zagros fold belt is that Hormuz salt is the main detachment level in the Fars arc and southeastern Izeh zone (e.g., Edgell, 1996; Talbot and Alavi, 1996; Sepehr et al., 2006). The late Proterozoic to Early Cambrian Hormuz salt is located at the base of the 10–12-km (6.2–7.4-mi)-thick Paleozoic to Cenozoic sedimentary succession of the Arabian margin, which deformed during the Zagros collision. The occurrence of more than 200 Hormuz salt diapirs in the Fars arc and southeast Persian Gulf (Callot et al., 2007; Jahani et al., 2007), and southeastern and central segments of the imbricate zone, constrain the minimum extent of this weak unit at depth (Figure 1). The lack of Hormuz salt diapirs in both the Dezful embayment (Khuzestan Province) and Pusht-e Kuh arc (Lurestan Province) has been commonly attributed to the absence or the insignificant thickness of this salt unit in the two provinces (Talbot and Alavi, 1996; Bahroudi and Koyi, 2003). Despite the lack of Hormuz salt diapirs in both the Lurestan and Khuzestan stratigraphic provinces, the geometry of large folds as well as their distribution in map view seems to indicate that these two provinces are likewise detached above a weak and efficient detachment level located either near to or at the base of the cover sequence.

The southwestern Zagros deformational front forms a relatively rectilinear boundary in front of both arcs and embayments (ZDF in Figure 1). The width of the Zagros fold belt, calculated from the trace of the high Zagros fault to the front of folds cropping out, is 250 km (155 mi) in the center of the Fars arc and only about 150 km (93 mi) in both the Pusht-e Kuh arc and Dezful embayment (Figure 1). This width, however, can reach about 210 km (130 mi) in the Pusht-e Kuh arc if we include the buried folds that lie beneath the Mesopotamian foreland basin in both Iran and Iraq (Figure 1). The noteworthy result is that the Zagros deformational front is roughly rectilinear and subparallel to the high Zagros fault, suggesting similar propagation of folding along the three main tectonic domains of the Zagros fold belt in Iran. The present distribution of the Zagros fold belt in salients and reentrants is a feature produced by late uplift of large already folded regions above basement thrusts that produced a regional flexure or monocline at their front (MFF defined by Falcon, 1961). Although

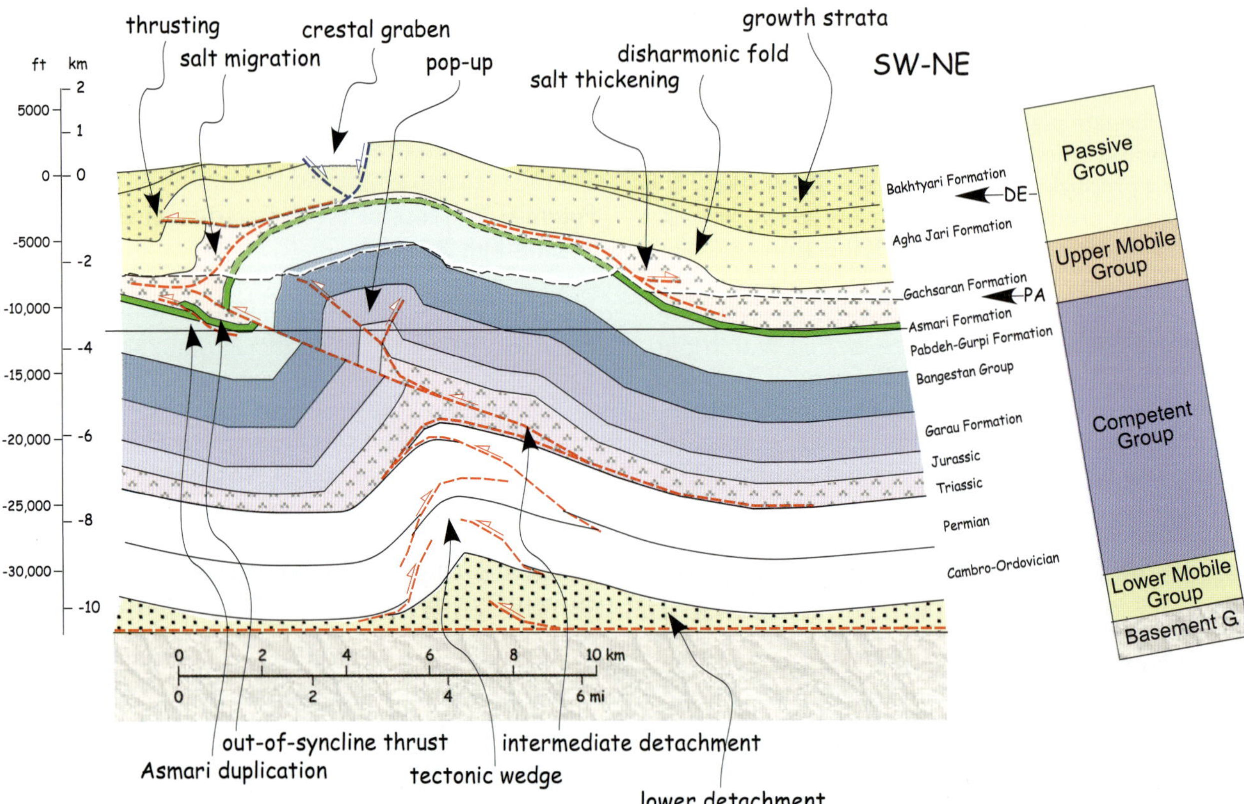

Figure 16. Conceptual model of folding integrating observations from the Pusht-e Kuh arc (Competent Group) and Dezful embayment (Passive Group) tectonic domains. The PA and DE represent the outcropping levels for the Pusht-e Kuh arc and Dezful embayment, respectively. G. = Group.

210–250 km (130–155 mi) wide, the Zagros fold belt has maximum shortening values of only 15% in the Pusht-e Kuh arc and Dezful embayment. Folding of wide areas with relatively low shortening values as well as low taper (e.g., Talbot and Alavi, 1996; Ford, 2004) may also indicate activation of a very efficient mechanism to propagate deformation forward. This is typical of fold and thrust belts detached above thick ductile layers as proved in detachment fold and thrust belts elsewhere (e.g., Salt Ranges, Pyrenees, and Jura Mountains; Butler et al., 1987; Vergés et al., 1992; Sommaruga, 1999, respectively) as well as in analog models (e.g., Costa and Vendeville, 2002). Although not conclusive, data presented in this chapter support the existence of an efficient detachment layer near to or at the base of the cover sequence above which folding could develop in Dezful embayment and Pusht-e Kuh arc domains in a similar way to the Fars arc.

Style of Folding in the Pusht-e Kuh Arc and Implications for Hydrocarbons

In the Zagros, folds are shaped by two major detachments decoupling the thick cover sequence in different compartments. Above the lower Mobile Group (Hormuz evaporites or equivalent), the stiff Competent Group is separated from the Passive Group by the upper Mobile Group at the level of the Gachsaran evaporites. This large-scale framework, however, can be completed by additional detachment levels that complicate the geometry of the anticlines at depth. Although other recent works deal with the same subject in the Zagros fold belt, no studies exist yet for the Pusht-e Kuh arc folds. In the following, a summary of our findings on the entire structure of this tectonic domain combined with the Passive Group structure from the western region of the Dezful embayment is given (Figure 16).

The decoupling between different rigid units is recorded in the Passive Group, above the mobile Gachsaran evaporites (Figure 16). These evaporites, sandwiched between the Competent and the Passive groups, migrate from the growing anticlines in the Competent Group toward the adjacent synclines. As a consequence of the remobilization of these evaporites, the amplitude of folding in the Passive Group may vary with respect to equivalent folds in the Competent Group. If folding amplitudes decrease, the consequent reduction in shortening is accommodated by thrusting with maximum displacements of few kilometers, especially in the footwall of the MFF. In addition, the gravitational collapse of the anticline crestal domains during their growth may increase the displacements along the thrusts that cut the Passive Group. Thrusts and back thrusts may develop in the Passive Group showing fishtail arrangements when linked together (Figure 16).

At deeper levels, below the Gachsaran evaporites, the Competent Group contains multiple detachments especially toward the northeast of the Kabir Kuh anticline that represents a paleogeographic boundary during the Mesozoic between platform and basin (Figure 16). The unique size and geometry of the Kabir Kuh anticline are due to the thick tectonic accumulation of Triassic Dashtak evaporites in the core of the fold. This accumulation increases fold amplitude and determines the box fold geometry of the Kabir Kuh anticline. Further shortening affecting the already evolved box fold produces propagation of one or more thrusts, bounding short and subvertical segments of the fold forelimbs, from the Dashtak detachment level through the forelimb. The upward propagation of these thrusts in the forelimb can be along already formed axial surfaces (fold hinges) displaying elevated angles with both hanging-wall and footwall domains. These thrusts might displace the culmination above the detachment to the southwest with respect to the anticline culmination at depth. In some cases, the reactivation of pairs of axial surfaces produces the uplift of the entire anticline culmination as a pop-up structure. At deeper levels, we inferred that low-angle uplimb thrusting accommodates shortening in the core of the anticline at the level of Paleozoic strata. This interpretation is based on the occurrence of folded deep reflectors imaged in some of the seismic lines across the Dezful embayment. These folded deep reflectors deform above an efficient detachment level that must correspond (or is mechanically equivalent) to Hormuz evaporites.

In summary, we present a model in which the thick heterogeneous cover sequences of the Arabian plate folded according to the vertical and lateral variations of the mechanical stratigraphy. Although few major thrusts are observed at surface, thrusting is an important mechanism to accommodate shortening during final stages of fold growth, particularly in tight and upright anticlines. Disharmonic folding across these detachment levels may produce shifts of anticline culminations that in some cases can be of few kilometers. This is an important issue for hydrocarbon exploration in both Pusht-e Kuh arc and Dezful embayment regions.

One of the principal characteristics of the highly productive Zagros fold belt hydrocarbon province is the ideal alternation of folded source rocks with reservoirs and seals in conjunction to the right sequence of oil generation and migration with the timing of deformation (e.g., Bordenave and Hegre, 2005) (Figure 3). Both hydrocarbons-rich basinal shales and evaporitic seals represent the mechanically weaker materials in the stratigraphic pile and thus form the main detachments during compression. In the Zagros fold belt, a close link between the relative timing of oil generation, migration, and entrapment, and extensive fracture development and fold growth is also observed (Bordenave and Hegre, 2005). According to these authors, oil from middle Cretaceous and early Tertiary source rocks generated during deposition of the lower part of the Agha Jari Formation at about 10 Ma. This age of hydrocarbon generation is slightly older than the onset of folding growth (at least in frontal domains of the Pusht-e Kuh arc). Nevertheless, oil generation possibly continued during fold amplification in agreement with the recognition of asphalt-bearing sandstones within the Agha Jari Formation (Elmore and Farrand, 1981).

Asmari and Sarvak reservoirs in the Zagros fold belt are highly fractured but also interconnected in many cases across the seals. Thrusts presented in this work are secondary structures, normally characterized by relatively small displacements. However, at the anticline scale, they can form a linked system connecting reservoirs across seal units and might enhance connectivity across different reservoir units. The potential behavior of source rocks as detachment levels and the approximate concurrence between the timing of oil generation and tectonic deformation has been interpreted as having a feedback in which initial tectonic stresses in combination with abnormally high pore pressures within source rocks during oil generation may produce the initiation of those levels as detachments (Cobbold, 2005).

The potential relationships among mechanical stratigraphy and folding and thrusting as well as the relative timing between oil generation and trap formation are of crucial significance when exploring a region such as the Pusht-e Kuh arc. This uplifted region is an excellent natural laboratory to explore these potential relationships. The results of this study on folding geometries are directly applicable to petroleum exploration of recently awarded exploration areas within the Pusht-e Kuh arc and may apply to the rest of the Zagros fold belt.

CONCLUSIONS

The folds of the Pusht-e Kuh arc, in the northwestern Zagros fold belt, exhibit whaleback geometries outlined by the resistant limestones of the two main hydrocarbon reservoirs of the most prolific oil and gas fold and thrust belt province on Earth. A geological balanced cross section across the central part of the Kabir Kuh anticline, based on field observations and well data, is constructed as the backbone to link folding to the mechanical stratigraphy of the southeastern region of the Pusht-e Kuh arc. Although the large-scale mechanical properties of the 10–12-km (6.2–7.4-mi)-thick cover sequence of the Zagros has already been proposed, the results of this work are specific for the Pusht-e Kuh arc (Lurestan Province).

The Kabir Kuh anticline in its central part displays a box folding geometry characterized by a wide and rounded crestal domain, which is slightly tilted to the southwest. This geometry developed above an intermediate detachment at the level of Triassic Dashtak evaporites within the Competent Group. Below this detachment, the geometry of the fold changes to more acute with a narrower crestal region. Although conjectural, we propose low-angle thrusting forming a tectonic wedge below the Dashtak detachment. These thrusts increase shortening in the core of the anticline but with relatively smooth structural relief that matches with interpretations of deep and fairly continuous folded reflectors from the northwestern Dezful embayment. Subsidiary thrusts may reactivate axial surfaces as well as detachment levels to propitiate displacements of the crest of the anticline above its forelimb as inferred along the syncline to the southeast of the presented section. To the northeast of the Kabir Kuh anticline, the Garau and Amiran formations represent local detachment levels, whereas the Kalhur Member represents the local intermediate detachment to the southwest of the anticline.

The Gachsaran evaporites (upper Mobile Group), sandwiched between two stiff units, represent the main detachment decoupling the Competent and Passive groups. Gachsaran evaporites may migrate from anticlines to adjacent synclines, which in conjunction with thrusts affecting the Passive Group may produce a significant disharmonic folding.

The control exerted by the mechanical stratigraphy on folding style is of prime order. The lateral changes in sedimentary facies constrain the extent of the local detachment levels as the Amiran Formation and the Kalhur Member to the northeast and southwest of the Kabir Kuh anticline, respectively. The close relationships among oil source rocks and seals with detachment levels and the potential concurrence of oil generation, fold growth, and local thrusting need to be investigated to determine the potential feedback among all of these different processes. These links are crucial to understand hydrocarbon systems in complex regions.

ACKNOWLEDGMENTS

This is a contribution of the Group of Dynamics of the Lithosphere (GDL), financed by a collaborative project between the Institute of Earth Sciences Jaume Almera, CSIC of Barcelona (Spain), and Hydro Zagros Oil and Gas Tehran and StatoilHydro Research Center, with the partial support of project Team Consolider-Ingenio 2010 no. CSD 2006-00041. We also thank the support in the field of StatoilHydro and Norwegian People's Aid (NPA) staff, and the National Iranian Oil Company (NIOC) for their collaboration during this project. We thank I. Sharp who kindly provided the helicopter picture of the Kabir Kuh anticline, and S. Sherkati, E. Blanc, E. Casciello, and G. Casini for fruitful discussions. We thank J. de Vera, T. Flottmann, and an anonymous reviewer for their excellent and constructive comments. Finally, we dedicate this paper to the memory of Mohammad Sepehr who passed away while working in the Fars.

REFERENCES CITED

Abdollahie Fard, I., A. Braathen, M. Mokhtari, and S. Ahmad Alavi, 2006, Interaction of the Zagros fold-thrust belt and the Arabian type, deep-seated folds in the Abadan Plain and the Dezful Embayment, SW Iran: Petroleum Geoscience, v. 12, p. 347–362.

Agard, P., J. Omrani, J. Jolivet, and F. Mouthereau, 2005, Convergence history across Zagros (Iran): Constraints from collisional and earlier deformation: International Journal of Earth Sciences (Geol Rundsch), v. 94, p. 401–419.

Alaei, B., and J. Pajchel, 2006, Single arrival Kirchhoff prestack depth migration of complex faulted folds from the Zagros Mountains, Iran: Canadian Society of Exploration Geophysicists Recorder, January, p. 41–48.

Alavi, M., 1994, Tectonics of the Zagros orogenic belt of Iran: New data and interpretations: Tectonophysics, v. 229, p. 211–238.

Alavi, M., 2004, Regional stratigraphy of the Zagros fold-thrust belt of Iran and its proforeland evolution: American Journal of Science, v. 304, p. 1–20.

Bahroudi, A., and H. A. Koyi, 2003, Effect of spatial distribution of Hormuz salt in deformation style in the Zagros fold-and-thrust belt: An analog modeling approach: Journal of the Geological Society (London), v. 160, p. 719–733.

Bahroudi, A., and H. A. Koyi, 2004, Tectono-sedimentary framework of the Gachsaran Formation in the Zagros foreland basin: Marine and Petroleum Geology, v. 21, p. 1295–1310.

Berberian, M., 1995, Master "blind" thrust faults hidden under the Zagros folds: Active basement tectonics and surface morphotectonics: Tectonophysics, v. 241, p. 193–224.

Berberian, M., and G. C. P. King, 1981, Toward a paleogeography and tectonic evolution of Iran: Canadian Journal of Earth Sciences, v. 18, p. 210–265.

Beydoun, Z. R., M. W. Hughes Clarke, and R. Stoneley, 1992, Petroleum in the Zagros Basin: A late Tertiary foreland basin overprinted onto the outer edge of a vast hydrocarbon-rich Paleozoic–Mesozoic passive-margin shelf, *in* R. W.

Maequeen and D. H. Lackie, eds., Foreland basins and fold belts: AAPG Memoir 55, p. 309–339.

Blanc, E. J.-P., M. B. Allen, S. Inger, and H. Hassani, 2003, Structural styles in the Zagros simple folded zone, Iran: Journal of the Geological Society (London), v. 160, p. 401–412.

Bordenave, M. L., and J. A. Hegre, 2005, The influence of tectonics on the entrapment of oil in the Dezful Embayment, Zagros fold belt, Iran: Journal of Petroleum Geology, v. 28, p. 339–368.

Butler, R. W. H., M. P. Coward, G. M. Harwood, and R. J. Knipe, 1987, Salt, its control on thrust geometry, structural style and gravitational collapse along the Himalayan mountain front in the Salt Range of northern Pakistan, *in* J. J. Obrien and I. Lesche, eds., The dynamical geology of salt and related structures: Austin, Texas, Academic Press, p. 399–418.

Callot, J. P., S. Jahani, and J. Letouzey, 2007, The role of pre-existing diapirs in fold and thrust belt development, *in* O. Lacombe, J. Lavé, F. Roure, and J. Vergés, eds., Thrust belts and foreland basins from fold kinematics to hydrocarbon systems: Frontiers in Earth Sciences: Heidelberg, Springer, p. 309–326.

Carruba, S., C. R. Perotti, R. Buonaguro, R. Calabró, R. Carpi, and M. Naini, 2006, Structural pattern of the Zagros fold-and-thrust belt in the Dezful Embayment (SW Iran): Styles of continental contraction: Geological Society of America Special Paper 414, p. 11–32, doi:10.1130/2006.2414(02).

Casciello, E., J. Vergés, E. Saura, G. Casini, N. Fernández, E. Blanc, S. Homke, and D. Hunt, 2009, Fold patterns and multilayer rheology of the Lurestan Province, Zagros simply folded belt (Iran): Geological Society (London), v. 166, p. 1–13, doi:10.1144/0016-7649.2008-138.

Cobbold, P., 2005, Hydrocarbon generation, a mechanism of detachment in thin-skinned thrust belts: International Meeting on Thrust Belts and Foreland Basins, Rueil-Malmaison, December 2005, Abstracts Volume: Rueil-Malmaison, Institut Français du Pétrole, p. 104–107.

Colman-Sadd, S. P., 1978, Fold development in Zagros simply folded belt, southwest Iran: AAPG Bulletin, v. 62, p. 984–1003.

Costa, E., and B. Vendeville, 2002, Experimental insights on the geometry and kinematics of fold-and-thrust belts above weak, viscous evaporitic decollement: Journal of Structural Geology, v. 24, p. 1729–1739.

Dahlstrom, C. D. A., 1969, The upper detachment in concentric folding: Bulletin of Canadian Petroleum Geology, v. 17, p. 326–346.

Dixon, J. M., 2004, Physical (centrifuge) modeling of fold-thrust shortening across carbonate bank margins— Timing, vergence, and style of deformation, *in* K. R. McClay, ed., Thrust tectonics and hydrocarbon systems: AAPG Memoir 82, p. 223–238.

Dunnington, H. V., 1968, Salt-tectonic features of northern Iraq: Geological Society of America Special Paper, v. 88, p. 183–227.

Edgell, H. S., 1996, Salt tectonism in the Persian Gulf Basin, *in* G. I. Alsop, D. Blundell, and I. Davison, eds., Salt tectonics: Geological Society (London) Special Publication 100, p. 129–151.

Elmore, R. D., and W. R. Farrand, 1981, Asphalt-bearing sediment in synorogenic Miocene–Pliocene molasse, Zagros Mountains, Iran: AAPG Bulletin, v. 65, p. 1160–1165.

Falcon, N. L., 1961, Major earth-flexuring in the Zagros Mountains of south-west Iran: Quarterly Journal of the Geological Society (London), v. 117, p. 367–376.

Falcon, N. L., 1974, Southern Iran: Zagros Mountains, *in* A. M. Spencer, ed., Mesozoic–Cenozoic orogenic belts, data for orogenic studies: Geological Society (London) Special Publication 4, p. 9–22.

Ford, M., 2004, Depositional wedge tops: Interaction between low basal friction external orogenic wedges and flexural foreland basins: Basin Research, v. 16, p. 361–375, doi:10.1111/j.1365-2117.2004.00236.x.

Harrison, J. C., 1995, Tectonics and kinematics of a foreland folded belt influenced by salt, Arctic Canada, *in* M. P. A. Jackson, D. G. Roberts, and S. Snelson, eds., Salt tectonics: A global perspective: AAPG Memoir 65, p. 379–412.

Harrison, J. V., and N. L. Falcon, 1934, Collapse structures: Geological Magazine, v. 71, p. 529–539.

Harrison, J. V., and N. L. Falcon, 1937, The Saidmarreh landslip— West Iran: Geographical Journal, v. 89, p. 42–47.

Haynes, S. J., and H. McQuillan, 1974, Evolution of the Zagros suture zone, southern Iran: Geological Society of America Bulletin, v. 85, p. 739–744.

Homke, S., J. Vergés, M. Garcés, H. Emami, and R. Karpuz, 2004, Magnetostratigraphy of Miocene–Pliocene Zagros foreland deposits in the front of the Push-e Kush Arc (Lurestan Province, Iran): Earth and Planetary Science Letters, v. 225, p. 397–410.

Homke, S., J. Vergés, J. Serra-Kiel, G. Bernaola, M. Garcés, R. Karpuz, I. Sharp, M. H. Goodarzi, and I. M. Verdú, 2009, Late Cretaceous–Paleocene formation of the early Zagros foreland basin: Biostratigraphy and magnetostratigraphy of the Amiran, Taleh Zang and Kashkan sequence in Lurestan Province, SW Iran: Geological Society of America Bulletin, v. 121, p. 963–978, doi:10.1130/B26035.1.

Hull, C. E., and H. R. Warman, 1970, Asmari oil fields of Iran: AAPG Bulletin, p. 428–437.

Ion, D. C., S. Elder, and A. E. Pedder, 1951, The Agha Jari oilfield, south-west Persia: 3rd World Petroleum Congress, sec. 1: The Hague, J. Brill Publishing House, p. 162–186.

Jahani, S., J. P. Callot, D. Frizon de Lamotte, J. Letouzey, and P. Leturmy, 2007, The salt diapirs of the eastern Fars province (Zagros, Iran): A brief outline of their past and present, *in* O. Lacombe, J. Lavé, F. Roure, and J. Vergés, eds., Thrust belts and foreland basins from fold kinematics to hydrocarbon systems: Frontiers in Earth Sciences: Heidelberg, Springer, p. 289–308.

James, G. A., and J. G. Wynd, 1965, Stratigraphic nomenclature of Iranian oil consortium agreement area: AAPG Bulletin, v. 49, p. 2182–2245.

Kashfi, M. S., 1980, Stratigraphy and environmental sedimentology of lower Fars Group (Miocene), south-southwest Iran: AAPG Bulletin, v. 64, p. 2095–2107.

Letouzey, J., B. Colletta, R. Vially, and J. C. Chermette, 1995, Evolution of salt-related structures in compressional settings, *in* M. P. A. Jackson, D. G. Roberts, and S. Snelson, eds., Salt tectonics: A global perspective: AAPG Memoir 65, p. 41–60.

MacLeod, J. H., 1970, Kabir Kuh geological compilation map: Iranian Oil Operating Companies, scale 1:100,000, sheet number 20812 W.

McQuarrie, N., 2004, Crustal scale geometry of the Zagros fold-thrust belt, Iran: Journal of Structural Geology, v. 26, p. 519–535.

McQuillan, H., 1974, Fracture patterns on Kuh-e Asmari anticline, southwest Iran: AAPG Bulletin, v. 58, p. 236–246.

Mitra, S., 2002, Fold-accommodation faults: AAPG Bulletin, v. 86, p. 671–693.

Mitra, S., 2003, A unified kinematic model for the evolution of detachment folds: Journal of Structural Geology, v. 25, p. 1659–1673.

Mohajjel, M., and C. L. Fergusson, 2000, Dextral transpression in Late Cretaceous continental collision, Sanandaj-Sirjan zone, western Iran: Journal of Structural Geology, v. 22, p. 1125–1139.

Mohammed, S. A. G., 2006, Megaseismic section across the northeastern slope of the Arabian plate, Iraq: GeoArabia, v. 11, p. 77–90.

Molinaro, M., J. C. Guezou, P. Leturmy, S. A. Eshraghi, and D. Frizon de Lamotte, 2004, The origin of changes in structural style across the Bandar Abbas syntaxis, SE Zagros (Iran): Marine and Petroleum Geology, v. 21, p. 735–752.

Molinaro, M., P. Leturmy, J.-C. Guezou, D. Frizon de Lamotte, and S. A. Eshraghi, 2005, The structure and kinematics of the south-eastern Zagros fold thrust belt; Iran: From thin-skinned to thick-skinned tectonics: Tectonics, v. 24, p. TC3007, doi:10.1029/2004TC001633.

Murris, R. J., 1980, Middle East: Stratigraphic evolution and oil habitat: AAPG Bulletin, v. 64, p. 597–618.

O'B Perry, J. T., and A. Setudehnia, 1967, Kūh-e Kamestān geological compilation map: National Iranian Oil Company (NIOC), scale 1:100,000, sheet number 20821E.

O'Brien, C. A. E., 1950, Tectonic problems of the oil field belt of southwest Iran: Proceedings of 18th International Geological Congress, London: Part 6, p. 45–58.

O'Brien, C. A. E., 1957, Salt diapirism in south Persia: Geologie en Mijnbouw, v. 19, p. 337–376.

Oveisi, B., J. Lavé, and P. van der Beek, 2007, Rates and processes of active folding evidenced by Pleistocene terraces at the central Zagros front (Iran), *in* O. Lacombe, J. Lavé, F. Roure, and J. Vergés, eds., Thrust belts and foreland basins from fold kinematics to hydrocarbon systems: Frontiers in Earth Sciences: New York, Springer, p. 267–288.

Price, N. J., and J. W. Cosgrove, 1990, Analysis of geological structures: Cambridge, Cambridge University Press, 520 p.

Sans, M., and J. Vergés, 1995, Fold development related to contractional salt tectonics, Southeastern Pyrenean thrust front, Spain, *in* M. P. A. Jackson, D. G. Roberts, and S. Snelson, eds., Salt tectonics: A global perspective: AAPG Memoir 65, p. 369–378.

Sattarzadeh, Y., J. W. Cosgrove, and C. Vita-Finzi, 2000, The interplay of faulting and folding during the evolution of the Zagros deformation belt, *in* J. W. Cosgrove and M. S. Ameen, eds., Forced folds and fractures: Geological Society (London) Special Publication 169, p. 187–196.

Sepehr, M., and J. W. Cosgrove, 2004, Structural framework of the Zagros fold-thrust belt, Iran: Marine and Petroleum Geology, v. 21, p. 829–843.

Sepehr, M., J. W. Cosgrove, and M. Moieni, 2006, The impact of cover rock rheology on the style of folding in the Zagros fold-thrust belt: Tectonophysics, v. 427, p. 265–281.

Setudehnia, A., and J. T. O'B Perry, 1967, Dehluran geological compilation map: Iranian Oil Operating Companies (IOOC), scale 1:100,000, sheet number 20816 W.

Sherkati, S., and J. Letouzey, 2004, Variation of structural style and basin evolution in the central Zagros (Izeh zone and Dezful Embayment), Iran: Marine and Petroleum Geology, v. 21, p. 535–554.

Sherkati, S., M. Molinaro, D. Frizon de Lamotte, and J. Letouzey, 2005, Detachment folding in the central and eastern Zagros fold-belt (Iran): Salt mobility, multiple detachments and late basement control: Journal of Structural Geology, v. 27, p. 1680–1696, doi:10.1016/j.jsg.2005.05.010.

Sommaruga, A., 1999, Decollement tectonics in the Jura foreland fold-and-thrust belt: Marine and Petroleum Geology, v. 16, p. 111–134.

Srivastava, D. C., and R. J. Lisle, 2004, Rapid analysis of fold shape using Bézier curves: Journal of Structural Geology, v. 26, p. 1553–1559.

Stöcklin, J., 1968, Structural history and tectonics of Iran: A review: AAPG Bulletin, v. 52, p. 1229–1258.

Stoneley, R., 1990, The Middle East basin: A summary overview, *in* J. Brooks, ed., Classic petroleum provinces: Geological Society (London) Special Publication 50, p. 293–298.

Talbot, C. J., and M. Alavi, 1996, The past of a future syntaxis across the Zagros, *in* G. I. Alsop, D. J. Blundell, and I. Davison, eds., Salt tectonics: Geological Society (London) Special Publication 100, p. 89–109.

Vergés, J., 2007, Drainage responses to oblique and lateral thrust ramps: A review, *in* G. Nichols, C. Paola, and E. Williams, eds., Sedimentary processes, environments and basins: A tribute to Peter Friend: International Association of Sedimentologists Special Publication 38, p. 29–47.

Vergés, J., J. A. Muñoz, and A. Martínez, 1992, South Pyrenean fold-and-thrust belt: Role of foreland evaporitic levels in thrust geometry, *in* K. R. McClay, ed., Thrust tectonics: London, Chapman and Hall, p. 255–264.

Wallace, W. K., and T. X. Homza, 2004, Detachment folds versus fault-propagation folds, and their truncation by thrust faults, *in* K. McClay, ed., Thrust tectonics and hydrocarbon systems: AAPG Memoir 82, p. 324–355.

Ziegler, M. A., 2001, Late Permian to Holocene Paleofacies evolution of the Arabian plate and its hydrocarbon occurrences: GeoArabia, v. 6, p. 445–504.

5

Hardy, Stuart, and Richard W. Allmendinger, 2011, Trishear: A review of kinematics, mechanics, and applications, *in* K. McClay, J. H. Shaw, and J. Suppe, eds., Thrust fault-related folding: AAPG Memoir 94, p. 95 – 119.

Trishear: A Review of Kinematics, Mechanics, and Applications

Stuart Hardy

Institució Catalana de Recerca i Estudis Avançats (ICREA) and Grup de Geodinàmica i Anàlisi de Conques (GGAC) Departament de Geodinàmica i Geofísica, Facultat de Geologia, Universitat de Barcelona, Barcelona, Spain

Richard W. Allmendinger

Department of Earth and Atmospheric Sciences, Cornell University, Ithaca, New York

ABSTRACT

Trishear is a kinematic model of fault-propagation folding in which the decrease in displacement along the fault is accommodated by deformation in a triangular shear zone radiating from the tip line. This model has garnered increasing acceptance, particularly for cases where parallel kink-fold models do not work (e.g., footwall synclines, lateral and vertical changes in bedding thickness, and orientation). The articulation of the model in terms of velocity fields has enabled systematic explorations of the parameters controlling the trishear geometry; rapid, objective application of trishear to the simulation of real structures; and application of the model in three dimensions. The model has highlighted the importance of a parameter not unique to trishear, the propagation to slip ratio, which has a profound effect on fold geometry and is fundamental to understanding all types of fault-related folds. The drive to understand the significance of such parameters has instigated the application of several mechanical modeling strategies. Block-motion viscous, finite-element, and discrete-element analyses have all provided insight into trishearlike fault-propagation folds. Clearly, from these models, trishear most successfully simulates fold geometries where significant layered anisotropy is absent and the material is incompressible. Despite these modeling efforts, the significance of the trishear apical angle remains elusive. Trishear has been applied to a variety of real-world problems, including growth strata analysis, potential fracture distribution, paleoseismology, and even seismic hazard analysis.

INTRODUCTION

The relationship between fold shape and the evolution of the underlying fault is of interest for reasons ranging from hydrocarbon exploration to seismic risk evaluation to fault mechanics. Folds that form at the tip of blind faults, commonly called tip-line or fault-propagation folds, are especially interesting because the fold is typically the only exposed evidence available bearing on the nature of the underlying fault. The importance of

DOI:10.1306/13251334M943429

fault-propagation folds was first emphasized in work by petroleum exploration geologists more than 50 yr ago (Gallup, 1951; Fox, 1959). Much more recently, geophysicists have come to appreciate the importance of such structures and their association with blind seismogenic faults such as that associated with the Northridge earthquake in the Los Angeles Basin of southern California.

John Suppe and students (Suppe, 1983; Suppe and Medwedeff, 1990) introduced quantitative rigor to the analysis of fault-propagation folds by describing them in terms of kink-band geometry. Adoption of this approach by Jamison (1987), Mitra (1990), and many others has resulted in considerable progress in understanding their nature and evolution. Many structures, however, are not particularly amenable to the kink-band approach. These include basement-cored anticlines, fold-fault geometries with prominent footwall synclines, structures with deformation-induced variable limb thickness, and folds with progressive rotation of the forelimbs.

In 1991, Eric Erslev introduced a radically different and completely novel kinematic model for fault-propagation folds, which addresses some of the shortcomings of the kink-band approach. During his studies of the Laramide Rocky Mountain foreland in Colorado and Wyoming, he noticed that discrete fault zones within the basement diffuse outward and upward in a triangular zone of deformation in the overlying sedimentary section (Figure 1). He called these triangular zones of deformation trishear. Erslev's trishear (nonparallel shear planes oblique to bedding) is a kinematic fold model at the same hierarchical level as parallel folding (shear planes parallel to bedding) and similar folding (parallel shear planes oblique to bedding). Additionally, trishear provides a richer description of heterogeneous strain distribution at the tips of propagating faults, which may ultimately prove useful, for example, in studies of fracture distribution and orientation. The concept of trishear received little attention initially. However, during the decade following the publication of Hardy and Ford (1997) and Allmendinger (1998), a considerable upsurge of interest in this alternative model of fault-related folding has been observed (e.g., Ford et al., 1997; Allmendinger and Shaw, 2000; Fischer and Wilkerson, 2000; Grelaud et al., 2000; Zehnder and Allmendinger, 2000; Champion et al., 2001; Carena and Suppe, 2002; Cristallini and Allmendinger, 2002; Johnson and Johnson, 2002a, b; Khalil and McClay, 2002; Cardozo et al., 2003; Finch et al., 2003; Cristallini et al., 2004).

This chapter provides a historical review of trishear fault-propagation folding from its inception to its current realization. In the process, we not only present its mathematical description and demonstrate its application to real structures, but also use it as an exemplar of the functions of kinematic and mechanical structural models and how they can complement one another.

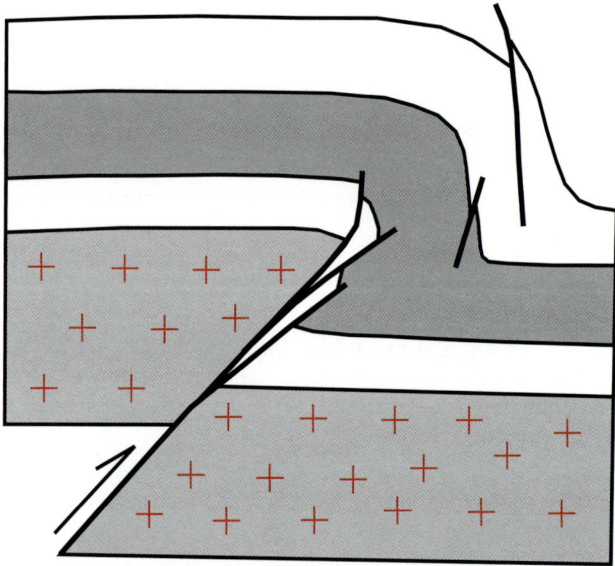

Figure 1. Diagrammatic illustration of qualitative deformation in a triangular shear zone as derived from observational evidence (redrawn from Erslev, 1991).

OBSERVATIONAL EVIDENCE AND PREVIOUS TERMINOLOGY

Prior to the development of trishear, most kinematic models for fault-propagation folding were based on kink-band migration (e.g., Suppe, 1983; Suppe and Medwedeff, 1990), which predicts uniform bed dips and homogeneous deformation in fold limbs. However, several observations suggest that simple kink-band migration kinematics may not be applicable in all cases of fault-propagation folding (Figure 2). Many of these features are obvious in one of the first fault-propagation folds ever described, the Turner Valley anticline (Gallup, 1951) (Figure 2a), and others are recognized in basement-involved provinces such as the Laramide Rocky Mountains (e.g., see articles and references in Schmidt et al., 1993) and the Sierras Pampeanas of western Argentina (Jordan and Allmendinger, 1986). First, we observe the common occurrence of asymmetric fold pairs verging in the direction of thrusting comprising a footwall syncline and hanging-wall anticline, with a thrust cutting through the common limb (Figure 2a). Second, many fault-propagation folds display limbs in which beds are not uniform in dip and where deformation is characteristically heterogeneous, particularly in the steep forelimbs where local tectonic thickening and thinning are observed. Third, in areas involving competent massive lithologies, such as crystalline basement, the application of kink-band kinematics based on flexural slip seems inappropriate (Figure 2b). Fourth, growth strata in many foreland fold and thrust belts display variations in dip and thickness across fault-related folds, which are interpreted to indicate progressive rotation of fold limbs (Figure 2b).

Fault-propagation folds with trishearlike geometries also accompany the formation of normal faults (Hardy and McClay, 1999). In these areas, characteristic features include normal-drag folding in both the hanging wall and the footwall and the formation of subsidiary high-angle reverse faults in the hanging wall (Withjack et al., 1990; Patton et al., 1994; Sharp et al., 2000; Gawthorpe and Hardy, 2002). These structures differ from the more commonly recognized reverse-drag roll over anticlines associated with listric normal faults (Figure 2c).

Structures such as these have commonly been described in the literature as forced folds or drape folds (e.g., Matthews and Work, 1978; Jamison and Stearns, 1982; Withjack et al., 1990; Couples et al., 1994; Couples and Lewis, 1998; Cosgrove and Ameen, 2000). We prefer not to use these terms because of their obvious genetic implications: the idea that the basement is somehow forcing the cover to fold. Basement and cover are inexorably linked in these structures, and the critical mechanical interfaces may not be at that unconformity at all. Furthermore, such terms implicitly suggest that the tip line does not propagate from the basement into the cover during deformation, something that clearly does happen in many structures.

The term drag fold is likewise problematic because it suggests that drag (friction and entrainment of layering in a broad zone of heterogeneous simple shear) along an existing fault produced the folds. Although this process may happen in nature, in trishear, all of the folding occurs in the triangular shear zone in front of the tip line. When, as a result of propagation, the fault breaches these rocks, all folding has ceased; no drag exists. Nevertheless, the tip-line folds subsequently cut by the propagating fault appear as if they were drag folds.

KINEMATIC MODELS AND THEIR DEVELOPMENT

Erslev's Trishear

Erslev's (1991) breakthrough article outlined a two-dimensional (2-D) geometric construction, which appeared to ensure conservation of area and produced geometries similar to those encountered in natural structures associated with both reverse and normal faults (Figure 3). A key kinematic component of the model is the trishear angle, the apex of which is focused on the fault tip line and which, Erslev (1991) argued, must be symmetric with respect to the fault plane. The algorithm, described in greater detail by Erslev and Rogers (1993), implied a flux of material from a rigid moving, hanging-wall block to a static footwall block, with displacement magnitude decreasing in a similar sense. Erslev recognized that the fault tip itself was a key component of the model and proposed two end members, hanging-wall-fixed and footwall-fixed trishear, in which the tip was fixed to the hanging wall or footwall, respectively. Erslev's algorithm, as well as all subsequent implementations by other authors, is realized numerically as no analytical or graphic solution exists. The program Erslev develop for Windows computers is available as shareware by writing to him.

Erslev (1991) also briefly made reference to a method for producing heterogeneous trishear in which strain is more concentrated toward the center of the shear zone. Because he did not have an analytical solution to the velocity field, heterogeneous trishear was implemented in Erslev's program by a progressive, cyclical narrowing of the trishear angle: a displacement increment would first occur at an open angle, and with each subsequent small increment, the angle would be reduced until it reached some minimum value. At the next displacement step, the angle would be reset at its maximum value, and the cycle of diminishing trishear angles would start over. Although Erslev's heterogeneous trishear is kinematically possible, there is no reason to believe that real fault zones actually evolve with such a progressive cyclical narrowing of the trishear angle.

In the years immediately following the publication of Erslev's (1991) article, the trishear concept did not really catch on, and the article received few citations (only seven prior to 1997). This may have been because Erslev's (1991) first article was very brief and gave the readers little idea of how the algorithm functioned or how to construct cross sections of real structures using trishear. Nonetheless, the 1991 article was key in that it introduced the concept of a triangular zone of distributed shear, which opened outward and upward from a fault tip line (Figures 1, 3).

Two-dimensional Velocity Description of Trishear

Hardy and Ford (1997) built on Erslev's (1991) original idea and in the process introduced two key concepts, which ultimately contributed to more widespread acceptance of the trishear model by the structural community. First, they derived their solution using a simple velocity description of deformation that contained explicit terms for the horizontal and vertical velocities within the shear zone (Figure 4). This allowed efficient,

Fault-propagation folds

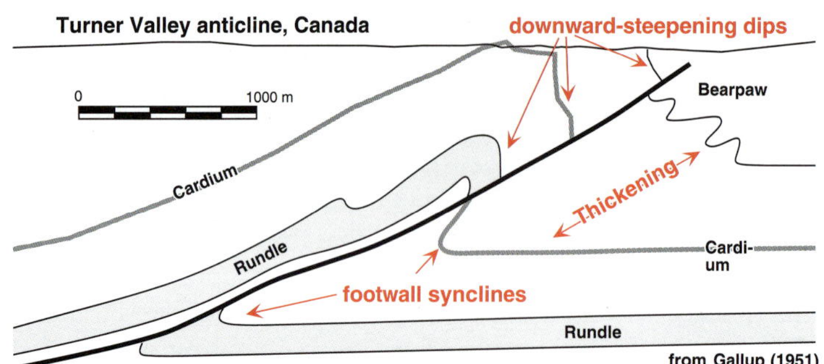

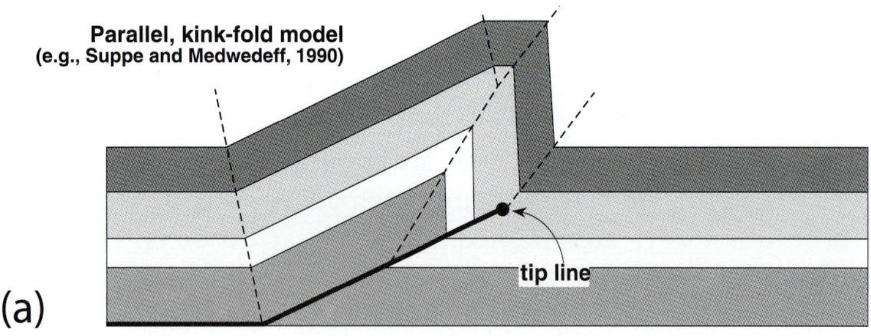

(a)

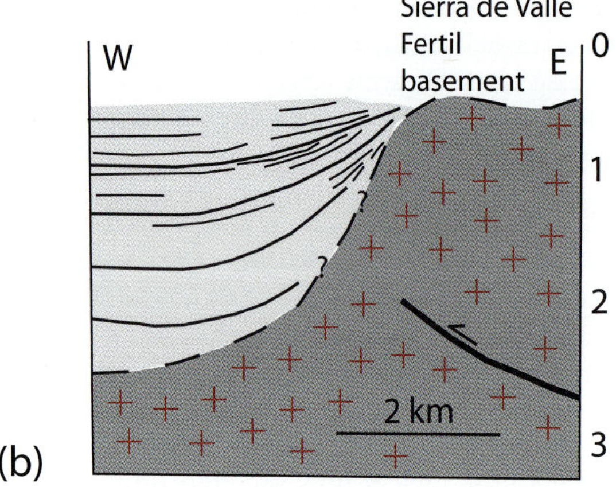

(b)

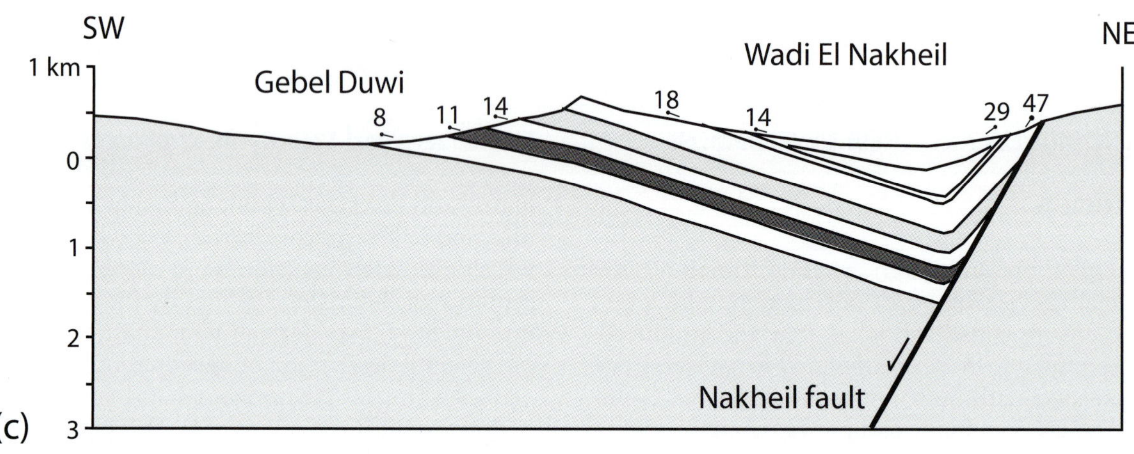

(c)

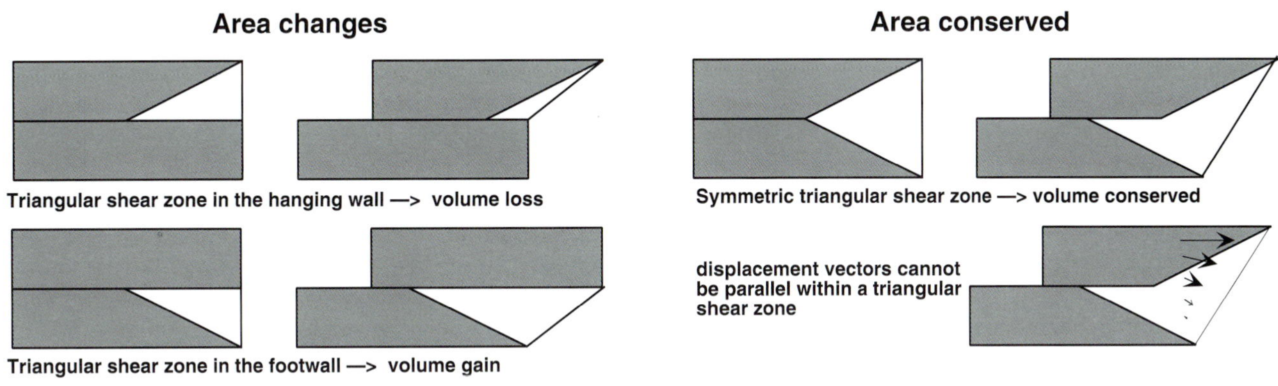

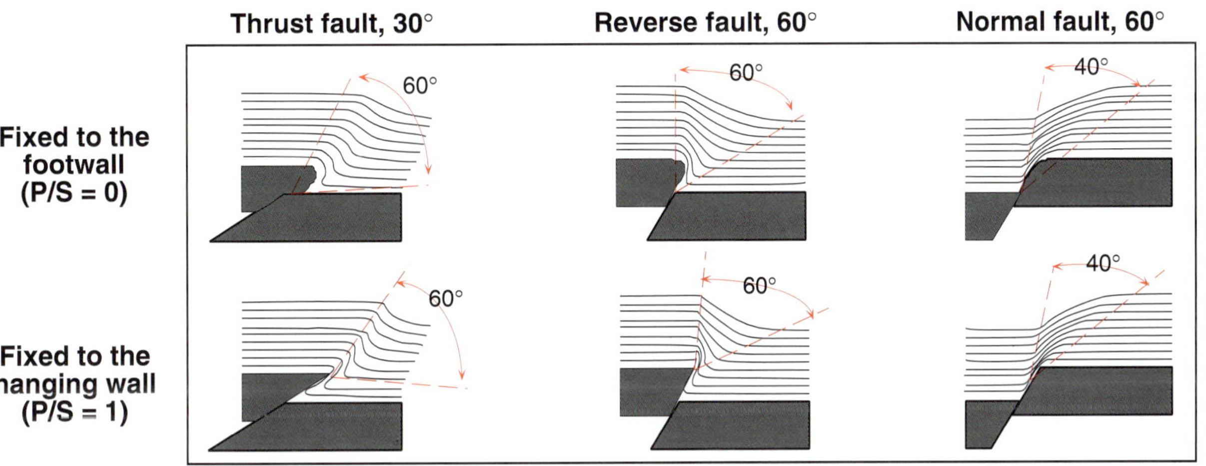

Figure 3. Erslev's geometric explanation and model of trishear (1991). P/S = propagation to slip ratio.

constrained, forward modeling to be undertaken as demonstrated by the rapid subsequent publication of Allmendinger (1998), which used Hardy and Ford's velocity description. Hardy and Ford (1997) also examined the implications of the trishear model for growth sedimentation and showed that the forelimb rotation inherent in trishear produced growth strata with characteristic progressive unconformities.

In current usage, most workers use the subsequent trishear velocity description by Zehnder and Allmendinger (2000) (Figure 5a). Their boundary conditions represent explicit statements of the assumptions used by previous authors: (1) the hanging wall is rigid so every point in the hanging wall moves with the same velocity, parallel to the fault zone; (2) the footwall is held fixed so that all points therein have a velocity of zero; and (3) that area must be conserved in the triangular shear zone between the hanging wall and footwall, or in precise mechanical terms, the flow is incompressible. Incompressible flow requires that the divergence of the velocity field is zero.

$$\mathrm{div}(v) \equiv \frac{\partial v_x}{\partial x} + \frac{\partial v_y}{\partial y} = 0 \tag{1}$$

With a coordinate system where the x-axis is parallel to the fault and the origin is centered on the fault tip line, boundary conditions 1 and 2 at the hanging-wall

Figure 2. Examples of natural fault-related folds exhibiting features difficult or problematic to explain using kink-band kinematics. (a) Smooth fold profiles with footwall synclines, complex deformation in forelimb with thickening and thinning; the Turner Valley anticline and comparison with a kink-band model. (b) Line drawing of a seismic line across the westward-verging Sierra de Valle Fertil, in the Precordilleran thrust belt of Argentina, showing a cumulative wedge on the forelimb with progressive limb rotation (modified from Zapata and Allmendinger, 1996). (c) Extensional fault-propagation fold from the Gulf of Suez (modified from Khalil and McClay, 2002).

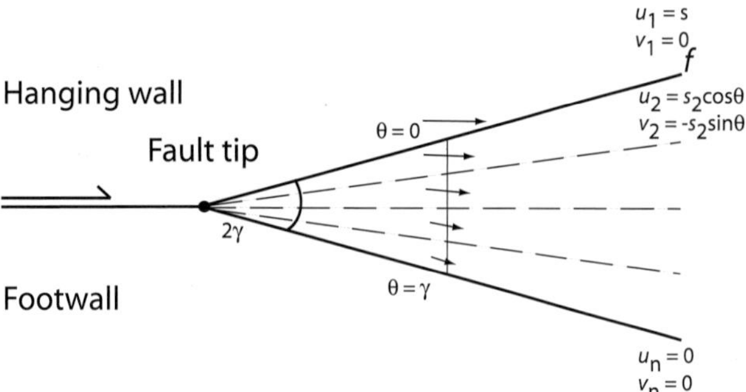

Figure 4. Hardy and Ford's (1997) kinematic model of trishear, velocity vectors within the hanging wall and shear zone, and sectors of equal velocity are illustrated schematically. The symmetric shear zone has an apical angle of 2γ (redrawn from Hardy and Ford, 1997).

and footwall margins, respectively, of the triangular shear zone can be written as

$$\begin{aligned} v_x = v_o \quad &\text{and} \quad v_y = 0 \quad \text{at} \quad y = x\tan\phi_1 \\ v_x = 0 \quad &\text{and} \quad v_y = 0 \quad \text{at} \quad y = -x\tan\phi_2 \end{aligned} \tag{2}$$

where v_o is the velocity of the hanging wall, x is the distance from the tip line, and ϕ_1 and ϕ_2 sum together to form the trishear angle. The solution from this point is to assume a distribution for v_x and use equations 1 and 2 to calculate the corresponding v_y. Because an infinite number of possible v_x distributions exist, an equally large number of v_y can be calculated, and unlike what was implied in Erslev's (1991) original proposal, no unique solution to the trishear velocity field exists. Zehnder and Allmendinger (2000) presented a formulation that, at its simplest, assumes a linear variation in v_x across the trishear zone but that also allows one to specify varying degrees of concentration of the deformation in the center of the trishear zone (Figure 5b). Written in vector format

$$\begin{aligned} v_{(x,y)} = \frac{v_o}{2}\Bigg\{ & \left[\operatorname{sgn}(y)\left(\frac{|y|}{x\tan\phi}\right)^{\frac{1}{c}} + 1\right]\hat{\imath} \\ & + \frac{\tan\phi}{1+c}\left[\left(\frac{|y|}{x\tan\phi}\right)^{\frac{(1+c)}{c}} - 1\right]\hat{\jmath}\Bigg\} \end{aligned} \tag{3}$$

where c is the center concentration factor; when $c = 1$, v_x varies linearly across the trishear zone. The linear version of Zehnder and Allmendinger (2000) is slightly different than Hardy and Ford's (1997) velocity field in that the latter assumed a linear variation in the magnitude of the total velocity vector, $|v(x,y)|$, instead of just a linear variation in v_x. Because $v_x >> v_y$, the two velocity fields are, however, very similar.

Although the center concentration, c in equation 3, is superficially similar to Erslev's (1991) heterogeneous trishear, the resulting finite strain is very different: the former maintains deformation more concentrated in the center at every single displacement step, whereas the latter does not. Equation 3 represents only one of many ways of achieving trishear deformation in which the displacement and velocity gradients and hence strain (rates) are higher in the center of the trishear zone. The sine velocity field of Zehnder and Allmendinger (2000) also accomplishes this (see Figure 5b). Because a typographical error in the equation for the sine field in Zehnder and Allmendinger is observed, we present the correct version.

$$\begin{aligned} v(x,y) &= v_o\left[\frac{1}{2}(\sin\beta + 1)\hat{\imath} + \frac{\tan\phi}{\pi}\left(\cos\beta + \beta\sin\beta - \frac{\pi}{2}\right)\hat{\jmath}\right], \\ \beta(x,y) &\equiv \frac{y\pi}{2x\tan\phi} \end{aligned} \tag{4}$$

This field is qualitatively more nearly similar to heterogeneous trishear, but recall that (1) it is only one of many ways to produce such an effect and (2) similar velocity fields have not commonly been observed in mechanical models (see review below). We suggest abandoning Erslev's (1991) ad-hoc heterogeneous trishear, and when one wishes to simulate greater deformation in the center of the triangular zone, use either equation 3 with a value of c greater than 1, equation 4, or derive some other equally precise mathematical formulation.

In addition to demonstrating an infinite number of solutions to the trishear velocity field, Zehnder and Allmendinger (2000) also showed that trishear zones need not be symmetric with respect to the fault. Although kinematically feasible, we have found few real examples or mechanical modeling results, which suggest that asymmetric trishear is important in nature. Zehnder and Allmendinger's (2000) algorithms, including asymmetric and center-concentrated trishear, are implemented in Allmendinger's (2007) computer program.

Propagation to Slip Ratio

The second key contribution of Hardy and Ford (1997) to the development of trishear was to relax some of the constraints inherent in Erslev's original model; in particular, they introduced the concept of independent fault tip propagation in the form of the propagation to slip ratio (P/S). Erslev's (1991) footwall-fixed and hanging-wall-fixed trishear are, in fact, special cases of Hardy

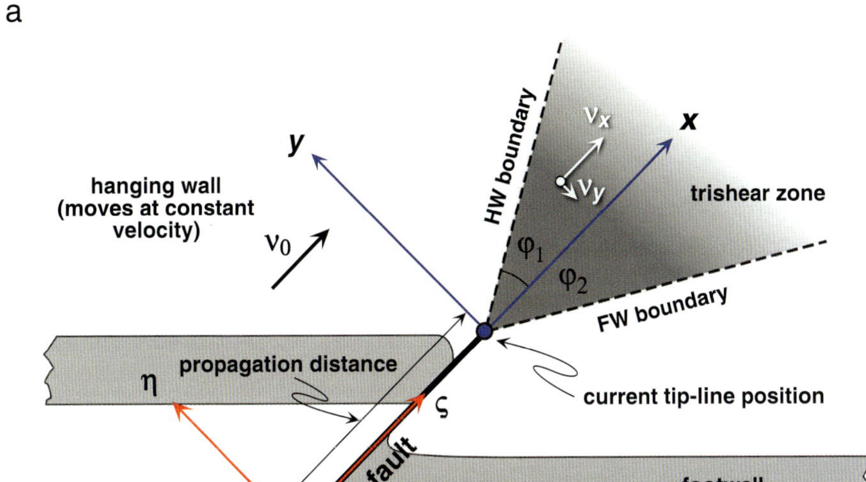

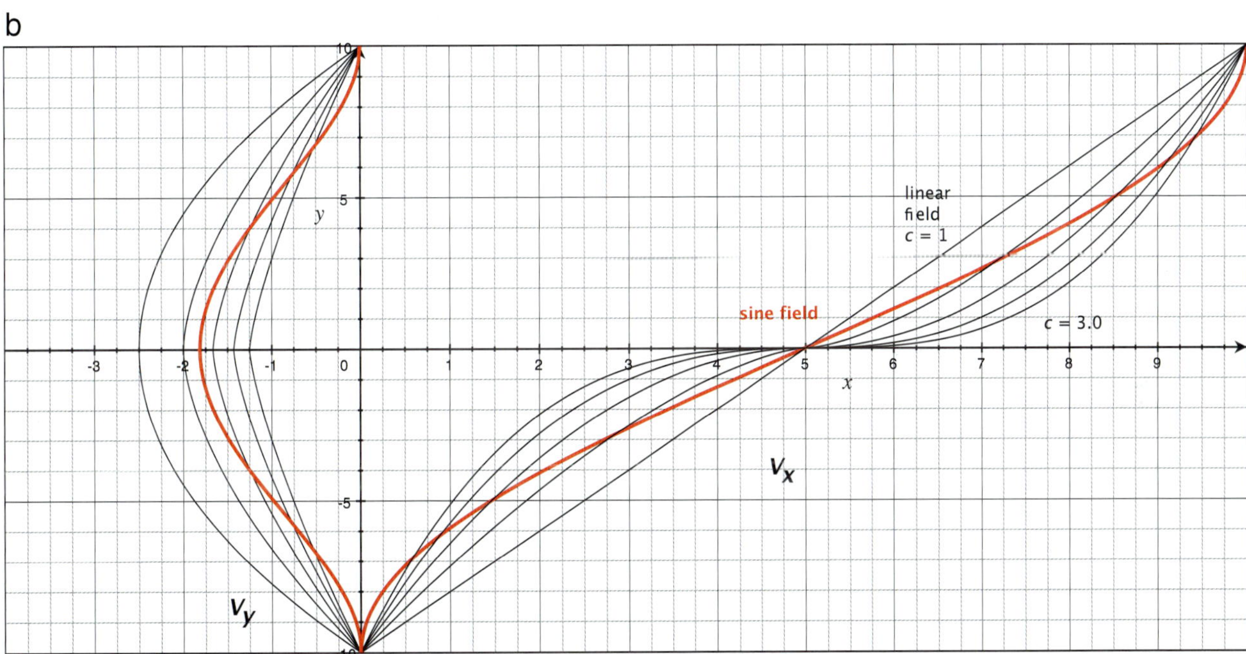

Figure 5. (a) Diagram of Zehnder and Allmendinger's (2000) concept velocity model of trishear; HW = hanging wall; FW = footwall; (b) illustration of typical velocity fields encompassed by Zehnder and Allmendinger's model. In this figure, the shallower the slope of the curve, the steeper the velocity gradient and therefore the greater the strain at that point in the trishear zone.

and Ford's trishear in which P/S = 0 and P/S = 1, respectively. The concept of P/S, which has its roots in the work by Elliott (1976) and Williams and Chapman (1983), is not unique to trishear kinematics. It is implicit, but fixed by the geometry, in kink fault-propagation folding where, for example, a ramp off a detachment has a P/S = 2 (see Suppe and Medwedeff, 1990; Hardy, 1997). In fact, all simple fault-related fold types described by thrust belt geologists, detachment, fault-propagation, and fault-bend folds, can be considered as points along a continuum of P/S variation (Allmendinger and Shaw, 2000; Allmendinger et al., 2004). As described below, P/S is the single most important factor determining the shape of a trishear fold.

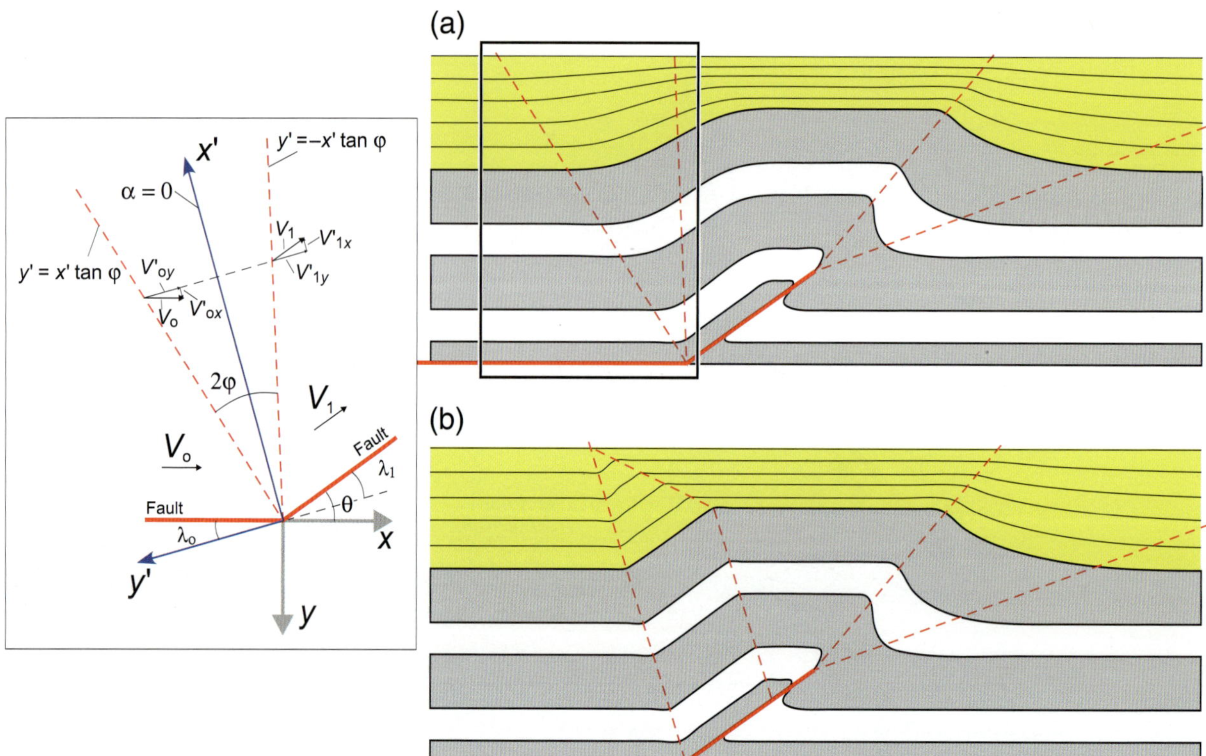

Figure 6. Comparison of (a) backlimb trishear modified from Cristallini and Allmendinger (2002) with (b) parallel kink fault-bend fold backlimb geometry. Strata in yellow are growth strata that accumulated during the evolution of the two structures. The inset box shows the velocity distribution and key angles for the backlimb trishear analysis.

Backlimb Trishear

Cristallini and Allmendinger (2002) suggested that trishear kinematics could also be applied to the backlimbs of ramp anticlines. They observed that analog models and some natural examples (also modes 1 and 2 of Serra, 1977) display smoothly curving backlimb fold hinges over sharp, angular bends in the underlying thrust surface. This geometry can be reproduced by defining a triangular zone focused on the fault bend (Figure 6). The results of the backlimb trishear numerical model compare well to analog experiments. The backlimb fold geometries produced by this method at least superficially resemble those produced by (1) similar folding over a curved footwall ramp and (2) shear fault-bend folding (Suppe et al., 2004). Thus, the choice of kinematic method used to simulate curving backlimbs will depend first on excellent imaging of the fault bend itself, and second on whether the thickening over the lower corner of the footwall ramp is restricted to one layer (shear fault-bend folding) or occurs in all layers (backlimb trishear).

Three-dimensional Trishear

Three-dimensional (3-D) analysis of trishear deformation was initially studied by Cooper and Hardy (1999), Fischer et al. (1999), and Fischer and Wilkerson (2000). These authors linked serial 2-D sections together, assuming dip-slip displacement, and thus their models were not strictly 3-D. The main purpose of Fischer and Wilkerson's (2000) work was not an investigation of trishear kinematics but an attempt to predict fracture orientations from an analysis of surface curvature. Cristallini and Allmendinger (2001) presented a pseudo–3-D trishear model in which various parameters were permitted to vary along strike and oblique slip could be modeled. They showed that these variations could be combined in an infinite number of ways, facilitating the simulation of many real structures. For example, they found that a thrust changing from blind to emergent can be produced by a change in the slip or P/S along strike and that folds with forelimbs changing from overturned to upright along strike can be modeled either by changing the slip, P/S, or trishear angle. Models, including growth strata, showed that distinguishing between growth and pregrowth strata using map patterns alone is practically impossible. As a field test, they modeled the oblique slip east Kaibab monocline, demonstrating a good fit between the field observations and model predictions.

As regards a 3-D velocity description of trishear, incompressibility is expressed in its 2-D form in equation 1

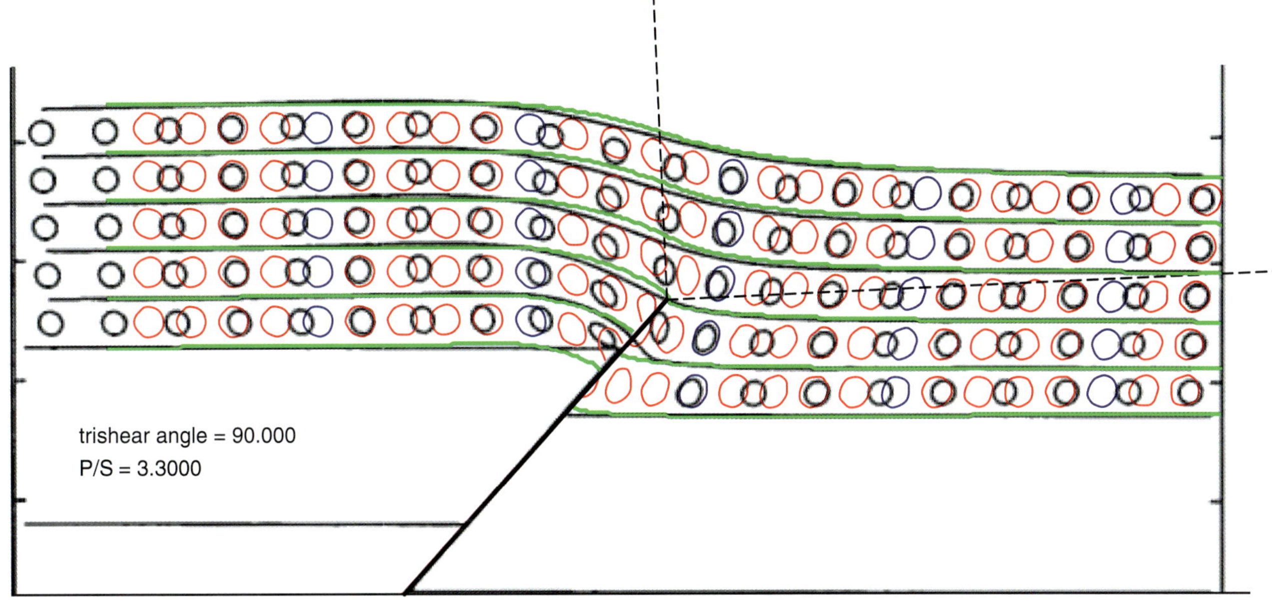

Figure 7. Inverse trishear modeling (in color) of the isotropic block-viscous model (in black) modified from Johnson and Johnson (2002b). Note the similarity of strain and bed geometry produced by the two models. P/S = propagation to slip ratio.

but can be extended easily to three dimensions. Of course, as in 2-D trishear, an infinite number of solutions to the velocity field in 3-D trishear exist. Cristallini et al. (2004) presented one possible solution and showed it to be a reasonable approximation to some real structures, although many other acceptable solutions probably exist as well. In their article, they suggested that 3-D trishear may be able to model flower structures in strike-slip fault zones.

FITTING THE TRISHEAR MODEL TO REAL STRUCTURES

Even after the publication of Hardy and Ford (1997), the problem remained: how does one apply this kinematic model to real structures? In kink-fold models, one can measure dip panels, kink angles, and cutoff angles and then use Suppe's (1983) graphics to determine the appropriate models; these graphics are arguably one of the great advantages of and prime reasons for the great success and widespread application of the kink-band method. In trishear, however, there is nothing obvious to measure in folded rocks to tell us what P/S and the trishear angle are.

Inverse Modeling: Identifying the Best-Fitting Model

Allmendinger (1998) realized that this problem could be solved because trishear deformation is reversible: a thrust fault-cored trishear fold can be unfolded by taking the final geometry and running the fault backward as if it were a normal fault. To do this in an automated fashion requires the identification of best-fit criteria; Allmendinger chose a least-squares linear regression to identify the point at which an unfolding keybed most closely approximates a straight line (in two dimensions). The key-bed approach of Allmendinger (1998) works very well where the geometry of one bed is well known in advance, as in the case of high-quality seismic reflection and borehole data, or where the structure is small enough or the structural relief great enough to provide a complete definition of a single bed (Figure 7).

Most field geologists, however, have transect data where multiple stratigraphic contacts and their dips are well known along a single line of section, but no one bed is completely known, geometrically. Cardozo (2005) showed that if stratigraphic thicknesses were well known, a best-fitting model could also be identified with a linear least-squares approach. Both Cardozo's (2005) fit to profile and Allmendinger's (1998) key-bed grid searches employ an χ^2 minimization scheme as described by Press et al. (1986).

Because the basic trishear forward model algorithm (Zehnder and Allmendinger, 2000) is computationally efficient, both Allmendinger (1998) and Cardozo (2005) employed a brute-force grid search across the parameter space. Three of those parameters, fault ramp angle and x and y positions of the tip line, describe the current, final geometry of the structure. With good exposure, these can commonly be measured directly, thus limiting the scope of the grid search. One kinematic parameter, fault slip, can sometimes be constrained by offset of planar beds outside of the folded region. The remaining kinematic parameters, P/S, trishear angle, center concentration, and asymmetry, can commonly be determined only by grid searching. Although more computationally time-consuming, the advantage of a comprehensive grid search is that local minimums in the solution field can be readily identified.

Inverse modeling via grid searching has proven to be remarkably good at simulating the geometry and finite strain of real structures, numerical and analog models as demonstrated by a variety of studies (e.g., Figure 7). However, a few limitations reduce its effectiveness for many real structures. Most notably both types of grid searches assume constant parameters during the evolution of the structure. Although a forward model can be constructed by changing the fault angle, P/S, or any other parameter during the evolution of the model, constructing an inverse model that would test not only all possible combinations of parameters, but also any possible change in those parameters during evolution, although hypothetically possible, would result in an intractable situation. Thus, the results of the grid search should be regarded only as the best-fit average values. Fine-tuning by hand is commonly necessary. The one exception to this limitation is where growth strata accumulate during changes in trishear parameters. As described in the Applications section, progressive restoration can detect changing trishear parameters.

Both Allmendinger et al. (2004) and Cardozo (2005) conducted sensitivity analyses by grid searching known forward models as if they were unknown structures. With reasonably fine-scale numerical integration (as represented by choosing a small displacement step size), the grid search finds the correct starting parameters of the forward model. An important result of these experiments is to show that P/S and trishear angle are approximately inversely correlated: similar final geometries can result from a small trishear angle and high P/S or vice versa. The lower the P/S or the smaller the trishear angle, the more strain accrues within the triangular shear zone. A change of 0.2 in P/S is approximately equivalent to a 15–20° change in trishear angle in terms of their effect on the final geometry of the structure.

Evaluating Uncertainty in Goodness of Fit

Because trishear is a model that is applied to real data, it is incumbent upon the geologist to specify not only the best-fitting model but also the uncertainty of the fit, just as one should give the uncertainty of a fit of a cylindrical fold axis to a set of observed bedding poles or a mean vector to a set of paleomagnetic poles. A rigorous estimate of uncertainty is particularly important for evaluating risk, whether it be the economic risk inherent in hydrocarbon exploration or the economic and human risk caused by seismic hazard.

In the case of the key-bed grid search, the uncertainty arises from errors in the digitized bed position. Where an exploration well penetrates the keybed, the error in position may be less than 10 m (33 ft) vertical, but where the bed is imaged only on seismic reflection data, the error in bed position depends on the conversion from time to depth (which depends on how well the velocities are known) and may be on the order of several hundred meters. In the case of the fit-to-profile grid search, the uncertainty comes from errors in position of stratigraphic contact along the line of section, the dips of those stratigraphic units, and the stratigraphic thicknesses. Intuitively, errors in dips are probably the most important. Note that all possible sources of uncertainty are not considered. For example, the approach does not consider possible errors in the choice of a kinematic model (e.g., whether a kink-fold approximation would give a better fit than trishear).

The numerical implementation of trishear makes it ideally suited to Monte Carlo simulation. We follow the implementation described by Press et al. (1986) and have found that about 200 simulations are sufficient to define the 1-sigma errors in the grid search fit of a model to the data. In the Applications section, we revisit a study by Allmendinger and Shaw (2000), demonstrating the application of Monte Carlo simulation to a situation that relates to seismic hazard.

MECHANICAL ANALYSIS OF TRISHEARLIKE FOLDS

A variety of different mechanical models have been applied in an effort to better understand the origins and the realms of applicability of the trishear kinematic model (and the physical meaning of the various parameters). The need for a thorough mechanical analysis, the asking of why and under what conditions a particular type of descriptive kinematics develops, is not unique to trishear but is particularly germane to many issues raised by trishear. Trishear illustrates the manner in which the two approaches are, in fact, complementary. We return to this point at the end of this section.

The first discussion of the mechanical questions that need to be asked regarding the trishear kinematic model was presented by Allmendinger (1998). He identified two key parameters that warranted some physical explanation: the P/S and the trishear apical angle (cf. Figure 5). Additionally, trishear's implicit assumption of a rigid hanging-wall translation needed justification. This was then followed up by Zehnder and Allmendinger

Table 1. Mechanical Modeling of Trishearlike Structures

Model	Pros	Cons	References
Block-motion viscous analysis	Analytically and computationally simple; good insight into processes	No fault-tip propagation; rigid substrate; limited heterogeneity	Patton and Fletcher (1995); Johnson and Johnson (2002a, b)
Finite-element analysis	Broad array of rheologies; prescribe tip propagation	Tip propagation artificial; distorted meshes at large strains (P/S < 3); resulting trishear angles always <20°	Cardozo et al. (2003)
Discrete-element analysis	Most geologically realistic; natural tip propagation; heterogeneity and layer anisotropy	Computationally intensive; rigid substrate; complex results not always easy to relate to process	Finch et al. (2003, 2004); Cardozo et al. (2005)

(2000) who, while deriving their (kinematic) velocity field for trishear, showed that the trishear model produces $1/r$ singular strain rates, consistent with results for cracks in elastic-plastic materials. These articles hinted at the need for a more thorough mechanical analysis of trishear.

Alternative Modeling Strategies, Their Implementation, and Results

Although there have been several mathematical and mechanical modeling studies investigating the deformation of overburden caused by slip on a buried fault (e.g., Sanford, 1959; Haneberg, 1992, 1993; Patton and Fletcher, 1995), none of these studies addressed, or was focused on, trishear. In this section, we review those studies that have used mechanical models to address features with a direct relationship to trishear (Table 1).

Block-Motion Viscous Modeling

The first explicit attempt at mechanical modeling of trishear folding (Table 1) was presented in a set of two companion articles by Johnson and Johnson (2002a, b), although a similar approach was previously applied by Patton and Fletcher (1995, 1998). They correctly recognized that although many authors had compared the geometry of folds generated by trishear with experimental and field data, the validity of the assumed trishear velocity fields had not been checked against theoretical (mechanical) models. This was particularly important given that Zehnder and Allmendinger (2000) had recently shown that an infinite number of possible velocity fields that satisfied the boundary conditions assumed in trishear existed.

To address this problem, they presented a series of mechanical models based on viscous folding theory. They solved a boundary value problem appropriate to folding of an incompressible viscous anisotropic fluid (Johnson and Fletcher, 1994; Patton and Fletcher, 1995, 1998). A velocity boundary condition was applied at the basement cover interface arising from rigid motion along a basement fault, and a traction-free boundary condition to the upper ground surface. The velocities and stresses throughout the body were then determined from the numerical solution.

Using this model, they examined the effects of several key variables on fold forms and velocities: shape of the basement fault, anisotropy of the sedimentary cover, and nature of the basement-cover interface. They found that, particularly in isotropic sedimentary cover above a planar fault, the velocity distributions in the mechanical model followed in general terms the velocity fields assumed in trishear kinematic models (Johnson and Johnson, 2002a). In these models, they also noted that the forelimb was thickened near the synclinal hinge and thinned between the anticlinal and synclinal hinges similar to predictions of the kinematic models (Johnson and Johnson, 2002b) (Figure 8). However, they found that anisotropic sedimentary cover produced more parallel, kinklike folds. In addition, the assumptions of a rigid hanging wall and a static footwall in the kinematic models were best approximated when the sedimentary cover was welded to the basement.

They concluded from their mechanical modeling that the assumption of a triangular zone of deformation in trishear kinematic models was reasonable, particularly when dealing with an isotropic sedimentary cover that was welded to the basement. Fold forms in such models closely resembled those produced by the kinematic model. They also noted that the vertex angle of the triangular zone of deformation in such cases was uniformly 110°, suggesting a constraint on one of the kinematic model parameters but offering no explanation as to why it should be so.

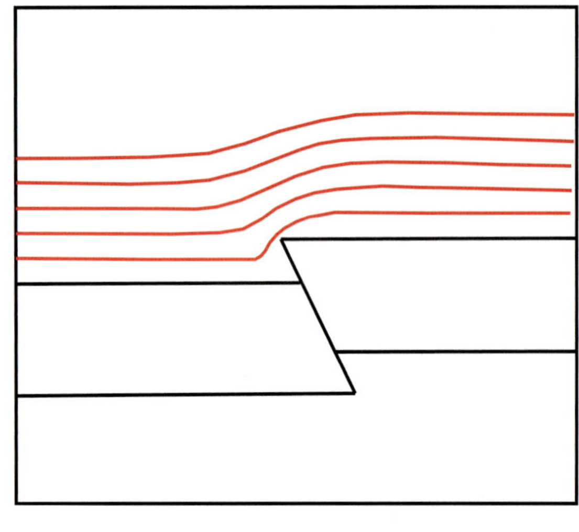

(a)

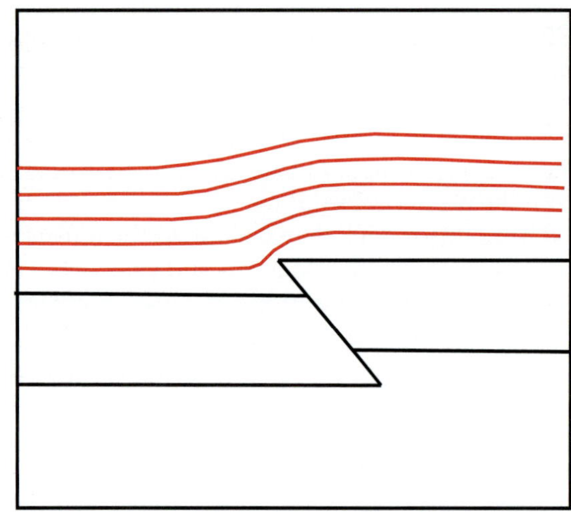

(b)

Figure 8. Examples of block-motion viscous modeling modified from Johnson and Johnson (2002a). Results are illustrated for an isotropic cover welded to a basement above a (a) 60° reverse fault and a (b) 45° reverse fault (redrawn from Johnson and Johnson, 2002a).

Although these modeling results for the first time gave support to the basic concepts of trishear, a triangular zone of deformation opening outward and upward from a fault tip, and the linear form of the velocity field used by many authors, they were limited in their general applicability because of one crucial factor: lack of faulting within, or basement fault propagation into, the cover (Table 1). In real structures, such fault propagation is a key process, and these viscous models could only really represent one particular end member of the spectrum of behavior encompassed by the P/S in the kinematic models.

Finite-Element Modeling

Cardozo et al. (2003), using a finite-element model (FEM), were motivated by concerns similar to those of Johnson and Johnson (2002a, b) but specifically addressed the issue of fault-tip propagation into the cover sequence. They presented a series of 2-D, large-deformation, finite-element simulations of faults propagating in elasto-plastic frictional and frictionless materials, with faults propagating at 3–3.5 times their slip rate. In their models, fault-tip propagation was included by specifying a mechanical discontinuity that developed during the deformation according to a specified history of fault growth.

They found that a fault propagating through an elasto-plastic material produced a wave of deformation ahead of its tip line. This triangular zone of deformation was found to be slightly asymmetric to the fault tip but migrated with it during deformation. Velocity fields within the zone of deformation were steady (Figure 9). Fold geometries, finite strain, and velocity fields in models with incompressible sedimentary cover materials most closely resembled those produced by the trishear kinematic model. Importantly, they confirmed the applicability of the ad-hoc linear velocity field used in many of the kinematic models. However, they found that inverse modeling of the mechanical models gave trishear apical angles in the range of 20–30°, much smaller than those commonly determined from analysis of natural structures and those derived from viscous folding models.

In summary, these modeling results gave further support to the basic concepts of trishear, a triangular zone of deformation opening outward and upward from a fault tip, and the linear form of the velocity field used by many authors. They also showed that the idea of a zone of deformation attached to a propagating fault tip had a sound mechanical basis. However, although fault propagation was included in the finite-element modeling scheme, it was implemented in a sense as part of the boundary conditions and thus could give no insight into what controlled fault propagation (and the P/S) in natural structures (Table 1).

Discrete-Element Modeling

In contrast to block-motion viscous models and FEMs, which are continuum techniques, discrete-element models (DEMs) are well suited for studying problems in which discontinuities (faults, joints, or fractures) are important because they allow deformations involving large relative motion of individual elements and do not require the complex remeshing at high strains that, e.g., finite-element techniques, typically require. These

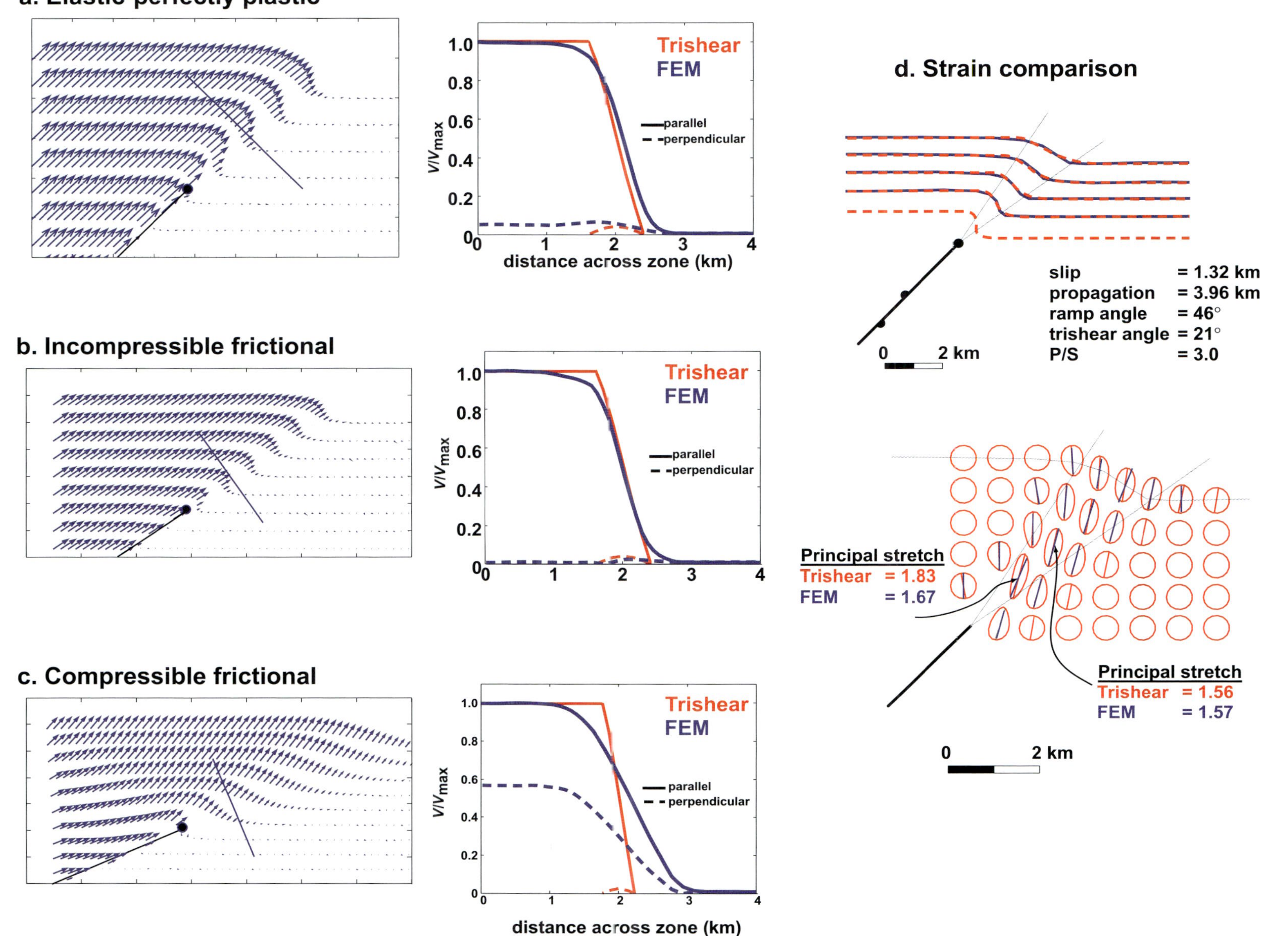

Figure 9. Summary of finite-element modeling results modified from Cardozo et al. (2003). In panels a and b, trishear provides an excellent description of the velocities observed in the finite-element analysis, whereas a compressible material yields a velocity field totally unlike that assumed for trishear because of the nonrigid behavior of the hanging wall. (d) Comparison of bedding geometries and finite strains produced by an incompressible elastic–perfectly plastic material and simulated by trishear. FEM = finite-element model; P/S = propagation to slip ratio.

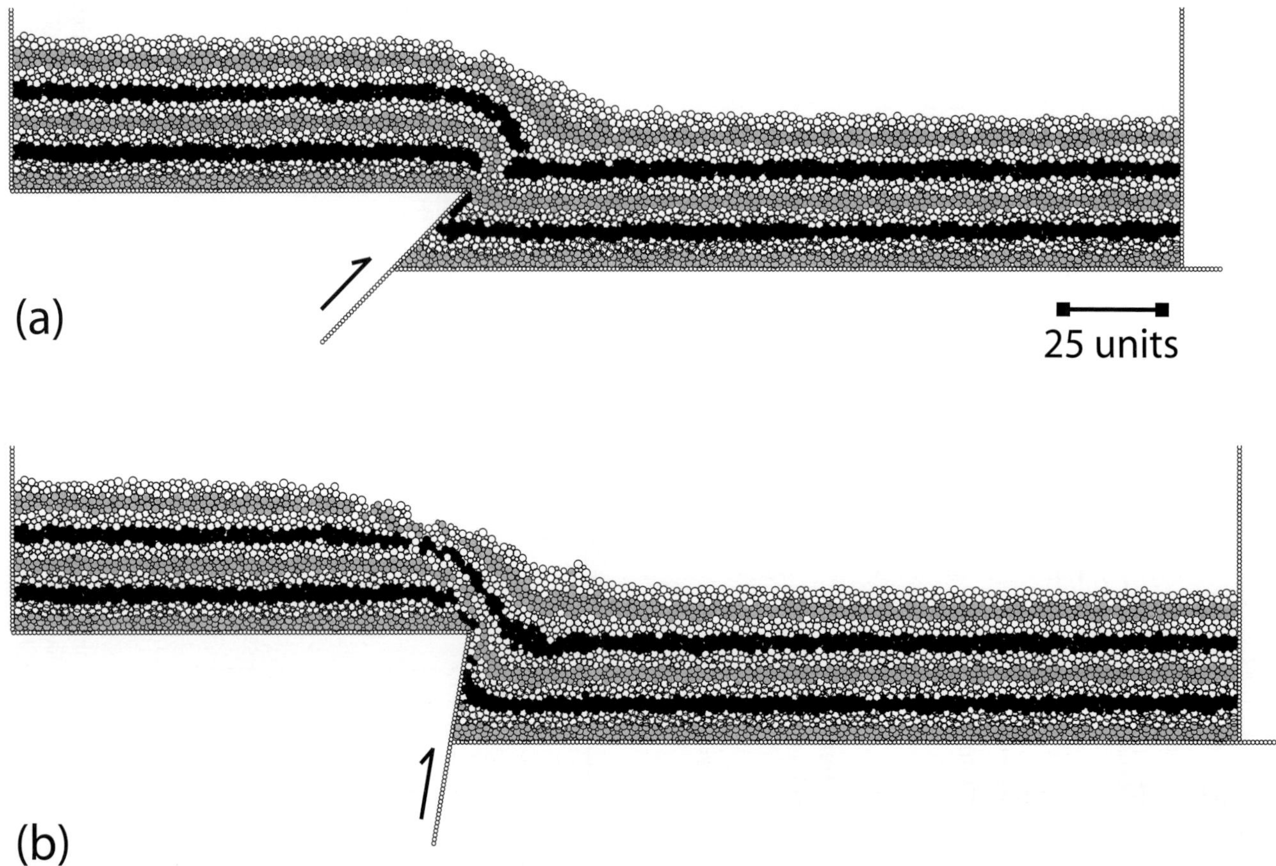

Figure 10. Discrete-element modeling results modified from Finch et al. (2003); a homogeneous, standard strength cover is deformed above a blind basement fault. (a) Basement fault dips at 45°, (b) basement fault dips at 80°. Displacement on the basement fault in both examples is 26.4 model units.

characteristics have led to their recent use in the mechanical investigation of trishear fault-propagation folding (Finch et al., 2003, 2004; Cardozo et al., 2005).

In the first application of discrete-element modeling to trishear, Finch et al. (2003) used a discrete-element technique to investigate the mechanical control of sedimentary cover strength upon fault propagation and trishear fold development. They examined boundary conditions very similar to those of Johnson and Johnson (2002a, b) and Cardozo et al. (2003): slip on a rigid basement fault block overlain by a weaker sedimentary cover (Figure 10). The DEM used consisted of a series of soft spheres that obey Newton's equations of motion and initially interact with elastic forces under the influence of gravity. Particles are bonded until the separation between them exceeds a defined breaking strength at which time the bond breaks. The model was used to investigate the influence of basement fault dip and sedimentary cover strength on the geometry of folds and the rate of fault propagation. In all cases, an upward-widening monocline developed above the basement fault. Shallow-basement fault dips produced homogeneous thickening of the monocline limb, whereas steeper dips produced contemporaneous thinning and thickening within the monocline (Figure 10). The key results of Finch et al. were that, with decreasing cover strength, the zone of deformation became wider, localization did not occur on a single fault and, fold geometries resembled trishear fold profiles with low propagation to slip ratios (P/S, 1). In contrast, a stronger cover produced a narrower zone of deformation, localization on a single fault, and more rapid fault propagation (similar to trishear fold profiles where P/S, 2–3; Figure 10).

Recently, Cardozo et al. (2005) also used discrete-element modeling to better understand the development of two contrasting structures: the Rip Van Winkle (southeast New York) and La Zeta (southwest Mendoza, Argentina) anticlines. In their model, the rock mass is represented by an assemblage of rigid, frictional, linear-elastic, circular particles that displace independently from one another and interact at contacts between the particles. In both anticlines, faults nucleating at distinct stratigraphic levels open upward into triangular zones of folding. Folding intensity and finite strain attenuate with distance from the fault tip. Cardozo et al. (2005) found that trishear reproduced the bulk geometry and finite strain of the relatively homogeneous limestone sequence of the Rip Van Winkle anticline and predicted an initial location of the fault tip consistent with the field observations. Folding of the heterogeneous sedimentary section of La Zeta anticline, however, could not be simulated by the trishear model alone. However, distinct-element modeling of the anticlines indicated that their contrasting kinematics could have resulted from differences in mechanical stratigraphy and initial stress state. The distinct-element modeling showed that folding in a homogeneous, normally consolidated assemblage is trishearlike and resembled the Rip Van Winkle anticline. However, folding in a mechanically layered, heterogeneous, overconsolidated assemblage departed from the trishear model and more closely resembled the La Zeta anticline.

Insights Gained from Mechanical Models

The mechanical modeling undertaken so far has clearly given us much insight into some of the key issues in trishear fault-related folding, and we attempt to summarize them below.

- The use in kinematic models of a broadly triangular, symmetric zone of deformation opening upward and outward from the fault tip line is a good description of the zone of deformation seen in mechanical models.
- It appears that the kinematic model of trishear is most appropriately applied to isotropic, incompressible materials, be they viscous, elastoplastic, frictional, or frictionless. The mechanical models have shown that trishear thus appears to encompass quite a wide range in material behavior, suggesting that some general behavior of upper crustal materials under the assumed boundary conditions is being modeled.
- A linear velocity field, commonly assumed in kinematic models, is a good first-order approximation of the velocity fields seen in mechanical models. The center-concentrated or asymmetric velocity fields of Zehnder and Allmendinger (2000) have not been observed in mechanical experiments.
- Trishear appears to be most appropriate for materials in which limited anisotropy exists, and that as the material becomes more anisotropic, more parallel folds form and kink-band kinematics would appear more appropriate.
- Broadly, the P/S reflects the strength of the cover relative to the assumed rigid basement (or forcing member).
- The apical angle appears to be the most enigmatic of trishear parameters, very different angles have arisen from finite-element (20–30°) and block-motion viscous or discrete-element studies (c. 100°), and no clear understanding as to its meaning is observed. In addition, forward and inverse modeling studies (Allmendinger et al., 2004) have indicated an approximately inverse correlation of P/S and trishear angle, so perhaps it is not an independent variable.

Complementary Nature of Kinematic and Mechanical Models

In the authors' opinion, kinematic modeling and mechanical modeling are complementary approaches: they are very different in their philosophy but have something to offer each other. Kinematic models (forward or inverse) deal with displacements (or velocities) as a result of an assumed model subject to physically and geologically reasonable constraints (incompressibility, preservation of line length, bed thickness, etc). Mechanical models, however, contain several basic equations (constitutive relations, strain or deformation rate compatibility, equations of equilibrium, etc.) and imposed boundary conditions from which the internal displacements (velocities) and stresses are, through numerical solution, determined.

Undoubtedly, mechanical modeling is more complete and amenable to rigorous analysis; yet kinematic models are more straightforward and commonly considerably faster. Mechanical models are excellent tools with which to evaluate the effects of lithology, loading history, and loading rates on fault-propagation folding; yet the calibration and upscaling of the models are not straightforward, and their complexity makes them sometimes impractical. This is particularly so in the case of DEMs in which a rich material behavior very reminiscent of geological examples is produced but at the expense perhaps of a deeper understanding of why particular features form. For example, precisely why faults propagate in the DEMs is still unclear. Additionally, mechanical models are sufficiently time-consuming to carry out that grid searching (or other forms of inversion) for a best-fit solution to a real structure is rendered impractical.

Kinematic models, however, are simplifications of nature designed with the nonexpert in mind and provide a systematic framework within which to construct balanced cross sections. In other words, their purpose

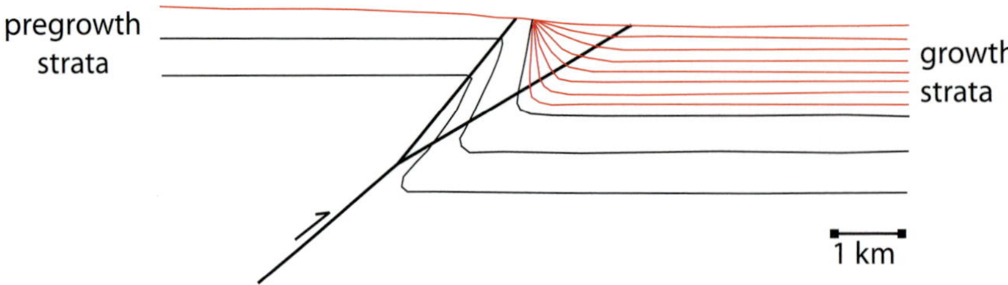

Figure 11. Trishear forward model with growth strata modified from Hardy and Ford (1997). The model illustrated is a hanging-wall-fixed example in which the base level rise was less than the uplift rate of the hanging-wall block. In addition to background sedimentation, the model included erosion of the uplifting hanging-wall block.

is to simulate a final geometry, thereby providing a reasonable estimate of finite strain and allowing one a framework for projecting a deformed geometry to depth. Because they are simple, one can test the fit of millions of different combinations of parameters to a particular structure within a matter of minutes on a modern desktop computer. Such comprehensive and objective probing of a parameter space cannot be accomplished with most mechanical models, where a single model run may take many tens of hours.

Trishear is perhaps an ideal example (exemplar) of where kinematic modeling and mechanical modeling are both complementary and individually necessary for some legitimate structural purposes. Because trishear has proved to be very effective at simulating real structures, it has stimulated a search for the mechanical significance, if any, of the parameters involved. The mechanical modeling has in turn validated (from a mechanical point of view) some of the key concepts embodied in the kinematic model, allowing a better understanding of when its use is appropriate and therefore providing greater confidence in the results produced.

APPLICATIONS

From its initial application to the large basement-cored anticlines of the Laramide Rocky Mountains (Erslev, 1991; Erslev and Rogers, 1993), the trishear kinematic model has found applications in a diverse range of geological settings. We review here some of those applications and then summarize some of the lessons learned.

Progressive Limb Rotation and Growth Strata

Trishear has proven appealing and useful to the structural geology community through its ability to simulate growth strata geometries that are difficult to simulate using kink-band migration (see Hardy and Ford, 1997). Growth strata in many foreland fold and thrust belts display variations in dip and thickness across fault-related folds that are interpreted to indicate progressive rotation of fold limbs (e.g., Riba. 1976; Nichols, 1987). However, in (simple) kink-band models, limb dips are attained instantaneously and do not change through time (e.g., Suppe et al., 1992). One of the most distinctive aspects of trishear kinematics is the ability to simulate progressive limb rotations. This can lead to a progressive change in limb dip from shallowly dipping at surface to subvertical to overturned at depth. Overturning of growth strata turns out to be a natural consequence of trishear fault propagation when a structure takes up a large amount of slip (Hardy and Ford, 1997). An important result of these studies is to show that this fanning of strata occurs not only in growth strata but also in pregrowth strata. We illustrate this in Figure 11 where a trishear fold and associated growth strata are shown. Changes in layer dip can also be produced by varying either the P/S or the trishear angle during the deformation (Allmendinger, 1998).

To demonstrate the utility of using trishear to conduct progressive restorations of growth strata and, in the process, deduce changing parameters during deformation, we offer the analysis of a synthetic example where we know a priori the history used to generate the structure (Figure 12). The original forward model was generated using a growth sequence where first the trishear angle was changed from 60 to 40° and then the P/S was changed from 1.5 to 2.5, as shown. Five pregrowth beds and four growth layers exist. If we simply do a bulk restoration using the highest pregrowth layer (Figure 12a), the grid search algorithm finds values of P/S and trishear angle that fall between the ones actually used to generate the forward model. Additionally, the total slip calculated from the grid search is too large, and the initial tip-line position is incorrect. A progressive restoration (Figure 12b) results in the correct calculation of slip and initial tip-line position, and the calculated intermediate values of P/S and trishear angle are much closer to those actually used to generate the forward model initially. The overall fits, as shown by the χ^2 values, are much better in the case of the progressive restoration. Thus, if trishear values change during evolution of the structure, they will be detected by progressive restoration using the trishear grid search.

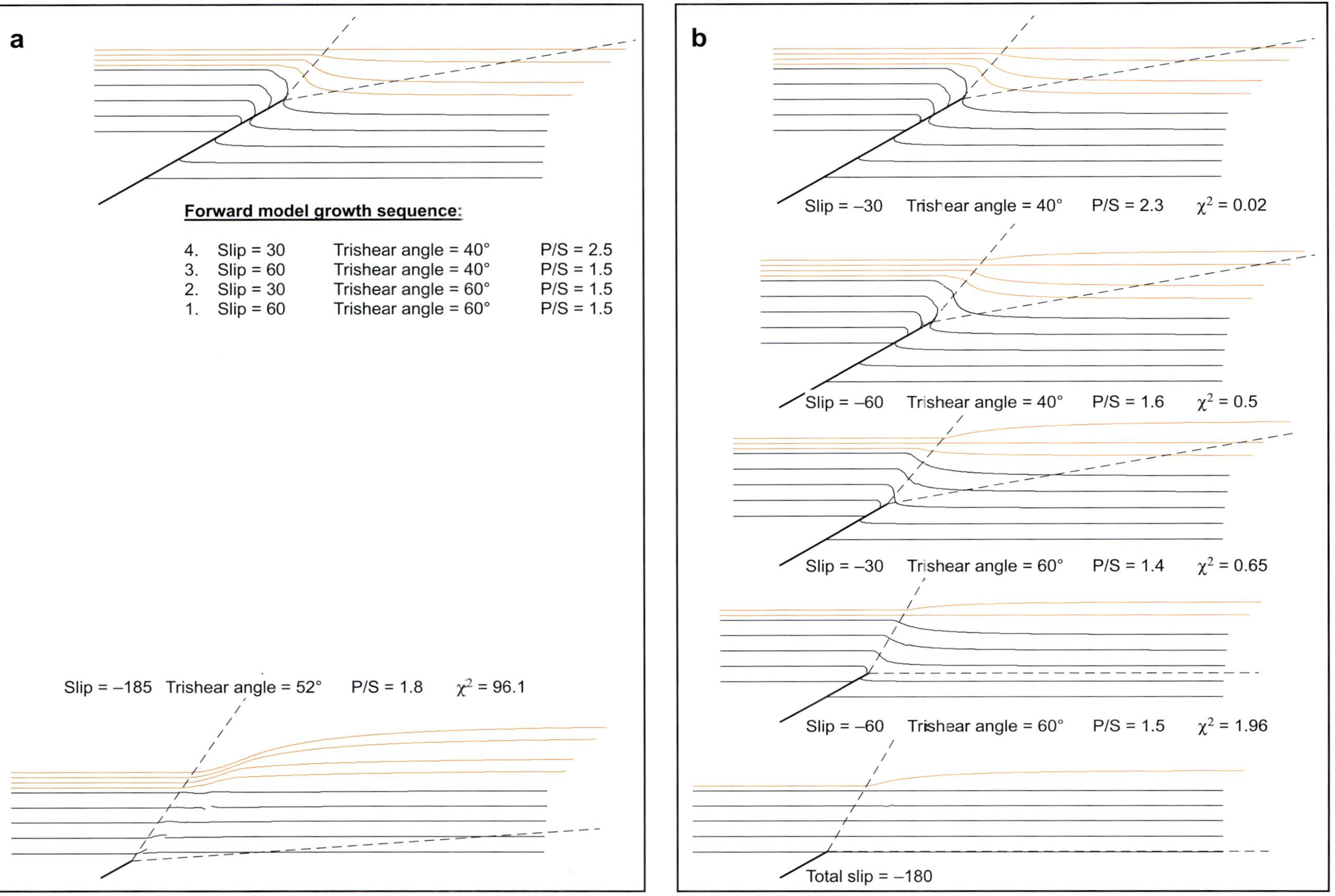

Figure 12. Illustration of the use of growth strata to enable progressive grid search restorations of a hypothetical structure produced with a known deformation history. (a) The left side shows the final geometry and steps to create it and at the bottom, the grid search restoration in a single step to an initial undeformed geometry. (b) The right side shows the progressive restoration of the same structure using growth strata, resulting in a much-improved overall solution. P/S = propagation to slip ratio.

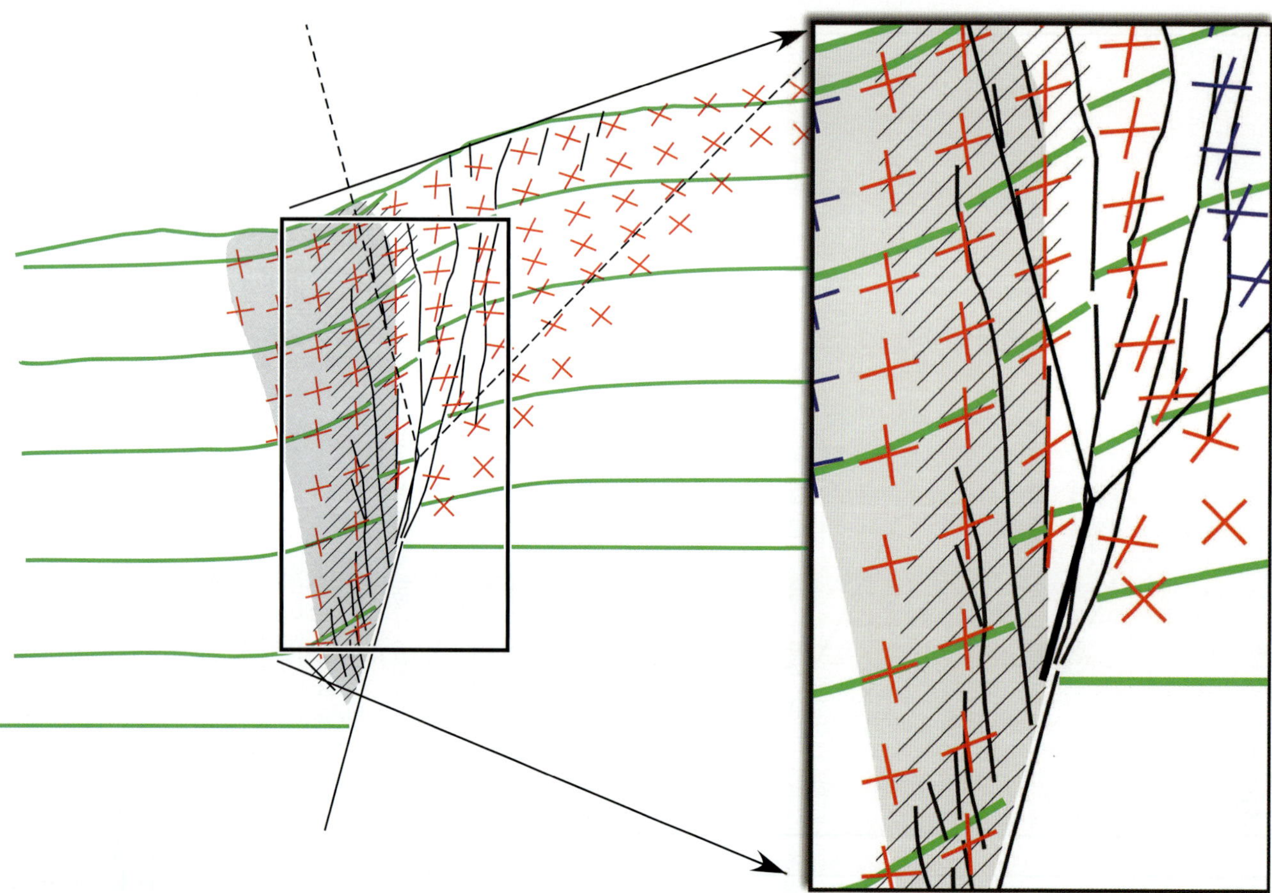

Figure 13. Trishear simulation (green lines) modified from Withjack et al.'s (1990) clay model of extensional forced folds (modified from Allmendinger, 1998). The small red and blue lines show the lines of no finite elongation, which are proxies for potential shear plane orientation. Note their parallelism with the observed fractures (black lines) in the clay model. The light-gray region shows the region where high-angle reverse faults are predicted by the trishear model, and the hachured region is where high-angle reverse faults were observed in the clay model.

Strain Distribution and Fracture Orientation

Bedding-thickness changes during deformation are a reflection of the strain that the rocks have experienced. Because trishear simulates bed geometries well under many conditions, it should be no surprise that it also reproduces well the finite strain in a structure. Various studies comparing trishear predictions to the strain distribution in natural structures have shown this to be the case (e.g., Allmendinger et al., 2004; Cardozo et al., 2005). Whether the strain path in trishear matches that of natural structures is less certain; trishear models contain a rich record of strain evolution through the history of the structure, but most natural examples have had an elusive and fragmentary record of incremental strains to compare with. Studies such as those by Fisher and Anastasio (1994) are needed to confirm the trishear incremental strain history.

As a kinematic model, trishear can say nothing directly about strain mechanisms and partitioning. Thus, one inevitably is faced with assuming that cleavages in natural example should be perpendicular to the short axis of the finite-strain ellipse in the trishear simulation, or joints should be perpendicular to the long axis. Fracture orientations are more problematic because their kinematic interpretation is less clear, particularly where large finite strains are involved. We have suggested that lines of no finite longitudinal extension in trishear models can be used as proxies for the orientation of faults. Comparison of trishear predictions (Allmendinger, 1998; Hardy and McClay, 1999) with analog clay models (Figure 13) of extensional folds (Withjack et al., 1990), as well as with the minor fault orientations in real structures, shows that lines of no finite longitudinal extension can successfully be used to map shear fracture orientations. One of the seemingly counterintuitive results of these studies is to show that high-angle reverse faults can be associated with folding over the tips of propagating normal faults. Likewise, small low-angle normal faults may be associated with large thrust faults.

Propagation Distance and Nucleation

The propagation distance can be used to determine where the fault tip line was when the trishear fold began to form. This tip-line initiation point may take one of three modes (Allmendinger and Shaw, 2000; Allmendinger et al., 2004): (1) where the fault ramps off a regional decollement, (2) the final tip-line position of a fault that has subsequently been reactivated, or (3) the point in the crust at which an isolated crack began to grow into a fault. Thus, the first mode may tell us the depth to the decollement, whereas the second mode leads to the interpretation of fault reactivation. The third mode may yield insight about stress concentrating asperities or where regional stress levels are sufficient to cause significant crack growth. The fault that results from this third mode may be truly isolated from other faults; although the usual structural interpretation is that all faults must be linked to other faults via decollements, there is no reason why this must be the case.

Allmendinger and Shaw (2000) studied the propagation of the Puente Hills thrust fault in the Los Angeles Basin by modeling the Santa Fe Springs anticline. In detail, the Santa Fe Springs anticline exhibits both trishear-like folding and late-stage kink folding. The propagation distance in the best-fitting trishear model suggests that the Puente Hills thrust nucleated about 18 km (11.1 mi) downdip from its final (current) position. This position coincides with the location of the 1987 magnitude 6.0 Whittier Narrows earthquake. They suggested that both the earthquake and the nucleation depth might be related to the position of maximum differential stress in the crust. Because of the possible implications of this study for seismic hazard assessment, an estimate of the uncertainty associated with the best-fitting model is highly desirable.

As mentioned above, the uncertainty in the best-fitting parameters for a trishear model can be obtained using Monte Carlo simulation. In this case, as in all trishear key-bed grid searches, the uncertainty arises from imperfect knowledge of the final bed geometry. In the case of the Santa Fe Springs anticline, two distinctly different types of information are used to determine bed geometry: (1) conversion from time to depth based on an assumed velocity model for the seismic reflection data and (2) uncertainty in picks of stratigraphic boundaries in the hydrocarbon exploration wells as well as their projection along strike onto the line of section. We lack sufficient information to evaluate the second source, but its uncertainty is assuredly (much) smaller than that arising from the seismic reflection data. Therefore, we model below only the uncertainty caused by the time-depth conversion; the resulting estimate will be a maximum for the structure.

The modeled bed is within the Tertiary stratigraphic sequence between 1- and 3-km (0.6- and 1.9-mi) depth. A 10% error in a velocity estimate of 4000 m/s (13,123 ft/s) at 1-s two-way traveltime would yield a positional uncertainty in the depth to bedding of ±200 m (±656 ft); a 5% error would yield a ±100-m (±328-ft) uncertainty. To run the Monte Carlo simulation, we generate 200 synthetic data sets where the entire bed is shifted uniformly within the range according to random numbers drawn from a Gaussian distribution (see Press et al., 1986). To take the data from the exploration wells into account, one would need a substantially more complicated scheme in which the uncertainty of individual points along the bed varied as a function of distance from the wells. The results of the simulation for both 100- and 200-m (656-ft) bed uncertainty are shown in Figure 14. A 100-m (328-ft) error in bed position results in about ±1100-m (±3609-ft) uncertainty in propagation and nearly ±200-m (±256-ft) in displacement at the 1-sigma level. A 200-m (256-ft) uncertainty in bed position would yield ±2000 m (±6562 ft) in propagation and nearly ±800 m (±2625 ft) in displacement. These results are consistent with the cruder method of estimating errors reported by Allmendinger and Shaw (2000).

Note that this exercise does not demonstrate that trishear is the best model or interpretation for this structure because we have only evaluated the uncertainties in the best-fitting trishear model. Hypothetically, one could calculate the misfit for the best-fitting kink models applied to the Santa Fe Springs anticline, which could then be compared to the trishear best fit. However, there is no formal way to do this at the present time.

Seismic Hazard and Paleoseismology

Fold scarps along seismically active faults are very common, a fact increasingly being recognized by the paleoseismological community (e.g., Champion et al., 2001; Ishiyama et al., 2004). The application of kinematic models to such structures is a relatively new development, and trishear kinematic modeling of these features has proved successful in several areas. Perhaps the most successful application of trishear in this context was that undertaken by Champion et al. (2001) with respect to the Reelfoot blind thrust in the New

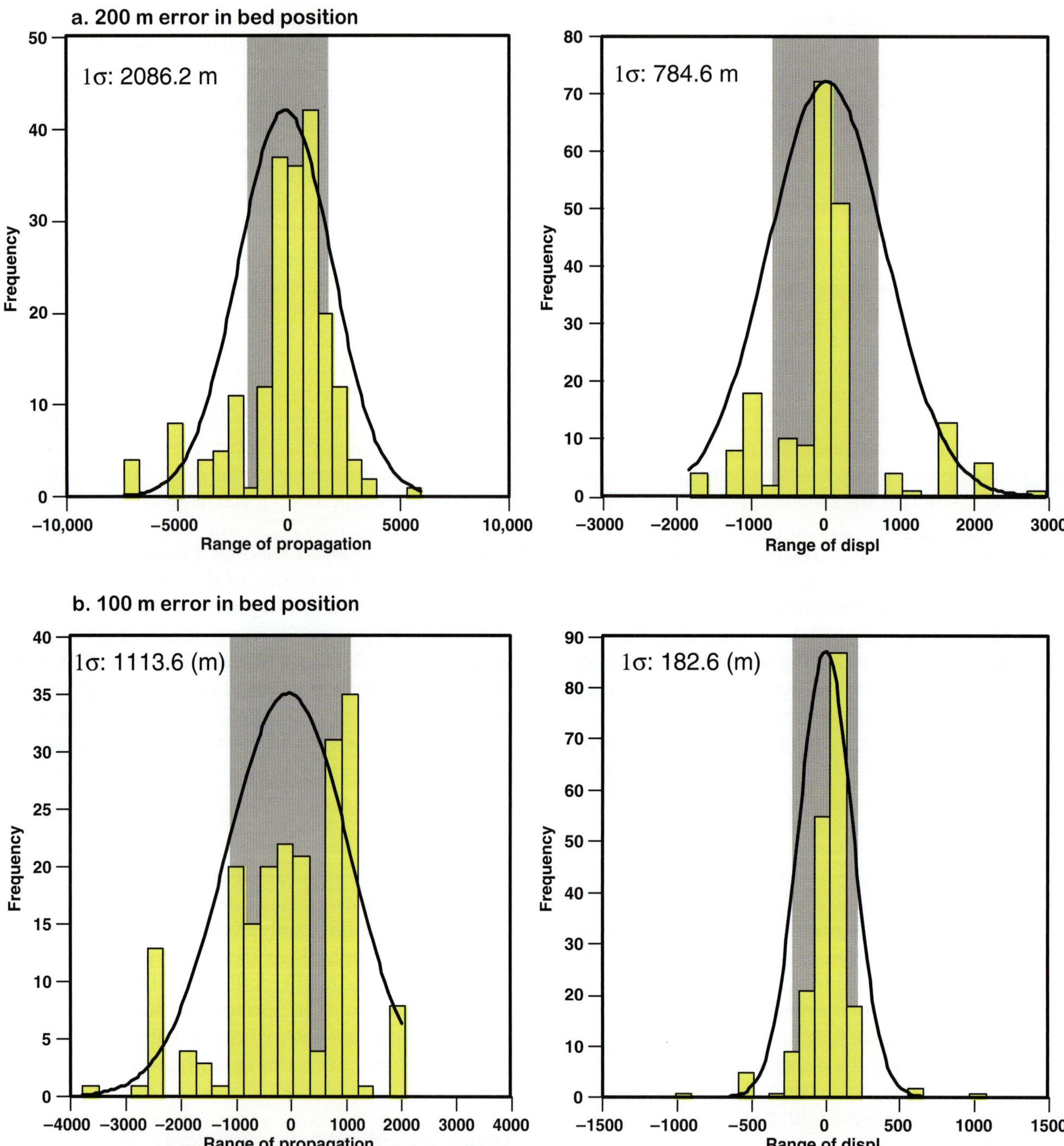

Figure 14. Monte Carlo statistics for the Santa Fe Springs anticline model modified from Allmendinger and Shaw (2000). The 1-sigma range of permissible values (i.e., 66% confidence interval) is shown in gray behind the histogram. See discussion in the text. displ = displacement.

Madrid seismic zone. Previous work using kink-band models (Mueller et al., 1999) had determined a slip rate of 6.1 ± 0.7 mm/yr (0.24 ± 0.03 in./yr) on the Reelfoot fault. Through a combination of trenching, boring, digital elevation models, and (trishear) structural modeling, they were able to revise the slip rate on the Reelfoot fault to 3.9 ± 0.1 mm/yr (0.15 ± 0.003 in./yr). This approach appears to offer some promise in the modeling of very shallow folds associated with deeper, blind faults.

SUMMARY AND CONCLUSIONS

Trishear has successfully developed into a mainstream kinematic model for simulating structural geometries on balanced cross sections. The numerical implementation of trishear has facilitated inverse modeling via grid searching, which allows an objective choice of best-fitting model, something that has been lacking from most balanced cross section strategies. Likewise, realistic uncertainty estimates, also lacking from most balanced sections, can also be made for trishear structures using Monte Carlo simulation. In this way, trishear represents the forerunner to a necessary new approach to balanced section construction wherein the community moves away from the old, graphical, intuitive, and subjective methodologies to one that more closely resembles quantitative forecasting of probable geometries.

Perhaps also because of its numerical realization, the development of the kinematic model has been accompanied by mechanical modeling of similar structures. In the span of a few years, most major mechanical modeling strategies, viscous, finite element, and discrete element, have been applied to trishearlike structures in an attempt to acquire a deeper understanding of their origins, probe the validity of commonly assumed trishear velocity fields, and understand the physical and rheological controls on key parameters such as P/S. This near simultaneous development reflects the complementary nature of the kinematic and mechanical approaches to modeling structures. Although each approach can exist on its own, the result is clearly richer when pursued together. Although trishear has been shown to be successful at simulating geometries and finite strains, it tells us nothing to explain how the structure results from a combination of boundary conditions, rheologies, and environmental conditions.

In the context of understanding the trishear kinematic model, the mechanical modeling conducted to date has provided some fairly robust insights. The most important conclusion is that the folded sequence lying above the tip line must be relatively isotropic in mechanical properties; anisotropy results in a more kinklike folding in front of the tip line. Likewise, the deformation must be incompressible; otherwise, the hanging wall undergoes significant deformation, violating a primary assumption of trishear kinematics that the hanging wall is rigid. Assuming an isotropic sequence, the trishear linear velocity field is an excellent first-order approximation to the velocities that result from a wide variety of modeling approaches and rheologies. That is, viscous, elastic-plastic, and frictional rheologies in analytical, finite-element, or discrete-element formulations all produce trishearlike geometries. Trishear is thus most appropriately applied in situations in which bedding-parallel heterogeneity is not a major factor in the cover stratigraphy or the cover is prefractured with a variety of orientations of fractures, shear planes, and faults (cf. Patton and Fletcher, 1998; Hardy and Finch, 2007). This situation corresponds to the isotropic viscous block models of Johnson and Johnson (2002a, b), the weak isotropic DEMs of Finch et al. (2003, 2004), and the incompressible FEMs of Cardozo et al. (2003). It clearly is inappropriate when one particular fracture set or bedding is the dominant plane of movement. In such situations, the deformation of the cover would be expected to depart from that predicted by the trishear kinematic model based as it is on an isotropic cover. This situation would correspond to the strongly anistropic block viscous models of Patton and Fletcher (1998) and Johnson and Johnson (2002a, b). Indeed, without explicitly referring to trishear, Patton and Fletcher (1998) found that the development of a wedge-shaped (triangular) region above a basement fault, in which deformation becomes more diffuse upward, was maximum when the cover was isotropic, and that when a strong anistropy was introduced, the cover deformed either in a more kink-band fashion or in discrete shear zones reminiscent of Coulomb materials.

To illustrate and amplify these points, we now show some recent discrete-element modeling results from Hardy and Finch (2007). In this study, they examined the availability, or lack, of bedding-parallel slip surfaces in sedimentary cover above a blind basement fault. In Figure 15, we show the change in fold form above a blind basement fault, in which the cover varies from being isotropic to strongly anisotropic. In all models, the weak layers have identical strengths; in Figure 15a, the strong and weak layers have the same strength; in Figure 15b, the strong layers are three times stronger; in Figure 15c, they are five times stronger; and in Figure 15d, they are seven times stronger. These results clearly show the transition from classic trishear fold forms in the isotropic cover (Figure 15a) to much more kinklike fold forms in the anisotropic cover (Figure 15d).

The concept of propagation to slip is fundamental to all types of fold models and is not unique to trishear. However, only trishear treats P/S as an independent parameter. In all other kinematic models, P/S is fixed by the geometry. The P/S is impossible to explore using existing viscous folding theory models. Although the parameter can be prescribed in FEMs by unzipping the nodes of the mesh in front of the fault at some arbitrary rate, only DEMs allow for natural tip-line propagation. Although the complexity of DEMs makes it more difficult to identify mechanical controls on various parameters, repeated experiments suggest that P/S is fundamentally related to the strength of the cover in front of the tip line.

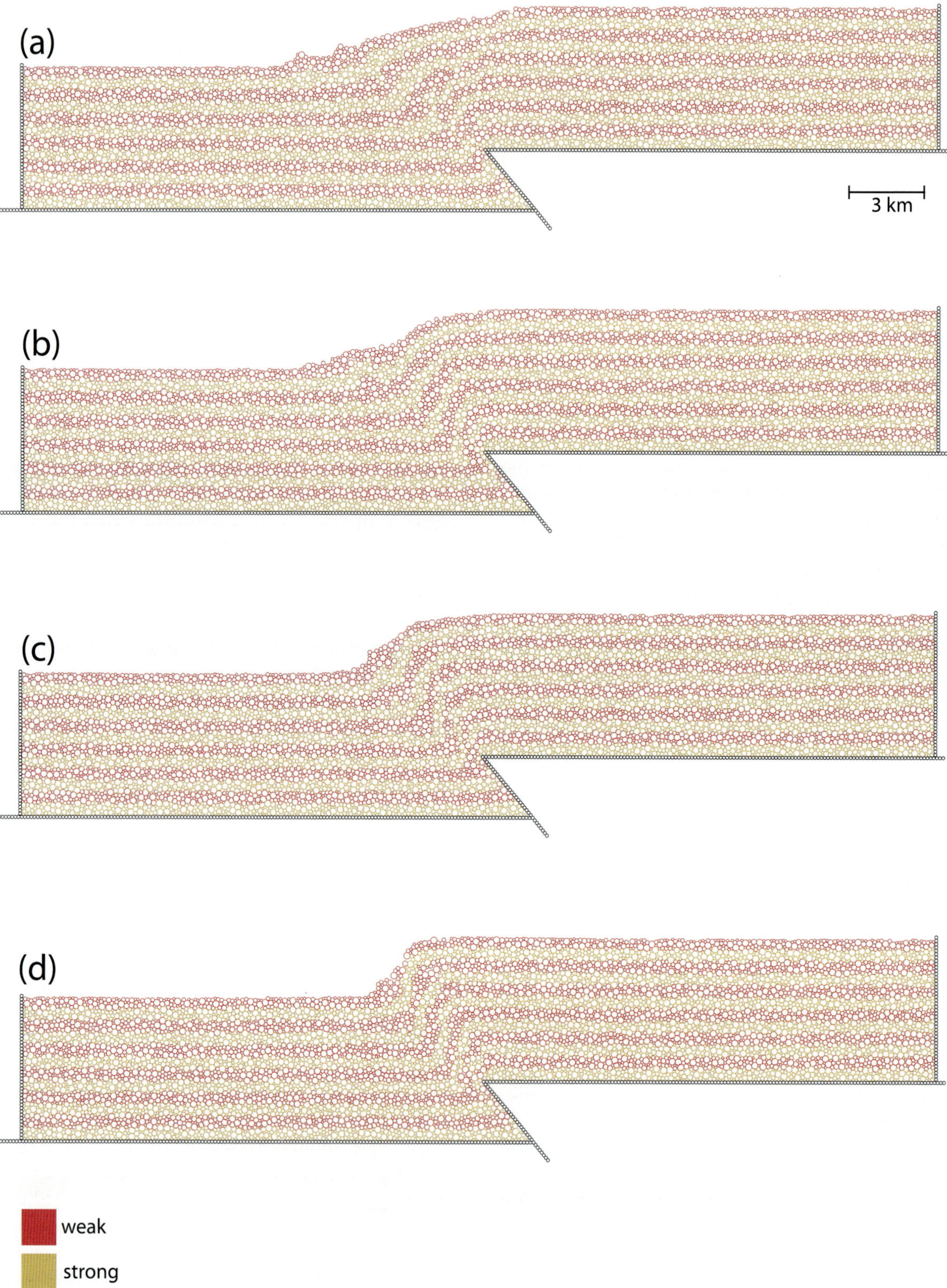

Figure 15. Discrete-element model results showing the change in fold form above a blind basement fault, in which the cover varies from being isotropic to strongly anisotropic (a–d). In all models, the weak layers have identical strengths; (a) the strong and weak layers have the same strength; (b) the strong layers are three times stronger; (c) the strong layers are five times stronger; (d) the strong layers are seven times stronger. These results clearly show the transition from classic trishear fold forms in the isotropic cover (a) to much more kinklike fold forms in the anisotropic cover (d).

The significance of the trishear apical angle remains most elusive. Different mechanical modeling strategies have yielded completely different and apparently incompatible results. Viscous folding theory always produces very open apical angles, whereas FEMs only yield apical angles less than 20°. Sensitivity analysis using only the kinematic model indicates that P/S and trishear apical angle are roughly inversely correlated, and thus the apical angle may not be a truly independent parameter.

We anticipate that future research will result in more sophisticated inverse modeling of trishear structures in two and three dimensions. More importantly, guided by results of ongoing mechanical modeling, more complicated hybrid kinematic models are likely to be developed in which different parts or different structural levels deform in distinctly different kinematic styles. We regard such feedback between more comprehensive mechanical modeling and simpler trishear-style kinematic modeling as very positive and constructive. The mechanical modeling provides a deeper understanding of physical processes, whereas trishear-style kinematics allows rapid, objective simulation of real structures. One can envision a day in the not too distant future when computers have become sufficiently powerful that mechanical models themselves can be used in an iterative fashion to simulate real structures, bypassing the simpler kinematic approach altogether.

ACKNOWLEDGMENTS

This review grew out of discussions during the International Conference on Theory and Application of Fault-related Folding held in Beijing, June 2005. The authors are grateful to the conference organizers for hosting such a stimulating meeting and for providing financial support to attend. We are most grateful to our many colleagues with whom we have explored trishear-like folding in the past, including Mary Ford, Ernesto Cristallini, Néstor Cardozo, Alan Zehnder, Emma Finch, Kate Cooper, Asdrubal Bernal, John Shaw, Kavi Bhalla, and Juli Morgan. Reviews by John Suppe and Ken McClay are greatly appreciated. Additionally, we thank several colleagues who have had considerable impact on our understanding of fault-related folds, including Ken McClay, John Suppe, Ray Fletcher, and Karl Mueller. Allmendinger acknowledges support for his trishear research from the U.S. National Science Foundation (EAR-9814348, EAR-0125557) and the donors to the Petroleum Research Fund of the American Chemical Society (ACS-PRF 34882-AC2). Hardy acknowledges support by Institució Catalana de Recerca i Estudis Avançats (ICREA), the Centre de Supercomputació de Catalunya (CESCA), the Geomod 3-D project (CGL2004-05816-C02-01/BTE), and the Geomodels program.

REFERENCES CITED

Allmendinger, R. W., 1998, Inverse and forward numerical modeling of trishear fault-propagation folds: Tectonics, v. 17, p. 640–656, doi:10.1029/98TC01907.

Allmendinger, R. W., 2007, FaultFold Computer Program, available for noncommercial download use at http://www.geo.cornell.edu/geology/faculty/RWA/FF_DL.html (accessed December 8, 2010).

Allmendinger, R. W., and J. H. Shaw, 2000, Estimation of fault propagation distance from fold shape: Implications for earthquake seismicity: Geology, v. 28, p. 1099–1102, doi:10.1130/0091-7613(2000)28<1099:EOFPDF>2.0.CO;2.

Allmendinger, R. W., T. R. Zapata, R. Manceda, and F. Dzelalija, 2004, Trishear kinematic modeling of structures, with examples from the Neuquén Basin, Argentina, *in* K. R. McClay, ed., Thrust tectonics and hydrocarbon systems: AAPG Memoir 82, p. 356–371.

Cardozo, N., 2005, Trishear modeling of fold bedding data along a topographic profile: Journal of Structural Geology, v. 27, p. 495–502, doi:10.1016/j.jsg.2004.10.004.

Cardozo, N., K. Bhalla, A. T. Zehnder, and R. W. Allmendinger, 2003, Mechanical models of fault propagation folds and comparison to the trishear kinematic model: Journal of Structural Geology, v. 25, p. 1–18, doi:10.1016/S0191-8141(02)00013-5.

Cardozo, N., R. W. Allmendinger, and J. K. Morgan, 2005, Influence of mechanical stratigraphy and initial stress state on the formation of two fault propagation folds: Journal of Structural Geology, v. 27, p. 1954–1972, doi:10.1016/j.jsg.2005.06.003.

Carena, S., and J. Suppe, 2002, Three-dimensional imaging of active structures using earthquake aftershocks: The Northridge thrust, California: Journal of Structural Geology, v. 24, p. 887–904, doi:10.1016/S0191-8141(01)00110-9.

Champion, J., K. Mueller, A. Tate, and M. Guccione, 2001, Geometry, numerical models and revised slip rate for the Reelfoot fault and trishear fault-propagation fold, New Madrid seismic zone: Engineering Geology, v. 62, p. 31–49, doi:10.1016/S0013-7952(01)00048-5.

Cooper, K., and S. Hardy, 1999, The interaction between fault-related folding and sedimentation in 3-D (abs.): Thrust Tectonics Conference, Royal Holloway, University of London, p. 156.

Cosgrove, J. W., and M. S. Ameen, eds., 2000, Forced folds and fractures: Geological Society (London) Special Publication 169, 225 p.

Couples, G. D., and H. Lewis, 1998, Lateral variations of strain in experimental forced folds: Tectonophysics, v. 295, p. 79–91, doi:10.1016/S0040-1951(98)00116-4.

Couples, G. D., D. W. Stearns, and J. W. Handin, 1994, Kinematics of experimental forced folds and their relevance to cross-section balancing: Tectonophysics, v. 233, p. 193–213, doi:10.1016/0040-1951(94)90241-0.

Cristallini, E. O., and R. W. Allmendinger, 2001, Pseudo 3-D modeling of trishear fault-propagation folding: Journal of Structural Geology, v. 23, p. 1883–1899, doi:10.1016/S0191-8141(01)00034-7.

Cristallini, E. O., and R. W. Allmendinger, 2002, Backlimb trishear: A kinematic model for curved folds developed over angular fault bends: Journal of Structural Geology, v. 24, p. 289–295, doi:10.1016/S0191-8141(01)00063-3.

Cristallini, E. O., L. Giambiagi, and R. W. Allmendinger, 2004, True three-dimensional trishear: A kinematic model for strike-slip and oblique-slip deformation: Geological Society of America Bulletin, v. 116, p. 938–952, doi:10.1130/B25273.1.

Elliott, D., 1976, The energy balance and deformation mechanisms of thrust sheets: Philosophical Transactions of the Royal Astronomical Society, v. A-283, p. 289–312.

Erslev, E. A., 1991, Trishear fault-propagation folding: Geology, v. 19, p. 617–620, doi:10.1130/0091-7613(1991)019<0617:TFPF>2.3.CO;2.

Erslev, E. A., and J. L. Rogers, 1993, Basement-cover geometry of Laramide fault-propagation folds, *in* C. J. Schmidt, R. B. Chase, and E. A. Erslev, eds., Laramide basement deformation in the Rocky Mountain foreland of the western United States: Geological Society of America Special Paper 280, p. 125–146.

Finch, E., S. Hardy, and R. L. Gawthorpe, 2003, Discrete element modeling of contractional fault-propagation folding above rigid basement fault blocks: Journal of Structural Geology, v. 25, p. 515–528, doi:10.1016/S0191-8141(02)00053-6.

Finch, E., S. Hardy, and R. L. Gawthorpe, 2004, Discrete element modeling of extensional fault-propagation folding above rigid basement fault blocks: Basin Research, v. 16, p. 489–506, doi:10.1111/j.1365-2117.2004.00244.x.

Fischer, M. P., and M. S. Wilkerson, 2000, Predicting the orientation of joints from fold shape: Results of pseudo-three-dimensional modeling and curvature analysis: Geology, v. 28, p. 15–18, doi:10.1130/0091-7613(2000)28<15:PTOOJF>2.0.CO;2.

Fischer, M. P., M. S. Wilkerson, and R. D. Christensen, 1999, Three-dimensional evolution of trishear fault-propagation folds: Relations between fold geometry, fault geometry and fault displacement (abs.): Thrust Tectonics 99, Royal Holloway, University of London, Abstracts, p. 198.

Fisher, D. M., and D. J. Anastasio, 1994, Kinematic analysis of a large-scale leading edge fold, Lost River Range, Idaho: Journal of Structural Geology, v. 16, p. 337–354, doi:10.1016/0191-8141(94)90039-6.

Ford, M., E. Williams, A. Artoni, J. Vergés, and S. Hardy, 1997, Progressive evolution of a fault propagation fold pair from growth strata geometries, Sant Llorenç de Morunys, SE Pyrenees: Journal of Structural Geology, v. 19, p. 413–441, doi:10.1016/S0191-8141(96)00116-2.

Fox, F. G., 1959, Structure and accumulation of hydrocarbons in southern Foothills, Alberta, Canada: AAPG Bulletin, v. 43, p. 992–1025.

Gallup, W. B., 1951, Geology of Turner Valley oil and gas field, Alberta, Canada: AAPG Bulletin, v. 34, p. 797–821.

Gawthorpe, R., and S. Hardy, 2002, Extensional fault-propagation folding and base-level change as controls on growth strata geometries: Sedimentary Geology, v. 146, p. 47–56.

Grelaud, S., D. Buil, S. Hardy, and D. Frizon de Lamotte, 2000, Trishear kinematic model of fault-propagation folding and sequential development of minor structures: The Oupia anticline (NE Pyrenees, France) case study: Geological Society of France Bulletin, v. 171, no. 4, p. 441–449, doi:10.2113/171.4.441.

Haneberg,W. C., 1992, Drape folding of compressible elastic layers: I. Analytical solutions for vertical uplift: Journal of Structural Geology, v. 14, p. 713–721, doi:10.1016/0191-8141(92)90128-J.

Haneberg, W. C., 1993, Drape folding of compressible elastic layers: II. Matrix solution for two-layer folds: Journal of Structural Geology, v. 15, p. 923–932, doi:10.1016/0191-8141(93)90185-D.

Hardy, S., 1997, A velocity description of constant thickness fault-propagation folding: Journal of Structural Geology, v. 19, p. 893–896, doi:10.1016/S0191-8141(97)00013-8.

Hardy, S., and E. Finch, 2007, Mechanical stratigraphy and the transition from trishear to kink-band fault-propagation fold forms above blind basement thrust faults: A discrete element study: Marine and Petroleum Geology, v. 42, p. 75–90, doi:10.1016/j.marpetgeo.2006.09.001.

Hardy, S., and M. Ford, 1997, Numerical modeling of trishear fault-propagation folding: Tectonics, v. 16, p. 841–854, doi:10.1029/97TC01171.

Hardy, S., and K. McClay, 1999, Kinematic modeling of extensional fault-propagation folding: Journal of Structural Geology, v. 21, p. 695–702, doi:10.1016/S0191-8141(99)00072-3.

Ishiyama, T., K. Mueller, M. Togo, A. Okada, and K. Takemura, 2004, Geomorphology, kinematic history, and earthquake behavior of the active Kuwana wedge thrust anticline, central Japan: Journal of Geophysical Research, v. 109, p. B12408, doi:10.1029/2003JB002547.

Jamison, W. R., 1987, Geometric analysis of fold development in overthrust terranes: Journal of Structural Geology, v. 9, p. 207–219, doi:10.1016/0191-8141(87)90026-5.

Jamison, W. R., and D. W. Stearns, 1982, Tectonic deformation of Wingate Sandstone, Colorado National Monument: AAPG Bulletin, v. 66, p. 2584–2608.

Johnson, A. M., and R. C. Fletcher, 1994, Folding of viscous layers: New York, Columbia University Press, 461 p.

Johnson, K. M., and A. M. Johnson, 2002a, Mechanical models of trishear-like folds: Journal of Structural Geology, v. 24, p. 277–287, doi:10.1016/S0191-8141(01)00062-1.

Johnson, K. M., and A. M. Johnson, 2002b, Mechanical analysis of the geometry of forced-folds: Journal of Structural Geology, v. 24, p. 401–410, doi:10.1016/S0191-8141(01)00085-2.

Jordan, T. E., and R. W. Allmendinger, 1986, The Sierras Pampeanas of Argentina: A modern analog of Rocky Mountain foreland deformation: American Journal of Science, v. 286, p. 737–764.

Khalil, S., and K. R. McClay, 2002, Extensional fault-related folding, northwestern Red Sea, Egypt: Journal of Structural Geology, v. 24, p. 743–762, doi:10.1016/S0191-8141(01)00118-3.

Matthews III, V., and D. F. Work, 1978, Laramide folding associated with basement block faulting along the northeastern flank of the Front Range, Colorado, *in* V. Matthews III, ed., Laramide folding associated with basement block faulting in the western United States: Geological Society of America Memoir 151, p. 101–123.

Mitra, S., 1990, Fault-propagation folds: Geometry, kinematic evolution, and hydrocarbon traps: AAPG Bulletin, v. 74, p. 921–945.

Mueller, K., J. Champion, M. Guccione, and K. Kelson, 1999, Fault slip rates in the modern New Madrid seismic zone: Science, v. 284, p. 619–622, doi:10.1126/science.284.5414.619.

Nichols, G. J., 1987, Syntectonic alluvial fan sedimentation, southern Pyrenees: Geological Magazine, v. 124, p. 121–133, doi:10.1017/S0016756800015934.

Patton, T. L., and R. C. Fletcher, 1995, Mathematical block-motion model for deformation of a layer above a buried fault of arbitrary dip and sense of slip: Journal of Structural Geology, v. 17, p. 1455–1472, doi:10.1016/0191-8141(95)00034-B.

Patton, T. L., and R. C. Fletcher, 1998, A rheological model for fractured rock: Journal of Structural Geology, v. 20, p. 491–502, doi:10.1016/S0191-8141(98)00002-9.

Patton, T. L., A. R. Moustafa, R. A. Nelson, and A. S. Abdine, 1994, Tectonic evolution and structural setting of the Suez Rift, *in* S. M. Landon, ed., Interior rift basins: AAPG Memoir 59, p. 9–55.

Press, W. H., B. P. Flannery, S. A. Teukolsky, and W. T. Vetterling, 1986, Numerical recipes: The art of scientific computing: Cambridge, United Kingdom, Cambridge University Press, 818 p.

Riba, O., 1976, Syntectonic unconformities of the Alto Cardener, Spanish Pyrenees: A genetic interpretation: Sedimentary Geology, v. 15, p. 213–233, doi:10.1016/0037-0738(76)90017-8.

Sanford, A. R., 1959, Analytical and experimental study of simple geologic structures: Geological Society of America Bulletin, v. 70, p. 19–52, doi:10.1130/0016-7606(1959)70[19:AAESOS]2.0.CO;2.

Schmidt, C. J., R. B. Chase, and E. A. Erslev, eds., 1993, Laramide basement deformation in the Rocky Mountain foreland of the western United States: Geological Society of America Special Paper, 280, 365 p.

Serra, S., 1977, Styles of deformation in the ramp regions of overthrust faults, *in* E. L. E. A. Heisey, ed., Rocky Mountain thrust belt geology and resources: Wyoming Geological Association Annual Field Conference, v. 29, p. 487–498.

Sharp, I. R., R. L. Gawthorpe, J. R. Underhill, and S. Gupta, 2000, Fault-propagation folding in extensional settings: Examples of structural style and synrift sedimentary response from the Suez rift, Sinai, Egypt: Geological Society of America Bulletin, v. 112, p. 1877–1899, doi:10.1130/0016-7606(2000)112<1877:FPFIES>2.0.CO;2.

Suppe, J., 1983, Geometry and kinematics of fault-bend folding: American Journal of Science, v. 283, p. 684–721.

Suppe, J., and D. Medwedeff, 1990, Geometry and kinematics of fault-propagation folding: Eclogae Geologicae Helvetiae, v. 83, p. 409–454.

Suppe, J., G. T. Chou, and S. C. Hook, 1992, Rates of folding and faulting determined from growth strata, *in* K. R. McClay, ed., Thrust tectonics: London, Chapman and Hall, p. 105–121.

Suppe, J., C. D. Connors, and Y. Zhang, 2004, Shear fault-bend folding, *in* K. R. McClay, ed., Thrust tectonics and hydrocarbon systems: AAPG Memoir 82, p. 303–323.

Williams, G., and T. Chapman, 1983, Strains developed in the hanging walls of thrusts due to their slip/propagation rate, a dislocation model: Journal of Structural Geology, v. 5, p. 563–571, doi:10.1016/0191-8141(83)90068-8.

Withjack, M. O., J. Olson, and E. Peterson, 1990, Experimental models of extensional forced folds: AAPG Bulletin, v. 74, p. 1038–1054.

Zapata, T. R., and R. W. Allmendinger, 1996, Growth stratal records of instantaneous and progressive limb rotation in the Precordillera thrust belt and Bermejo Basin, Argentina: Tectonics, v. 15, p. 1065–1083, doi:10.1029/96TC00431.

Zehnder, A. T., and R. W. Allmendinger, 2000, Velocity field for the trishear model: Journal of Structural Geology, v. 22, p. 1009–1014, doi:10.1016/S0191-8141(00)00037-7.

6

Tavani, S., and F. Storti, 2011, Layer-parallel shortening templates associated with double-edge fault-propagation folding, *in* K. McClay, J. H. Shaw, and J. Suppe, eds., Thrust fault-related folding: AAPG Memoir 94, p. 121 – 135.

Layer-parallel Shortening Templates Associated with Double-edge Fault-propagation Folding

S. Tavani[1]
Dipartimento di Scienze Geologiche, Università degli studi "Roma Tre", Rome, Italy

F. Storti
Dipartimento di Scienze Geologiche, Università degli studi "Roma Tre", Rome, Italy

ABSTRACT

Pressure-solution cleavage is frequently among the most abundant mesostructures in carbonate thrust wedges. It can exert a primary function in fluid migration, and consequently, understanding its time-space evolution can significantly impact the reservoir modeling and performance. The evidence that pressure-solution cleavage is commonly at a high angle to bedding and, in many cases, displays a frequency distribution relating to the host-fold geometry indicates a partial synfolding development driven by fault-fold kinematics. Double-edge fault-propagation folding assumes layer-parallel shortening during fold evolution. Accordingly, this model can provide a tool for inferring the distribution of pressure-solution cleavage within thrust-related folds that, under appropriate stress conditions, can significantly improve secondary porosity and permeability in reservoirs.

We summarize the pressure-solution cleavage pattern in three anticlines that have possibly developed by double-edge fault-propagation folding, and then we analyze the deformation patterns associated with double-edge fault-propagation folding, investigating the influence of different model parameters (i.e., ramp propagation history, shape, and initial length) onto the cross-sectional deformation pattern in fault-propagation anticlines. Modeling results indicate that, in carbonate thrust wedges, forelimb panels and footwall sectors close to the thrust ramp can provide promising targets for hydrocarbon exploration.

INTRODUCTION

The deformation structure pattern strongly influences hydrocarbon migration and accumulation in foreland fold and thrust belts (e.g., Nelson, 1985; Mitra, 1988; Cooper, 1992; Gholipour, 1998; Aydin, 2000; Van Dijk et al., 2000; Florez-Nino et al., 2005). Its development relates to several factors, including the environmental

[1]*Present address*: Departament de Geodinamica i Geofisica, Facultat de Geologia, Universitat de Barcelona, Barcelona, Spain.

DOI:10.1306/13251335M943120

Figure 1. (a) Layer-perpendicular, stratabound, constantly spaced pressure-solution cleavages in the Maiolica limestone, Fiastrone anticline, Italy. (b) Tar-filled pressure-solution cleavage in the Bolognano limestone, Majella anticline, Italy. In both (a) and (b), a 10-cm-wide compass provides a reference for the scale.

conditions of deformation (e.g., Chester et al., 1991; Stewart and Alvarez, 1991; Jamison, 1992; Lemiszki et al., 1994; Woodward, 1999), the mechanical stratigraphy of the folded multilayer (e.g., Corbett et al., 1987; Woodward and Rutherford, 1989; Protzman and Mitra, 1990; Gross, 1995; Couzens and Wiltschko, 1996; Fischer and Jackson, 1999; Chester, 2003), and the folding mechanism (e.g., De Sitter, 1956; Ramsay, 1974; Sanderson, 1982; Dahlstrom, 1990; Srivastava and Engelder, 1990; Fischer et al., 1992; Storti and Salvini, 1996; Salvini and Storti, 2001). In many cases, pressure-solution cleavages are among the most abundant mesoscale structures associated with folding, particularly in carbonate multilayers (Figure 1) (e.g., Marshak and Engelder, 1985; Mitra and Yonkee, 1985; Mitra, 1988; Holl and Anastasio, 1995; Dunne and Caldanaro, 1997; Davidson et al., 1998; Storti and Salvini, 2001; Tavani et al., 2004). Cleavage development reduces rock permeability, particularly in the direction perpendicular to the cleavage surface (e.g., Nelson, 1985). However, once formed, it can provide a mechanical discontinuity able to be reopened under appropriate stress conditions, and consequently, it can eventually produce additional pathways for oil migration and accumulation (Figure 1b). It follows that understanding the time-space evolution of syntectonic solution cleavage networks can significantly impact the reservoir modeling and performance (e.g., Leythaeuser et al., 1995; Aydin, 2000; Van Geet et al., 2002).

In well-bedded carbonate multilayers deforming at shallow depth, tectonic pressure-solution cleavage is frequently represented by surfaces at a high angle to bedding. This observation frequently leads to a prefolding interpretation. However, a deeper analysis of pressure-solution cleavage attribute variability shows that, in many cases, both its angle to bedding and, particularly, its frequency relate with the host-fold geometry, thus supporting a partial synfolding origin. In these cases, pressure-solution cleavage is an active factor during folding by reducing the cross-sectional area, which because of its orientation (i.e., at a high angle to bedding) occurs at a rather constant bed thickness. In these situations, those models assuming line-length and/or area preservation (e.g., Suppe, 1983; Jamison, 1987; Chester and Chester, 1990; Mitra, 1990; Suppe and Medwedeff, 1990; Erslev, 1991; Epard and Groshong, 1995; Wickham, 1995; Poblet and McClay, 1996) cannot provide accurate predictions of fold kinematics and the associated deformation pattern. Double-edge fault-propagation folding (Tavani et al., 2006a) is a geometric and kinematic model that can help to overcome this limitation because it assumes synfolding line-length variation and bed-thickness preservation in multilayers deforming by flexural slip.

In this work, we summarize synfolding pressure-solution cleavage distributions of three anticlines that provided insights for the formalization of the double-edge fault-propagation folding model, and then we discuss the cross-sectional layer-parallel shortening distributions predicted for double-edge fault-propagation anticlines. Implication for hydrocarbon exploration and development is discussed.

PRESSURE-SOLUTION CLEAVAGE DISTRIBUTION IN THRUST-RELATED ANTICLINES

In this section, we summarize the pressure-solution cleavage distributions in three thrust-related anticlines: the Chaudrons anticline (Tavani et al., 2004), the Añisclo

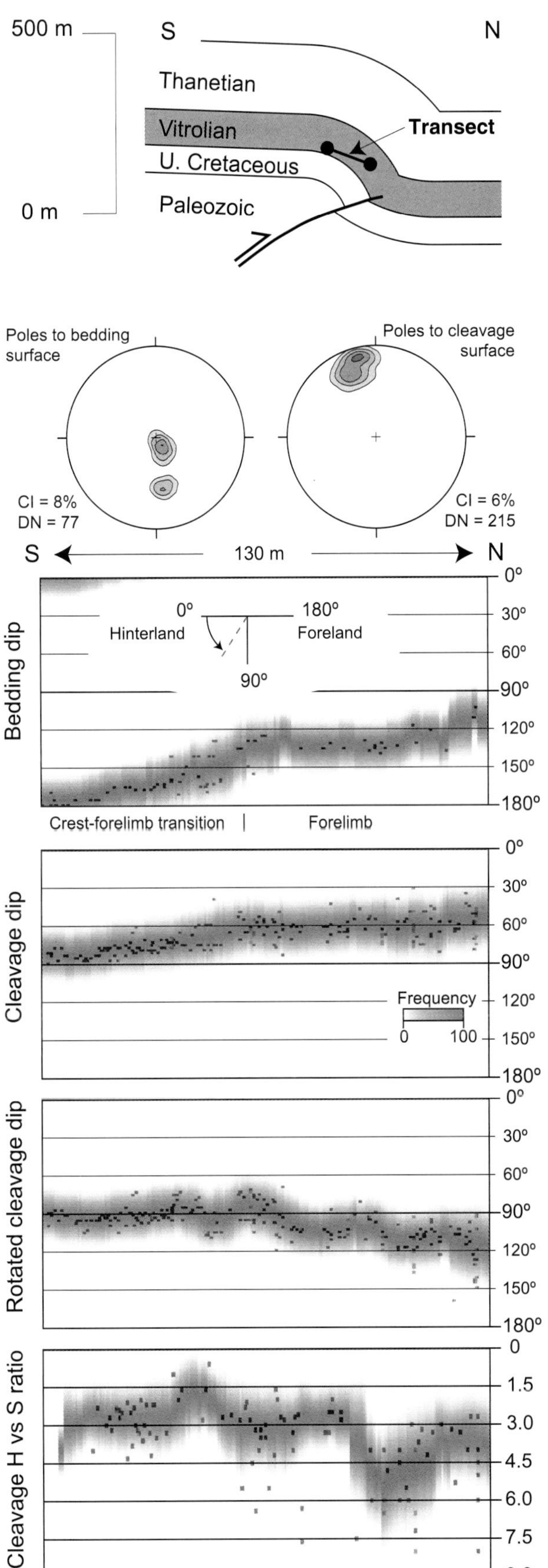

Figure 2. Pressure-solution cleavage distribution in the Chaudrons anticline. Geological cross section with transect trace (light gray indicates the analyzed unit). Cumulative contouring of both poles to bedding and poles to pressure-solution cleavage and transect analysis showing the variability of bedding dip, cleavage dip, rotated cleavage dip, and cleavage frequency (i.e., bed thickness versus cleavage spacing ratio) across the fold strike. The smoothed and filtered parameter frequency is shown in gray; points correspond to original data (see Salvini et al., 1999, for details). In the stereoplots, CI is the contouring interval and DN is the data number. In the transects, H is the bed thickness and S is the cleavage spacing. U = Upper.

anticline (Tavani et al., 2006b), and the Mt. Catria anticline (Tavani et al., 2008). In these anticlines, we collected data on faults, tectonic pressure-solution cleavages, joints, and veins. The description and analysis of the complete data sets are provided in the above-cited works, and the methodology for structural data collection and analysis is illustrated by Tavani et al. (2006b). In this work, we analyze the across-strike variability of pressure-solution cleavage attributes using structural transects (e.g., Wise and McCrory, 1982; Salvini et al., 1999; Tavani et al., 2004) oriented perpendicularly to the corresponding fold axis. In the transect analysis, we project the data position onto a section line and plot various structural parameters as the y variable along the transect (Figures 2–4). For each anticline, the data sets are analyzed in a single mechanical unit to reduce the influence of the mechanical stratigraphy onto the deformation pattern analyses (e.g., Corbett et al., 1987; Protzman and Mitra, 1990; Gross, 1995; Fischer and Jackson, 1999).

Chaudrons Anticline

The Eocene Chaudrons anticline is an east-west–trending anticline located in the Corbières Area, northern Pyrenees (France). The exposed multilayer includes carbonatic and siliciclastic units, ranging in age from Upper Cretaceous to upper Paleocene. The back of the anticline is offset by normal faults formed during a late Oligocene–Miocene event, and the backlimb is not exposed. We collected cleavage data in the Vitrolian formation, which includes alternating sandstones, silts, and clays. Data were collected along a transect whose starting and ending points correspond to the crest-forelimb transition and to the steeper forelimb sector, respectively (Figure 2). Pressure-solution cleavages strike parallel to the fold axis. Their dip, both in the present orientation and after bedding-dip removal, and its frequency, computed as bed thickness versus cleavage spacing ratio (e.g., Tavani et al., 2006b), vary along the transport direction following the progressive variation of the bedding dip.

Añisclo Anticline

The north-south–trending middle Eocene Añisclo anticline is located in the Spanish Pyrenees. The involved multilayer includes alternating limestones and marls of Upper Cretaceous to middle Eocene age overlying the Keuper Triassic evaporites (Figure 3). The backlimb of the Añisclo anticline in the study area is characterized by a bedding dip ranging from 20 to 45° toward the east and by a gentle transition to a nearly flat-lying crest. The crest-forelimb transition in the oldest exposed rocks is commonly coincident with an intensely deformed zone, which is interpreted to be associated with a steep splay of the main Añisclo thrust. The forelimb is characterized by a curvilinear shape that progressively reaches an overturned attitude of 65° to the east and then curves to a moderate westward dip before transitioning westward to the flat-lying foreland. In the Añisclo anticline, pressure-solution cleavage data were collected in marly limestone units. Here pressure-solution cleavages strike parallel to the fold axis (Figure 3). Average cleavage dip, after bedding-dip removal, passes from about 80° in the backlimb to about 100° in the forelimb. In the foreland sectors, its value is about 90°. However, both in the forelimb and in the foreland sectors close to the leading syncline, a second cleavage set oblique to bedding occurs, with an opposite dip in the two sectors. The cleavage frequency, computed only for the cleavage set at high angle to bedding, relates with the structural position. It is rather constant in the backlimb and in the crest and increases in the forelimb and in the foreland sector close to the leading syncline. Note that, in the transect analysis, due to the noncylindrical geometry of the anticline, the data position in the foreland does not exactly correspond to the distance from the leading syncline.

Mt. Catria Anticline

The Mt. Catria anticline is located in the external sector of the northern Apennines (Italy) and formed in the Messinian. The exposed stratigraphic succession consists of a Lower Jurassic, poorly layered platform limestone; a Lower Jurassic–Miocene pelagic sequence, including mainly well-bedded limestones, marly limestones,

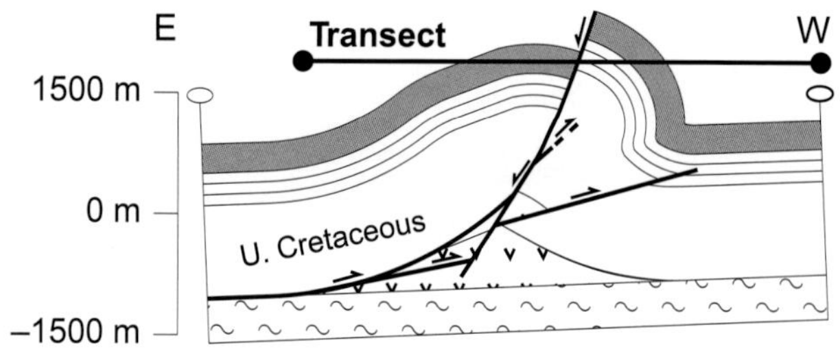

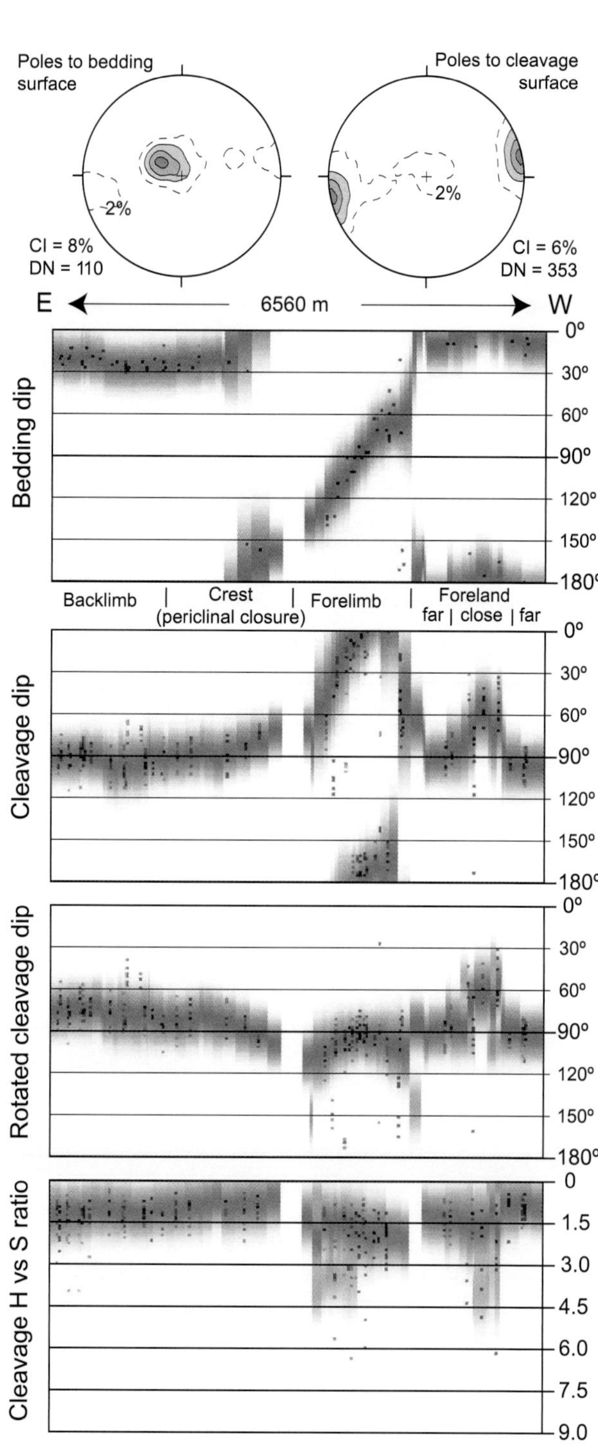

Figure 3. Pressure-solution cleavage distribution in the Añisclo anticline. Balanced cross section with transect trace (modified from Tavani et al., 2006b). Cumulative contouring of both poles to bedding and poles to pressure-solution cleavage and transect analysis showing the variability of bedding dip, cleavage dip, rotated cleavage dip, and cleavage frequency. In the stereoplots, CI is the contouring interval and DN is the data number. In the transects, H is the bed thickness and S is the cleavage spacing. U = Upper.

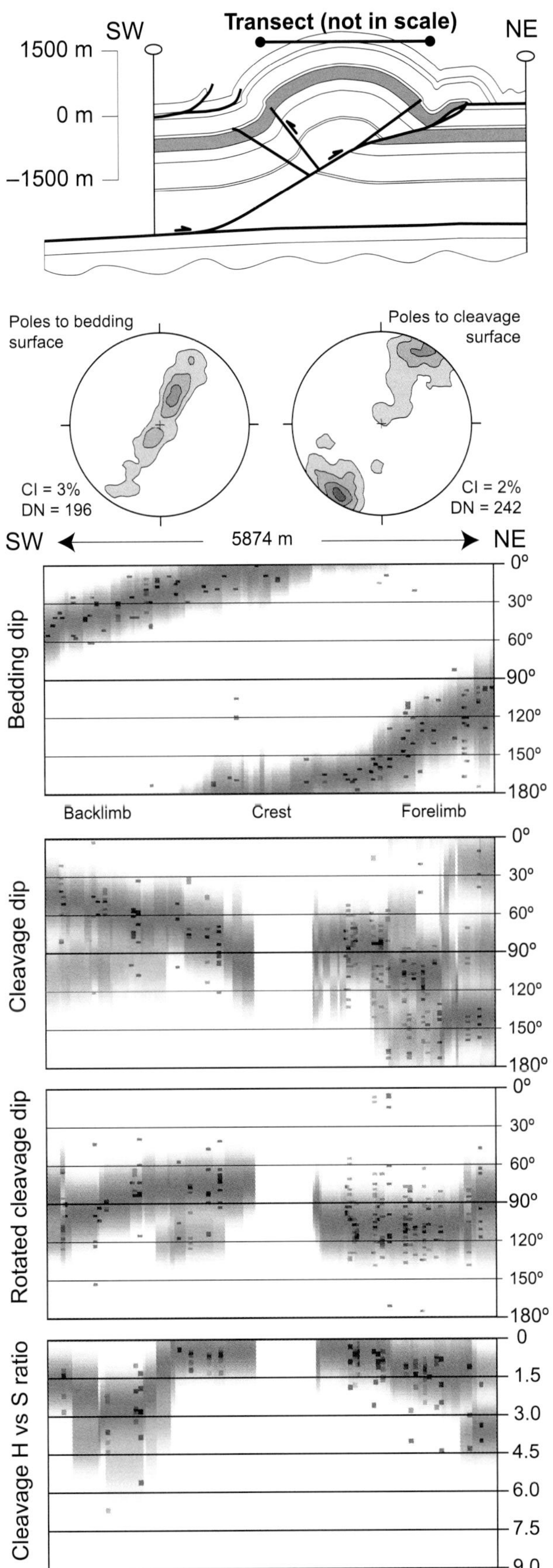

and marls; and a Miocene to Pliocene siliciclastic sequence consisting mainly of sandstones and marls. In the study area, the Mt. Catria anticline trends northwest–southeast and exhibits a concentric geometry. Bedding dip across the fold strike progressively passes from 60° toward the southwest in the southwestern sectors and up to 90° in the northeastern sector (Figure 4). We present cleavage data collected in the Maiolica formation (Upper Jurassic–Lower Cretaceous), which consists of well-bedded limestones with cherty layers and nodules. Pressure-solution cleavages strike parallel to the fold axis. Cleavage dip varies along the transport direction following the progressive variation of the bedding dip. However, rotated cleavage dip is rather scattered around an average value of 90°, being slightly hinterland dipping in the backlimb and foreland dipping in the forelimb. Cleavage frequency reduces from the fold limb toward the crestal sector.

Insights for Double-edge Fault-propagation Folding

In all three anticlines, pressure-solution cleavages strike parallel to the fold axis and are at high angle to bedding. The angle between cleavage and bedding surfaces varies with the structural position. This, coupled with the observation that cleavage frequency relates with the structural position, supports a partial synfolding origin. Rotated cleavage dip variation is quantitatively limited, and deviations from the vertical mostly do not exceed 10–15°. For these angular values, a bed thinning always lower than 2% is expected for shortening values (perpendicular to the cleavage surface) lower than 20%. A bed thinning of 4% is expected for a shortening of about 30%. These observations support the use of the layer-parallel shortening assumption for pressure-solution cleavage development.

OUTLINE OF DOUBLE-EDGE FAULT-PROPAGATION FOLDING

The initial configuration of double-edge fault-propagation folding consists of a basal decollement accommodating

Figure 4. Pressure-solution cleavage distribution in the Mt. Catria anticline. Balanced cross section with transect trace (modified from Tavani et al., 2008). Cumulative contouring of both poles to bedding and poles to pressure-solution cleavage and transect analysis showing the variability of bedding dip, cleavage dip, rotated cleavage dip, and cleavage frequency. Because of the slightly plunging geometry of the anticline, the transect length in the cross section is not in scale. In the stereoplots, CI is the contouring interval and DN is the data number. In the transects, H is the bed thickness and S is the cleavage spacing.

layer-parallel contraction, and a thrust ramp nucleating somewhere within the multilayer (Figure 5a) and propagating by the outward migration of both the upper (Ut) and lower (Lt) fault tip (Tavani et al., 2006a). A hanging-wall translation above a straight immature ramp (i.e., nonconnected to a layer-parallel decollement) symmetrically propagating outward at a constant rate and with a low upper-ramp tip slip vs. propagation ratio (S/P) produces an anticlinal fold geometry consisting of five panels named hinterland (HL), downward-propagation panel (DpP), crestal panel (CP), forelimb panel (FP), and foreland (FL), respectively (Figure 5b). Two straight hinges (γ_1 and γ_2) are pinned at the lower and upper ramp tips and divide the HL from the DpP panels and the FP from the FL panels, respectively. The outward propagation of the ramp tips causes the migration of the corresponding hinges γ_1 and γ_2 (active hinges; Suppe et al., 1992) and, consequently, the progressive widening of the DpP and FP panels by migration of material from the hinterland and the foreland, respectively (Figure 5b). Contemporarily, the CP panel is passively translated along the ramp, thus preserving its width during fold evolution.

Material removal is imposed in the sectors of the multilayer stratigraphically underlying and overlying Lt_0 and Ut_0, respectively, to compensate for excess forelandward and hinterlandward shear (Figure 5b). The amount of removed material increases during fold evolution and depends on the residence time above or below the ramp. Layers crosscut by the advancing fault tip terminate any further incremental variation of line length. When the lower tip (Lt) joins the basal decollement, a new panel (BP) develops in the backlimb and material removal terminates in the lower multilayer sector. However, if the upper tip, Ut, joins a decollement layer, the FP panel is translated above the upper decollement and, consequently, material removal terminates in the upper multilayer sector. Eventually, both the ramp tips may join a decollement layer forming a staircase fault trajectory along which the entire anticline is translated onto the foreland by fault-bend folding (Figure 5c).

A characteristic feature of double-edge fault-propagation folding is the dependence of the cutoff angle in the DpP and FP panels on the ramp S/P other than the dip. High S/P in the upper ramp tip cause the development of a footwall syncline in the model (Figure 5d). In this case, a footwall syncline panel (SP) adds to the standard configuration. Progressive fold growth imposes material removal in the layers stratigraphically overlaying point C_3 and material addition in the underlying layers. The upward migration of the upper ramp tip and, consequently, of point C_3 imposes the superimposition of constant thickness stretching in those

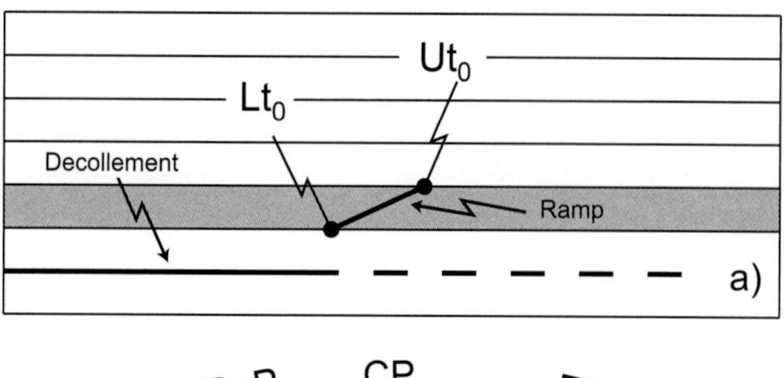

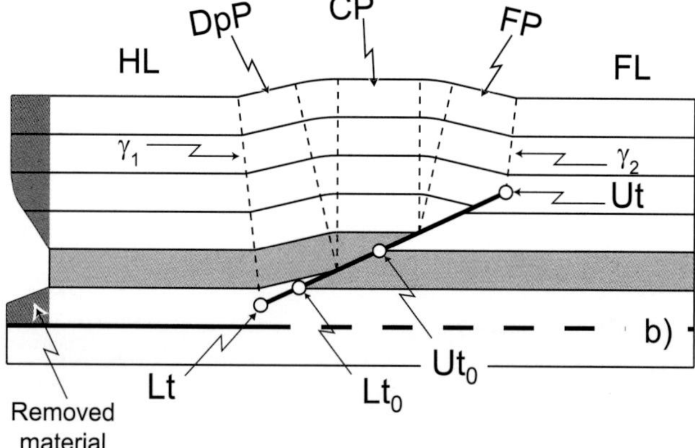

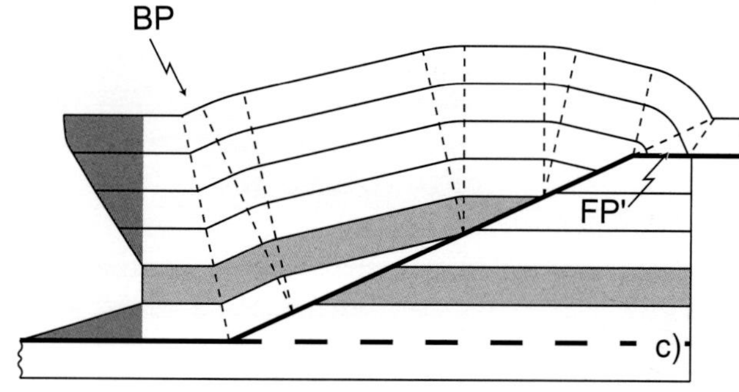

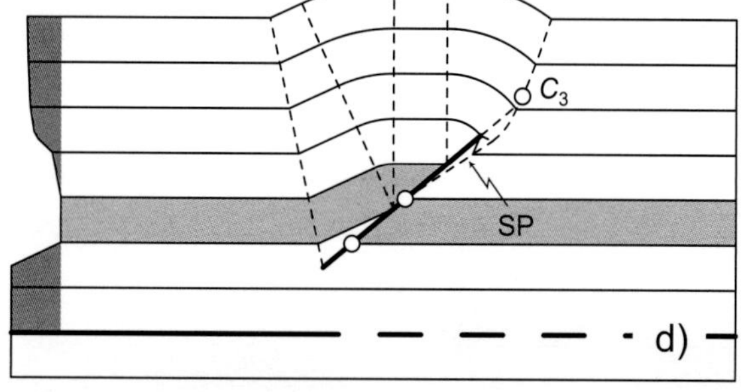

Figure 5. Kinematic evolution of double-edge fault-propagation folding. (a) Prefolding configuration; (b) fold geometry when the ramp tips propagate outward; (c) transported double-edge fault-propagation folding; (d) internal architecture of double-edge fault-propagation folding when a footwall syncline develops. The light gray area indicates the layers faulted before the hanging-wall motion. The patterned area indicates removed material. Ut = upper fault tip; Lt = lower fault tip; HL = hinterland panel; DpP = downward-propagation panel; CP = crestal panel; FP = forelimb panel; FL = foreland panel; BP = backlimb panel; SP = footwall syncline panel.

layers previously affected by contraction. A comprehensive quantitative description of double-edge fault-propagation folding is provided by Tavani et al. (2006a).

MATERIAL REMOVAL BY LAYER-PARALLEL SHORTENING

The geometrical model described above allows us to quantify the amount of material that can be removed from the section by pressure solution to compensate excess shear when second-order faulting and folding do not occur. To constrain where this material removal occurs, we again use field data. In the Chaudrons anticline, cleavage frequency reduces from the steeper forelimb sector toward the crest. In the Añisclo anticline, cleavage frequency is high in the close foreland and in the forelimb and reduces away from these sectors. In the Mt. Catria anticline, we observe a frequency reduction from the fold limbs toward the crestal sector. In all cases, a cleavage frequency reduction away from the leading syncline is observed. This reduction is fast at first but progressively reduces until cleavage frequency attains a regional trend. A quantitative correspondence between cleavage frequency and layer-parallel shortening cannot be established. However, areas with a greater cleavage frequency are expected to be more shortened than areas displaying less penetrative cleavage (e.g., Alvarez et al., 1978). The final distribution of layer-parallel shortening depends on both its incremental distribution and on the particle path during folding. We propose that, at each step, the amount of layer-parallel shortening relates to the distance from the leading syncline through the following exponential equation

$$\mathrm{LPS} = \mathrm{LPS}_0 \times e^{-\eta_1 x} \quad (1)$$

where x is the distance from the leading syncline, η_1 is the attenuation coefficient, and LPS_0 is the amount of layer-parallel shortening at the origin (i.e., the leading syncline).

Equation 1 can be derived also with the following reasoning. In a multilayer undergoing layer-parallel translation, an amount of the total displacement is expected to convert into layer-parallel shortening (Figure 6). This process can be described by the following equation

$$\frac{\Delta \mathrm{St}}{\mathrm{St}} = -\eta_2 \Delta x \quad (2)$$

where St is the displacement applied at the edge of a segment of length Δx, ΔSt is the amount of St converted into layer-parallel shortening, and η_2 is the attenuation coefficient. In the continuum space, equation 2 can be written as

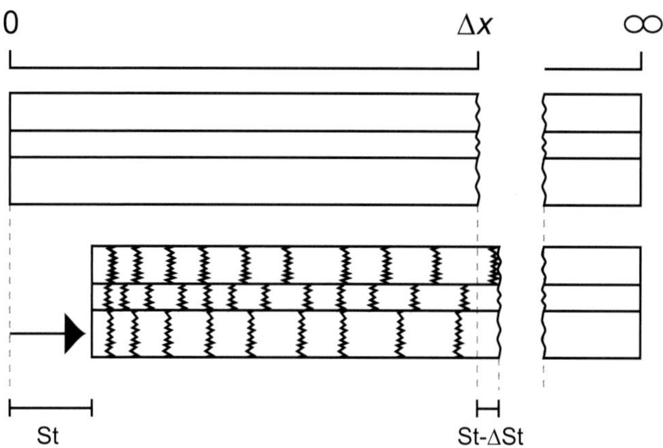

Figure 6. Multilayer converting layer-parallel translation (St) in layer-parallel shortening at constant bed thickness by pressure-solution cleavage development. See the text for details.

$$\frac{\mathrm{dSt}}{\mathrm{d}x} = -\eta_2 \mathrm{St} \quad (3)$$

and consequently

$$\mathrm{St} = S_0 e^{-\eta_2 x} \quad (4)$$

where S_0 is the initial displacement and x is the distance from the origin (i.e., the point at which S_0 is applied). The amount of layer-parallel shortening at a certain distance will be given by

$$\mathrm{LPS} = \frac{([x + \mathrm{d}x + \mathrm{St} + \mathrm{dSt}] - [x + \mathrm{St}]) - ([x + \mathrm{d}x] - x)}{([x + \mathrm{d}x] - x)} \times 100 \quad (5)$$

Simplifying and introducing equation 4, we obtain

$$\mathrm{LPS} = 100 \times \eta_2 S_0 e^{-\eta_2 X} \quad (6)$$

By comparing equations 1 and 6, we obtain that $\eta_1 = \eta_2 = \eta$, $\mathrm{LPS}_0 = 100 \times \eta S_0$.

NUMERICAL IMPLEMENTATION

Equation 6 was implemented in a numerical model of double-edge fault propagation. The numerical multilayer consisted of 64,000 points lying along 160 layers and was deformed according to the equations of double-edge fault-propagation folding, i.e., the distance between

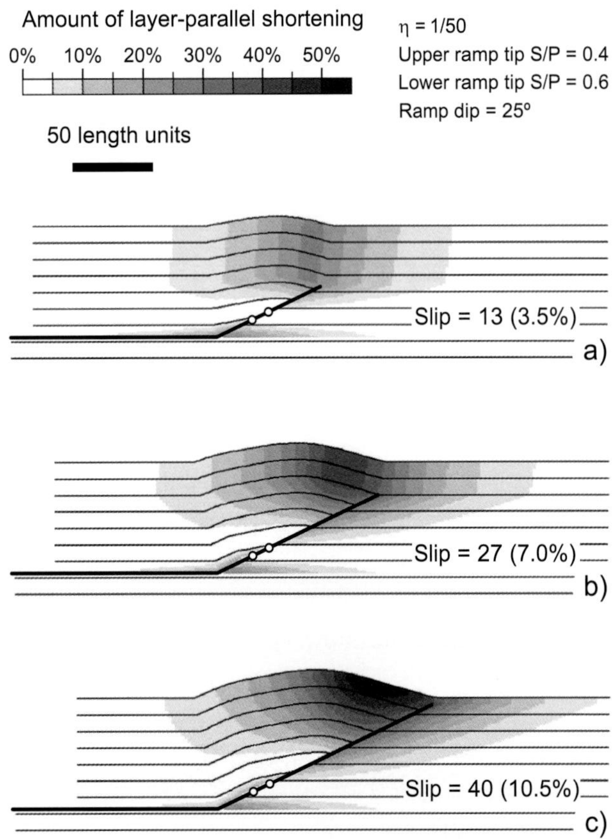

Figure 7. Progressive evolution of a double-edge fault-propagation anticline and corresponding layer-parallel shortening value distribution. White circles indicate upper and lower ramp tip nucleation sites. The gray scale indicates the amount of layer-parallel shortening. S/P = fault slip to propagation ratio.

adjacent layers was assumed constant, whereas the distance between neighboring points lying along the same layer modified during folding. In particular, for each step, the modeling allows to define the amount of material that has to be removed. For the upper ramp tip propagation, the starting point ($x = 0$) from which the layer-parallel translation is applied is assumed to be located at the intersection between a given layer and the axial surface dividing the forelimb from the foreland. For the propagation of the lower ramp tip, the $x = 0$ point of a given layer is assumed to have the X coordinate of the lower ramp tip. With this configuration, layer-parallel shortening decreases both forelandward and hinterlandward from these points.

Figure 7 shows the evolution of a double-edge fault-propagation anticline associated with a ramp nucleated in the lower half of the multilayer (circles lying along the fault represent the upper and lower tip initial position, respectively) and propagating upward and downward at constant rates. The S/P associated with the lower and upper ramp tip was 0.60 and 0.4, respectively. The value of η is 1/50 in the entire multilayer. Layer-parallel shortening affects those layers stratigraphically located below the lower ramp tip and above the upper ramp tip, respectively, and its amount is represented in gray scale. The intensity of deformation produced by the propagation of the upper ramp tip increases toward the forelimb. Moreover, in the layers stratigraphically overlying the initial position of the upper ramp tip, deformation intensity increases upward. Layer-parallel shortening associated with the propagation of the lower ramp tip increases downward and decreases away from the ramp. With increasing shortening and ramp propagation, the cross-sectional width of the area affected by layer-parallel shortening increases, as well as deformation intensity. In the upper multilayer sector, the central part of the forelimb is predicted to experience the highest amount of layer-parallel shortening (Figure 7).

This basic deformation pattern at constant S/P is controlled by (1) the attenuation coefficient (η), (2) the position of the ramp nucleation site within the multilayer, and (3) the ramp step-up angle. The influence of η on the spatial distribution of deformation is illustrated in Figure 8. The two models are characterized by the same fault-fold architecture and amount of total displacement. Increasing η causes the narrowing of the area affected by layer-parallel shortening and the increase of the maximum intensity of deformation. High values of η cause the localization of layer-parallel shortening in the forelimb where deformation intensity is

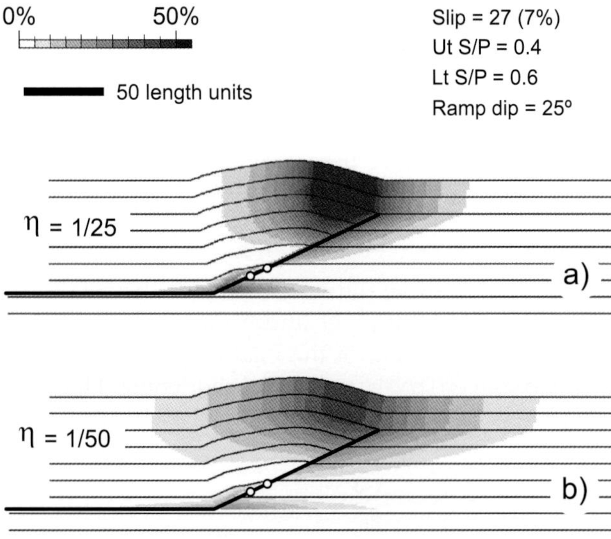

Figure 8. Layer-parallel shortening distribution produced by double-edge fault-propagation folding at different attenuation coefficients and identical fault-fold shapes. S/P = fault slip to propagation ratio; Ut = upper fault tip; Lt = lower fault tip.

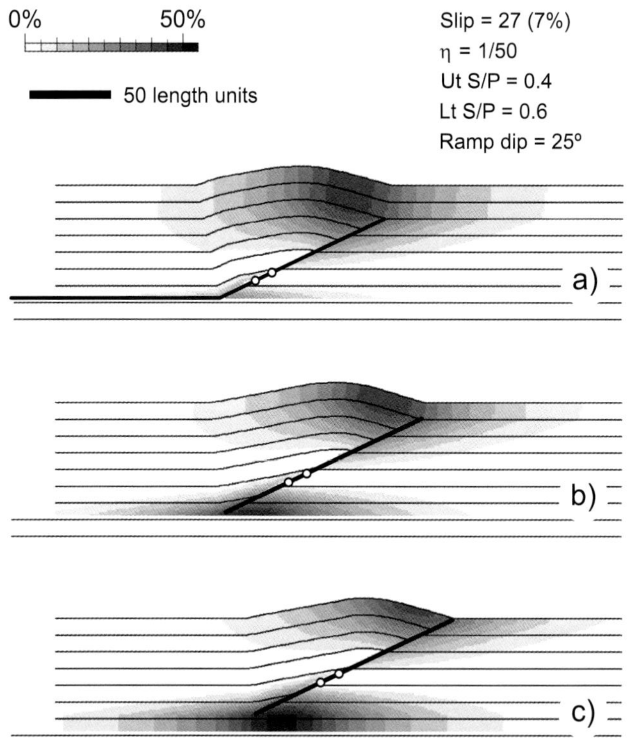

Figure 9. Layer-parallel shortening distribution produced by double-edge fault-propagation folding at different ramp nucleation sites. S/P = fault slip to propagation ratio; Ut = upper fault tip; Lt = lower fault tip.

highest (Figure 8a). The function of the ramp nucleation site is investigated in models illustrated in Figure 9. Shallowing the nucleation site implies that the upper multilayer sectors have a shorter resident time above the ramp tip. As a consequence, they undergo lower amounts of layer-parallel shortening compared with the corresponding layers deformed in a fault-propagation anticline that developed above a ramp nucleated in the lower part of the multilayer. The same reasoning applies to layers underlying the lower ramp tip. It follows that anticlines that developed ahead thrust ramps nucleated in the lower sector of the multilayer are predicted to have experienced intense layer-parallel shortening in the upper part of the multilayer. Layer-parallel shortening is expected to be negligible in the fold core (Figure 9a). Thrust ramps nucleated at higher stratigraphic elevations within the multilayer produce anticlines characterized by less intense layer-parallel shortening in the outer layers. A broad and heavily deformed region develops in the lower part of the multilayer and affects both the hanging wall and the footwall (Figure 9c).

The influence of the thrust ramp step-up angle on the pattern of layer-parallel shortening associated with double-edge fault-propagation folding is investigated in Figure 10. The three models show that (1) the highest amounts of layer-parallel shortening systematically occur in the forelimb and have comparable intensities and (2) decreasing the ramp step-up angle causes broadening of the deformation area.

Another factor influencing the cross-sectional pattern of layer-parallel shortening in double-edge fault-propagation anticlines is the fault S/P. For a given ramp step-up angle and nucleation site, the increasing of the S/P implies the increase of the fold limb dip, the increase of removed material, and consequently, the increase of the layer-parallel shortening value (Figure 11).

Symmetrical partitioning of the layer-parallel shortening accommodation area with respect to the trailing syncline is a particular case of a wide spectrum of possible situations where these areas are different in the opposite sides of the leading syncline. Models in Figure 12

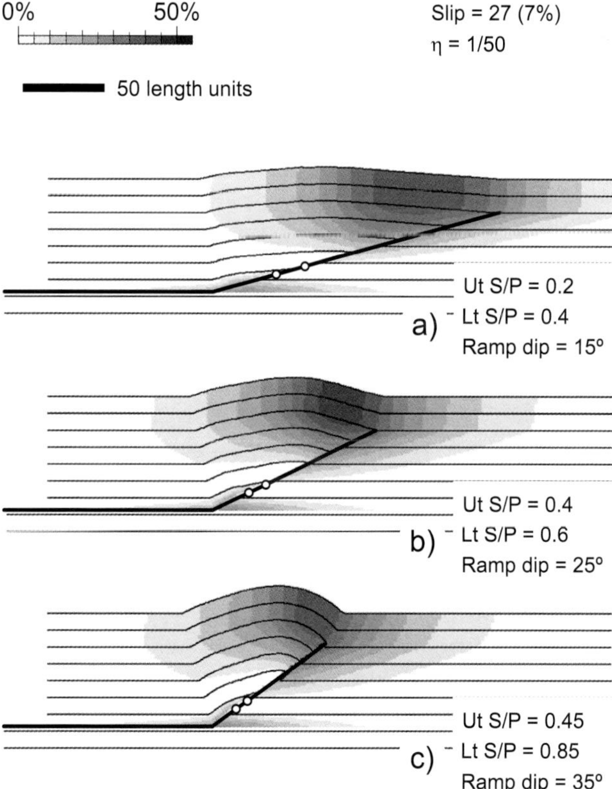

Figure 10. Layer-parallel shortening distribution produced by double-edge fault-propagation folding at different ramp step-up angles. The initial stratigraphic elevation of the upper and lower ramp tip is constant. Notice that, in the three examples, the propagation rates of the upper and lower ramp tips are different but the vertical component is constant for both, i.e., the final stratigraphic elevation of the upper ramp tip is the same. S/P = fault slip to propagation ratio; Ut = upper fault tip; Lt = lower fault tip.

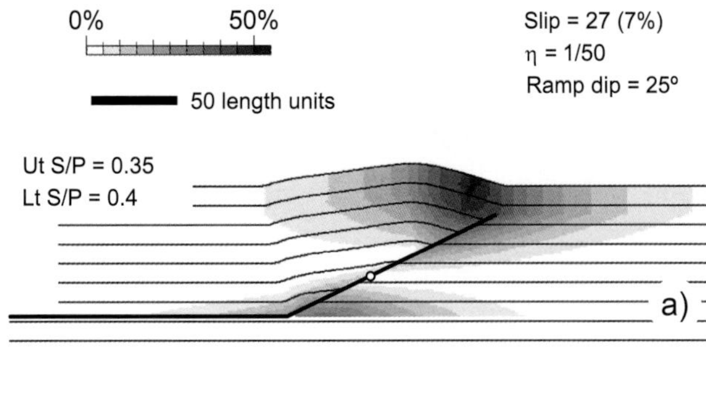

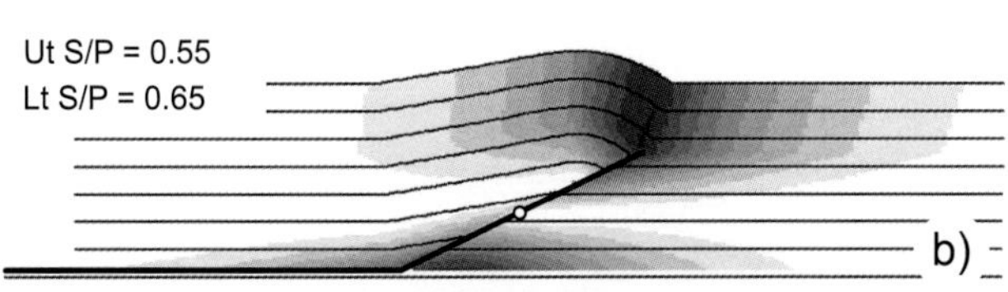

Figure 11. Layer-parallel shortening distribution produced by double-edge fault-propagation folding at different ramp S/P. The ramp nucleated as a single point. S/P = fault slip to propagation ratio; Ut = upper fault tip; Lt = lower fault tip.

show the effect of such different partitioning. Dominant hinterlandward accommodation produces a near-symmetrical distribution of layer-parallel shortening intensity with respect to the anticlinal crest (Figure 12a). Decreasing this partitioning causes the migration of the most intensely shortened area toward the forelimb (Figure 12b, c). Eventually, the dominant forelandward accommodation of layer-parallel shortening results in the near-symmetrical distribution of the most intensely shortened area about the leading syncline.

In summary, in our two-dimensional simulations, layer-parallel shortening varies across strike and within the multilayer. Its amount increases with increasing S/P associated to both the upper and lower ramp tips, i.e., slow ramp propagation rates imply high deformation intensities. Layer-parallel shortening values in the upper multilayer sector increase with increasing the distance from the upper ramp tip nucleation point. Similarly, in the lower multilayer sector, layer-parallel shortening increases with increasing the distance from the lower ramp tip nucleation point. The cross-sectional area affected by layer-parallel shortening increases with decreasing both the attenuation coefficient and the ramp dip. Accommodation of layer-parallel shortening with respect to the axial surface dividing the forelimb from the foreland can be either symmetrical or asymmetrical, influencing the location of the maximum amount of layer-parallel shortening. Despite this, the maximum layer-parallel shortening values are always located within the forelimb and the surrounding areas.

DISCUSSION

The amounts of layer-parallel shortening associated with double-edge fault-propagation folding are consistent with values observed in many thrust and fold belts where they commonly range between 5 and 30% (e.g., Engelder and Engelder, 1977; Fisher and Coward, 1982; Marshak and Engelder, 1985; Ferrill and Dunne, 1989;

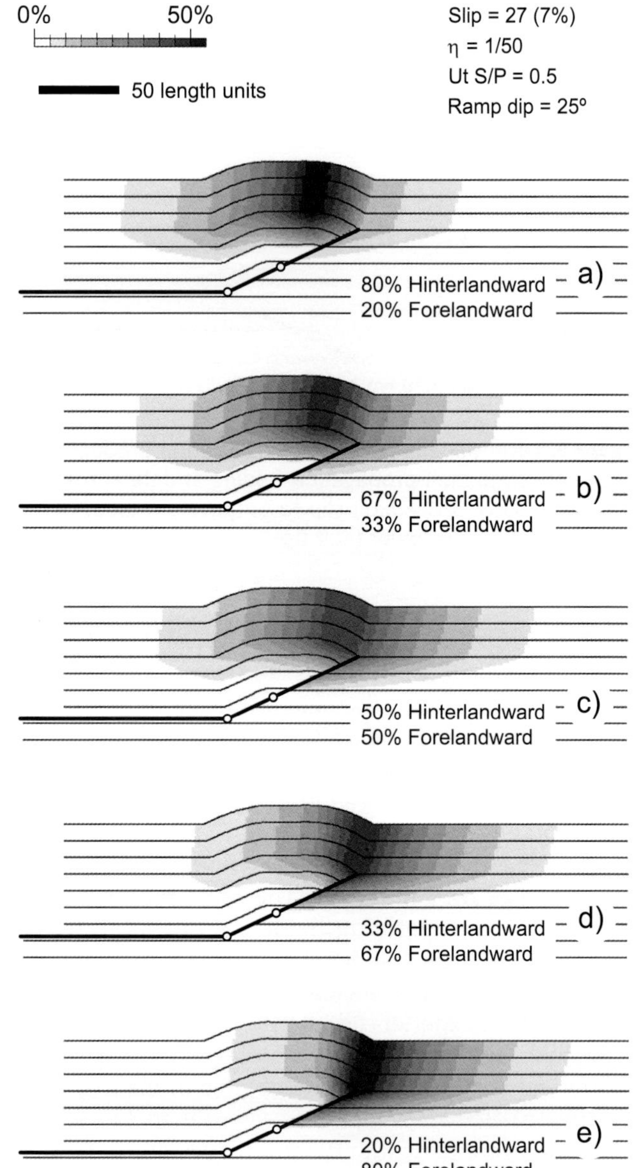

Figure 12. Layer-parallel shortening distribution associated with different partitioning of the layer-parallel shortening accommodation area with respect to the trailing syncline. In all the models, the ramp lower tip initial position is located at the intersection between the ramp and the lower flat. S/P = fault slip to propagation ratio; Ut = upper fault tip.

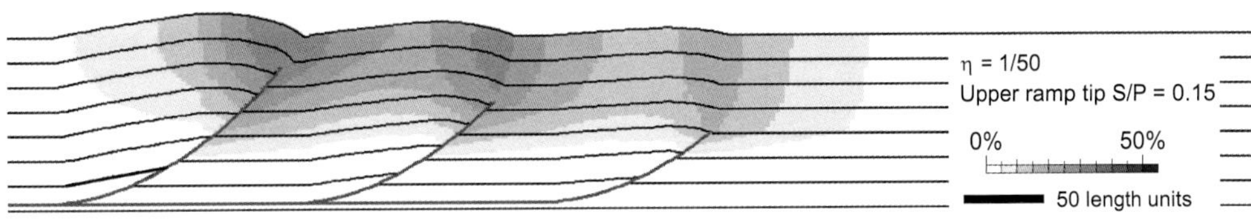

Figure 13. Layer-parallel shortening distribution produced by three double-edge fault-propagation anticlines developed in a piggyback sequence above a natural curved fault shape. S/P = fault slip to propagation ratio.

Protzman and Mitra, 1990; Onasch, 1993; Holl and Anastasio, 1995; Harris and Van Der Pluijm, 1998; Whitaker and Bartholomew, 1999; Sans et al., 2003; among others). Values higher than those observed in thrust and fold belts (i.e., greater than 30%) are obtained for low values of the attenuation coefficient. This parameter is expected to depend on several factors that influence the capability of the rock multilayer to attenuate layer-parallel shortening. These factors include the rock type and initial porosity, fluid flow, strain rate, and temperature, which control the effectiveness of rock dissolution (e.g., Rutter, 1983; Marshak and Engelder, 1985; Holl and Anastasio, 1995). Accordingly, double-edge fault-propagation folding can provide an effective folding mechanism when the deforming multilayer mechanical properties and the environmental conditions of deformation imply low η values. In poorly lithified siliciclastic rocks undergoing layer-parallel shortening, tectonic layer-parallel compaction is expected. In carbonate multilayers, layer-parallel shortening is ensured by layer-perpendicular pressure-solution cleavage development. In the latter case, cleavage development is facilitated in an open-water circulation system where the possibility for fast calcite removal enhances the pressure-solution process (Engelder and Marshak, 1985). Both the above-mentioned conditions imply rather shallow depths within thrust and fold belts. At greater depths, or in close hydraulic systems, a lower amount of layer-parallel shortening is expected. In these conditions, double-edge fault-propagation folding can operate only in the initial stages of fold amplification, then likely triggering fault breakthrough and foreland translation of the anticline.

Modeling results indicate that the deformation associated with the propagation of the upper ramp tip has a spatial distribution closely resembling that associated with trishear fault-propagation folding (e.g., Erslev and Mayborn, 1997; Cardozo et al., 2005). As in trishear fault-propagation folding, in double-edge fault-propagation folding, the S/P and the ramp dip influence the final width of the deformation area. These analogies suggest that the two folding mechanisms can be regarded as end-member models of fault-propagation folding. Double edge is expected to operate in a well-layered material undergoing layer-parallel shortening at constant bed thickness; trishear operates when the folded multilayer consists of poorly layered material and/or when the multilayer does not allow layer-parallel shortening at constant bed thickness. The two models can also operate together in multilayers characterized by variable mechanical properties.

A cautionary note concerning the deformation pattern in double-edge fault-propagation anticlines has to be introduced. Models illustrated in this work simulate isolated anticlines, whereas natural thrust wedges consist of multiple anticlines located at a given distance. The growth of an anticline by double-edge fault-propagation folding leads to the deformation of its foreland that, in a piggyback thrust propagation sequence, will be redeformed by the development of the next anticline, and consequently, the rock material will be reshortened (Figure 13). This implies that cumulative deformation by synfolding layer-parallel shortening of a given anticline is influenced by the growth of the anticline itself and by the growth of the neighboring ones. The pressure-solution cleavage distribution in the backlimb of both the Mt. Catria and the Añisclo anticlines could be related to the development of the associated inner structures.

IMPLICATION FOR HYDROCARBON RESEARCH AND DEVELOPMENT

The development of tectonic pressure-solution cleavage is a primary factor in modifying the pretectonic rock permeability and porosity, particularly in carbonate multilayers. Pressure-solution cleavage development can imply (1) porosity reduction by tectonic compaction induced by grain-to-grain solution (e.g., Beutner, 1978; Engelder and Marshak, 1985; Holl and Anastasio, 1995); (2) porosity reduction induced by mass transfer from the solution surfaces to both the pores and the extensional features in the neighboring areas of the solution surfaces (e.g., Elliott, 1973; Fletcher and Pollard, 1981; Mitra et al., 1984); and (3) permeability increase by increasing rock connectivity (e.g., Alsharhan and Whittle, 1994; Davidson et al., 1998; Van Geet et al., 2002).

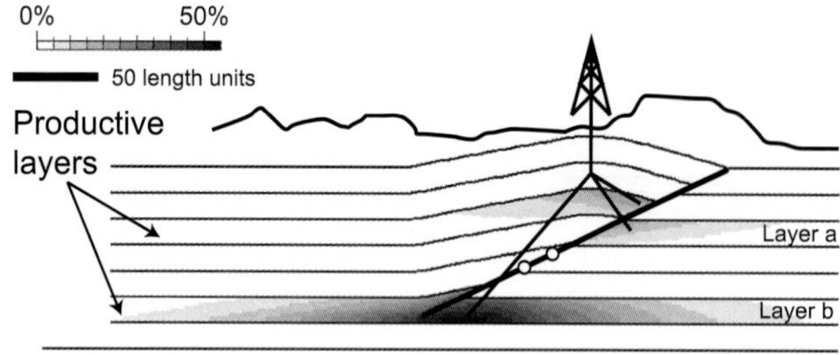

Figure 14. Example of the influence of the predicted distribution of layer-parallel shortening along two key layers on well plan optimization. The most intensely cleaved areas are in the forelimb (layer a) and in the footwall of the main ramp (layer a, b).

In the case that pressure-solution cleavage networks are under appropriate stress conditions favoring their dilation and their behavior as additional pathways for fluid migration and accumulation, the basic models presented here provide useful templates for simulating pressure-solution cleavage distribution at depth. In particular, our results indicate that reservoir layers are expected to increase their connectivity in the forelimb and in the footwall sectors close to the thrust ramp (Figure 14). An opposite path is expected in siliciclastic reservoirs, where layer-parallel shortening induces layer-parallel compaction, and consequently, these sectors are expected to be characterized by porosity reduction. Deformation localization in the forelimb of fault-propagation anticlines is commonly expected in fault-propagation anticlines (e.g., Erslev and Mayborn, 1997; Salvini and Storti, 2001). Conversely, the occurrence of a potential reservoir layer in the footwall sectors close to the thrust ramp represents a possible new play.

CONCLUSIONS

The presented models simulated the distribution and timing of layer-parallel shortening expected to develop during the growth of double-edge fault-propagation anticlines. Modeling results can be summarized as follows.

1) Layer-parallel shortening is virtually zero in the layer interval between the initial position of the upper and lower ramp tip, respectively, and increases both upward and downward with increasing the distance from the upper and lower ramp tip initial position.
2) At a given point, the layer-parallel shortening intensity increases with proceeding folding until the corresponding stratigraphic level is cut by the fault.
3) The amount of layer-parallel shortening increases with increasing the ramp S/P.
4) The width of the cross-sectional area affected by layer-parallel shortening increases with decreasing both the attenuation coefficient and the ramp dip.
5) Forelimb sectors are expected to exhibit the higher values of layer-parallel shortening.

In double-edge fault-propagation folds, development of pressure-solution cleavage networks in carbonate multilayers can provide additional porosity and permeability when the stress conditions at the time of hydrocarbon migration are appropriate. In particular, the forelimb panel and the footwall sectors close to the ramp can increase their pretectonic permeability, thus providing suitable areas for hydrocarbon exploration. On the contrary, in siliciclastic multilayers, these sectors are expected to be the less suitable areas for hydrocarbon exploration because of preferential tectonic compaction.

ACKNOWLEDGMENTS

This work was conducted with the financial support of the Italian Ministero dell'Università e della Ricerca (MIUR), the MODES-4D (CGL2007-66431-C02-02/BTE) project, and the Grup de Recerca de Geodinàmica i Anàlisi de Conques (2001SRG-000074). We gratefully acknowledge constructive reviews from A. Amibilia Cabeza, S. Hardy, and an anonymous reviewer, which helped us to improve an early version of the manuscript. We thank F. Salvini for useful suggestions.

REFERENCES CITED

Alsharhan, A. S., and G. L. Whittle, 1994, Sedimentary diagenetic interpretation and reservoir characteristics of the Middle Jurassic (Araej Formation) in the southern Arabian Gulf: Marine and Petroleum Geology, v. 12, p. 615–628, doi:10.1016/0264-8172(95)98088-M.

Alvarez, W., T. Engelder, and P. Geiser, 1978, Classification of solution cleavage in pelagic limestones: Geology, v. 6, p. 263–266, doi:10.1130/0091-7613(1978)6<263:COSCIP>2.0.CO;2.

Aydin, A., 2000, Fractures, faults, and hydrocarbon entrapment, migration, and flow: Marine and Petroleum Geology, v. 17, p. 797–814, doi:10.1016/S0264-8172(00)00020-9.

Beutner, E. C., 1978, Slaty cleavage and related strain in Martinsburg slate, Delaware Water Gap, New Jersey: American Journal of Science, v. 278, p. 1–23.

Cardozo, N., R. W. Allmendinger, and J. K. Morgan, 2005, Influence of mechanical stratigraphy and initial stress state on the formation of two fault propagation folds: Journal of Structural Geology, v. 27, p. 1954–1972, doi:10.1016/j.jsg.2005.06.003.

Chester, J. S., 2003, Mechanical stratigraphy and fault-fold interaction, Absaroka thrust sheet, Salt River Range, Wyoming: Journal of Structural Geology, v. 25, p. 1171–1192, doi:10.1016/S0191-8141(02)00151-7.

Chester, J. S., and F. M. Chester, 1990, Fault-propagation folds above thrusts with constant dip: Journal of Structural Geology, v. 12, p. 903–910, doi:10.1016/0191-8141(90)90063-5.

Chester, J. S., J. M. Logan, and J. H. Spang, 1991, Influence of layering and boundary conditions on fault-bend and fault-propagation folding: AAPG Bulletin, v. 103, p. 1059–1072.

Cooper, M., 1992, The analysis of fracture systems in subsurface thrust structures from the foothills of the Canadian Rockies, *in* K. R. McClay, ed., Thrust tectonics: London, Chapman & Hall, p. 105–121.

Corbett, K., M. Friedman, and J. Spang, 1987, Fracture development and mechanical stratigraphy of Austin chalk, Texas: AAPG Bulletin, v. 71, p. 17–28.

Couzens, B. A., and D. V. Wiltschko, 1996, The control of mechanical stratigraphy on the formation of triangle zones: Bulletin of Canadian Petroleum Geology, v. 44, p. 165–179.

Dahlstrom, C. D. A., 1990, Geometric constraints derived from the law of conservation of volume and applied to evolutionary models for detachment folding: AAPG Bulletin, v. 74, p. 336–344.

Davidson, S. G., D. J. Anastasio, G. E. Bebout, J. E. Holl, and C. A. Hedlund, 1998, Volume loss and metasomatism during cleavage formation in carbonate rocks: Journal of Structural Geology, v. 20, p. 707–726, doi:10.1016/S0191-8141(98)00008-X.

De Sitter, L. V., 1956, Structural geology: New York, McGraw-Hill, 552 p.

Dunne, W. M., and A. J. Caldanaro, 1997, Evolution of solution structures in deformed quartz arenite: Geometric change related to permeability changes: Journal of Structural Geology, v. 19, p. 663–672, doi:10.1016/S0191-8141(96)00105-8.

Elliott, D., 1973, Diffusion flow laws in metamorphic rock: Geological Society of America Bulletin, v. 84, p. 2645–2664, doi:10.1130/0016-7606(1973)84<2645:DFLIMR>2.0.CO;2.

Engelder, T., and P. Engelder, 1977, Fossil distortion and decollement tectonics of the Appalachian Plateau: Geology, v. 5, p. 457–460, doi:10.1130/0091-7613(1977)5<457:FDADTO>2.0.CO;2.

Engelder, T., and S. Marshak, 1985, Disjunctive cleavage formed at shallow depths in sedimentary rocks: Journal of Structural Geology, v. 7, p. 327–342, doi:10.1016/0191-8141(85)90039-2.

Epard, J.-L., and R. H. Groshong Jr., 1995, Kinematic model of detachment folding including limb rotation, fixed hinges and layer-parallel strain: Tectonophysics, v. 247, p. 85–103, doi:10.1016/0040-1951(94)00266-C.

Erslev, E. A., 1991, Trishear fault-propagation folding: Geology, v. 19, p. 617–620, doi:10.1130/0091-7613(1991)019<0617:TFPF>2.3.CO;2.

Erslev, E. A., and K. R. Mayborn, 1997, Multiple geometries and modes of fault-propagation folding in the Canadian thrust belt: Journal of Structural Geology, v. 19, p. 321–335.

Ferrill, D. A., and W. M. Dunne, 1989, Cover deformation above a blind duplex: An example from West Virginia, U.S.A.: Journal of Structural Geology, v. 11, p. 421–431, doi:10.1016/0191-8141(89)90019-9.

Fischer, M. P., and P. B. Jackson, 1999, Stratigraphic controls on deformation patterns in fault-related folds: A detachment fold example from the Sierra Madre Oriental, northeast Mexico: Journal of Structural Geology, v. 21, p. 613–633, doi:10.1016/S0191-8141(99)00044-9.

Fischer, M. P., N. B. Woodward, and M. M. Mitchell, 1992, The kinematics of break-thrust folds: Journal of Structural Geology, v. 14, p. 451–460, doi:10.1016/0191-8141(92)90105-6.

Fisher, M. W., and M. P. Coward, 1982, Strain and folds within thrusts sheets: An analysis of the Heilam sheet, northwest Scotland: Tectonophysics, v. 88, p. 291–312, doi:10.1016/0040-1951(82)90241-4.

Fletcher, R. C., and D. D. Pollard, 1981, Anticrack model for pressure solution surfaces: Geology, v. 9, p. 419–424, doi:10.1130/0091-7613(1981)9<419:AMFPSS>2.0.CO;2.

Florez-Nino, J.-M., A. Aydin, G. Mavko, M. Antonellini, and A. Ayaviri, 2005, Fault and fracture systems in a fold and thrust belt: An example from Bolivia: AAPG Bulletin, v. 89, p. 471–493, doi:10.1306/11120404032.

Gholipour, A. M., 1998, Patterns and structural position of productive fractures in the Asmari reservoirs, Southwest Iran: Journal of Canadian Petroleum Technology, v. 37, p. 44–50.

Gross, M. R., 1995, Fracture partitioning: Failure mode as a function of lithology in the Monterey Formation of coastal California: Geological Society of America Bulletin, v. 107, p. 779–792, doi:10.1130/0016-7606(1995)107<0779:FPFMAA>2.3.CO;2.

Harris, J. H., and B. A. Van Der Pluijm, 1998, Relative timing of calcite twinning strain and fold-thrust belt development; Hudson Valley fold-thrust belt, New York, U.S.A.: Journal of Structural Geology, v. 20, p. 21–31, doi:10.1016/S0191-8141(97)00093-X.

Holl, J. E., and D. J. Anastasio, 1995, Cleavage development within a foreland fold and thrust belt, southern Pyrenees, Spain: Journal of Structural Geology, v. 17, p. 357–369, doi:10.1016/0191-8141(94)00062-5.

Jamison, W. R., 1987, Geometric analysis of fold development in overthrust terranes: Journal of Structural Geology, v. 9, p. 207–219, doi:10.1016/0191-8141(87)90026-5.

Jamison, W. R., 1992, Stress controls of fold thrust style, *in* K. R. McClay, ed., Thrust tectonics: London, Chapman & Hall, p. 155–164.

Lemiszki, P. J., J. D. Landes, and R. D. Hatcher, 1994, Controls on hinge-parallel extension fracturing in single-layer tangential-longitudinal strain folds: Journal of Geophysical Research, v. 99, p. 22,027–22,042.

Leythaeuser, D., O. Borromeo, F. Mosca, R. Di Primo, M. Randke, and R. G. Shaefer, 1995, Pressure solution in carbonate source rocks and its control on petroleum generation and migration: Marine and Petroleum Geology, v. 12, p. 717–733, doi:10.1016/0264-8172(95)93597-W.

Marshak, S., and T. Engelder, 1985, Development of cleavage in limestones of a fold-thrust belt in eastern New York: Journal of Structural Geology, v. 7, p. 345–359, doi:10.1016/0191-8141(85)90040-9.

Mitra, S., 1988, Effects of deformation mechanisms on reservoir potential in central Appalachian overthrust belt: AAPG Bulletin, v. 72, p. 536–554.

Mitra, S., 1990, Fault-propagation folds: Geometry, kinematic evolution, and hydrocarbon traps: AAPG Bulletin, v. 74, p. 921–945.

Mitra, G., and W. A. Yonkee, 1985, Relationship of spaced cleavage to fold and thrust in the Idaho-Utah-Wyoming thrust belt: Journal of Structural Geology, v. 7, p. 361–373, doi:10.1016/0191-8141(85)90041-0.

Mitra, G., W. A. Yonkee, and D. J. Gentry, 1984, Solution cleavages and its relationship to major structures in the Idaho-Utah-Wyoming thrust belt: Geology, v. 12, p. 354–358, doi:10.1130/0091-7613(1984)12<354:SCAIRT>2.0.CO;2.

Nelson, R. A., 1985, Geological analysis of naturally fractured reservoir: Houston, Texas, Gulf Publishing, 320 p.

Onasch, C. M., 1993, Determination of pressure solution shortening in sandstones: Tectonophysics, v. 227, p. 145–159, doi:10.1016/0040-1951(93)90092-X.

Poblet, J., and K. McClay, 1996, Geometry and kinematics of single layer detachment folds: AAPG Bulletin, v. 80, p. 1085–1109.

Protzman, G. M., and G. Mitra, 1990, Strain fabric associated with the Meade thrust sheet: Implications for cross-section balancing: Journal of Structural Geology, v. 12, p. 403–417, doi:10.1016/0191-8141(90)90030-3.

Ramsay, J. G., 1974, Development of chevron folds: Geological Society of America Bulletin, v. 85, p. 1741–1754.

Rutter, E. H., 1983, Pressure solution in nature, theory, and experiment: Journal of the Geological Society, v. 140, p. 725–740, doi:10.1144/gsjgs.140.5.0725.

Salvini, F., and F. Storti, 2001, The distribution of deformation in parallel fault-related folds with migrating axial surfaces: Comparison between fault-propagation and fault-bend folding: Journal of Structural Geology, v. 23, p. 25–32, doi:10.1016/S0191-8141(00)00081-X.

Salvini, F., A. Billi, and D. U. Wise, 1999, Strike-slip fault-propagation cleavage in carbonate rocks: The Mattinata fault zone, southern Apennines, Italy: Journal of Structural Geology, v. 21, p. 1731–1749, doi:10.1016/S0191-8141(99)00120-0.

Sanderson, D. J., 1982, Models of strain variation in nappes and thrust sheets: A review: Tectonophysics, v. 88, p. 201–233, doi:10.1016/0040-1951(82)90237-2.

Sans, M., J. Vergés, E. Gomis, J. M. Parés, M. Schiattarella, A. Travé, F. Calvet, P. Santanach, and A. Doulcet, 2003, Layer parallel shortening in salt-detached folds: Constraint on cross-section restoration: Tectonophysics, v. 372, p. 85–104, doi:10.1016/S0040-1951(03)00233-6.

Srivastava, D. C., and T. Engelder, 1990, Crack-propagation sequence and pore-fluid circulations during fault-bend folding in the Appalachian Valley and Ridge, central Pennsylvania: Geological Society of America Bulletin, v. 102, p. 116–128, doi:10.1130/0016-7606(1990)102<0116:CPSAPF>2.3.CO;2.

Stewart, K. G., and W. Alvarez, 1991, Mobile-hinge kinking in layered rocks and models: Journal of Structural Geology, v. 13, p. 243–259, doi:10.1016/0191-8141(91)90126-4.

Storti, F., and F. Salvini, 1996, Progressive rollover fault-propagation folding: A possible kinematic mechanism to generate regional-scale recumbent folds in shallow foreland belts: AAPG Bulletin, v. 80, p. 174–193.

Storti, F., and F. Salvini, 2001, The evolution of a model trap in the central Apennines, Italy: Fracture patterns, fault reactivation and development of cataclastic rocks in carbonates at the Narni anticline: Journal of Petroleum Geology, v. 24, p. 171–190, doi:10.1111/j.1747-5457.2001.tb00666.x.

Suppe, J., 1983, Geometry and kinematics of fault-bend folding: American Journal of Sciences, v. 283, p. 684–721.

Suppe, J., and D. A. Medwedeff, 1990, Geometry and kinematics of fault-propagation folding: Ecoglae Geologicae Helvetiae, v. 83, p. 409–454.

Suppe, J., G. T. Chou, and S. C. Hook, 1992, Rates of folding and faulting determined from growth strata, *in* K. R. McClay, ed., Thrust tectonics: London, Chapman & Hall, p. 105–121.

Tavani, S., L. Louis, C. Souque, P. Robion, F. Salvini, and D. Frizon de Lamotte, 2004, Folding related fracture pattern and physical properties of rocks in the Chaudrons ramp-related anticline (Corbières, France), *in* R. Swennen, F. Roure, and J. Granath, eds., Deformation, fluid flow and reservoir appraisal: AAPG Hedberg Series 1, p. 257–275.

Tavani, S., F. Storti, and F. Salvini, 2006a, Double-edge fault-propagation folding: Geometry and kinematics: Journal of Structural Geology, v. 28, no. 1, p. 19–35.

Tavani, S., F. Storti, O. Fernández, J. A. Muñoz, and F. Salvini, 2006b, 3-D deformation pattern analysis and evolution of the Añisclo anticline, southern Pyrenees: Journal of Structural Geology, v. 28, p. 695–712, doi:10.1016/j.jsg.2006.01.009.

Tavani, S., F. Storti, F. Salvini, and C. Toscano, 2008, Stratigraphic versus structural control on the deformation pattern associated with the evolution of the Mt. Catria anticline, Italy: Journal of Structural Geology, v. 30, no. 5, p. 664–681.

Van Dijk, J. P., M. Bello, C. Toscano, A. Bersani, and S. Nardon, 2000, Tectonic model and three-dimensional fracture network analysis of Monte Alpi (southern Apennines): Tectonophysics, v. 324, p. 203–237, doi:10.1016/S0040-1951(00)00138-4.

Van Geet, M., R. Swennen, C. Durmishi, F. Roure, and P. H. Muchez, 2002, Paragenesis of Cretaceous to Eocene carbonate reservoirs in the Ionian fold and thrust belt (Albania): Relation between tectonism and fluid flow: Sedimentology, v. 49, p. 697–718, doi:10.1046/j.1365-3091.2002.00476.x.

Whitaker, A. E., and M. J. Bartholomew, 1999, Layer parallel shortening: A mechanism for determining deformation timing at the junction of the central and southern

Appalachians: American Journal of Science, v. 299, p. 238–254, doi:10.2475/ajs.299.3.238.

Wickham, J., 1995, Fault displacement-gradient folds and the structure at Lost Hills, California (U.S.A.): Journal of Structural Geology, v. 17, p. 1293–1302, doi:10.1016/0191-8141(95)00029-D.

Wise, D. U., and T. A. McCrory, 1982, A new method of fracture analysis: Azimuth versus traverse distance plots: Geological Society of America Bulletin, v. 93, p. 889–897, doi:10.1130/0016-7606(1982)93<889:ANMOFA>2.0.CO;2.

Woodward, N. B., 1999, Competitive macroscopic deformation processes: Journal of Structural Geology, v. 21, p. 1209–1218, doi:10.1016/S0191-8141(99)00076-0.

Woodward, N. B., and E. Rutherford Jr., 1989, Structural lithic units in external orogenic zones: Tectonophysics, v. 158, p. 247–258, doi:10.1016/0040-1951(89)90327-2.

7

Alonso, J. L., F. Colombo, and O. Riba, 2011, Folding mechanisms in a fault-propagation fold inferred from the analysis of unconformity angles: The Sant Llorenç growth structure (Pyrenees, Spain), *in* K. McClay, J. H. Shaw, and J. Suppe, eds., Thrust fault-related folding: AAPG Memoir 94, p. 137–151.

Folding Mechanisms in a Fault-propagation Fold Inferred from the Analysis of Unconformity Angles: The Sant Llorenç Growth Structure (Pyrenees, Spain)

J. L. Alonso

Departamento de Geología, Universidad de Oviedo, Oviedo, Spain

F. Colombo

Departament d'Estratigrafía, Paleontología i Geociences Marines, Universitat de Barcelona, Barcelona, Spain

O. Riba

Departament d'Estratigrafía, Paleontología i Geociences Marines, Universitat de Barcelona, Barcelona, Spain

ABSTRACT

This chapter shows, from the theoretical point of view, that growth strata geometries above angular unconformities depend on the modification of initial unconformity angles due to different folding mechanisms operating in underlying layers. This theory is applied to the Sant Llorenç de Morunys growth structure, where the analysis of the unconformity angles shows that classic folding mechanisms, such as layer-parallel simple shear in both angular and curved folds, and tangential-longitudinal strain operated in the forelimb of a fault-propagation fold. These folding mechanisms, inferred from changes in unconformity angles, are consistent with the minor structures and lithological features of the involved stratigraphic units. Forward modeling, including limb rotation, hinge migration, and the above-mentioned folding mechanisms, explains the main structural features of the syntectonic sediments in the Sant Llorenç de Morunys fault-propagation fold.

INTRODUCTION

The kinematics of fold amplification is a subject of enduring interest in structural geology. In particular, concerning fault-related folds, two basic models of fault-propagation folds, developed by kink-band migration (Suppe and Medwedeff, 1990) or involving limb rotation (Erslev, 1992; Erslev and Mayborn, 1997), have been proposed.

Sant Llorenç de Morunys, Spanish Pyrenees, is a classic example of a growth structure located in the forelimb of a fault-propagation fold (Vergés, 1993; Ford

DOI:10.1306/13251336M943430

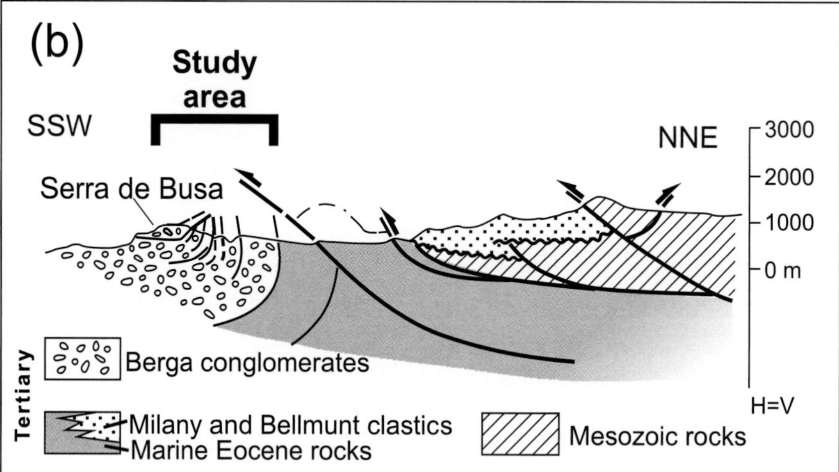

Figure 1. (a) Location of the study area within the Pyrenees. (b) Regional section across the study area (modified from Vergés, 1993). The location is in panel a.

et al., 1997). This classic example is of great interest because a thick sequence of well-exposed Tertiary conglomerates records the progressive evolution of the forelimb of a fault-propagation fold (Figure 1). However, it is a controversial structure whose fold kinematics has been explained invoking both limb rotation (Riba, 1976; Ford et al., 1997) and models involving beds rolling through axial surfaces (Suppe et al., 1997; Novoa et al., 2000; Rafini and Mercier, 2002).

In this study, the Sant Llorenç de Morunys area has been mapped on 1:5000 topographic and orthophoto maps, paying special attention to unconformity angles. Cross sections have been built using dip data and structure contours. The deep valley incision and the nonplanar bedding surfaces of the poorly sorted conglomerates cause the structure contours constructed from bed traces to be more reliable in some places than dip data obtained with a compass. In fact, our map and cross sections (Figures 2–4) are very close to those produced by previous workers (Riba, 1976; Ford et al., 1997; Suppe et al., 1997) but for some details around angular unconformities. The main features of the growth structure are local angular unconformities that disappear toward the syncline hinge and a dip change from overturned to flat-lying beds in the forelimb (Figures 1b, 2–4). Two wedges displaying major thickness changes (Figures 4, 5) associated with two major angular unconformities evidence two main growth stages. These growth wedges occur where the sedimentary rocks are more finely grained, as in Jordan et al.'s (1988) tectonostratigraphic models (Figures 2, 4). The first growth wedge in fact displays two angular unconformities near the tip of the principal angular unconformity: the youngest one is located at the bottom of the Pont de Les Cases formation and the oldest one within the Camps de Vall-Llonga formation (Figures 3, 4; 5b).

Unlike previous models, our kinematic model has been constructed changing unconformity angles during deformation as a result of layer-parallel shear in underlying layers. This mechanism is recorded by flexural slip

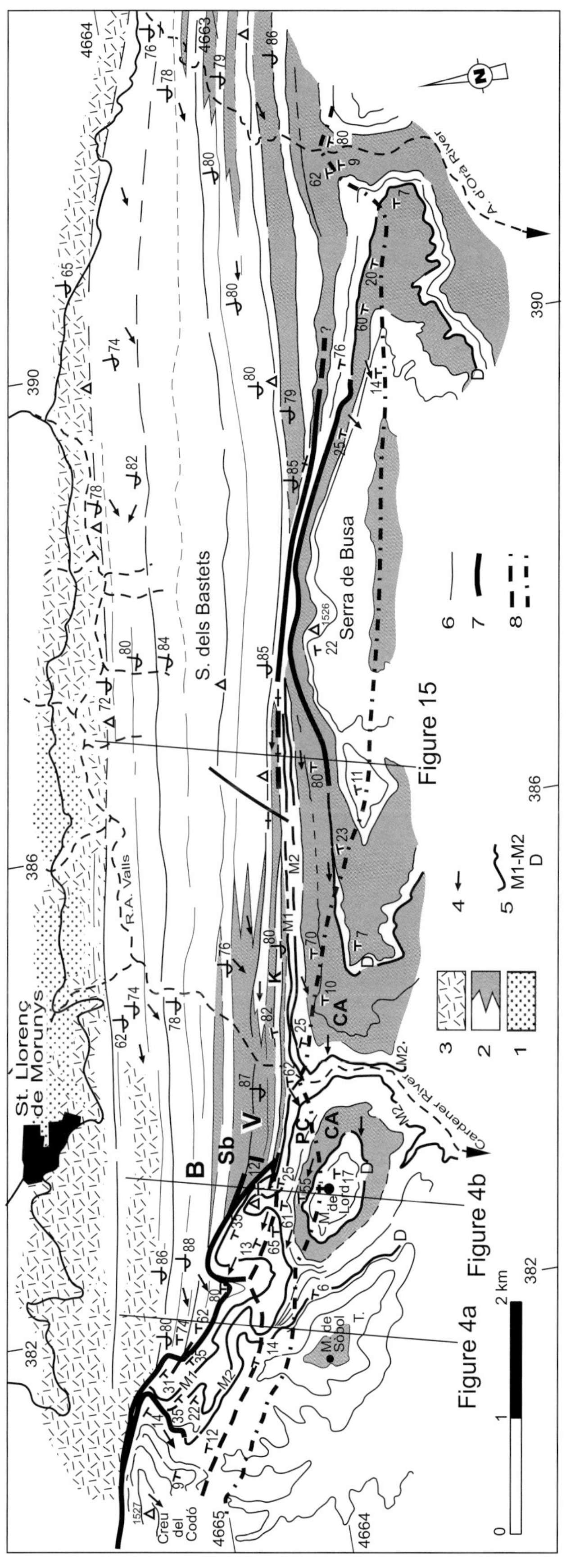

Figure 2. Geologic map of the Sant (St.) Llorenç growth structure located in Figure 1a (compiled from Riba, 1976; Suppe et al., 1977; Ford et al., 1997, and our own data). (1) Marine marls. (2) Berga group, left white = conglomerate-rich succession; right gray = sandstone-rich succession. Capital letters in the map label formations of the Berga group: B = El Bastets formation; Sb = Sobirana formation; V = Camps de Vall-Llonga formation; K = El Castell formation; PC = Pont de Les Cases formation; CA = Les Cases Altes formation (Ford et al., 1997). (3) Quaternary deposits. (4) Paleocurrent data. (5) Key horizons: M1 – M2 = lower and upper marker horizons of Ford et al. (1997); D = conglomerate horizon from the tip of the third angular unconformity. (6) Bed traces. (7) Angular unconformities. (8) Anticline and syncline axial traces.

faults, which are common in the forelimb as documented by Ford et al. (1997), and by bedding-parallel shear zones deflecting a weak cleavage developed in the most incompetent beds (Figure 6).

The aims of this study are first, from the theoretical point of view, to show the function of the modification of initial unconformity angles in the geometry of overlying growth strata, as a result of layer-parallel simple shear in both angular and concentric folds, and second, to apply this theory to the Sant Llorenç example, showing that the analysis of the unconformity angles provides significant data on the folding mechanisms that operated in the forelimb of a fault-propagation fold.

PREVIOUS MODELS OF THE SANT LLORENÇ DE MORUNYS GROWTH STRUCTURE

One of the first attempts to explain the Sant Llorenç structure is that of Riba (1976), who emphasized the close relationship between the so-called progressive unconformities and the tips of local angular unconformities, and explained this lateral change in terms of acceleration and subsequent deceleration of the limb rotation rate (Figure 7). Subsequently, Ford et al. (1997) proposed basically a limb rotation mechanism within a trishear fault-propagation fold model. At the same time, Suppe et al. (1997) used a curved-hinge fold model involving both hinge migration and limb rotation. In our view, all former models applied to the Sant Llorenç structure have some weaknesses. The model by Suppe et al. (1997) explains successfully the geometry of the second growth wedge, above the main unconformity, showing that its development required hinge migration, but fails to explain the dip changes below the principal unconformity, particularly the overturned attitude of the oldest beds between 700- and 1000-m (2296- and 3281-ft) height above sea level (Figure 4). However,

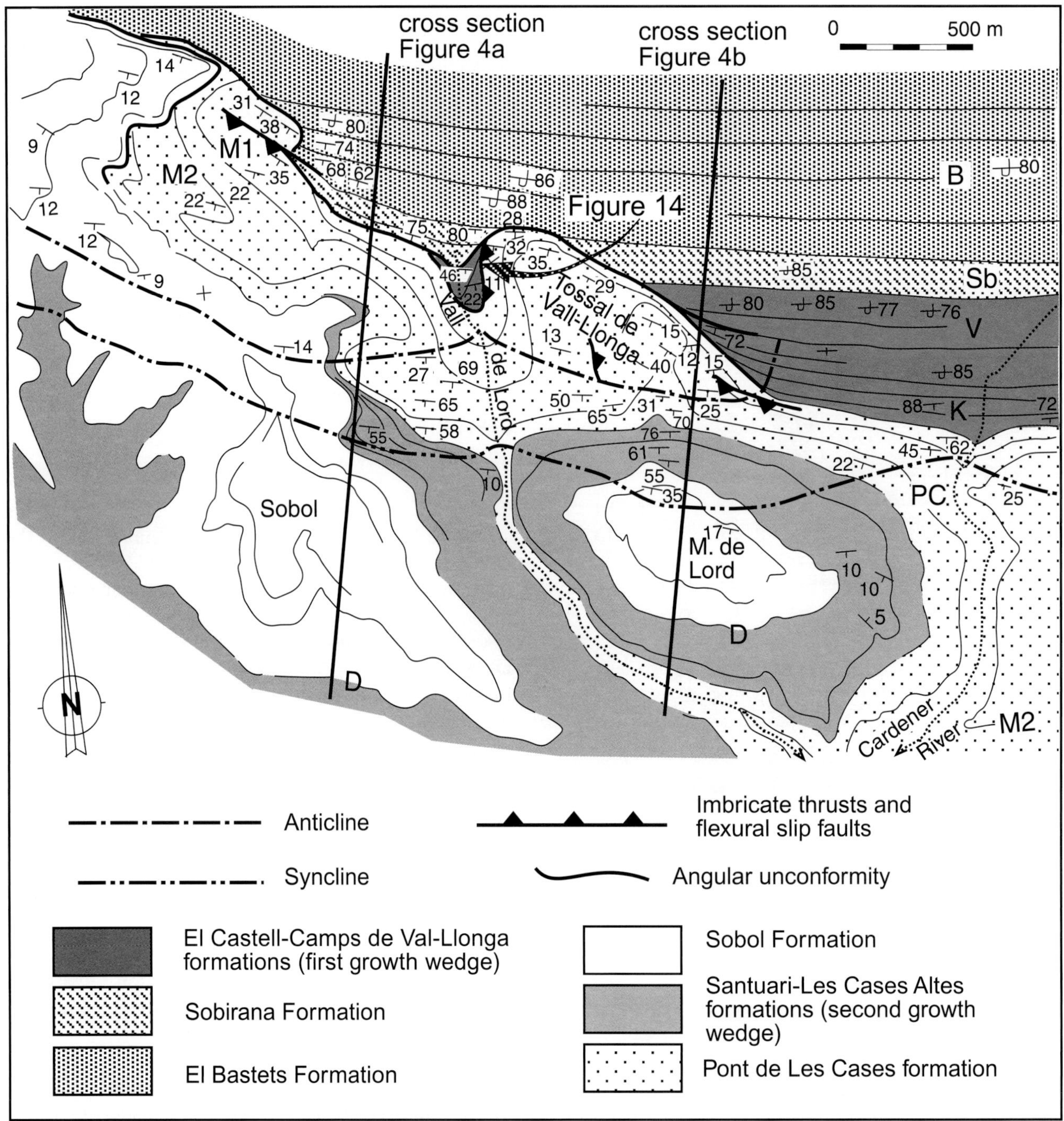

Figure 3. Detailed geologic map of the western part of the map in Figure 2, showing the two growth wedges (gray) linked to two main angular unconformities. Stratigraphic names are from Ford et al. (1997). Capital letters in the map are the same as in the legend of Figure 2.

limb rotation models explain the dip changes in the forelimb, from overturned to flat lying, but involve some inconsistencies. For instance, if a layer located just above the principal unconformity is restored, maintaining the unconformity angle, we obtain unreasonable results (Figure 8). Thus, if we restore unconformable layers to a horizontal attitude, the oldest layers remain with a 75° dip, but if we restore the same layer ahead of the unconformity tip, the dip of the oldest layers becomes about 30°. A simple rotation model (Figure 7)

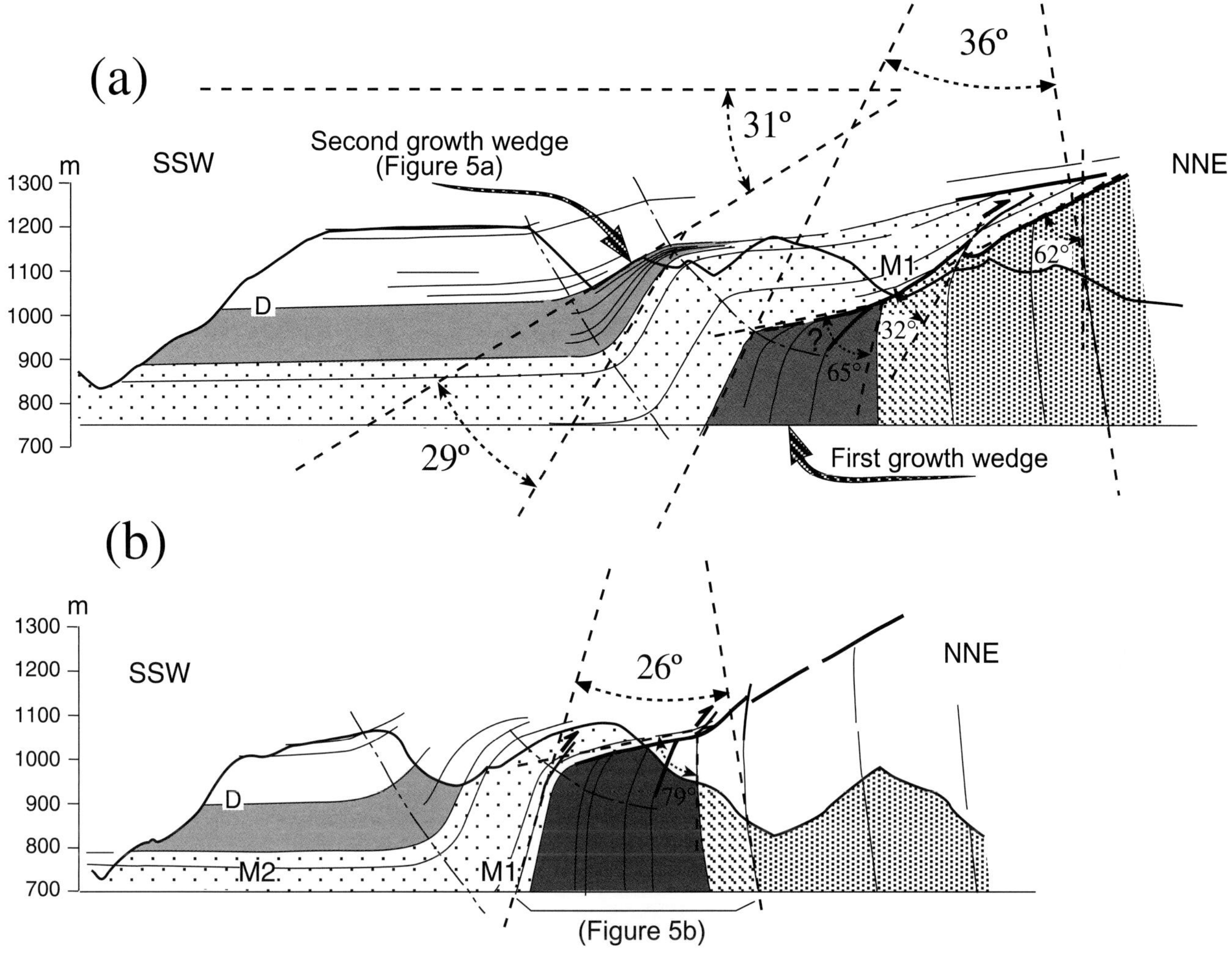

Figure 4. Geologic cross sections of the Sant Llorenç growth structure in the western slope of the Cardener River. The location is in Figures 2 and 3. Stratigraphic legend is the same as in Figure 3. Unconformity angles and external rotation angles for the forelimb are shown with dashed lines.

predicts that unconformity angles increase toward the oldest beds. However, this is not what we observe in the Sant Llorenç angular unconformity. The unconformity angle is greater above the Camps de Vall-Llonga formation (65–79°) than above the Bastets formation (62°) and much greater than above the Sobirana conglomerates (32°) (Figure 4a, b), with the Camps de Vall-Llonga formation being younger than the Sobirana conglomerates.

To solve these problems, we propose a model in which the unconformity angles change during fold growth as a result of different folding mechanisms in underlying layers. Accordingly, we show that the main geometrical features of the Sant Llorenç de Morunys growth structure can be explained considering the modification of unconformity angles due to classic folding mechanisms in the forelimb.

GROWTH STRATA GEOMETRIES RESULTING FROM MODIFICATION OF UNCONFORMITY ANGLES BY FLEXURAL AND CHEVRON MECHANISMS IN UNDERLYING LAYERS

Growth strata frequently display local angular unconformities as a result of fluctuations in sedimentation and uplift rates across growing structures (Rocco and Jaboli, 1958; Riba, 1976; Anadón et al., 1986; Poblet et al., 1998). If the folds underlying these unconformities keep on developing during or after the deposition of overlying sediments, these angular unconformities will be deformed depending on the folding mechanisms across the fold limbs (Figure 9).

To understand the function of folding mechanisms in the changes of unconformity angles and their consequences on growth strata geometry, we first analyze a

Figure 5. Photographs of the Sant Llorenç de Morunys (a) upper or second and (b) lower or first growth wedges looking west. The location scale and orientation are in Figure 4. The stratigraphic units are labeled with capital letters as in the legend of Figure 2.

simple growth symmetric fold and then we model an asymmetric monoclinal fold similar to the Sant Llorenç structure. The modification of the initial unconformity angle, β_0, depends on the folding mechanisms in the underlying layers across the limbs (Alonso, 1989). The modification of unconformity angles due to layer-parallel simple shear in both angular and concentric folds is shown in Figure 9. The modification of the unconformity angle, β, is that of a line deformed by simple shear. The shearing strain, γ, is given by the angle of dip of the layer $\gamma = \theta$ in curved parallel folds and $\gamma = \tan\theta$ in angular folds (Ramsay, 1967), assuming constant bed length during folding. Ramsay (1967) refers to both types of folding as flexural and chevron folds, respectively. In the case of reactivated folds (Figure 9), the relationship between the initial unconformity angle, β_0, and the

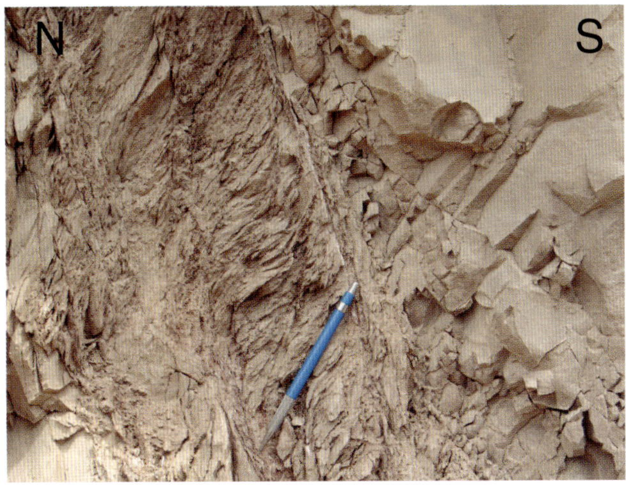

Figure 6. Bedding-parallel shear zones deflecting a weak cleavage developed in the most incompetent beds of the El Castell formation; Solsona-Sant Llorenç de Morunys road.

modified unconformity angle (β' in curved-parallel folds and β'' in angular folds) is

$$\mathrm{Cot}\beta' = \cot\beta_0 - \Delta\theta$$

$$\mathrm{Cot}\beta'' = \cot\beta_0 - \Delta\tan\theta$$

If we model discontinuous layer-parallel slip, an additional mechanism in individual layers, tangential-longitudinal strain, for instance, will be necessary (Figure 9, left limb). This mechanism is developed in nonstratified stratigraphic units and produces layer-parallel extension in the outer arc of the folded layer and layer-parallel shortening in its inner arc (Ramsay, 1967, his figure 7.63). Strain values using this mechanism depend on the fold curvature so that planar limbs of layers folded by tangential-longitudinal strain remain unstrained during fold development. If planar limbs are unstrained, the unconformity angle will remain constant, and consequently, the unconformable layers will rotate together with the underlying layers as a rigid body, at the same rate, during fold tightening. In folds reactivated by layer-parallel simple shear, the unconformity angle increases, and therefore, the unconformable layers rotate at a lesser rate than the preunconformity beds (Figure 9). Angular folding gives rise to a greater modification of unconformity angles than curved parallel folds because it implies higher bed-parallel shearing strain. Figure 10 shows the growth strata geometries developed on an angular unconformity as a result of underlying fold tightening in angular and concentric folds. During the last stages of angular fold growth, the unconformable layers may rotate even in a sense opposite to that of the fold limb because angle β increases at a higher rate than limb rotation (Figure 10e, f), giving rise to seemingly paradoxical structures such as antiforms over synforms (Figure 10e). In this way, growth strata may thicken toward the syncline hinge during the initial stages of fold growth and, later, younger growth strata may thin toward the syncline hinge, reaching their maximum thickness at the point where the folding mechanism changes (Figure 10).

FORWARD MODELING OF THE SANT LLORENÇ DE MORUNYS GROWTH STRUCTURE

When modeling asymmetric monoclinal folds, like the Sant Llorenç growth syncline, if constant bed thickness and limb rotation are assumed, deformation compatibility problems arise around the hinge zone. If the hinge is fixed, a fault along the axial plane will be necessary (Figure 11a). Other solutions are hinge migration plus rotation (limited activity axial surfaces of Poblet et al., 1997) (Figure 11b) or shearing along the horizontal limb (Erslev and Mayborn, 1997) (Figure 11c). If solely kink-band migration is assumed, then the unconformity angle will remain constant during fold amplification (Figure 11d) (Hubert-Ferrari et al., 2007).

In the Sant Llorenç syncline, no fault along the axial surface and no significant deformation of the horizontal limb are observed and the unconformity angle displays great changes along the unconformity surface. Therefore, we will apply the hinge migration plus rotation model in our forward models. In this case, the shear strain is given by $\gamma = 2\Delta\tan(\theta/2)$ (Suppe, 1985).

Two balanced forward models have been constructed to simulate the evolution of the Sant Llorenç de Morunys structure. A first simple model of a monoclinal fold amplified by limb rotation and subsidiary hinge migration (Figure 12) may explain the main features of the angular unconformable relationships (compare Figures 4a, b

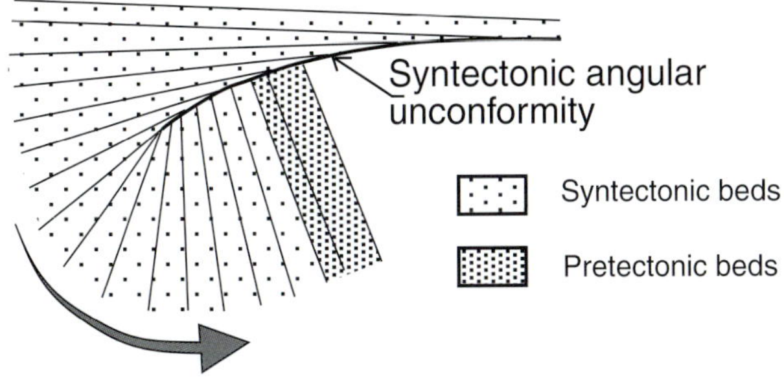

Figure 7. Model of syntectonic unconformity developed by rigid limb rotation (modified from Riba, 1976).

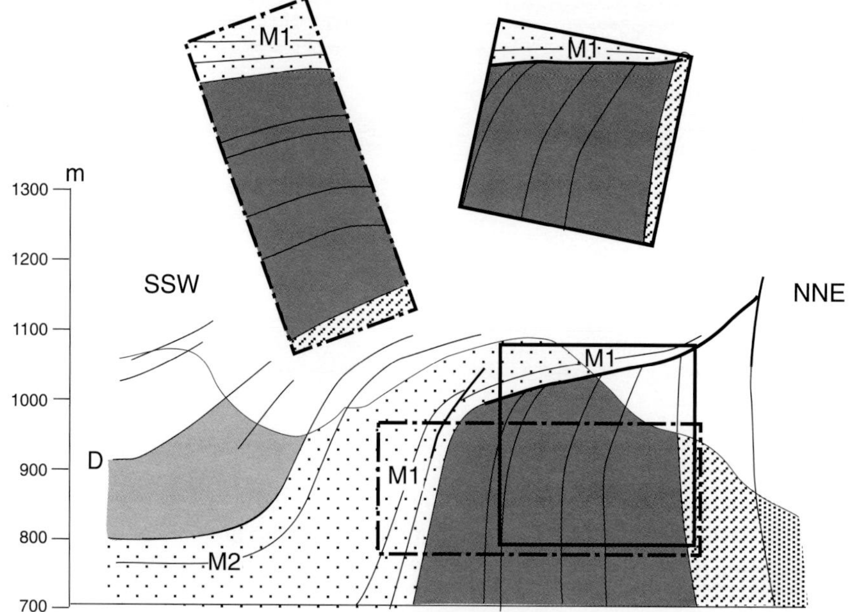

Figure 8. Rigid body restoration of the different limbs of the Sant Llorenç de Morunys growth anticline, in the cross sections of Figure 4, provides inconsistent dips for preunconformity beds. The M1 has been restored to a horizontal attitude.

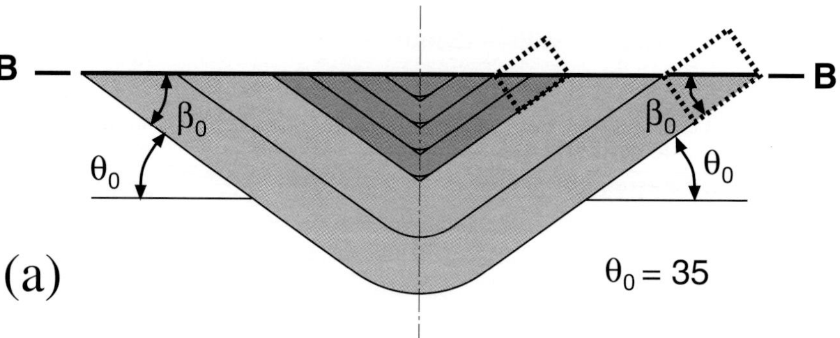

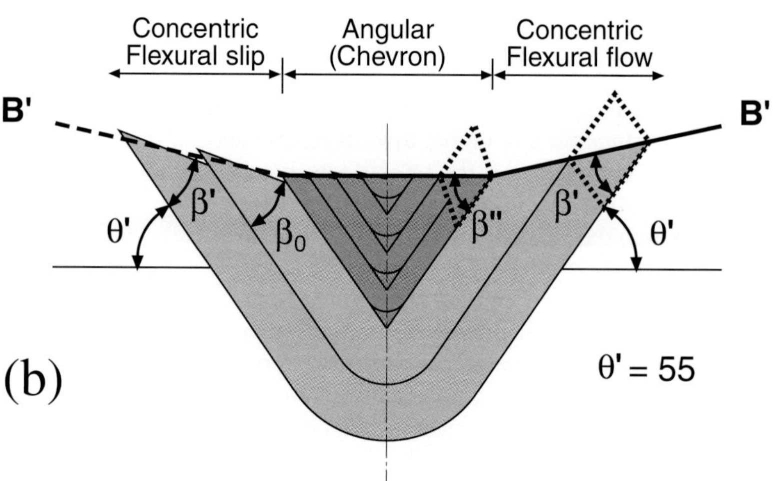

Figure 9. Modification of unconformity angles by different folding mechanisms. (a) Initial stage of a syncline truncated by an angular unconformity B. (b) The syncline with both angular and concentric geometries shown in panel a is amplified by layer-parallel simple shear (modified from Alonso, 1989).

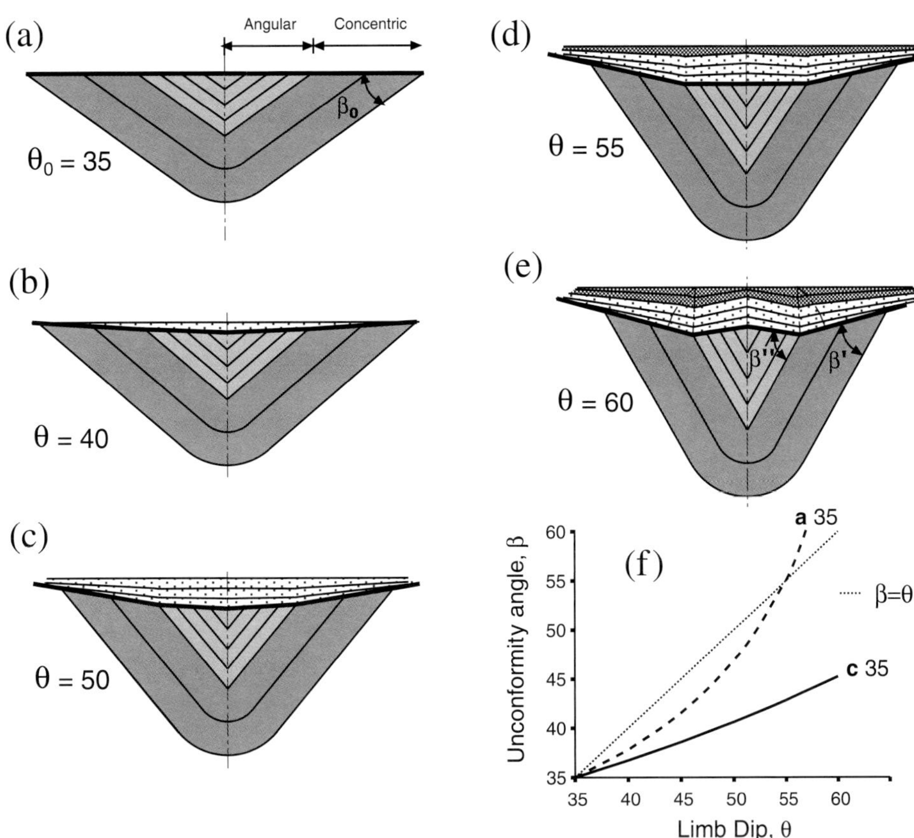

Figure 10. (a–e) Geometries of growth strata developed above an underlying fold tightening with angular and concentric geometries. (f) Modification of the unconformity angle β' during the fold growth stages (a–e) from an initial unconformity angle of 35° in angular (dashed line a) and curved parallel (line c) folds.

and 12), but it does not explain the dip changes observed across the forelimb. To fit better the actual geometry of the Sant Llorenç structure, shown in Figure 4a and b, three wedges located beneath the principal unconformity have been included in the model (Figure 13). Both forward models presume layer-parallel simple shear related to angular folds in preunconformity strata for the Camps de Vall-Llonga and El Castell formations, which display the maximum unconformity angle, whereas for the Bastets-Sobirana formations, concentric folding has been assumed. These assumptions are consistent with field data because flexural slip faults and bedding-parallel shear zones can be found at El Castell, Camps de Vall-Llonga, and Bastets formations (Figure 6). These folding mechanisms are also consistent with the lithological features of the formations involved. Thus, the Camps de Vall-Llonga and El Castell formations display thinner conglomerate beds and a higher proportion of incompetent beds than the Bastets-Sobirana formations.

In the models of Figures 12 and 13, an initial unconformity angle of 35° has been assumed. Two methods determine the initial unconformity angles. If the unconformity angle has been modified, the dip change across the forelimb from the pregrowth strata up to the oldest bed overlying the angular unconformity supplies the original angle (36–26° in Figure 4). The second method consists of finding the minimum unconformity angle, which is always associated with the most competent and poorly stratified units (Alonso, 1989), folded by tangential-longitudinal strain because, in this mechanism, planar limbs undergo rigid body rotation and no modification of the unconformity angle occurs. At Sant Llorenç, the minimum angle, β (32°), is associated with the most competent and poorly stratified formation (lower part of the Sobirana conglomerates).

The model in Figure 13 reproduces well the dip changes across the forelimb, the angular unconformity features, and the parallel beds in the lower part of the upper growth wedge (Figure 4a) as a result of syncline hinge migration. The meaning of the upper angular unconformity, developed when the forelimb dip was 65°, is well illustrated too. However, the model yields an anticline tighter than of the cross section as a result of stacking of wedges above the anticline hinge (Figure 13g, i), where they were translated because of layer-parallel shear. This anticline has been flattened (Figure 13h, j) by decreasing shear strain, assuming layer-parallel shear in the horizontal limb (Figure 13g) (Erslev and Mayborn, 1997). Although we have not found layer-parallel shear evidences in this limb along the Solsona-Sant Llorenç road, the bad exposures of the Les Cases Altes formation (Figures 3, 4) do not allow us to discard this mechanism.

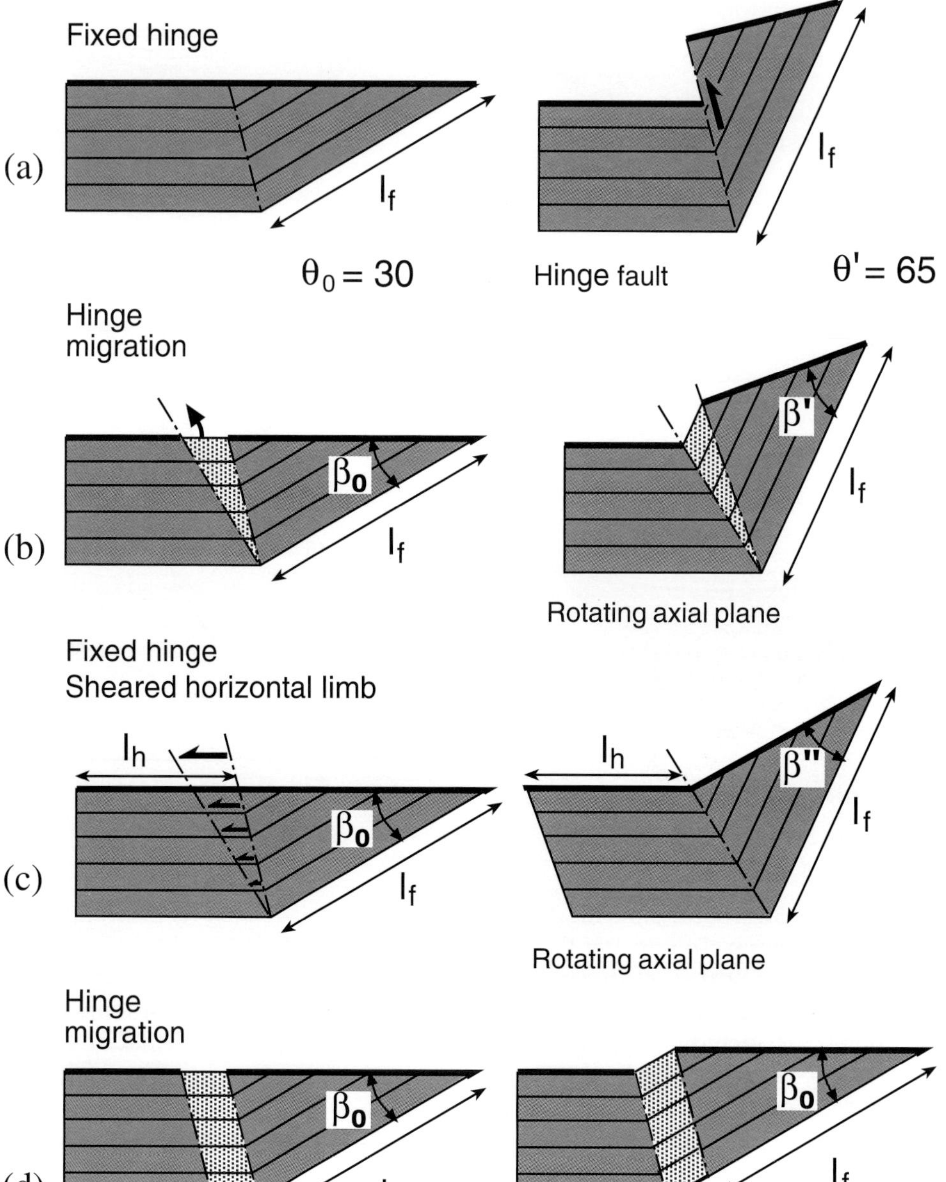

Figure 11. (a–d) Different kinematic models of amplification of angular monoclinal folds with an unconformity truncating the dipping limb. lf and lh = initial lengths of the forelimb and horizontal limb, respectively. Models a, b, and c involve limb rotation, whereas model d is a self-similar model.

The Sant Llorenç natural example has been quantitatively analyzed to compare the unconformity angles shown in the actual cross section (Figure 4) for the principal unconformity with the theoretical values obtained from simple shear and illustrated in the forward models. Figure 14 shows the good agreement between the theoretically predicted angles (curves) and the observed values of β'. Deviations from the field data may be attributed to inaccuracies in the initial unconformity angle assumed or to mixed mechanisms. The variations of the unconformity angle may be linked to changes from tangential-longitudinal strain to layer-parallel slip in both angular and curved parallel folds across the stratigraphic sequence of the fold limb. The folding mechanisms inferred from the analysis of unconformity angles make it possible to predict fold shapes at depth. Thus, concentric shapes are likely to occur in the subsurface part of the Sant Llorenç syncline at the Sobirana-Bastets conglomerates, whereas angular hinges or a mixed shape between concentric and angular is probable for the Camps de Vall-Llonga formation.

The unconformable layers rotated in a different way above and ahead of the unconformity tip, and this is the reason why the anticline hinge zone is located at the tip of the angular unconformity (Figure 4a, b). In our view, the anticline developed in the Sant Llorenç growth structure is not a fault-propagation fold but resulted from deformation of the angular unconformity. Although the

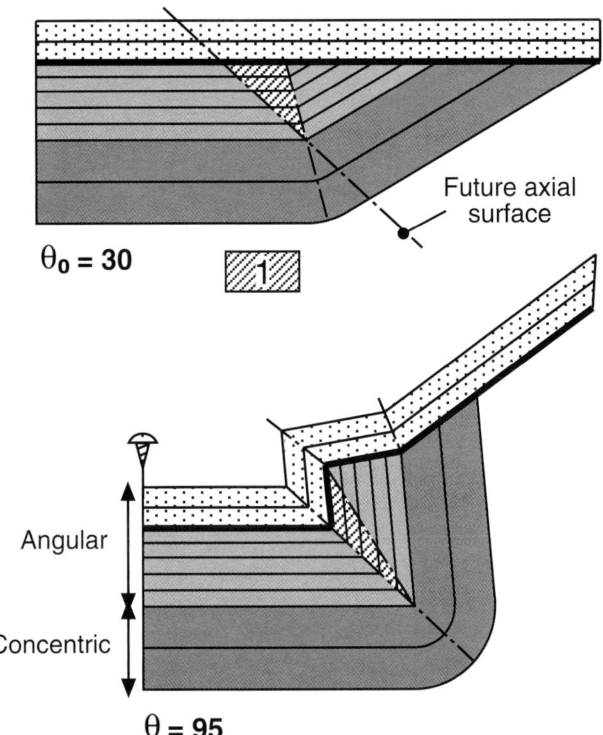

Figure 12. Forward model of a monoclinal fold amplified by limb rotation and subsidiary hinge migration to maintain bed thickness. The angular unconformity above the dipping limb is deformed by layer-parallel simple shear in angular and concentric folds. (1) Part of the horizontal limb migrating into the dipping limb.

beds situated ahead of the unconformity tip rotate at the same rate as the fold limb, the beds above the angular unconformity rotate more slowly because the unconformity angle increases because of layer-parallel shear (Figure 13). The resulting geometry is an anticline involving the unconformable layers, with its hinge zone typically located at the tip line of the angular unconformity. This anticline does not involve beds located below the angular unconformity. A similar structure can be obtained by parallel migration of axial planes (Hubert-Ferrari et al., 2007), but in that case, the unconformity angle does not change (Figure 11d) because the unconformity surface is not deformed by layer-parallel shear.

We agree with Suppe et al. (1997) that most of the forelimb youngest beds came from the horizontal limb as a result of hinge migration because no thickness changes occur in the lower part of the second growth wedge (Figure 4a). However, unlike hinge migration models, the anticline and syncline axial traces only display a slight upward convergence (Figure 4a). Moreover, the upper part of the growth wedges could have developed as a result of limb rotation (Figure 13). Small thickness changes in the anticline hinge for the lower part of the Pont de Les Cases formation (Figure 4b), similar to the kink-band migration model (Suppe et al., 1997), have been avoided in our forward model. However, these changes could be easily reproduced between stages d and e in Figure 13, considering that the level of the unconformity surface in the cross section of Figure 4b (dashed line in Figure 13d) implies a close location of the syncline and anticline hinges.

During fold amplification, the unconformity surface commonly becomes a surface of decollement because of the structures that have to be developed in the unconformable beds to accommodate the changes in length of the unconformity surface (Figure 13). In this case, we can expect detached structures accommodating layer-parallel shortening in unconformable beds. Imbricate thrusts detaching from different bedding surfaces and shortening the unconformable layers can be seen right above the principal unconformity surface along the Sant Llorenç-Sanctuary road (Figure 15). A fault gouge along the unconformity surface has also been found in that locality. Regarding the first angular unconformity, the problem of excess length in the unconformable beds, inferred from the forward model (Figure 13), has been avoided in this model. In our view, this excess length was accommodated downward, beneath the principal unconformity surface, where not only reverse faults (Ford et al., 1997) but also minor normal faults implying layer-parallel shortening have been observed in the subvertical beds of the Camps de Vall-Llonga and the Castell formations. This is the result of the difficulty of deforming the thick and massive sequence of the Pont de Les Cases unconformable conglomerates.

What is the meaning of the folds underlying the principal unconformity that look like drag folds? Why does the anticline axial trace die out at the base of the Camps de Vall-Llonga formation? The folds below the main unconformity, which can be seen to the east of Tossal de Vall-Llonga (Figure 4b), could be interpreted as drag folds related to a detachment along the unconformity surface (Figures 4b, 13), but in our view, their significance is different (Figure 13). They are the remnants of the first growth wedge, perhaps preserved in Vall de Lord (Figure 4a) (Ford et al., 1997) and truncated to the east of Tossal de Vall-Llonga (Figures 4b, 13d).

NONCYLINDRICAL MODEL FOR THE SANT LLORENÇ STRUCTURE: LATERAL CHANGES IN FOLDING MECHANISMS

Although the syncline axis is subhorizontal (compare hinge height for the same stratigraphic horizon in different cross sections, for instance, for the bottom of the Pont

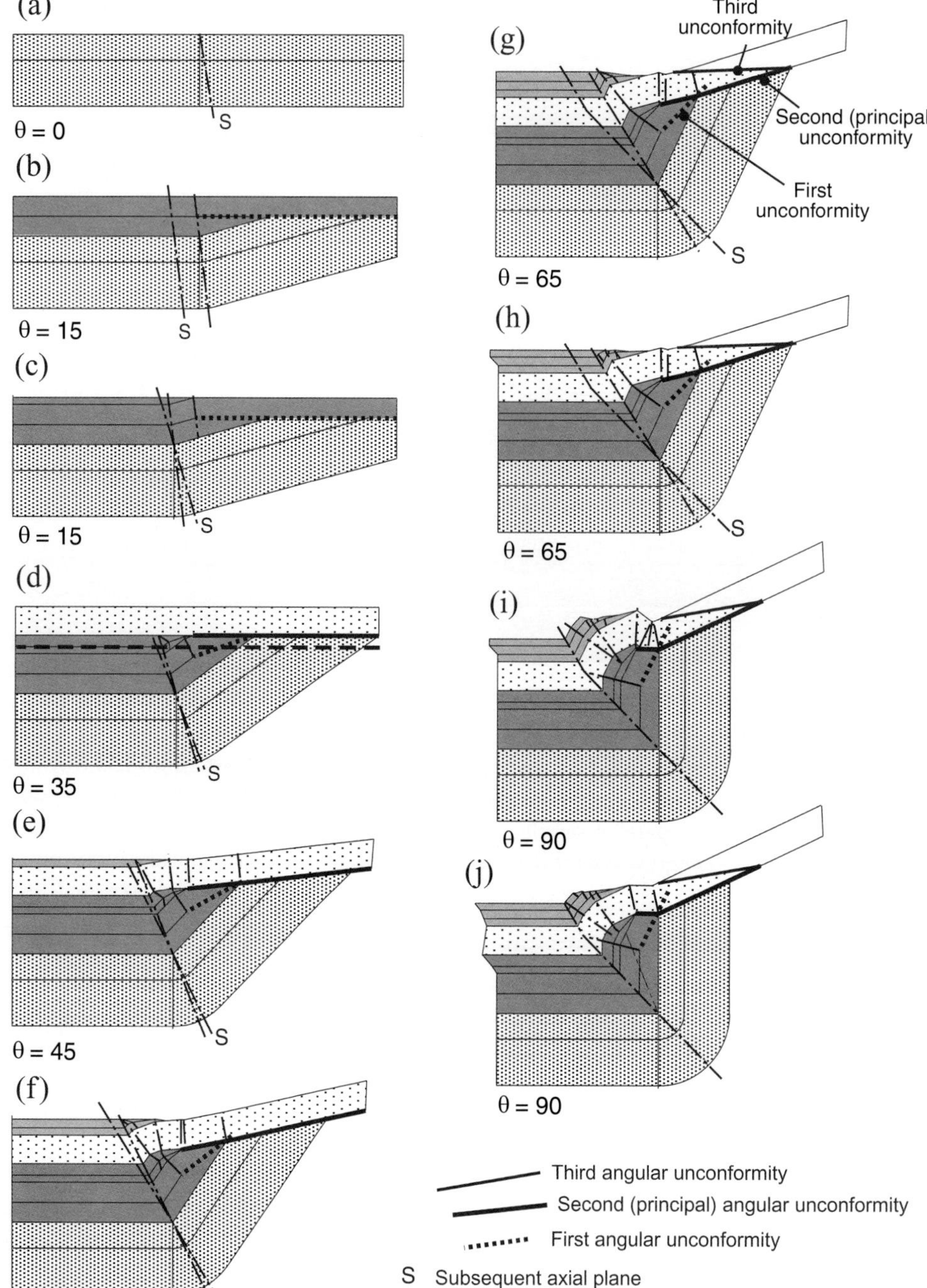

Figure 13. Balanced forward model of a monoclinal fold amplified by limb rotation and subsidiary hinge migration, simulating the sequential development of the cross section in Figure 4a. Stratigraphic legend is the same as in Figure 3. (a–d) Development of the first growth wedge. The dashed line in panel d shows the level of the principal unconformity in the cross section of Figure 4b. (e–h) Development of the second growth wedge. (j) Present structure.

de Les Cases formation), different dips of the same beds along strike and lateral changes in the forelimb length imply a noncylindrical structure. Changes in height and dip of the angular unconformities along strike also occur (Figures 2, 4, 16). All these structural variations can be explained as a result of lateral changes in folding mechanisms and limb rotation rate and are also consistent with the expected rheology of the layers involved. Bedding dips increase eastward, which may result from an initially slower rotation on the eastern side that is in the Busa section of Figure 16. Although the finite rotation of the oldest layers (El Bastets formation) was around 100–105° in both western and eastern areas, limb rotation before the deposition of the Pont de Les Cases formation is around 36° on the westernmost section (Figure 4a), 26° in section of Figure 4b, and 17° at the Busa section (Figure 16), perhaps because the thicker and more massive conglomerates in the eastern section (Figure 2)

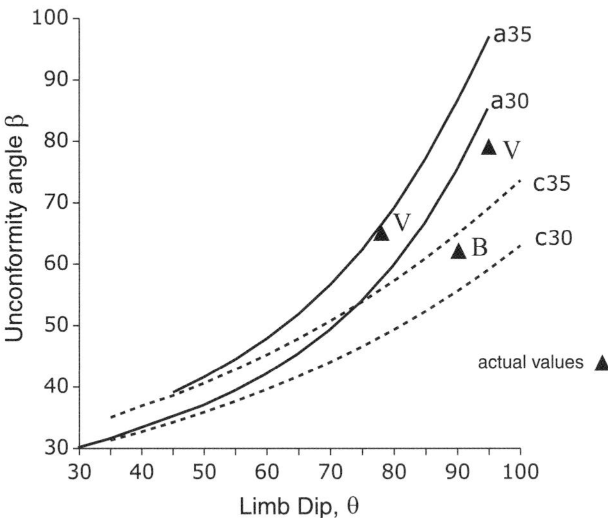

Figure 14. Modification of initial unconformity angles of 30 and 35° as a result of increasing limb dip for (c) growth curved parallel folds and (a) angular folds. Actual values for the Camps de Vall-Llonga (V) and El Bastets (B) formations in the cross sections of Figure 4 are shown with triangles.

resisted deformation more strongly during the initial stages of fold amplification. From the Pont de Les Cases formation, the up-section rotation was faster in the Busa section than in the western areas. The dip of the layers right above the main angular unconformity above the El Castell-Camps de Vall-Llonga formations also increases eastward. It displays a flat-lying attitude in the section depicted in Figure 4a, 15° south dipping in Figure 4b, and a near-vertical attitude in the section depicted in Figure 16. This structural change along strike can be attributed to a change in folding mechanism from layer-parallel simple shear in the western sections to tangential-longitudinal strain in the eastern section, and that is why the initial unconformity angle would have changed from 35 to 60–80° in the western sections and remained constant (about 20°) in the eastern section. This change is consistent with the expected rheology of the layers involved. The Bastets and Camps de Vall-Llonga formations change from massive and thick conglomerate beds in the eastern section (proximal part of alluvial fans) to thinner beds of conglomerates westward, where intercalations of sandstone and siltstone-clay beds occur. This lateral change is also recorded by the westward plunge of the anticline axis and the lateral variation of the forelimb length (Ford et al., 1997). At the lower marker horizon (M1), the forelimb length is 300 m (984 ft) in Figure 4a, 350 m (1148 ft) in Figure 4b, and more than 700 m (2296 ft) in Figure 16. The higher level of erosion in the eastern sector during the development of the principal unconformity is in agreement with the paleocurrent data, indicating sedimentary transport from the northeast for the successions underlying and overlying that unconformity (Figure 2).

CONCLUSIONS

The geometry of growth sediments above angular unconformities should be interpreted considering the changes of unconformity angles due to layer-parallel simple shear during fold growth.

When growth strata develop a local angular unconformity and the underlying beds rotate with bedding-parallel shear related to angular or curved parallel folds, the beds above and ahead of the unconformity surface will rotate in different ways, giving rise to a typical anticline with its hinge zone located at the tip of the angular unconformity.

In the Sant Llorenç de Morunys growth structure, eastern Pyrenees, the variations in the unconformity angles provide information on the folding mechanisms involved in the forelimb of a fault-propagation fold. The degree of modification of the initial unconformity

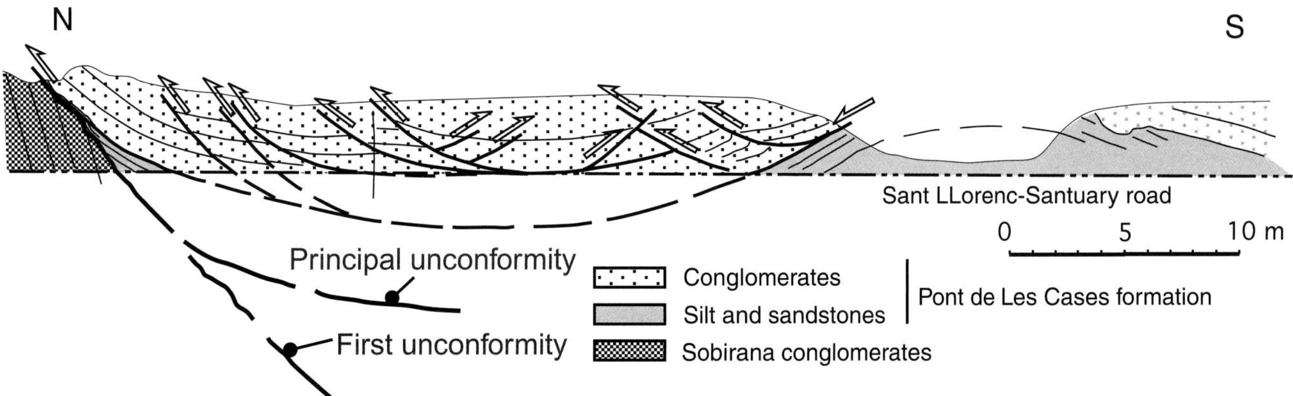

Figure 15. Imbricate thrusts developed above the principal angular unconformity. The location is in Figure 3.

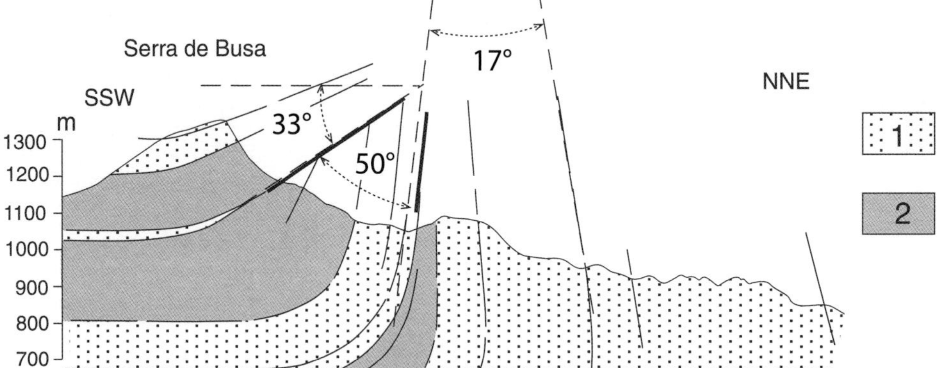

Figure 16. Geologic cross section of the Sant Llorenç growth structure in the eastern slope of the Cardener River. (1) Conglomerates. (2) Sandstone-rich succession. The location is in Figure 2.

angles shows that layer-parallel simple shear in both angular and curved folds and tangential-longitudinal strain operated in the forelimb, and these mechanisms are consistent with the mechanical properties of the stratigraphic units involved. Unconformable layers were detached and shortened as a result of changes in length of the unconformity surface.

Both limb rotation and hinge migration occurred during fold development, but most of the hinge migration could be a consequence of asymmetric limb rotation to maintain bed thickness.

The above-mentioned conclusions allow us to infer fold geometries at depth. Thus, buried angular or rounded fold hinges can be predicted by estimating the degree of modification of the initial unconformity angle across fold limbs.

ACKNOWLEDGMENTS

We thank Josep Poblet and Mayte Bulnes for their encouragement and critical reading of a previous version of the manuscript. We are grateful to John Suppe and Josep Anton Muñoz who suggested to us working in the lower growth wedge of the Sant Llorenç structure. Three anonymous referees helped to improve the manuscript. We also thank Ana Ojanguren for helping with the English translation. Projects BTE2002-04316-C0303 and CGL2006-12415-C03-02 of the Spanish Ministry of Science and Technology, and the Grup de Qualitat SRG2002-2004, Department d'Universitats Recerca y Societat de l'Informació, Generalitat de Catalunya, partially funded this work.

REFERENCES CITED

Alonso, J. L., 1989, Fold reactivation involving angular unconformable sequences: Theoretical analysis and natural examples from the Cantabrian zone (northwest Spain): Tectonophysics, v. 170, p. 57–77, doi:10.1016/0040-1951(89)90103-0.

Anadón, P., L. Cabrera, F. Colombo, M. Marzo, and O. Riba, 1986, Syntectonic intraformational unconformities in alluvial fan deposits, eastern Ebro Basin margins NE Spain, *in* P. A. Allen and P. Homewood, eds., Foreland basins: International Association of Sedimentologists Special Publication 8, p. 259–271.

Erslev, E., 1992, Trishear fault-propagation folding: Geology, v. 19, p. 617–620, doi:10.1130/0091-7613(1991)019<0617:TFPF>2.3.CO;2.

Erslev, E., and K. R. Mayborn, 1997, Multiple geometries and modes of fault-propagation folding in the Canadian thrust belt: Journal of Structural Geology, v. 19, p. 321–335, doi:10.1016/S0191-8141(97)83027-1.

Ford, M., E. A. Williams, A. Artoni, J. Vergés, and S. Hardy, 1997, Progressive evolution of a fault-related fold pair from growth strata geometries, Sant Llorenç de Morunys, SE Pyrenees: Journal of Structural Geology, v. 19, p. 413–441, doi:10.1016/S0191-8141(96)00116-2.

Hubert-Ferrari, A., J. Suppe, R. Gonzalez-Mieres, and X. Wang, 2007, Mechanisms of active folding of the landscape (southern Tian Shan, China): Journal of Geophysical Research, v. 112, p. B03S09, doi:10.1029/2006JB004362.

Jordan, T. E., P. B. Flemings, and J. A. Beer, 1988, Dating of thrust-fault activity by use of foreland basin strata, *in* K. Kleinspehn and C. Paola, eds., New perspectives in basin analysis: New York, Springer-Verlag, p. 307–330.

Novoa, E., J. Suppe, and J. H. Shaw, 2000, Inclined shear restoration of growth folds: AAPG Bulletin, v. 84, p. 787–804.

Poblet, J., K. McClay, F. Storti, and J. A. Muñoz, 1997, Geometries of syntectonic sediments associated with single-layer detachment folds: Journal of Structural Geology, v. 19, p. 369–381.

Poblet, J., J. A. Muñoz, A. Travé, and J. Serra-Kiel, 1998, Quantifying the kinematics of detachment folds using 3D geometry: The example of the Mediano anticline (Pyrenees, Spain): Geological Society of America Bulletin, v. 110, p. 111–125, doi:10.1130/0016-7606(1998)110<0111:QTKODF>2.3.CO;2.

Rafini, S., and E. Mercier, 2002, Forward modeling of foreland basins progressive unconformities: Sedimentary Geology, v. 146, p. 75–89, doi:10.1016/S0037-0738(01)00167-1.

Ramsay, J. G., 1967, Folding and fracturing of rocks: New York, McGraw-Hill Inc., 568 p.

Riba, O., 1976, Syntectonic unconformities of the Alto Cardener, Spanish Pyrenees: A genetic interpretation: Sedimentary Geology, v. 15, p. 213–233, doi:10.1016/0037-0738(76)90017-8.

Rocco, T., and D. Jaboli, 1958, Geology and hydrocarbons of the Poo Basin, *in* L. G. Weeks, ed., Habitat of oil: A symposium: Tulsa, Oklahoma, AAPG, p. 1153–1167.

Suppe, J., 1985, Principles of structural geology: Englewood Cliffs, New Jersey, Prentice Hall, 537 p.

Suppe, J., and D. A. Medwedeff, 1990, Geometry and kinematics of fault-propagation folding: Eclogae Geologicae Helvetiae, v. 83, p. 409–454.

Suppe, J., F. Sabat, J. A. Muñoz, J. Poblet, E. Roca, and J. Vergés, 1997, Bed-by-bed fold growth by kink-band migration: Sant Llorenç de Morunys, eastern Pyrenees: Journal of Structural Geology, v. 19, p. 443–461, doi:10.1016/S0191-8141(96)00103-4.

Vergés, J., 1993, Estudi geologic del vessant sud del Pirineu oriental i central, Evolució cinematica en 3D: Ph.D. thesis, Universitat de Barcelona, Barcelona, Spain, 193 p.

8

Yue, Li-Fan, John Suppe, and Jih-Hao Hung, 2011, Two contrasting kinematic styles of active folding above thrust ramps, western Taiwan, *in* K. McClay, J. H. Shaw, and J. Suppe, eds., Thrust fault-related folding: AAPG Memoir 94, p. 153–186.

Two Contrasting Kinematic Styles of Active Folding Above Thrust Ramps, Western Taiwan

Li-Fan Yue[1]
Department of Geosciences, Princeton University, Princeton, New Jersey, U.S.A.

John Suppe[2]
Department of Geosciences, National Taiwan University, Taipei, Taiwan

Jih-Hao Hung
Department of Earth Sciences, National Central University, Jhongli, Taiwan

ABSTRACT

Two adjacent active thrust ramps in western Taiwan show contrasting hanging-wall structural geometries that suggest different kinematics, although they involve the same stratigraphic section and basal detachment. The Chelungpu thrust shows a classic fault-bend folding geometry, which predicts folding by kink-band migration without limb rotation, whereas the hanging wall of the Changhua thrust shows the characteristic geometry of a shear fault-bend folding, which predicts a progressive limb rotation with minor kink-band migration. We test the kinematic predictions of classical and shear fault-bend folding theories by analyzing deformed flights of terraces and coseismic displacements in the 7.6 moment magnitude scale Chi-Chi earthquake. In particular, differences in terrace uplift across active axial surfaces are used to show that the assumptions of classical fault-bend folding are closely approximated, including constant fault-parallel displacement, implying conservation of bed length, and hanging-wall uplift rates that are proportional to the sine of the fault dip. This provides a basis for precise determination of total fault slip because the formation of each terrace, combined with terrace dating, gives long-term fault-slip rates for the Chelungpu thrust system. Even the coseismic displacements of 3 to 9 m (10 to 29 ft) in the Chi-Chi earthquake are approximately fault parallel but have additional transient components that are averaged out over the time scale of terrace deformation, which represents 10–100 large earthquakes. In contrast, terrace deformation in the hanging wall of the Changhua thrust ramp shows progressive limb rotation, as predicted from its shear fault-bend folding geometry, which combined with terrace dating allows an estimation of the long-term fault-slip rate of 21 mm/yr (0.83 in./yr) over the last 31 ka. A combined shortening rate of 37 mm/yr (1.46 in./yr) is obtained for this part of the western Taiwan thrust belt, which is about 45% of

[1]*Present address*: Chevron Energy Technology Company, Houston, Texas, U.S.A.
[2]*Also at*: Department of Geosciences, Princeton University, Princeton, New Jersey, U.S.A.

DOI:10.1306/13251337M943431

the total plate-tectonic shortening rate across Taiwan. The Changhua shear fault-bend fold ramp is in the early stages of its development with only 1.7 km (1.06 mi) total displacement, whereas the Chelungpu classical fault-bend fold ramp in the same stratigraphy has nearly an order of magnitude more displacement (~14 km [8.7 mi]). We suggest that shear fault-bend folding may be favored mechanically at low displacement, whereas classical fault-bend folding would be favored at large displacement.

INTRODUCTION

Thrust ramps have long been recognized as one of the characteristic structural elements of fold and thrust belts, and their associated folds are key structural traps for thrust-belt petroleum exploration (e.g., Macqueen and Leckie, 1992; Shaw et al., 2005), for example, the Zagros fold and thrust belt (McQuarrie, 2004; Sherkati et al., 2006), Niger Delta (Bilotti and Shaw, 2005; Corredor et al., 2005a), Canadian thrust belt (Erslev and Mayborn, 1997; Fermor, 1999; Larson and Price, 2006), both north and south flanks of the Chinese Tianshan orogenic belt (Jungger Basin and Tarim Basin; Yin et al., 1998; Allen et al., 1999; He et al., 2005; Hubert-Ferrari et al., 2007), and the eastern Venezuela foreland basin (Parnaud et al., 1995). The existence of diverse folding styles associated with thrust ramps suggests that there must be a significant variety to the mechanisms of ramp propagation as well as to the subsequent displacement along them and deformation around them. Here we document a fundamental contrast in styles of thrust-ramp kinematics in two adjacent active ramps in the western Taiwan thrust belt, both of which step up from the same detachment horizon.

Several simple kinematic models have been developed to describe the diverse folding mechanisms that have been recognized in thrust belts, including detachment folding, fault-bend folding, and fault-propagation folding models, together with their more complex combinations (e.g., Suppe, 1983; Suppe and Medwedeff, 1990; Erslev, 1991, 1993; Jordan and Noack, 1992; Suppe et al., 1992, 2004; Narr and Suppe, 1994; Epard and Groshong, 1995; Homza and Wallace, 1995; Poblet and McClay, 1996; Hardy and Ford, 1997; Medwedeff and Suppe, 1997; Poblet et al., 1997; Storti and Poblet, 1997; Allmendinger, 1998; Zehnder and Allmendinger, 2000; Atkinson and Wallace, 2003; Mitra, 2003; Gonzalez-Mieres and Suppe, 2006; Hardy and Connors, 2006). These models make very different predictions for the distribution of instantaneous uplift relative to total uplift in these folds and for the resulting geometries of growth strata (Suppe and Medwedeff, 1990; Suppe et al., 1992, 1997, 2004; Erslev, 1993; Poblet and McClay, 1996; Poblet et al., 1997; Storti and Poblet, 1997; Allmendinger, 1998; Gonzalez-Mieres and Suppe, 2006). If these models are reasonable approximations to some actual structures, then we expect to see a rich variety of contrasting kinematics in folding. In this chapter, we test the predictions of fault-related fold models by comparing the kinematics constrained by deformed flights of terraces and coseismic displacements in the 7.6 moment magnitude scale Chi-Chi earthquake with the fault and finite fold geometries.

Kinematic Models of Thrust Ramps

By far the simplest kinematic models of folding above thrust ramps are the classical fault-bend folding and shear fault-bend folding mechanisms (Suppe, 1983; Suppe et al., 2004), which are illustrated in Figure 1 using models composed of straight thrust ramps of constant dip, stepping up from a bedding-parallel detachment. These models might be called toy models because they attempt to reduce the end-member modes of fault-bend folding to their essential behavior from a balancing and kinematic perspective. Our examples from Taiwan are somewhat more complex than these toy models because they show variable fault dips and variable pregrowth stratigraphic thicknesses; nevertheless, these models are sufficient to introduce the essence of these end-member structural styles.

The practical qualitative distinction between these end-member folding types is easily made with seismic images (Suppe et al., 2004; Shaw et al., 2005). Classical fault-bend folding produces backlimb dips that are parallel to the fault ramp, with limb lengths that are equal to the fault slip (Figure 1A). In contrast, shear fault-bend folding produces ramp anticlines characterized by very long gentle backlimbs that dip much less than the fault ramp (Figure 1B, C). The length of the backlimb in shear fault-bend folds is typically much greater than the fault slip. The pure-shear and simple-shear end members are distinguished by differences in the geometry of the back syncline in the weak basal layer (see Figure 1). However, in what follows, we will focus less on the finite shape of the fold than on the instantaneous kinematics of these structures.

Classical fault-bend folding (Suppe, 1983) has the theoretical property that all motion of the hanging wall is parallel to the adjacent fault segment; therefore, the

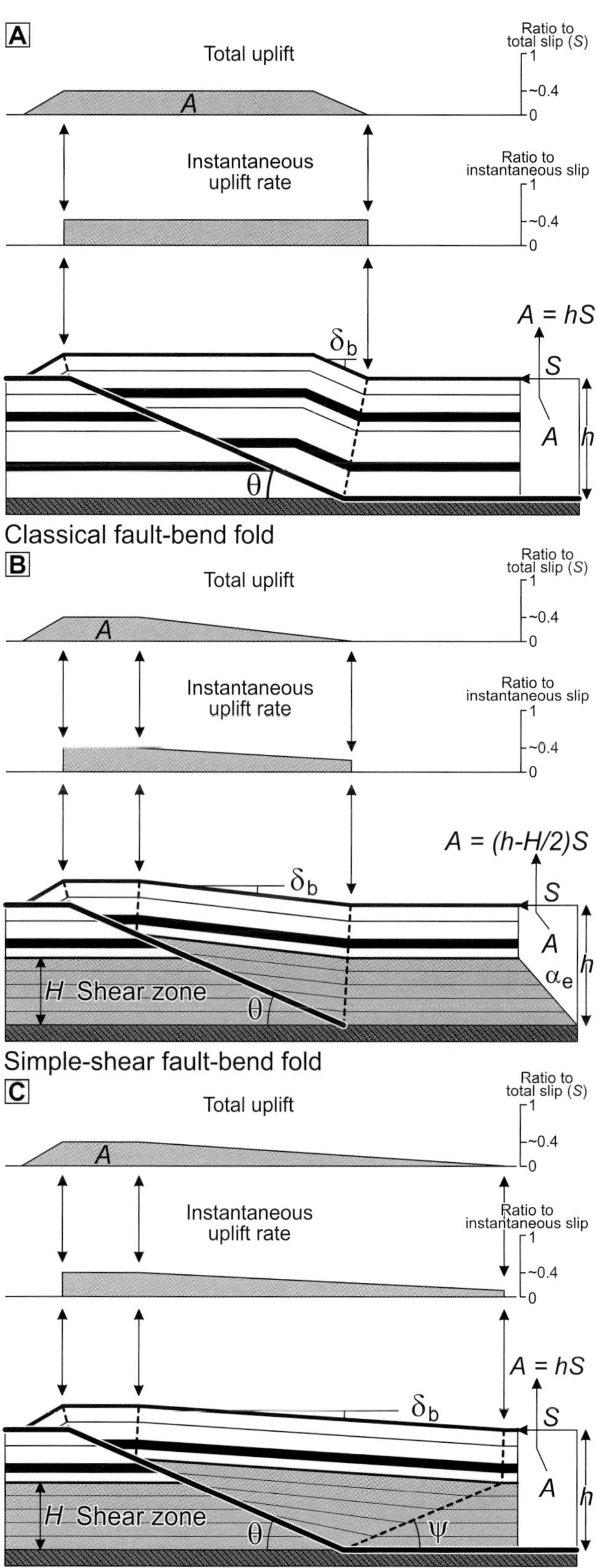

Figure 1. Kinematic models of classical no-shear fault-bend folding (fbf) (A) and two end members of shear fault-bend folding (panel B is a simple-shear fault-bend fold and panel C is a pure-shear fault-bend fold) with the same ramp dip showing different fold geometries on the fault ramps, different instantaneous uplift rate, and total uplift profiles (Suppe, 1983; Suppe et al., 2004). Note that the no-shear fault-bend fold displays a uniform instantaneous uplift profile, and the shear fault-bend folds show backward-tilted instantaneous uplift profiles on the backlimb. Here *A* is the total deformed area, *H* is the thickness of the weak decollement layer, and *h* is the original total thickness of the strata. Parameters of examples are the ramp dip, $\theta = 23^\circ$, for all three cases; the backlimb dip, $\delta_b = \theta = 23^\circ$, for the no-shear fault-bend fold example; the backlimb dip, $\delta_b = 6.5^\circ$, and the shear, $\alpha_e = 42^\circ$, for the simple-shear fault-bend fold example; and the backlimb dip, $\delta_b = 3.7^\circ$, and the dip of the back syncline, $\psi = 22.5^\circ$, for the pure-shear fault-bend fold example.

instantaneous uplift rate, h, is everywhere equal to the fault-slip rate, S, times the sine of the fault dip θ

$$h = S \sin \theta \tag{1}$$

However, in the case of oblique-slip faults, which include our examples from Taiwan, the equation is written in terms of the dip-slip component of the fault-slip rate.

$$h = S \cos \alpha \sin \theta \tag{2}$$

where α is the difference in azimuth of the slip vector and the dip direction. In either case, it follows that the long-term uplift rate will be constant over regions in which the fault dip is constant and abrupt discontinuities in uplift rate will exist across active axial surfaces that bound regions of different fault dip (Figure 1). In essence, the deformation associated with moving through fault bends is kink-band migration that results from the discontinuity in uplift rate.

In contrast, shear fault-bend folding (Suppe et al., 2004; Hardy and Connors, 2006) is characterized by a combination of limb rotation and kink-band migration in the hanging walls of thrust ramps (Figure 1B, C). The progressive rotation of the hanging wall above the fault ramp is caused by a progressive straining of the hanging wall by either pure shear or simple shear or by some combination of the two end members. Typically, this shear of the hanging wall is confined to a weak basal layer of finite thickness as shown in Figure 1, but in other cases, the entire hanging wall may strain heterogeneously (e.g., Shaw et al., 2005, p. 40). In addition, a discontinuity in uplift rate across the back synclinal axial surface is observed, similar to classical fault-bend folding (Figure 1A) associated with kink-band migration.

Notice in Figure 1 that the total finite uplift of these ramp models is different from their instantaneous uplift. For this reason, we expect the deformed shapes of young surfaces, such as terraces, to be different in location and shape from the finite structure. This noncollocated (not spatially coincident) uplift behavior is in contrast with pure-shear detachment folds that have an instantaneous uplift pattern that is linearly proportional to the finite uplift (Hubert-Ferrari et al., 2007).

Thrust Ramps in Western Taiwan

We document the contrasting kinematic styles of folding above the two frontal thrust ramps of the active fold and thrust belt of the western Taiwan Pliocene–Pleistocene clastic foreland basin (Figure 2). The westernmost structure is the Pakuashan-Tatushan anticline, which lies above the Changhua blind thrust ramp (Delcaillau et al., 1998; Wang et al., 2003) and forms the western limit of the current Taichung piggyback depositional basin. The next interior structure is the Chelungpu thrust sheet, which forms the current eastern depositional limit of the Taichung Basin. Seismic imaging, drilling, and surface geology indicate that both structures involve the same Pliocene and younger stratigraphy and lie above a detachment in the Pliocene Chinshui Shale, which is near the base of the foreland-basin sequence (Teng, 1990; Chen et al., 2001; Lin et al., 2003). Both the Changhua and Chelungpu thrusts step up from the Chinshui Shale detachment (Figure 3A) (see also Yue et al., 2005).

The Changhua and Chelungpu thrust ramps are both active, which allows us to study their kinematic behavior on several time scales (coseismic, geomorphic, and total structure). The Chelungpu thrust ruptured in the 1999 Chi-Chi earthquake (7.6 moment magnitude scale) with surface displacements typically in the range of 3–9 m (10–29 ft) (Yu et al., 2001; Dominguez et al., 2003; Lee et al., 2003). Much of the approximately 90-km (56-mi)-long surface break is shown in red in Figure 2. This earthquake is by far the best-instrumented historic thrust-belt earthquake (Ma et al., 1999, 2006). Coseismic surface displacements of the hanging wall are very well constrained by geodetic data (Yu et al., 2001; Dominguez et al., 2003; Lee et al., 2003). A longer term record of surface deformation in the hanging wall of the Chelungpu thrust is provided by flights of uplifted and folded terraces along the Wu and Tachia rivers (Figure 2B).

The Pakuashan-Tatushan anticlines are also actively deforming as recorded by extensive sequences of well-preserved flights of deformed and uplifted fluvial terraces (Figure 2B) (Shih and Yang, 1985; Delcaillau et al., 1998; Mouthereau et al., 1999; Delcaillau, 2001; Chen et al., 2003a; Sung and Chen, 2004). The Changhua thrust is presumed to be seismically active and could be the source of one or more of the large preinstrumental historic earthquakes in the region (Figure 2), such as the 1845 6–6.5 magnitude Taichung earthquake or the 1848 6.7–7.1 magnitude Changhua earthquake (Tsai, 1985; Cheng and Yeh, 1989).

In the following sections, we show that these two active thrust ramps, stepping up from the same Chinshui Shale detachment, have fundamentally different hanging-wall kinematics based on analysis of deformed terraces and coseismic displacements in the Chi-Chi earthquake. This contrast in kinematics should be expected from qualitative inspection of the structural styles of the finite structures seen in cross section (Figure 3A). The Changhua structure shows a long gentle backlimb and narrow steep front limb typical of shear fault-bend folding (compare Figure 1B, C); therefore, we should expect to see evidence for progressive backlimb rotation. In contrast, the Chelungpu thrust sheet is bedding parallel in the hanging wall; therefore, we expect to see classic fault-bend fold kinematics, with uplift proportional to the sine of the fault dip (equations 1, 2) and no limb rotation (Figure 1A).

Relatively little subsurface data are available over most of the length of the Changhua structure; therefore, our analysis is limited to the southern end of the Pakuashan anticline where several seismic lines exist together with well-developed flights of deformed terraces. In contrast, the three-dimensional (3-D) geometry of the Chelungpu thrust sheet is well known based on seismic images and surface geology (Yue et al., 2005), but well-developed flights of deformed terraces are only developed locally, particularly along the Wu and Tachia rivers (Figure 2B). We begin with the Chelungpu thrust ramp.

CHELUNGPU THRUST RAMP

Structural Geometry and Predicted Kinematics

The Chelungpu thrust shows a nearly classic ramp-flat subsurface geometry. For example, in map view, the Chelungpu thrust runs nearly parallel to bedding in the hanging wall in the Pliocene Chinshui Shale for nearly 50 km (31 mi) in classic thrust-belt fashion (Figure 2). This observation suggests that the fault at depth flattens at the base of the ramp to a nearly flat footwall detachment in the Chinshui Shale, which is confirmed by seismic profiles (B, D, E, and F in Figure 2A and Figures 3, 4), as discussed in detail by Yue et al. (2005) as part of their development of a detailed 3-D model of the Sanyi-Chelungpu thrust system (Figure 5). This parallelism between the thrust and bedding is also shown by the general parallelism of the 1999 Chi-Chi surface rupture, the

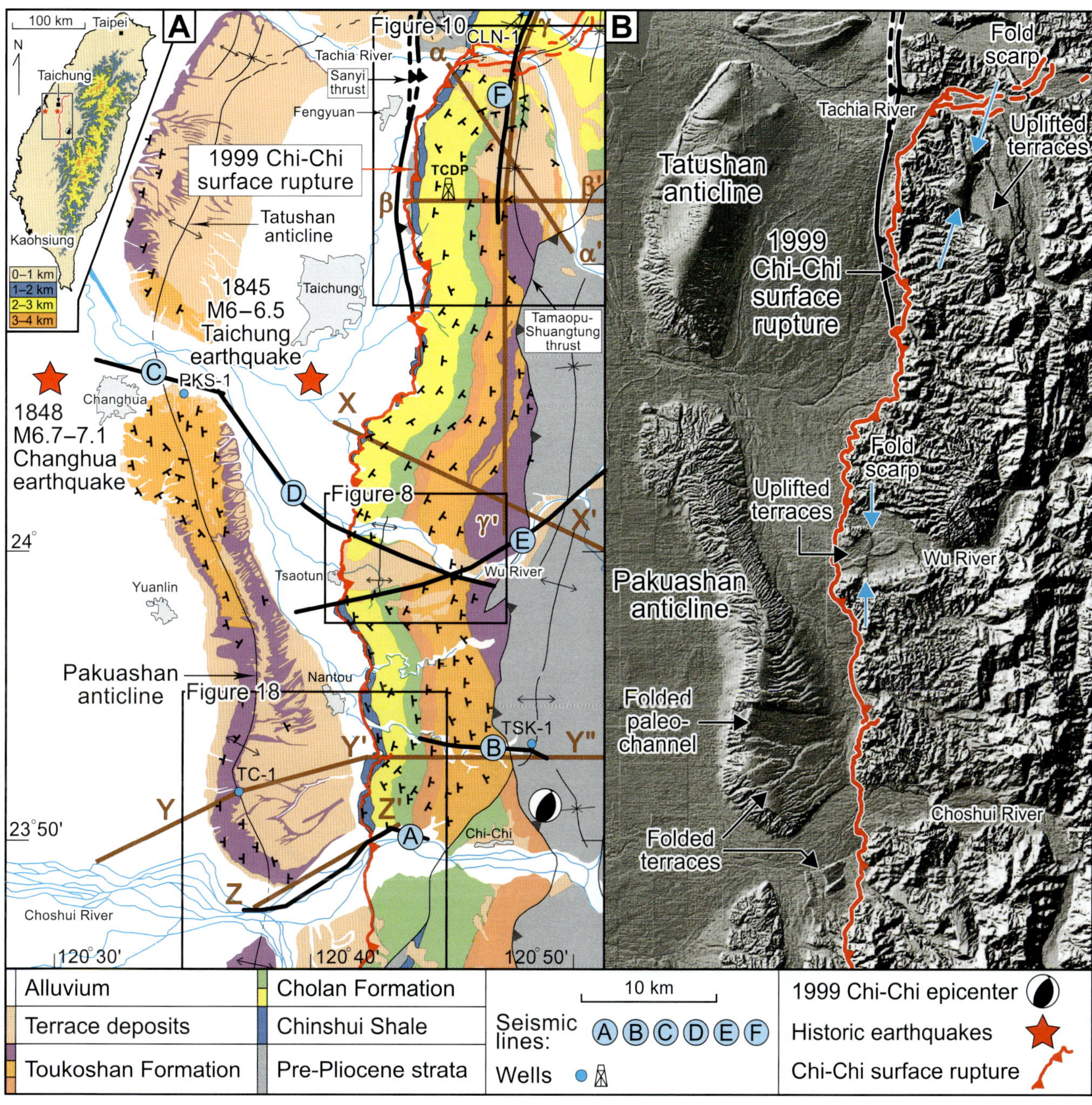

Figure 2. (A) Geologic map of the Chelungpu thrust sheet and the Pakuashan anticline with the surface rupture of the 1999 Chi-Chi earthquake shown in red and main shocks of three large earthquakes: 1845 6–6.5 magnitude Taichung earthquake, 1848 6.7–7.1 magnitude Changhua earthquake (Tsai, 1985; Cheng and Yeh, 1989), and 1999 7.6 moment magnitude scale Chi-Chi earthquake. The Chelungpu thrust runs parallel to the hanging-wall bedding in the Pliocene Chinshui Shale displaying the so-called bacon structure (Yue et al., 2005). Note that terrace deposits shown in pink cover mostly the southern Pakuashan anticline and also the east-dipping monoclinal strata of the Chelungpu thrust sheet along three main rivers in this area, Tachia, Wu, and Choshui rivers, which are used to demonstrate different folding behaviors of these two structures. See the text for details. The geologic map is modified from the one of the Chinese Petroleum Corporation (1982). For references for seismic lines and wells, see Yue et al. (2005). (B) Shaded relief map of the study area. The Pakuashan anticline is clearly folded up from the regional topography with a folded paleochannel on top of it. Also, uplifted terraces are found extensively along those three major rivers in this area. The marked lineations denoted by light blue arrows are fold scarps. See the text for details. TCDP = Taiwan Chelungpu Fault Drilling Project; M = magnitude.

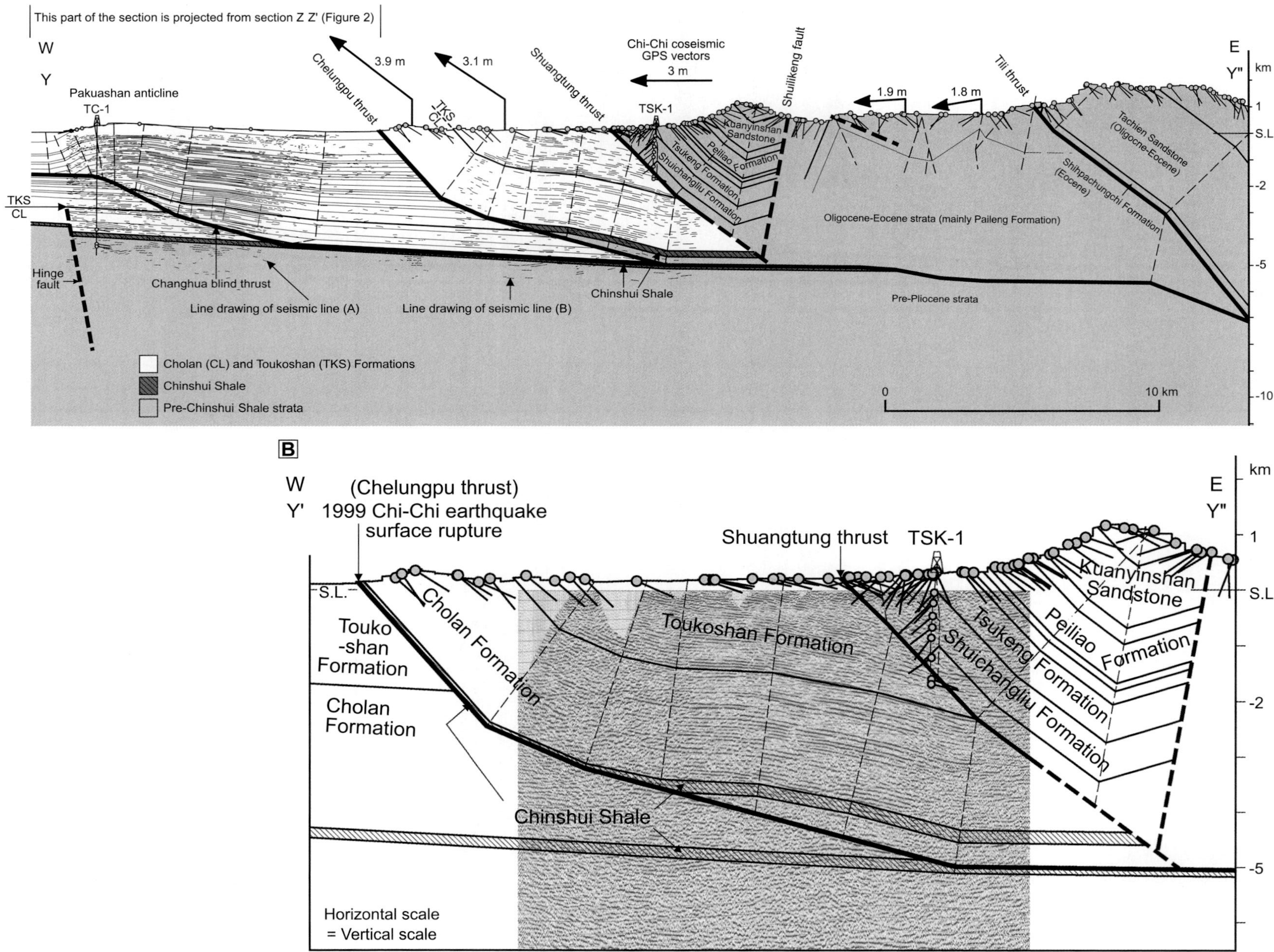
A
This part of the section is projected from section Z Z' (Figure 2)
W
Y
Pakuashan anticline
TC-1
Chelungpu thrust
3.9 m
3.1 m
TKS
CL
Shuangtung thrust
Chi-Chi coseismic
GPS vectors
3 m
TSK-1
Kuanyinshan Sandstone
Peiliao Formation
Tsukeng Formation
Shuichangliu Formation
Shuilikeng fault
1.9 m
1.8 m
Tili thrust
Tachien Sandstone (Oligocne-Eocene)
Shihpachungchi Formation (Eocene)
Oligocene-Eocene strata (mainly Paileng Formation)
E
Y"
km
1
S.L.
-2
-5
-10
TKS
CL
Hinge fault
Changhua blind thrust
Line drawing of seismic line (A)
Line drawing of seismic line (B)
Chinshui Shale
Pre-Pliocene strata
Cholan (CL) and Toukoshan (TKS) Formations
Chinshui Shale
Pre-Chinshui Shale strata
0
10 km
B
W
Y'
(Chelungpu thrust)
1999 Chi-Chi earthquake
surface rupture
Shuangtung thrust
TSK-1
E
Y"
km
1
S.L.
-2
-5
S.L.
Touko-shan Formation
Cholan Formation
Cholan Formation
Toukoshan Formation
Tsukeng Formation
Shuichangliu Formation
Peiliao Formation
Kuanyinshan Sandstone
Chinshui Shale
Horizontal scale
= Vertical scale

hanging-wall measurements of bedding strike, and the formation boundaries on the geologic map (Figure 2A), which display the so-called bacon structure typical of the ramp-flat structural style. Finally, the hanging-wall Chinshui Shale detachment is confirmed at depth by several post-Chi-Chi scientific boreholes that penetrated the fault, including the approximately 2-km (1.2-mi)-deep Taiwan Chelungpu Fault Drilling Project (TCDP) borehole (Figures 2, 4C) (Ma et al., 2006; Hung et al., 2007, 2009).

Note that the Chelungpu thrust is part of the larger Sanyi-Chelungpu thrust system (Figure 5), which is composed of a single ramp-flat Sanyi-Chelungpu thrust to the south but branches northward along a well-defined branch line into the Sanyi ramp-flat thrust and the bedding-parallel north Chelungpu Chinshui detachment in the hanging wall of the Sanyi thrust (Figure 4A–C). The Sanyi thrust also runs parallel to bedding in its hanging wall but at a deeper stratigraphic level in the Pliocene and Miocene Kueichulin to Kuanyinshan formations (Figures 2, 4A–C) (see also Yue et al., 2005). Note that the surface rupture in the Chi-Chi earthquake was confined to the Sanyi-Chelungpu thrust and the north Chelungpu Chinshui detachment, with no surface rupture on the Sanyi thrust farther to the north (Figures 2, 5).

These observations indicating the nearly fault-parallel bedding in the hanging wall are important for our present purposes because of their kinematic implications. They imply that the long-term and geomorphic displacement of the hanging wall is not only fault parallel (Figure 1A) but also parallel to bedding in the hanging wall, which is a property we will apply in the following sections to constrain the kinematics of the Chelungpu thrust sheet.

Coseismic Displacements of the Chelungpu Thrust Sheet

The 1999 Chi-Chi earthquake (7.6 magnitude) is important because it is one of the best-instrumented historic large earthquakes on a near-surface fault (<5–6 km [3.1–3.7 mi]) (Ma et al., 1999, 2006). It is particularly important for our purposes because it has excellent geodetic data on the coseismic displacement at several locations in the hanging wall (Yu et al., 2001; Dominguez et al., 2003; Lee et al., 2003), combined with a well-determined geometry on the immediately underlying fault (Yue et al., 2005). The locations of these measured coseismic displacements are in the region where the underlying fault lies almost entirely along the Chinshui Shale in the hanging wall (Figures 2–4) and the fault geometry is well determined in 3-D (Figure 5).

Fault-bend folding theory predicts that the displacement vectors, summed over several earthquake cycles to remove elastic transients, are everywhere parallel to the underlying fault shape (Suppe, 1983). These displacement vectors measured at the land surface are also predicted to be parallel to nearby bedding measurements because of the bedding-parallel nature of the fault. In the simple model in Figure 6A, the surface-displacement vectors are shown to be inclined and parallel to the local ramp dip and to adjacent surface-bedding dip measurements. Furthermore, whereas the displacement magnitudes are observed to be heterogeneous in a single earthquake (Figure 5A), the displacement magnitudes that accumulate over several earthquake cycles are expected to become constant in the slip direction, such that $S_0 = S_1 = S_2 = S_3$ in Figure 6A, if bed length and layer thickness are conserved in a bedding-parallel thrust (Suppe, 1983). However, lateral variation in slip is still possible, with associated distortion of the thrust sheet. We evaluate the extent to which the Chelungpu thrust sheet approximates this fault-parallel constant-slip behavior.

Remarkably, the fault-parallel displacement that is expected to accumulate over several earthquake cycles is closely approximated in a single large earthquake as shown by the orientations of coseismic surface displacements of the Chi-Chi earthquake based on Global positioning System (GPS) measurements (Yue et al, 2005). As summarized in Figure 6B, the plunges of the coseismic displacement vectors, δ_{GPS}, are statistically equal to the apparent dip, δ'_B, of the adjacent bedding in the direction of displacement. This demonstrates that the coseismic surface displacements are statistically parallel

Figure 3. (A) Regional east–west section YY″ with the line drawing of reflectors in seismic lines A and B (they are separated about 6 km [3.7 mi], locations in Figure 2A), showing that both the Changhua blind thrust and the Chelungpu thrust have developed their ramps from the same Chinshui detachment at a depth of 5.5–6 km (3.4–3.7 mi). See the text for details. The Global Positioning System (GPS) vectors are projected showing the component of the Chi-Chi earthquake coseismic displacement in the plane of the section (Yue et al., 2005). (B) Structural interpretation of the depth-converted seismic line B (Wang et al., 2000) with dip measurements from the TSK-1 well and geologic maps, showing that the Chelungpu thrust has a classic ramp-flat geometry with a multibend fault-bend fold on top. The Chelungpu thrust flattens to the Chinshui detachment level to the east of the TSK-1 well. Two dip measurements near the bottom of the TSK-1 well and surface dip measurements agree with seismic reflections and together display an east-dipping monoclinal sequence. The little stratigraphic wedge along the base of the Chelungpu thrust is completely based on the seismic interpretation, which can be compensated in the restoration phase (Yue et al., 2005). S.L. = sea level.

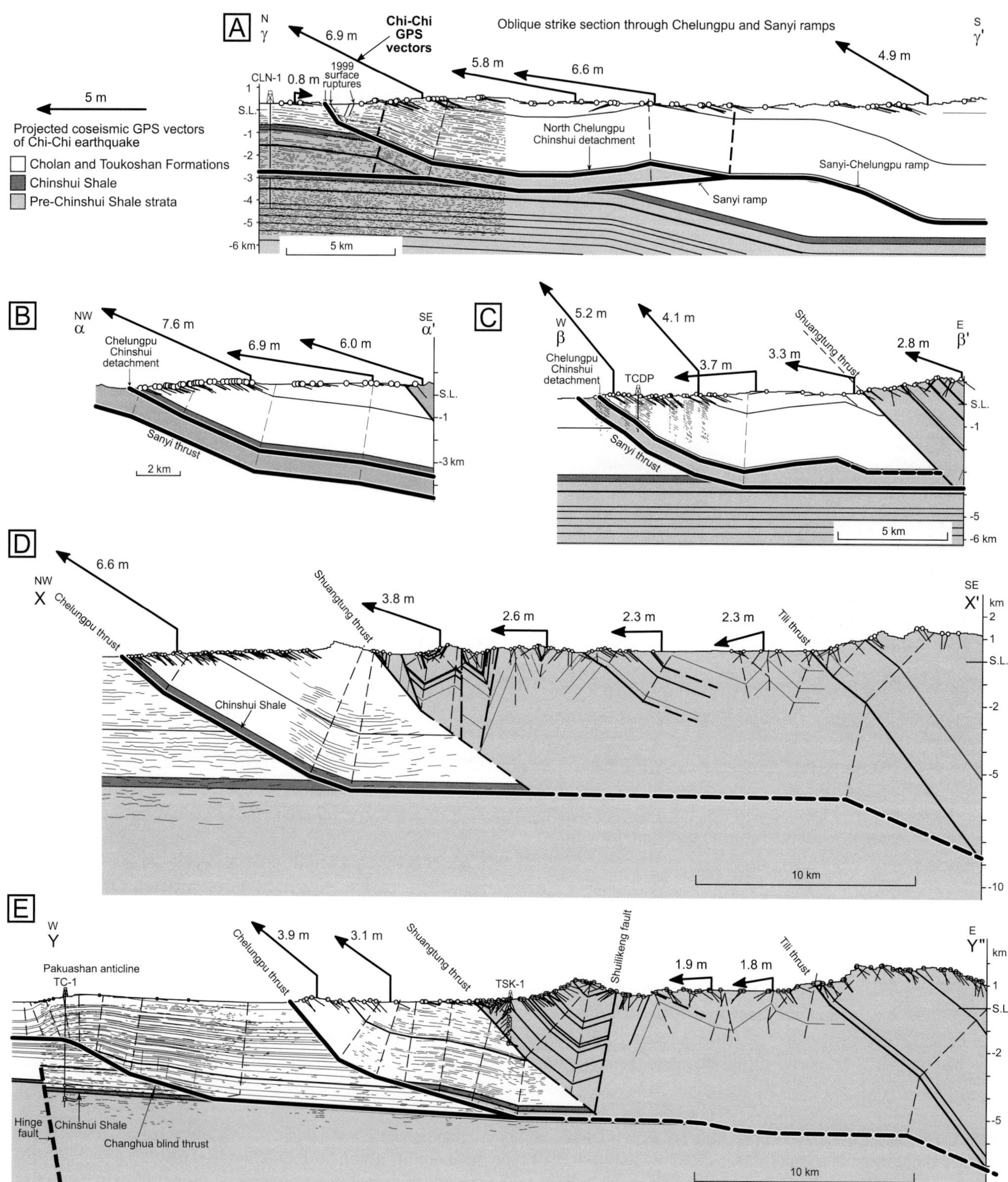

Figure 4. Five cross sections of the Chelungpu thrust showing parallelism between the Chelungpu thrust plane (thick black lines) and coseismic vectors. Sections A to E are named sections γγ′, αα′, ββ′, XX′, and YY″ in Figure 2A and Yue et al. (2005). Those vectors turn to subhorizontal as the thrust turns to flat in sections C to E. Sections A, C, D, and E are constrained by seismic lines and wells (for details, see Yue et al., 2005). Note that the vectors are approximately fault parallel, suggesting that the surface coseismic displacement is dominated by net structural growth. GPS = Global Positioning System; S.L. = sea level; TCDP = Taiwan Chelungpu Fault Drilling Project.

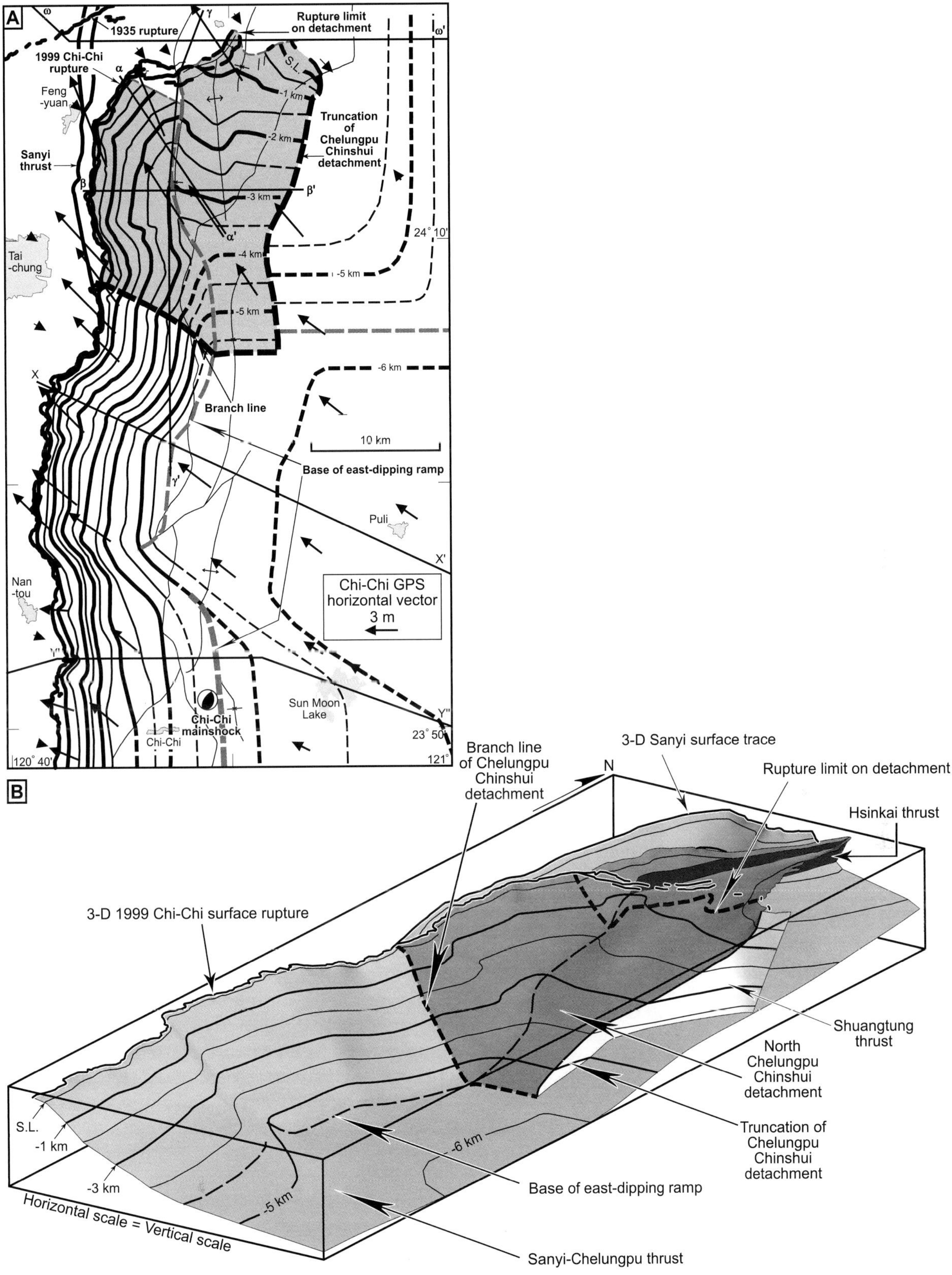

Figure 5. (A) Chelungpu fault model (500-m [1640-ft] contours) together with the horizontal component of the GPS coseismic displacements of Yu et al. (2001). The north Chelungpu Chinshui Shale detachment, shown in gray, takes off northward from the Sanyi-Chelungpu thrust along the branch line (see sections ββ′ and γγ′ in Figure 4). The north Chelungpu Chinshui Shale detachment is truncated to the east by the Shuangtung thrust. (B) Oblique 3-D view of the Sanyi-Chelungpu fault model showing its ramp-flat geometry. See Yue et al. (2005) for details. GPS = Global Positioning System; S.L. = sea level.

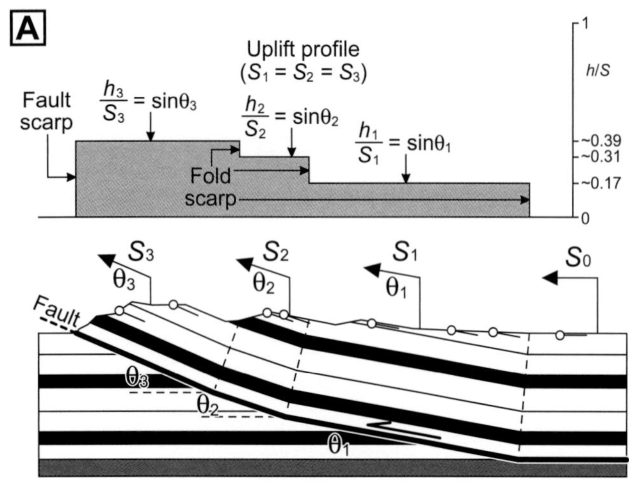

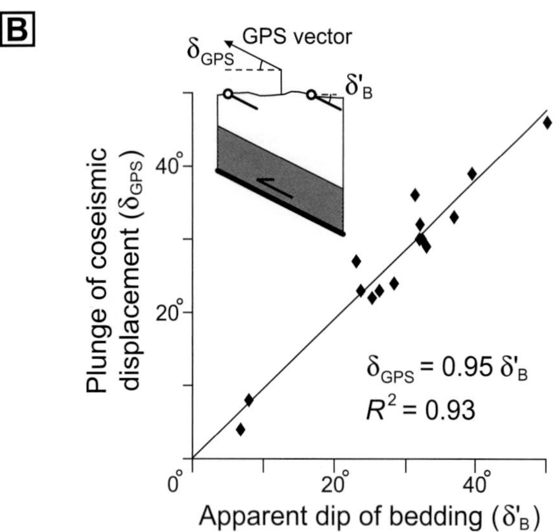

Figure 6. (A) Multibend thrust-ramp model showing that (1) coseismic slip vectors are predicted to be parallel to the thrust-ramp plane because of the bedding-parallel nature of the fault, which is observed in Figure 4; (2) uplifts of the terraces are equal to the vertical components of the corresponding slip vectors. (B) Observed parallelism of local orientation of bedding above the Chelungpu ramp and coseismic surface displacements of the 1999 Chi-Chi earthquake. This shows that the coseismic displacements in the Chi-Chi event are dominantly fault parallel and mostly reflect net structural and geomorphic growth (Yue et al., 2005). GPS = Global Positioning System.

to the adjacent bedding-parallel fault ramp, in agreement with fault-bend folding predictions. This approximate but not exact parallelism of individual data points is shown in the cross sections, which plot the components of the coseismic displacements that lie in the plane of the sections (Figures 3A, 4).

These observations suggest that the surface coseismic displacements are dominated by net structural growth; however, the fact that the displacement magnitudes measured at the surface show substantial gradients in the direction of slip not only reflects the heterogeneous slip on the fault surface in a single earthquake but also requires a transient component to the hanging-wall deformation. This transient component may be a combination of elastic effects and fracture dilatancy. Part of this transient deformation is predicted by elastic-dislocation modeling to appear in the geodetic data as nonfault-parallel displacement vectors, especially near the rupture terminations (Johnson et al., 2001; Dominguez et al., 2003; Johnson and Segall, 2003; Loevenbruck et al., 2004; Wang et al., 2004). For example, downward-pointing vectors are expected in the hanging wall near the eastern limit of the Chi-Chi rupture, which are observed, as seen in the eastern part of Figure 4D and E.

Thus, we see that the long-term kinematic predictions of fault-bend folding for the Chelungpu thrust ramp are approximated by the coseismic displacement orientations in the Chi-Chi earthquake (Figure 6B). However, the heterogeneous displacement magnitudes in the slip direction and the nonfault-parallel components to the displacement (Figures 4, 5A) indicate that several earthquake cycles may be required to accumulate the constant fault-parallel displacement in the slip direction that is predicted for the finite structure. The flights of terraces deposited over the deforming hanging wall will provide insight into the kinematics over a time scale that represents 10–100 large earthquakes.

Deformed Terraces of the Chelungpu Thrust Sheet

The deformation of hanging-wall fluvial terraces above thrust ramps of the sort modeled in Figure 6A can be expected to provide a relatively accessible record of the vertical component of the cumulative fault-parallel displacement, which is expected to be proportional to the sine of the fault dip (equations 1, 2). Discrete changes in ramp dip are expected to be reflected in discrete discontinuities in uplift rate across active axial surfaces, as shown at the top of Figure 6A. Therefore, the total uplift of a deformed terrace on the two sides of an active axial surface is predicted to be $h_1 = S_1 \sin\theta_1$ and $h_2 = S_2 \sin\theta_2$, where θ_1 and θ_2 are the bedding and fault dips on the two sides (Figure 7A). It follows that

$$\frac{h_2}{h_1} = \frac{S_2 \sin\theta_2}{S_1 \sin\theta_1} \tag{3}$$

However, in the cases of oblique slip or changes in dip direction across the axial surface (equation 2)

$$\frac{h_2}{h_1} = \frac{S_2 \cos\alpha_2 \sin\theta_2}{S_1 \cos\alpha_1 \sin\theta_1} \tag{4}$$

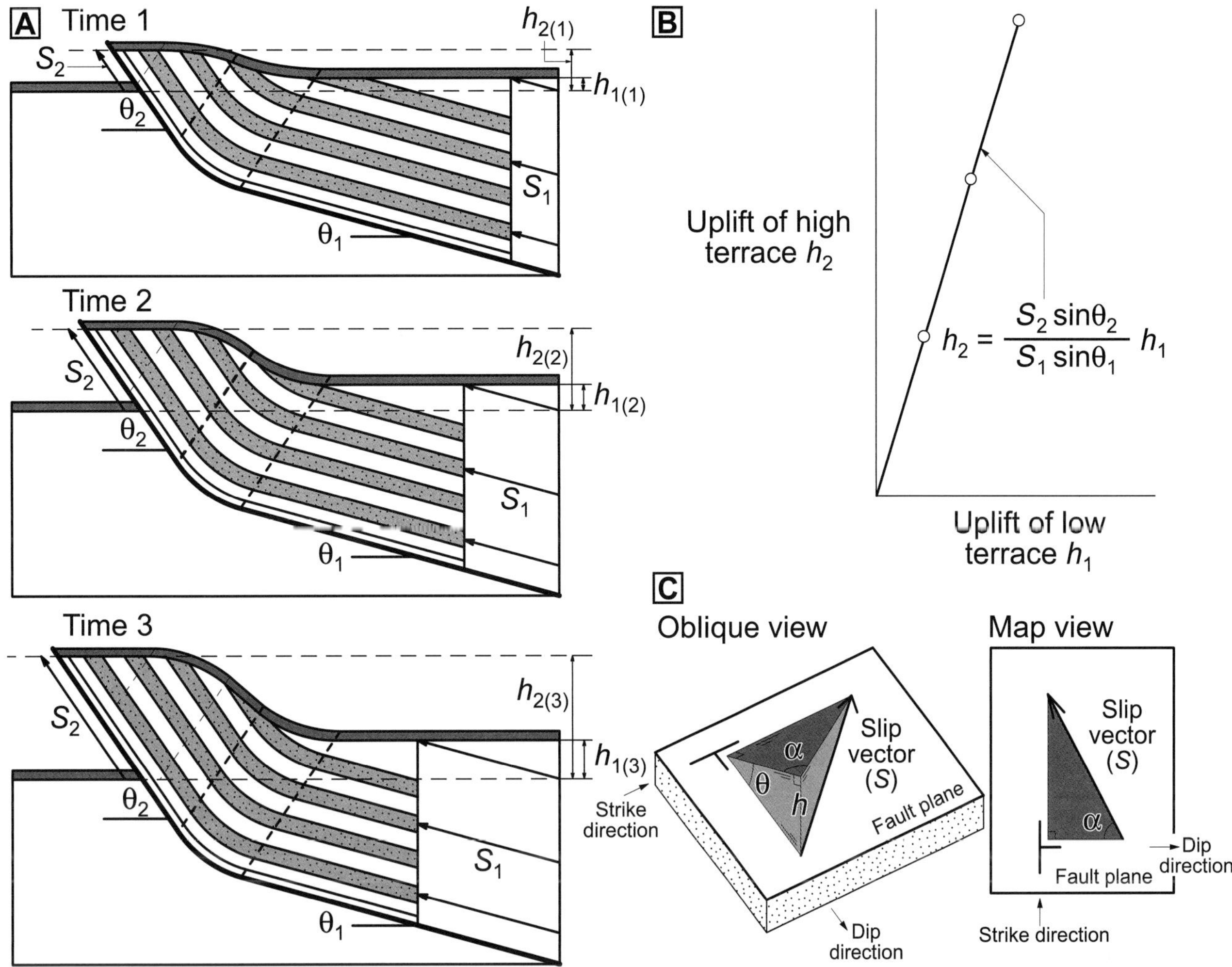

Figure 7. (A) Simple fault-bend folding model of progressive terrace uplift and kink-band migration across a discrete bend in a thrust ramp. (B) The ratios of uplift, h_1/h_2, of a flight of deformed terraces are linearly related by a slope that is $\sin\theta_1/\sin\theta_2$. We use such graphs in Figure 12 to show that terrace uplift in the Chelungpu thrust sheet is consistent with the predictions of fault-bend folding. (C) Correction of the net vertical component of an oblique slip vector to the strike of a given fault plane. See the text for details.

where α_1 and α_2 are the differences in azimuth between the slip vector and the dip direction (Figure 7C). As the transient components of displacement become negligible after the accumulation of a moderate number of large earthquakes, the ratio of slip is expected to approach $S_2/S_1 = 1$ if the bed length is conserved (Suppe, 1983). We use equations 3 and 4 to test the simple kinematic model of Figure 6 over the time scale of formation of flights of fluvial terraces (10–100 large earthquakes) based on measurements of uplifts, $\langle h_1, h_2 \rangle$, for a set of terraces cut by an active axial surface with associated bedding dips, $\langle \theta_1, \theta_2 \rangle$. In general, the measured uplifts, $\langle h_1, h_2 \rangle$, may need to be corrected for changes in base level produced by a variety of causes, as discussed by Lavé and Avouac (2000). The differences in azimuth, $\langle \alpha_1, \alpha_2 \rangle$, are taken from the coseismic displacement vectors in the Chi-Chi earthquake and the 3-D fault model (Figure 5A) (see also Yue et al., 2005; Yue, 2007). The predicted behavior in the simplest case is shown in Figure 7, in which the slope of the graph is given by equations 3 and 4.

In addition to the vertical displacement recorded by terraces, the motion of particles through the active hinge zone produces fold scarps that grow progressively in width with each successive earthquake by a process of kink-band migration (Suppe et al., 1997, 2000; Lai et al., 2006; Chen et al., 2007; Hubert-Ferrari et al., 2007). Detailed analysis of these fold scarps can in some cases provide a complete history of the horizontal components of displacement as shown by Hubert-Ferrari et al. (2007). Such fold scarps have been documented in the Chelungpu thrust sheet, and these grew coseismically in the Chi-Chi earthquake (Suppe et al., 2000; Lai et al., 2006; Chen et al., 2007).

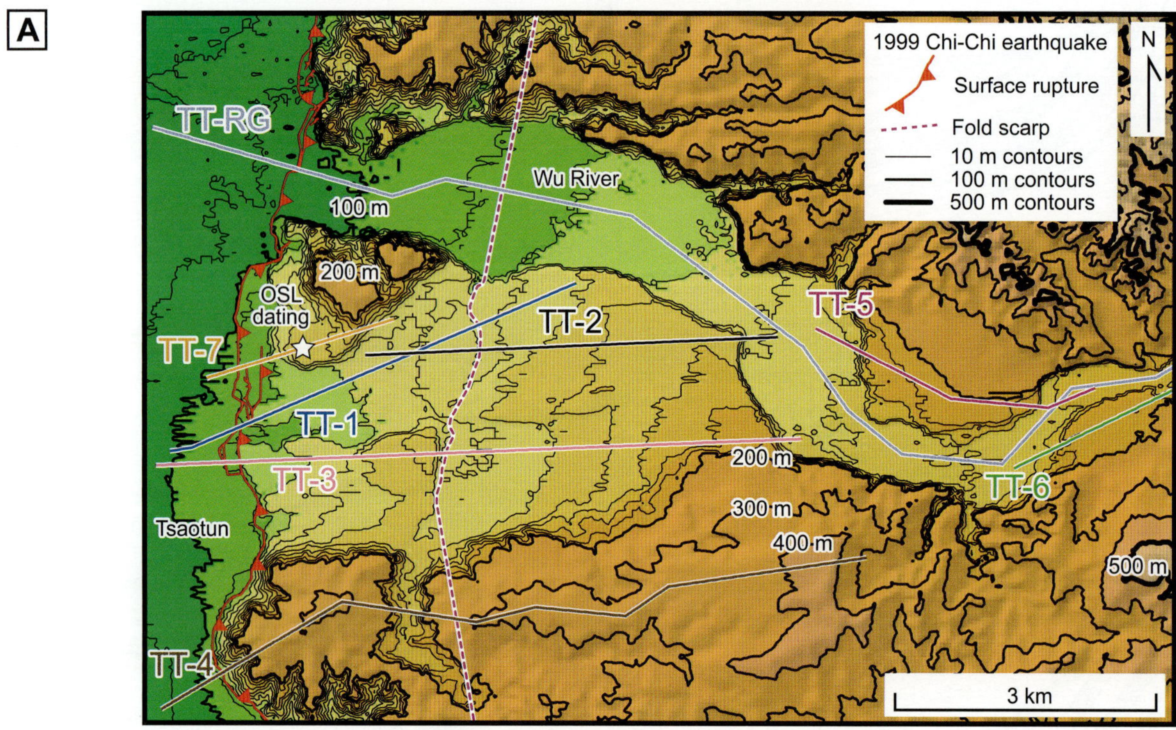

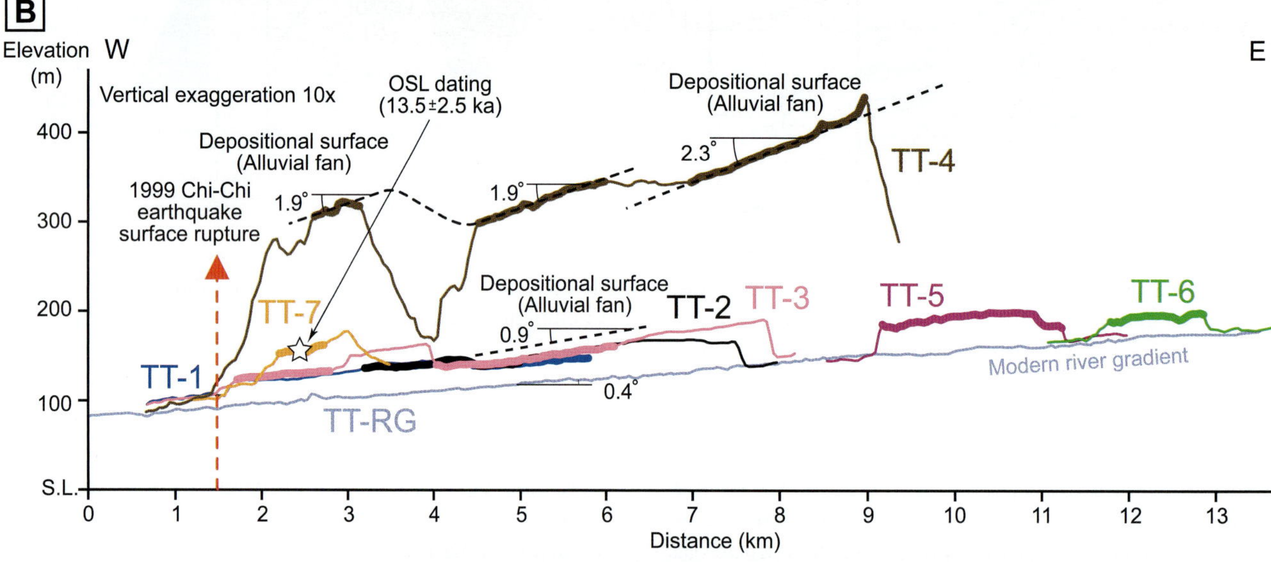

Figure 8. (A) Ten-meter (33-ft)-contour topographic map of the Tsaotun Terrace with locations of seven profiles and a modern river profile. The north–south trend of fold scarps shown in red dashed lines implies that the underlying Chelungpu thrust contains one major fault bend here (see Figure 4D, section XX′, and Figure 5A for the location). One dated sample collected on the high terrace by Chen et al. (2003b) is shown by a star. (B) Seven terrace profiles and a modern river profile of the Tsaotun Terrace with a vertical exaggeration of 10×. Those terraces have slopes close to the modern river gradient, which demonstrates that they have been uplifted homogeneously. S.L. = sea level; OSL dating = optically stimulated luminescence dating; TT = Tsaotun; RG = river gradient.

Terrace Uplift Along the Wu and Tachia Rivers

In this section, we determine the vertical components of displacements from elevations of deformed flights of terraces that are intersected by active axial surfaces in the hanging wall of the Chelungpu thrust sheet (Figure 2B). We study two areas of extensive terrace development, along the Wu River east of Tsaotun in the south and

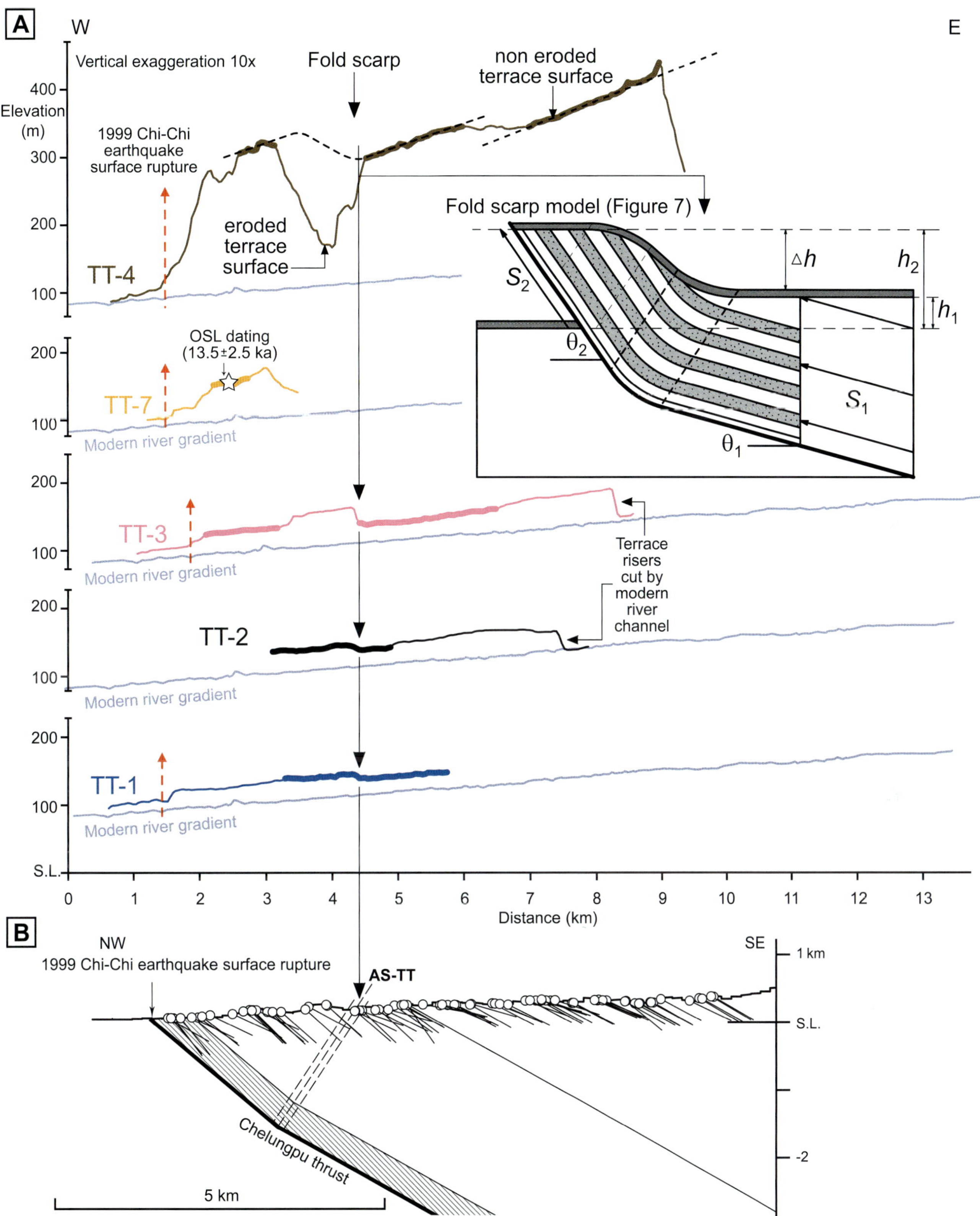

Figure 9. (A) Five terrace profiles of the Tsaotun Terrace with a vertical exaggeration of 10×, showing that (1) the gradient of terrace surfaces is similar to the modern river gradient except profile TT-4 (see text for details) and (2) a distinct fold scarp can be traced from north to south in the Tsaotun area. (B) Cross section D in Figure 4 (also see Yue et al., 2005). Note that the fault scarps are associated with axial surface AS-TT. TT = Tsaotun Terrace; S.L. = sea level; OSL dating = optically stimulated luminescence dating.

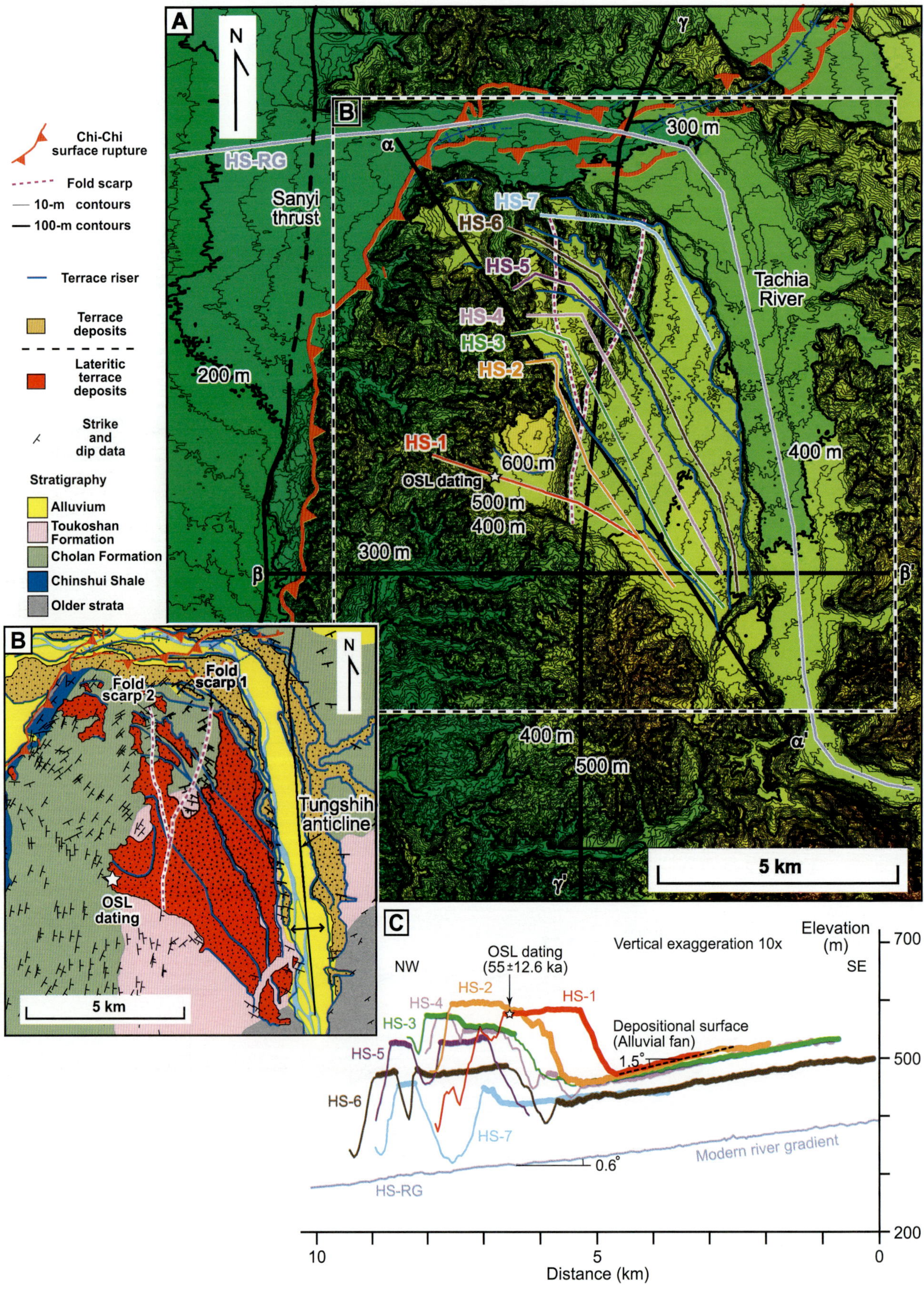
A
N
Chi-Chi surface rupture
Fold scarp
10-m contours
100-m contours
Terrace riser
Terrace deposits
Lateritic terrace deposits
Strike and dip data
Stratigraphy
Alluvium
Toukoshan Formation
Cholan Formation
Chinshui Shale
Older strata
HS-RG
Sanyi thrust
300 m
200 m
HS-7
HS-6
HS-5
HS-4
HS-3
HS-2
HS-1
600 m
OSL dating
500 m
400 m
Tachia River
5 km
B
Fold scarp 2
Fold scarp 1
Tungshih anticline
C
NW
SE
OSL dating (55±12.6 ka)
Vertical exaggeration 10x
Elevation (m)
Depositional surface (Alluvial fan)
1.5°
0.6°
Modern river gradient
700
500
200
10
5
0
Distance (km)

along the Tachia River near Hsinshe (Figure 2B). In the next section, we will analyze the structural implications of these terrace uplift measurements.

The terrace profiles (Figures 8–11) were extracted from a 40-m (131-ft) digital elevation model (DEM) that is based on 1:5000 topographic maps with a 5-m (16-ft) contour interval; therefore, the vertical resolution is much better than the horizontal. The topographic gradients of long terrace profiles are generally similar to the modern Wu and Tachia river gradients (0.4–0.6°) but are slightly higher, with the older terraces having progressively higher gradients, reaching as high as 1.4–1.9°. These steeper gradients could possibly be structural; however, all of the terrace gradients (0.6–2.3°) are within the range of observed depositional gradients for more proximal rivers and fans (Lin, 1957; Shih et al., 1986; see also Blair and McPherson, 1994). Therefore, we interpret the 1–2° steeper gradients of the highest terraces as primary depositional gradients reflecting progressively more proximal sources for higher terraces. These more proximal gradients suggest that the presently throughgoing major Wu and Tachia rivers are a relatively recent development due to the deactivation of the Shuangtung thrust, immediately to the east. Prior to deactivation, uplift above the Shuangtung thrust ramp formed the barrier that allowed deposition of the Puli piggyback basin to the east of the Shuangtung thrust. This basin is now being incised by the Wu and Tachia river systems.

The uplift of each terrace segment (h_x) was measured as its relief relative to an assumed paleoriver profile, which for the lower terraces is similar to the modern river profile (Rockwell et al., 1984; Molnar, 1987; Burbank et al., 1996; Lavé and Avouac, 2000; Burbank and Anderson, 2001). Higher more proximal gradients are used for the higher terraces. We remove the present river gradient from the terrace profiles, which is a partial removal of the original gradient that allows for better measurement of the uplift. In addition, we require uplifts relative to the fault; therefore, we have made small corrections (~+10%) for temporal changes in base level, constrained in the footwall by depths of dated horizons in the Wufeng borehole about 10 km (6.2 mi) to the north of Tsaotun (Lai et al., 2006). We measured the uplift of the upper surfaces of the terraces, which may introduce some error because there may be some difference in the age of the end of deposition on the two sides of a fold scarp with uplift ratios, h_2/h_1, that are therefore too small. This effect may be significant for terrace HS-6, which shows anomalously low h_2/h_1. We note that the observed terrace gradients are similar for the two sides of the fold scarp, which is consistent with fault-bend folding.

Wu River (Tsaotun) Terraces. We begin with the Wu River terraces, which are deformed by a fold scarp in the hanging wall of the Chelungpu thrust about 2.5 km (1.5 mi) east of its surface trace near the city of Tsaotun (Figures 8, 9). This fold scarp was activated in the Chi-Chi earthquake (Chen et al., 2007). Profiles of the principal terraces are labeled TT-1 to TT-7 in Figure 8B. Terrace TT-7 has an optically stimulated luminescence (OSL) date of 13.5 ± 2.5 Ma (Chen et al., 2003b). All but the highest terrace (TT-4) show gradients similar to the modern profile of the Wu River, whereas profile TT-4 shows a substantially higher gradient, which is consistent with a proximal source, suggesting that a throughgoing major Wu River may not have existed at the time of this terrace. The long terrace profiles, TT-1 to TT-4, cross the fold scarp (Figure 9A). Bedding dip data collected north of the Wu River show a steeper mean dip ($\bar{\theta}_3 = 42°$, n = 37) west of the Chi-Chi fold scarp where it crossed the Wu River in the Chi-Chi earthquake and lower mean dip to the east ($\bar{\theta}_1 = 28°$, n = 34) (Figure 9B), with identical bedding strike. This axial surface is associated with an upward increase in dip of the Chelungpu fault (see also Figures 3, 4).

Tachia River (Hsinshe) Terraces. The folding and uplift of the terraces near Hsinshe along the Tachia River are more complex because the fold scarp branches northward into two fold scarps (Figure 10). Furthermore, the strikes of bedding and the associated fault are quite different across each fault bend and are variable along strike, reflecting a complex three dimensionality to the Sanyi-Chelungpu fault system in this region as shown in Figure 5 (see also Yue et al., 2005). Therefore, the coseismic displacement vectors from the Chi-Chi earthquake, which are relatively constant in orientation in this region (Figure 5A) (Dominguez et al., 2003), have a variable orientation, α, relative to the local bedding or fault dip direction, requiring the use of

Figure 10. (A) Ten-meter (33-ft)-contour topographic map of the Hsinshe Terrace with locations of seven profiles and a modern river profile. Note that, for elevations higher than 300 m (984 ft), the contour density has been reduced in some places. (B) Geologic map of the Hsinshe Terrace. Note that the fold scarp shown in red dashed lines splits into two branches to the north, becoming fold scarp 1 and fold scarp 2. This implies that the underlying fault ramp here changes from single bend to two bends to the north. One dated sample collected on the highest terrace by Chen et al. (2003b) is shown by a star. (C) Seven terrace profiles and a modern river profile of the Hsinshe Terrace with a vertical exaggeration of 10×. Profiles are arranged with the modern river profile to show the parallelism, which demonstrates that they have been uplifted homogeneously (Figures 1, 6). OSL dating = optically stimulated luminescence dating; HS = Hsinshe; RG = river gradient.

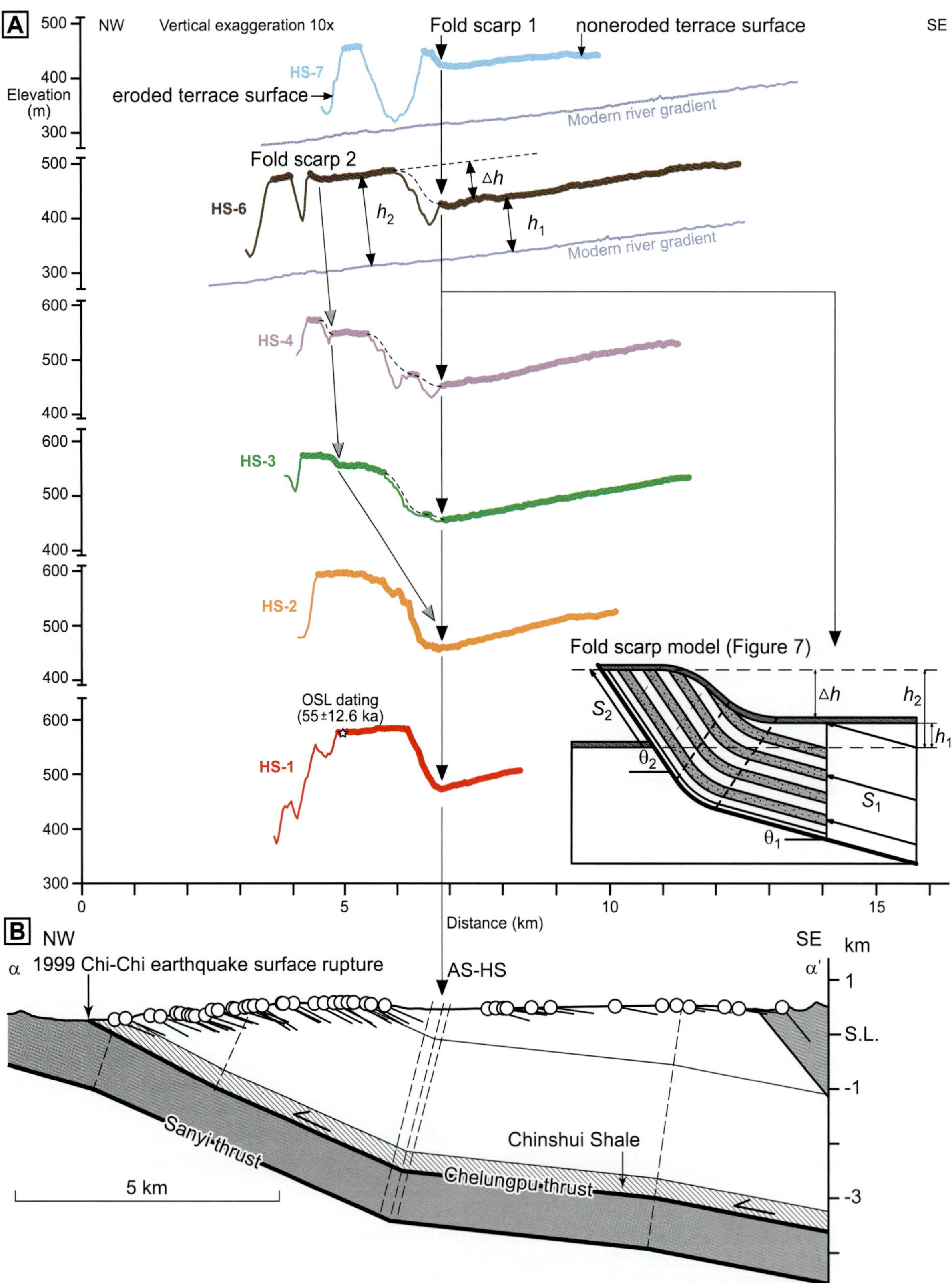

Figure 11. (A) Six along-strike profiles across the Hsinshe fold scarp of folded terraces of different ages and a modern river profile with a vertical exaggeration of 10×. Profiles are lined up along the base of fold scarp 1 for comparison. Note that fold scarp 2 branches from scarp 1 north of profile HS-2. (B) Cross section B in Figure 4 (see also Yue et al., 2005). Note that fault scarp 1 is associated with axial surface AS-HS. HS = Hsinshe; S.L. = sea level; OSL dating = optically stimulated luminescence dating.

equations 2 and 4. Profiles of the principal through-going terraces that are cut by the fold scarps are shown in Figure 11 together with a representative cross section drawn parallel to the coseismic displacement vector. The highest terrace shows a substantially steeper gradient than the modern Tachia River, suggesting a more proximal source (Figures 10B, 11A), similar to the highest terrace in the Tsaotun area discussed above. The estimated terrace uplifts are given by Yue (2007).

Structural Implications of Terrace Uplift Data

The terrace profiles of the Tsaotun and Hsinshe areas (Figures 8–11) are qualitatively consistent with the predictions of classical fault-bend folding, showing progressively larger differential uplift across the fold scarps for older terraces, with similar terrace gradients on the two sides of the fold scarps and no evidence for back tilting of the sort predicted by shear fault-bend folding (Figure 1). We interpret the 1–2° steeper gradients of the highest terraces as primary depositional gradients because they are within the range expected for more proximal sources, as discussed above, although we recognize that a very small homogeneous simple shear within the Chelungpu thrust sheet would produce a 1–2° forward tilting of the terraces, as would a modest growth of subthrust folds.

We now use the uplift data to test the predictions of classical fault-bend folding using the simple uplift model of Figure 7. The uplift of the terraces on the two sides of an axial surface should be linearly proportional, with the slope of the regression given by equations 3 and 4. The data for Tsaotun approximate this linearly proportional relationship (Figure 12A). The data for Hsinshe, although in general agreement, show significant scatter (Figure 12B–D). We attribute this scatter to the observed variable dip magnitude, θ, and dip direction, α, relative to the slip direction in the Hsinshe area (Yue, 2007) based on the following analysis.

Each observed pair of terrace uplifts (h_1, h_2) and associated bedding dips (θ_1, θ_2) predicts a slip ratio of

$$\frac{S_1}{S_2} = \frac{h_1 \sin\theta_2 \cos\alpha_2}{h_2 \sin\theta_1 \cos\alpha_1} \qquad (5)$$

where the displacement azimuths (α_1, α_2) are the coseismic azimuths of the Chi-Chi earthquake (Figure 5A), which are expected to approximate the long-term displacement azimuth because they are parallel to large-scale corrugations of the Chelungpu fault surface (Yue et al., 2005). Therefore, if we plot the ratio of fault dip and slip azimuth, $\sin\theta_2 \cos\alpha_2/\sin\theta_1 \cos\alpha_1$ as a function of the ratio of terrace heights, h_2/h_1, the data should lie along a line of $S_1/S_2 = 1$ based on the prediction of fault-bend folding that slip is conserved if the bedding is fault parallel (Suppe, 1983). This comparison for the combined data of the Tsaotun and Hsinshe areas (Figure 13) shows a strong linear relationship with a slope close to $S_1/S_2 = 1$. These results confirm that the terrace uplift data closely approximate the predictions of classical fault-bend folding for a bedding-parallel hanging wall, given by equation 2. Therefore, over the time scale of terrace uplift, which amounts to 10–100 large earthquakes, the transient deviations from the predictions of classical fault-bend folding observed in the Chi-Chi coseismic displacements (Figures 4, 6) are removed. Figure 13 implies that the displacement field, on the time scale of terrace uplift and folding, is constant in magnitude in the slip direction, as assumed in equations 3 and 4, and is fault and bedding parallel, as assumed in equations 1 and 2.

The combined analysis based on terrace uplift and fault and bedding dip data of Figure 13 is important beyond being a test of fault-bend folding theory because it demonstrates that reliable estimates of fault slip can be obtained by application of equation 2, as proposed previously by Lavé and Avouac (2000). These fault-slip magnitudes are retrieved from both fold-scarp heights and terrace heights, which yield similar results in this case (Yue, 2007). The advantage of using fold-scarp heights instead of terrace heights to estimate the fault slip is that we can avoid the difficulty of determining the absolute height of regional base level. Furthermore, with suitable dating of terraces, slip rates and slip histories can be obtained. One terrace in each area has been dated using optically stimulated luminescence (OSL) measurements (Chen et al., 2003b); therefore, we can estimate a fault-slip rate for each area as well as the ages of the undated terraces, assuming constant long-term slip rate for each area, as shown in Figure 14 (Yue, 2007). Remarkably, the estimated slip rates differ by about a factor of two, with a slip rate of 7.8 + 1.8–1.2 mm/yr (0.31 + 0.07–0.05 in./yr) (6.8 + 1.6–1.0 mm/yr [0.27 + 0.06–0.04 in./yr] from terrace-height measurement) over the last 13.5 ± 2.5 ka for the Tsaotun area and 14.5 + 4–2.7 mm/yr (0.57 + 0.16–0.11 in./yr) (13.8 + 4.1–2.6 mm/yr [0.54 + 0.16–0.10 in./yr] from terrace-height measurement) over the last 55 ± 12.6 ka for the Hsinshe area. This factor-of-two difference in slip rate needs to be confirmed by dating of other terraces; nevertheless, it is large relative to the stated errors. Therefore, we briefly consider this difference.

Chen et al. (2003b) also noted that the difference in terrace uplift between the Tsaotun and Hsinshe areas agrees with regional differences in slip magnitude in the Chi-Chi earthquake based on fault-scarp heights. The mean Chi-Chi scarp heights (Lee et al., 2003) are 1.6 m (5.2 ft) to the south of the branch line of the

A Tsaotun Terrace

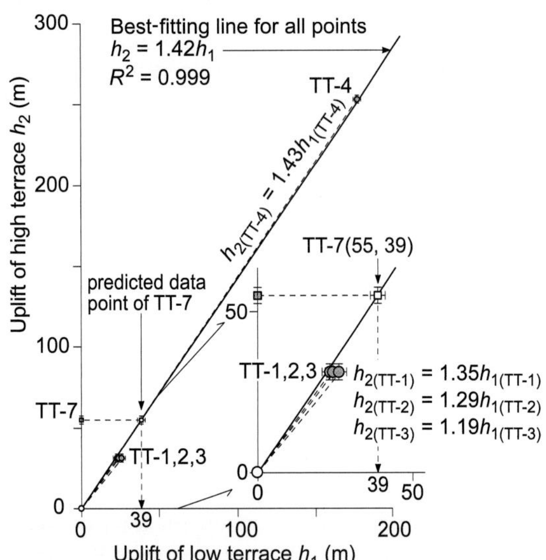

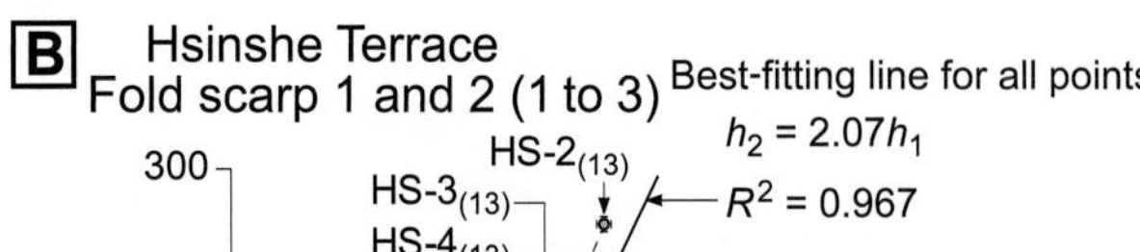

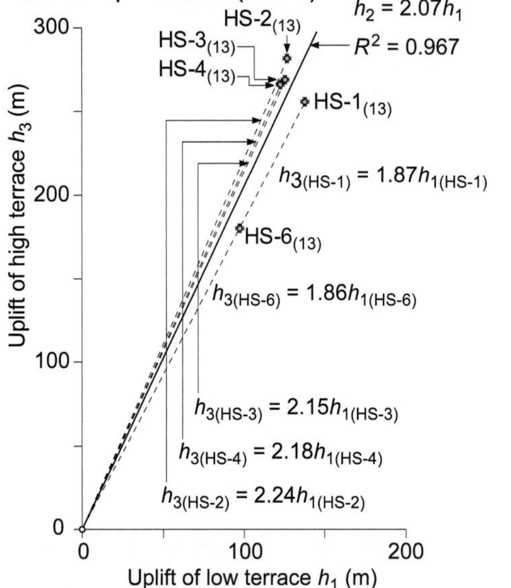

C Hsinshe Terrace
Fold scarp 1 (1 to 2)

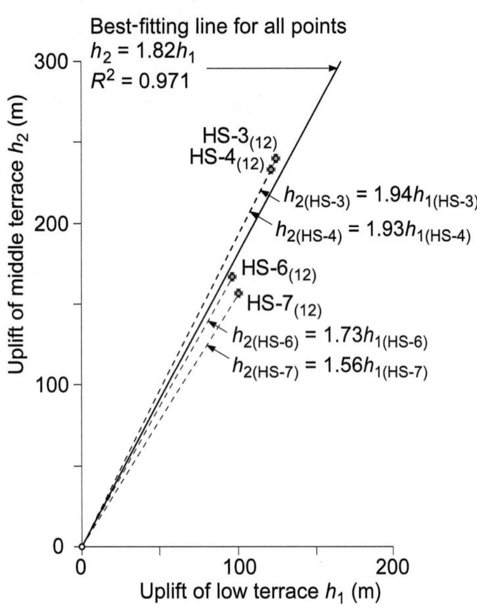

D Hsinshe Terrace
Fold scarp 2 (2 to 3)

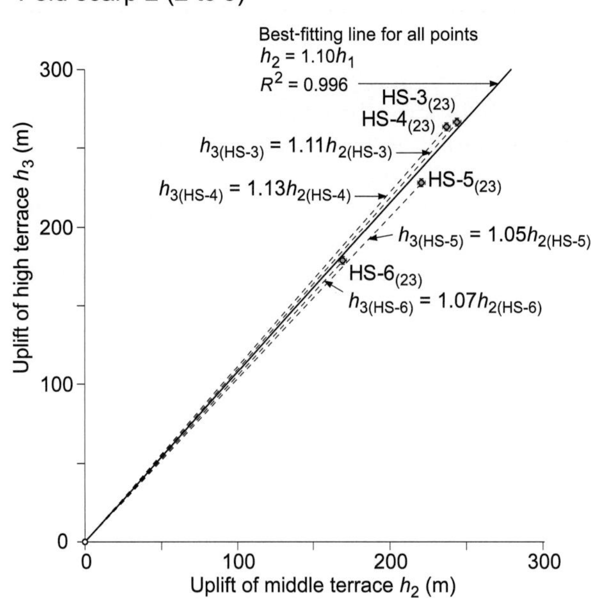

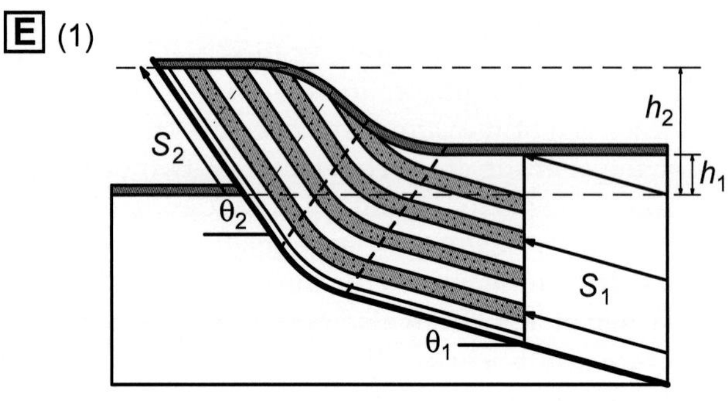

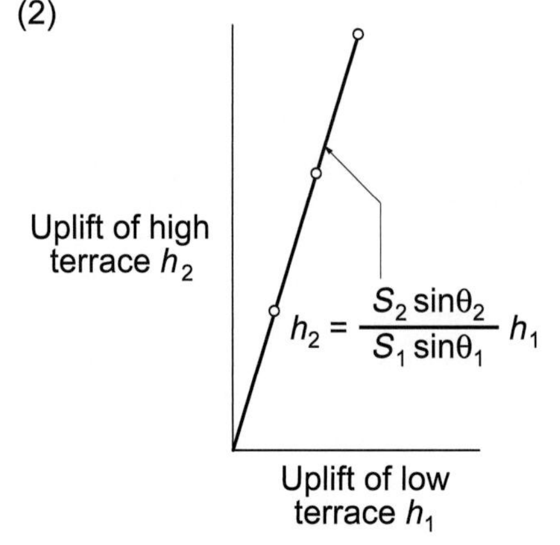

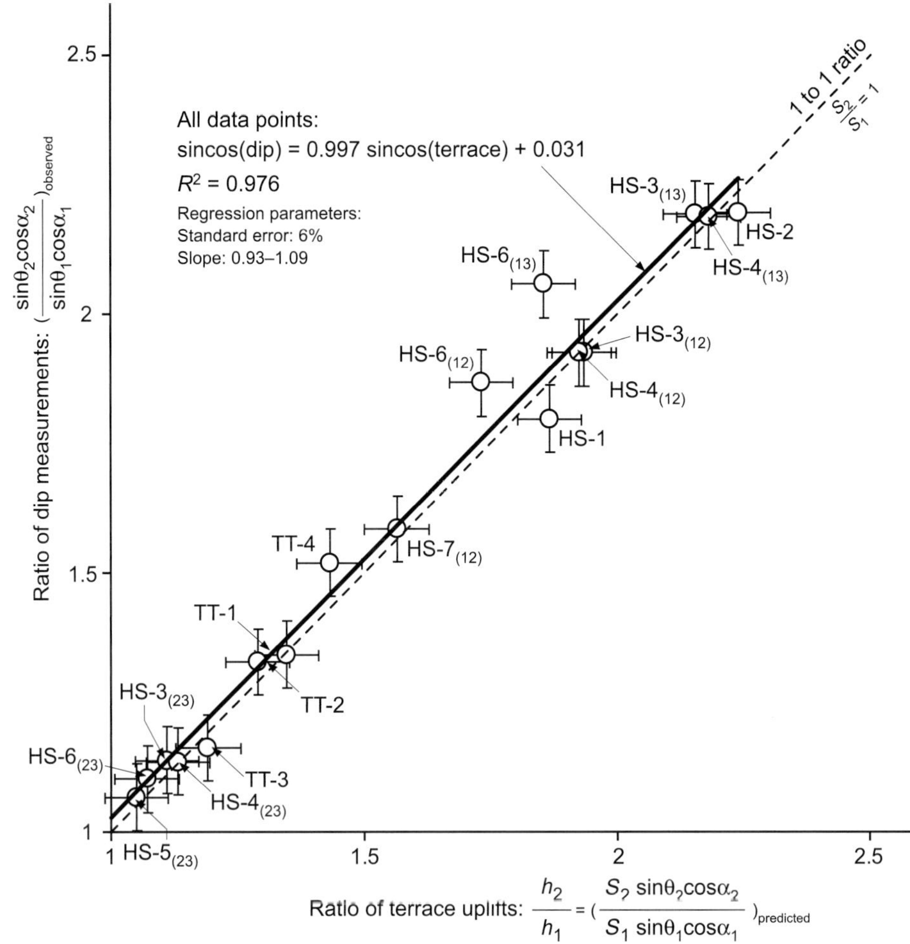

Figure 13. Relationship between observed terrace uplifts, h_1/h_2, across axial surfaces in the Chelungpu thrust sheet and the uplift predicted from fault-bend folding theory (equation 4), $h_1/h_2 = \cos\alpha_1 \sin\theta_1 \cos\alpha_2 \sin\theta_2$, where θ is the fault dip or dip of hanging-wall bedding and α is the difference in azimuth between the dip direction and the slip vector, which are derived from coseismic displacements in the Chi-Chi earthquake. The data, within their uncertainties, are in agreement with the fault-bend folding theory, which is given by the dashed line representing equation 4. TT = Tsaotun; HS = Hsinshe.

Sanyi-Chelungpu fault system, including the Tsaotun area and 3.8 m (12.5 ft) to the north of the branch line (Figure 15). The coseismic surface displacements within the interior of the Chelungpu thrust sheet (Yu et al., 2001; Dominguez et al., 2003) also show an approximate factor-of-two difference north and south of the branch line (Figure 15). Therefore, it seems possible that the Chi-Chi earthquake approximates a characteristic earthquake for the ramp system of the Chelungpu thrust, as suggested by Chen et al. (2003b). If the cumulative long-term slip north and south of the branch line differs by a factor of two, then we would expect significantly higher regional topography to the north than in the south, which is observed as shown in Figure 15. The mean elevation above the Chelungpu thrust ramp north of the branch line is 368 and 233 m (1207 and 764 ft) to the south (Figure 15). Even with a maximum correction to the regional base level (~42 m [138 ft] above the sea level in the Taichung Basin), it still shows a significant difference between mean elevation in the north and south (Figure 15). Therefore, it seems possible that the long-term slip rate north and south of the branch line has been substantially different over the last approximately 55 ka.

The Chelungpu thrust south of the branch line has a total slip of about 14 km (8.7 mi). North of the branch line, the Sanyi thrust has similarly large slip, but the north Chelungpu Chinshui detachment, which slipped in the Chi-Chi earthquake, shows a total displacement of only about 0.3 ± 0.1 km (0.19 ± 0.06 mi) (Yue et al.,

Figure 12. Observed terrace uplift relationships across fold scarps in comparison with model predictions, (A) Tsaotun Terrace (the solid best-fitting line with the data point of section TT-4 and the dashed best-fitting line without it) and (B, C, D) Hsinshe Terrace. The slopes of graphs are the ratios between the sine of ramp dips. Note each data point with an error bar of ±5 m (16 ft). (E) Simple fault-bend folding model (1) and the predicted linear relationship between uplifts, h_1/h_2, of a flight of deformed terraces. TT = Tsatun; HS = Hsinshe.

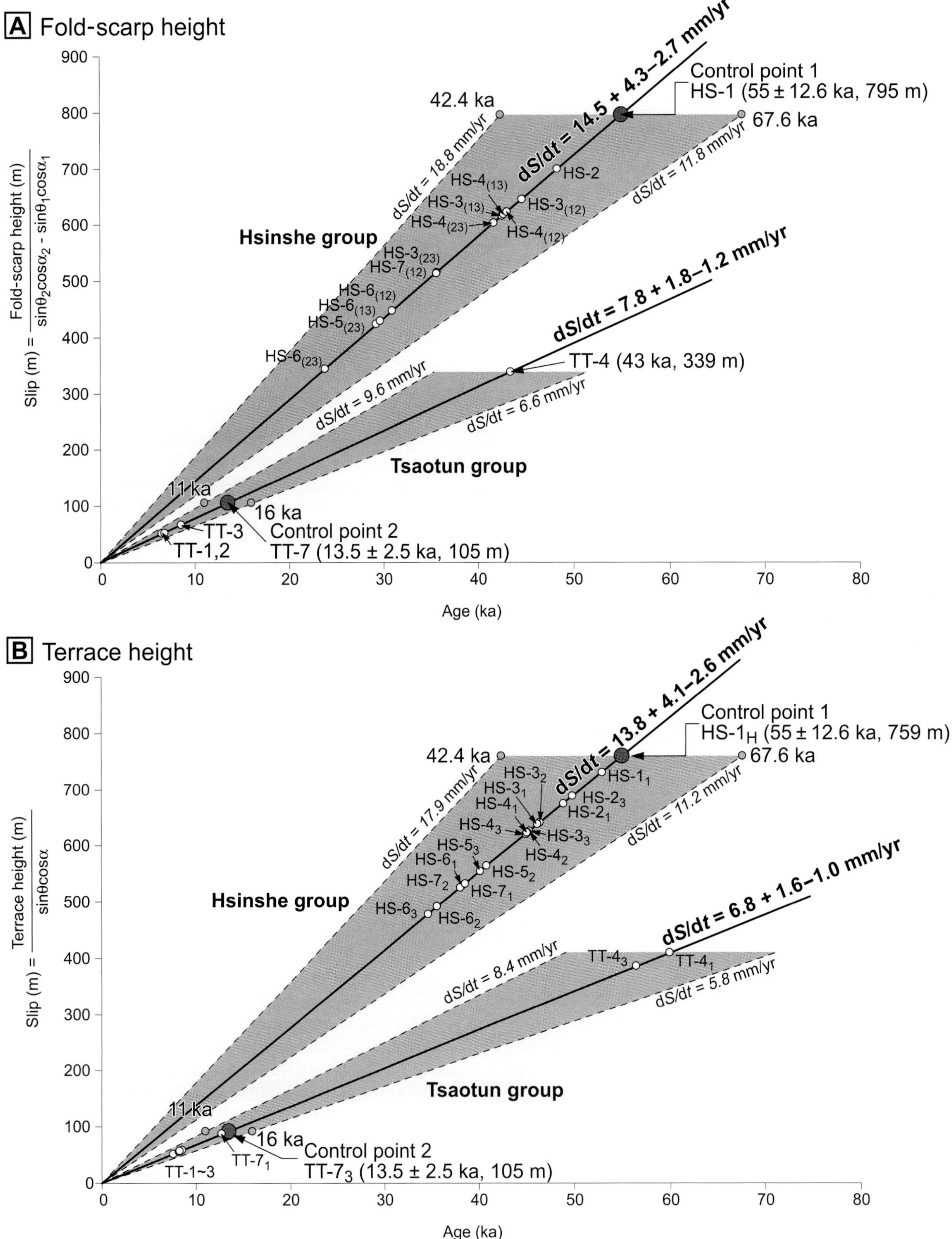

Figure 14. Relationship between age and fault slip derived for the Chelungpu thrust sheet in the Hsinshe and Tsaotun areas, (A) from the fold-scarp height and (B) from the terrace height. See the text for details.

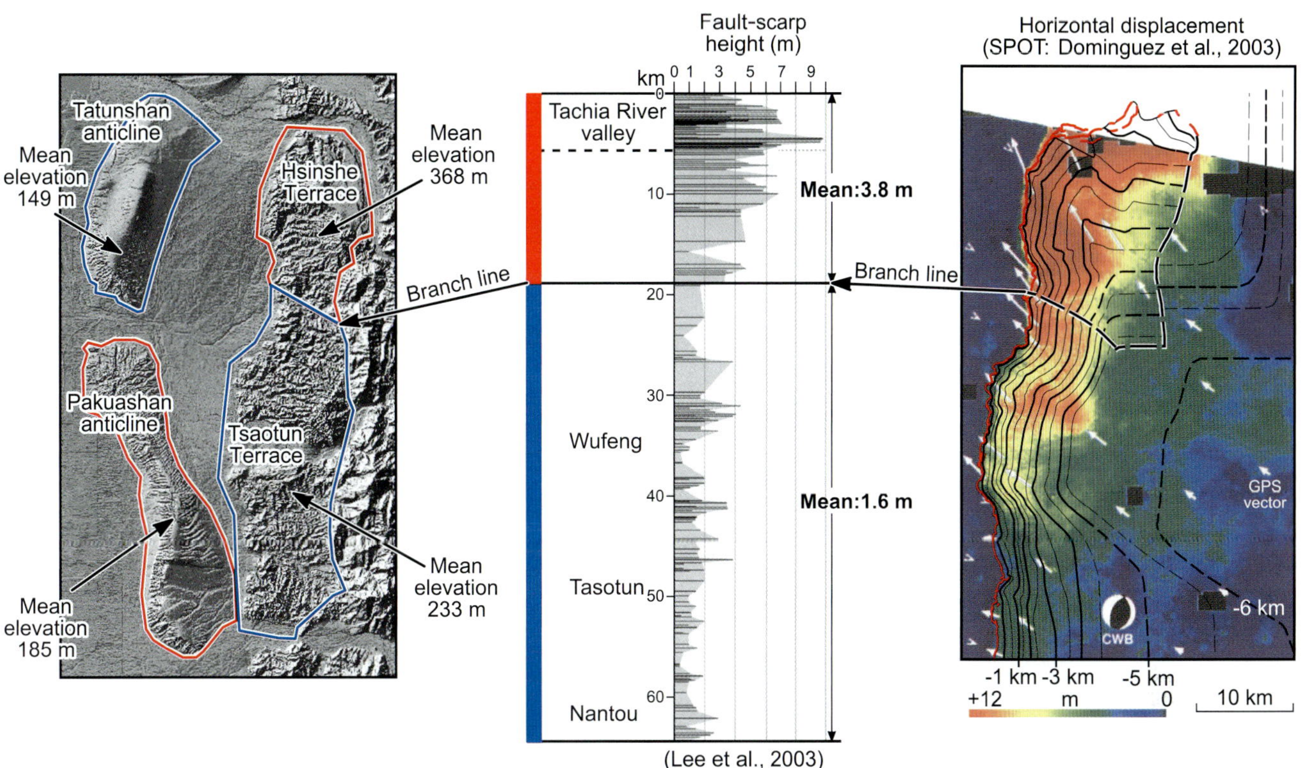

Figure 15. Comparison between coseismic displacement fields of the Chi-Chi earthquake (Dominguez et al., 2003; Lee et al., 2003) and topography of the Chelungpu and also Changhua thrust sheets, which may reflect the cumulative uplift in many earthquakes. See the text for discussion. GPS = Global Positioning System; SPOT = Satellite pour Observation de la Terre; CWB = Central Weather Bureau.

2005) (Figure 5). The Sanyi thrust did not slip at the surface in the Chi-Chi earthquake or in the 1935 Tuntzuchiao earthquake (7.1 local magnitude scale); however, topographic evidence of Holocene activity south of the Tachia River exists. The total fault slip of 600–800 m (1968–2625 ft) recorded by the higher Hsinshe terraces is significantly larger than the total slip on the north Chelungpu Chinshui detachment. Therefore, at least half of the Hsinshe terrace uplift and folding must be associated with slip on the Sanyi thrust.

CHANGHUA THRUST RAMP

Structural Geometry and Total Shortening

The Pakuashan-Tatushan anticline is the frontal structure of the western Taiwan thrust belt, forming the western limit of the current Taichung depositional basin (Figures 2, 3A). It appears to be young geomorphically because much of its upper surface in the north and the south is composed of folded fluvial terraces as well as exposures of the Pleistocene Toukoshan Formation in areas of incision, particularly along the western limb of the fold (Figure 2) (Shih and Yang, 1985; Delcaillau et al., 1998; Mouthereau et al., 1999; Delcaillau, 2001; Chen et al., 2003a; Sung and Chen, 2004). The surface of the Pakuashan anticline, at its southern end, shows a very long backlimb (~6–7 km [3.7–4.3 mi]) defined by fluvial terraces dipping 1.4–4.2° to the east. In contrast, its front limb is narrow (1–2 km [0.6–1.2 mi]) and exposes the Toukoshan Formation dipping 4–17° to the west. To the north, the topographic form of the Pakuashan-Tatushan anticline is more symmetric, suggesting a somewhat different subsurface structure (Figure 2); nevertheless, the mean elevations are similar (Figure 15). In any case, relatively few subsurface data are available over most of the length of the anticline; therefore, our analysis is limited to the southern end of the Pakuashan anticline where several seismic lines exist (Hung and Suppe, 2002; Wang et al., 2002; see also Yue et al., 2005) together with well-developed flights of deformed terraces.

We have constructed a composite structural cross section for the southern Pakuashan anticline based on seismic line A along the Choshui River, the TC-1 well about 7–8 km to the north and surface geology (Figures 2, 3A). The seismic image shows a strongly

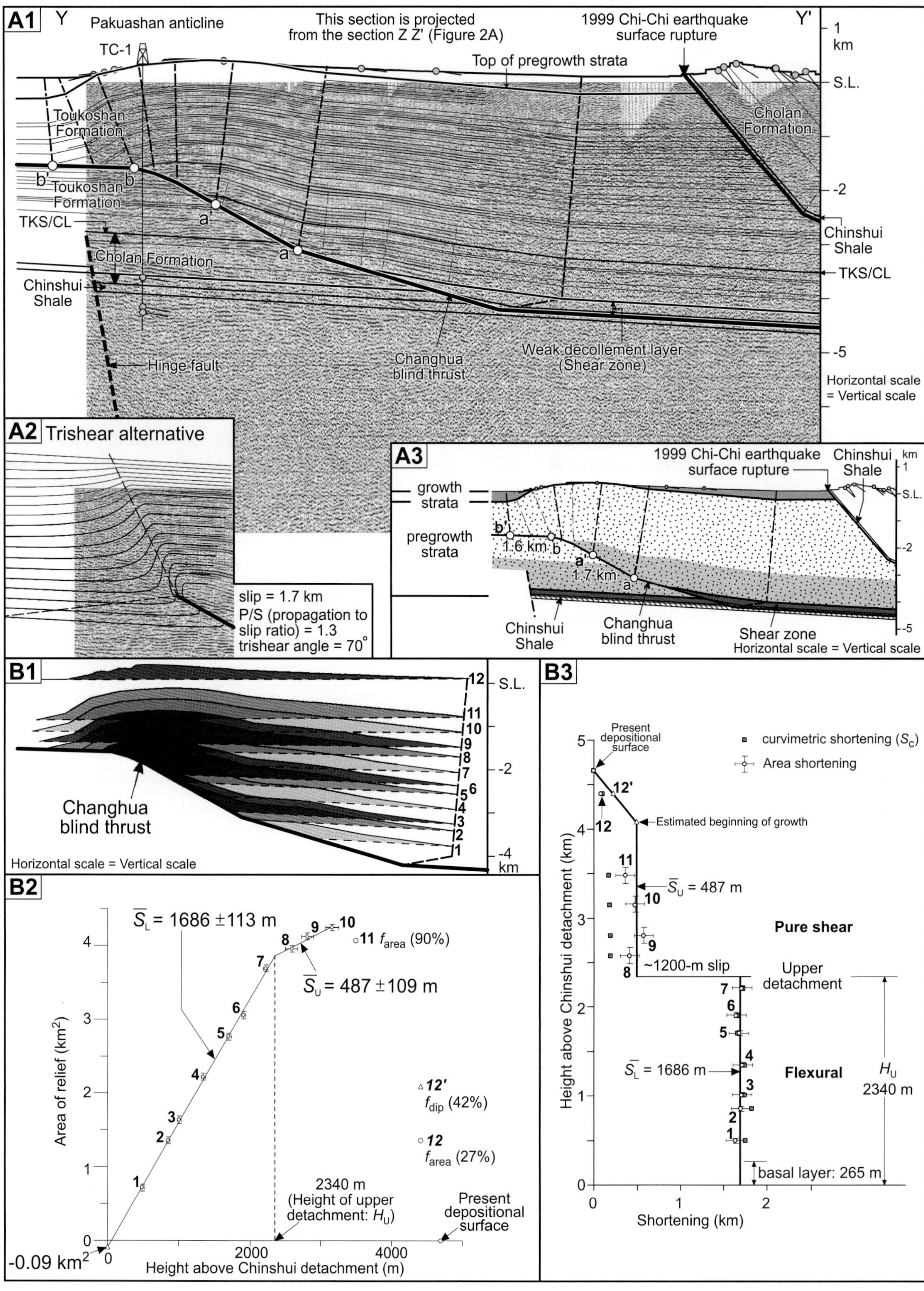
A1
Y
Pakuashan anticline
This section is projected
from the section Z Z' (Figure 2A)
1999 Chi-Chi earthquake
surface rupture
Y'
TC-1
Top of pregrowth strata
1
km
S.L.
Toukoshan
Formation
Cholan
Formation
b'
b
Toukoshan
Formation
a'
-2
TKS/CL
a
Chinshui
Shale
Cholan Formation
TKS/CL
Chinshui
Shale
Weak decollement layer
(Shear zone)
Changhua
blind thrust
Hinge fault
-5
Horizontal scale
= Vertical scale
A2 Trishear alternative
slip = 1.7 km
P/S (propagation to
slip ratio) = 1.3
trishear angle = 70°
A3
1999 Chi-Chi earthquake
surface rupture
Chinshui
Shale
km
growth
strata
pregrowth
strata
1.6 km
1.7 km
Chinshui
Shale
Changhua
blind thrust
Shear zone
Horizontal scale = Vertical scale
B1
Changhua
blind thrust
Horizontal scale = Vertical scale
B2
$\bar{S}_L$ = 1686 ±113 m
$\bar{S}_U$ = 487 ±109 m
11 f_{area} (90%)
12' f_{dip} (42%)
12 f_{area} (27%)
2340 m
(Height of upper
detachment: H_U)
Present
depositional
surface
-0.09 km²
Area of relief (km²)
Height above Chinshui detachment (m)
B3
Present
depositional
surface
curvimetric shortening (S_c)
Area shortening
Estimated beginning of growth
$\bar{S}_U$ = 487 m
Pure shear
~1200-m slip
Upper
detachment
$\bar{S}_L$ = 1686 m
Flexural
H_U
2340 m
basal layer: 265 m
Height above Chinshui detachment (km)
Shortening (km)

asymmetric anticline, in agreement with surface geology and morphology (Figure 16A). The subsurface fold has a very long (~6–7 km [3.7–4.3 mi]) backlimb with a gentle dip, about 5° near the top of the section, and a narrow front limb (~2 km [1.2 mi]) that is less well imaged but nevertheless shows near-surface dips similar to the surface dips. Flat reflectors are imaged below the front limb and the crest of the anticline at depths of 2–5 km (1.2–3.1 mi), which is confirmed by the TC-1 well. The Changhua thrust ramp marks the boundary between the flat reflectors of the footwall and the gently east-dipping reflectors of the long east limb of the fold (Figure 16A). The thrust ramp is constrained by the reflectors to be shallower dipping about 16° at depths of 3.5–4.5 km (2.2–2.8 mi) and steeper dipping about 28° at depths of 2–3.5 km (1.2–2.2 mi). The base of the ramp is constrained by the location of the back hanging-wall syncline to be near the level of the Chinshui Shale (Figure 16A), as discussed below, which agrees with the detachment level of the Chelungpu thrust to the east (Figure 3A) (Yue et al., 2005).

The general form of the Pakuashan anticline is similar to models of shear fault-bend folds (Figure 1) and to better seismic images elsewhere in the world (Suppe et al., 2004; Corredor et al., 2005a, b; Shaw et al., 2005). In particular, Pakuashan shows a long backlimb that dips more gently (~5°) than the ramp dip (16–28°) and a very narrow front limb; therefore, a shear fault-bend fold interpretation is suggested but one that is somewhat more complex than the simple end-member models (Figure 1) because of the upward-steepening bend in the ramp (Figure 16A). The distinction between pure-shear and simple-shear end-member models depends on the geometry of the back syncline within the weak basal shear layer (Figure 17). The simple-shear end member is precluded for the Pakuashan structure because the synclinal axial surface does not project to the base of the ramp as a bisecting axial surface. We adopt a pure-shear solution. The dip angle of the synclinal axial surface is determined to be nearly equal to the ramp dip from the pure-shear fault-bend folding theory (Suppe et al., 2004). The top of the weak basal layer is located at the bend in the back syncline. This solution correctly predicts the location of the anticlinal axial surface about 2.8 km (1.7 mi) up the ramp, which marks the hanging-wall cutoff of the top of the basal layer (compare Figures 1C and 16A).

The fault ramp steepens rather abruptly from 16 to 28°, producing a synclinal fold with a shape that fits the predictions of fault-bend folding and indicates a ramp slip of 1.7 km (1.06 mi) based on the cutoff length aa′ in Figure 16A. This slip agrees with slip obtained from the independent area of relief and bed-length shortening measurements discussed below. Above the bend, the fault ramp projects upward into the narrow front limb of the fold, which is not well imaged, but apparently it does not reach the surface. This observation is consistent with regional geomorphic and shallow seismic data, suggesting that the Changhua thrust does not reach the surface (see also Delcaillau et al., 1998; Wang et al., 2003). Two solutions have been explored for this narrow front limb. (1) We have fit the surface dip data and shallow seismic reflections of 5 to 15° to a trishear fault-propagation fold model (Erslev, 1991; Allmendinger, 1998; Zehnder and Allmendinger, 2000). This solution shown in Figure 16A3 predicts that all the 1.7 km (1.06 mi) of slip on the ramp is consumed in folding, which disagrees with area-of-relief measurements that indicate a mean shortening of only about 0.5 km (0.3 mi) for the uppermost horizons of the Pakuashan fold (Figure 16B). (2) Therefore, we adopt a fault-bend folding solution (Figure 16A), which satisfies

Figure 16. (A1) Structural interpretation of the Pakuashan anticline in seismic line A (Hung and Suppe, 2002) as a shear fault-bend fold, showing a characteristic gently dipping (~5°), very long (~6–7km) backlimb; narrow front limb (~2 km, ~4–17°); rather steep ramp dip (16° and 28°); and a 0.3-km (0.19-mi)-thick basal weak decollement layer (shear zone) (compare with simpler shear fault-bend fold models in Figures 1, 17). The input fault slip is about 1.7 km (1.06 mi) (see the text for discussion). (A2) An alternative trishear model of the front limb (Allmendinger, 1998). (A3) Simplified structural interpretation showing basic elements of the Pakuashan structure. Note that the kink-band width (aa′), 1.7 km (1.06 mi), agrees with the slip (1.69 km) derived from an area-relief analysis in panel B. (B1) Simplified cross section showing area-of-relief measurements for the hanging wall of the Changhua thrust ramp. (B2) Graph of the area-of-relief measurements as a function of height. The lower data (horizons 1–7) fit a linear regression with good statistics ($R^2 = 0.998$) with a homogeneous shortening of $\bar{S}_L = 1686 \pm 113$ m (371 ft). The upper horizons (8–10) show a lower shortening of $\bar{S}_U = 487 \pm 109$ m (1598 ± 358 ft), which is consistent with a fault-bend fold interpretation for horizons above the upper detachment. The height of the upper detachment, H_U, is estimated at 2340 m (7677 ft) from the intersection of the two linear regressions. Horizon 12 is the topographic surface, which shows a low area of relief, indicating that it is a growth horizon. (B3) Graph of shortening components as a function of height. The mean shortening for individual data points below the upper detachment is computed as $\bar{S} = Ah_b$ and above the upper detachment is $\bar{S} = (A - \bar{S}_L H_U)/(h_b - H_U)$. The curvimetric or bed-length shortening, S_c, is nearly equal to the mean shortening in the lower horizons; therefore, bed length is mostly conserved and deformation is flexural. In contrast, $S_c < \bar{S}$ in the upper horizons, which therefore shows significant pure shear. S.L. = sea level; TKS/CL = Toukoshan Formation/Cholan Formation; P/S = propagation to slip ratio.

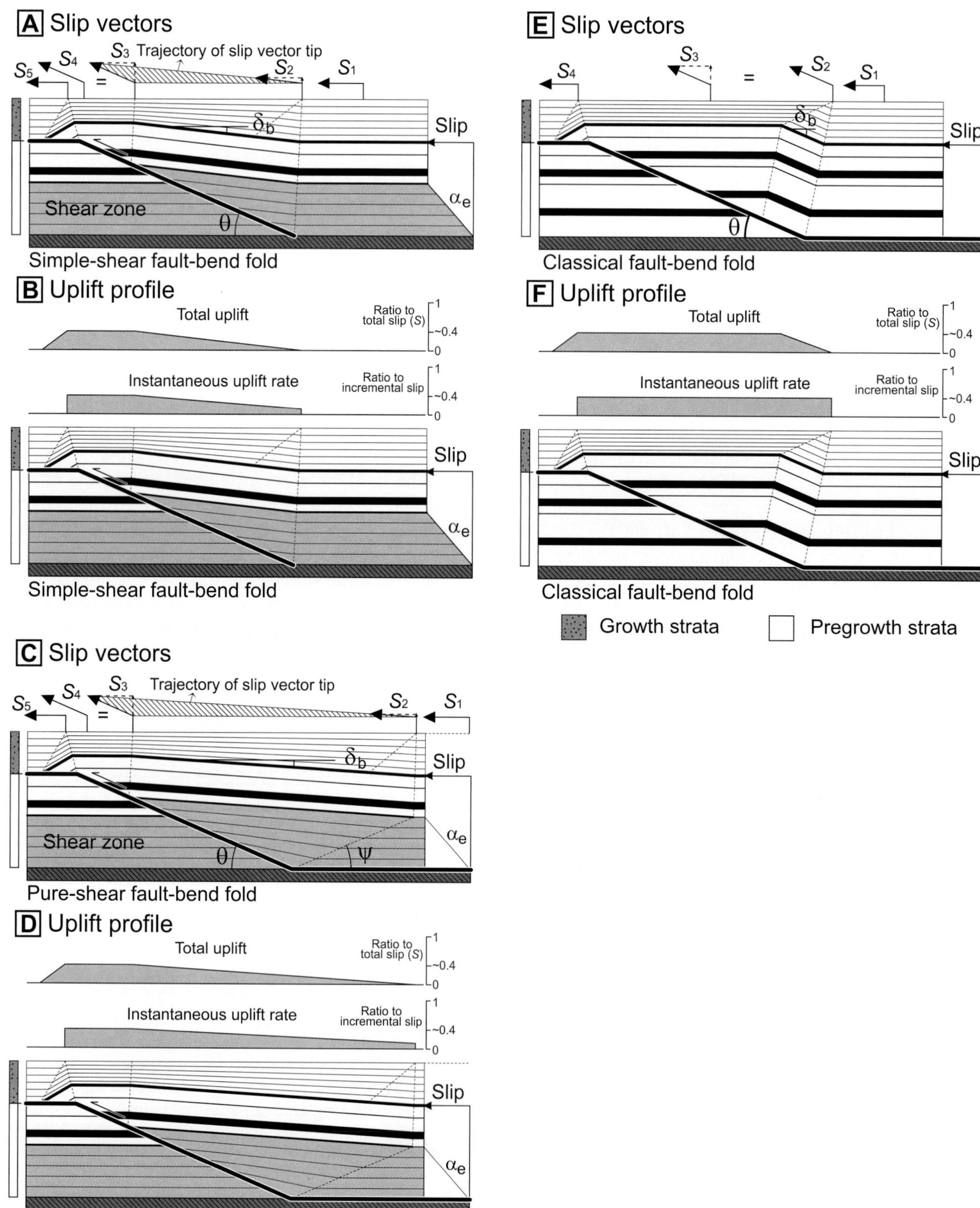

Figure 17. Instantaneous and total uplift profiles and total slip vectors for three end-member fault-bend folding (fbf) models (for parameters, see Figure 1). Note that the inclination of slip vectors varies linearly across the backlimb of the shear fault-bend fold models, varying between horizontal at the back syncline to parallel to the ramp at the top of the backlimb. The limb rotates progressively as shown by the growth strata. The progressive backward tiling is similar to observed terrace profiles for the Pakuashan anticline shown in Figures 18 and 19.

the shortening constraints, sending about 1.2 km (0.7 mi) of slip out along the upper detachment.

Measurements of area of structural relief as a function of height (Epard and Groshong, 1993; Gonzalez-Mieres and Suppe, 2006) for lower stratigraphic horizons abutting the ramp (horizons 1–7, Figure 16B) show a linear relationship indicating an excellent least-squares shortening of $\overline{S}_L = 1686 \pm 113$ m (5531 ± 371 ft) (Figure 16B2), which agrees with the fault slip of approximately 1700 m (5577 ft) shown by the fault-bend folding above the upward-steepening bend in the ramp (aa′ in Figure 16A). Furthermore, the intercept is close to the origin (-0.09 km^2 [-0.03 mi^2]); therefore, a solution with negligible simple shear in the basal layer is indicated. The upper horizons (8–10) show a lower shortening magnitude of $\overline{S}_U = 487 \pm 109$ m (1598 ± 358 ft), indicating a slip on the upper detachment of about 1200 m (3937 ft). The height of the upper detachment H_U is estimated at 2340 m (7677 ft) from the intersection of the two linear regressions. The slip ratio, $R = 1200$ m/1700 m (3937 ft/5577 ft) = 0.7, is smaller than predicted by fault-bend folding theory ($R \geq 0.8$), assuming conservation of bed length and area, which indicates that additional slip (150–200 m [492–656 ft]) is consumed in the fold crest above the upper detachment. This additional pure-shear shortening in the fold crest is seen in the bed-length shortening discussed below (Figure 16B3). Horizon 12, which is the topographic (terrace) surface, shows a much lower area of relief, indicating that it is a growth horizon.

The components of shortening as a function of height are shown in Figure 16B3. The mean shortening for individual data points is computed as the area divided by the height above the detachment, $\overline{S} = A/h_b$, for horizons below the upper detachment. In contrast, the shortening of horizons above the upper detachment is computed as the area of relief corrected for the contribution of the area of shortening between the two detachments, $\overline{S}_L H_U$, divided by the height above the upper detachment, $\overline{S} = (A - \overline{S}_L H_U)/(h_b - H_U)$. These computed mean shortenings for individual measurements below the upper detachment are close to constant as a function of height, indicating a lack of simple shear within the thrust sheet except for the possibility of a small component in the basal layer. The curvimetric or bed-length shortening, S_c, is nearly equal to the mean shortening in the lower horizons (Figure 16B3); therefore, bed length is mostly conserved and deformation is flexural. In contrast, the curvimetric shortening, S_c, is only about 41% of the mean shortening, $\overline{S}_U$, in the upper horizons, which implies that approximately 200 m (656 ft) of constant-volume pure-shear shortening exists. We presume that the remaining approximately 1200 m (3937 ft) of shortening is consumed in horizontal compaction, as is observed in other frontal structures at shallow depths (cf. Gonzalez-Mieres and Suppe, 2006). The slightly lower area of relief of horizon 11 relative to the trend of the underlying horizons (Figure 16B1) may be ascribed to either the beginning of growth or a higher proportion of horizontal compaction at that level. Horizon 12, based on its area of relief, shows a fraction, $f_a = 27\%$, of the pregrowth shortening based on $A = f_a[\overline{S}_L H_U + \overline{S}_U(h_b - H_U)]$, which assumes an equal fraction for both levels of the structure.

A similar shear fault-bend fold structural form with a long gentle backlimb is observed 40 km (25 mi) to the north in an older seismic line (line C, Figure 2) (Chen, 1978). The Changhua blind thrust can also be traced on this seismic line, and the limb dips in the hanging-wall strata are indeed gentler than the fault ramp dip, indicating that a shear fault-bend folding interpretation may be appropriate farther north than our area of study.

Shortening History from Deformed Terraces

Shear fault-bend folding models indicate a hanging-wall kinematics that is quite different from classical fault-bend folding (Figure 17). The long backlimb is marked by progressive limb rotation plus minor kink-band migration, as recorded in growth strata. Because of this rotation, the particle displacement vectors in the backlimb above the basal shear zone vary linearly between being parallel to the fault ramp, S_3, on the fold crest (Figure 17A, C) to being close to horizontal, S_2, at the back syncline. In this section, we make use of the nearly linear relationship between backlimb dip and shortening given by shear fault-bend folding theory (Suppe et al., 2004, their figures 10, 11) to infer the shortening history and slip rate of the Changhua fault based on the history of limb rotation of the Pakuashan anticline.

The predicted progressive limb rotation does not appear to be imaged as growth strata on the seismic image (Figure 16A). The shallowest backlimb reflectors are probably pregrowth strata because they are parallel to the uppermost well-imaged throughgoing reflectors. Therefore, the direct constraints on progressive fold growth are limited to surface observations, which fortunately consist of well-developed flights of deformed fluvial terraces, including a 3–4-km (1.9–2.5-mi)-wide folded paleoriver channel (Figures 2B, 18), similar in size to the present Choshui River channel, which is the location of seismic line A (Figure 18).

We need to combine the constraints on terrace folding with the seismic interpretation and its constraints on total fault slip. We do this by tracing the main fold axial surfaces identified in the cross section (Figure 16A) in map view based on surface bedding strikes in the

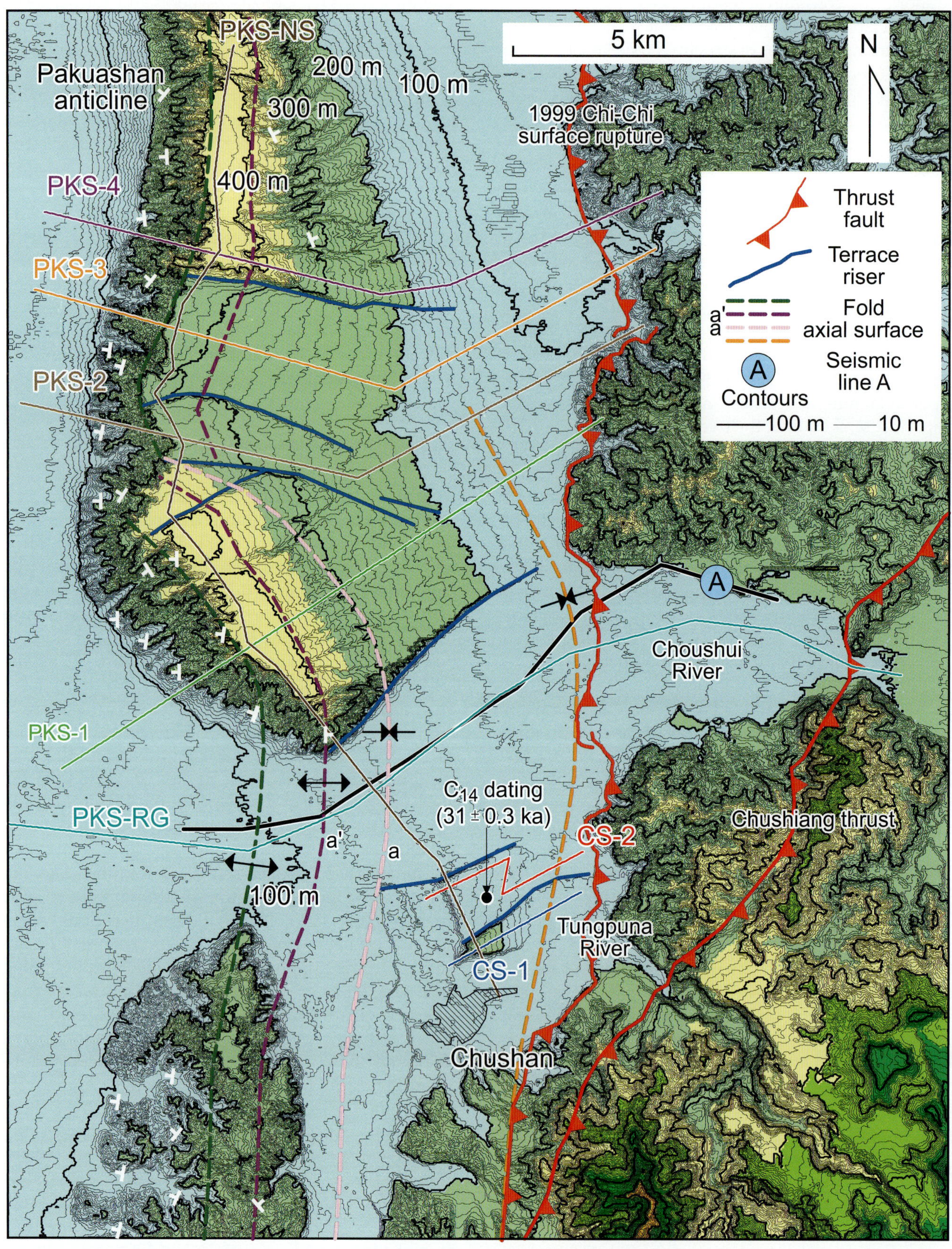
PKS-NS
Pakuashan anticline
200 m
100 m
300 m
400 m
5 km
N
1999 Chi-Chi surface rupture
PKS-4
PKS-3
PKS-2
PKS-1
PKS-RG
Thrust fault
Terrace riser
a'
a
Fold axial surface
A
Seismic line A
Contours
100 m
10 m
Choushui River
C14 dating (31 ± 0.3 ka)
a'
a
100 m
CS-2
CS-1
Tungpuna River
Chushiang thrust
Chushan

Toukoshan Formation and topographic contours on the folded terraces, as shown in Figure 18. This exercise shows that terrace profiles CS-1 and CS-2 south of the seismic line and the eastern part of terrace profile PKS-1 north of the seismic line lie above the shear fault-bend fold and are east of the kink-band aa′ associated with the synclinal bend in the ramp (Figure 16A), which folds the upper part of the PKS-1 terrace (Figures 18, 19). Farther north, in the area of terrace profiles PKS-2, PKS-3, and PKS-4, kink-band aa′ is missing; therefore, the fault geometry has changed. A substantial change in fault geometry is also indicated by the changes in strike of the topographic contours between PKS-2 and PKS-4 (Figure 18). Nevertheless, these northern terraces appear to be over the shear fault-bend fold based on the axial-surface mapping. The terrace risers indicate relative ages of CS-1 > CS-2, PKS-1 > PKS-2 > PKS-3, and PKS-4 > PKS-3 (Figure 18), which agree with their differences in dip (Figure 18).

We focus our analysis on the terraces adjacent to the seismic cross section (CS-1, CS-2, PKS-1) to estimate the shortening history and rate based on two C_{14} dates of 30,400 ± 200 yr before present (BP) (NTU-3279) and 30,950 ± 290 yr BP (NTU-3509) for terrace CS-2 from Ota et al. (2002). We measure the apparent dips of the terraces in the direction of the seismic line (CS-1 = 4.2°, PKS-1 = 2.3°, CS-2 = 1.5°), which can be compared with the apparent dip of the uppermost reflectors of the seismic line (4.6 ± 0.1°, Figure 16A). We assume an initial dip for all these surfaces equal to the present river gradient of −0.4°. Shear fault-bend folding theory indicates for a fault dip of 16° and limb dips of 2–5° that fault slip and change in dip are linearly proportional (Suppe et al., 2004), which allows us to assign an input fault slip, S, for each change in terrace or reflector dip based on

$$S = \frac{\delta_{\text{terrace}}}{\delta_{\text{pregrowth}}} S_{\text{pregrowth}} \qquad (6)$$

where the total input fault slip, $S_{\text{pregrowth}}$, is 1686 ± 113 m (5531 ± 371 ft) based on the analysis of area of relief. Applying this to the dated terrace CS-2, we obtain a fault slip of 649 ± 113 m (2129 ± 371 ft), which is 38 ± 9% of the total shortening. Combining with the C_{14} dates, we estimate a mean slip rate of 21 ± 3.8 mm/yr (0.83 ± 0.15 in./yr) over the last 31 ka in the direction of the seismic line (Figure 20B). If this rate is representative of the entire history, we then estimate a time of inception of slip on the Changhua thrust at 80 ± 5 ka (Figure 20A). This age estimation could be the upper bound here because the paleodepositional surface might be steeper than −0.4° (e.g., alluvial fan surface) (Figure 20A). In addition, we can estimate fractional slip and ages for the other terraces, which are given in Figure 20.

The slip direction on the Changhua thrust is not directly known. Here we assume it is parallel to the interseismic geodetic shortening azimuth of approximately 300 ± 5° (e.g., stations I007 and CPUL; Yu et al., 2001) (Figure 20). These directions make a very minor correction to the total displacement and slip rate over the last 31 ka for the Changhua blind thrust because the slip azimuth and seismic line are symmetric about the dip direction (Figure 20C). The true slip rate over the last 31 ka in the southern Pakuashan anticline is therefore 21 ± 5.8 mm/yr (0.83 ± 0.23 in./yr). This result is similar to the shortening rate estimated by Simões et al. (2007b), which based on modeling the growth strata of the fold yields an average shortening rate of 16.3 ± 4.1 mm/yr (0.64 ± 0.16 in./yr) for the whole Pakuashan anticline (15.2 ± 3.7 mm/yr [0.60 ± 0.14 in./yr] for the southern Pakuashan anticline) along an azimuth of 298° and the time of initiation of the fold to 62.2 ± 9.6 ka.

Based on the vertical offset of a strath terrace and C_{14} dates, a determination of slip rates on the Chelungpu and Chushiang thrusts about 5 km (3.1 mi) to the southeast along the Dungpuna River (Figure 20C) by Simões et al. (2007a) gives a combined rate of 15.8 ± 5.1 mm/yr (0.62 ± 0.20 in./yr) with a similar azimuth of 300° over the last 12 ka. Therefore, the total shortening rate for these three frontal structures near the Choshui River of 36.8 ± 10.9 mm/yr (1.45 ± 0.43 in./yr) over the last 12–31 ka is indicated.

This total shortening rate of 37 mm/yr (1.46 in./yr) is about 45% of the total plate-tectonic shortening rate

Figure 18. Topographic map of the southern Pakuashan anticline showing 5–7-km (3.1–4.3-mi)-wide tilted fluvial terraces on the east limb and a narrow approximately 2-km (1.2-mi)-wide front limb north of the Choshui River and two smaller tilted terrace fragments preserved south of the river (north of Chushan). The relative ages of the terraces are constrained by the terrace risers and by the progressive tilting of the terrace surfaces. Traces of axial surfaces that fold the terraces are shown; axial surfaces of kink-band aa′ are imaged in seismic line A (Figure 16A1). Note that the PKS-3 profile lies within the 3-km (1.9-mi)-wide paleochannel marked in Figure 2B. The PKS-3 paleochannel is similar in width to the present Choshui River channel (see also strike profile PKS-NS, Figure 19E). The location of two dated samples reported by Ota et al. (2002) on the Chushan Terrace is shown by a black circle labeled C_{14} dating. PKS = Pakuashan; CS = Chushan; RG = river gradient; NS = north–south.

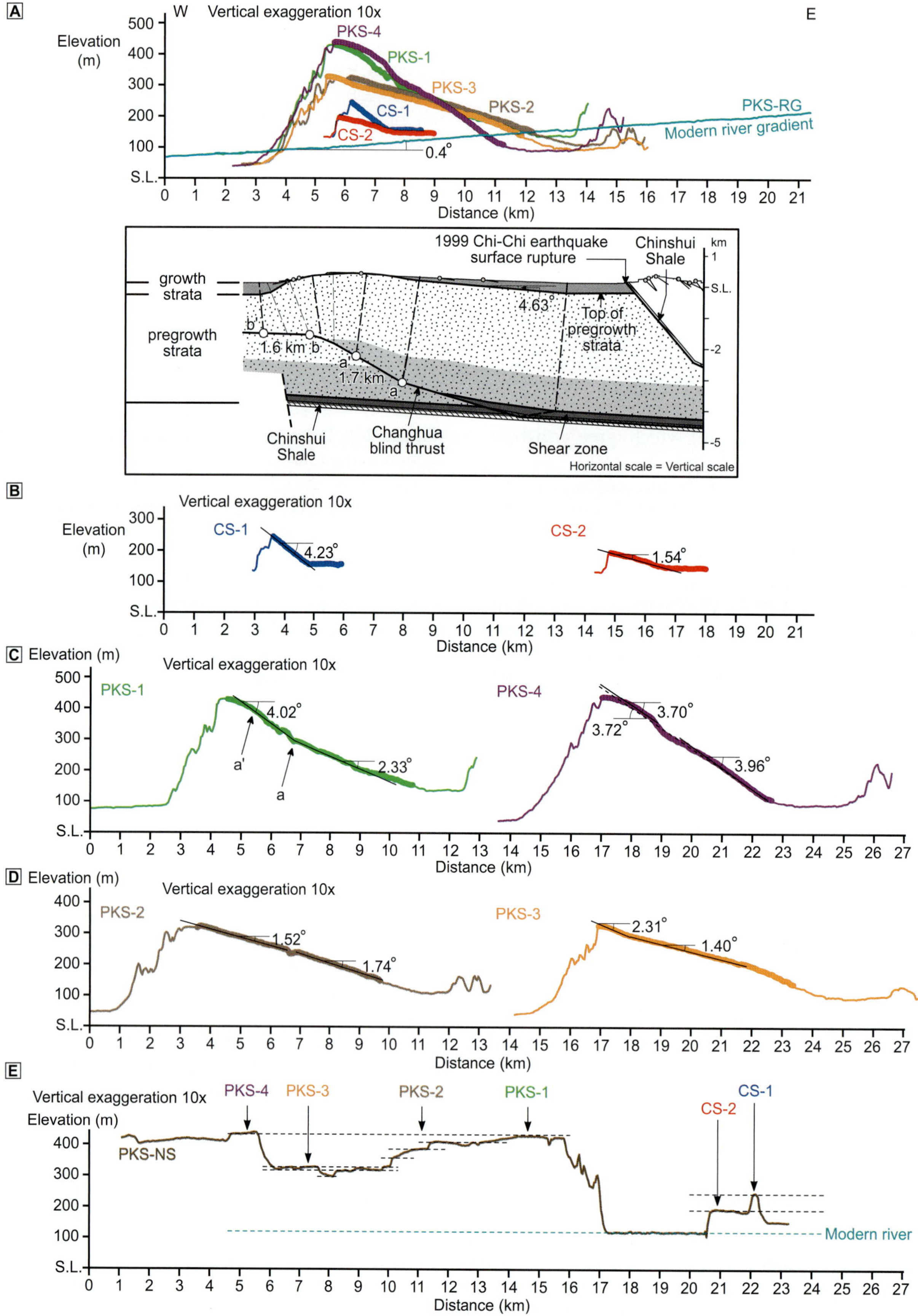
A
W
Vertical exaggeration 10x
E
Elevation (m)
PKS-4
PKS-1
PKS-3
PKS-2
CS-1
CS-2
PKS-RG
Modern river gradient
0.4°
Distance (km)
1999 Chi-Chi earthquake surface rupture
Chinshui Shale
growth strata
pregrowth strata
4.63°
Top of pregrowth strata
1.6 km
1.7 km
Chinshui Shale
Changhua blind thrust
Shear zone
Horizontal scale = Vertical scale
B
Vertical exaggeration 10x
Elevation (m)
CS-1
4.23°
CS-2
1.54°
Distance (km)
C
Elevation (m)
Vertical exaggeration 10x
PKS-1
4.02°
2.33°
PKS-4
3.70°
3.72°
3.96°
Distance (km)
D
Elevation (m)
Vertical exaggeration 10x
PKS-2
1.52°
1.74°
PKS-3
2.31°
1.40°
Distance (km)
E
Vertical exaggeration 10x
Elevation (m)
PKS-NS
Modern river
Distance (km)

across Taiwan (Station S063 at Lutao, 81 ± 0.5 mm/yr (3.1 ± 0.02 in./yr), Azimuth 305°; Yu and Kuo, 2001), suggesting that these three frontal structures are the most important in western Taiwan. This rate is similar to the estimated rate of westward propagation of the foreland flexure of 39.5–44.5 mm/yr (1.55 ± 1.75 in./yr) over the last 2 Ma (Simões and Avouac, 2006). Therefore, we expect the Changhua thrust to be a substantial seismic hazard. By applying a structural interpretation similar to our cross section to the entire length of the Pakuashan and Tatushan anticlines, we estimate an area of the thrust ramp about 600 km^2 (232 mi^2), suggesting a nominal maximum characteristic magnitude of 6.8 based on empirical relationships of Wells and Coppersmith (1994). The estimated area of the combined ramp and lower detachment of the Changhua blind thrust is approximately 1700 km^2, indicating a nominal maximum characteristic Wells and Coppersmith magnitude of 7.2. These estimated nominal maximum magnitudes of 6.8–7.2 may be compared with two historical earthquakes in central Taiwan, the 1845 6–6.5 magnitude Taichung earthquake and the 1848 6.7–7.1 magnitude Changhua earthquake (Tsai, 1985; Cheng and Yeh, 1989) (Figure 2). In contrast, the estimated area of slip in the 1999 (7.6 moment magnitude scale) Chi-Chi earthquake (Chelungpu thrust) is about 4000–4300 km^2 (1544–1660 mi^2), nearly twice the possible rupture area of the Changhua blind thrust. Therefore, in light of the higher slip rate of the Changhua thrust, a shorter mean recurrence interval may be expected. The region of the Changhua thrust is a highly populated area with approximately 25% of the population of Taiwan (~5 million people); therefore, significant seismic risk may exist.

DISCUSSION

We have shown that the active folding above two adjacent thrust ramps involving the same detachment and hanging-wall stratigraphy has substantially different kinematics, as recorded by deformed flights of fluvial terraces and coseismic displacement in the 1999 Chi-Chi earthquake. The Chelungpu thrust ramp, which has a total displacement of about 14 km (8.7 mi), shows classic fault-bend folding by kink-band migration, whereas the adjacent and younger Changhua thrust ramp, with only 1.7-km (1.06-mi) total displacement, shows a close agreement to pure-shear fault-bend folding with progressive limb rotation. It is tempting to suggest that these two thrust ramps in their early stages of development both formed as shear fault-bend folds and the Chelungpu thrust underwent a transition to classical fault-bend folding as large slip accumulated. This is essentially a mechanical conjecture that cannot be tested here; however, kinematic models (e.g., Figures 1, 17) show that, with increasing limb dip, continuously increasing strain rates are required for maintaining the pure-shear fault-bend folding mode of deformation. The increase in total strain is seen in the thickness change across the synclinal axial surface within the basal layer of a pure-shear fault-bend fold, indicating substantial bulk deformation (Figures 1, 17C, D). In contrast, classical fault-bend folding requires much less bulk deformation. Pure-shear fault-bend folding may be favored over classical fault-bend folding at initiation of the thrust ramp because there may be a smaller bending resistance. In contrast, after large displacement (20–30 times the basal layer thickness), the limb dips are quite similar in the two modes of folding, so classical fault-bend folding might be favored (Suppe et al., 2004). The Changhua thrust has a slip (1.7 km [1.06 mi]) that is about six times the basal layer thickness of 265 m (869 ft) (Figure 16A), whereas the Chelungpu thrust has a slip (~14 km) that is about 50 times the basal layer thickness. Therefore, it seems a plausible conjecture that the Chelungpu thrust ramp began as a pure-shear fault-bend fold such as the Changhua ramp and later was transformed to its present classical mode of deformation.

Our study of uplifted terraces in the hanging-wall of the Chelungpu thrust provides a rather strong test of the assumptions of classical fault-bend folding. The ratio of terrace uplift across fold scarps, in comparison with changes in fault dip, shows a close agreement with the assumptions of conservation of bed length, layer thickness, and fault slip across fault bends (Figure 13). Therefore, we can use measurements of terrace uplift to obtain robust determinations of fault slip since terrace formation, which we have done for several terraces. Most of these terraces remain undated, but based on two terrace dates, we obtain estimates of

Figure 19. (A) Seven terrace profiles of the southern Pakuashan anticline and the modern Choshui river profile (Figure 18); vertical exaggeration 10×. A progressive backlimb tilting toward the east is observed, ranging from 1.5 to 4.2° (B–D) in comparison with the regional gradient of the modern Choshui river profile of 0.4° to the west, which is in agreement with the expectations of limb rotation for shear fault-bend folding (Figure 17). The kink-band aa′ in profile PKS-1 is produced by slip through the bend of the fault ramp (Figure 6). (E) The north–south section shows the relationship between six terrace profiles and a modern river channel. S.L. = sea level; PKS = Pakuashan; CS = Chushan; RG = river gradient; NS = north–south.

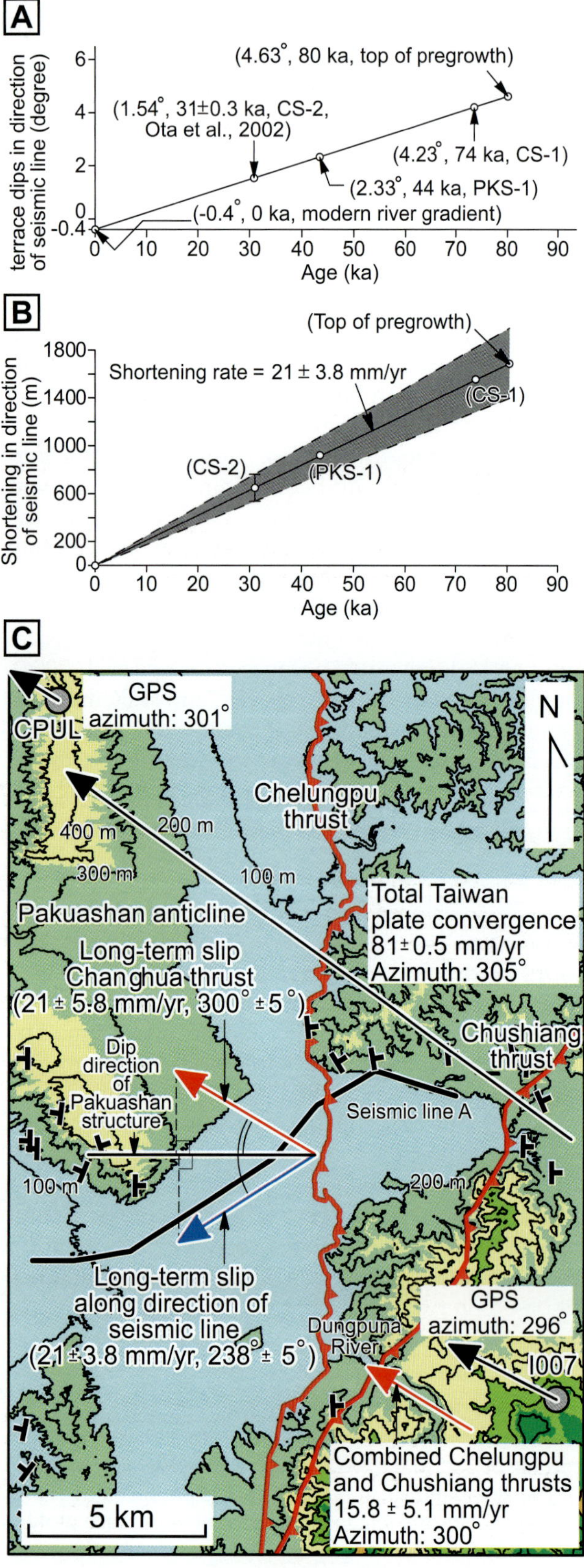

the long-term slip rate on the Chelungpu thrust in two areas, which differ by a factor of two (Figure 14). These two areas show a similar difference in slip in the 1999 Chi-Chi earthquake (Figure 15).

Subsurface structural analysis of the Pakuashan anticline above the Changhua thrust shows a close agreement of the pregrowth structure to pure-shear fault-bend folding theory, with a total fault slip of 1.7 km (1.06 mi) (Figure 16). The area of structural relief and bed-length shortening measurements proved to be powerful aids in structural interpretation (Figure 16). Flights of fluvial terraces above the thrust ramp show progressive rotation, which is used to determine the fault slip since deposition of each terrace. One dated terrace yields an estimate of the shortening rate over the last 31 ka of 2.1 cm/yr (0.83 in./yr).

These results show that, with appropriate subsurface structural analysis, strongly constrained estimates of fault slip can be obtained from deformed fluvial terraces and other growth horizons in the hanging walls of thrust ramps. With robust age control, detailed deformation histories and well-constrained fault-slip rates are obtainable, which are important for active-tectonic and seismic-hazard analysis.

ACKNOWLEDGMENTS

We are grateful to Yue-Gau Chen of the National Taiwan University for providing us with DEM data. We are extremely grateful to PetroChina for hosting the International Conference on Theory and Application

Figure 20. (A) Graph of expected terrace age as a function of terrace dip for the backlimb of the Pakuashan anticline. A zero age corresponds to a −0.4° dip, which is the modern river gradient. Only one terrace, CS-2, is currently dated, which has an age of about 31 ka (Ota et al., 2002). According to seismic line A, the limb dip of the top of the pregrowth strata is 4.6°, which yields an expected age about 80 ka for the beginning of growth. See the text for discussion. (B) Shortening-age graph of the Pakuashan anticline in growth strata with a constant shortening rate of 21 ± 3.8 mm/yr (0.83 ± 0.15 in./yr) in the direction of the seismic line. (C) The long-term slip of the Changhua blind thrust is assumed to be parallel to the GPS shortening azimuth at 300 ± 5°, which, given the dip of the structure near the seismic line, gives a long-term slip rate of 21 ± 5.8 mm/yr (0.83 ± 0.23 in./yr). The combined long-term slip rate of the Changhua blind thrust, Chelungpu thrust, and Chushiang thrust is estimated at 36.8 ± 10.9 mm/yr (1.45 ± 0.43 in./yr) over the last 12–31 ka, which is about 45% of the total plate shortening across Taiwan. See the text for details. GPS = Global Positioning System; PKS = Pakuashan; CS = Chushan; CPUL = GPS station.

of Fault-Related Folding in Foreland Basins, Beijing, China, 2005, which stimulated the publishing of this chapter and the volume. Yue thanks Princeton University for providing support during 2001–2007. Special thanks to Ken R. McClay for the effort to edit the volume for publication. The trishear interpretation in Figure 16A2 was produced using the Fault/Fold program of Rick Allmendinger.

REFERENCES CITED

Allen, M. B., S. J. Vincent, and P. J. Wheeler, 1999, Late Cenozoic tectonics of the Kepingtage thrust zone: Interactions of the Tien Shan and Tarim Basin, northwest China: Tectonics, v. 18, no. 4, p. 639–654, doi:10.1029/1999TC900019.

Allmendinger, R. W., 1998, Inverse and forward numerical modeling of trishear fault-propagation folds: Tectonics, v. 17, no. 4, p. 640–656, doi:10.1029/98TC01907.

Atkinson, P. K., and W. K. Wallace, 2003, Competent unit thickness variation in detachment folds in the northeastern Brooks Range, Alaska: Geometric analysis and a conceptual model: Journal of Structural Geology, v. 25, no. 10, p. 1751–1771, doi:10.1016/S0191-8141(03)00003-8.

Blair, T. C., and J. G. McPherson, 1994, Alluvial fans and their natural distinction from rivers based on morphology, hydraulic processes, sedimentary processes, and facies assemblages: Journal of Sedimentary Research, v. 64, no. 3, p. 450–489.

Bilotti, F., and J. H. Shaw, 2005, Deep-water Niger Delta fold and thrust belt modeled as a critical-taper wedge: The influence of elevated basal fluid pressure on structural styles: AAPG Bulletin, v. 89, no. 11, p. 1475–1491, doi:10.1306/06130505002.

Burbank, D. W., and R. S. Anderson, 2001, Tectonic geomorphology: Oxford, Blackwell Scientific, 270 p.

Burbank, D. W., J. Leland, E. Fielding, R. S. Anderson, N. Brozovic, M. R. Reid, and C. Duncan, 1996, Bedrock incision, rock uplift and threshold hillslopes in the northwestern Himalayas: Nature, v. 379, no. 6565, p. 505–510, doi:10.1038/379505a0.

Chen, J. S., 1978, A comparative study of the refraction and reflection seismic data obtained on the Changhua plain to the Peikang shelf, Taiwan: Petroleum Geology of Taiwan, v. 15, p. 199–217.

Chen, W. S., K. D. Ridgway, C. S. Horng, Y. G. Chen, K. S. Shea, and M. G. Yeh, 2001, Stratigraphic architecture, magnetostratigraphy, and incised-valley systems of the Pliocene–Pleistocene collisional marine foreland basin of Taiwan: Geological Society of America Bulletin, v. 113, no. 10, p. 1249–1271, doi:10.1130/0016-7606(2001)113<1249:SAMAIV>2.0.CO;2.

Chen, Y. C., Q. C. Sung, and K. Y. Cheng, 2003a, Along-strike variations of morphotectonic features in the western foothills of Taiwan: Tectonic implications based on stream-gradient and hypsometric analysis: Geomorphology, v. 56, no. 1–2, p. 109–137, doi:10.1016/S0169-555X(03)00059-X.

Chen, Y. G., Y. W. Chen, W. S. Chen, J. F. Zhang, H. Zhao, L. P. Zhou, and S. H. Li, 2003b, Preliminary results of long-term slip rates of 1999 earthquake fault by luminescence and radiocarbon dating: Quaternary Science Reviews, v. 22, no. 10–13, p. 1213–1221, doi:10.1016/S0277-3791(03)00037-4.

Chen, Y. G., K. Y. Lai, Y. H. Lee, J. Suppe, W. S. Chen, Y. N. N. Lin, Y. Wang, J. H. Hung, and Y. T. Kuo, 2007, Coseismic fold scarps and their kinematic behavior in the 1999 Chi-Chi earthquake Taiwan: Journal of Geophysical Research, v. 112, p. B03S02, doi:10.1029/2006JB004388.

Cheng, S. N., and Y. T. Yeh, 1989, Catalog of the earthquakes in Taiwan from 1604 to 1988: Research report of project IES-R-661 of the Institute of Earth Sciences of Academia Sinica, Taiwan, 253 p.

Chinese Petroleum Corporation, 1982, Geological map of Tai-Chung: Taiwan Petroleum Exploration Division, Chinese Petroleum Corporation, ROC, scale 1:100,000.

Corredor, F., J. Shaw, and F. Bilotti, 2005a, Structural styles in the deep-water fold and thrust belts of the Niger Delta: AAPG Bulletin, v. 89, no. 6, p. 753–780, doi:10.1306/02170504074.

Corredor, F., J. Shaw, and J. Suppe, 2005b, Shear fault-bend folding, deep water Niger Delta, *in* J. Shaw, C. Connors, and J. Suppe, eds., Seismic interpretation of contractional fault-related folds: AAPG Studies in Geology 53, p. 87–92.

Delcaillau, B., 2001, Geomorphic response to growing fault-related folds: Example from the foothills of central Taiwan: Geodinamica Acta, v. 14, no. 5, p. 265–287, doi:10.1016/S0985-3111(01)01071-3.

Delcaillau, B., B. Deffontaines, L. Floissac, J. Angelier, J. Deramond, P. Souquet, H. T. Chu, and J. F. Lee, 1998, Morphotectonic evidence from lateral propagation of an active frontal fold; Pakuashan anticline, foothills of Taiwan: Geomorphology, v. 24, no. 4, p. 263–290, doi:10.1016/S0169-555X(98)00020-8.

Dominguez, S., J. P. Avouac, and R. Michel, 2003, Horizontal coseismic deformation of the 1999 Chi-Chi earthquake measured from SPOT satellite images: Implications for the seismic cycle along the western foothills of central Taiwan: Journal of Geophysical Research, v. 108, no. B2, p. 2083, doi:10.1029/2001JB000951.

Epard, J. L., and R. H. Groshong, 1993, Excess area and depth to detachment: AAPG Bulletin, v. 77, no. 8, p. 1291–1302.

Epard, J. L., and R. H. Groshong, 1995, Kinematic model of detachment folding including limb rotation, fixed hinges and layer-parallel strain: Tectonophysics, v. 247, no. 1–4, p. 85–103, doi:10.1016/0040-1951(94)00266-C.

Erslev, E. A., 1991, Trishear fault-propagation folding: Geology, v. 19, no. 6, p. 617–620, doi:10.1130/0091-7613(1991)019<0617:TFPF>2.3.CO;2.

Erslev, E. A., 1993, Thrusts, back-thrusts, and detachment of Laramide foreland arches, *in* C. J. Schmidt, R. Chase, and

E. A. Erslev, eds., Laramide basement deformation in the Rocky Mountain foreland of the western United States: Geological Society of America Special Paper 280, p. 339–358.

Erslev, E. A., and K. R. Mayborn, 1997, Multiple geometries and modes of fault-propagation folding in the Canadian thrust belt: Journal of Structural Geology, v. 19, no. 3–4, p. 321–335, doi:10.1016/S0191-8141(97)83027-1.

Fermor, P., 1999, Aspects of the three-dimensional structure of the Alberta foothills and front ranges: Geological Society of America Bulletin, v. 111, no. 3, p. 317–346, doi:10.1130/0016-7606(1999)111<0317:AOTTDS>2.3.CO;2.

Gonzalez-Mieres, R., and J. Suppe, 2006, Relief and shortening in detachment folds: Journal of Structural Geology, v. 28, no. 10, p. 1785–1807, doi:10.1016/j.jsg.2006.07.001.

Hardy, S., and C. D. Connors, 2006, A velocity description of shear fault-bend folding: Journal of Structural Geology, v. 28, no. 3, p. 536–543, doi:10.1016/j.jsg.2005.12.015.

Hardy, S., and M. Ford, 1997, Numerical modeling of trishear fault propagation folding: Tectonics, v. 16, no. 5, p. 841–854, doi:10.1029/97TC01171.

He, D., J. Suppe, G. Yang, S. Guan, S. Huang, X. Shi, X. Wang, and C. Zhang, 2005, Guidebook for fieldtrip in south and north Tianshan foreland basin, Xinjiang Uygur Autonomous Region, China: International Conference on Theory and Application of Fault-Related Folding in Foreland Basins, Beijing, China, 78 p.

Homza, T. X., and W. K. Wallace, 1995, Geometric and kinematic models for detachment folds with fixed and variable detachment depths: Journal of Structural Geology, v. 17, no. 4, p. 575–588, doi:10.1016/0191-8141(94)00077-D.

Hubert-Ferrari, A., J. Suppe, R. Gonzalez-Mieres, and X. Wang, 2007, Mechanisms of active folding of the landscape (southern Tian Shan, China): Journal of Geophysical Research, v. 112, B03S09, doi:10.1029/2006JB004362.

Hung, J., and J. Suppe, 2002, Subsurface geometry of the Sani-Chelungpu faults and fold scarp formation in the 1999 Chi-Chi Taiwan earthquake: Eos Transactions, American Geophysical Union, v. 83, suppl. 47, T61B-1268, p. F1279.

Hung, J. H., Y. H. Wu, E. C. Yeh, J. C. Wu, and TCDP Scientific Party, 2007, Subsurface structure, physical properties, and fault-zone characteristics in the scientific drill holes of Taiwan Chelungpu-Fault Drilling Project: Terrestrial, Atmospheric, and Oceanic Sciences, v. 18, no. 2, p. 271–293, doi:10.3319/TAO.TCDP0706S06.

Hung, J. H., K. F. Ma, Y. H. Wu, H. Y. Wu, H. Ito, W. Lin, and E. C. Yeh, 2009, Structure geology, physical properties, fault zone characteristics and stress state in scientific drill holes of Taiwan Chelungpu Fault Drilling Project: Tectonophysics, v. 466, p. 307–321, doi:10.1016/j.tecto.2007.11.014.

Johnson, K. M., and P. Segall, 2003, Imaging the ramp-decollement geometry of the Chelungpu fault using coseismic GPS displacements from the 1999 Chi-Chi, Taiwan earthquake: Tectonophysics, v. 378, no. 1–2, p. 123–139, doi:10.1016/j.tecto.2003.10.020.

Johnson, K. M., Y. J. Hsu, P. Segall, and S. B. Yu, 2001, Fault geometry and slip distribution of the 1999 Chi-Chi, Taiwan earthquake imaged from inversion of GPS data: Geophysical Research Letters, v. 28, no. 11, p. 2285–2288, doi:10.1029/2000GL012761.

Jordan, P., and T. Noack, 1992, Hanging wall geometry of overthrusts emanating from ductile decollements, *in* K. McClay, ed., Thrust tectonics: London, United Kingdom, Chapman & Hall, p. 311–318.

Lai, K. Y., Y. G. Chen, J. H. Hung, J. Suppe, L. F. Yue, and Y. W. Chen, 2006, Surface deformation related to kink-folding above an active fault: Evidence from geomorphic features and co-seismic slips: Quaternary International, v. 147, p. 44–54, doi:10.1016/j.quaint.2005.09.005.

Larson, K. P., and R. A. Price, 2006, The southern termination of the western main ranges of the Canadian Rockies, near Fort Steele, British Columbia: Stratigraphy, structure, and tectonic implications: Bulletin of Canadian Petroleum Geology, v. 54, no. 1, p. 37–61, doi:10.2113/54.1.37.

Lavé, J., and J. P. Avouac, 2000, Active folding of fluvial terraces across the Siwalik Hills (Himalaya of central Nepal): Journal of Geophysical Research, v. 105, no. B3, p. 5735–5770, doi:10.1029/1999JB900292.

Lee, Y. H., M. L. Hsieh, S. D. Lu, T. S. Shih, W. Y. Wu, Y. Sugiyama, T. Azuma, and Y. Kariya, 2003, Slip vectors of the surface rupture of the 1999 Chi-Chi earthquake, western Taiwan: Journal of Structural Geology, v. 25, no. 11, p. 1917–1931, doi:10.1016/S0191-8141(03)00039-7.

Lin, A. T., A. B. Watts, and S. P. Hesselbo, 2003, Cenozoic stratigraphy and subsidence history of the South China Sea margin in the Taiwan region: Basin Research, v. 15, p. 453–478, doi:10.1046/j.1365-2117.2003.00215.x.

Lin, C. C., 1957, Topography of Taiwan: Taiwan, Taiwan Provincial Documentary Committee, 424 p.

Loevenbruck, A., R. Cattin, X. Le Pichon, S. Dominguez, and R. Michel, 2004, Coseismic slip resolution and post-seismic relaxation time of the 1999 Chi-Chi, Taiwan, earthquake as constrained by geological observations, geodetic measurements and seismicity: Geophysical Journal International, v. 158, no. 1, p. 310–326, doi:10.1111/j.1365-246X.2004.02285.x.

Ma, K. F., C. T. Lee, Y. B. Tsai, T. C. Shin, and J. Mori, 1999, The Chi-Chi Taiwan earthquake: Large surface displacements on inland thrust fault: Eos Transactions, American Geophysical Union, v. 80, p. 605–611.

Ma, K. F., et al., 2006, Slip zone and energetics of a large earthquake from the Taiwan Chelungpu-fault Drilling Project: Nature, v. 444, no. 7118, p. 473–476, doi:10.1038/nature05253.

Macqueen, R. W., and D. A. Leckie, 1992, Introduction, *in* R. W. Macqueen and D. A. Leckie, eds., Foreland basins and fold belts: AAPG Memoir 55, p. 1–8.

McQuarrie, N., 2004, Crustal-scale geometry of Zagros fold-thrust belt, Iran: Journal of Structural Geology, v. 26, no. 3, p. 519–535, doi:10.1016/j.jsg.2003.08.009.

Medwedeff, D. A., and J. Suppe, 1997, Multibend fault-bend folding: Journal of Structural Geology, v. 19, no. 3–4, p. 279–292, doi:10.1016/S0191-8141(97)83026-X.

Mitra, S., 2003, A unified kinematic model for the evolution of detachment folds: Journal of Structural Geology, v. 25, no. 10, p. 1659–1673, doi:10.1016/S0191-8141(02)00198-0.

Molnar, P., 1987, Inversion of profiles of uplift rates for the geometry of dip-slip faults at depth, with examples from the Alps and the Himalaya: Annales Geophysicae Series B-Terrestrial and Planetary Physics, v. 5 no. 6, p. 663–670.

Mouthereau, F., O. Lacombe, B. Deffontaines, J. Angelier, H. T. Chu, and C. T. Lee, 1999, Quaternary transfer faulting and belt front deformation at Pakuashan (western Taiwan): Tectonics, v. 18, no. 2, p. 215–230, doi:10.1029/1998TC900025.

Narr, W., and J. Suppe, 1994, Kinematics of basement-involved compressive structures: American Journal of Science, v. 294, no. 7, p. 802–860.

Ota, Y., J. B. H. Shyu, Y. G. Chen, and M. L. Hsieh, 2002, Deformation and age of fluvial terraces south of the Choushui River, central Taiwan, and their tectonic implications: Western Pacific Earth Sciences, v. 2, no. 3, p. 251–260.

Parnaud, F., Y. Gou, J. C. Pascual, I. Truskowski, O. Gallango, and H. Passalacqua, 1995, Petroleum geology of the central part of the eastern Venezuelan Basin: AAPG Memoir 62, p. 741–756.

Poblet, J., and K. McClay, 1996, Geometry and kinematics of single-layer detachment folds: AAPG Bulletin, v. 80, no. 7, p. 1085–1109.

Poblet, J., K. McClay, F. Storti, and J. A. Munoz, 1997, Geometries of syntectonic sediments associated with single-layer detachment folds: Journal of Structural Geology, v. 19, no. 3–4, p. 369–381, doi:10.1016/S0191-8141(96)00113-7.

Rockwell, T. K., E. A. Keller, M. N. Clark, and D. L. Johnson, 1984, Chronology and rates of faulting of Ventura River terraces, California: Geological Society of America Bulletin, v. 95, no. 12, p. 1466–1474, doi:10.1130/0016-7606(1984)95<1466:CAROFO>2.0.CO;2.

Shaw, J. H., C. Connors, and J. Suppe, 2005, Seismic interpretation of contractional fault-related folds— An AAPG seismic atlas: AAPG Studies in Geology 53, 156 p.

Sherkati, S., J. Letouzey, and D. Frizon de Lamotte, 2006, Central Zagros fold-thrust belt (Iran): New insights from seismic data, field observation, and sandbox modeling: Tectonics, v. 25, p. TC4007, doi:10.1029/2004TC001766.

Shih, T. T., and G. S. Yang, 1985, The active faults and geomorphic surfaces of Pakua Tableland in Taiwan (in Chinese): Geographical Research, v. 11, p. 173–186.

Shih, T. T., K. S. Teng, K. S. Yang, and M. Y. Hsu, 1986, Active faults and geomorphic surfaces of the Hsinshe terraces: Geomorphic Bulletin, v. 5, p. 29–39.

Simões, M., and J. P. Avouac, 2006, Investigating the kinematics of mountain building in Taiwan from the spatiotemporal evolution of the foreland basin and western foothills: Journal of Geophysical Research, v. 111, p. B10401, doi:10.1029/2005JB004209.

Simões, M., J. P. Avouac, and Y. G. Chen, 2007a, Slip rates on the Chelungpu and Chushiang thrust faults inferred from a deformed strath terrace along the Dungpuna river, west-central Taiwan: Journal of Geophysical Research, v. 112, p. B03S10, doi:10.1029/2005JB004200.

Simões, M., J. P. Avouac, Y. G. Chen, A. K. Singhvi, C. Y. Wang, M. Jaiswal, Y. C. Chan, and S. Bernard, 2007b, Kinematic analysis of the Pakuashan fault-tip fold, west-central Taiwan: Shortening rate and age of folding inception: Journal of Geophysical Research, v. 112, p. B03S14, doi:10.1029/2005JB004198.

Storti, F., and J. Poblet, 1997, Growth stratal architectures associated to decollement folds and fault-propagation folds: Inferences on fold kinematics: Tectonophysics, v. 282, no. 1–4, p. 353–373, doi:10.1016/S0040-1951(97)00230-8.

Sung, Q. C., and Y. C. Chen, 2004, Geomorphic evidence and kinematic model for Quaternary transfer faulting of the Pakuashan anticline, central Taiwan: Journal of Asian Earth Sciences, v. 24, no. 3, p. 389–404, doi:10.1016/j.jseaes.2004.02.003.

Suppe, J., 1983, Geometry and kinematics of fault-bend folding: American Journal of Science, v. 283, no. 7, p. 684–721.

Suppe, J., and D. A. Medwedeff, 1990, Geometry and kinematics of fault-propagation folding: Eclogae Geologicae Helvetiae, v. 83, no. 3, p. 409–454.

Suppe, J., G. T. Chou, and S. C. Hook, 1992, Rates of folding and faulting determined from growth strata, *in* K. McClay, ed., Thrust Tectonics: London, United Kingdom, Chapman & Hall, p. 105–121.

Suppe, J., F. Sabat, J. A. Muñoz, J. Poblet, E. Roca, and J. Verges, 1997, Bed-by-bed fold growth by kink-band migration: Sant Llorenç de Morunys, eastern Pyrenees: Journal of Structural Geology, v. 19, no. 3–4, p. 443–461, doi:10.1016/S0191-8141(96)00103-4.

Suppe, J., Y. H. Lee, Y. G. Chen, and J. H. Hung, 2000, Coseismic fault-bend folding: Fold scarp formation in the 1999 Chi-Chi earthquake (M 7.6), Taiwan fold-and-thrust belt: Eos, v. 81, no. 48, p. 874.

Suppe, J., C. D. Connors, and Y. K. Zhang, 2004, Shear fault-bend folding, *in* K. R. McClay, ed., Thrust tectonics and hydrocarbon systems: AAPG Memoir 82, p. 303–323.

Teng, L. S., 1990, Geotectonic evolution of late Cenozoic arc-continent collision in Taiwan: Tectonophysics, v. 183, no. 1–4, p. 57–76, doi:10.1016/0040-1951(90)90188-E.

Tsai, Y. B., 1985, A study of disastrous earthquakes in Taiwan, 1683–1895: Bulletin of the Institute of Earth Sciences, Academica Sinica, v. 5, p. 1–44.

Wang, C. Y., C. H. Chang, and H. Y. Yen, 2000, An interpretation of the 1999 Chi-Chi earthquake on Taiwan based on the thin-skinned thrust model: Terrestrial, Atmospheric and Oceanic Sciences, v. 11, no. 3, p. 609–630.

Wang, C. Y., C. L. Li, F. C. Su, M. T. Leu, M. S. Wu, S. H. Lai, and C. C. Chern, 2002, Structural mapping of the 1999 Chi-Chi earthquake fault, Taiwan by seismic reflection methods: Terrestrial, Atmospheric and Oceanic Sciences, v. 13, no. 2, p. 211–226.

Wang, C. Y., S. Y. Kuo, W. L. Shyu, and J. W. Hsiao, 2003, Investigating near-surface structures under the Changhua fault, west-central Taiwan by the reflection seismic method: Terrestrial, Atmospheric and Oceanic Sciences, v. 14, no. 3, p. 343–367.

Wang, W. M., Y. M. He, and Z. X. Yao, 2004, Complexity of the coseismic rupture for 1999 Chi-Chi Earthquake (Taiwan) from inversion of GPS observations: Tectonophysics, v. 382, no. 3–4, p. 151–172, doi:10.1016/j.tecto.2004.01.005.

Wells, D. L., and K. J. Coppersmith, 1994, New empirical relationships among magnitude, rupture length, rupture width, rupture area, and surface displacement: Bulletin of the Seismological Society of America, v. 84, no. 4, p. 974–1002.

Yin, A., S. Nie, P. Craig, T. M. Harrison, F. J. Ryerson, X. L. Qian, and G. Yang, 1998, Late Cenozoic tectonic evolution of the southern Chinese Tian Shan: Tectonics, v. 17, no. 1, p. 1–27, doi:10.1029/97TC03140.

Yu, S. B., and L. C. Kuo, 2001, Present-day crustal motion along the longitudinal valley fault, eastern Taiwan: Tectonophysics, v. 333, no. 1–2, p. 199–217, doi:10.1016/S0040-1951(00)00275-4.

Yu, S. B., et al., 2001, Preseismic deformation and coseismic displacements associated with the 1999 Chi-Chi, Taiwan, earthquake: Bulletin of the Seismological Society of America, v. 91, no. 5, p. 995–1012.

Yue, L. F., 2007, Active structural growth in central Taiwan in relationship to large earthquakes and pore-fluid pressures: Ph.D. thesis, Princeton University, Princeton, New Jersey, 262 p.

Yue, L. F., J. Suppe, and J. H. Hung, 2005, Structural geology of a classic thrust belt earthquake: The 1999 Chi-Chi earthquake Taiwan (M_w = 7.6): Journal of Structural Geology, v. 27, no. 11, p. 2058–2083, doi:10.1016/j.jsg.2005.05.020.

Zehnder, A. T., and R. W. Allmendinger, 2000, Velocity field for the trishear model: Journal of Structural Geology, v. 22, no. 8, p. 1009–1014, doi:10.1016/S0191-8141(00)00037-7.

9

Rivero, Carlos, and John H. Shaw, 2011, Active folding and blind thrust faulting induced by basin inversion processes, inner California borderlands, *in* K. McClay, J. H. Shaw, and J. Suppe, eds., Thrust fault-related folding: AAPG Memoir 94, p. 187–214.

Active Folding and Blind Thrust Faulting Induced by Basin Inversion Processes, Inner California Borderlands

Carlos Rivero[1]

Department of Earth and Planetary Sciences, Harvard University, Cambridge, Massachusetts, U.S.A.

John H. Shaw

Department of Earth and Planetary Sciences, Harvard University, Cambridge, Massachusetts, U.S.A.

ABSTRACT

The present bathymetry, basin geometries, and spatial earthquake distribution in the inner California borderlands reflect complex basin inversion processes that reactivated two low-angle Miocene extensional detachments as blind thrust faults during the Pliocene to Holocene. The Oceanside and the Thirtymile Bank detachments comprise the inner California blind thrust system. These low-angle detachments originated during Neogene crustal extension that opened the inner California borderlands, creating a rift system that controlled the deposition of early to late Miocene sedimentary units and the exhumation of the metamorphic Catalina schist. During the Pliocene, a transpressional regime induced by oblique convergence between the Pacific and the North American plates reactivated the Oceanside and the Thirtymile Bank detachments as blind thrust faults. This reactivation generated regional structural wedges cored by faulted basement blocks that inverted the sedimentary basins in the hanging wall of the Miocene extensional detachments and induced contractional fold trends along the coastal plain of Orange and San Diego counties. Favorably oriented high-angle normal faults were also reactivated, creating zones of oblique and strike-slip faulting and folding such as the offshore segments of the Rose Canyon, San Diego, and the Newport-Inglewood fault zones. We evaluate several different styles of geometric and kinematic interactions between these high-angle strike-slip faults and the low-angle detachments, and favor interpretations where deep oblique slip is partitioned at shallow crustal levels into thrusting and right-lateral strike-slip faulting.

Analyses of seismic reflection profiles, well data, earthquake information, and sea-floor geology indicate that the Oceanside and the Thirtymile Bank blind thrust faults are active and represent important sources of earthquakes in this region. Restored balanced cross sections provide a minimum southwest-directed slip of 2.2–2.7 km (1.4–1.8 mi) on the Oceanside thrust and illustrate the function of this detachment in controlling the processes of basin inversion and the development of the overlying fold and thrust belt.

[1]*Present address*: Structural Geology Team, Chevron Exploration Technology Company, Houston, Texas.

DOI:10.1306/13251338M943432

INTRODUCTION

The southern margin of the inner California borderland province (sICB) comprises the San Pedro shelf and the Oceanside-Capistrano Basin (Figure 1). This region is limited to the north by the Los Angeles Basin, to the east by the coastal plain of the Orange and San Diego counties, to the west by the Santa Catalina Island, and to the south by the United States-Mexico maritime boundary (Vedder, 1976; Drewry and Victor, 1997).

Geologic data from nearby islands, sea-floor samples, and oil industry wells indicate that the sedimentary rocks deposited in the sICB are mainly Quaternary to lower Miocene turbiditic and fluvial sand-shale deposits overlying Cretaceous to Paleogene marine conglomerate, sandstone to siltstone formations, and Mesozoic Catalina schist (Legg, 1980; Vedder et al., 1986; Clarke et al., 1987, among others).

The main physiographic characteristic of the sICB is a series of elongated basins and ridges trending subparallel to the coast and to the relative slip vector between the Pacific and the North American plates (Luyendyk, 1991; Atwater and Stock, 1998). Based on the presence of these physiographic features, geophysical data, and coastal geology, several authors have attributed the earthquake activity in this part of the inner California borderlands to a regional system of active strike-slip faults similar to that in the onshore region around the Peninsular Ranges (e.g., Legg and Ortega, 1978; Clarke et al., 1987; Legg, 1989). However, the inner California borderlands do not display the apparent spatial correlation between earthquake activity and regional strike-slip fault zones that is observed around the onshore region of the Peninsular Ranges (Figure 1). In contrast, seismicity in this area is diffuse and scattered (Ziony and Jones, 1989; Astiz and Shearer, 2000; Richards-Dinger and Shearer, 2000). Poorer offshore earthquake locations because of limited station coverage presumably contribute to this pattern; however, they alone are insufficient to explain why seismicity is not localized along a few major offshore strike-slip faults. Moreover, the focal mechanism of the 1986 (5.6 local magnitude scale) Oceanside earthquake, the largest recorded event in the region, indicates faulting dominated by thrust motion (Hauksson and Jones, 1988; Pacheco and Nábêlek, 1988). Thus, we interpret these observations to reflect a complex mixture of strike-slip and blind thrust faulting in the inner borderlands that is similar to the style of deformation in the onshore Los Angeles Basin (Hauksson, 1990; Wright, 1991; Shaw and Suppe, 1996).

During the past decade, several authors have described the function of a Miocene system of low-angle normal faults in the Neogene opening of the inner borderland province and in the clockwise rotation and translation of the Transverse Ranges from the south to its present location (Yeats, 1976; Crouch and Suppe, 1993; Nicholson et al., 1993; Bohannon and Geist, 1998; Ingersoll and Rumelhart, 1999). Rivero et al. (2000) suggested that the seismogenic inner borderland blind thrust system originated when two of these detachments were reactivated by basin inversion processes initiated in the late Pliocene, during the onset of the modern transpressional regime (Figure 2). Here we confirm and elaborate on these results, showing that large-scale thrust faulting and folding are driven by structural wedging (Medwedeff, 1992) and argue that basin inversion processes are the primary tectonic mechanisms controlling the physiography and geology of the southern margin of the inner borderlands.

To document these basin inversion mechanisms, we analyze more than 10,000 km (6214 mi) of industry seismic reflection profiles, well data, seismicity, and sea-floor geologic maps. We perform a structural analysis that involves kinematic and forward modeling techniques based on quantitative structural relationships between fold and fault shapes (Suppe, 1983; Mount et al., 1990; Erslev, 1991; Suppe and Medwedeff, 1992; Allmendinger, 1998). We also use advanced three-dimensional (3-D) modeling techniques to generate precise representations of fault surfaces and key stratigraphic markers (Plesch et al., 2007). The lateral extent and geometry of the active blind thrust ramps and fold trends are determined by mapping of direct fault-plane reflections and folded reflections throughout the basin areas covered by the seismic grid (Figures 2, 3). The 3-D modeling was also used to quantify the distribution of dip slip on the active fault system and to further constrain the geometric analysis.

To illustrate our findings, we present new regional structural interpretations of the active submarine fold and thrust belt located offshore Dana Point, and new geometric representations of the offshore Newport-Inglewood, Rose Canyon, and San Diego Trough fault zones. These new structural interpretations are consistent with basin inversion processes and the presence of both active blind thrust and strike-slip faults in the southern inner California borderlands.

BASIN INVERSION

Formation of sedimentary basins commonly involves extensional deformation of the crust, with concomitant development of rifting and normal faulting (Bally and Snelson, 1980; Allen and Allen, 1990). Subsequent compressional-transpressional tectonic phases generally induce the contraction of the basins in a process called basin inversion or inversion tectonics (Gleinner

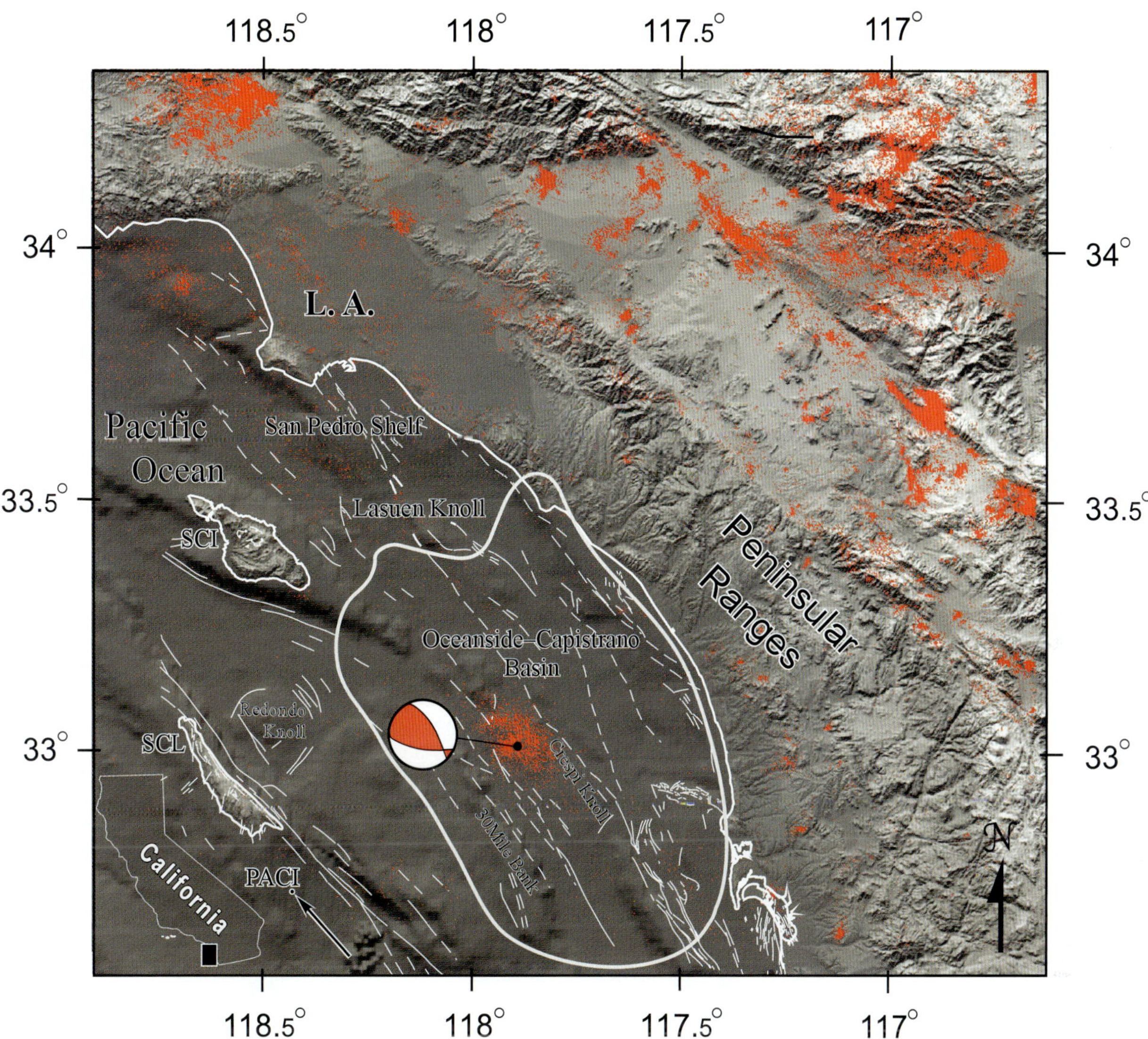

Figure 1. Map of the inner California borderlands and the study area. Major surface fault traces and earthquake locations (1977 to 2000) are shown (from Richards-Dinger and Shearer, 2000). The earthquake focal mechanism represents the 1986 (5.3 local magnitude scale) Oceanside earthquake location from Astiz and Shearer (2000). Pacific NOAM slip vector (PACI) from McCaffrey (2005). L. A. = Los Angeles Basin; SCI = Santa Catalina Island; SCL = San Clemente Island. The bathymetric contour interval is 200 m (656 ft). Digital southern California topography was generated from digital elevation data provided by the U.S. Geological Survey.

and Boegner, 1981; Bally, 1984), characterized by uplifting, flexure, and folding of the basin floor and the sedimentary infill (Ziegler, 1983; Williams et al., 1989; Letouzey, 1990) (Figure 4).

As the tectonic phase of compression or transpression proceeds, favorably oriented normal faults reverse their movement and become totally or partially inverted (Bally, 1983). The onset of reverse slip normally occurs along the deepest parts of the reactivated fault, typically where the fault surface bounds a preexistent graben or half graben (Figures 4, 5). The upward movement of the hanging-wall block due to the reverse slip continuously deforms shallower strata and older structures and may induce deep-seated structural wedging and back thrusting (Roure et al., 1990). At shallow levels, the propagation of the deformation eventually leads to the development of new fault structures (blind or emergent shortcuts) in either the footwall or hanging-wall blocks (McClay, 1992, 1995).

The reactivation of the faults produces broad anticlines located directly on top of extensional rollovers and synextensional stratigraphic wedges (Bally, 1984; McClay, 1989; Letouzey, 1990), and tighter folds localized above the reactivated fault tips (Figure 5). As a

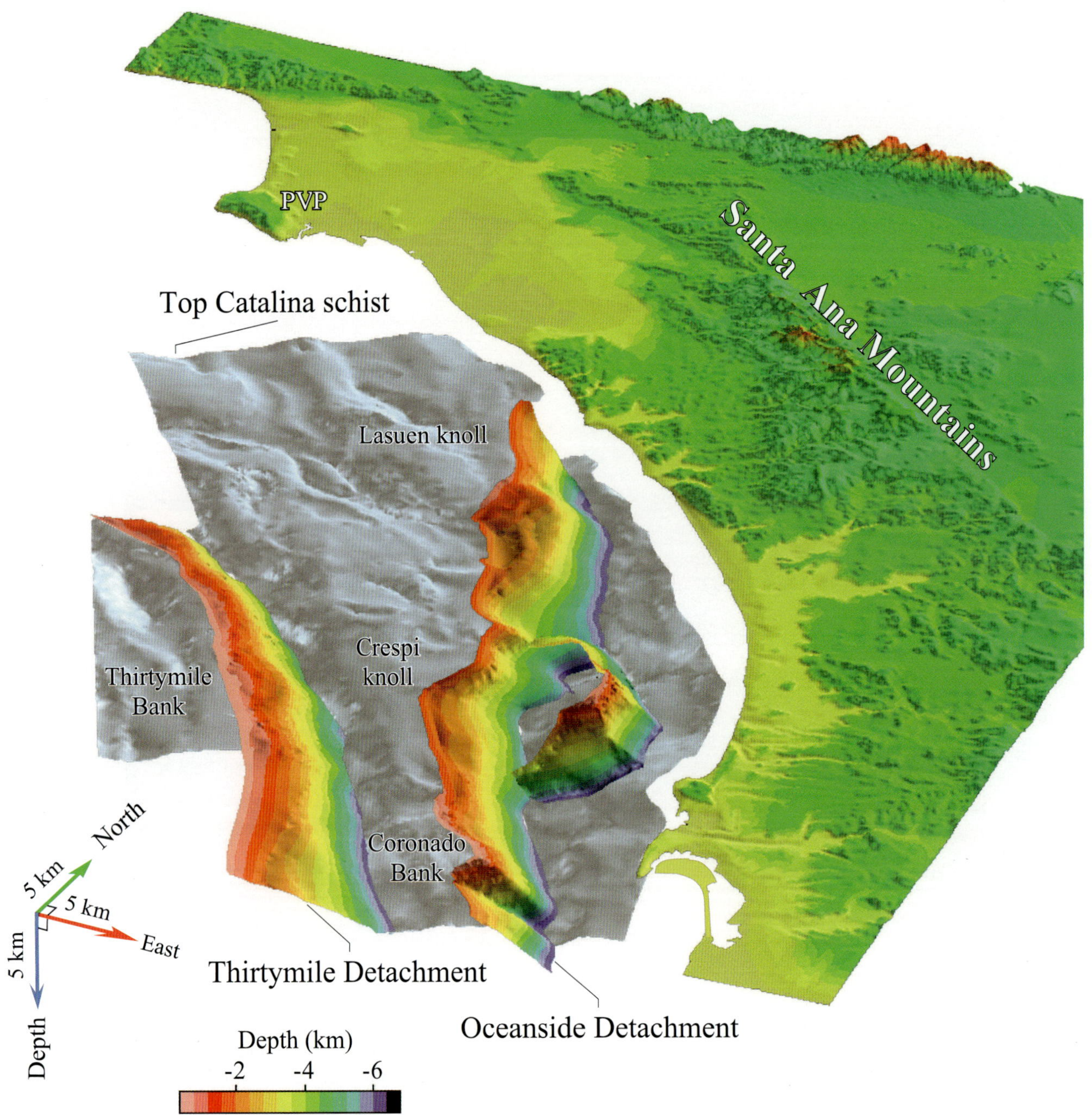

Figure 2. Perspective view of the inner California blind thrust system: the Oceanside and the Thirtymile Bank thrusts. The blue surface represents the top of the basement (Catalina schist). Digital shaded relief of the southern California topography was generated from digital elevation data provided by the U.S. Geological Survey. PVP = Palos Verdes peninsula.

consequence, synextensional units in the hanging-wall block are commonly preserved within the core of contractional folds at higher structural levels than their correlative footwall and basin regional elevations (Figure 5) (Williams et al., 1989). Commonly, the contractional folding is also recorded by syntectonic growth strata that thicken away from the contractional folds, providing a record of the kinematic evolution of the structure (Shaw and Suppe, 1996; Suppe et al., 1997). As a result, we commonly observe that slip along an inverted normal fault changes from net extension at deep stratigraphic levels to net contraction in the contractional growth strata and shallow parts of the postrift sequence (Cooper et al., 1989; Hayward and Graham, 1989).

General observations suggest that inversion processes do not always reactivate the entire suite of preexistent

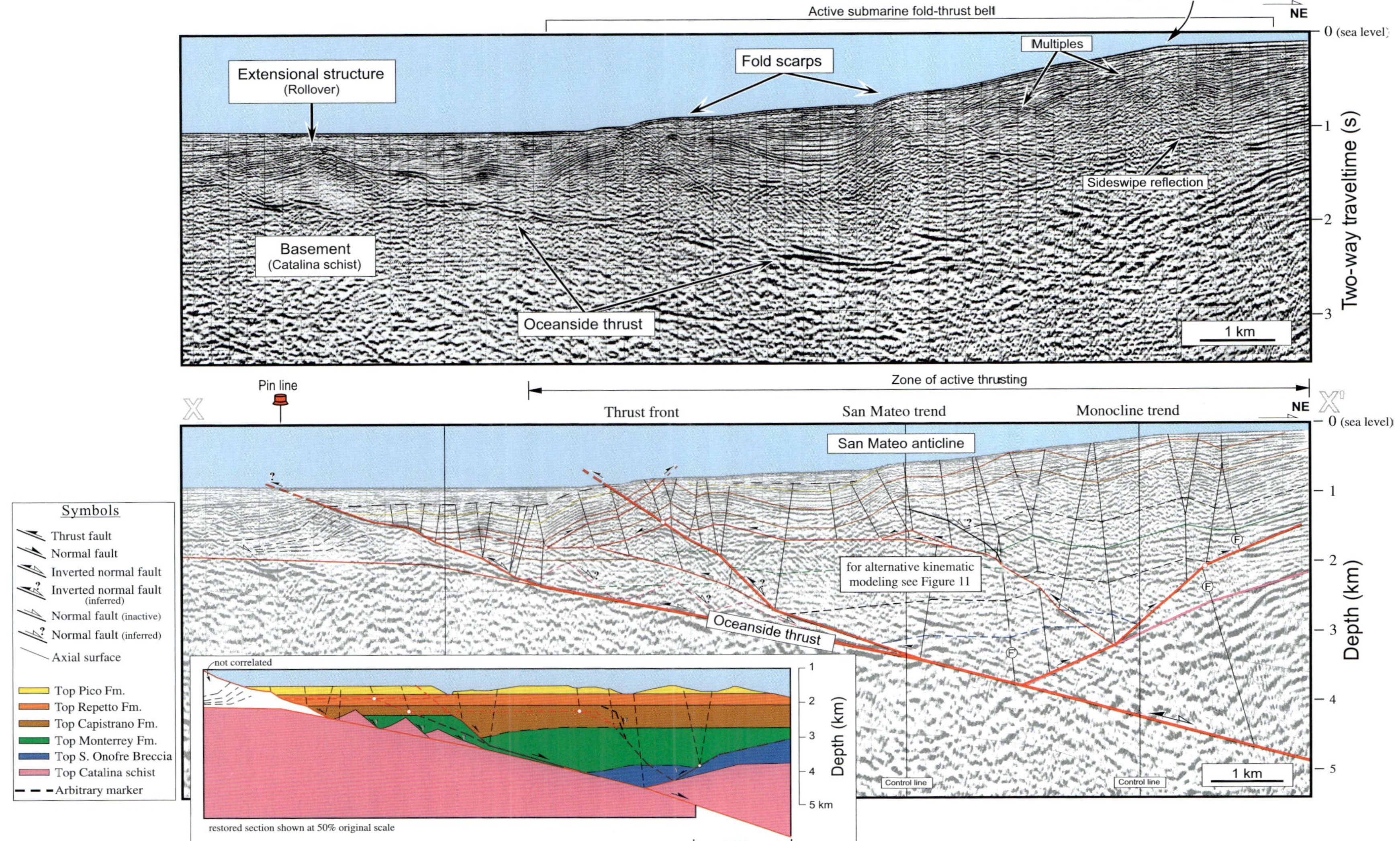

Figure 3. (top) Migrated seismic reflection profile across the active submarine fold and thrust belt located to the south of Lasuen Knoll. The seismic profile images the thrust front and a regional backthrust zone (see Figure 6 for the location). (bottom) Balanced geologic cross section XX′ derived from the seismic profile, lateral correlation of well information, direct fault-plane reflections, and fault-related folds theories. The cross section depicts the complex structural contractional trends produced by early Pliocene to Holocene reactivation of a series of northeast-dipping Miocene normal faults. The section also illustrates the role that basin inversion processes and associated structural wedging play to control the location of the thrust front, the San Mateo trend, and the monocline trend. Fm. = Formation; S. Onofre = San Onofre.

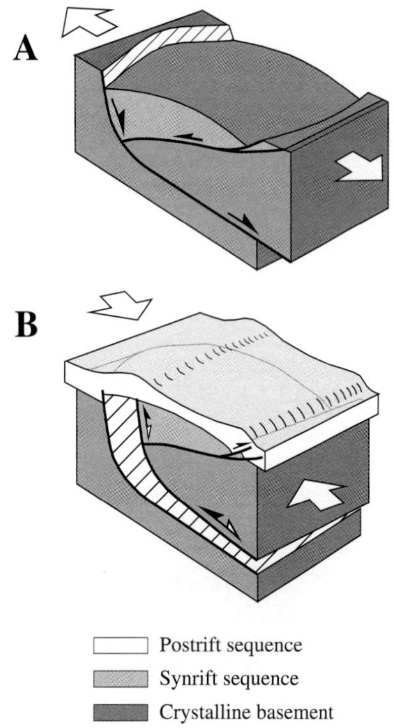

Figure 4. Conceptual model of basin inversion (modified from Bally, 1984). (A) Development of the extensional half graben and associated rollover structure. (B) Basin inversion phase characterized by the development of asymmetric contractional folds above the tip line of the old normal fault and structural wedging involving back thrusts within the old half graben.

normal faults; thus, we commonly observe isolated extensional faults in close proximity with inverted and reverse faults (Lowell, 1995; McClay, 1995). Depending on the coupling between the reactivated fault surfaces and the direction of the contractional slip, a varied suite of blind or emergent structures such as back thrusts, structural wedges, and out-of-sequence thrusts may be present. All these elements give a complex picture to the settings where basin inversion operates, making them prone to be confused with flower structures (Roure and Colletta, 1996; Colletta et al., 1997; Roure et al., 1997).

In this study, we document basin inversion processes in the inner California borderlands (Figures 1, 6) that share many of the characteristics documented in other inverted basins (Bally, 1984; Badley et al., 1989; Lowell, 1995), in analog (sand-box) experiments (McClay and Buchanan, 1992; Schreurs and Colletta, 1998), and in numerical models (Huyghe and Mugnier, 1991; Buiter and Pfiffner, 2003). Many of these previous studies, however, consider reactivation of steeply dipping planar and listric normal fault geometries, with little or no reference to low-angle normal systems because activation of upper crustal detachments is rare (Butler, 1989). Thus, the southern inner California borderlands offers a unique opportunity to investigate basin inversion processes originated by the reactivation of Miocene low-angle detachments. High-resolution seismic reflection profiles across this region illuminate and document active growth folding above blind thrust structures at Lasuen Knoll, Crespi Knoll, Coronado Bank, and the San Diego Trough–Thirtymile Bank region (Figures 3, 5–7). In these structures, contractional Pliocene sediments and younger units overlay Miocene rocks deposited during the rifting phase that opened the inner California borderlands. These Neogene rocks commonly defined local asymmetric anticlines sitting on top of regional rollovers. The normal faults associated with these rollovers are commonly synthetic to regional basal detachments. We interpret these offshore trends as the bathymetric expression of folding and faulting processes generated by the onset of the present transpressional regime in the southern inner California borderlands in the Pliocene. This transpressional regime induced the inversion of the Miocene depocenters (Figure 2) and reactivated a pair of regional low-angle detachments as the inner California borderlands blind thrust system (Rivero et al., 2000). In the next section, we integrate fault-bend fold theory and forward modeling techniques with seismic reflection profiles, well information, and seismicity to construct balanced cross sections and kinematic models that describe our interpretations of basin inversion and associated structural wedging across the major active trends of the southern inner California borderlands (Figure 6).

Our results are compatible with strike-slip displacement along the offshore segments of the Newport-Inglewood and Rose Canyon fault zones, and we provide new constraints on the geometries and kinematics of these fault systems. The results also provide insight into the subsurface geometries of complex zones where coeval strike slip and thrust fault interact.

ACTIVE SUBMARINE FOLD AND THRUST BELT

High-resolution seismic reflection profiles offshore Dana Point show the presence of a shallow west-vergent submarine fold and thrust belt defining a broad zone of active thrusting and related folding sitting on top of the well-illuminated Oceanside thrust (Figures 3, 6). The main characteristics of the submarine fold and thrust belt are the development of a large-scale structural wedge, a series of smaller forethrusts, and associated contractional folds located in the hanging-wall block of the Oceanside thrust. The thrust sheets are emplaced on, and structurally controlled by, the regional Miocene rift

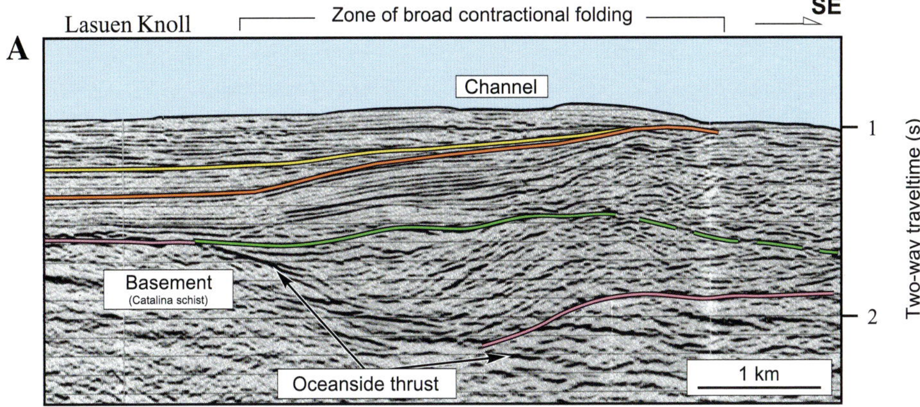

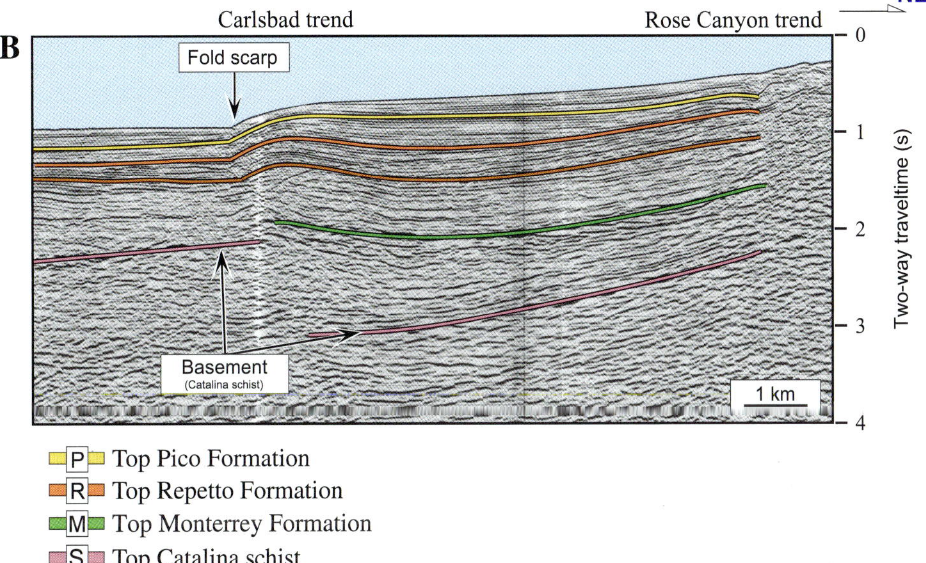

Figure 5. Seismic examples of basin inversion structures associated with the activity of the Oceanside thrust. (A) Half-graben reactivation along a lateral ramp of the Oceanside thrust. (B) Inversion structure developed by contractional reactivation of the Carlsbad fault. In both cases, seismic reflections within the Monterrey Formation define the stratigraphic expansion of the synrift sequence. Similarly, the phase of basin inversion is recorded by the contractional folds and thinning of the syncontractional Pico Formation on the crest of the anticlines. The location of seismic lines is shown in Figure 6.

system that is expressed in the seismic data as extensional rollovers and normal faults (Figures 2, 5). Although locally deformed in tight anticlines, the basic elements of the rift system are still clearly recognizable in the seismic data (Figure 2) (Yeats, 1976; Crouch and Suppe, 1993; Bohannon and Geist, 1998).

Several of these contractional and extensional structures were previously interpreted as wrench-related thrust and folds and as flower structures produced by active offshore segments of the Newport-Inglewood strike-slip fault (Legg and Kennedy, 1979; Crouch and Bachman, 1989; Fisher and Mills, 1991). In contrast, we observe that the contractional trends sole into the Oceanside thrust (Figures 3, 5, 7). This implies that the Oceanside thrust is a regional detachment level for the contractional deformation observed in the southern inner California borderlands (Figures 3, 7). Folded strata and regional near-sea-floor unconformities delineate several active piggyback basins preserved on top of the thrusts. Stratigraphic control provided by the nearby Mobil MSCH-1 San Clemente well and the Shell Oceanside well (Figure 6) dates the onset of the basin inversion phase as Pliocene (Figures 3, 5).

Motion on the Oceanside thrust generated four prominent contractional fold trends. Three of these trends are foreland-directed structures, namely the San Mateo, the San Onofre, and the Carlsbad trends (Figures 3, 7, 8). These active structures are characterized by thrust sheets that extend laterally for several kilometers and produce prominent fold scarps in the sea floor. The fourth trend is characterized by hinterland thrusting, which is manifested in a laterally continuous monocline that controls the relief and bathymetric expression of the coastal shelf (Figure 6). This monocline is the result of the interaction between a shallow west-dipping backthrust system and the deep-seated, east-dipping Oceanside thrust. In the following section, we present detailed structural evaluations of each of these trends.

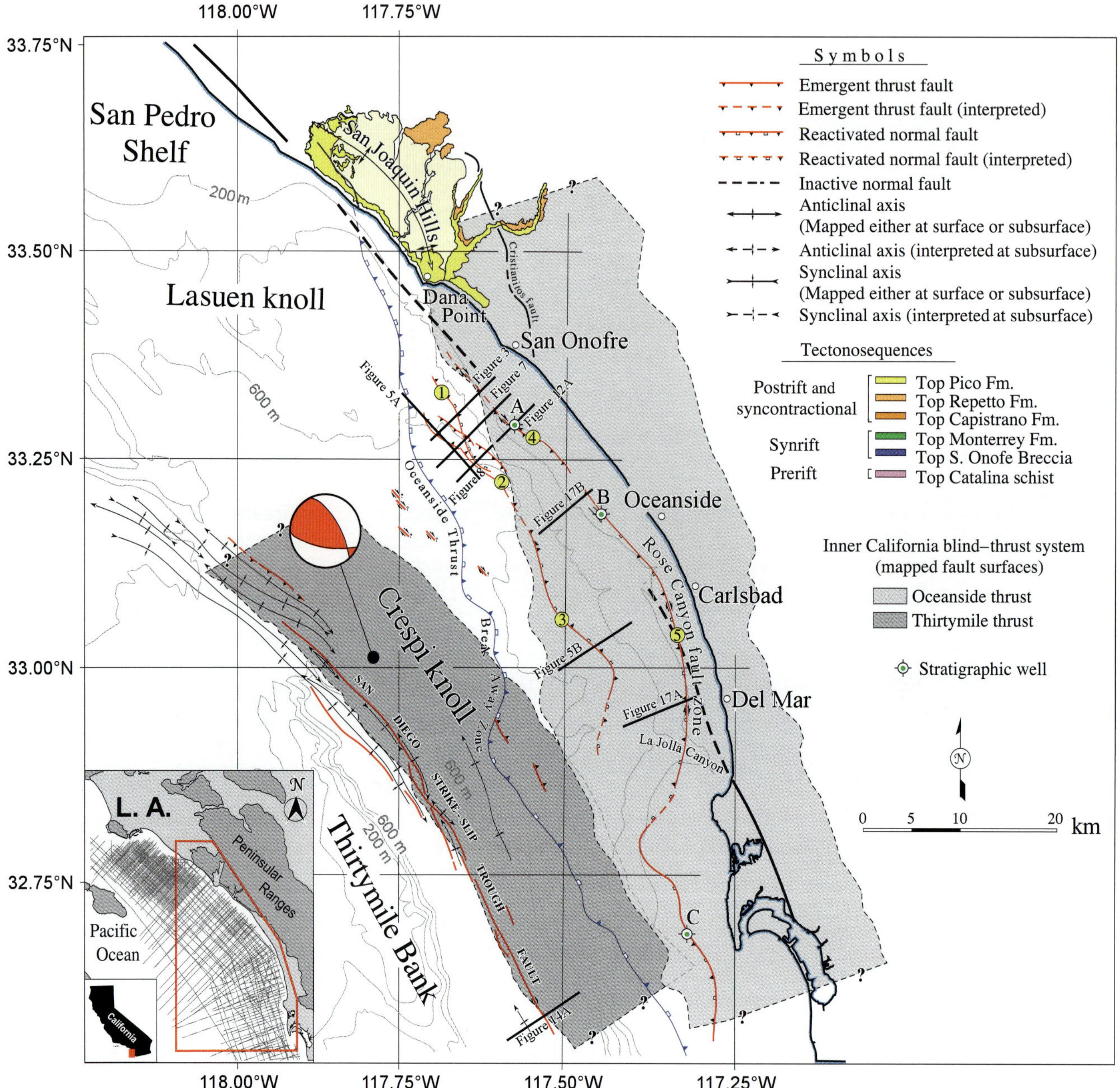

Figure 6. Map of major structural trends discussed in this study. The San Mateo (1), San Onofre (2), and Carlsbad trends (3) are the three major forethrust structures; the shelf monocline trend (4) is a regional backthrust system that may extend into the onshore San Joaquin Hills. Bathymetric contour lines illustrate the wide shelf offshore Orange County, which is produced by a regional southwest-propagating structural wedge. The locations of the mapped parts of the offshore Rose Canyon fault are shown (5). The breakaway zone of the Oceanside thrust is also indicated. Well locations A, B, and C correspond to the Mobil MSCH-1 San Clemente well, the Shell Oceanside well, and the Point Loma well, respectively. Inset is the regional grid of seismic reflection data used in this study. L. A. = Los Angeles.

SAN MATEO TREND

The San Mateo trend is located offshore Orange County at the southern margin of Lasuen Knoll (Figure 6). This trend extends for more than 12 km (7.4 mi), parallel to the coastline between Dana Point and Oceanside. Crouch and Bachman (1989), Fisher and Mills (1991), and Mills (1991) originally mapped the trend as an active frontal thrust that is part of an outer complex zone associated with the dextral motion of the Newport-Inglewood fault.

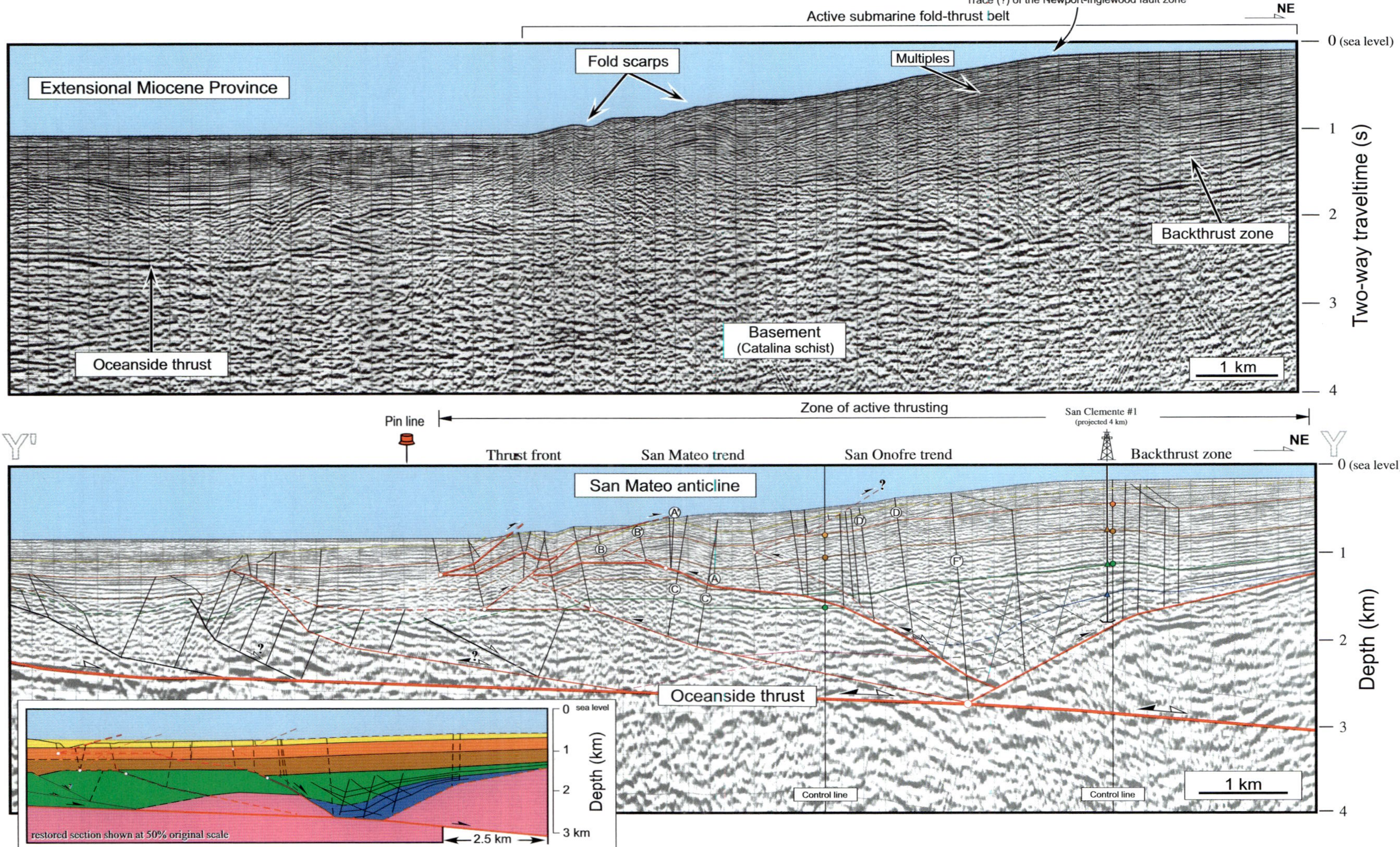

Figure 7. (top) Migrated seismic reflection profile across the San Mateo trend, offshore Dana Point (see Figure 6 for the location). Extensional Miocene structures are observable next to the thrust front position. (bottom) Balanced geologic cross section YY′ developed from the seismic profile, well information, and direct fault-plane reflections. The structural interpretation highlights the relationship of the contractional structures to a regional structural wedge and the reactivation of the preexistent Miocene normal faults during the phase Pliocene basin inversion. Restoration of the proposed structural scenario suggests 2.5 km (1.5 mi) as a minimum amount of Pliocene to Holocene shortening. San Clemente 1 = Mobil MSCH-1 San Clemente well.

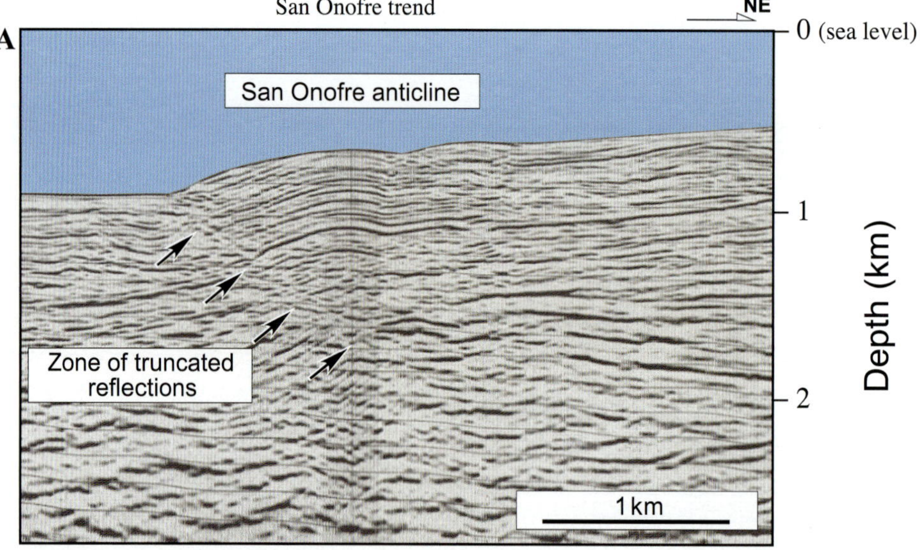

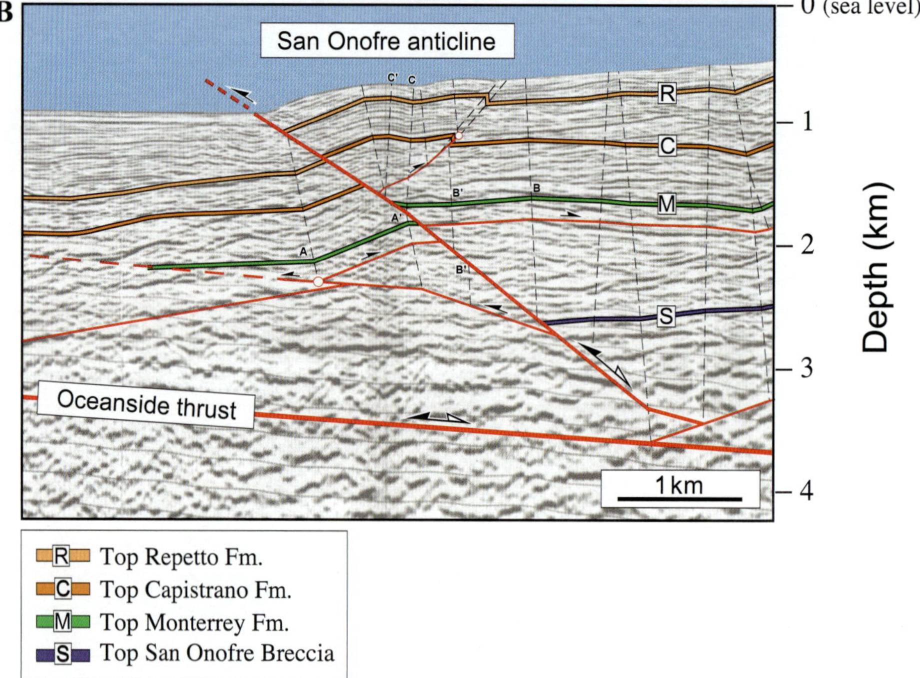

Figure 8. Detail of a seismic reflection profile across the San Onofre trend (see Figure 6 for the location). (A) Sea-floor scarps, shallow folding, and offset of seismic reflections define the location of the thrust fault. (B) Structural interpretation describing the geometry of the San Onofre anticline and the origin of this trend as a breakthrough thrust produced during the reactivation of a Miocene normal fault. Fm. = Formation.

Fisher and Mills (1991) inferred that Quaternary wrench deformation produced by a broad basement anomaly at a left step of the Newport-Inglewood–Rose Canyon fault zone was responsible for this and other nearby contractional trends. In contrast, Rivero et al. (2000) suggested that this deformation results from thrusting on the Oceanside detachment, which may be independent from strike-slip motion on the Newport-Inglewood–Rose Canyon fault zone.

Time and depth-converted seismic reflection profiles across the San Mateo trend used in our analysis indicate that the structure consists of a shallow and relatively symmetric faulted anticline that produces a series of pronounced bathymetric scarps on the sea floor (Figures 3, 7). The backlimb of the structure is defined by a syncline that contains Quaternary and older sediments (Figures 3, 7). At the thrust front, two additional small anticlines deform the uppermost sedimentary section (Figure 7). The

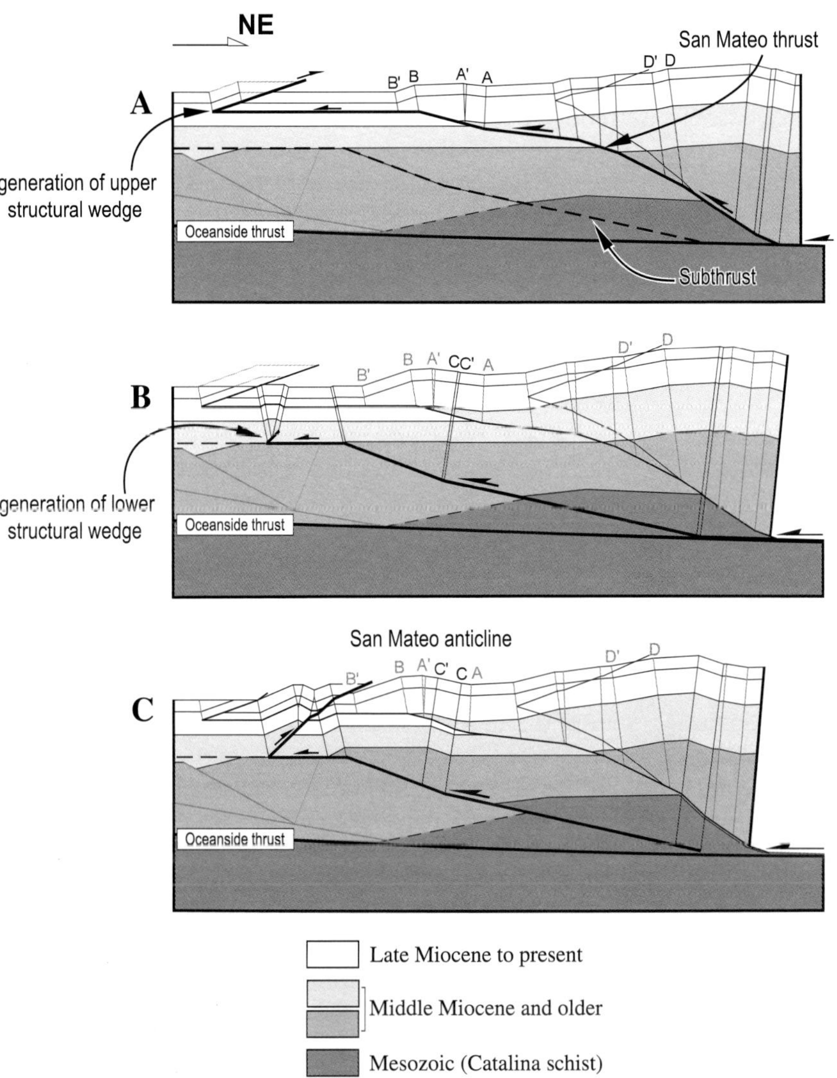

Figure 9. Kinematic modeling of the development of the San Mateo trend. (A) Minor displacement on the Oceanside thrust produces the reactivation of a synthetic normal fault and the incipient development of the San Mateo thrust. Simultaneously, an upper structural wedge is also developed in the thrust front position, which dissipates contractional slip as emergent back thrusting. (B) Generation of a subthrust (footwall shortcut), which imbricates and translates the San Mateo anticline. A coeval lower structural wedge related with the subthrust is formed at the thrust front position, in a similar way as for the San Mateo thrust. (C) Final configuration of the San Mateo trend and the thrust front domain.

observed folding of Pliocene and younger strata in the syncline and the related sea-floor fold scarps above the anticlines suggest Quaternary activity of the San Mateo trend.

Direct fault-plane reflections and stratigraphic cutoffs constrain the location of a 25° north-dipping fault plane that produces the anticline. We refer to this as the San Mateo thrust fault and apply fold-related fault theories (Suppe, 1983) to estimate the downdip projection of the fault to the northeast (Figure 7). This downward projection corresponds to the location of a preexistent half-graben structure situated in the hanging wall of the basal Oceanside thrust (Figures 3, 7). The half graben is defined by a westward-thickening sedimentary wedge in which the beds dip progressively steeper with depth, a typical style of extensional half grabens. Stratigraphic control provided by the nearby Mobil MSCH-1 San Clemente well indicates a Miocene age for the half graben, with the synrift sequence corresponding to the middle Miocene Monterrey Formation. This rollover structure is laterally persistent along the entire mapped part of the San Mateo thrust and shows evidence of structural inversion (Figure 7).

We combine these observations with fold-related fault theories and balanced forward modeling techniques (Mount et al., 1990; Shaw and Suppe, 1996) to interpret the San Mateo anticline as a multibend fault-bend fold developed by the reactivation of the underlying Miocene normal fault associated with the described half graben (Figure 9). The amount of contractional slip derived from panel BB′ is 0.58 km (0.36 mi). This value is consistent with the estimated 0.55 km (0.34 mi) of contractional slip recorded by kink-band DD′, which affected the Miocene rift basin, refolding the rollover and the younger units deposited in shallower levels.

We interpret that the gently dipping Oceanside thrust is not cut or deformed by the San Mateo fault, which is limited to its hanging-wall block (Figures 3, 7). This interpretation is based on the listric geometry of the San Mateo thrust with depth. This listric geometry is observed along the lateral extension of the trend (Figure 7) and is required to explain the rollover panel observed in reflections of the Monterrey Formation (Figure 7). The interpretation implies at least approximately 0.55 km (0.34 mi) of slip on the San Mateo thrust soles into the Oceanside thrust.

Structural analysis of the backlimb reflections of the San Mateo anticline shows a kink-band panel (CC′) that extends downward into the footwall block of the San Mateo thrust. Thus, the thrust fault appears to be folded by this kink band based on the geometric criteria proposed by Shaw et al. (1999). This implies that the San Mateo trend is imbricated by an underlying younger thrust (Figures 7, 9). This deeper thrust emerges to the east, where its slip is transferred to shallow back thrusts and dissipated in contractional folding. This fault extends to depth beneath and imbricates the San Mateo thrust.

Details of the structural geometry observed in the San Mateo trend appear to vary along strike. To the north, at location XX′ (Figures 3, 6), for example, the forelimb of the anticline expresses a stepper geometry that can be explained by imbricated fault-related folding. To the south, the thrust geometry becomes a multibend geometry ending in a buried tip wedge. Despite this variation, shortening across the trend changes very little, ranging from 0.5 to 0.7 km (0.3 to 0.4 mi).

SAN ONOFRE TREND

The San Onofre trend is a large, northwest-striking faulted anticline with pronounced bathymetric expression located offshore San Onofre and Oceanside (Figure 6). The structure was previously interpreted as an active, near-vertical strand of the Newport-Inglewood fault zone that potentially represented the reactivated offshore extension of the Miocene Cristianitos normal fault (Fisher and Mills, 1991; Mills, 1991). Based on several observations, Fisher and Mills (1991) proposed a recent right-lateral strike-slip movement on the onshore Cristianitos normal fault. However, Shlemon (1992) and Shlemon and Rockwell (1992) identified a regressive sequence of undisturbed marine deposits of Quaternary age (stage 5e ~125 k.y.) concealing the surface trace of the Cristianitos fault at San Onofre beach. This indicates that the Cristianitos fault is inactive and therefore unlikely to have produced the San Onofre trend, which is generally considered active. Moreover, our analysis suggests that the San Onofre trend is formed mostly by slip on a northeast-dipping thrust fault above the Oceanside thrust, and thus, it is not a simple vertically dipping extension of the Cristianitos or Newport-Inglewood trends.

High-resolution seismic reflection profiles across the San Onofre trend image a complex ramp anticline with pronounced bathymetric relief (Figure 8). Folded and laterally persistent unconformities and stratigraphic reflectors of syncontractional Repettian–Venturian and younger strata suggest the late Pliocene to Quaternary activity of the trend (Fisher and Mills, 1991). A dipping zone of shallow reflection truncations observed in the seismic images also indicates that the San Onofre trend is produced by an emergent thrust fault (Figure 8). Seismic reflection terminations as well as offset axial surfaces are used to interpret the shallow location and dip (32–34° northeast) of the thrust fault, which we defined as the San Onofre thrust (Figures 6, 8). Using fault-bend fold theories (Suppe, 1983), we estimate that the thrust steepens to a northeast-dipping angle of 38 to 40°, which is consistent with the dipping value of the fault reflections observed in depth-converted seismic images.

Similar to the San Mateo anticline, thickening of the Monterrey Formation toward the fault in the hanging-wall block of the San Onofre thrust suggests that this young contractional structure reactivated a Miocene normal fault. Using forward modeling techniques (Mount et al., 1990; Shaw and Suppe, 1996) and 3-D structural analysis (Rivero and Shaw, 2005), we interpret the San Onofre trend as a faulted decollement wedge. This wedge developed above a south-plunging segment of the underlying San Mateo thrust (Figures 8, 10). The decollement wedge is defined by the presence of a southwest-dipping kink-band AA′ located below the San Onofre thrust-plane reflections. The wedge and the kink band extend upward into the hanging wall of the San Onofre fault but are clearly offset by the thrust. Structural relief of the postrift top Repetto and Pico formations, which is controlled by lateral seismic correlation, confirms the fault offset.

On the basis of these geometric and stratigraphic relationships, we interpret the San Onofre trend as a late Pliocene to Quaternary fault-related fold originated by inversion of a half graben and thrusting that extends to the sea floor (Figure 8). Our structural scenario for the origin of the San Onofre anticline does not require the presence of a major strike-slip fault in the San Onofre trend nor its connection with other onshore trends. In contrast, the proposed solution describes a series of northeast-dipping thrust ramps beneath the trend with a minimum foreland-directed slip of 1 km (0.6 mi) for the San Mateo thrust and of 0.25 km (0.15 mi) for the San Onofre thrust. These thrusts appear to sole at depth with the Oceanside thrust.

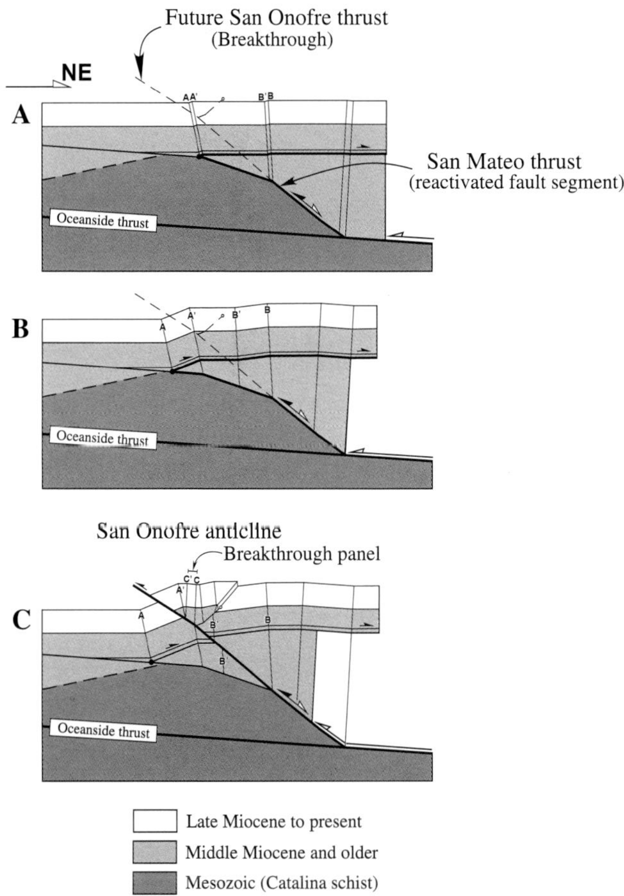

Figure 10. Balanced sequential development of the San Onofre and the San Mateo structures. (A) Activation of the San Mateo thrust as a reactivated Miocene normal fault synthetic to the Oceanside thrust. (B) Development of a structural wedge and folding of the future San Onofre thrust sheet. (C) Slip on the young breakthrough structure cutting across the northern limb of the structural wedge generates the San Onofre anticline.

CARLSBAD TREND

Evidence for active thrusting associated with the inner California borderlands blind thrust system is also present at Crespi Knoll, adjacent to San Diego Bay (Figure 6). At this location, the north-south–striking Carlsbad trend extends for more than 25 km (15.5 mi) offshore between Carlsbad and Del Mar. The trend was originally recognized by Crouch and Bachman (1989) and by Mills (1991) as a compressional thrust-related fold. Bohannon and Geist (1998) identified the trend as the offshore extension of the Newport-Inglewood–Rose Canyon fault zone and noted the active character of the structure and its potential downward merging with the Oceanside detachment.

In the industry seismic reflection profiles, the Carlsbad trend is illuminated as a wide, asymmetric, and laterally continuous contractional anticline overlying a Miocene half graben located in the hanging wall of the Oceanside thrust (Figures 5B, 11). Reflection truncations also illuminate a northeast-dipping thrust fault situated beneath the trend, consistent with the observed southwest vergence of the anticline. We refer to this thrust fault as the Carlsbad thrust, following original terminology from Fisher and Mills (1991).

Downdip projection of the Carlsbad thrust coincides with the location of a high-angle normal fault synthetic to the regional Oceanside detachment (Figure 11). The normal fault offsets the top of the regional acoustic basement defined by the Mesozoic Catalina schist. This fault also controlled the formation of a middle to late Miocene asymmetric half graben in which the synrift Monterrey Formation thickens toward the fault (Figures 5B, 11). Highly reflective units within the Monterrey Formation and the underlying San Onofre Breccia highlight the geometry of the growth sequence within the broad rollover. The high reflectivity levels may be related with Miocene volcanic rocks extruded and exposed in highly extended regions of the inner borderlands during the peak of the Neogene rifting phase (Vedder et al., 1986; Clarke et al., 1987; Crouch and Suppe, 1993). Similar seismic facies have been reported elsewhere within the Topanga Formation of the Los Angeles Basin and some other areas of the inner California borderlands (Luyendyk, 1991; Wright, 1991).

Regional stratigraphic correlation indicates that contractional growth sediments of the late Pliocene (Venturian) to Quaternary(?) age are present atop the Carlsbad trend (Figure 11). These growth strata are folded and truncated at the sea floor in a way consistent with contractional tip-line folding above the reactivated normal fault. The structural relief of these and older horizons can be explained in terms of basin inversion processes induced by the reactivation of the Oceanside thrust and by reverse motion on the described synthetic normal fault. At depth, the basal section of the Monterrey Formation still preserves the original dip orientation produced by the extensional phase. This suggests a limited amount of reverse motion on the inverted Carlsbad fault because the null point is somewhere close to this stratigraphic level (Williams et al., 1989).

Our structural analysis indicates that the Carlsbad trend is a contractional anticline formed above the tip of an inverted normal fault. The geometry of the structure suggests that the Carlsbad anticline may have originated by trishear fault-propagation folding (Figure 11) (Erslev, 1991; Hardy and Ford, 1997; Allmendinger, 1998, 2000); although triple-junction kinematics (Narr and Suppe, 1994) are also possible (Rivero, 2004). Ideally,

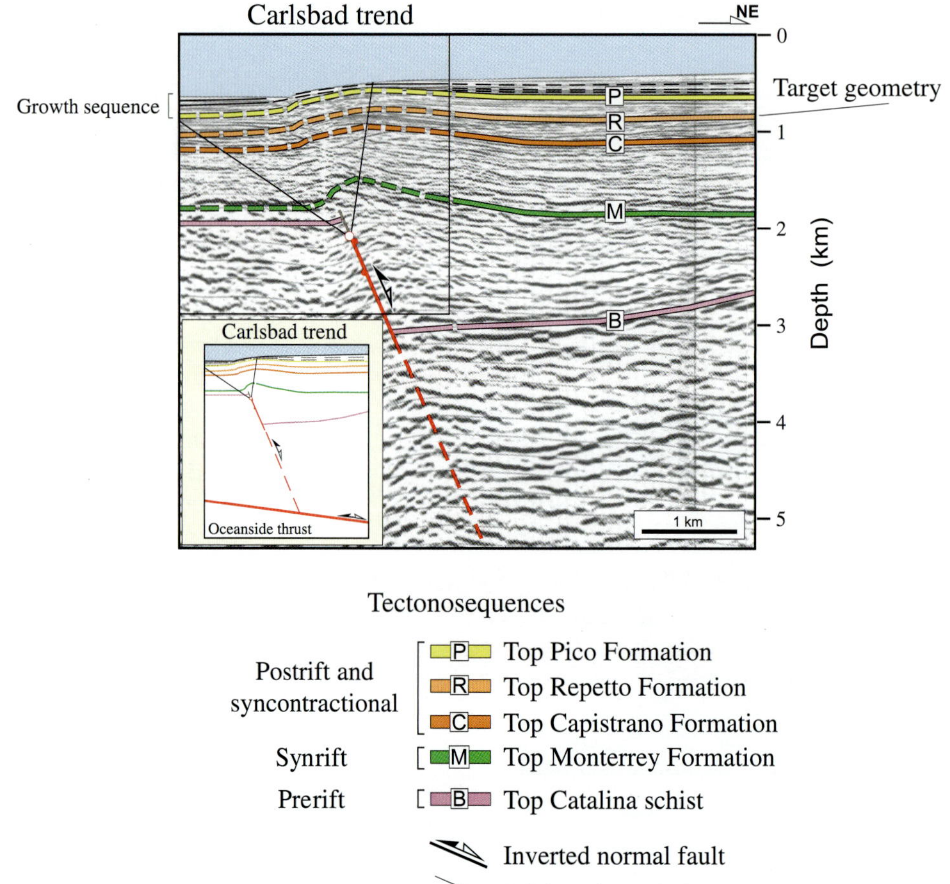

Figure 11. Structural interpretation of the Carlsbad trend in a depth-converted seismic reflection profile. A best-fitted model computed with the Trishear software (Allmendinger, 1998, 2000) for the Carlsbad anticline assuming trishear fault-tip propagation kinematics resolved the frontal limb of the Carlsbad anticline by reverse slip of a steeply northeast-dipping thrust, consistent with that observed in the depth-converted seismic image. The same solution accounts for rounded-hinge geometries and an estimated 0.4 to 0.6 km (0.2 to 0.4 mi) of displacement in the Carlsbad fault. The value of the correlation coefficient for this fitted model is $R = 0.98932$.

a geometric analysis of the contractional growth strata would be sufficient to distinguish between the two alternatives (Hardy and Poblet, 1994; Ford et al., 1997; Suppe et al., 1997). However, contractional growth strata atop the Carlsbad trend define a very thin and shallow sedimentary sequence that is very difficult to analyze in the seismic images.

In summary, our structural analysis indicates that the Carlsbad trend is a propagating tip fold produced by the structural inversion of a Miocene normal fault soled into the Oceanside thrust. Onlapping and modestly folded near-sea-floor sediments of shallow Quaternary(?) age are interpreted as evidence of Quaternary(?) activity on the Carlsbad and Oceanside thrusts. Inversion of the folded shape of the Carlsbad anticline using trishear estimates between 0.4 and 0.6 km (0.2 and 0.4 mi) of total thrust slip in the Carlsbad thrust.

SHELF MONOCLINE TREND

The offshore coastal shelf between Dana Point and Carlsbad exhibits a pronounced widening observed in bathymetric contours of the sea floor (Figure 6). This widening displaces the shelf break seaward, from an average distance of about 2 km (1.2 mi) in front of the San Joaquin Hills and Carlsbad, to a maximum of 10 km (3.2 mi) offshore San Onofre. Previous neotectonic studies in the region attributed this variation to recent activity in the offshore trace of the Newport-Inglewood strike-slip fault zone (Legg, 1980; Fisher et al., 1988; Fisher and Mills, 1991; Mills, 1991, among others). These authors suggested that a complex and continuous wrench fault system associated with the active trace of this strike-slip fault produced a flower structure with bathymetric expression that controls the sea-floor bathymetry of the area.

In contrast, industry seismic reflection profiles in this region illuminate a broad and well-expressed monocline dipping at about 15 to 20° to the southwest, which underlies and forms the wide shelf (Figures 3, 7, 12). In the seismic images, the monocline is defined by northeast-dipping axial surfaces F and F′ (Figures 3, 7, 12). South of San Onofre, the coastal shelf narrows again where the monocline is not present (Figure 6). We interpret this monocline as a rift shoulder of the Oceanside detachment, which has been reactivated as a structural

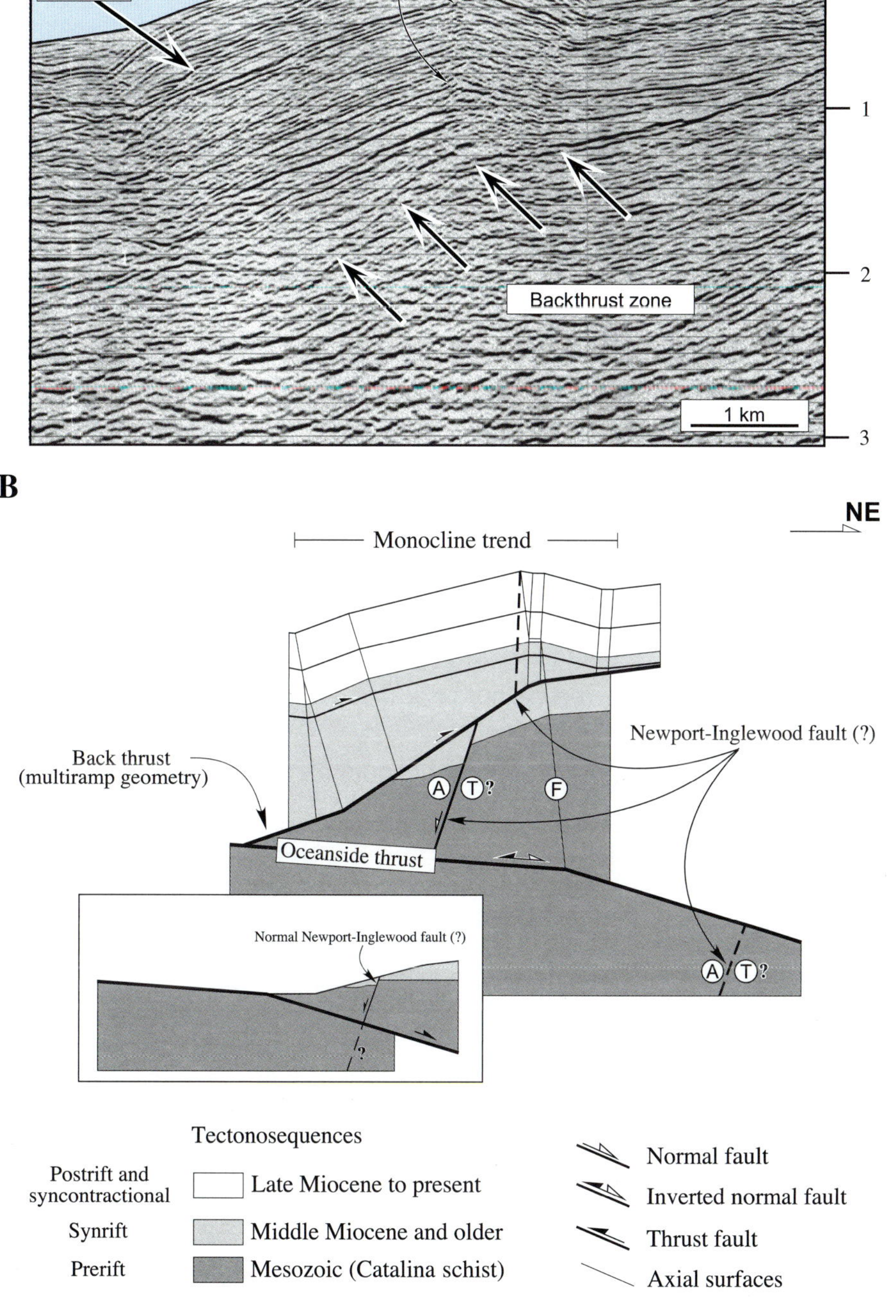

Figure 12. (A) High-resolution seismic reflection profile illustrating the geometry of the backthrust system in a region located offshore Oceanside (see Figure 6 for the location). The seismic data illuminate a well-developed zone of fault-related folding located in the hanging-wall block of the Oceanside thrust. Compare with the shelf domain in Figure 7. (B) Retrodeformable structural solution for the backthrust system and the monocline structure located at the coastal shelf of Orange County. The solution also considers the interaction of the monocline trend and the offshore part of the Newport-Inglewood fault zone. In this solution, the strike-slip fault is offset by the structural wedge, yielding offset fault segments in different structural blocks. This solution allows the coeval activity of the low-angle and the strike-slip faults but reflects a high degree of fault complexity. This type of complexity, induced by the Oceanside thrust, is likely responsible for the along-strike geometric segmentation of the Newport-Inglewood and the Rose Canyon strike-slip faults south of Dana Point.

wedge during tectonic inversion. The rift shoulder developed on the hanging-wall block of the Oceanside detachment by east-directed normal motion induced during the Miocene extension that affected the inner California borderlands.

Seismic correlation across the monocline provided by the Mobil MSCH-1 San Clemente and the Shell Oceanside wells reveals shallow folding involving the Repetto and the Pico formations. These units appear continuous across the monocline trend but form a set of asymmetric, northeast-vergent folds consistent with hinterland-directed structural transport during tectonic inversion of the underlying rift (Figures 3, 7, 12). The seismic images indicate that the style of deformation of this shallow folding changes laterally from single fault-bend and fault-propagation fold anticlines to more complex imbricated systems (Figure 12). Normally, the crests of these shallow anticlines appear truncated by the modern sea floor.

Direct fault-plane reflections and stratigraphic cutoffs of Miocene units at middle levels of the monocline reveal the location of a backthrust fault dipping 21 to 23° to the southwest (Figures 3, 7, 12). This and similar back thrusts underlie the shallow hinterland-directed folds. In a stratigraphic sense, the position of the back thrust commonly coincides with the tilted contact between the shaly Monterrey Formation and the underlying San Onofre Breccia.

We apply analytical techniques provided by fault-related fold theories (Suppe, 1983; Medwedeff, 1992) and forward modeling strategies (Mount et al., 1990) to interpret the observed fault-related fold and its relation with the monocline geometry (Figure 12). Figures 3, 7, and 12 show three different structural cross sections located across the continental platform between San Joaquin Hills and Carlsbad. The monocline trend is well defined at the right side of the structural cross sections, where a southwest-dipping panel is limited by axial surfaces F and F′. This panel extends continuously throughout the wide platform and is expressed in the low-slope gradient observed along this region (Figure 13). The southwest-dipping panel is not observed anywhere below the structural level defined by the Oceanside detachment. Thus, we infer that the monocline structure is limited at depth by the Oceanside detachment, consistent with the proposed origin of the monocline as a Miocene rift shoulder and the presence of fore thrusts restricted to the hanging-wall block of the Oceanside thrust.

The structural and stratigraphic relations observed in the seismic data between the monocline trend, the shallow thrust-fold structures, and the Oceanside thrust suggest the presence of a wedge geometry (Medwedeff, 1988) involving the metamorphic basement (Figure 13). In this scenario, a component of slip on the Oceanside thrust is directed offshore, leading to the development of the previously described submarine fold and thrust belt composed of the San Mateo, the San Onofre, and the Carlsbad trends. The development of a foreland-propagating wedge tip within the monocline, however, also allowed some of the contractional slip to be transferred to a shallow backthrust system ramping up from the Oceanside thrust (Figure 13). Thus, slip may have been partitioned between foreland- and hinterland-directed structures during the more recent evolution of this active submarine fold and thrust belt. Similar crustal wedging processes have been reported in other areas of California (Medwedeff, 1992; Shaw and Suppe, 1996; Novoa, 1997) and represent a common mechanism of deformation observed in many sedimentary basins and fold and thrust belts (Ziegler, 1983; Cooper and Williams, 1989; Roure and Colletta, 1996; Colletta et al., 1997).

Rivero et al. (2000) and Rivero (2004) noted the spatial correlation of this structural wedge with Quaternary uplift observed in adjacent coastal areas east of the anomalously wide region of the coastal shelf (Figures 2, 6). Previous neotectonic studies based on Quaternary marine terraces and strand lines located along the coastal plain of Orange and San Diego counties indicate continuous uplift of this region during the last 120–80 k.y. (Lajoie et al., 1979, 1992; Barrie and Gath, 1992; Kern and Rockwell, 1992; Grant et al., 1999; Kier and Mueller, 1999). Moreover, some of the local high rates of tectonic uplift and folding observed in the San Joaquin Hills have been recently associated with an active blind thrust fault dipping to the southwest (Mueller et al., 1998; Grant et al., 1999) and with activity in the Oceanside thrust itself (Rivero et al., 2000).

Figure 6 shows the bathymetric expression of the monocline structure at several locations offshore the San Joaquin Hills. We use axial surface mapping techniques (Shaw et al., 1994; Rowan and Linares, 2000) and 3-D structural modeling to control the extent of the synclinal structure and back thrusts related with the structural wedge (axial surface F′). The results suggest that these trends extend toward the onshore region into the vicinities of the San Joaquin Hills. If the structural wedge extends into the onshore domain, as suggested by the map patterns, then it is possible that an onshore segment of the backthrust fault observed in the seismic images is responsible for the Quaternary uplifting and folding documented by coastal tectonics studies in the San Joaquin Hills (Lajoie et al., 1979; Grant et al., 1999; Kier and Mueller, 1999) (Figure 6). If so, this implies that the backthrust fault is active, which in turn suggests that the basal Oceanside thrust should be also active (Rivero et al., 2000).

STRIKE-SLIP AND BLIND THRUST FAULT INTERACTIONS

The inner California borderlands is cut by several right-lateral strike-slip fault zones that have been defined based on sea-floor geomorphology, shallow-marine geophysical data, and seismicity (Wilcox et al., 1973; Yeats, 1973; Crowell, 1974; Sylvester, 1988, among others). The two most conspicuous of these structures are the San Diego Trough strike-slip fault, located on the western margin of the inner California borderlands (Vedder et al., 1986), and the eastern Newport-Inglewood–Rose Canyon fault zone, which runs parallel to the coastline of the Orange and San Diego counties (Harding, 1973; Vedder et al.,

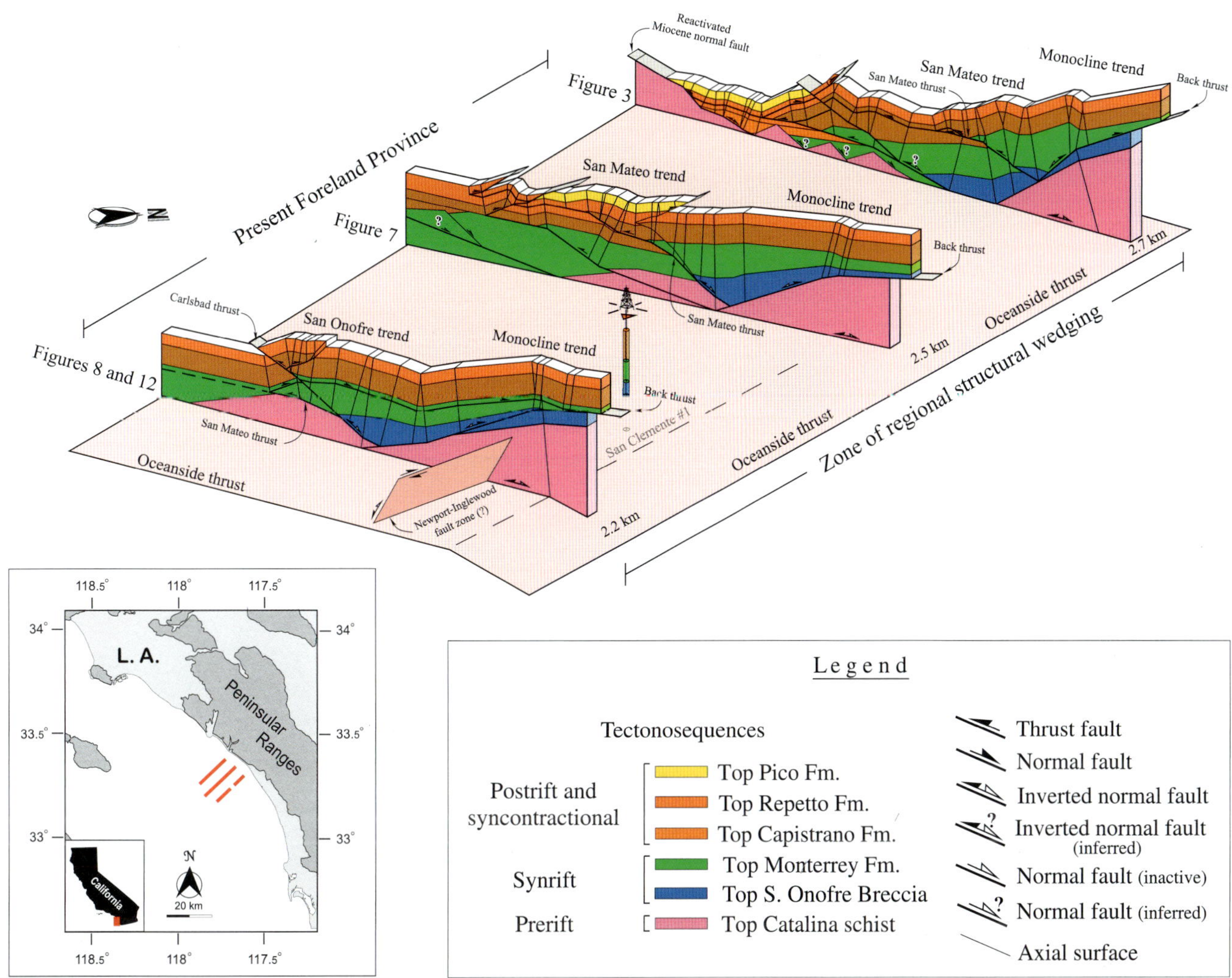

Figure 13. Diagrammatic representation of the main structural elements and trends observed along three transects shown in Figures 3, 7, 8, and 12. The representation illustrates the lateral continuity of the offshore-dipping monocline and the role of the Oceanside thrust as a regional basal detachment level. The model also highlights the complex arrangements of the modern contractional trends within the active submarine fold and thrust belt, and the control induced by the Miocene normal fault system and the propagating structural wedge in their locations. Note that the most important fold structures in a profile do not necessarily correlate between sections (i.e., observe the transition of the San Mateo and San Onofre anticlines between locations YY′ and ZZ′). Fm. = Formation; S. Onofre = San Onofre.

1986). Here we consider the geometric and kinematic interactions of these strike-slip fault zones with the low-angle Oceanside and Thirtymile Bank detachment surfaces.

The San Diego Trough strike-slip fault is a continuous wrench fault zone that extends for more than 50 km (31 mi) along a bathymetric low known as the San Diego Trough (Legg et al., 1991). This fault zone has been described as a characteristic dextral strike-slip system that dissects low-angle Miocene detachment faults, including the Thirtymile Bank detachment (Legg and Nicholson, 1993). These authors also suggest that the San Diego Trough strike-slip fault is active on the basis of fault scarps, modern slump and flow deposits, and offset of submarine channels observed at the sea floor.

The San Diego Trough strike-slip fault is well imaged to depths of about 5 km (3.1 mi) in the industry seismic reflection data available for this study (Figures 6, 14A). The fault consists of one or more steeply dipping splays defined by near vertically aligned truncations of reflections within the sedimentary fill of the basin (Figure 14A). Individual fault splays are somewhat discontinuous, but the fault system follows the central location of the triangular San Diego Trough. The traces of these faults

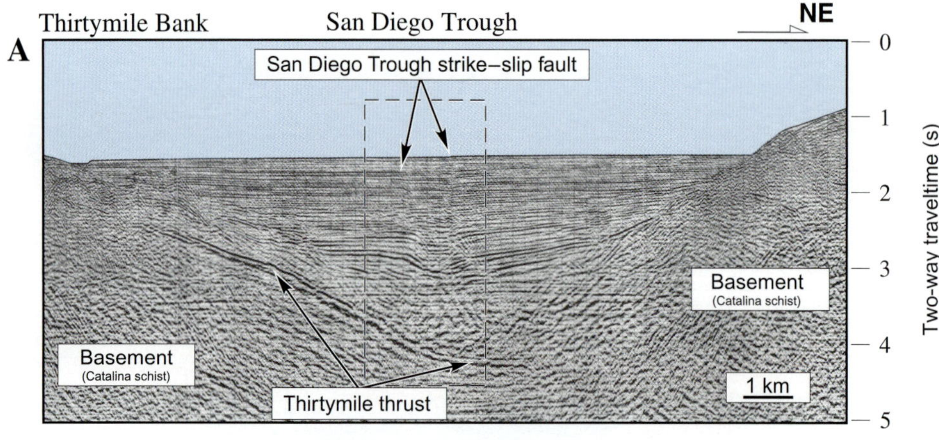

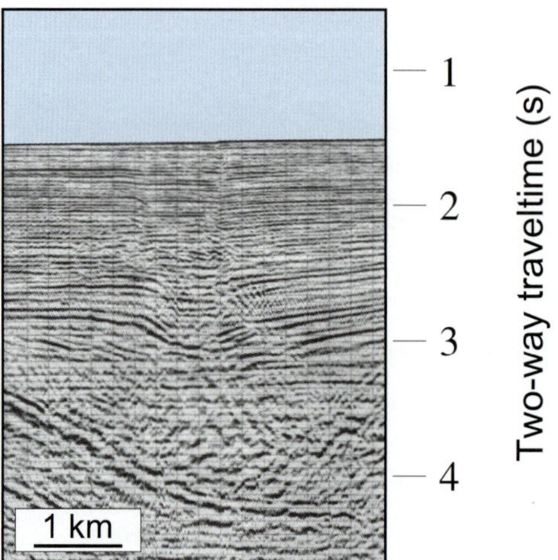

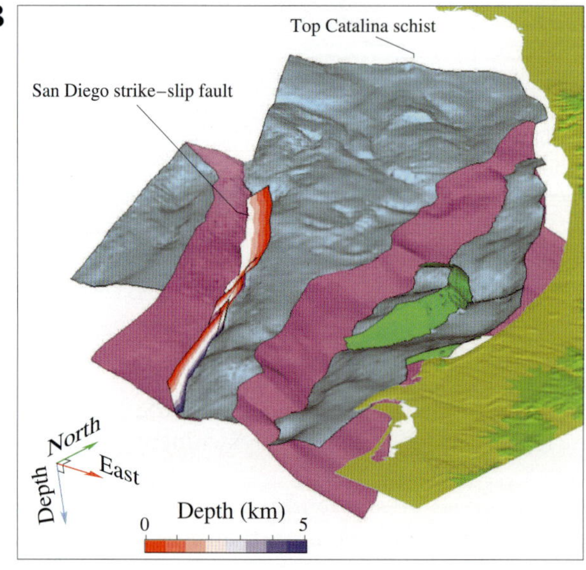

Figure 14. Seismic expression of the San Diego Trough strike-slip Fault at Thirtymile Bank. (A) Medium segment of the San Diego Trough strike-slip fault composed of two nearby fault branches. (B) Three-dimensional representation of the San Diego Trough strike-slip fault built from seismic reflection profiles similar to panel A, showing the relation between the Thirtymile Bank thrust and the San Diego Trough strike-slip fault.

produce small scarps at the sea floor that coincide with the location of the San Diego Trough strike-slip fault as mapped by Vedder et al. (1986).

We modeled the 3-D geometry of the San Diego Trough strike-slip fault using the grid of high-resolution seismic data, as well as bathymetric data and seismicity

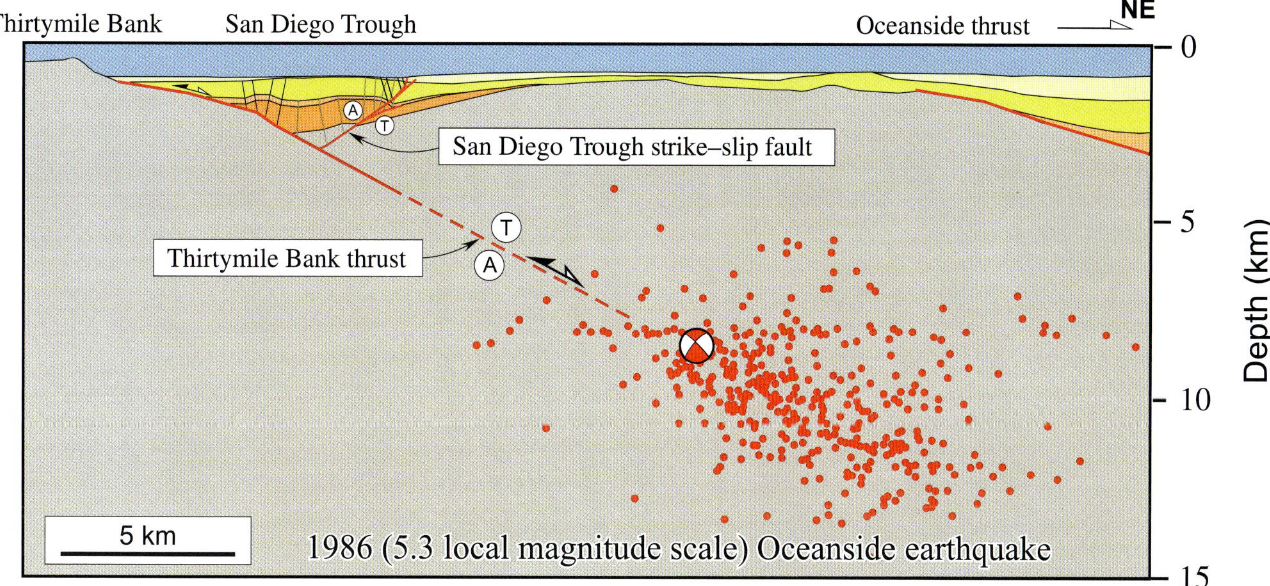

Figure 15. Structural cross section across the Thirtymile Bank thrust and the San Diego Trough strike-slip fault zone at the epicentral location of the 1986 (5.3 local magnitude scale) Oceanside sequence. The earthquake appears to have ruptured the downdip extent of the Thirtymile Bank detachment and not the San Diego Trough strike-slip fault. The association of the earthquake and the Thirtymile Bank fault is consistent with the hypocentral locations and main-shock focal mechanism.

(Figure 14B). Our analysis indicates that the strike-slip faulting and folding are confined to a zone along the center of the San Diego Trough. The zone is well expressed next to Coronado Banks where the faults are mostly defined by discrete vertical segments, several of which reach and offset the sea floor (Figure 14B). In contrast, the fault zone is more diffuse toward the northern Santa Catalina Basin (Figure 6). The basin in this area is shallower, and the geometry of the San Diego Trough fault becomes more complex, with a southwesterly dip to the main fault splay and evidence of oblique slip manifest in contractional folds adjacent to the fault. North of this region, the strike-slip fault seems to terminate into a zone of active thrusting and folding overlying the Thirtymile Bank thrust (Figure 15). Near-vertical fault splays are no longer observed; instead, deformation is accommodated by north-northwest–trending contractional folds with bathymetric expression (Figure 6).

Rivero et al. (2000) and Rivero (2004) described four possible modes of interaction between high-angle strike-slip faults and low-angle thrusts (Figure 16). Here we evaluate in detail each one of these modes on the basis of the results from our structural analysis.

The first two scenarios describe cases where the low-angle Thirtymile Bank detachment either dies out or is offset by the active San Diego Trough strike-slip fault zone (Legg et al., 1991; Legg and Nicholson, 1993). These interactions imply that the strike-slip faults cut down through the entire seismogenic crust, offsetting and rendering inactive the low-angle thrust faults, as it is assumed by most current hazard assessments (Figure 16A,

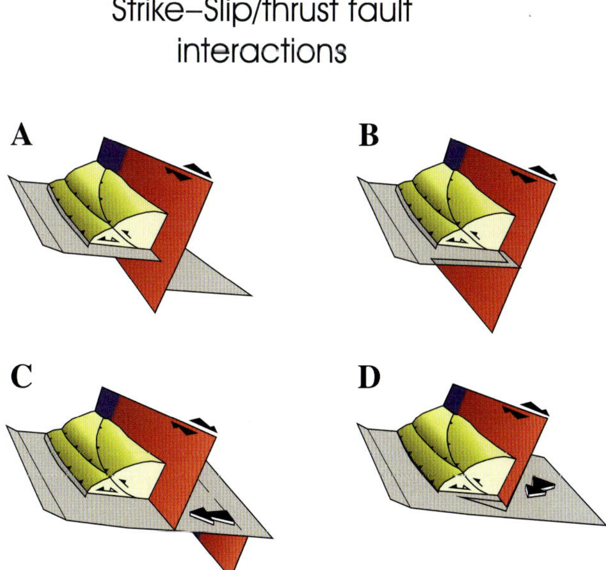

Figure 16. Schematic representation of different structural scenarios considered in this study for strike-slip and blind thrust fault interactions (modified from Rivero et al., 2000).

B). In these configurations, some thrust motion might still be possible but only as part of a highly oblique slip derived from motion on the strike-slip faults. An alternative set of solutions considers that the San Diego strike-slip fault is either offset by or merges with the active low-angle detachment. For these solutions, both fault systems are potentially active and possibly independent earthquake sources (Rivero, 2004).

Seismic reflection data and hypocentral locations of the 1986 (5.3 local magnitude scale) Oceanside earthquake sequence suggest that the Thirtymile Bank detachment is not truncated by the San Diego Trough fault but extends to the northeast beyond the strike-slip zone (Figures 6, 15). Previous authors (Hauksson and Jones, 1988; Pacheco and Nábêlek, 1988) attributed the Oceanside earthquake to dextral motion along the San Diego Trough strike-slip fault rupturing at depths of about 8 km (5 mi) in an unmapped restraining bend. In contrast, Rivero et al. (2000) related the origin of the 1986 (5.3 local magnitude scale) Oceanside earthquake sequence to the activity on the gently dipping Thirtymile Bank thrust fault based on spatial correlation between mapped parts of the detachment fault and the earthquake cluster as relocated by Astiz and Shearer (2000) (Figure 15). Through our mapping, we note that the earthquake epicenter was located 6 km (3.7 mi) east of the closest mapped segments of the San Diego fault (Figure 6), further supporting the association of the earthquake with the Thirtymile Bank fault. This interpretation is consistent with scenarios where the San Diego Trough strike-slip fault is either truncated by or merges down into the gently dipping Thirtymile Bank detachment (Figure 16C, D). As the Thirtymile Bank thrust manifests little reverse slip at this location, we would expect only minor offset of the San Diego Trough strike-slip fault if it extends into the footwall of the Thirtymile detachment.

The seismic data do not have enough resolution to rule out the presence of a deep segment of the San Diego strike-slip fault located in the footwall block of the Thirtymile Bank fault. Therefore, the mode of fault interaction depicted in Figure 16C remains viable. However, systematic mapping of the San Diego Trough fault and the Thirtymile Bank detachment indicates that the merging point between these two faults generally coincides with the shallow hanging-wall cutoff of the crystalline basement on the Thirtymile Bank fault (Figures 14, 15). This relation reflects that the strike-slip fault has broken through the deepest part of the sedimentary basin and hence the weakest part of the upper crust. Moreover, the orientations of the two faults, with the strike-slip fault trending parallel to the strike of the detachment, are consistent with a system where oblique motion at depth is fully partitioned into shallow zones of strike-slip and dip-slip faulting (Fitch, 1970; McCaffrey, 1996; Colletta et al., 1997; Roure et al., 1997). Thus, we consider that the mode of fault interaction depicted in Figure 16D with shallow-slip partitioning is the most likely description of the structural relationship between the Thirtymile Bank and the San Diego Trough strike-slip faults.

The Newport-Inglewood–Rose Canyon fault zone and the Oceanside thrust are other examples of strike-slip and thrust faults that interact at depth in the inner California borderlands. The Newport-Inglewood–Rose Canyon fault zone is a near-shore zone of active right-lateral faulting along the coast line of Orange and San Diego counties in southern California (Harding, 1973; Wilcox et al., 1973). Some of these authors defined a continuous strike-slip fault zone that extends from more than 100 km (62 mi) between Newport Beach and the San Diego Peninsula (Harding, 1973; Legg and Kennedy, 1979; Ziony and Jones, 1989). However, these and other authors have proposed many different locations for the surface traces of the Newport-Inglewood–Rose Canyon fault zone (Fisher and Mills, 1991).

Although recent activity of the Newport-Inglewood–Rose Canyon fault zone is well expressed onshore in the Los Angeles and San Diego regions, seismicity along the offshore segments of this fault zone is scattered and diffuse with no clear correlation between earthquake locations and the proposed offshore traces of the fault zone (Figure 1). Nevertheless, these offshore trends are considered active strike-slip systems based on sets of positive flower structures with complex bathymetric traces identified from offshore seismic data sets (Clarke et al., 1987; Fisher et al., 1988; Ziony and Jones, 1989; Mills, 1991). Comparison of the proposed fault traces with high-resolution bathymetric data and seismic reflection profiles reveals the systematic alignment between the locations of the shelf break and the mapped strike-slip fault zones south of Dana Point (Figures 3, 7, 12). In the Carlsbad region offshore from the San Diego Peninsula, the shelf break is discrete and traditionally attributed to the offshore segment of the onshore Rose Canyon strike-slip fault (Vedder et al., 1986; Ziony and Jones, 1989; Rockwell et al., 1992). Crouch and Suppe (1993) reported sedimentary rocks of Eocene and Miocene ages exposed in this narrow bathymetric slope and the coastal shelf. Stratigraphic control provided by the nearby Oceanside well, combined with seismic reflection data, indicates that these rocks are at higher structural elevation than younger Pliocene (Venturian–Repettian) strata deposited farther offshore. We interpret much of this structural relief to reflect Neogene extension or transtensional movement along the offshore Rose Canyon fault zone (Figure 17).

The offshore Rose Canyon fault zone is clearly recognized in seismic reflection profiles offshore Oceanside

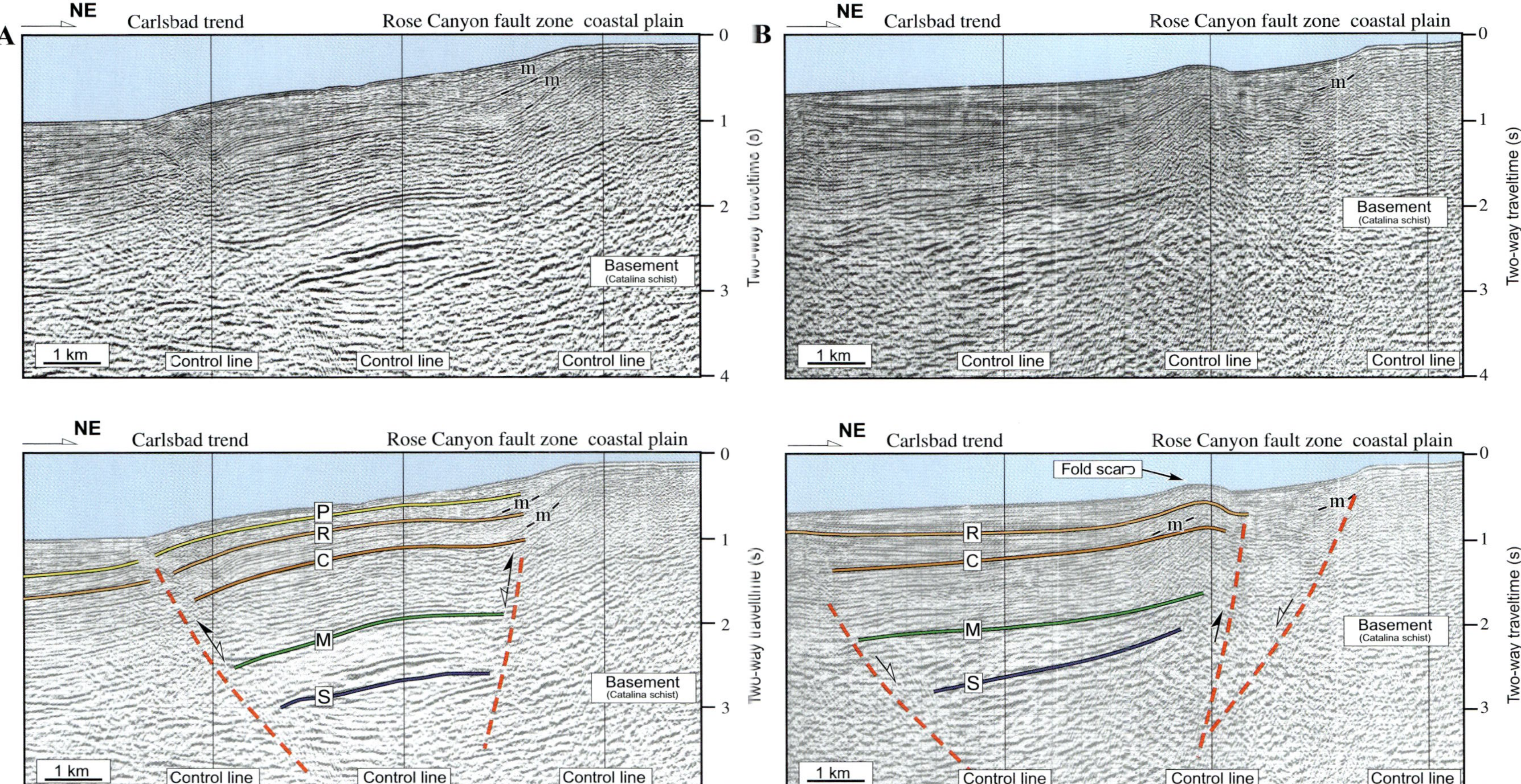

Figure 17. Seismic reflection profiles illustrating the variation in the geometry of the offshore Rose Canyon fault along the strike (see Figure 6 for the locations). In all the profiles, the location of the shelf break is structurally controlled by the presence of a fault, with normal separation and southwest dip. Contractional faulting occurs along faults that lie to the west of the shelf break. (A) Northern segment of the Rose Canyon fault. Pliocene contractional folding is associated with the northeast-dipping Carlsbad thrust. (B) Asymmetric contraction developed on a fault splay in the hanging wall of the Rose Canyon fault. The contraction affects Miocene and Pliocene–Holocene(?) sedimentary units. m = sea floor multiples; P = Top Pico Formation; R = Top Repetto Formation; C = Top Capistrano Formation; M = Top Monterrey Formation; S = Top San Onofre Formation.

and Carlsbad (Figure 17A). The broad fault zone extends continuously for more than 20 km (12.4 mi) between Oceanside and Del Mar, dipping steeply to the southwest at about 55 to 65°. Several splays related with the mapped fault zone show evidences of recent activity. On some of these splays, east-vergent contractional anticlines trending parallel to the coastline deform sedimentary rocks of Repettian and younger ages (Figure 6). Occasionally, the contractional trends define prominent fold scarps in the sea floor (Figure 17A). Other fault splays, particularly along the shelf break, appear to exhibit normal offset. Presumably, these contractional and extensional structures represent local restraining and releasing bends along the offshore extension of the Rose Canyon strike-slip fault.

Several authors have defined the geometry of the Rose Canyon strike-slip fault in the San Diego Peninsula on the basis of its surface expression between San Diego and Point Loma (Legg, 1980; Ziony and Jones, 1989; Fisher and Mills, 1991; Rockwell et al., 1992; Lindwall and Rockwell, 1995). The projection of this onshore segment north of Point Loma (and thus its connection with our mapped Rose Canyon fault zone located offshore) is uncertain because of poor fault expression and lack of seismic coverage at the La Jolla Canyon (Figure 6). At La Jolla Canyon, the mapped Rose Canyon fault zone changes its orientation to a southwest-trending direction, away from the San Diego Peninsula (Figures 6, 18), whereas the projection of the onshore Rose Canyon strike-slip fault continues to the north. This implies that the modern Rose Canyon fault zone has a pronounced bend or other form of geometric segment boundary in this area (Figures 6, 18). North of Oceanside, the offshore Rose Canyon fault zone extends into an area of active thrusting and wedging associated with the submarine fold and thrust belt located above the Oceanside thrust (Figure 6).

The offshore segment of the Newport-Inglewood fault zone is present to the north of the submarine fold and thrust belt that marks the northern extent of the Rose Canyon fault. The structural character of the Newport-Inglewood fault is well defined in the Los Angeles area, mostly because of its surface expression, and by the prolific amount of subsurface geologic and geophysical data obtained from oil fields located along its trace (Harding, 1973; Yeats, 1973; Wright, 1991). In this area, the fault geometry of the Newport-Inglewood fault zone has been interpreted as a simple vertical fault zone that changes to a more complex arrangement of segmented en echelon faults toward the south (Harding, 1973; Petersen and Wesnousky, 1994). Recent tectonic activity of the Newport-Inglewood fault zone in the Los Angeles region is documented by the 1920 (4.9 local magnitude scale) Inglewood earthquake and the 1933 (6.3 local magnitude scale) Long Beach earthquake (Hauksson and Gross, 1991; Wright, 1991). These events showed a right-lateral movement of a northwest-trending fault plane, dipping at around 70° to the east (Hauksson and Gross, 1991).

Offshore in the Dana Point and Carlsbad areas, the proposed traces of the offshore Newport-Inglewood strike-slip fault commonly lie above a monocline structure that we previously described in this study as having formed by structural wedging above the Oceanside thrust (Figure 12). Geologic interpretations of seismic reflection data generally show a continuous stratigraphic correlation of the Neogene units across the monocline trend. In many cases, the seismic data indicate that previously interpreted strike-slip fault splays correspond to active hinges of contractional anticlines produced by Pliocene to Holocene backthrust motion on a deep structural wedge (Figures 3, 7, 12). Nevertheless, a set of apparently truncated reflections within the monocline delineate a steep south-dipping zone that may correspond to the offshore trace of the Newport-Inglewood fault (Figure 12A). These truncated reflections seem to define an active strike-slip fault that may cut down into the structural wedge and must intersect the Oceanside thrust at a depth of 4 km (2.5 mi).

The offshore Newport-Inglewood fault may extend northward to link with the onshore Newport-Inglewood fault segment (Harding, 1973; Wilcox et al., 1973; Crowell, 1974; Legg and Kennedy, 1979; Sylvester, 1988; California Geological Survey, 2002). However, the presence and activity of the Oceanside thrust in this region preclude such a simple geometric linkage, at least at shallow levels. As a consequence, we speculate that the offshore San Joaquin Hills area most likely represents a geometric segment boundary between the onshore and offshore traces of the Newport-Inglewood strike-slip fault.

To the south of Dana Point, the offshore segment of the Newport-Inglewood fault zone may extend south into the Oceanside area, where the offshore Rose Canyon fault zone and the submarine fold and thrust belt converge (Figure 18). The direct linkage of the Newport-Inglewood fault zone and the offshore Rose Canyon strike-slip fault in this area of active faulting and thrusting, however, is not clear (Figure 6). If the linkage exits, as it has been proposed elsewhere (Harding, 1973; Wilcox et al., 1973; Crowell, 1974; Legg and Kennedy, 1979; Sylvester, 1988; California Geological Survey, 2002, among others), then it is through a highly complicated zone of fault splays and active folding observable in map view and cross sections. This zone of complexity arises because of the presence and activity of the Oceanside thrust. Thus, Rivero (2004) inferred that the offshore Newport-Inglewood strike-slip fault and the Rose Canyon

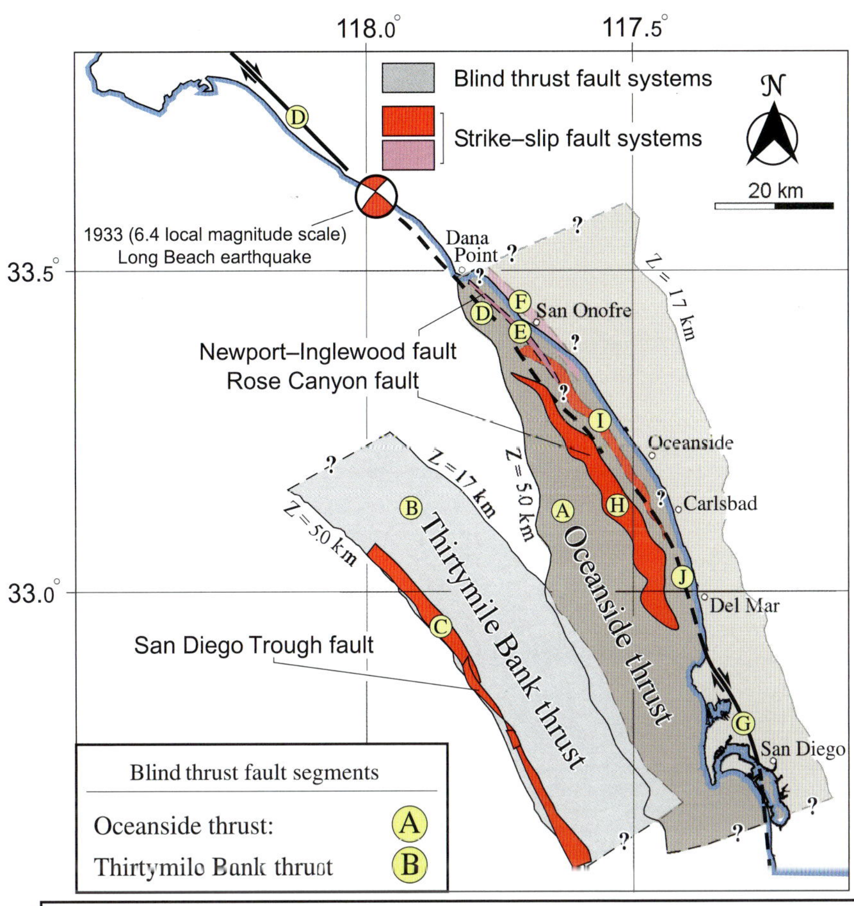

Strike–slip fault segments	Location	Code
San Diego Trough strike–slip fault (mapped)	HW	C
Newport–Inglewood strike–slip fault		D
Offshore Newport–Inglewood strike–slip fault	HW	E
	FW	F
Onshore Rose Canyon fault		G
Offshore Rose Canyon strike–slip fault (mapped)	HW	H
	FW	I
Offshore Rose Canyon strike–slip fault		J

Figure 18. Regional map view of the Oceanside and the Thirtymile Bank thrusts and the mapped and inferred offshore segments of the strike-slip San Diego Trough, Newport-Inglewood, and the Rose Canyon faults. Suggested onshore and offshore traces of the Newport-Inglewood and the Rose Canyon faults by the California Geological Survey (2002) are also shown for comparison (black and dashed lines). The footwall segment of the San Diego Trough fault is not represented. HW = hanging wall; FW = footwall.

fault zones are geometrically segmented from one another in this region and thus may represent independent sources of earthquakes (Figure 18).

At depth, the Newport-Inglewood and the Rose Canyon strike-slip fault zones intersect with the Oceanside thrust. This intersection occurs at relatively shallow levels of about 4 km (2.5 mi) to the north and deeper approximately 10 km (6.2 mi) in the south. Data are insufficient to uniquely define the manner in which these two fault systems interact. However, late Tertiary to Holocene activity on both systems and our documentation of several kilometers of west-directed shortening on the Oceanside thrust imply that the thrust fault is not truncated by the Newport-Inglewood and the Rose Canyon strike-slip fault zones (cases A or B, Figures 16, 18). Instead, scenarios where the two fault systems interact at depth in a manner consistent with their coeval activity (cases C and D) are favored.

CONCLUSIONS

Structural interpretations of high-resolution seismic reflection data, 3-D geologic modeling, well data, and earthquake information confirm the presence of a large blind thrust system underlying the coastal plain and offshore region of southern California. Our study indicates that the inner California blind thrust system consists of the Oceanside and the Thirtymile Bank thrusts, two Miocene low-angle detachments that were reactivated as blind thrusts by Pliocene basin inversion processes. The basin inversion processes were initiated during the Pliocene onset of the modern transpressional regime.

Our analysis suggests that active contractional folding and faulting are driven by regional structural wedging and basin inversion mechanisms. These processes partially reactivated ramp segments of the detachments and induced the generation of northwest–southeast contractional trends on the hanging walls of the Oceanside and Thirtymile Bank thrusts. Many of these contractional trends have bathymetric and topographic expression in the form of prominent fold scarps that control the morphology of the modern sea floor at the shelf and coastal plain of Orange County and the offshore Thirtymile Bank region.

The characteristic structural style of the contractional trends changes in the study region from complex arrays of imbricated fold and thrust systems in the north to discrete and individual broad fault-related folds in the south. These imbricated thrust systems are well expressed offshore Dana Point, San Onofre, and Carlsbad, which include foreland- and hinterland-directed thrust ramps of Pliocene and younger age. In the south, oblique contractional motion induced the reactivation of Miocene high-angle normal faults antithetic to the Oceanside thrust as broad contractional anticlines. We attribute the origin of the offshore segment of the Rose Canyon strike-slip fault to Pliocene to the Holocene structural reactivation of one of these high-angle normal faults.

Seismic reflection data and 3-D geologic modeling also help define dextral strike-slip faulting in the offshore San Diego Trough region previously identified as the San Diego strike-slip fault. We interpret this strike-slip fault as a young vertical structure developed by shallow strain partitioning of transpressional (southwest-directed) contractional slip transfer from the underlying Thirtymile Bank thrust.

Analysis of the potential fault interactions between the strike-slip faults and the blind thrusts suggests that the vertical faults either merge at depth with or are offset by the low-angle thrusts. Thus, both types of fault systems are deemed likely to be active and should be considered in the context of regional earthquake hazards assessment.

ACKNOWLEDGMENTS

This research benefited from valuable contributions by M. Peter Suess and Frank Bilotti. Freddy Corredor, Chris Guzofski, Andreas Plesch, James Dolan, and Chris Sorlien provided numerous discussions that improved many of the ideas presented in this study. We also thank Peter Shearer for assistance in integrating relocated earthquake hypocenters into our analysis. This research was partially funded by the National Science Foundation Grant EAR 0087648, Harvard University, and the Southern California Earthquake Center (SCEC). Texaco, the Minerals Management Service (MMS), and other industry sponsors provided the well and seismic data used in this research. Financial support to Carlos Rivero provided by the Fulbright Grant "Energy for the XXI Century" is deeply appreciated.

REFERENCES CITED

Allen, P. A., and J. R. Allen, 1990, Basin analysis: Principles and applications, 2d ed.: New York, Wiley–Blackwell, 560 p.

Allmendinger, R. W., 1998, Inverse and forward numerical modeling of trishear fault-propagation folds: Tectonics, v. 17, p. 640–656, doi:10.1029/98TC01907.

Allmendinger, R. W., 2000, Trishear 4.5 Computer Program for Windows: http://www.geo.cornell.edu/geology/faculty/RWA/programs.html (accessed October 1, 2010).

Astiz, L., and P. M. Shearer, 2000, Earthquake relocation in the Inner California Borderland: Bulletin of the Seismological Society of America, v. 90, p. 425–449, doi:10.1785/0119990022.

Atwater, T., and J. Stock, 1998, Pacific-North America plate tectonics of the Neogene southwestern United States: An update: International Geology Review, v. 40, p. 375–402.

Badley, M. E., J. D. Price, and L. C. Backshall, 1989, Inversion, reactivated faults and related structures: Seismic examples from the southern North Sea, *in* M. A. Cooper and G. D. Williams, eds., Inversion tectonics: Geological Society Special Publication 44, p. 201–219.

Bally, A. W., 1983, Seismic expression of structural styles: AAPG Studies in Geology 15, v. 3, 29 p.

Bally, A. W., 1984, Tectogénèse et sismique réflexion: Bulletin Société Géologique de France, v. 7, p. 279–285.

Bally, A. W., and S. Snelson, 1980, The realm of subsidence, *in* A. Miall, ed., Facts and principles of world petroleum occurrence: Canadian Society of Petroleum Geology Memoir 6, p. 9–94.

Barrie, D., and E. Gath, 1992, Neotectonic uplift and ages of Pleistocene marine terraces, San Joaquin Hills, Orange county, California, *in* E. Heath and L. Lewis, eds., The regressive Pleistocene shoreline in southern California: Santa Ana, California: South Coast Geological Society Annual Field Trip Guidebook 20, p. 115–121.

Bohannon, R., and E. Geist, 1998, Upper crustal structure and Neogene tectonic development of the California continental borderland: Geological Society of America Bulletin, v. 110, p. 779–800, doi:10.1130/0016-7606(1998)110<0779:UCSANT>2.3.CO;2.

Buiter, S., and O. Pfiffner, 2003, Numerical models of the inversion of half-graben basins: Tectonics, v. 22, no. 5, p. 1057–1073, doi:10.1029/2002TC001417.

Butler, R. W. H., 1989, The influence of pre-existing basin structure on thrust system evolution of the western Alps, *in* M. A. Cooper and G. D. Williams, eds., Inversion tectonics: Geological Society Special Publication 44, p. 105–122.

California Geological Survey, 2002, California fault parameters-Interactive fault parameter map of California: www.consrc.ca.gov/cgs/rghm/psha/fault_parameters/htm/index.htm (accessed January 30, 2003).

Clarke, S. H., H. G. Greene, M. P. Kennedy, and J. G. Vedder, 1987, Geologic map of the inner-southern California continental margin, *in* G. H. Greene and M. P. Kennedy, eds., Geology of the mid-southern California continental margin: U.S. Geological Survey, Department of Conservation, Mines and Geology, State of California, California Continental Margin Map Series, scale 1:250,000, 7 sheets.

Colletta, B., F. Roure, B. De Toni, D. Loureiro, H. Passalacqua, and Y. Gou, 1997, Tectonic inheritance, crustal architecture, and contrasting structural styles in the Venezuela Andes: Tectonics, v. 16, p. 777–794, doi:10.1029/97TC01659.

Cooper, M. A., and G. D. Williams, 1989, Inversion tectonics: Geological Society Special Publication 44, 375 p.

Cooper, M. A., G. D. Williams, P. C. De Graciansky, R. W. Murphy, T. Needham, D. DePaor, R. Stoneley, S. P. Todd, J. P. Turner, and P. A. Ziegler, 1989, Inversion tectonics—A discussion, *in* M. A. Cooper and G. D. Williams, eds., Inversion tectonics: Geological Society Special Publication 44, p. 335–347.

Crouch, J. K., and S. Bachman, 1989, Exploration potential of Offshore Newport-Inglewood fault zone: AAPG Bulletin, v. 73, p. 536.

Crouch, J. K., and J. Suppe, 1993, Late Cenozoic tectonic evolution of the Los Angeles basin and inner California borderland: A model for core complex-like crustal extension: Geological Society of America Bulletin, v. 105, p. 1415–1434, doi:10.1130/0016-7606(1993)105<1415:LCTEOT>2.3.CO;2.

Crowell, J. C., 1974, Origin of Late Cenozoic basins in southern California, *in* W. R. Dickinson, ed., Tectonics and sedimentation: SEPM Special Publication 22, p. 190–204.

Drewry, S. D., and F. W. Victor, 1997, Inner Borderland province in Minerals Management Service-Pacific Outer Continental Shelf Region: Ocean Continental Shelf (OCS) Report MMS 97-0019, p. 125–128.

Erslev, E., 1991, Trishear fault-propagation folding: Geology, v. 19, p. 617–620, doi:10.1130/0091-7613(1991)019<0617:TFPF>2.3.CO;2.

Fisher, P. J., and G. I. Mills, 1991, The offshore Newport–Inglewood Rose Canyon fault zone, California: Structure, segmentation and tectonics, *in* P. L. Abbott and W. J. Elliott, eds., Environmental perils San Diego region: San Diego, California, San Diego Association of Geologists, p. 17–36.

Fisher, P. J., G. I. Mills, and G. Simila, 1988, The offshore Newport-Inglewood fault zone, southern California (abs.): Geological Society of America Abstracts with Programs, v. 20, no. 7, p. A91.

Fitch, T. J., 1970, Focal mechanism along inclined earthquake zones in the Indonesia-Philippines region: Journal of Geophysical Research, v. 75, p. 1431–1444, doi:10.1029/JB075i008p01431.

Ford, M., E. Williams, A. Artoni, A. Vergés, and S. Hardy, 1997, Progressive evolution of a fault-related fold pair from growth strata geometries, Sant Llorenç de Morunys, SE Pyrenees: Journal of Structural Geology, v. 19, p. 413–441, doi:10.1016/S0191-8141(96)00116-2.

Gleinner, K. W., and P. L. Boegner, 1981, Sole pit inversion tectonics, *in* L. V. Illing and G. D. Hobson, eds., Petroleum geology of the continental shelf of northwest Europe: London, Institute of Petroleum, p. 110–120.

Grant, L. B., K. L. Mueller, E. M. Gath, and R. Munro, 1999, Late Quaternary uplift and earthquake potential of the San Joaquin Hills, southern Los Angeles Basin, California: Geology, v. 27, p. 1031–1034, doi:10.1130/0091-7613(1999)027<1031:LQUAEP>2.3.CO;2.

Harding, T. P., 1973, Newport-Inglewood trend, California—An example of wrench style of deformation: AAPG Bulletin, v. 60, p. 97–116.

Hardy, S., and M. Ford, 1997, Numerical modeling of trishear fault propagation folding: Tectonics, v. 16, p. 841–854, doi:10.1029/97TC01171.

Hardy, S., and J. Poblet, 1994, Geometric and numerical model of progressive limb rotation in detachment folds: Geology, v. 22, p. 371–374, doi:10.1130/0091-7613(1994)022<0371:GANMOP>2.3.CO;2.

Hauksson, E., 1990, Earthquakes, faulting and stress in the Los Angeles Basin: Journal of Geophysical Research, v. 95, no. B10, p. 15,365–15,394, doi:10.1029/JB095iB10p15365.

Hauksson, E., and S. Gross, 1991, Source parameters of the 1933 Long Beach earthquake: Bulletin of the Seismological Society of America, v. 81, p. 81–91.

Hauksson, E., and L. Jones, 1988, The July 1986 Oceanside (M_L 5.3) earthquake sequence in the continental Borderland, southern California: Bulletin of the Seismological Society of America, v. 78, p. 1885–1906.

Hayward, A. B., and R. H. Graham, 1989, Some geometrical characteristic of inversion, *in* M. A. Cooper and G. D. Williams, eds., Inversion tectonics: Geological Society Special Publication 44, p. 17–40.

Huyghe, P., and J. L. Mugnier, 1991, Short-cut geometry during structural inversions: competition between faulting and reactivation: Bulletin Societé Géologique de France, v. 163, p. 691–700.

Ingersoll, R. V., and P. E. Rumelhart, 1999, Three-stage evolution of the Los Angeles Basin, southern California: Geology, v. 27, p. 593–596, doi:10.1130/0091-7613(1999)027<0593:TSEOTL>2.3.CO;2.

Kern, J. P., and T. K. Rockwell, 1992, Chronology and deformation of Quaternary marine shorelines, San Diego County, California, *in* E. Heath and L. Lewis, eds., The regressive Pleistocene shoreline in southern California: Santa Ana, California: South Coast Geological Society Annual Field Trip Guidebook 20, p. 1–7.

Kier, G., and K. Mueller, 1999, Evidence for active shortening in the offshore Borderlands and its implications for blind-thrust hazards in the Coastal Orange and San Diego counties: 1999 Southern California Earthquake Center (SCEC) Annual Meeting, Los Angeles, California, Proceedings and Abstracts, p. 71.

Lajoie, K. R., J. P. Kern, and J. F. Wehmiller, 1979, Quaternary marine shorelines and crustal deformation, San Diego to Santa Barbara, California, *in* P. Abbott, ed., Geologic excursions in the southern California area: San Diego, California, San Diego State University, p. 3–15.

Lajoie, K. R., D. J. Ponti, C. L. Powell, S. A. Mathieson, and A. M. Sarna-Wojcicki, 1992, Emergent marine strandlines and associated sediments, coastal California: A record of Quaternary sea-level fluctuations, vertical tectonic movements, climatic changes and coastal processes, *in* E. Heath and L. Lewis, eds., The regressive Pleistocene shoreline in southern California: South Coast Geological Society Annual Field Trip Guidebook 20, p. 81–104.

Legg, M., and M. Kennedy, 1979, Faulting offshore San Diego and northern Baja California, *in* P. Abbott and W. Elliott, eds., Earthquakes and other perils, San Diego region: San Diego Association of Geologist, Geological Society of America Field Trip Guidebook, p. 29–46.

Legg, M., and C. Nicholson, 1993, Geologic structure and tectonic evolution of the San Diego Trough and vicinity, California continental borderland: AAPG Bulletin, v. 77, p. 705.

Legg, M., and V. W. Ortega, 1978, New evidences for major faulting in the inner borderland off northern Baja California, Mexico (abs.): Transactions, American Geophysical Union, v. 59, p. 1134.

Legg, M., V. Wong, and F. Suarez, 1991, Geologic structure and tectonics of the inner continental borderland of northern Baja California, *in* J. P. Dauphin and B. R. Simoneit, eds., The gulf and Peninsular province of the Californias: AAPG Memoir 47, p. 145–177.

Legg, M. R., 1980, Seismicity and tectonics of the inner continental borderland of southern California and northern Baja California, Mexico: M.S. thesis, University of California at San Diego, 60 p.

Legg, M. R., 1989, Faulting and seismotectonics of the inner continental borderland west of San Diego, *in* G. Roquemore and S. Tanges, eds., Workshop on "The Seismic Risk in the San Diego Region: Special Focus on the Rose Canyon Fault System," June 29–30, 1989, San Diego, California, p. 50–70.

Letouzey, J., 1990, Fault reactivation, inversion and fold-thrust belt, *in* J. Letouzey, ed., Petroleum and tectonics in mobile belts: Paris, Editions Technip, p. 101–128.

Lindwall, S. C., and T. K. Rockwell, 1995, Holocene activity of the Rose Canyon zone in San Diego, California: Journal of Geophysical Research, v. 100, p. 24,121–24,132.

Lowell, J. D., 1995, Mechanics of basin inversion from worldwide examples, *in* J. Buchanan and P. Buchanan, eds., Basin inversion: Geological Society Special Publication 88, p. 39–57.

Luyendyk, B. P., 1991, A model for the Neogene crustal rotation, transtension, and transpression in southern California: Geological Society of America Bulletin, v. 103, p. 1528–1536, doi:10.1130/0016-7606(1991)103<1528:AMFNCR>2.3.CO;2.

McCaffrey, R., 1996, Slip partitioning at the convergent plate boundaries of SE Asia, *in* R. Hall and D. J. Blundell, eds., Tectonic evolution of Southeast Asia: Geological Society (London) Special Publication 106, p. 3–18.

McCaffrey, R., 2005, Block kinematics of the Pacific-North America plate boundary in the southwestern United States from inversion of GPS, seismological, and geological data: Journal of Geophysical Research, v. 110, p. B07401, doi:10.1029/2004JB003307.

McClay, K. R., 1989, Analog models on inversion tectonics, *in* M. A. Cooper and G. D. Williams, eds., Inversion tectonics: Geological Society Special Publication 44, p. 41–59.

McClay, K. R., 1992, Glossary of thrust tectonics terms, *in* K. R. McClay, ed., Thrust tectonics: London, Chapman and Hall, p. 419–433.

McClay, K. R., 1995, The geometries and kinematics of inverted fault systems: A review of analog models studies, *in* J. Buchanan and P. Buchanan, eds., Basin inversion: Geological Society Special Publication 88, p. 97–118.

McClay, K. R., and P. G. Buchanan, 1992, Thrust faults in inverted extensional basins, *in* K. R. McClay, ed., Thrust tectonics: London, Chapman and Hall, p. 93–104.

Medwedeff, D., 1988, Structural analysis and tectonic significance of late Tertiary and Quaternary, compressive growth-folding, San Joaquin Valley, California: Ph.D. thesis, Princeton University, Princeton, New Jersey, 184 p.

Medwedeff, D., 1992, Geometry and kinematics of an active, laterally propagating wedge thrust, Wheeler California Ridge, *in* S. Mitra and G. W. Fisher, eds., Structural geology of fold and thrust belts: Baltimore, Johns Hopkins University Press, p. 3–28.

Mills, G. I., 1991, The structure and activity of the offshore Newport-Inglewood-Rose Canyon zone of deformation: M.S. thesis, California State University, Northridge, 207 p.

Mount, V., J. Suppe, and S. Hook, 1990, A forward modeling strategy for balancing cross sections: AAPG Bulletin, v. 74, p. 521–531.

Mueller, K., L. B. Grant, and I. Gath, 1998, Late Quaternary growth of the San Joaquin Hills anticline: A new source of blind-thrust earthquakes in the Los Angeles Basin: Seismological Research Letters, v. 69, p. 161.

Narr, W., and J. Suppe, 1994, Kinematics of basement-involved compressive structures: American Journal of Science, v. 294, p. 802–860.

Nicholson, C., C. Sorlien, and M. Legg, 1993, Crustal imaging and extreme Miocene extension of the Inner California

Continental Borderland: Geological Society of America Abstracts with Programs, v. 25, p. A-418.

Novoa, E., 1997, Two and three-dimensional analysis of structural trends in the Santa Barbara Channel, California, U.S.A.: Ph.D. thesis, Princeton University, Princeton, New Jersey, 154 p.

Pacheco, J., and J. Nábêlek, 1988, Source mechanisms of three moderate California earthquakes of July 1986: Bulletin of the Seismological Society of America, v. 78, p. 1907–1929.

Petersen, M. D., and S. G. Wesnousky, 1994, Fault slip rates and earthquake histories for active faults in southern California: Bulletin of the Seismological Society of America, v. 84, p. 1608–1649.

Plesch, A., et al., 2007, Community fault model (CFM) for southern California: Bulletin of the Seismological Society of America, v. 97, no. 6, p. 1793–1802.

Richards-Dinger, K. B., and P. M. Shearer, 2000, Earthquake locations in Southern California obtained using source-specific station terms: Journal of Geophysical Research, v. 105, p. 10,939–10,960.

Rivero, C., 2004, Origin of active blind-thrust faults in the southern Inner California Borderlands: Ph.D. thesis, Harvard University, Cambridge, Massachusetts, 146 p.

Rivero, C., and J. H. Shaw, 2005, Fault-related folding in reactivated offshore basins, California, *in* J. H. Shaw, C. Connors, and J. Suppe, eds., Seismic interpretation of contractional fault-related folds: AAPG Studies in Geology 53, 156 p.

Rivero, C., J. H. Shaw, and K. Mueller, 2000, Oceanside and Thirtymile Bank blind thrusts: Implications for earthquake hazards in coastal southern California: Geology, v. 28, no. 10, p. 891–894, doi:10.1130/0091-7613(2000)28<891:OATBBT>2.0.CO;2.

Rockwell, T. K., S. C. Lindvall, C. Haraden, C. Kenji, and E. Baker, 1992, Minimum Holocene slip rate for the Rose Canyon fault, *in* E. Heath and L. Lewis, eds., The regressive Pleistocene shoreline in southern California: Santa Ana, California: South Coast Geological Society Annual Field Trip Guidebook 20, p. 55–64.

Roure, F., and B. Colletta, 1996, Cenozoic foreland inversions between the Alps and Pyrenees, *in* P. A. Ziegler and F. Horwath, eds., Structure and prospects of the Alpine foreland and basins: Peri-Tethys Memoir 2, p. 173–209.

Roure, F., D. W. Howell, S. Guellec, and P. Casero, 1990, Shallow structures induced by deep-seating thrusting, *in* J. Letouzey, ed., Petroleum and tectonics in mobile belts: Paris, Editions Technip, p. 15–30.

Roure, F., B. Colletta, B. De Toni, D. Loureiro, H. Passalacqua, and Y. Gou, 1997, Within-plate deformations in the Maracaibo and east Zulia basins, western Venezuela: Marine and Petroleum Geology, v. 14, p. 139–163, doi:10.1016/S0264-8172(96)00063-3.

Rowan, M. G., and R. Linares, 2000, Fold-evolution matrices and axial-surface analysis of fault-bend folds: Applications to the Medina anticline, eastern Cordillera, Colombia: AAPG Bulletin, v. 84, p. 741–764.

Schreurs, G., and B. Colletta, 1998, Analog modeling of faulting in zones of continental transpression and transtension, *in* R. E. Holdsworth, R. A. Strachan, and J. F. Dewey, eds., Continental transpressional and transtensional tectonics: Geological Society (London) Special Publication 135, p. 59–79.

Shaw, J. H., and J. Suppe, 1996, Earthquake hazards of active blind-thrust faults under the central Los Angeles Basin, California: Journal of Geophysical Research, v. 101, no. B4, p. 8623–8642, doi:10.1029/95JB03453.

Shaw, J. H., S. C. Hook, and J. Suppe, 1994, Structural trend analysis by axial surface mapping: AAPG Bulletin, v. 78, p. 700–721.

Shaw, J. H., F. Bilotti, and P. Brennan, 1999, Patterns of imbricated thrusting: Geological Society of America Bulletin, v. 111, p. 1140–1154, doi:10.1130/0016-7606(1999)111<1140:POIT>2.3.CO;2.

Shlemon, R. J., 1992, The Cristianitos fault and Quaternary geology, San Onofre State Beach, California, *in* E. Heath and L. Lewis, eds., The regressive Pleistocene shoreline in southern California: South Coast Geological Society Annual Field Trip Guidebook 20, p. 9–12.

Shlemon, R. J., and T. Rockwell, 1992, Field trip itinerary San Onofre to San Diego-September, 1992, *in* E. Heath and L. Lewis, eds., The regressive Pleistocene shoreline in southern California: South Coast Geological Society Annual Field Trip Guidebook 20, p. vi–ix.

Suppe, J., 1983, Geometry and kinematics of fault-bend folding: American Journal of Science, v. 283, p. 684–721.

Suppe, J., and D. A. Medwedeff, 1992, Geometry and kinematics of fault-propagation folds: Eclogae Geologicae Helvetiae, v. 83, p. 409–454.

Suppe, J., F. Sàbat, J. A. Munoz, J. Poblet, E. Roca, and J. Vergés, 1997, Bed-by-bed growth by kink-band migration: San Llorenç de Morunys, eastern Pyrenees: Journal of Structural Geology, v. 19, p. 443–461, doi:10.1016/S0191-8141(96)00103-4.

Sylvester, A. G., 1988, Strike-slip faults: Geological Society of American Bulletin, v. 100, p. 1666–1703, doi:10.1130/0016-7606(1988)100<1666:SSF>2.3.CO;2.

Vedder, J. G., 1976, Precursors and evolution of the name California Continental Borderland, *in* D. G. Howell, ed., Aspects of the geologic history of the California Continental Borderland: Pacific Section of the AAPG, Miscellaneous Publication 24, p. 6–11.

Vedder, J. G., H. G. Greene, S. H. Clarke, and M. P. Kennedy, 1986, Geologic map of the mid-southern California continental margin, *in* G. H. Greene and M. P. Kennedy, eds., Geology of the mid-southern California continental margin: U.S. Geological Survey, Department of Conservation, Mines and Geology, State of California, California Continental Margin Map Series, scale 1:250,000, 7 sheets.

Wilcox, R. E., T. P. Harding, and D. R. Seeley, 1973, Basic wrench tectonics: AAPG Bulletin, v. 57, p. 74–96.

Williams, G. D., C. M. Powel, and M. A. Cooper, 1989, Geometry and kinematics of inversion tectonic, *in* M. A. Cooper and G. D. Williams, eds., Inversion tectonics: Geological Society Special Publication 44, p. 3–15.

Wright, T. L., 1991, Structural geology and tectonic evolution of the Los Angeles Basin, *in* K. T. Biddle, ed., Active margin basins: AAPG Memoir 52, p. 35–134.

Yeats, R. S., 1973, The Newport-Inglewood fault zone, Los Angeles Basin, California: AAPG Bulletin, v. 57, p. 117–135.

Yeats, R. S., 1976, Extension versus strike-slip origin of the southern California Borderland, *in* D. G. Howell, ed., Aspects of the geology history of the California continental borderland: Pacific Section of the AAPG, Miscellaneous Publication 24, p. 455–485.

Ziegler, P. A., 1983, Inverted basins in the alpine foreland, *in* A. W. Bally, ed., Seismic expression of structural styles—A picture and work atlas: AAPG Studies in Geology 15, Tulsa, p. 3.3(3)–3.3(12).

Ziony, J. I., and L. M. Jones, 1989, Map of late Quaternary faults and 1978–1984 seismicity of the Los Angeles region, California: U.S. Geological Survey Map MF-1964, scale 1:250,000, 1 sheet.

10

Wang, Xin, John Suppe, Shuwei Guan, Aurelia Hubert-Ferrari, Ramon Gonzalez-Mieres, and Chengzao Jia, 2011, Cenozoic structure and tectonic evolution of the Kuqa fold belt, southern Tianshan, China, *in* K. McClay, J. H. Shaw, and J. Suppe, eds., Thrust fault-related folding: AAPG Memoir 94, p. 215–243.

Cenozoic Structure and Tectonic Evolution of the Kuqa Fold Belt, Southern Tianshan, China

Xin Wang
Department of Geosciences, Zhejiang University, Hang Zhou, China

John Suppe[1]
Department of Geosciences, National Taiwan University, Taipei, Taiwan

Shuwei Guan
Research Institute of Petroleum Exploration and Development, PetroChina, Beijing, China

Aurelia Hubert-Ferrari
Géomorphologie, Université de Liège, Liège, Belgium

Ramon Gonzalez-Mieres[2]
Department of Geosciences, Princeton University, Princeton, New Jersey, U.S.A.

Chengzao Jia
Research Institute of Petroleum Exploration and Development, PetroChina, Beijing, China

ABSTRACT

The east–west-trending late Cenozoic Kuqa fold belt is a part of the compressive southern margin of the Tianshan Mountains in western China. Approximately 20,000 km (12,000 mi) of two-dimensional seismic reflection profiles are integrated with surface geology and well data to examine the deformation style and structural evolution of the Kuqa fold belt. Mesozoic through Holocene strata in the northern Tarim Basin have been deformed in a thrust system that roots northward into the Paleozoic basement of the southern Tianshan. The south-vergent deformation is characterized by a series of forward-breaking thrust faults, fault-related folds, and detachment folds. Two major decollement levels exist: an upper detachment in salt-gypsum lithologies in the Paleogene–Miocene Kumgeliem, Suweiyi, and Jidike formations, and the lower detachment mostly within Jurassic coal and mudstone strata. Fault-propagation folds

[1]*Also at*: Department of Geosciences, Princeton University, Princeton, New Jersey, U.S.A.
[2]*Present address*: Saudi Arabian Chevron, Chevron Global Upstream and Gas, Houston, Texas, U.S.A.

DOI:10.1306/13251339M94389

developed above both detachments and have been refolded in some cases by displacement on the lower thrust faults. Imbricate thrust faults and duplex structures linking the two detachments developed with salt that apparently flowed into the cores of the duplex structure. Near the high Tianshan mountain front, Mesozoic and Cenozoic strata are involved in deformation that began at approximately 25–26 Ma as documented by growth strata north of Kuqa. Toward the southward limit of the fold belt, Miocene through Holocene strata are folded in the Quilitage and Yaken anticlines, which began growing above a thrust system that propagated at about 5.5 Ma. The Yaken anticline at the south edge of the eastern Kuqa fold belt has only emerged as a topographic anticline in the last 0.2–0.3 Ma associated with an acceleration of the Quilitage-Yaken thrust system. Structural restoration suggests a shortening of 15–20 km (9–12 mi) across the eastern Kuqa fold belt. Considering that this shortening began about 25 Ma, the average shortening rate was about 0.7 mm/yr (0.03 in./yr). Because the frontal thrust system underlying the Quilitage and Yaken anticlines has a shortening of 6 km (3.7 mi) that began approximately 5.5 Ma, their average shortening rate is about 1.1 mm/yr (0.04 in./yr). However, the shortening rate on this frontal system from about 5.5 Ma to about 0.2–0.3 Ma is approximately 0.6 mm/yr (0.02 in./yr) followed by an acceleration to about 4–5 mm/yr (0.16–0.19 in./yr) at approximately 0.2–0.3 Ma, causing the topographic emergence of these structures. These results indicate that shortening rates in the Kuqa fold belt have increased in the late Pleistocene, which is consistent with more regional present-day geodetic shortening rates of about 9 mm/yr (0.35 in./yr) across the southern Tianshan, which also indicate a substantial acceleration relative to Neogene shortening rates.

INTRODUCTION

The Kuqa thrust and fold belt along the front of southern Tianshan contains a record of the interactions between the Tianshan Mountains and the Tarim Basin during the Neogene to Holocene (Lu et al., 1994; Jia, 1997; Yin et al., 1998; Allen et al., 1999; Burchfiel et al., 1999; Hubert-Ferrari et al., 2007) (Figure 1). Cenozoic compressional deformation exists on both the northern and southern flanks of the Tianshan, which is part of the broad India-Eurasia collision (Molnar and Tapponnier, 1975; Tapponnier and Molnar, 1979; Allen et al., 1991; Hendrix et al., 1994; Sobel, 1995; Sobel and Dumitru, 1997) (Figure 1). In the Kuqa area, only a few of active faults reach the surface and offset the youngest sediments; these have been investigated in previous work, and the rates and processes of crustal shortening and thickening in the Tianshan are documented (Brown et al., 1998; Yin et al., 1998; Burchfiel et al., 1999). However, many faults are blind. Similarly, the active faults on the northern flank of the Tianshan have been investigated (Avouac et al., 1993). Despite these substantial research efforts, discussing the subsurface structure of the Cenozoic deformation along the southern and northern flanks of the Tianshan has been difficult in the past because only few seismic reflection profiles have been published (Lu et al., 1994; Jia, 1997; Hubert-Ferrari et al., 2005b, 2007).

In this chapter, we present data on the structure and evolution of the Cenozoic Kuqa fold belt based on analysis of the surface geology, seismic reflection profiles, and well data. We interpreted almost 20,000 km (12,000 mi) of two-dimensional (2-D) seismic profiles in the Kuqa fold belt obtained by the Tarim Oilfield Company. In addition, 59 wells (most over 4000 m [13,000 ft] deep) are used to control the subsurface structure. The locations of these wells, with abbreviated names such as YH1 or DQ5, are shown on the maps and cross sections. Surface geological maps at scales of 1:100,000 and 1:50,000 were integrated into this project (Zhong and Xia, 1998a). These maps, along with new structural field measurements, provide important surface constraints on the subsurface structural interpretation. Detailed stratigraphic profiles of the stratigraphic reference sections are published at a scale of 1:2500 (Zhong and Xia, 1998b) to which maps, seismic horizons, and wells are tied throughout the Kuqa Basin. Furthermore, strike and dip data in remote and inaccessible regions along seismic lines were obtained from stereo Corona satellite images using techniques described by Hubert-Ferrari et al. (2007, their appendix A).

We first review the geology and stratigraphy of the Kuqa fold belt, which is an important component in understanding the structure of the area; then major geological structures of the Kuqa fold belt are described. The timing and rate of Cenozoic deformation of the Kuqa fold belt are constrained by key seismic lines that show unconformities and growth strata that allow us, combined with magnetostratigraphic data tied to the mapped seismic horizons, to estimate the timing of deformation and

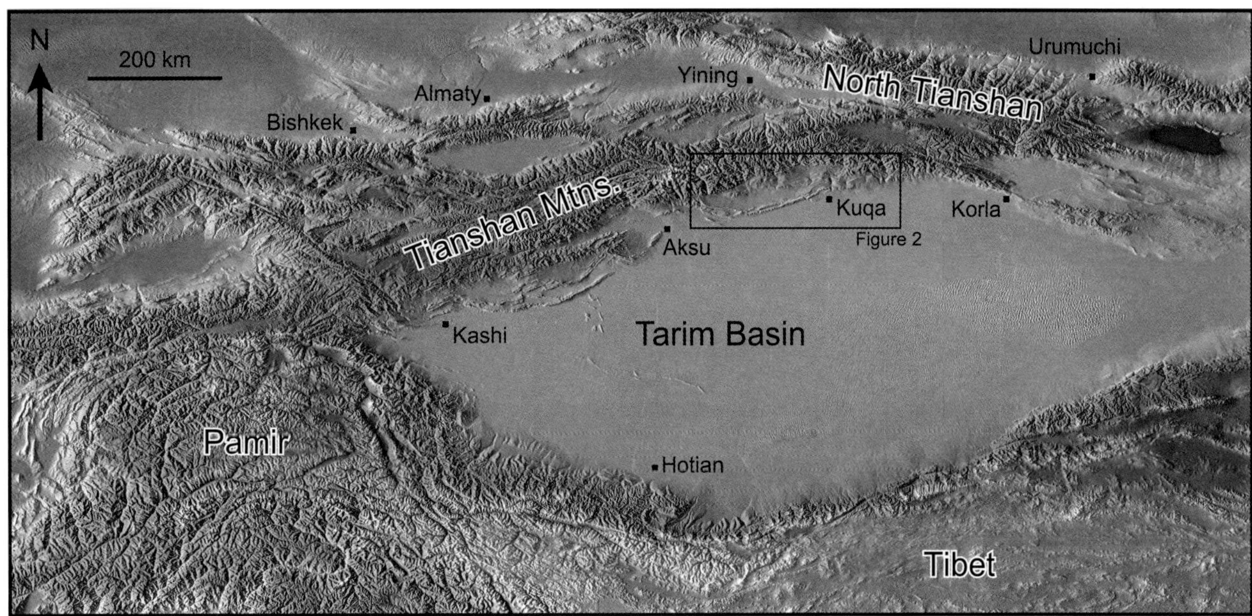

Figure 1. Regional location of the Kuqa fold belt (Figure 2) at the southern edge of the Tianshan Mountains (Mtns.) and the northern edge of the Tarim Basin.

slip rate of the thrust faults. These results suggest a substantial acceleration in shortening in the late Pleistocene.

REGIONAL SETTING

Overview of the Study Area

The Kuqa fold belt is located in the northern part of the Tarim Basin at the southern edge of the Tianshan, extending 300–400 km (200–250 mi) west to east from Aksu to Korle with a width of 40–80 km (25–50 mi) (Jia, 1997) (Figures 1, 2). The Kuqa fold belt is a part of the Cenozoic southern thrust belt of the Tianshan, with the Kalping-Wushi thrust zone to the west and the Korla transfer system to the east (Yin et al., 1998). The Kuqa fold belt is composed almost entirely of terrestrial Triassic, Jurassic, Cretaceous, late Paleogene, Neogene, and Quaternary strata with a total Mesozoic–Cenozoic thickness of up to 12,000 m (40,000 ft) in the Baicheng basin and Cenozoic strata exceeding 8000 m (25,000 ft). The series of south-vergent thrust and fold belts in the Kuqa fold belt developed sequentially from north to south and can be subdivided into four parts according to deformation styles: the northern Kuqa fold belt, Baicheng syncline, Quilitage anticline, and Yaken anticline (Lu et al., 1994; Jia, 1997) (Figure 2). Several oil and gas fields have been discovered in the Kuqa fold belt, including the Kela, Dina, Yinan, Dabei, and Tubei gas fields and the Dawanqi oil field, with sufficient production to justify completion of a gas pipeline to eastern China.

Stratigraphy

The stratigraphy of the Kuqa fold belt has been extensively investigated (Hao et al., 1982; Ye and Huang, 1990; Zhong and Xia, 1998a, b). Mesozoic strata are well established from outcrop studies in the northern Kuqa fold belt (Hendrix et al., 1992; Zhong and Xia, 1998a, b), and the Cenozoic sequence is constrained by surface outcrops and wells in the Quilitage anticline (Yin et al., 1998; Zhong and Xia, 1998a, b). Mesozoic and Paleogene biostratigraphy of the Kuqa fold belt has been extensively investigated based on studies of bivalves, algae, Ostracoda, pollens, and gastropods (Lu and Luo, 1990; Ye and Huang, 1990; Zhong and Xia, 1998a, b). The Neogene fluvial sequence, which is deposited in an arid environment, is of questionable biostratigraphic age because fossils are rare. Nevertheless, detailed magnetostratigraphic studies in the Kuqa fold belt provide important and detailed timing constraints for Cenozoic sediments in the area. The nomenclature of the Mesozoic and Cenozoic stratigraphic units used in this chapter follows that of Yin et al. (1998) and Zhong and Xia (1998a, b), who divided the sedimentary units principally based on their lithology, fossil content, and magnetostratigraphic data (Figure 2).

Assigning absolute ages to formation boundaries requires some attention to detail and is important for

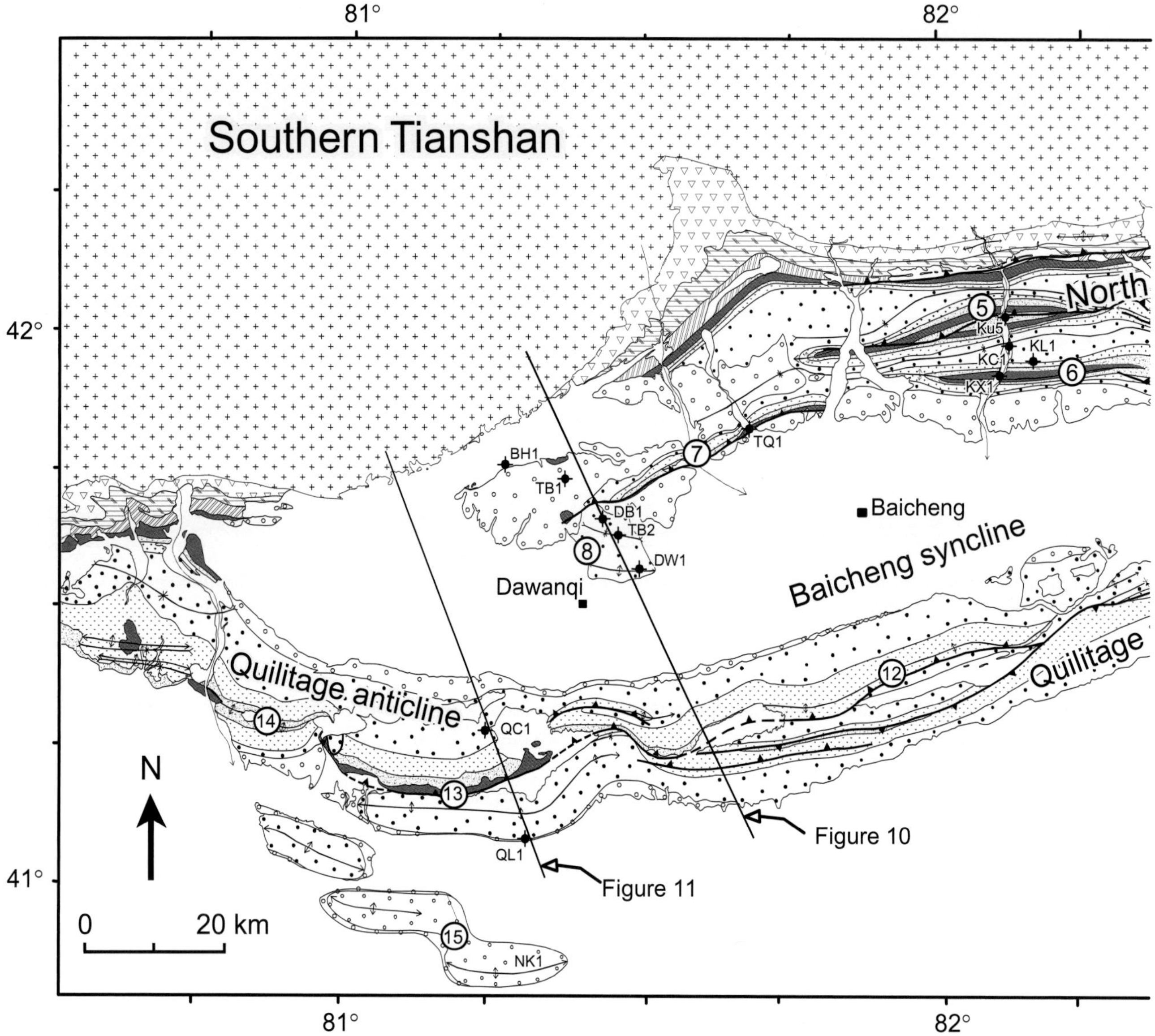

Figure 2. Geological map of the Kuqa thrust and fold belt (geology simplified with reinterpretation from Zhong and Xia, 1998a). The belt is divided into four segments: the northern Kuqa thrust and fold belt, the Baicheng syncline, the Quilitage anticline, and the Yaken anticline (Lu et al., 1994; Jia, 1997). (1) Tuger anticline, (2) Yiqikelike anticline, (3) Jidike anticline, (4) Bashijiqike anticline, (5) Kumgeliem anticline, (6) Kasangtuokai anticline, (7) Tuzimaza anticline, (8) Dawanqi anticline, (9) eastern Quilitage anticline, (10) Kuqadawu anticline, (11) southern Quilitage anticline, (12) northern Quilitage anticline, (13) Miskantag anticline, (14) eastern Awate anticline, (15) Kalayuergun anticline, (16) Yaken anticline. Locations of the cross sections (Figures 6–11, 13) are shown. Fm = formation.

determining tectonic timing and rates. Most of the Cenozoic magnetostratigraphic data are from the eastern Kuqa Basin (Yin et al., 1998; Meng, 1999; Charreau et al., 2006, 2008; Huang et al., 2006, 2008) and along or close to the cross section 40 km (25 mi) northeast of Kuqa (labeled Figure 7 in Figure 2), which is also the location of the detailed (1:2500) stratigraphic reference sections of Zhong and Xia (1998b) that define the formation boundaries used in the present chapter. The Meng (1999) magnetostratigraphy was collected along the Zhong and Xia reference sections and uses their formation boundaries and thicknesses. The high-resolution magnetostratigraphy of Charreau et al. (2006) does not use formation boundaries but coincides with Zhong and Xia's (1998b) and Meng's (1999) reference sections to which we have correlated it. The high-resolution magnetostratigraphic study of Huang et al. (2006) uses different formation boundaries, but we have been able to

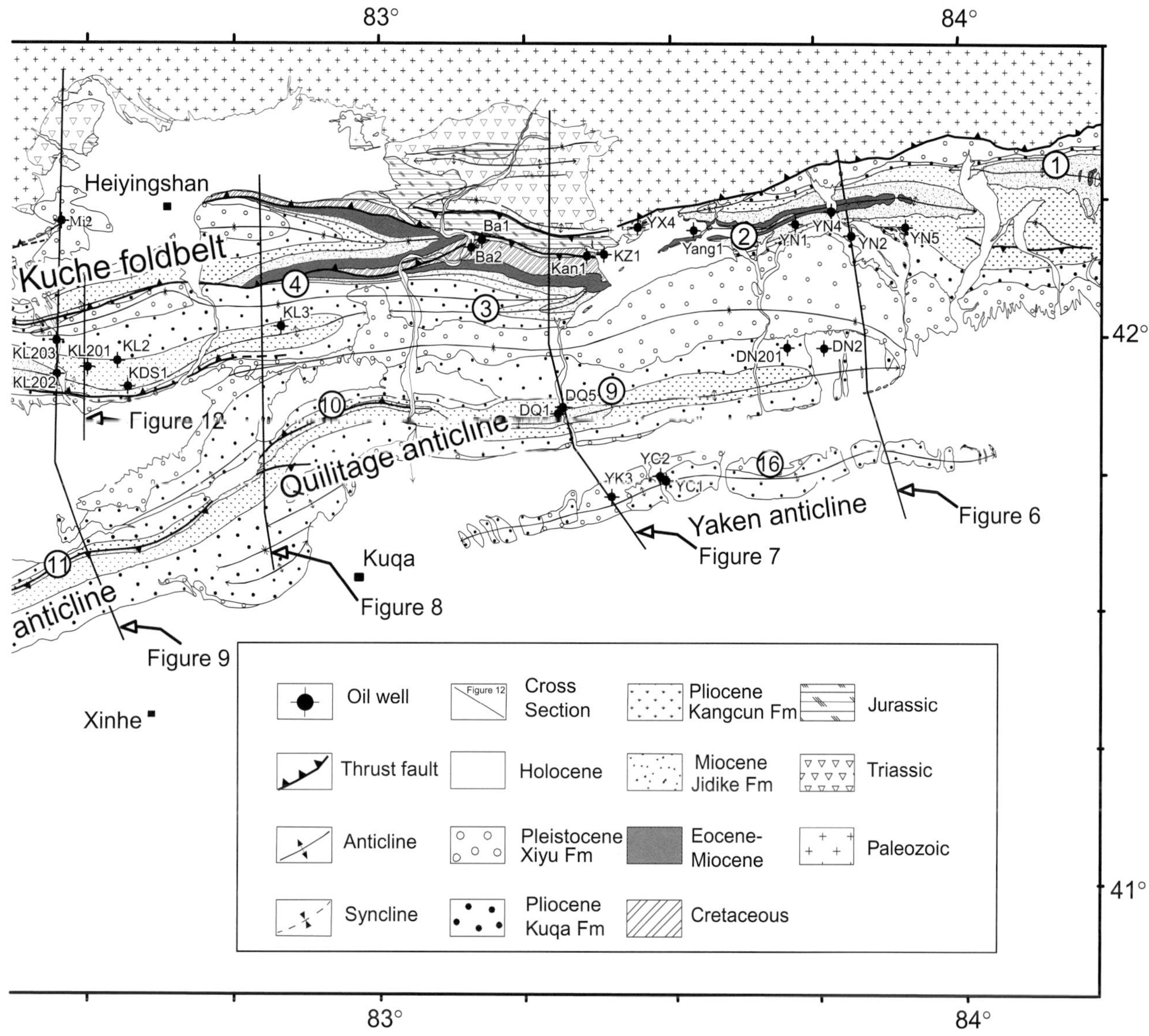

Figure 2. (cont.).

correlate it unambiguously to the nearby Zhong and Xia (1998b) and Meng (1999) stratigraphy, making use of the coordinates of the Huang sections (B. J. Huang, 2006, personal communication), tracing key horizons at the surface and correlating key reversal signatures of the magnetostratigraphic sections. The ages of Cenozoic formations in the present chapter therefore are based on Zhong and Xia's (1998b) stratigraphy, which has been tied throughout the Kuqa Basin in seismic lines, surface mapping, and well picks. However, we emphasize that the late Cenozoic formation boundaries are regionally time transgressive.

The Mesozoic stratigraphy is well constrained by outcrop exposures along the northern foothills of the southern Tianshan, deep wells, and seismic profiles. The Triassic sequence, composed of the Ehaobulake, Karamay, Huangshanjie, and Taliqik formations, is up to 2000 m (6562 ft) thick (Zhong and Xia, 1998a, b). Fluvial conglomerate, sandstone, siltstone, and shale unconformably overlie Permian rhyolitic rocks. During the Early and Middle Triassic, the Kuqa region had a relatively arid climate, which by late Triassic became more humid (Hendrix et al., 1992). The Jurassic, including the Ahe, Yangxia, Kezilenuer, Qakemake, Qigu, and Kalaza formations (up to 2100 m [6890 ft] thick), is in conformable contact with the underlying Triassic (Hendrix et al., 1992; Zhong and Xia, 1998a, b). It is composed of massive gray sandstones grading upward into fine greenish sandstone, siltstone, and shale. Coal horizons with thickness of several tens of meters have been found in the Lower to Middle Jurassic, which are important detachment horizons. The Jurassic has been interpreted as meandering fluvial sediments with local lacustrine deltaic influence (Hendrix et al., 1992; Zhong and Xia, 1998a). The Triassic and Jurassic shales are considered to be the major source rocks that generated the bulk of the reservoired oil in the Kuqa fold belt (Jia, 1997). The preserved Cretaceous strata in the Kuqa region are composed of the Yageliemu, Shusanhe, Baxigai, and Basjiqike formations (up to 1600 m [5000 ft]), which lie unconformably on the Jurassic (Hendrix et al., 1992; Zhong and Xia, 1998a). A recent magnetostratigraphic study (Peng et al., 2006) shows that there is a 45-Ma hiatus between the Yageliemu, Shusanhe, Baxigai formations formed during the early–middle stage of the Early Cretaceous (142–124 Ma) and the Basjiqike Formation deposited during the Late Cretaceous (79–66 Ma). The Cretaceous sequence begins with a 10–20-m (30–70-ft)-thick basal conglomerate grading into coarse- to medium-grained sandstone interbedded with fine-grained red siltstone and shale. Isopach maps show a thick sequence of Mesozoic strata thinning southward in the northern and central Kuqa fold belt, with depocenters located along the southern Tianshan front (Figure 3).

Late Paleogene–early Miocene strata unconformably overlie the Mesozoic along the northern Kuqa fold belt and are up to 1200 m (4000 ft) thick (Zhong and Xia, 1998a, b). They are divided into the Talake, Xiaokuzibai, and Avate formations in the western region of the Kuqa fold belt and the Kumgeliem and Suweiyi formations in the eastern part of the Kuqa fold belt. The basal part of the late Paleogene strata (the Kumgeliem formation in the east and the Talake Formation in the west) is composed of a brown-gray 4–20-m (10–70-ft)-thick conglomerate. Overlying strata are a sequence of red and reddish-brown mudstone, siltstone, and fine-grained sandstone interbedded with gypsum and thin dolomite. This basal Cenozoic sequence in the Kuqa area, deposited in a marginal marine environment, has been considered to be Oligocene to early Miocene in age (Lu and Luo, 1990; Ye and Huang, 1990; Yin et al., 1998; Zhong and Xia, 1998a). In contrast, Eocene spores are reported from the Talake formation in the western Kuqa Basin (Zhong and Xia, 1998a, b). Magnetostratigraphic data show that the Kumgeliem and Suweiyi formations of the Kuqa River and Bestantuogela sections in the north of Kuqa belong to Oligocene–early Miocene (Yin et al., 1998; Meng, 1999; Charreau et al., 2006, Huang et al., 2006), and that an approximately 30–35 Ma sedimentary hiatus with the underlying Cretaceous strata exists (Peng et al., 2006). The age of the base of the Kumgeliem is not well constrained but is greater than 32 Ma, using the Zhong and Xia (1998b) formation boundary. The age of the Suweiyi formation is better constrained magnetostratigraphically, with a base at 25–26 Ma and the top at 20–21 Ma, using Zhong and Xia's (1998b) formation boundaries. The depocenter in the late Paleogene–early Miocene trended west–east along the present axis of the Quilitage anticline, which demonstrates that the Quilitage anticline was folded in the area where the basal Cenozoic sequence reached a maximum thickness (Figure 4).

The Neogene is divided into the Jidike, Kanchung, and Kuqa formations (Yin et al., 1998; Zhong and Xia, 1998a, b). The Jidike formation consists of yellowish-red and gray mudstone interbedded with gypsum, siltstone, and medium-grained sandstone (Zhong and Xia, 1998a, b). Gypsum beds are present in the lower Jidike in the eastern Kuqa fold belt, indicating that brackish water conditions may have only existed in the eastern region during the early Jidike. We favor the interpretation that the Jidike formation represents a transition from a lacustrine to braided-fluvial depositional setting (Yin et al., 1998). Several depocenters of the Jidike formation (Figure 4) may indicate sites of shallow basins, with the deepest lake located at the eastern part of the Kuqa fold belt where the Jidike formation is up to 1600 m (5249 ft) thick. Fossils found in the Jidike formation suggest ages ranging from Oligocene to Miocene (Lu and Luo, 1990; Ye and Huang, 1990; Zhong and Xia, 1998a). These ages overlap with the Miocene magnetostratigraphic ages of 21–15 Ma from the Jidike formation in the northern Kuqa area by Huang et al. (2006) using Zhong and Xia's (1998b) formation boundaries. Magnetostratigraphic data 15 km (9 mi) southward in the same sequence by Charreau et al. (2006) give an age for the top of the Jidike formation of 10.5 Ma, indicating that the formation is time transgressive and has prograded south. Evaporite of the Jidike formation is one of the decollement horizons in the eastern part of the Kuqa fold belt. The Kanchung and Kuqa formations are composed of thick sandstones interbedded with siltstone and conglomerate. The entire unit is characterized by an increase in the abundance of coarse-grained sandstone and conglomerate beds upward, probably reflecting a dramatic increase in topographic relief of the southern Tianshan and Kuqa fold belt from Kanchung to Kuqa times. Analyses of the lithologic character and fossil assemblages of the Kanchung and Kuqa groups indicate that the Kanchung formation was deposited in a lacustrine and alluvial-fan environment and that the Kuqa formation was deposited in a fluvial environment (Ye and Huang, 1990; Yin et al., 1998).

Recent high-resolution magnetostratigraphy of Huang et al. (2006, 2008) and Charreau et al. (2006, 2008) in two locations separated by about 15 km (9 mi) in the eastern Kuqa fold belt shows that the Kanchung formation is time transgressive. The Kanchung formation

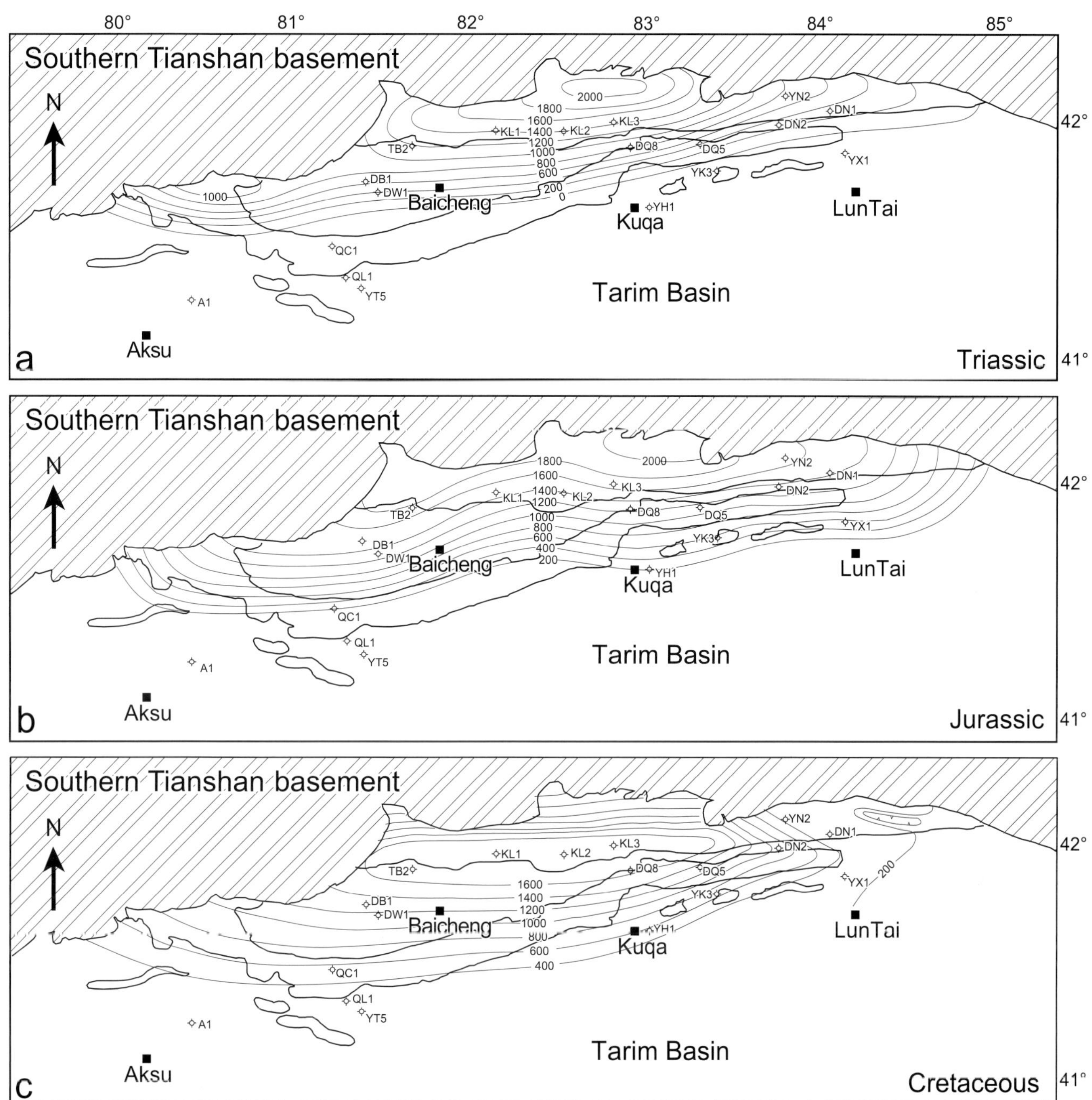

Figure 3. Isopach map of Mesozoic basins in the Kuqa area. (a) Thickness of Triassic strata with a 200-m (656-ft) contour interval. (b) Thickness of Jurassic strata with a 200-m (656-ft) contour interval. (c) Thickness of Cretaceous strata with a 200-m (656-ft) contour interval (modified from Ye and Huang, 1990).

is 15 to 6 Ma old in the northern section of Huang et al. (2006) and 10.5 to 7 Ma old in the southern Quilitage section of Charreau et al. (2006) based on Zhong and Xia's (1998b) formation boundaries. Therefore, the base of the Kanchung formation has prograded southward, whereas the base of the overlying Kuqa formation has prograded rapidly northward, which agrees with the rising base level suggested by the termination of the time-transgressive growth unconformity in Figure 8b. The Kanchung and Kuqa formations are up to 1200 and 4000 m (4000 and 13,000 ft) thick, respectively (Figure 4). This implies that the sedimentation rate increased greatly in the last 10 Ma. During Kanchung deposition, the sedimentation rate was about 0.2–0.3 mm/yr (0.007–0.012 in./yr) (Zhong and Xia, 1998b; Charreau et al., 2006, 2008; Huang et al., 2006, 2008). The rate increased

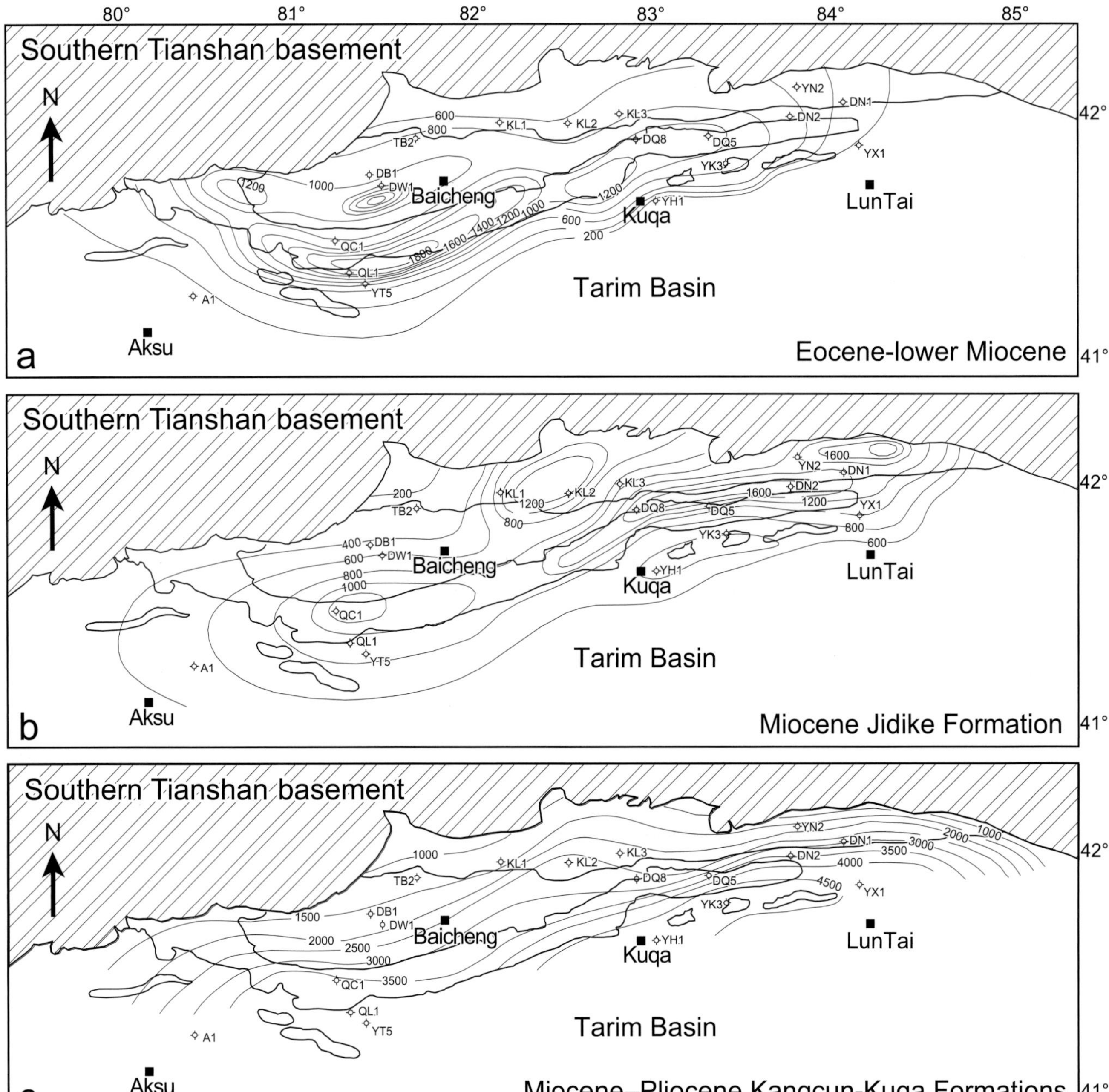

Figure 4. Isopach map of Tertiary basins in the Kuqa area. (a) Thickness of Eocene–lower Miocene strata with a 200-m (656-ft) contour interval. (b) Thickness of the Miocene Jidike formation with a 200-m (656-ft) contour interval. (c) Thickness of the Miocene Kanchung-Kuqa formations with a 500-m (1640-ft) contour interval (modified from Ye and Huang, 1990).

to 0.43 mm/yr (0.01y in./yr) 10–11 m.y. (Charreau et al., 2006, 2008). The late Miocene and Pliocene depocenters thin northward and southward, respectively, and were located in the central part of the Kuqa fold belt extending parallel to the trend of the Baicheng syncline (Figure 4). This suggests that the Kanchung and Kuqa formations accumulated during the uplift of the northern and southern Kuqa fold belt.

The younger Xiyu formation is a dark gray sequence of conglomerate with deposition starting 5 Ma ago in the northern Quilitage area (Charreau et al., 2006). It reaches a thickness of 600 m (1938 ft), interfingers with the Kuqa formation, and is locally unconformable on older formations because of fold growth (Hubert-Ferrari et al., 2005b, 2007). In the westernmost southern Tianshan, the Xiyu formation has been shown to be highly time transgressive (Heermance et al., 2007). The Xiyu formation represents the conglomeratic mountain-front alluvial facies, whereas the Kuqa formation represents the more distal sandy fluvial facies. The Xiyu-Kuqa facies boundary is currently just south of the thrust belt but was formerly within the area of the Yaken anticline just before it emerged topographically at about 0.2–0.3 ka (Hubert-Ferrari et al., 2005b, 2007).

MAJOR STRUCTURAL GEOLOGY

The major south Tianshan thrust fault that marks the topographic edge of the high Tianshan places diverse Paleozoic basement rocks of the high Tianshan Mountains over the Mesozoic and younger sedimentary cover of the adjacent foreland in both the easternmost Kuqa Basin extending east to Korla and in the southern Tianshan north and west of Aksu (Figures 1, 2). In contrast, in the central and westernmost Kuqa Basin, the basement-cover contact that marks the edge of the high Tianshan is depositional in outcrop (Figure 2) (also Hubert-Ferrari et al., 2005a). Nevertheless, the major south Tianshan basement thrust system is present in the subsurface as the fundamental structure of the northern Kuqa fold belt, as shown by a synthesis of seismic, outcrop, and well data (Figures 6–11). The southern front of this allochthonous Paleozoic in the subsurface is approximately 20–30 km (10–20 mi) south of the southernmost Paleozoic in outcrop and underlies a relatively narrow south-dipping monocline of Mesozoic and late Paleogene strata at the southern edge of the north Kuqa fold belt (Figure 2).

The north Kuqa fold belt is characterized by several east-west–striking thrust faults and folds and also confines the Heiyingshan piggyback basin on its north side, which is a present-day surface topographic depositional basin (Figure 2). The north Kuqa fold belt underwent multiple phases of deformation and uplift, so that Mesozoic and Tertiary strata are exposed along the belt. The main structure in the southern Kuqa fold belt is the active Quilitage anticline, which has 1200 m (4000 ft) of structural relief and is 20–30 km (10–20 mi) wide at the surface. The Quilitage anticline runs 280 km (174 mi) along strike and confines the great Baicheng topographic and sedimentary synclinal basin to the north, with only the Weigan River exiting the Baicheng basin across the Quilitage anticline, about 30 km (20 mi) west of Kuqa. Neogene and Quaternary strata are exposed within the Quilitage anticline. The Yaken anticline, 10 km (6 mi) south of the Quilitage anticline in the eastern Kuqa fold belt, is an active fold with beds dipping only a few degrees on both sides of the Quaternary surface (Hubert-Ferrari et al., 2005b, 2007). This anticline for about 100 km (60 mi) along strike marks the southern edge of the eastern Kuqa fold belt (Figures 2, 4).

Evaporites (gypsum and halite) are widely exposed in the western Kuqa fold belt along the eastern Awate, Tuzimaza, Kumgeliem, and Kasangtuokai anticlines (Figure 5). Exploration wells have also penetrated massive salt in the cores of many anticlines in the Kuqa fold belt, including the Dawanqi, Miskantag, Kasangtuokai, Bashijiqike, and western Quilitage anticlines. Two major evaporitic horizons in the Kuqa fold belt are observed. One is located in the Miocene Jidike formation in the eastern Kuqa fold belt, and the other is located in the Paleogene Kumgeliem formation in the western Kuqa fold belt (Figure 5). These formations represent the main shallow decollement horizon in the southern Kuqa fold belt and are consistent with the interpreted seismic sections (Figures 6–11). The lower decollement horizon developed in the northern part of the Kuqa fold belt is mostly within the Middle Jurassic mudstone and coal beds. The interpretation of seismic profiles shows that thrusts in the northern part of the Kuqa fold belt are located at this Jurassic level and that the rocks exposed in the cores of these anticlines in this area are Cretaceous and older (Figures 6–11).

Several regional profiles have been constructed across strike (Figures 6–11) constrained by two-dimensional seismic reflection profiles, well data, and surface geology. They provide the simple interpretations of the locally complex Cenozoic deformation of the Kuqa fold belt based on available surface and subsurface constraints. In the Kuqa fold belt, lateral velocity variations are anticipated due to the thrusting of high-velocity rocks from deeper to shallower positions and facies changes along strike. Interval velocities, determined from sonic logs and seismic stacking velocities, were used to depth convert seismic profiles by techniques of Suss and Shaw (2003). These velocities were used along with well data to determine the tops and thicknesses of different stratigraphic units of the Kuqa fold belt.

Eastern Part of the Kuqa Fold Belt

The Tuger, Yiqikelike, Jidike, eastern Quilitage, and Yaken anticlines are located in the eastern area of the Kuqa fold belt (Figure 2). We present two cross sections through the area. The eastern section (Figure 6) passes through the Yiqikelike anticline, the eastern tip of the eastern Quilitage anticline, and the eastern Yaken anticline. The western section (Figure 7) passes through the Jidike, eastern Quilitage, and western Yaken anticlines.

Along the eastern section (Figure 6), the Yiqikelike anticline is interpreted as a fault-propagation fold above

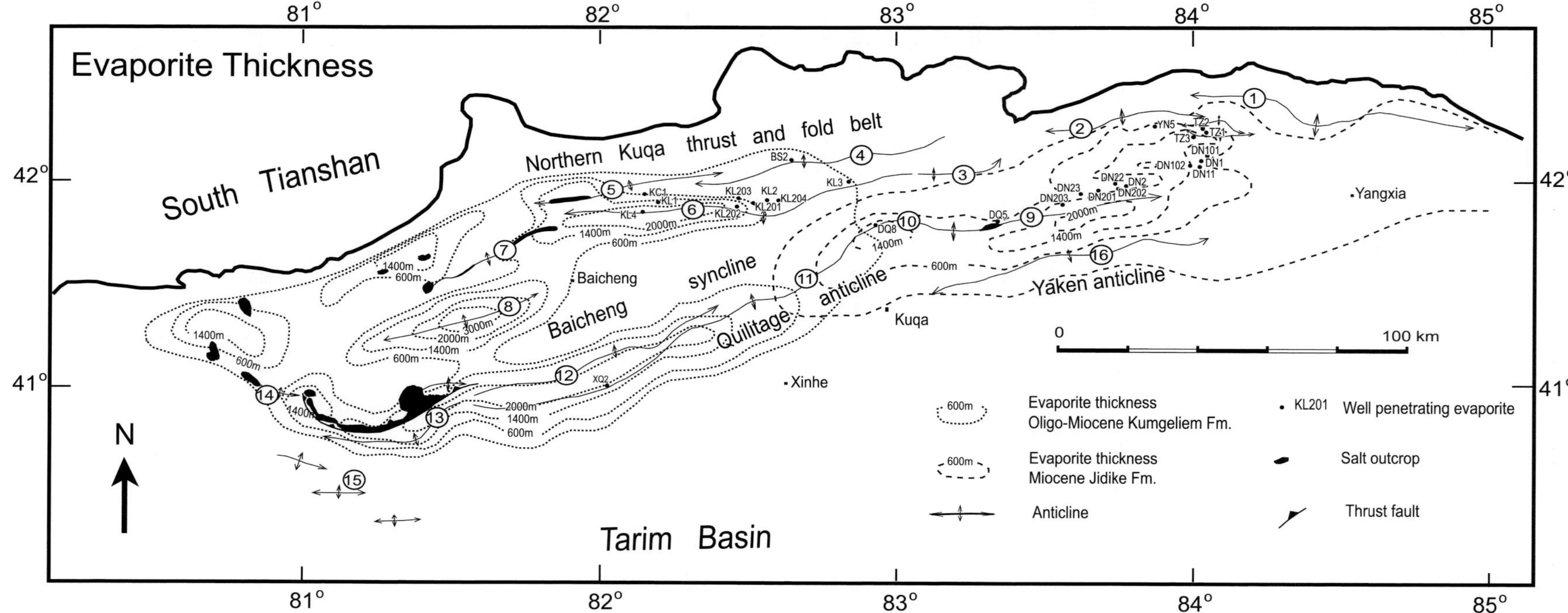

Figure 5. Isopach map of evaporites in the Miocene Jidike formation and in the Oligocene–Miocene Kumgeliem formation in the Kuqa area. Thickness of evaporites with an 800–600-m (2625–1968-ft) contour interval. (1) Tuger anticline, (2) Yiqikelike anticline, (3) Jidike anticline, (4) Bashijiqike anticline, (5) Kumgeliem anticline, (6) Kasangtuokai anticline, (7) Tuzimaza anticline, (8) Dawanqi anticline, (9) eastern Quilitage anticline, (10) Kuqadawu anticline, (11) southern Quilitage anticline, (12) northern Quilitage anticline, (13) Miskantag anticline, (14) eastern Awate anticline, (15) Kalayuergun anticline, (16) Yaken anticline (modified from Ye and Huang, 1990). Fm. = formation.

three imbricate thrust sheets that are stacked vertically by break-forward propagation. The steeply north-dipping imbricate faults are detached along the Middle Jurassic mudstone and coal beds and were encountered at measured depths of 1600, 2450, and 3239 m (5249, 8038, and 10,627 ft) in the YS4 well. Middle Jurassic strata, Yangxi and Kezilenuer formations, were repeated three times in the core of the Yiqikelike anticline where the Yiqikelike oil field was discovered in 1964. A buried anticline, here called the south Yiqikelike anticline, and the eastern tip of the eastern Quilitage anticline are located south of the Yiqikelike anticline (Figure 6). The south Yiqikelike anticline is a basement-involved fold, which is indicated by penetration of Carboniferous strata at the bottom of the YN2 well. The region of poor seismic data quality at and below the bottom of the well is interpreted as Paleozoic basement rocks that move upward along a north-dipping thrust ramp. The YN2 well penetrated Tertiary, Cretaceous, Jurassic, Triassic, and Carboniferous strata and encountered two minor faults at depths of 2850 and 4100 m (9350 and 13,451 ft). The eastern Quilitage anticline is interpreted as an asymmetric breakthrough fault-propagation fold with a steep and narrow forelimb and a long, planar backlimb. The forelimb was cut by a north-dipping reverse fault, which soles into evaporates of the Miocene Jidike formation, which detach this fold from deeper structure. Below the Jidike detachment, older Jurassic, Triassic, and Eocene strata have been folded, creating the structure of the Dina gas field discovered in 2001 (Figure 6). The Yaken anticline, which is the southernmost active structure, is a detachment fold above the Jidike evaporates (Hubert-Ferrari et al., 2005b; Gonzalez-Mieres and Suppe, 2006, 2011).

The interpretation of the east Kuqa profile (Figure 7) shows three north-dipping reverse faults rooting in the Paleozoic basement in the northern part of the profile. The thrust faults give rise to a broad anticline and a large amount of structural relief in the near-surface section (Figure 7). The Paleozoic basement rocks moved upward, and Triassic, Jurassic, and Cretaceous strata are exposed at an elevation of 2000 m (6500 ft) above sea level. The Jidike anticline is a hanging-wall fold above the south-dipping backthrust faults where Cretaceous and Tertiary strata were highly dissected and developed a large monocline structure (Figure 7). A major buried thrust ramp, now folded to horizontal, accounts for the large change in stratigraphic thickness across the syncline between the Jidike and Quilitage anticlines. The eastern Quilitage anticline has a box shape at the surface. The south limb of the anticline is narrow and steep, with a dip of 70 to 80°, developing several breaking thrusts in it, with horizontal strata on its crest. The north limb of the anticline is long and gentle, with a 50 to 60° dip. The subsurface structure of the eastern Quilitage is composed of two structural levels bounded by detachments: (1) a deep thrust ramp that steps up southward from a 9-km (5.5-mi)-deep lower detachment level in Upper Jurassic coal horizons and flattens to a 5-km (3-mi)-deep upper detachment in evaporates of the Miocene Jidike formation, producing a fault-bend fold anticline with limb dips of 10–15° above the deep thrust ramp; and (2) an overlying complex wedging system that produces the steep dips of the surface anticlinal core (Figure 7). The Yaken anticline, which occurred at the south of this section, is a detachment fold; the decollement fault underlying the anticline is nearly horizontal, without a ramp of thrust, sliding along the gypsum and shale layers of the Miocene Jidike formation (Figure 7).

The Quilitage upper detachment level extends to the south and corresponds to the Yaken basal detachment. Slip on the deep ramp (~5 km [3 mi]), plus deeper folding of the fault, has generated a 7- to 8-km (4- to 5-mi)-wide north-dipping kink band imaged in the seismic profile (Figure 7). The shallower structure of Quilitage is necessarily complex because Yaken, which is located above the same Jidike detachment, has consumed only 1.2 km (0.7 mi) of the 5-km (3.1-mi) slip coming from the deep ramp of Quilitage. The remaining 3.8 km (2.4 mi) is wedged back in the anticlinal core of Quilitage. Above the deep ramp, additional wedging occurs at a shallower level in the anticlinal core, located at the intersection between the back thrust and an upper bedding-parallel detachment level. At this shallower wedge tip, the thrust system splits into two main faults. The upper thrust reaches the surface in the south limb of Quilitage. Slip on the lower thrust is consumed in folding. Additional deep detachment folding and associated wedging in the north syncline completes the slip budget of Quilitage.

Central Part of the Kuqa Fold Belt

North-dipping imbricate thrust faults and roughly west-east–striking folds are interpreted in the northern part of the center of the Kuqa fold belt (Figures 2, 8, 9). The long profile in Figure 8a has been interpreted as three imbricate thrust faults and duplex wedges near the southern Tianshan deformation front. This area underwent multiple phases of deformation and experienced more uplift than the southern part of the Kuqa fold belt, exposing Mesozoic and Tertiary strata along the belt. To the south, Tertiary strata are preserved within the cores and limbs of Bashijiqike and Kasangtuokai anticlines. These anticlines are interpreted as fault-propagation folds with the detachment being along the gypsum and shale of the Paleogene Kumgeliem formation. The KL3

Eastern Yaken anticline

Eastern Quilitage a

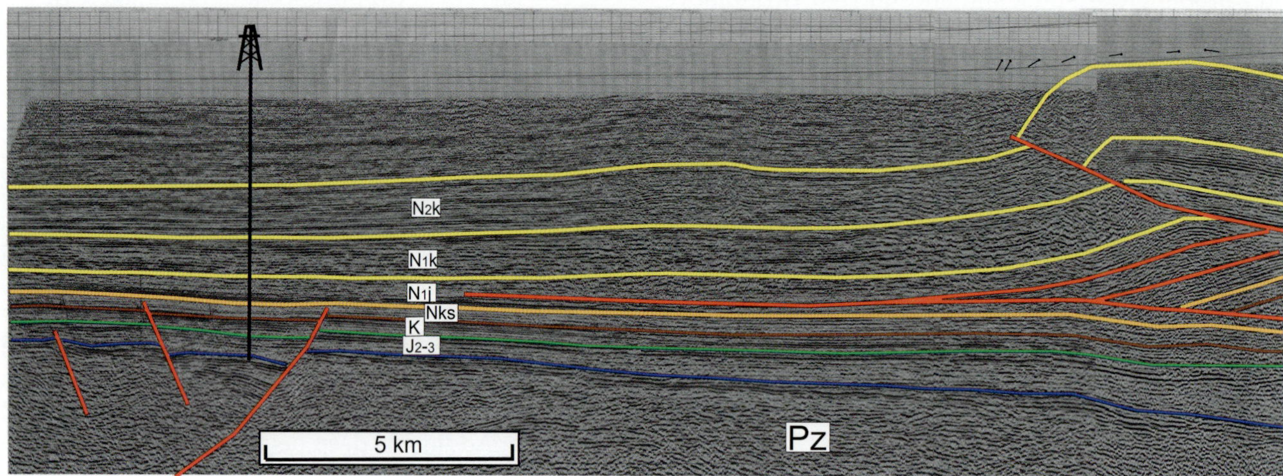

Figure 6. (a) Uninterpreted and (b) interpreted depth-converted 2-D seismic profile across the Yiqikelike, south Yiqikelike, eastern Quilitage, and Yaken anticlines. Horizontal scale equals vertical scale. The profile is located on Figure 2. N_2k = Pliocene–Pleistocene Kuqa formation; N_1k = Miocene Kanchung formation; N_1j = Miocene Jidike formation; Nks = Oligocene–Miocene Kumgeliem and Suweiyi formations; K = Cretaceous; J_{2-3} = Middle–Upper Jurassic; J_3 = Upper Jurassic; J_2 = Middle Jurassic; J_1 = Lower Jurassic; T = Triassic; Pz = Paleozoic; s.l. = sea level; S = Silurian; D = Devonian; C = Carboniferous; S. = Southern; ant. = anticline.

well penetrates the detachment fault (Figure 8a). The Kuqatawu anticline, which represents a transverse jog in the Quilitage anticline belt, is a tight fold, consisting of steep strata in the core and gentle dip strata in the limbs. From seismic reflection data, the anticline is dominated by the south-dipping back thrust in the forelimb, developing in the Miocene Jidike formation, similar to other sections through the central Quilitage anticline (Figure 9) (Hubert-Ferrari et al., 2007, their figures 14, 16). The deeper levels of the anticline are interpreted to have two fault-bend folds, whose lower detachment fault is in the coal bed at the base of the Jurassic and whose upper detachment fault is located in the gypsum layer of the Paleogene Kumgeliem formation. The ramps of the faults cut upward through the Jurassic, Cretaceous, and Paleogene strata. Slips on the two deep ramps have generated 4–5-km (2.5–3-mi)-wide north-dipping kink bands imaged in the seismic profile. The Tukelaketan anticline, one of the two deeper anticlines, has surface expression.

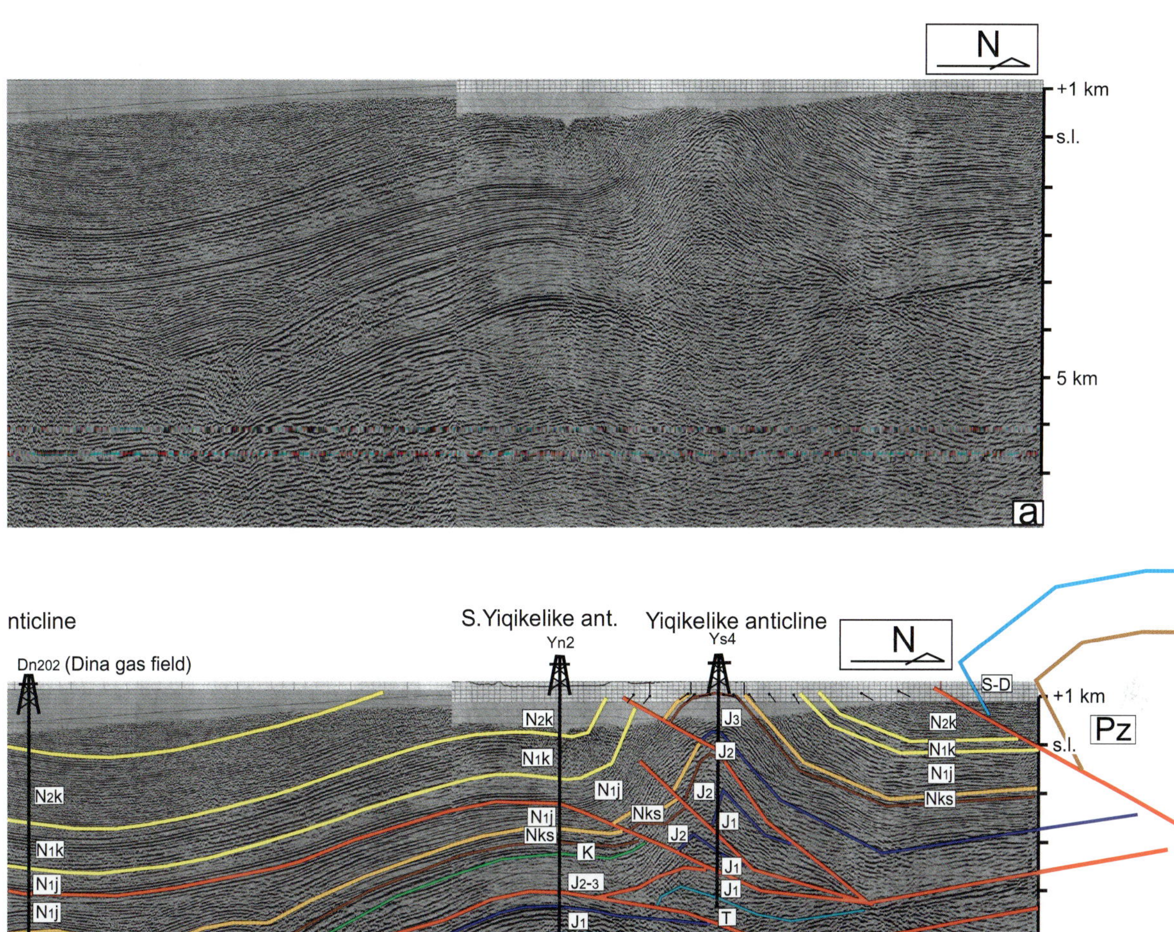

Figure 6. (cont.).

The structure of profile in Figure 9 has been interpreted to be composed of three imbricate thrust faults and duplex wedges near the southern Tianshan deformation front. The thrust faults slip southward along the gypsum and shale of the Paleogene Kumgeliem formation and cut upward through the Tertiary strata. The Kumgeliem and Kasangtuokai anticlines, developed on the ramps of the thrust faults, are interpreted as fault-propagation folds. The salt wall is interpreted to have developed in the footwall of the shallow thrust faults. The imbricate system beneath the salt wall was formed by break-forward propagation of thrust sheets that are suggested to terminate in the salt wall. Several deeper folds have been interpreted to develop in the Mesozoic strata. A pop-up structure is expressed as an anticline bounded by forward and back thrust underlying the Kasangtuokai anticline (Figure 9). A huge natural gas field (Kela gas field) with more than 200 billion m^3 (7.06 tcf) of natural gas was discovered in this structure in 1998. The south Quilitage anticline is a salt-cored detachment fold with a high-angle reverse fault on the crest. The fold is interpreted to be cored by the autochthonous Paleogene Kumgeliem salt. Thickened salt was penetrated by the XQ2 well, with the thickness of salt under the fold crest estimated at 3000–4000 m (10,000–13,000 ft) (Figure 9). The crest of the fold has been cut by a northern vergent bedding-parallel reverse fault, similar to other sections through the central Quilitage anticline (Figure 8a) (also see Hubert-Ferrari et al., 2007, their figures 14, 16). Growth strata (yellow dashed lines on Figure 9) are interpreted to be deposited on both flanks of fold, which shows the salt-cored detachment

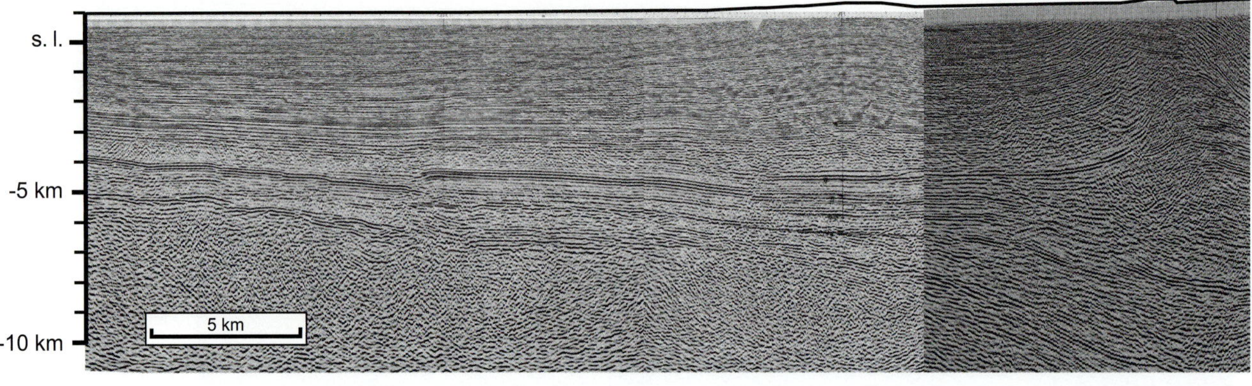

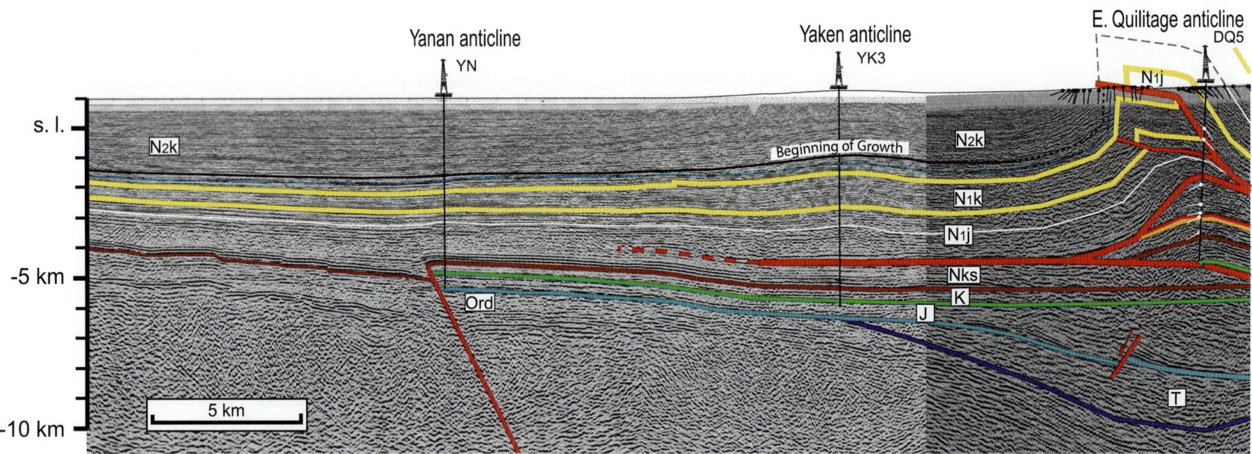

Figure 7. (a) Uninterpreted and (b) interpreted depth-converted 2-D seismic profile across the Jidike, eastern Quilitage, and Yaken anticlines. Interpretation is based on Hubert-Ferrari and Suppe (He and Suppe, 2005, p. 12) and Hubert-Ferrari et al. (2007). Horizontal scale equals vertical scale. The profile is located on Figure 2. N_2k = Pliocene–Pleistocene Kuqa formation; N_1k = Miocene Kanchung formation; N_1j = Miocene Jidike formation; Nks = Oligocene–Miocene Kumgeliem and Suweiyi formations; K = Cretaceous; J = Jurassic; s.l. = sea level; T = Triassic; Pz = Paleozoic; Ord = Ordovician; E. = Eastern.

fold and associated reverse faults of the south Quilitage anticline developed principally during the Pliocene and Quaternary.

Western Part of the Kuqa Fold Belt

The dominant structure north of the western Kuqa fold belt is a south-dipping monocline that overlies the north-dipping reverse faults in the Paleozoic basement (Figures 10, 11). The southern part of the western Kuqa fold belt is interpreted to have developed massive salt structures. Salt can be seen at the surface along the thrust faults that cut the backlimb (north limb) of the Miskantag anticline and adjacent to the eastern Awate and Tuzimaza anticlines (Figure 5). The massive salt has been drilled in the cores of these anticlines by exploration wells (Figures 11, 12). An east-west–striking compressed salt wall involving the Paleogene Kumgeliem formation is interpreted to develop underlying the Tuzimaza anticline. Diapiric ridges are exposed along the Tuzimaza anticline and between the Tuzimaza anticline and the south Tianshan mountain front (Figures 5, 10). Several north-dipping reverse faults are interpreted to develop beneath the diapirs. The Dawanqi, southern Quilitage, northern Quilitage, and Miskantag anticlines are interpreted to be cored by Paleogene Kumgeliem salt (Figures 10, 11). The Dawanqi anticline is a symmetric salt-cored fold. However, the north limb of the fold is interpreted to be uplifted by an underlying structure in the Paleozoic basement. The imbricate system beneath the fold was formed by break-forward propagation of the thrust sheets consisting of Mesozoic strata. The growth

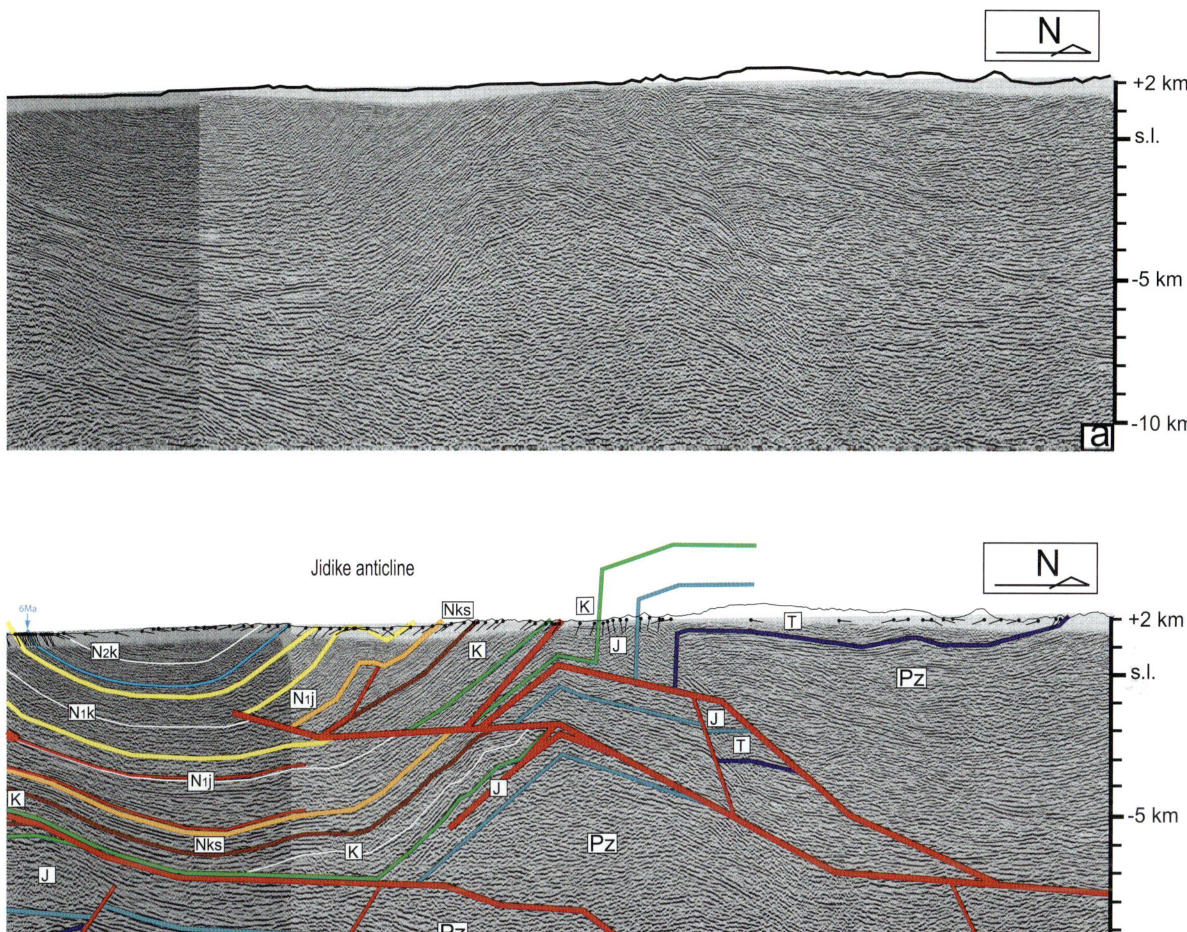

Figure 7. (cont.).

strata (shown by dashed lines) thin onto the fold crest fanning out on both flanks of the fold (Figure 10). The characteristics of the growth strata imply that the Dawanqi anticline was formed by limb rotation (Poblet et al., 1997). Growth strata, consisting of Pliocene and Quaternary strata, constrain the timing of formation of the Dawanqi anticline. Exploration wells discovered oil and gas in the shallow beds of the Dawanqi anticline (DW1 well) and the thrust sheets underlying the salt bed at the bottom of Tertiary strata (DB2 well).

The southern and northern Quilitage anticlines are two adjacent tight folds at surface. Seismic lines show that the two anticlines are linked at depth by a broad, salt-cored fold consisting of Paleogene Kumgeliem salt (Figure 10). The anticline has a narrow steep forelimb dipping 70 to 80° and a long planar backlimb consisting of Neogene and Quaternary strata dipping 40 to 20°. The crest of the anticline has been cut by a south-vergent high-angle reverse fault zone. Syntectonic growth strata (yellow dashed lines on Figure 10) have deposited on both flanks of the fold in the upper Pliocene and Pleistocene section. The deeper growth strata are thinned and deformed on the forelimb.

The structural profile of Figure 11 is located at the west end of the Kuqa fold belt, showing the Miskantag anticline and a major thrust fault that cuts the backlimb of the Miskantag anticline. These structures are part of the Quilitage anticline belt (Figure 2). The Miskantag anticline, consisting of the Tertiary strata, is a broad, salt-cored asymmetric fold with a long, gentle backlimb and a steeper, faulted forelimb. The continuous relatively flat basement that underlies the fold and the lack of an obvious thrust ramp beneath the fold indicate that the structure is a detachment fold dominated by ductile thickening of the Paleogene Kumgeliem salt in the core of the fold, with withdrawal from the adjacent syncline to the north. The growth strata (yellow dashed lines on Figure 11) deposited and deformed on the forelimb of the fold. The thrust fault that cut the backlimb of the Miskantag anticline is a north-dipping

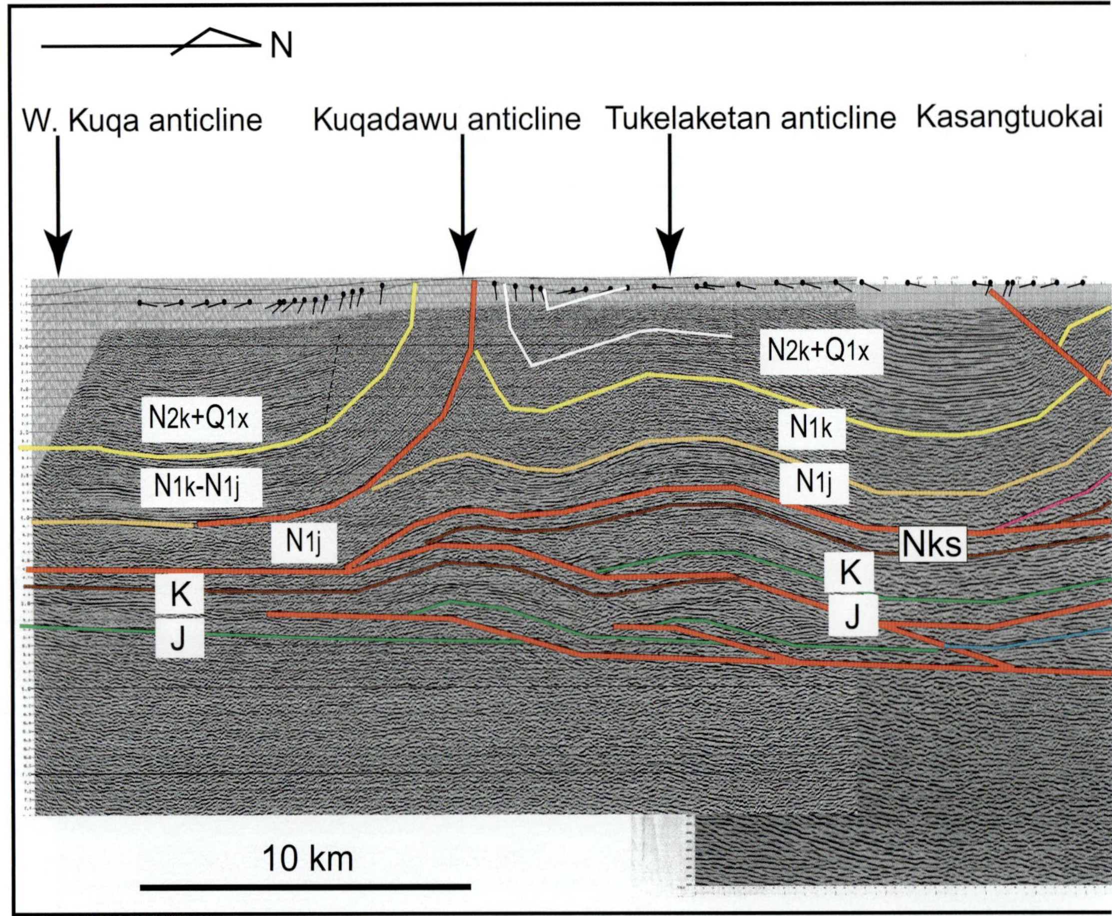

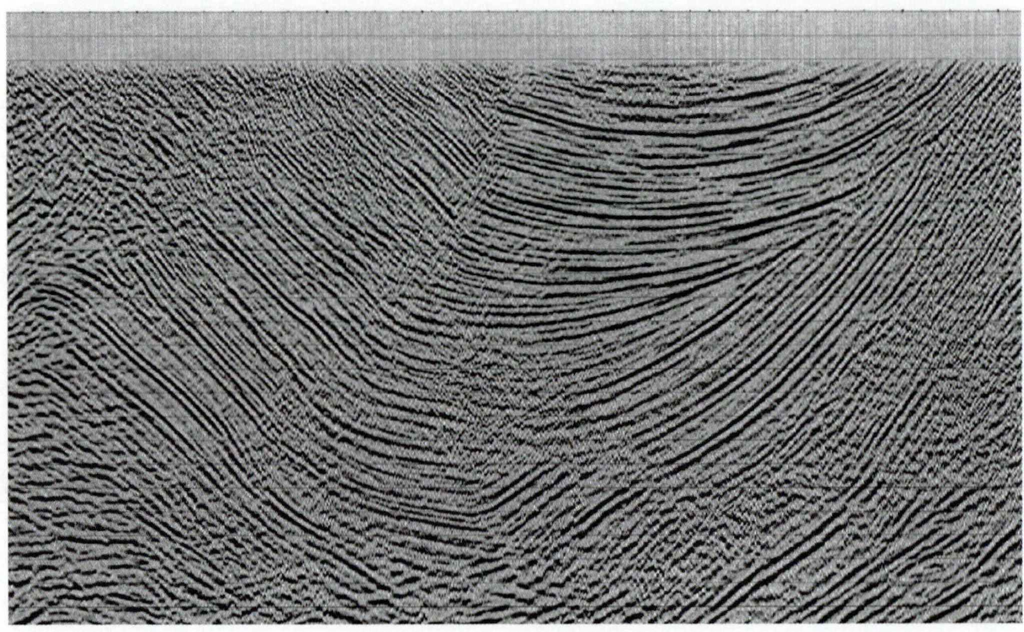

Figure 8. (a) Interpreted depth-converted 2-D seismic profile across the Bashijiqike, Kasangtuokai, Tukelaketan, and Kuqadawu anticlines. Horizontal scale equals vertical scale. The profile is located on Figure 2. (b) Seismic profile shows unconformities and growth strata developed in syncline (from Suppe in He and Suppe, 2005, p. 12). Q_1x = Pleistocene Xiyu formation; N_2k = Pliocene–Pleistocene Kuqa formation; N_1k = Miocene Kanchung formation; N_1j = Miocene Jidike formation; Nks = Oligocene–Miocene Kumgeliem and Suweiyi formations; K = Cretaceous; J = Jurassic; T = Triassic; s.l. = sea level; Pz = Paleozoic; F1 – F3 = faults.

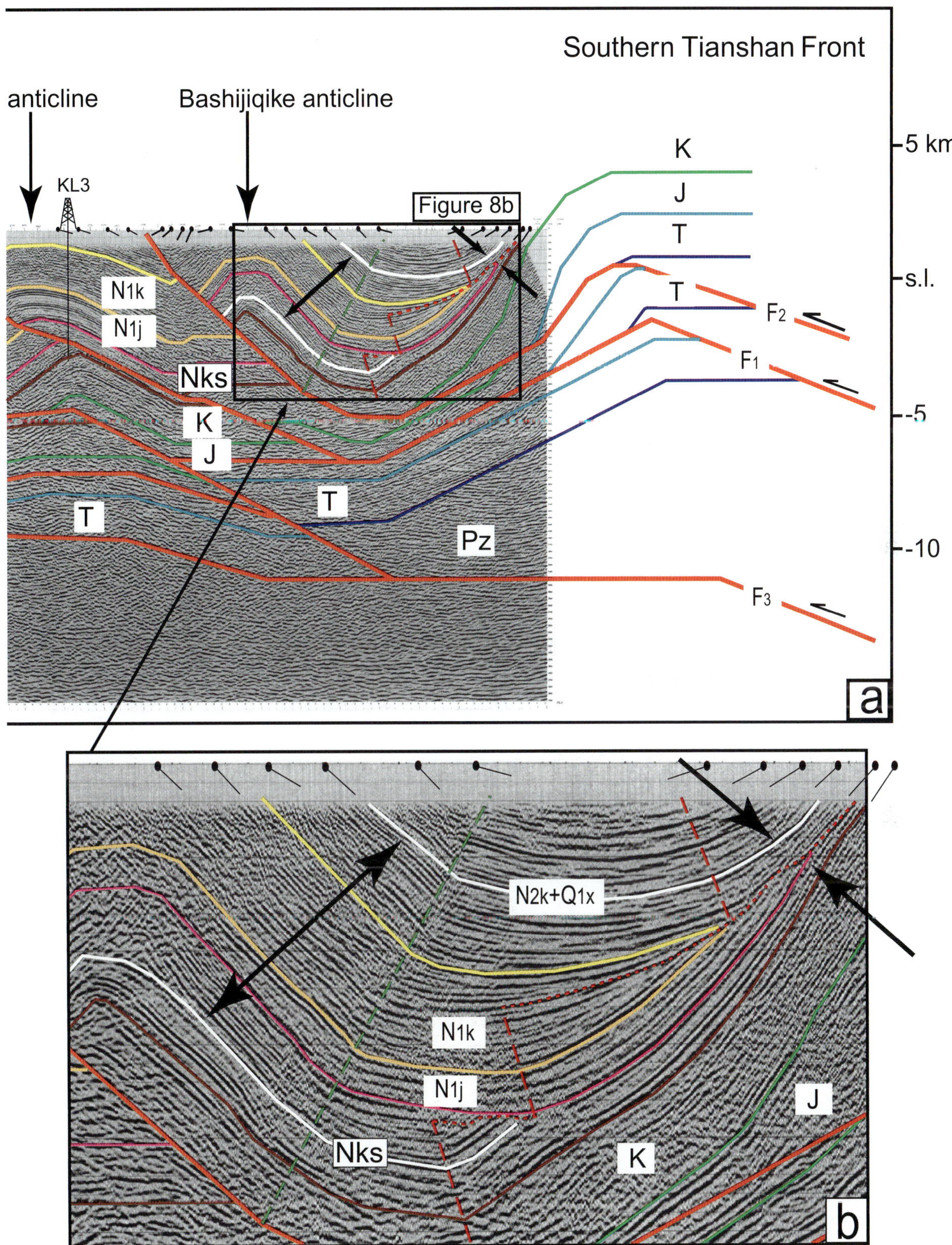

Figure 8. (cont.).

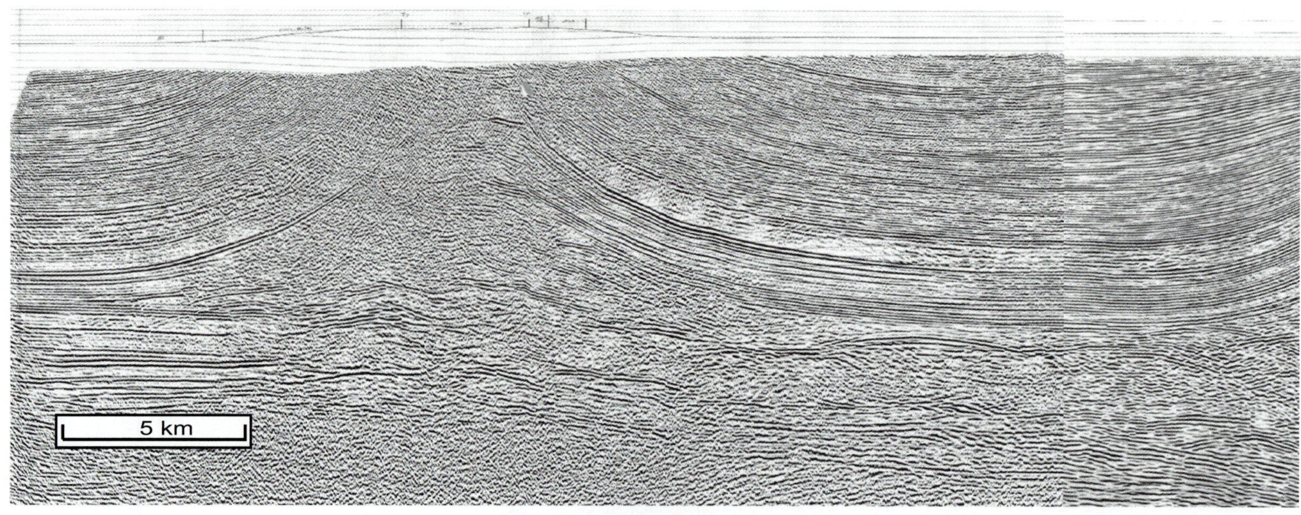

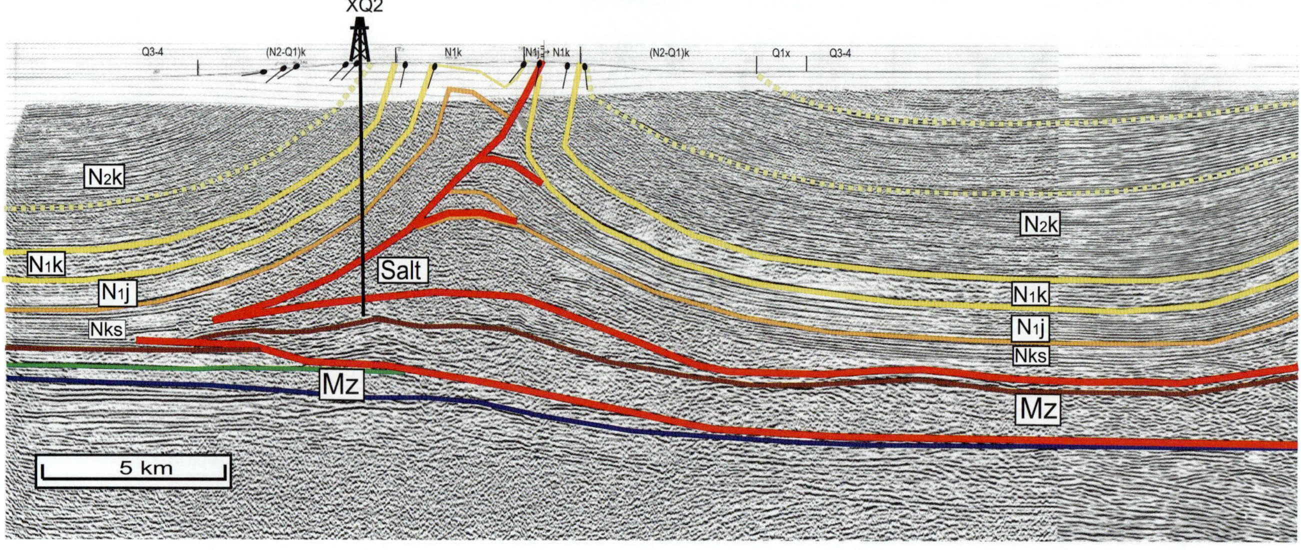

Figure 9. (a) Uninterpreted and (b) interpreted depth-converted 2-D seismic profile across the Bashijiqike, Kasangtuokai, and south Quilitage anticlines. Horizontal scale equals vertical scale. The profile is located on Figure 2. Yellow dashed lines show growth strata. N_2k = Pliocene–Pleistocene Kuqa formation; N_1k = Miocene Kanchung formation; N_1j = Miocene Jidike formation; Nks = Oligocene–Miocene Kumgeliem and Suweiyi formations; K = Cretaceous; J = Jurassic; T = Triassic; Mz = Mesozoic; Pz = Paleozoic.

fault with at least a displacement of 22 km (14 mi) (aa′, Figure 11). The fault is detached above a Paleogene Kumgeliem allochthonous salt nappe. Massive salt crops out along the fault (Figures 5, 11).

TIMING OF STRUCTURAL EVENTS

A late Oligocene or Miocene age of initiation of Cenozoic structuring in the southern Tianshan and Kuqa fold belt has been proposed previously based on several indirect indications (e.g., Hendrix et al., 1994; Lu et al., 1994; Sobel and Dumitru, 1997; Yin et al., 1998; Charreau et al., 2006, 2008; Huang et al., 2006, 2008; Sobel et al., 2006). These indications include the appearance of detritus that record the uplifting and eroding basement of the high Tianshan, including fission track and magnetic susceptibility anisotropy data and stratigraphic facies transitions. Here we report on observations of syntectonic growth unconformities and growth strata that allow us to determine ages of specific early structural events north of Kuqa, making use of magnetostratigraphic data discussed previously to constrain the ages of syntectonic growth strata. Later structural timing in

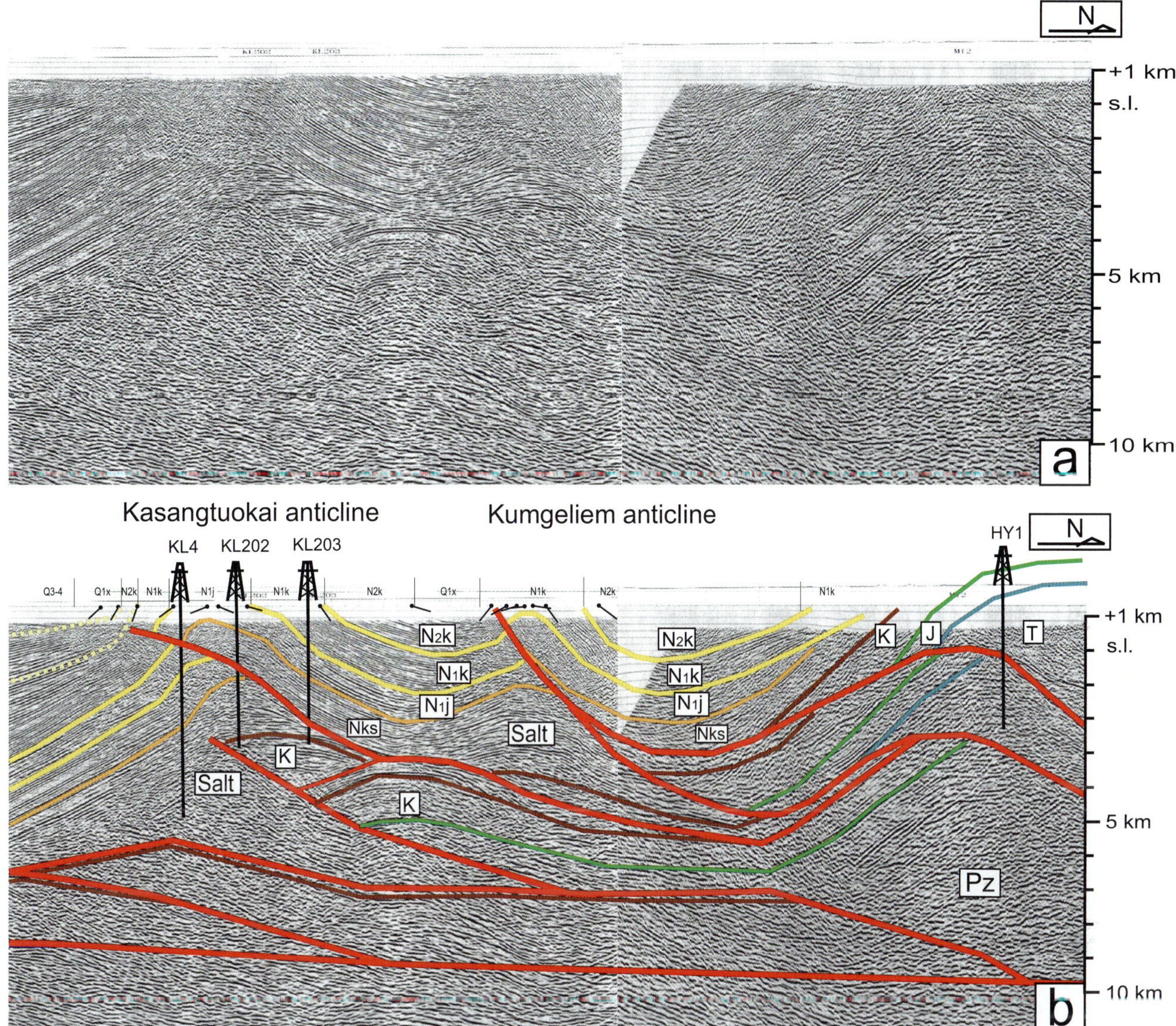

Figure 9. (cont.).

the Quilitage and Yaken anticlines is documented by Hubert-Ferrari et al. (2007). Elsewhere in the Kuqa fold belt, structural timing is constrained by growth strata imaged in the regional profiles (Figures 6–11).

In the northern Kuqa River area, time-transgressive growth angular unconformities have been found in association with growth strata in a fold limb south of the southern Tianshan front. Such unconformities have fold axial surfaces terminating at the two ends of an unconformity, indicating deposition during kink-band migration at a low sedimentation rate relative to uplift rate. At higher sedimentation rates, growth axial surfaces appear as shown in Figure 12. The general principles of such growth structures are given by Suppe et al. (1991, 1997).

The growth structures are shown in a progressively deformed syncline shown in Figure 8b. The thickness of the strata decreases northward, as indicated by the thickness variation of the late Paleogene and Neogene sequence, which is 3–4 km (1.9–2.5 mi) in the southern limb of the syncline and only several hundred meters on the northern limb (Figure 8b). The oldest unconformity observed to date in the Kuqa thrust belt records growth during deposition of the late Oligocene–early Miocene Suweiyi formation (Figure 8b). A hiatus in deformation during the Jidike (20–21 Ma to ~15 Ma) within this structure above this unconformity is observed, as shown by the constant thickness of the Jidike strata across the structure (Figure 9b). Growth resumed and was continuous during deposition of the Kanchung and lower Kuqa formations (~15 to <6 Ma) as shown by the development of a synclinal growth axial surface across which a large change in thickness is observed (Figure 8b).

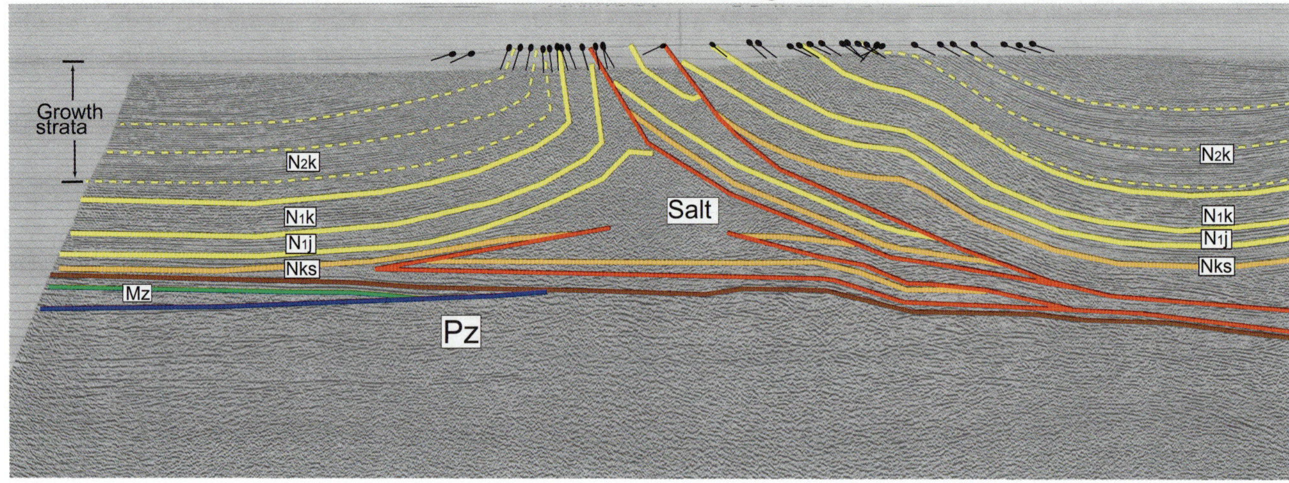

Figure 10. (a) Uninterpreted and (b) interpreted depth-converted 2-D seismic profile across the Tuzimaza, Dawanqi, and southern and northern Quilitage anticlines. Horizontal scale equals vertical scale. The profile is located on Figure 2. N_2k = Pliocene–Pleistocene Kuqa formation; N_1k = Miocene Kanchung formation; N_1j = Miocene Jidike formation; Nks = Oligocene–Miocene Kumgeliem and Suweiyi formations; K = Cretaceous; J = Jurassic; Mz = Mesozoic; Pz = Paleozoic; T = Triassic; S. = Southern; N. = Northern. Yellow dashed lines show the fan-dipping growth strata in the Kuqa formation.

Within this growth sequence, a time-transgressive growth angular unconformity that offsets two segments of the synclinal growth axial surface is observed (Figure 8b). This unconformity indicates a period of lower sedimentation rate relative to deformation rate, followed by a relatively higher sedimentation rate. This increased relative sedimentation rate that terminates the unconformity may correlate with the increase in sedimentation rate in the east Kuqa Basin at 10–11 Ma (Charreau et al., 2006, 2008). The uppermost stratigraphy records a hiatus or near-hiatus in deformation as evidenced by the nearly constant stratal thicknesses. These youngest strata are folded, recording a more recent resumption of deformation in this structural trend. The synclinal axial surface to the south deforms terrace gravels and is therefore presently active. Thus, we see a very long history of deformation recorded in this single complex synclinal structure, extending from the Oligocene to the present with two pronounced hiatuses in deformation that are at least local.

Growth strata containing a synclinal growth axial surface have been imaged in the front limb of the Kasangtuokai anticline 30 km (19 mi) to the southwest (Figures 9, 12). The boundary between growth and pre-

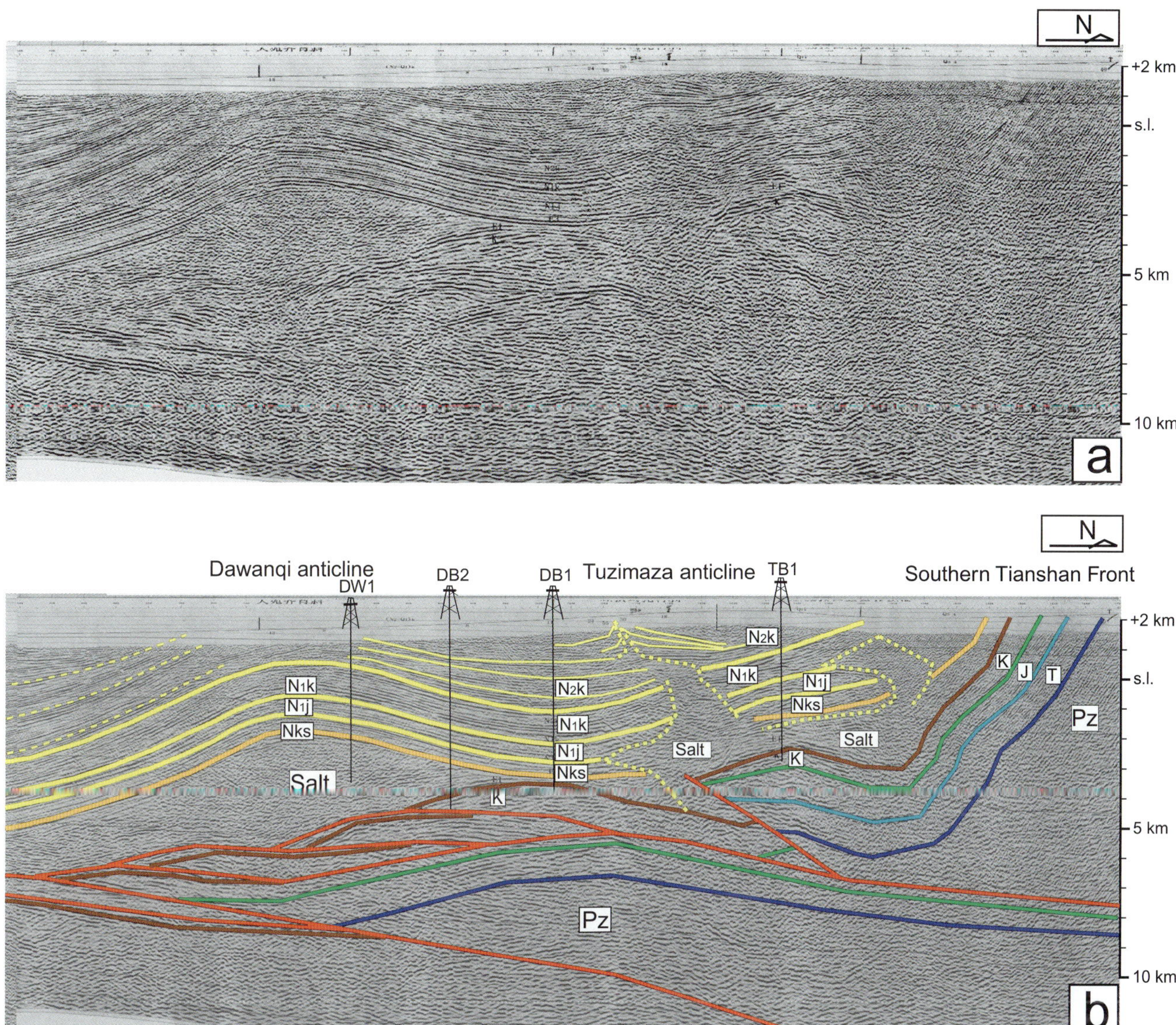

Figure 10. (cont.).

growth strata is located within the Miocene Kanchung formation and extends well up into the Kuqa formation (Figure 12), which implies that the Kasangtuokai anticline grew from approximately 8–10 to less than 3 Ma. The anticlinal growth may correlate with the upper unconformity in the northern syncline (Figure 8b). Growth strata developing in the northern and southern limbs of the Dawanqi anticline imply that the Dawanqi anticline developed during the post approximately 3-Ma deposition of the upper Kuqa formation (Figure 10b).

Growth strata of the eastern Quilitage anticline and the Yaken anticline are within the Pliocene–Pleistocene Kuqa formation with an age of about 5.5 Ma for the boundary between the pregrowth and growth strata, corresponding to the time of propagation and initiation of slip on the thrust system that underlies both the Yaken and Quilitage anticlines (Hubert-Ferrari et al., 2007) (Figure 7). The present topographic emergence of the Quilitage and Yaken anticlines is quite recent (~0.2–0.3 Ma) associated with a large acceleration of shortening rate in this system, as documented by Hubert-Ferrari et al. (2007). The growth history of the Yaken anticline is also discussed by Gonzalez-Mieres and Suppe (2011).

We demonstrate that the emplacement of the thrusts and fault-related folds propagated southward in the Kuqa fold belt. Near the mountain front, Mesozoic and Paleogene strata were involved in deformation related to thrusting and folding since approximately 25 Ma. Toward the southern edge of the Kuqa fold belt, strata in the Quilitage and Yaken anticlines began folding above a newly propagated thrust system about 5.5 Ma. The river network and alluvial fans are also deformed in this area.

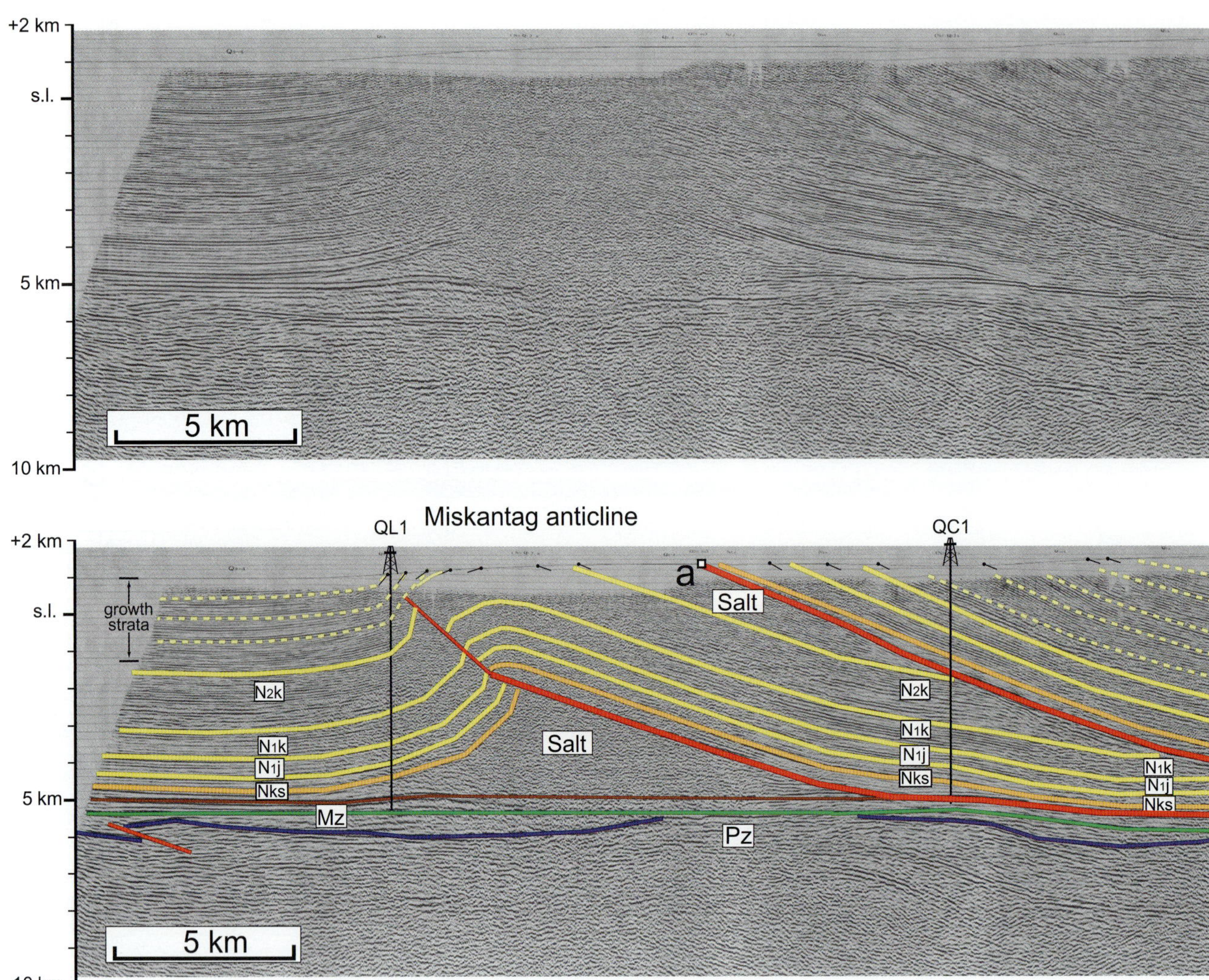

Figure 11. (a) Uninterpreted and (b) interpreted depth-converted 2-D seismic profile across the northern Kuqa fold belt and the Miskantag anticline. Horizontal scale equals vertical scale. The profile is located on Figure 2. N_2k = Pliocene–Pleistocene Kuqa formation; N_1k = Miocene Kanchung formation; N_1j = Miocene Jidike formation; Nks = Oligocene–Miocene Kumgeliem and Suweiyi formations; J = Jurassic; T = Triassic; K = Cretaceous; Mz = Mesozoic; Pz = Paleozoic. Yellow dashed lines show the growth strata in the Kuqa formation.

TECTONIC MODEL OF THE KUQA FOLD BELT

A forward kinematic model shows the interpreted evolution of the Cenozoic structure in the Kuqa river area of the central Kuqa fold belt (Figure 13). This model represents a simplified approximation to the structure in the seismic profile in Figure 8.

Stage 1

Fault 1 forms, and slip produces the kink band that defines the early Suweiyi growth in Figure 8b below a time-transgressive growth unconformity of the sort described by Suppe et al. (1991, 1997). Fault 1 is a multibend fault with two north-dipping ramps and two decollement zones; the upper decollement zone is the Oligocene Kumgeliem formation, and the lower zone is near the top of the Jurassic mudstone and coal sequence. Based on the width of the kink bands, displacement on fault 1 is less than 1 km (0.6 mi) at this stage. A hiatus in growth during Jidike deposition, prior to stage 2, is observed.

Stage 2

Following a hiatus in slip on fault 1 during Jidike deposition, fault 2 developed above fault 1 to form two ramp folds, the incipient anticline of the southern

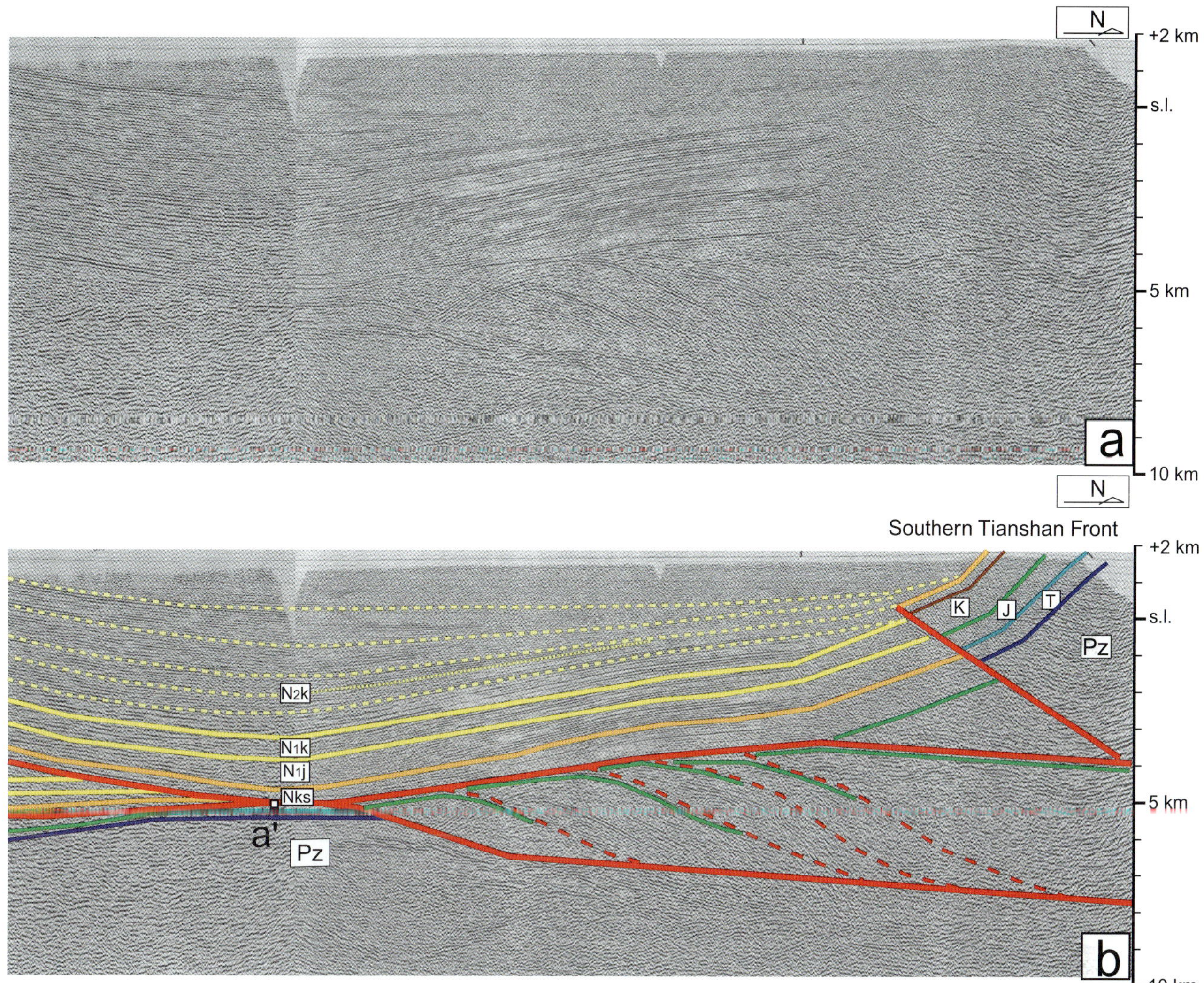

Figure 11. (cont.).

Tianshan front and the Bashijiqike anticline in the hanging wall. The thinning of Miocene and Pliocene rocks northward in the syncline between these anticlines shows that the structural growth of the southern Tianshan front was synchronous with Kangchung and Kuqa deposition. We find two time-transgressive angular unconformities (shown by wavy lines) in the growth strata wedge, an older one forming during the Suweiyi growth and a younger one during a period of lower sedimentation-to-uplift rate within a continuous growth interval during Kanchung and Kuqa deposition (see Suppe et al., 1991, 1997, for an explanation of such growth structures). Displacement on fault 2 is about 4 km (2.5 mi) (Figure 8a).

Stage 3

Fault 1 was reactivated in the late Miocene. Displacement on the lower fault 1 refolded the shallower fault 2 and the beds in its hanging wall. The Kasangtuokai anticline and the unnamed anticline, underlying the Bashijiqike anticline, develop in the hanging wall of fault 1. Part of the slip in the fault 1 system exits to the surface in the back thrust within the Kuqadawu anticline to the south. Displacement on fault 1 is about 5 km (3.1 mi) (Figure 8a).

Stage 4

In the Pliocene, slip on the fault 3 ramp generated a large uplift composed of Paleozoic and Triassic rocks, which refolded and uplifted the shallower early faults and anticlines. Fault 3 joins the fault 1 system updip at this stage, stepping up to a shallower detachment in the Jidike formation and amplifying a previous fault-bend fold in the hanging wall. Displacement on fault 3 is about 4 km (2.5 mi) (Figure 8a).

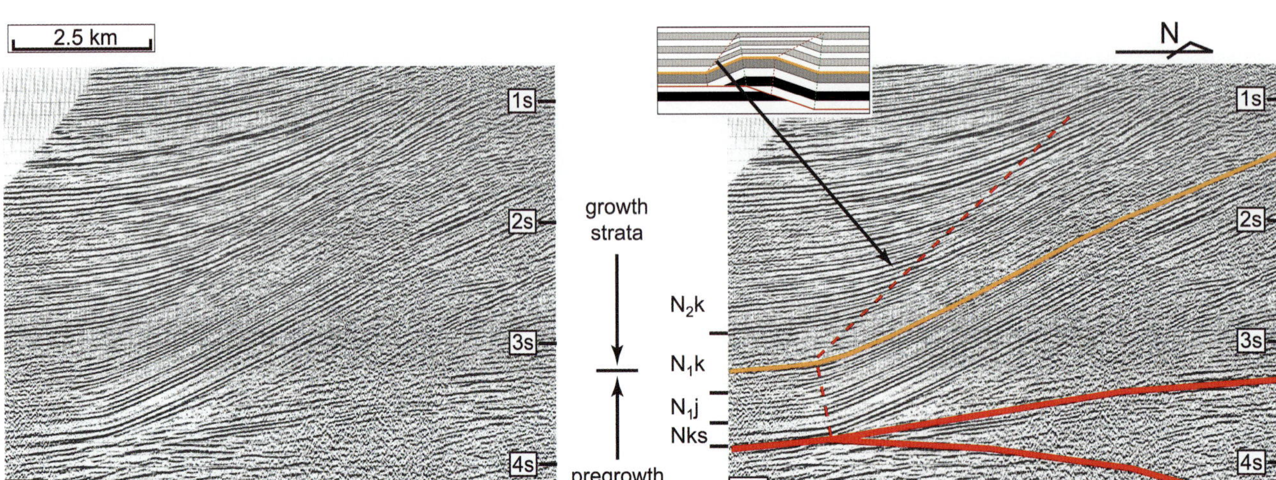

Figure 12. (a) Uninterpreted and (b) interpreted seismic profile showing growth strata developed in the front limb of the Kasangtuokai anticline. The boundary between pregrowth and growth strata at the middle of the Miocene Kanchung formation. The location of the profile is shown in Figure 2. N_2k = Pliocene–Pleistocene Kuqa formation; N_1k = Miocene Kanchung formation; N_1j = Miocene Jidike formation; Nks = Oligocene–Miocene Kumgeliem and Suweiyi formations. s = seconds.

Stage 5

At this stage, slip on the deep fault 3 transfers to a newly propagated fault 4 system and slip produces the lower fault-bend folds in the hanging wall that also refold shallow fault 1. The Tukelaketan anticline is an imbricate fault-bend fold north of a deep fault-bend fold below the Kuqadawu anticline. Slip on these two fault-bend folds terminates southward within the frontal west Kuqa anticline. Displacement on fault 4 is about 6 km (3.7 mi) (Figure 8a).

SHORTENING RATE OF THE KUQA FOLD BELT

The minimum shortening across the Kuqa fold belt estimated along our six regional sections (Figure 2) ranges from about 12 km (7.4 mi) in the east to 26 km (16 mi) in the west (Figures 6–11). We assume that deformation regionally began in the late Oligocene near the beginning of Suweiyi deposition (25–26 Ma), as locally documented by the growth strata in Figure 8b. This yields an average minimum long-term shortening rate of about 0.5 mm/yr (0.019 in./yr) in the east, 0.6 mm/yr (0.023 in./yr) near Kuqa, and 1.0 mm/yr (0.039 in./yr) in the westernmost Kuqa fold belt. Available data in the east Kuqa fold belt and from regional geodetic constraints indicate that the shortening rate has accelerated substantially toward the present, as outlined below. We find that the average shortening rate of 0.5–1.0 mm/yr (0.019–0.039 in./yr) since 25–26 Ma represents the sum of a much slower early average minimum shortening rate and a much faster recent rate.

The basal thrust system that underlies the Quilitage and Yaken anticlines, with a total slip of 6 km (3.7 mi), propagated about 5.5 Ma as shown by analysis of growth strata (Hubert-Ferrari et al., 2005b, 2007; Gonzalez-Mieres and Suppe, 2011), yielding an average slip rate of 1.1 mm/yr (0.043 in./yr). However, from 5.5 and 0.16–0.21 Ma, the shortening rate in the Yaken anticline was close to constant at 0.16 mm/yr (0.006 in./yr), which is about an order of magnitude slower than the average rate. During this period, the Yaken anticline was continuously buried by deposition. The shortening rate accelerated by about an order of magnitude to 1.2–1.6 mm/yr (0.047–0.062 in./yr) about 0.16–0.21 Ma, leading to topographic emergence. The Quilitage anticline underwent a similar acceleration and topographic emergence as shown by geomorphic analysis (Hubert-Ferrari et al., 2007). The total shortening rate of the Yaken-Quilitage system was about 0.6 mm/yr (0.023 in./yr) prior to acceleration and about 4–5 mm/yr (0.157–0.196 in./yr) since emergence. Thus, about half of the shortening has occurred in the late Quaternary since emergence.

Shortening of the western Kuqa fold belt is less constrained because of the flow of salt that contributes significantly to structural relief in this part of the Kuqa fold belt. It is likely that the evaporite deformation cannot be correctly restorable by 2-D analysis and is less clearly related to horizontal shortening. However, the thrust fault imaged on the backlimb of the Miskantag anticline shows that the minimum slip of the fault must be

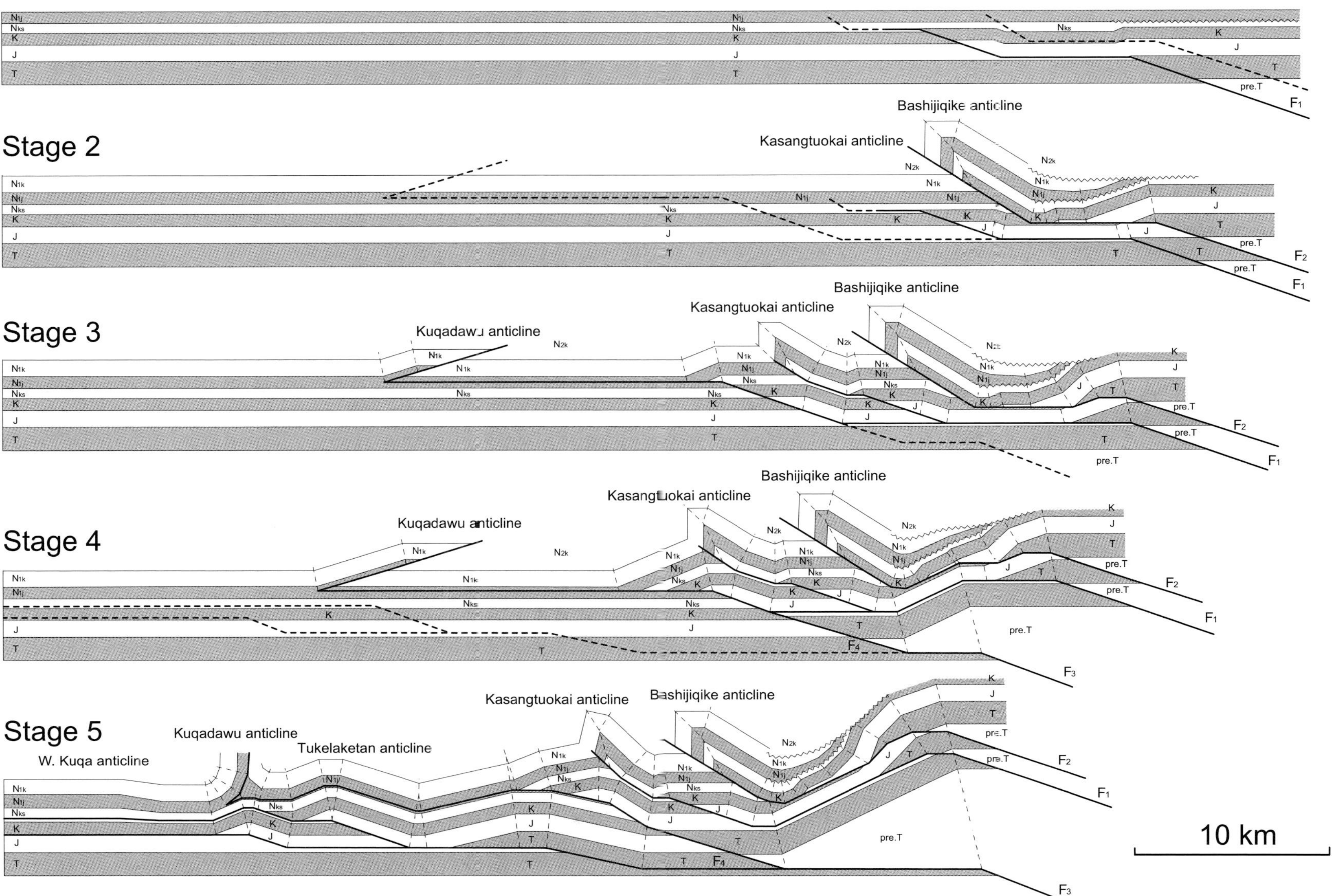

Figure 13. Simplified sequential model of the Kuqa profile (Figure 8) showing the progressive development of Cenozoic structures in the Kuqa fold belt (see Figure 2 for the location of the profile. N_2k = Pliocene–Pleistocene Kuqa formation; N_1k = Miocene Kanchung formation; N_1j = Miocene Jidike formation; Nks = Oligocene–Miocene Kumgeliem and Suweiyi formations; K = Cretaceous; J = Jurassic; T = Triassic; pre.T = pre-Triassic; ant. = anticline; W. = Western; F1–F4 = faults. See the text for discussion.

greater than the preserved length of the hanging-wall sheet, at least 22 km (14 mi) (aa′, Figure 11b). The growth strata, as indicated by the systematic decrease in dip angle as the strata become younger, were deposited during thrusting on the northern limb of the syncline that developed in the hanging wall of this thrust fault (Figure 11b). The oldest age of the growth is within the Kuqa formation (starting 7 Ma ago) placing constraints on the age of this thrust faulting. The minimum shortening rate of this fault, which is the principal structure of the southern part of the western Kuqa fold belt, is estimated to be approximately 3 mm/yr.

The observations from the eastern Kuqa fold belt, as discussed above, of a late Quaternary acceleration shortening rate to about 5 mm/yr (0.196 in./yr) in the frontal structures are in agreement with more regional geodetic measurements of shortening across the southern Tianshan. Wang et al. (2001, their appendix) presented data that indicate a present shortening rate of about 10 mm/yr (0.393 in./yr) between four stations in the Aksu region and two stations in the upper Yili River Basin near Yining (Figure 1), of which approximately 9 mm/yr (0.354 in./yr) is the north–south component and 2–3 mm/yr (0.078–0.118 in./yr) is the eastward component. This present geodetic rate, which corresponds to about 10 km/Ma (6.2 mi/Ma), contrasts strongly with the approximately 25 Ma beginning of deformation in the southern Tianshan and a long-term shortening rate of about 1 km/Ma (0.6 mi/Ma) until the late Quaternary. The present geodetic rate could not have been sustained for 25 Ma because the implied shortening would be excessive.

STRUCTURAL ASPECTS OF PETROLEUM POTENTIAL

Numerous gas fields have been discovered in Kuqa fold belt anticlines, including the Kela gas field, which is a relatively deep pop-up structure developed underlying the Kasangtuokai anticline in the central Kuqa fold belt; the Dabei gas field, which is a deeper structure beneath the Tuzimaza anticline in the western Kuqa fold belt; and the Dina gas field, which is a relatively deep fold underlying the shallow fold of the eastern Quilitage anticline. Based in part on the interpretations presented here, the Tuzimaza, Kasangtuokai, and eastern Quilitage area is considered to possess the most hydrocarbon potential, with relatively deep anticlines underlying shallow thrust sheets being the primary targets for future exploration.

In the Tuzimaza, Kasangtuokai, and eastern Quilitage area, the deeper structure exhibits imbricate thrust sheets beneath the salt wall (diapir). Such complex structures offer substantial and difficult challenges to determine the geometry and kinematics of thrust faults and folds. Some difficult questions such as the effect of the ductile deformation of salt layers in an imbricate system, multiple deformations of thrust faulting and folding, cross sections balancing, and three-dimensional model restoration of complex imbricate thrust and fold with salt cores require further investigation. Different models of the deeper folds would influence any petroleum prospects in the Kuqa fold belt.

DISCUSSION AND CONCLUSIONS

In this chapter, we describe the Cenozoic deformation and structural evolution of the Kuqa fold belt based on 20,000 km (12,000 mi) of 2-D seismic reflection profiles, detailed field measurements, and well data for several tens of exploration wells, which provide a more accurate quantitative picture of the geometry and kinematics of thrusts and folds along the southern piedmont of Tianshan than was available before. We defined the timing and rate of deformation of Miocene, Pliocene, and Pleistocene structures and discussed the tectonic evolution of the Kuqa fold belt.

The Kuqa fold belt is overridden from the north by a thrust system of Paleozoic rocks of the southern Tianshan that developed in the form of the basement thrust sheets and wedges that telescoped and thrust southward along the north-dipping reverse faults. Triassic cover rocks are extensively exposed in the hanging walls of these structures. Mesozoic and Paleogene strata are involved in the southward-dipping monocline that marks the southern limit of the basement duplication in the subsurface.

The Kuqa fold belt is characterized by a series of east-west–striking thrust faults and folds of Mesozoic–Cenozoic rocks. The thrusts are localized at two major decollement horizons with the shallow one located at salt-gypsum layers of the Oligocene Kumgeliem formations in the western part of the Kuqa fold belt and early Miocene Jidike formations in the eastern part of the Kuqa fold belt and the lower decollement in the Middle Jurassic mudstone and coal beds. Break-forward thrusts and duplex structure developed in the northern Kuqa fold belt, just in front of the southern Tianshan. At the surface, fault-propagation folds are exposed and Mesozoic and Tertiary are involved in deformation. Deeper blind thrust faults, formed by break-forward propagation of the thrust sheets consisting of Mesozoic strata, develop underlying the shallow thrust faults. Salt had apparently flowed into the core of the folds. Syntectonic growth strata have been deposited in the late Oligocene (25–26 Ma) and Neogene section.

The 230-km (140-mi)-long Quilitage anticline is a complex structure. The eastern Quilitage anticline shows an imbricate system consisting of a shallow box fold with a wedging core, and a fault-bend fold developed on a deeper thrust ramp. Westward, the Quilitage anticline developed to be cored by the autochthonous Paleogene Kumgeliem salt. Growth strata on the two limbs of the Quilitage anticline have deposited in the Pliocene and Quaternary sections. The Yaken anticline, 10 km (6 mi) south of the Quilitage anticline, is an active fold that began growth at about 5.5 Ma, with beds dipping only a few degrees on both sides of the Holocene surface.

The emplacement of the thrusts and fault-related folds propagated southward gradually and episodically in the Kuqa fold belt. Near the mountain front, Mesozoic and Paleogene strata are involved in the deformation starting at 25–26 Ma. Toward the south, Miocene and Pliocene strata are folded in the Quilitage and Yaken anticlines starting at about 5.5 Ma.

Restoration of balanced cross section suggests that the minimum shortening across the central Kuqa fold belt was more than 16–19 km (10–12 mi). Shortening began locally at approximately 25–26 Ma, yielding a long-term minimum average shortening rate of about 0.6 mm/yr (0.023 in./yr). The starting time of growth of the Quilitage and Yaken anticlines is about 5.5 Ma with a total shortening of about 6 km (3.7 mi), which yields an average shortening rate of about 1.2–1.6 mm/yr (0.047–0.062 in./yr). Studies of growth strata and landscape folding show that slip on the Quilitage-Yaken thrust system accelerated in the late Quaternary to about 4–5 mm/yr (0.157–0.196 in./yr) (Hubert-Ferrari et al., 2007). Calculated shortening rates indicate that the deformation became faster and that the crust underwent massive shortening since the late Pliocene in the Kuqa area. This acceleration is in agreement with more regional present-day geodetic shortening of approximately 10 mm/yr (0.393 in./yr) across the southern Tianshan just west of the Kuche basin, which could not have been sustained for long given the approximately 25–26 Ma beginning of deformation.

ACKNOWLEDGMENTS

This research was funded by the Tarim Oilfield Company of PetroChina, the China National Science Fund (grants 49972077 and 40372090), the U.S. National Science Foundation (grant EAR-0073759), Princeton University 3-D Structure Project, and the Swiss National Science Foundation (grant SNSF20020-101781). This research would have been impossible without the cooperation of the Tarim Oilfield Company that provided seismic and well data and supported our field work. We are particularly grateful to Lu Huafu of Nanjing University and Wang Zhaoming, Pi Xuejun, Li Qiming, Xie Huiwen, Pen Genxin, and Lei Ganlin of Tarim Oilfield Company for their hospitality and support of this work. Suppe is grateful for the hospitality of the Ludwigs-Maximillian University Munich and support of the Alexander Von Humboldt Foundation during the preparation of this manuscript.

REFERENCES CITED

Allen, M. B., B. F. Windley, C. Zhang, Z. Y. Zhao, and G. R. Wang, 1991, Basin evolution within and adjacent to the Tien Shan range, NW China: Journal of the Geological Society (London), v. 148, p. 369–378, doi:10.1144/gsjgs.148.2.0369.

Allen, M. B., S. J. Vincents, and P. J. Wheeler, 1999, Late Cenozoic tectonics of the Kepingtoge thrust zone: Interaction between the Tian Shan and the Tarim Basin, northwest China: Tectonics, v. 18, p. 639–654, doi:10.1029/1999TC900019.

Avouac, J. P., P. Tapponnier, and M. Bai, 1993, Active thrusting and folding along the northern Tien Shan and late Cenozoic rotation of the Tarim relative to Dzungaria and Kazakhstan: Journal of Geophysical Research, v. 98, p. 6755–6804, doi:10.1029/92JB01963.

Brown, E. T., et al., 1998, Estimation of slip rates in the southern Tian Shan using cosmic ray exposure dates of abandoned alluvial fans: Geological Society of America Bulletin, v. 110, p. 377–386, doi:10.1130/0016-7606(1998)110<0377:EOSRIT>2.3.CO;2.

Burchfiel, B. C., E. T. Brown, Q. Deng, F. Xianyue, L. Jun, P. Molnar, S. Jianbang, W. Zhangming, and Y. Huichuan, 1999, Crustal shortening on the margins of the Tienshan, China, Xinjiang: International Geology Review, v. 41, p. 665–700.

Charreau, J., S. Gilder, Y. Chen, S. Dominguez, J.-P. Avouac, S. Sen, M. Jolivet, Y. Li, and W. Wang, 2006, Magnetostratigraphy of the Yaha section, Tarim Basin (China): 11 Ma acceleration in erosion and uplift of the Tian Shan mountains: Geology, v. 34, p. 181–184, doi:10.1130/G22106.1.

Charreau, J., Y. Chen, S. Gilder, and L. Barier, 2008, Comment on "Magnetostratigraphic study of the Kuche depression, Tarim Basin, and Cenozoic uplift of the Tian Shan Range, western China" by B. Huang, J. D. A. Piper, S. Peng, T. Liu, Z. Li, Q. Wang, and R. Zhu, [Earth and Planetary Science Letters, 2006, doi:10.1016/j.espl.2006.09.020]: Earth and Planetary Science Letters, v. 268, p. 325–329.

Gonzalez-Mieres, R., and J. Suppe, 2006, Relief and shortening in detachment folds: Journal of Structural Geology, v. 28, p. 1785–1807, doi:10.1016/j.jsg.2006.07.001.

Gonzalez-Mieres, R., and J. Suppe, 2011, Shortening histories in active detachment folds based on area-of-relief methods, *in* K. McClay, J. H. Shaw, and J. Suppe, eds., Thrust fault-related folding: AAPG Memoir 94, p. 39–67.

Hao, Y., X. Zeng, Y. Qiu, and X. He, 1982, Miocene foraminifera of the Tarim Basin, Xinjiang and their geological significance: Bulletin of the Chinese Academy of Geological Science, v. 4, p. 69–79.

He, D., and J. Suppe, 2005, Guidebook for fieldtrip in south and north Tianshan foreland basin, Xinjiang Uygur Autonimous Region, China: International Conference on Theory and Application of Fault-Related Folding in Foreland Basins: Beijing, PetroChina, 78 p.

Heermance, R. V., J. Chen, D. W. Burbank, and C. Wang, 2007, Chronology and tectonic controls of late Tertiary deposition in the southwestern Tian Shan foreland, NW China: Basin Research, v. 19, p. 599–632, doi:10.1111/j.1365-2117.2007.00339.x.

Hendrix, M. S., S. A. Graham, A. R. Carroll, E. R. Sobel, C. L. McKnight, B. J. Schulein, and Z. Wang, 1992, Sedimentary record and climatic implications of recurrent deformation in the Tianshan: Evidence from Mesozoic strata of the north Tarim, south Junggar, and Turpan basins, northwest China: Geological Society of America Bulletin, v. 104, p. 53–79, doi:10.1130/0016-7606(1992)104<0053:SRACIO>2.3.CO;2.

Hendrix, M. S., T. A. Dumitru, and S. A. Graham, 1994, Late Oligocene–early Miocene unroofing in the Chinese Tian Shan: An early effect of the India-Asia collision: Geology, v. 22, p. 487–490, doi:10.1130/0091-7613(1994)022<0487:LOEMUI>2.3.CO;2.

Huang, B., J. D. A. Piper, S. Peng, T. Liu, Z. Li, Q. Wang, and R. Zhu, 2006, Magnetostratigraphic study of the Kuche depression, Tarim Basin, and Cenozoic uplift of the Tian Shan Range, western China: Earth and Planetary Science Letters, v. 251, p. 346–364.

Huang, B., J. D. A. Piper, and R. Zhu, 2008, Reply to comment by J. Charreau et al. on "Magnetostratigraphic study of the Kuche depression, Tarim Basin, and Cenozoic uplift of the Tian Shan Range, western China" [Earth and Planetary Science Letters, doi:10.1016/j.epsl.2008.01.025]: Earth and Planetary Science Letters, v. 275, p. 404–406, doi:10.1016/j.epsl.2008.06.053.

Hubert-Ferrari, A., J. Suppe, J. Van Der Woerd, X. Wang, and H. Lu, 2005a, Irregular earthquake cycle along the southern Tianshan front, Aksu area, China: Journal of Geophysical Research, v. 108, 18 p., B06402, doi:10.1029/2003JB002603.

Hubert-Ferrari, A., J. Suppe, X. Wang, and C. Jia, 2005b, The Yaken detachment fold, China, *in* J. Shaw, C. Connors, and J. Suppe, eds., Seismic interpretation of contractional fault-related folds: AAPG Studies in Geology 53, p. 110–113.

Hubert-Ferrari, A., J. Suppe, R. Gonzalez-Mieres, and W. Wang, 2007, Mechanisms of active folding of the landscape (southern Tian Shan, China): Journal of Geophysical Research, v. 112, 39 p., B03S09, doi:10.1029/2006JB004362.

Jia, C., 1997, The features of structure and petroleum geology on Tarim Basin of China: Beijing, Petroleum Industry Press, 438 p.

Lu, H., and Q. Luo, 1990, Fossil charophytes from the Tarim Basin, Xinjiang: Beijing, Science and Technology Literature Publishing House, 113 p.

Lu, H., D. Howell, D. Jia, D. Cai, S. Wu, C. Chen, Y. Shi, and Z. C. Valin, 1994, Rejuvenation of the Kuqa foreland basin, northern flank of the Tarim Basin, northwestern China: International Geology Review, v. 36, p. 1151–1158, doi:10.1080/00206819409465509.

Molnar, P., and P. Tapponnier, 1975, Cenozoic tectonics of Asia: Effects on a continental collision: Science, v. 189, p. 419–426, doi:10.1126/science.189.4201.419.

Meng, Z. F., 1999, The Cenozoic magnetic-strata of Kuqa river area, China (in Chinese): Research report of Tarim Oilfield Company, Xinjiang, 120 p.

Peng, S., Z. Li, B. Huang, T. Liu, and Wang, 2006, Magnetostratigraphic study of Cretaceous depositional succession in the northern Kuqa depression, northwest China: Journal Chinese Science Bulletin, v. 51, p. 97–107, doi:10.1007/s11434-005-0340-5.

Poblet, J., K. McClay, F. Storti, and J. A. Munoz, 1997, Geometries of syntectonic sediment associated with single layer detachment fold: Journal of Structural Geology, v. 19, p. 369–381, doi:10.1016/S0191-8141(96)00113-7.

Sobel, R. E., 1995, Basin analysis of the Jurassic–lower Cretaceous southwest Tarim Basin, northwest China: Geological Society of America Bulletin, v. 111, p. 709–724, doi:10.1130/0016-7606(1999)111<0709:BAOTJL>2.3.CO;2.

Sobel, R. E., and T. A. Dumitru, 1997, Thrusting and exhumation around the margins of the western Tarim Basin during the India-Asia collision: Journal of Geophysical Research, v. 102, p. 5043–5063.

Sobel, E. R., J. Chen, and R. V. Heermance, 2006, Late Oligocene–early Miocene initiation of shortening in the southwestern Chinese Tian Shan: Implications for Neogene shortening rate variations: Earth and Planetary Science Letters, v. 247, p. 70–81.

Suppe, J., G. T. Chou, and S. P. Hook, 1991, Rates of folding and faulting determined from growth strata, *in* K. R. McClay, ed., Thrust tectonics: London, Chapman & Hall, p. 105–121.

Suppe, J., F. Sabat, J. A. Muños, J. Poblet, E. Roca, and J. Verges, 1997, Bed-by-bed fold growth by kink-band migration: Sant Llorenç de Morunys, eastern Pyrenees: Journal of Structural Geology, v. 19, p. 443–461, doi:10.1016/S0191-8141(96)00103-4.

Suss, M. P., and J. H. Shaw, 2003, P wave seismic velocity structure derived from sonic logs and industry reflection data in the Los Angeles Basin, California: Journal of Geophysical Research, v. 108, p. 2170, doi:10.1029/2001JB001628.

Tapponnier, P., and P. Molnar, 1979, Active faulting and Cenozoic tectonics of the Tien Shan, Mongolia and Baykal regions: Journal of Geophysical Research, v. 84, p. 3425–3459, doi:10.1029/JB084iB07p03425.

Wang, Q., P. Z. Zhang, J. T. Bilham, R. Larson, K. M. Lai, X. Wu, J. Y. Li, J. Liu, Z. Yang, and Q. Chen, 2001, Present-day crustal deformation in Chana constrained by global positioning system measurements: Science, v. 294, p. 571–577, doi:10.1126/science.1063647.

Ye, C. H., and R. J. Huang, 1990, Tertiary stratigraphy of the Tarim Basin, *in* Z. Zhou and P. Chen, eds., Stratigraphy of the Tarim Basin: Beijing, Sciences Press, p. 308–363.

Yin, A., S. Nie, P. Craig, T. M. Harrison, F. J. Ryerson, X. L. Qian, and G. Yang, 1998, Late Cenozoic tectonic evolution of the southern Chinese Tian Shan: Tectonics, v. 17, p. 1–27, doi:10.1029/97TC03140.

Zhong, D., and W. S. Xia, 1998a, The investigation report of Mesozoic–Cenozoic strata, structure, sedimentary faces, and petroleum potential of Kuche foreland basin outcrop area (with 1:100,000 Kuche foreland basin geologic map) (in Chinese with English abstract): Research report of Tarim Oilfield Company, Xinjiang, 462 p.

Zhong, D., and W. S. Xia, 1998b, The investigation report of Mesozoic–Cenozoic strata, structure, sedimentary faces, and petroleum potential of Kuche foreland basin outcrop area (in Chinese): Research report of Tarim Oilfield Company, Xinjiang, supplemental plates, 156 p.

11

Kraemer, Pablo, José Silvestro, Federico Achilli, and Walter Brinkworth, 2011, Kinematics of a hybrid thick-thin-skinned fold and thrust belt recorded in Neogene syntectonic wedge-top basins, southern central Andes between 35° and 36°S, Malargüe, Argentina, *in* K. McClay, J. H. Shaw, and J. Suppe, eds., Thrust fault-related folding: AAPG Memoir 94, p. 245–270.

Kinematics of a Hybrid Thick-thin-skinned Fold and Thrust Belt Recorded in Neogene Syntectonic Wedge-top Basins, Southern Central Andes Between 35° and 36°S, Malargüe, Argentina

Pablo Kraemer
Wintershall Noordzee B.V., Rijswijk, Netherlands

José Silvestro
YPF S. A., Buenos Aires, Argentina

Federico Achilli
Total E&P Libye, Tripoli, Libya

Walter Brinkworth
Total Austral, Buenos Aires, Argentina

ABSTRACT

In this chapter, we study the geometry and evolution of the Andean fold and thrust belt located between 35° and 36°S. The main goal of this chapter is twofold; the first is to propose a new Neogene chronostratigraphic scheme based on outcrop studies and a new 39Ar/40Ar and K-Ar dating of volcanic intervals. The second is to integrate the structural kinematics deduced from the age and internal architecture of Neogene strata with two balanced cross sections to define the main pulses of fold-belt activity during the Miocene and Pliocene. Fold-belt kinematics is interpreted applying the Coulomb critical wedge model, and a succession of at least two cycles of activity is proposed. The hybrid thick-thin-skinned structural style of the fold belt is characterized by deep basement structures transferring shortening to the sedimentary cover, which is deformed into forethrust, back thrusts, and duplexes complicated by local salt tectonics. A net shortening of 30 km (19 mi) was calculated via restoration of two balanced cross sections. Clastic, pyroclastic, and volcanic synorogenic deposits form two megasequences called the Pincheira-Ventana of Miocene age (16–7 Ma) and the Malargüe of Pliocene age (5–1 Ma) composed of four tectonosequences, S1, S2, S3, and S4, respectively. Active structures acted as barriers bounding local independent synorogenic depocenters as is shown by palinspastic and paleogeographic

DOI:10.1306/13251340M943099

reconstructions. Two main pulses of deformation and a general foreland migration of deformation are evidenced by the spatial and temporal arrangement of synorogenic depocenters. Fold-belt kinematics is interpreted as a succession of two cycles of activity applying the Coulomb critical wedge model. Each cycle included a supercritical stage with out-of-sequence reactivation of basement structures and foreland migration of deformation followed by a subcritical stage with relative low rates of structural uplift and consequent high relative depositional rates in synorogenic basins. The results of this work agree with a progressive north to south deactivation of the Neogene deformation front between 33 and 37° being still active to the north but fossilized during the Pliocene and Miocene in the central and southern parts, respectively.

INTRODUCTION

The foreland deformation of the active margin of South America is characterized by strong along-strike changes in structural style and a variable eastward penetration of the Neogene deformation front (Figure 1). The chronology of deformation has been widely studied during recent years and suggests that the Neogene deformation front is actively deforming north of 34° but is mostly inactive south of 37°. The intermediate region between 34° and 37° has been comparatively less studied in terms of a detailed chronology of events, and this is the main focus of this chapter.

The studied area is located between 35° and 36° and combines the presence of excellent outcrops of Neogene deposits with abundant subsurface information generated by the extensive hydrocarbon exploration of the fold belt. The main goal of this chapter is twofold; the first is to analyze the stratigraphy of Neogene deposits to propose a new chronostratigraphic scheme based on outcrop studies and a new 39Ar/ 40Ar and K-Ar dating of volcanic intervals. The second is to integrate the structural kinematics deduced from the age and internal architecture of Neogene strata with two balanced cross sections to define the palegeography and chronology of the two main pulses of fold-belt activity during the Miocene and Pliocene.

TECTONIC SETTING

The major tectonic features of the Andes such as morphostructural changes and volcanism distribution define two major segments north and south of 33° (Jordan et al., 1983). This tectonic segmentation has been explained as the consequence of the change in subduction angle of the Nazca plate from a low angle north of 33° to a normal higher angle south of 34° (Baranzaghi and Isacks, 1976; Cahill and Isacks, 1992) (Figure 1).

To the north of 33° on the segment of horizontal subduction, the main morphostructural units include from east to west the Pampean ranges, Precordillera, Frontal Cordillera, and Cordillera principal, and the Neogene deformation front registers neotectonic and seismic activity (Jordan et al., 1983; Ramos, 1999). In the transitional zone between 33° and 34°, only the Frontal Cordillera and Cordillera principal are developed (Giambiagi et al., 2003) due to the plunge of the Pampean ranges and Precordillera toward the south. South of 34°, the Cordillera principal is the main morphostructural unit of the central Andes, and the passage to a normal subduction angle is accompanied by a westward retreat and progressive deactivation of the Neogene deformation front. The Malargüe region between 35° and 36° is located in the southern tip of the Cordillera principal (Yrigoyen, 1972) and shares a common stratigraphy and structural style. The main stratigraphic units include a Permian–Triassic basement associated to normal faulting, covered by Jurassic, Cretaceous, and Tertiary syntectonic deposits related to the Andean orogeny (Groeber, 1946; Kozlowski et al.,1989; Kozlowski et al., 1993).

The structure of the fold belt is characterized by a hybrid style with basement thrust sheets that transfer the shortening to the sedimentary cover, which is deformed in a typical thin-skinned style with well-developed triangular zones associated to halokinetic processes. The Tertiary syntectonic deposits were accumulated along a regional syncline axis, and the age and internal architecture of this interval registered the kinematics of the surrounding structures (Kozlowski et al., 1993; Kraemer and Zulliger, 1994; Manceda and Figueroa, 1995; Nocioni, 1996; Kraemer et al., 2000) (Figure 1).

MESOZOIC STRATIGRAPHY

The Mesozoic stratigraphy with its several detachment levels and the early normal faults produced during the Late Triassic–Early Jurassic rifting stage exerted a fundamental influence on the structural style and kinematic evolution of the fold belt. For such reason, a correlation panel of the Mesozoic interval was defined based on three stratigraphic sections and well data (Figure 2A).

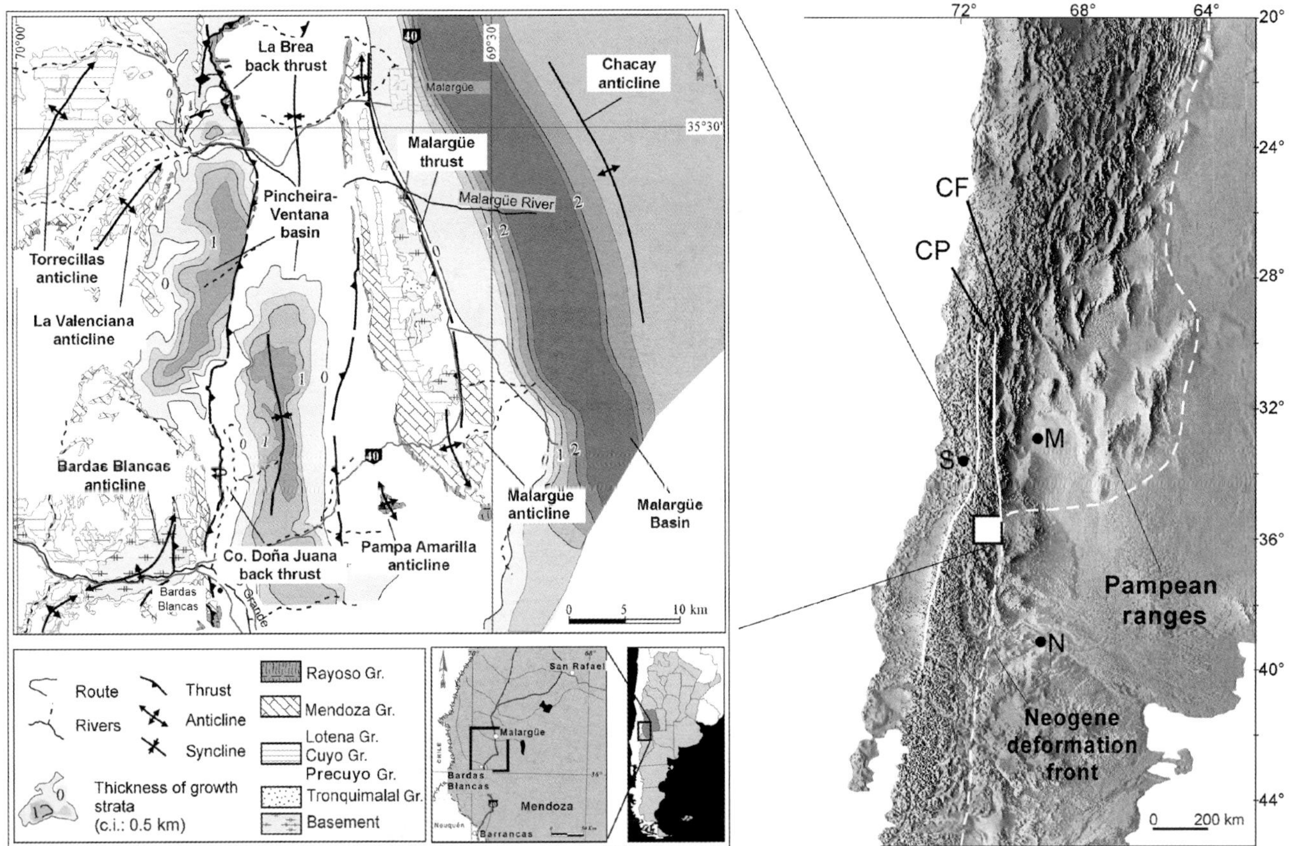

Figure 1. The studied area is located at the southern prolongation of the Cordillera principal (CP) between 35 and 36°, where the Neogene deformation front abruptly migrates toward the west. This coincides with the disparition of the Pampean ranges and Frontal Cordillera (CF) south of 33°. The main structural features of the studied area include the Torrecillas, La Valenciana, Bardas Blancas, and Malargüe basement folds and the La Brea-Dona Juana back thrust that bounds two syntectonic Pincheira-Ventana and Malargüe basins. c.i. = contour interval; Gr. = Group; Co. = Cerra (hill); M = Mendoza (city); S = Santiago (city); N = Neuquén.

Basement Rocks

In the studied area, the Permian–Triassic Choiyoi Group is the oldest unit and the one with greater relative competence; therefore, it is considered the structural and stratigraphic basement of the region. The volcanic breccias, tuffs, and andesite layers of this unit outcrop along the core of the main basement anticlines of the fold belt.

Synrift Sequences (Late Triassic–Early Jurassic)

The oldest sedimentary unit outcrops along the core of the Malargüe anticline and includes coarse conglomerates of the Chihuiu formation and black shales and sandstones of the Llantenes formation. These two fluviolacustrine formations attain a thickness of 450 m (1476 ft) and constitute the Tronquimalal Group that was deposited in local depocenters produced by Triassic half grabens (Figure 2B, C) (Artabe et al., 1998).

The Jurassic Precuyo Group crops out along the Malargüe anticline and rests unconformably over the Triassic and Early Jurassic, reaching a local thickness of 235 m (771 ft). It is composed of siltstone and reddish claystones with interbedded conglomerates and coarse tuffaceous sandstones. In the Bardas Blancas anticline, the synrift sequences show a reduced thickness of 45 m (148 ft) and rest unconformably over the Triassic units with a similar lithology.

The clastic marine deposits of the Cuyo Group reach 150 m (492 ft) in the studied area (Gulisano and Gutierrez Pleimling, 1995). In the Bardas Blancas anticline, the section includes sandstones with hummocky cross stratification and some bioclastic levels, whereas toward the Malargüe anticline, the conglomerate fraction increases. Along the eastern flank of the Torrecillas anticline, the Cuyo Group thickens up to 900 m (2953 ft) compared to the Bardas Blancas and Malargüe anticlines (Figure 2A). The basal levels are conglomerates and sandstones that gradually pass into a fining-upward sequence of sandstones

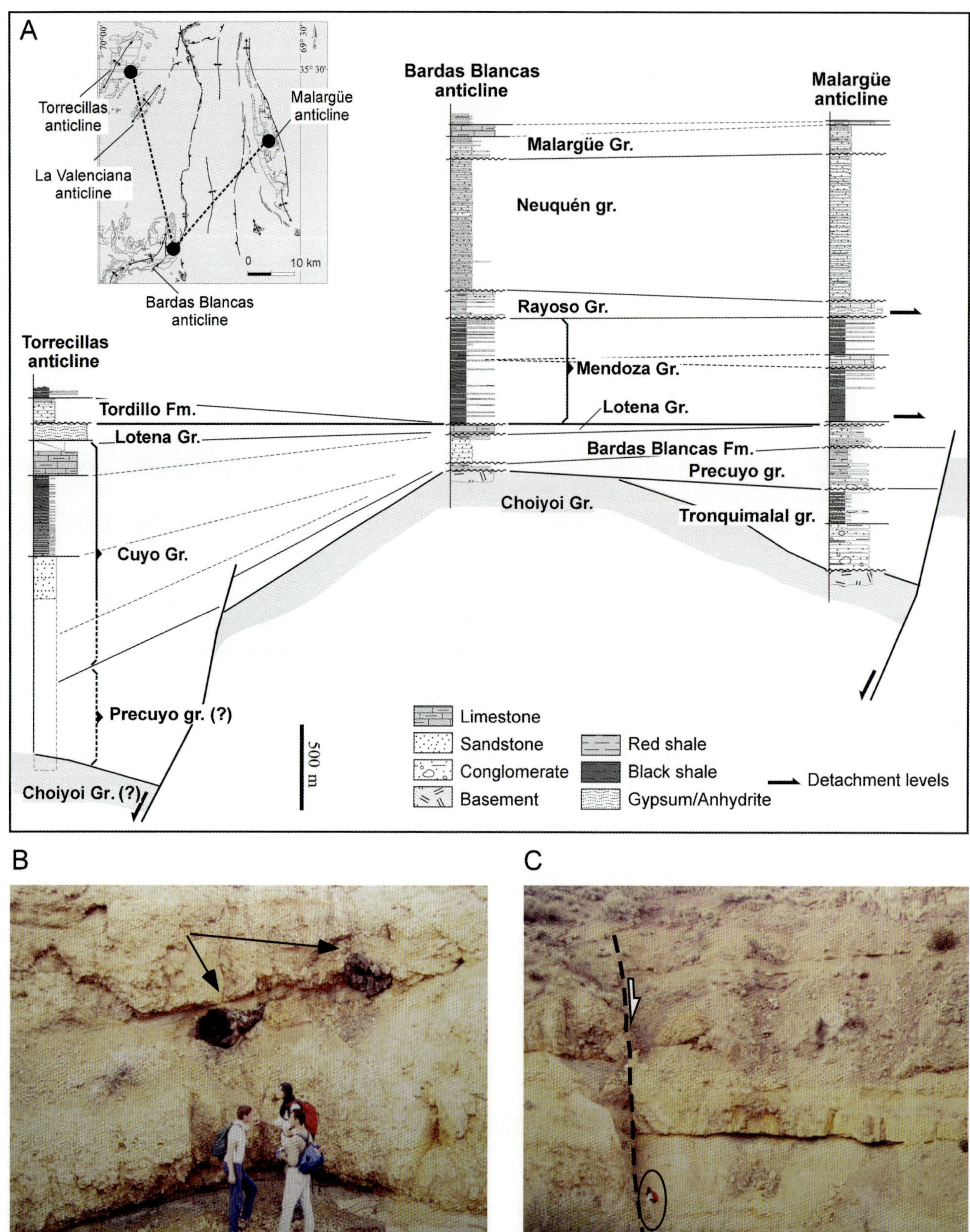

Figure 2. Stratigraphic panel showing the main facies and thickness changes of strata deposited in Triassic and Jurassic half grabens. (A) Triassic continental deposits with carbonized trunk remains outcropping in the core of the Malargüe anticline; (B) preserved normal faults originated during the early rifting stage; (C) (person inside the circle as reference). Gr. = Group; Fm. = Formation.

and an upper level of black shales covered by limestone and gypsum. The thickness increase of these units evidences the existence of two pulses of rifting and half-graben formation during the Triassic and Early Jurassic represented by the Malargüe and Torrecillas half grabens, respectively (Manceda and Figueroa, 1995).

Postrift and Sag Sequences (Late Jurassic, Cretaceous, and Paleogene)

The basal layers of the Lotena Group are represented by conglomerates in the Torrecillas anticline and covered by 55 m (180 ft) of limestones in the Bardas Blancas anticline. In the Malargüe anticline, however, the conglomerates and limestones were not deposited but some sparse evaporites of the Auquilco Formation crop out at the southern culmination of the anticline. This variable thickness and facies distribution suggest that the Torrecillas half graben was still subsident during this time.

The Mendoza Group begins with basal conglomerates of the Tordillo Formation that reaches a maximum thickness in the Torrecillas half graben but thins laterally toward the Bardas Blancas and Malargüe anticlines. The overlying unit is composed of 305 m (1001 ft) of black shales with an upward increase in carbonates, covered by 75 m (246 ft) of bioclastic limestones and 210 m (689 ft) of marls and limestones at the top. The Mendoza Group shows a more uniform thickness that suggests the predominance of long-wavelength thermal subsidence instead of local fault-controlled subsidence evidenced by older units.

The Rayoso Group shows a similar composition either in the Bardas Blancas or the Malargüe anticline and includes 25 to 50 m (82 to 164 ft) of evaporites covered by 110 m (361 ft) of alternating layers of gypsum and green-reddish mudstones and 40 m (131 ft) of gray limestones at the top of the unit.

The Neuquén Group consists of 850 m (2789 ft) of a monotone alternation between red sandstones and mudstones covered in turn by the Malargüe Group, which is composed of sandstones, mudstones, and limestones covered by continental red sandstones with a total thickness of 260 m (853 ft) (Legarreta et al., 1989).

The two evaporitic levels and shales of the Lotena and Mendoza groups are the three main detachment levels identified in the postrift sequences.

TERTIARY SYNTECTONIC STRATIGRAPHY

The Malargüe and Pincheira-Ventana basins are two north-south–trending syntectonic depocenters developed to the east and west of the Malargüe anticline, respectively (Figure 1). Fourteen stratigraphic sections were measured and correlated using pyroclastic bed and radiometric dating of volcanic levels. Surface and seismic stratigraphy complemented with radiometric dating of igneous rocks allowed the defining of four syntectonic sequences, S1 to S4, and one posttectonic sequence, S5. The internal architecture, facies, and thickness changes record the kinematics of active fold-belt structures.

Pincheira-Ventana Basin

This basin is limited to the west by the Torrecillas and Valenciana anticlines and to the east by the Malargüe anticline (Figure 1). The Pincheira-Ventana megasequence comprises two tectonosequences, S1 and S2, and reaches a maximum measured thickness of 1500 m (4921 ft). The La Brea-Dona Juana back thrust constitutes an internal high associated to a strong thickness change in syntectonic sediments that divides the basin into an eastern and a western zone (Figure 3).

Western Zone (Castillos De Pincheira-Cerro Butamallín)

The western zone is developed to the west of the La Brea-Dona Juana back thrusts (Figure 3), and the best outcrops are located along the margins of the Malargüe River where six stratigraphic sections were measured (Figure 4A). Three main lithostratigraphic units were identified, called the Molle, Butaló, and Pincheira formations. The correlation between different sections was achieved following very laterally continuous pyroclastic levels (Figure 4B) and radiometric dating of volcanic beds, and three unconformity-bounded sequences, S1, S2, and S3, were defined (Figure 5) (Kraemer and Zulliger, 1994; Kraemer et al., 2000).

Sequence S1 rests unconformably over a variably eroded Neuquén Group and is covered unconformably by sequence S2. Sequence S1 reaches a thickness of 300 m (984 ft), and the basal section is dominated by andesite pyroclastic breccias (Figure 4A), whereas the upper part consists of tuffaceous paraconglomerates and sandstones. This sequence includes the Mollelitense of Groeber (1946) and the Molle formation of Kozlowski et al. (1987). This sequence thins to the east and west because of basal and internal onlap associated to progressive unconformities. The 40Ar/36Ar dating of basal andesites support a middle Miocene age for S1 (15.8 ± 0.25 Ma, 14.38 ± 0.10 Ma, 14.78 ± 0.24 Ma, and 14.84 ± 0.3 Ma) (Table 1).

Sequence S2 rests unconformably over S1 and reaches a thickness of more than 500 m (1640 ft). In the central part of the basin, the basal levels start with 20 to 30 m (66 to 98 ft) of mudstones with intercalated fine sandstones with plant remains equivalent of the Butaló formation defined by Criado Roque (1950) (Figure 4A). Toward

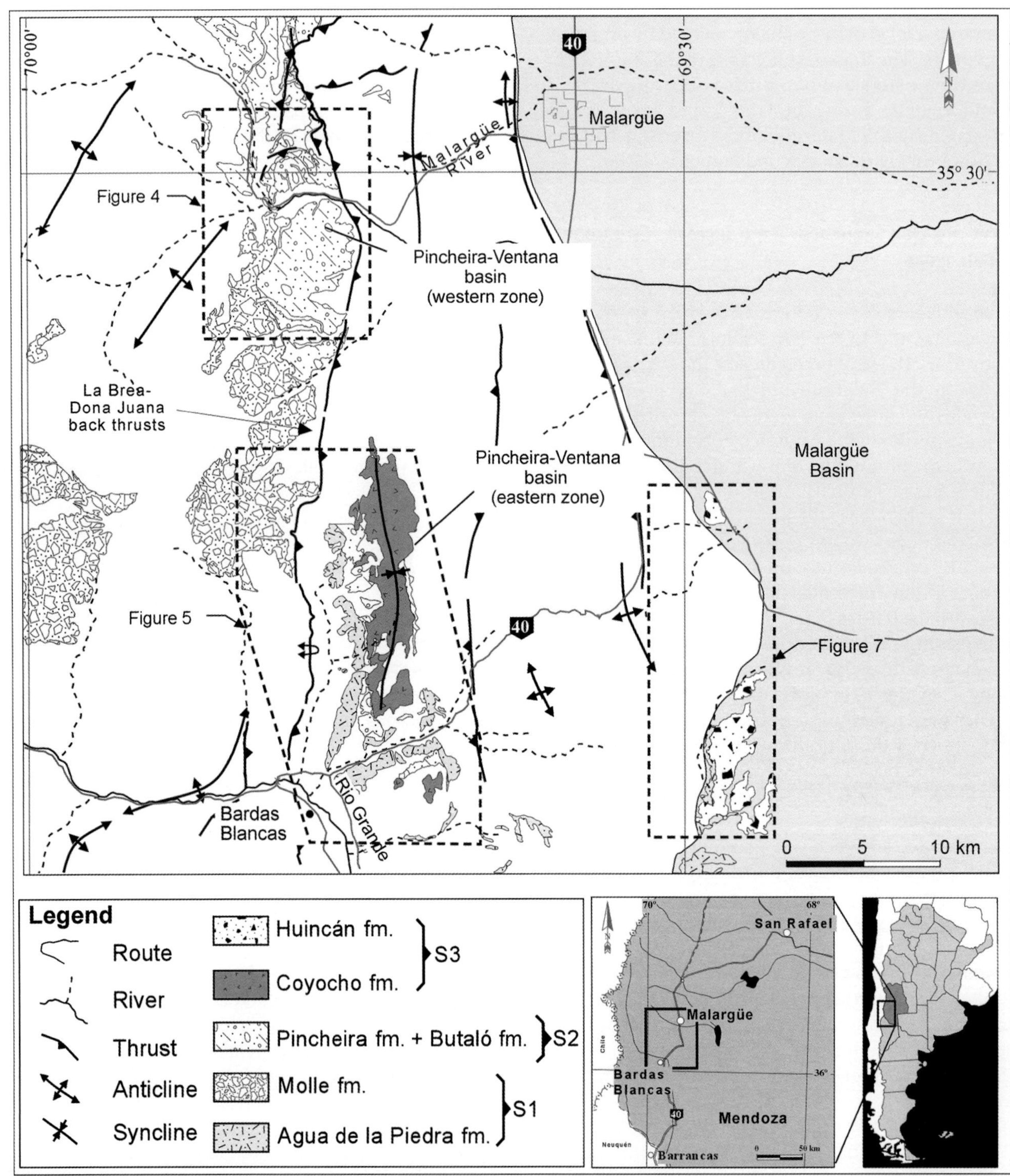

Figure 3. Geologic map of the main outcropping sedimentary sequences recognized at the Pincheira-Ventana Basin to the west and the Malargüe Basin to the east. Insets show the areas where detailed stratigraphic sections were measured. fm. = formation; S1–S3 = sequences.

the basin margins, the basal levels disappear laterally, and S2 is dominated by thick paraconglomerates with volcanic clasts changing into a mixed volcanic-sedimentary clast composition upward. This conglomeratic upper section of S2 was called Estratos de Pincheira by Criado Roque (1950), which we call here the Pincheira formation (Figure 4A), and measured paleocurrents indicate a source located to the south-southwest for this interval.

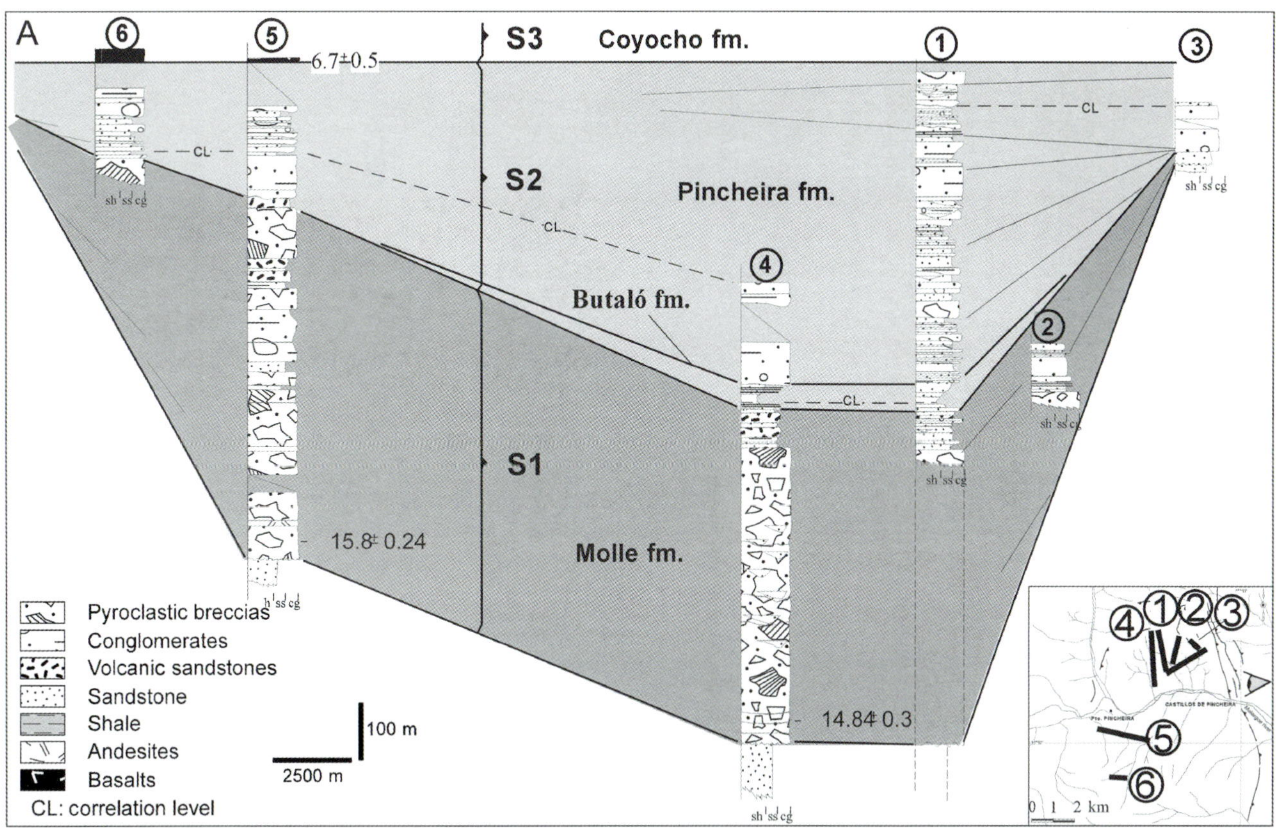

B

Figure 4. (A) Stratigraphic correlation of the western Pincheira-Ventana Basin showing the main lithostratigraphic units called the Molle, Butaló, and Pincheira formations and correlation levels (CL). The Pincheira-Ventana megasequence is formed by two sequences, S1 and S2, and radiometric dating of basal and overlying sequences defines a middle–upper Miocene age for this interval. (B) View toward the west of the typical outcrops of the Pincheira and Molle formations dipping to the southwest at the Castillos de Pincheira location (picture view direction indicated by the eye in the location map; see the location of the Pincheira-Ventana Basin in Figure 3). Mountain area depicted in photo is approximately 1.5 km (.9 mi) wide. fm. = formation; sh ss cg = shale-sandstone-conglomerate; S1–S3 = sequences.

The outcrops along the eastern margin of the basin next to the La Brea-Dona Juana back thrusts show a sudden increase in clast and slumped blocks of sedimentary composition, suggesting a local eastern backthrust sheet source for this upper section (Figure 6A). Sequence S2 is unconformably covered by upper Miocene–Pliocene

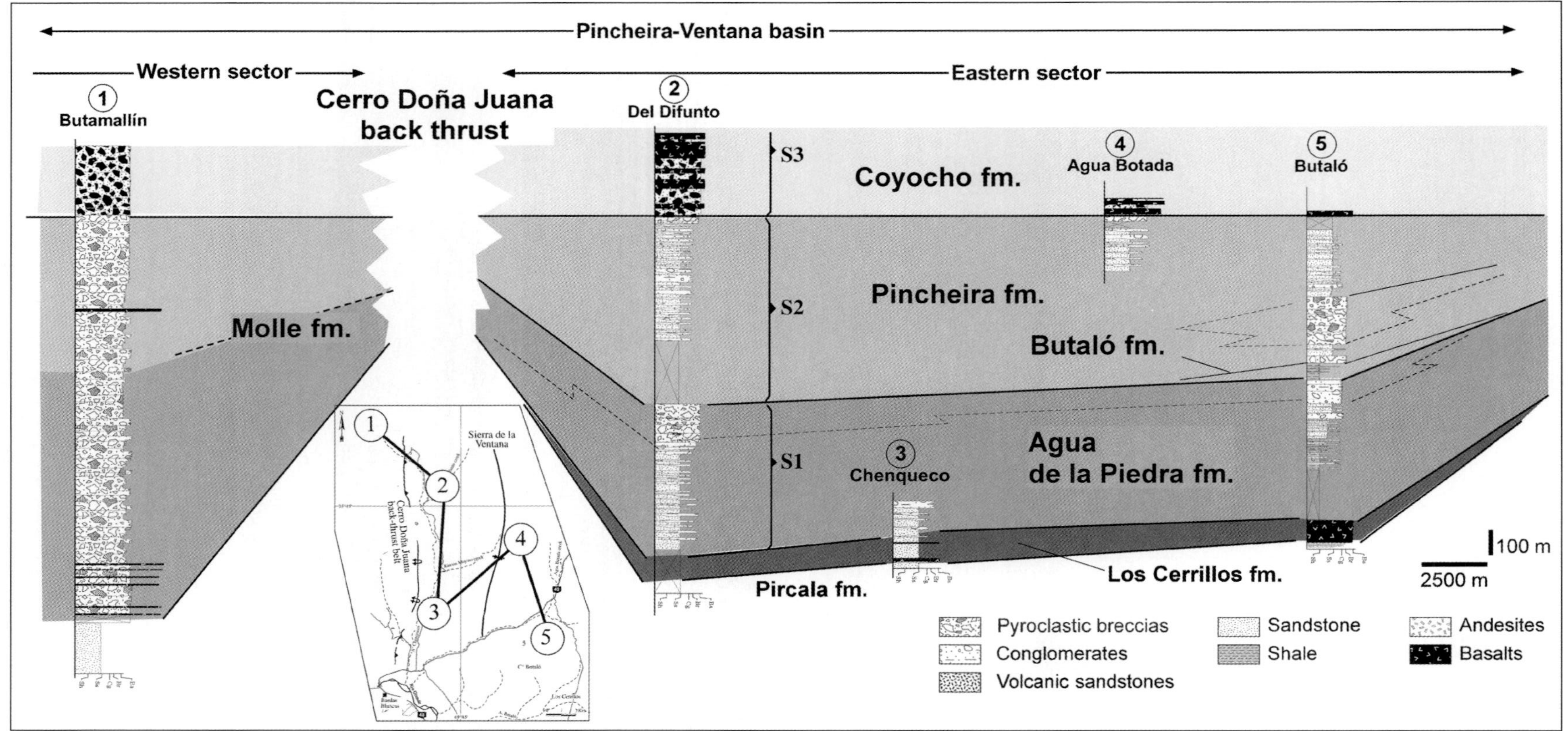

Figure 5. Stratigraphic sections measured at the southern extension of the Pincheira-Ventana Basin. The Pincheira-Ventana megasequence includes two sequences, S1 and S2, and the Malargüe megasequence includes only S3 (see the location in Figure 3). S1–S3 = sequences.

Table 1. Ar/Ar isotopic determinations.

Formation	Location	Analyzed Material	Step	^{39}Ar (%)	Plateau Age (Ma)	(^{40}Ar/^{36}Ar)i	Isocrone Age (Ma)	MSWD*	±2?
Molle	Castillos de Pincheira	Plagioclase				366.3 ± 4.8	14.38**	7.33	0.10
Molle	Castillos de Pincheira	Plagioclase		75.5	14.78**				0.24
Molle	Castillos de Pincheira	Plagioclase		74.4	15.18**				0.25
Molle	Castillos de Pincheira	Plagioclase		100.0	14.84**				0.30

*Mean sum weighted deviates.
**New Mexico Geochronological Research Laboratory.

basalts, breccias, and tuffs with K-Ar ages of 6.7 ± 0.5 Ma (Linares, 2001) belonging to the Malargüe megasequence.

Eastern Zone (Sierra De La Ventana-Cerro Butaló)

The eastern zone is developed to the southeast of the La Brea-Cerro Doña Juana back thrust (Figure 3). Five stratigraphic sections were measured (Figure 5), and three syntectonic sequences were defined, S1 and S2 of the Pincheira-Ventana megasequence and S3 of the Malargüe megasequence.

Sequence S1 rests unconformably over conglomerates, tuffs, and basalts of the Los Cerrillos formation (Groeber, 1946; Kozlowski et al., 1987) and is covered unconformably by the Pincheira formation along the eastern flank of the La Ventana syncline. In the Arroyo del Difunto section (Figure 5), the basal levels of S1 are reddish sandstones and conglomerates with some coarse channeled intervals. In the Cerro Butaló section, the basal section is composed of fine reddish sandstones and mudstones interbedded with light gray sandstones and conglomerates with tuffaceous matrix. In both sections (Figure 5), the clast composition of the basal part shows 50% of volcanic material and a sedimentary fraction with a clear prevalence of red sandstones sourced from the Neuquén Group as well as a minor participation of limestones. The volcanic clast participation increases upward up to 90 to 100% as well as the maximum clast size. The uppermost interval of S1 is composed of nearly 100 m (328 ft) of basaltic breccias with tuffaceous matrix. Paleocurrents have a dominant southwest to northeast flow direction, evidencing a dominant southwest source area during deposition of S1. The reddish conglomerates, sandstones, and mudstones of the lower interval of S1 are equivalent of the Agua de la Piedra formation defined by Criado Roque (1950), whereas the basalt breccias are included in the Molle formation as was defined by Kozlowski et al. (1987).

The basal section of sequence S2 rests unconformably over conglomerates of sequence S1 in the Cerro Butaló section, and it is composed of reddish mudstones with plant remains, bivalves, and gastropods intercalated with fine massive tabular sandstone beds with flow marks. This basal section is covered by massive basaltic agglomerates and breccias and greenish sandstones. In the Arroyo del Difunto section, the whole column is a monotonous succession of 1.5- to 3-m (4.9- to 10-ft)-thick massive coarse conglomerates (Figure 5).

The mudstones and tabular sandstones of the basal part are interpreted as lacustrine low-energy deposits alternating with episodic detritic flows from shoreline zones and are included in the Butaló formation. The agglomerates and breccias are interpreted as detritic flows and hyperconcentrated deposits originated from massive discharge of coarse sediments sourced from the nearest effusive centers and are included in the Molle formation (Kozlowski et al., 1987) (Figure 5).

Basalt layers and breccias of sequence S3 cover unconformably the conglomerates of sequence S2. These rocks are equivalent of the Coyocho formation, and K-Ar dating of S3 basalts yielded ages of 6.7 ± 0.5 Ma (Linares, 2001). Available radiometric dating supports a middle–upper Miocene age for the Pincheira–Ventana megasequence.

Malargüe Basin

This basin is located to the east of the Malargüe anticline and recorded more than 2000 m (6562 ft) of Tertiary sediments (Figure 1). Seismic sections complemented with surface stratigraphic sections allowed to define two syntectonic sequences, S3 and S4, included in the Malargüe megasequence and one posttectonic sequence, S5.

Outcrops

The oldest stratigraphic section of the Malargüe Basin outcrops along the southeast flank of the Malargüe anticline (Figure 3). In the Cerro Morado section located at the southeastern flank of the Malargüe anticline, 860 m (2821 ft) of volcaniclastic sediments was measured (Figure 7). The lower part consists of 140 m (459 ft) of

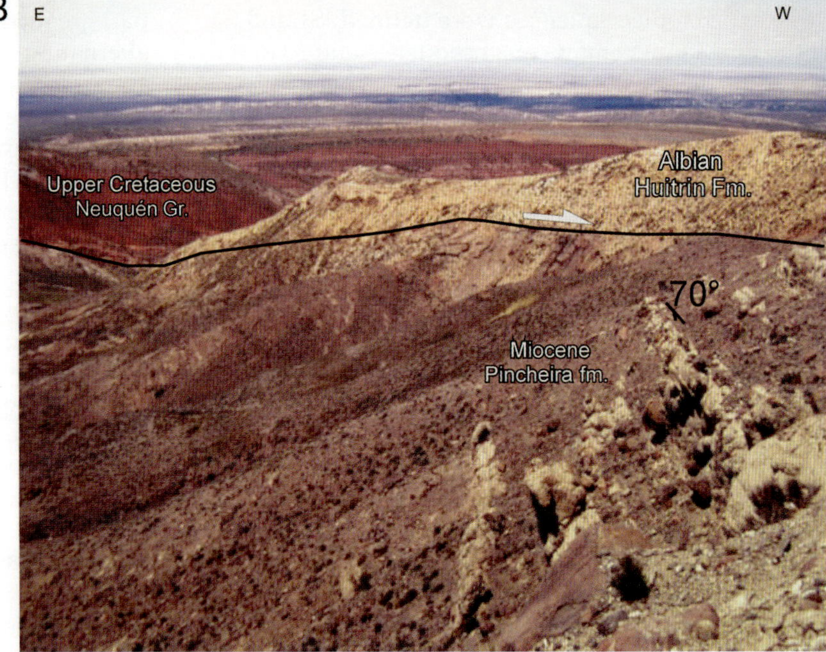

Figure 6. (A) Outcrops of the Miocene Pincheira formation showing upward-decreasing dips typical of growth strata associated to fold-limb rotation produced in this case by subsurface salt movement. The dashed lines outline the Miocene clastic wedges produced by the local erosion of thrust sheets carrying red sandstones of the upper Cretaceous Neuquén Group. (B) The overthrusted evaporitic rocks of the Huitrin Formation and Neuquén Group above the folded Pincheira formation demonstrate a post-Miocene late reactivation of the La Brea-Dona Juana back thrust. fm. = formation; gr. = group.

basaltic layers of the Los Cerrillos formation followed by mudstones, sandstones, and conglomerates of the Agua de la Piedra formation covered by conglomerates and tuffs of the Pincheira formation. This basal section is tentatively correlated with S1 and S2 of the Pincheira-Ventana megasequence until radiometric dating of this level becomes available.

In the Cerro Chachao section, 330 m (1083 ft) of basaltic and andesitic breccias rests unconformably over the pretectonic Malargüe Group. The K-Ar radiometric dating of andesite layers yields an age of 5.04 Ma (Table 2), suggesting the correlation of this section with the late Miocene–early Pliocene Coyocho formation of the Pincheira-Ventana Basin with ages of 6.7 ± 0.5 Ma (Figure 8).

The upper interval comprises 720 m (2362 ft) of basaltic layers, breccias, and conglomerates with maximum block diameters between 0.2 and 0.5 m (0.6 and 1.6 ft) of volcanic composition that evidence the proximity of effusive centers. Based on lithology similitudes, the upper levels of the Cerro Morado and Cerro Chachao sections are correlated to the Coyocho formation of sequence S3 (Figure 7).

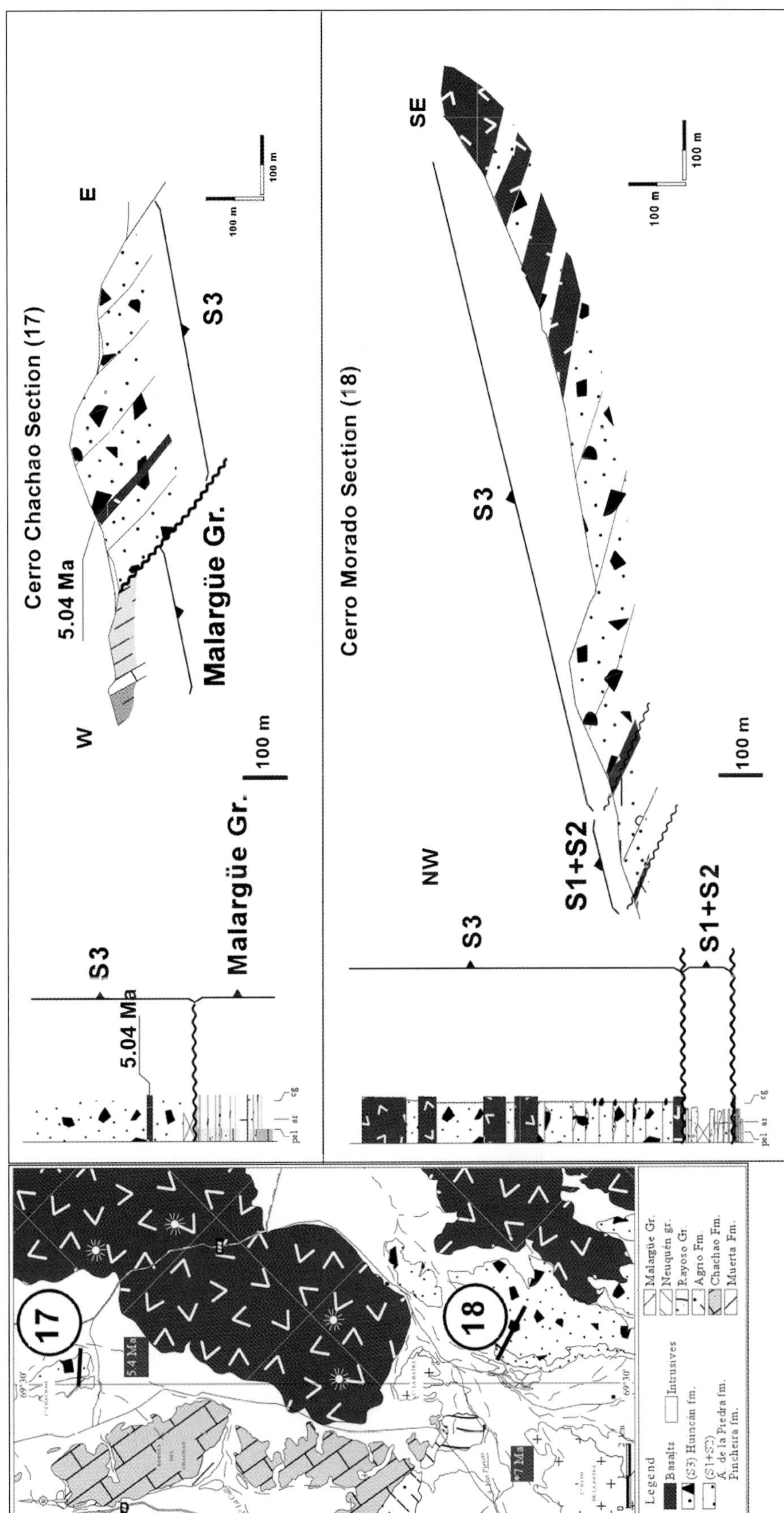

Figure 7. Stratigraphic sections measured along the southeastern flank of the Malargüe anticline showing the results of andesite K-Ar dating (see the location in Figure 3). A = Agua; S1–S3 = sequences.

Table 2. K-Ar isotopic determinations.

Formation	Location	Analyzed Material	K (%)	K error (%)	ccSTP*/g (10^{-6})	Ar^{40}Rad (nL/g)	Total (%)	Ar^{40}Atm (%)	Age (Ma)	Max Error (Ma)
Molle	Arroyo Chenqueco	Total rock	1.2908	0.5000	0.50			48.63	10.04**	0.4
Huincán	Cerro Chachao	Total rock	1.2400			0.2433	47.7		5.04[†]	0.1

*Cubic centimeters standard temperature and pressure.
**Centro de Pesquisas Geocronológicas, Instituto de Geociências, Universidade de Sao Paulo, Brazil.
[†]Institute of Geological and Nuclear Sciences Limited, New Zealand.

Seismic Interpretation

A seismic section located along the Malargüe River shows a lower thin basal interval that is correlated with the Chachao and Cerro Morado sections outcropping along the flank of the Malargüe anticline (Figure 9). This wedgelike interval is bounded at the base by the D1 + D2 unconformity, includes sequences S1 and S2 of the Pincheira-Ventana megasequence, and is bounded at its top by the D3 unconformity. The maximum local thickness ranges between 200 and 300 m (656 and 984 ft) and decreases toward the Malargüe anticline due to the erosion associated to the D3 unconformity (Figure 9).

Sequence S3 is well defined along the western margin of the basin and is bounded by the D3 and D4 unconformities at the base and top, respectively. The lower section of S3 displays wedgelike geometry with basal onlap and divergent reflections at the flank of the Malargüe anticline, suggesting a first pulse of structural growth. Toward the basin center, the lower section shows a plane-parallel arrangement with a paraconcordant relationship with the underlying unit. Sequence S3 develops an internal unconformity with basal onlap that separates the lower and upper sections. The upper section shows the same internal divergent reflection pattern as the lower section, suggesting a second pulse of structural growth of the Malargüe anticline. Sequence S3 reaches a maximum local thickness of 1100 m (3609 ft), and the upper limit is interpreted as an erosive truncation related to the D4 unconformity, which is the base of sequence S4. This later sequence reaches a thickness of 280 m (919 ft), onlaps toward the flank of the anticline, and is subparallel to prograding to the east, suggesting a third pulse of structural anticline growth.

A seismic section located along the eastern margin of the Malargüe Basin shows sequences S3 and S4 folded in a narrowing-upward geometry interpreted as a growth triangle bounded by two axial surfaces (Suppe et al., 1992) related to a basement structure. The last posttectonic depositional sequence, S5, rests unconformably over S4 and is formed by 350 m (1148 ft) of Pleistocene sediments and basalts recognized by low-frequency and high-amplitude seismic reflections (Figure 10).

STRUCTURE

The hybrid structural style that characterizes the Malargüe fold belt is defined by the general association of basement and cover structures developed during the Tertiary. Basement folds are first-order structures with wavelengths between 10 and 15 km (9 mi) and a north–south to northeast–southwest general trend (Figures 1, 8). The thickness increase and lateral facies changes observed in basement fold cores suggest that these folds were nucleated in Triassic–Jurassic half grabens (Figure 2) linked by extensional transfer zones inverted during the Tertiary compression (Manceda and Figueroa, 1995).

When basement sheets are thrusted and emplaced in higher structural levels, the shortening is transferred to thin-skinned structures. At least three main evaporitic and shale decollements with very low frictional resistance favored the development of cover fore- and backthrust and low-angle duplexes related to low frictional salt intervals (Ploszkiewicz and Gorroño, 1988; Manceda and Figueroa, 1995; Kozlowski et al., 1997).

La Valenciana and Torrecillas anticlines are interpreted as inverted half grabens where the basement blocks are thrusted over the sedimentary cover following the Huitrin Formation evaporites. Shortening transferred from basement to cover levels produces a La Brea-Dona Juana triangular zone, which is characterized by duplex structures of Agrio Formation limestones, a strong tectokinetic thickening of Huitrin Formation evaporates, and a passive behavior of the Neuquén and Malargüe groups (Figure 8A).

The uplift produced by blind duplexes and salt movement along the triangular zone is recorded in the formation of progressive unconformities, whereas the erosion of emergent Neuquén Group sandstone backthrust sheets sourced the syntectonic detritic wedges preserved in the Pincheira formation (Figure 6A). The late back thrusting of Neuquén Group and Aptian evaporites of the Huitrin Formation thrust sheets over folded conglomerates of the Pincheira formation (Figure 6B) confirms the continued structural reactivation of the triangular zone. The structural relationships evidence that although the period of main structural activity of the La Valenciana

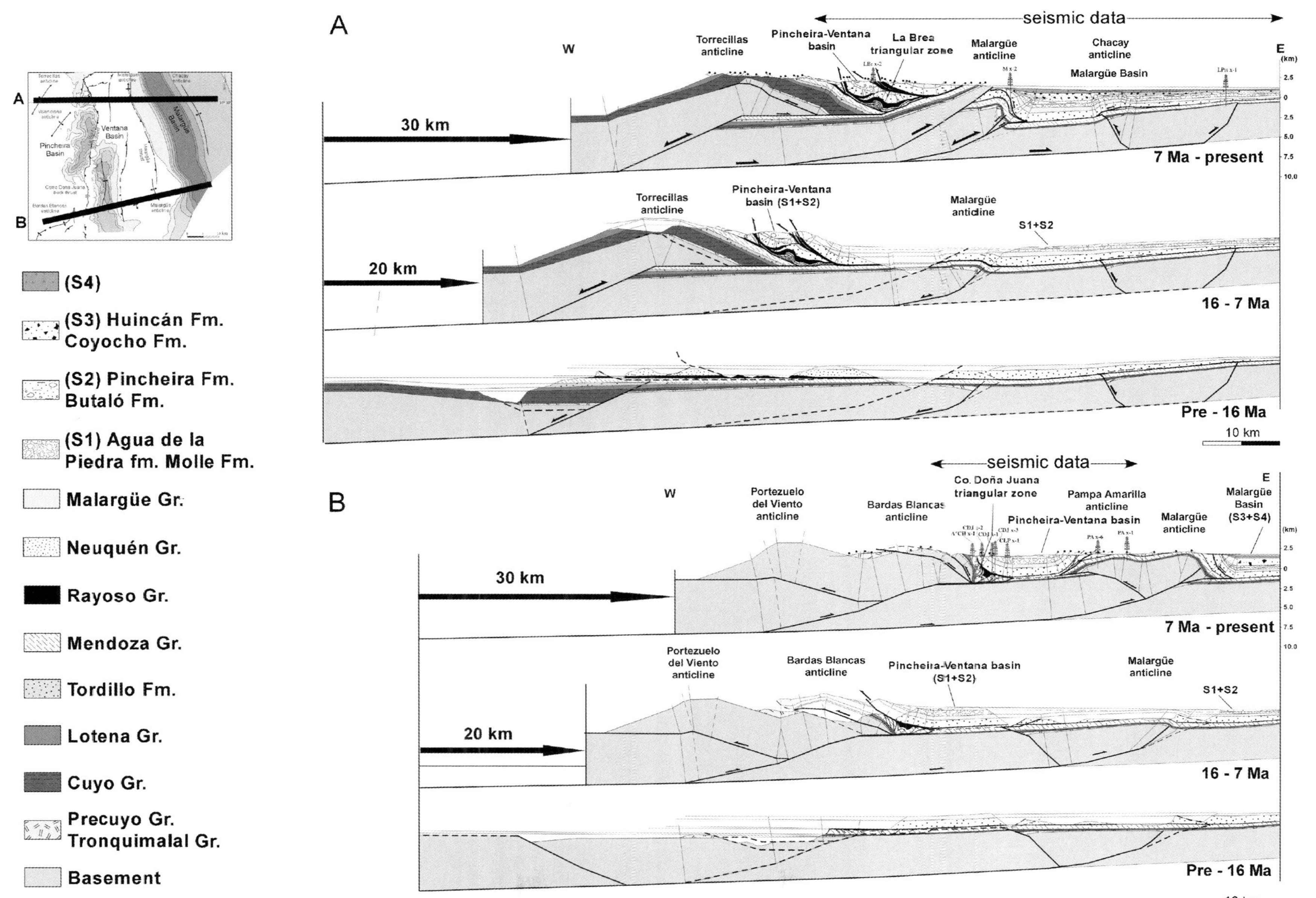

Figure 8. Balanced cross sections along the (A) Malargüe River and (B) Bardas Blancas. The restoration shows the main evolutionary stages of the fold belt during the Miocene and Pliocene, and the syntectonic perched basin developed during each stage typical of subaerial wedge-top basins. Fm. = Formation; Gr. = Group; Co. = Cerro (hill).

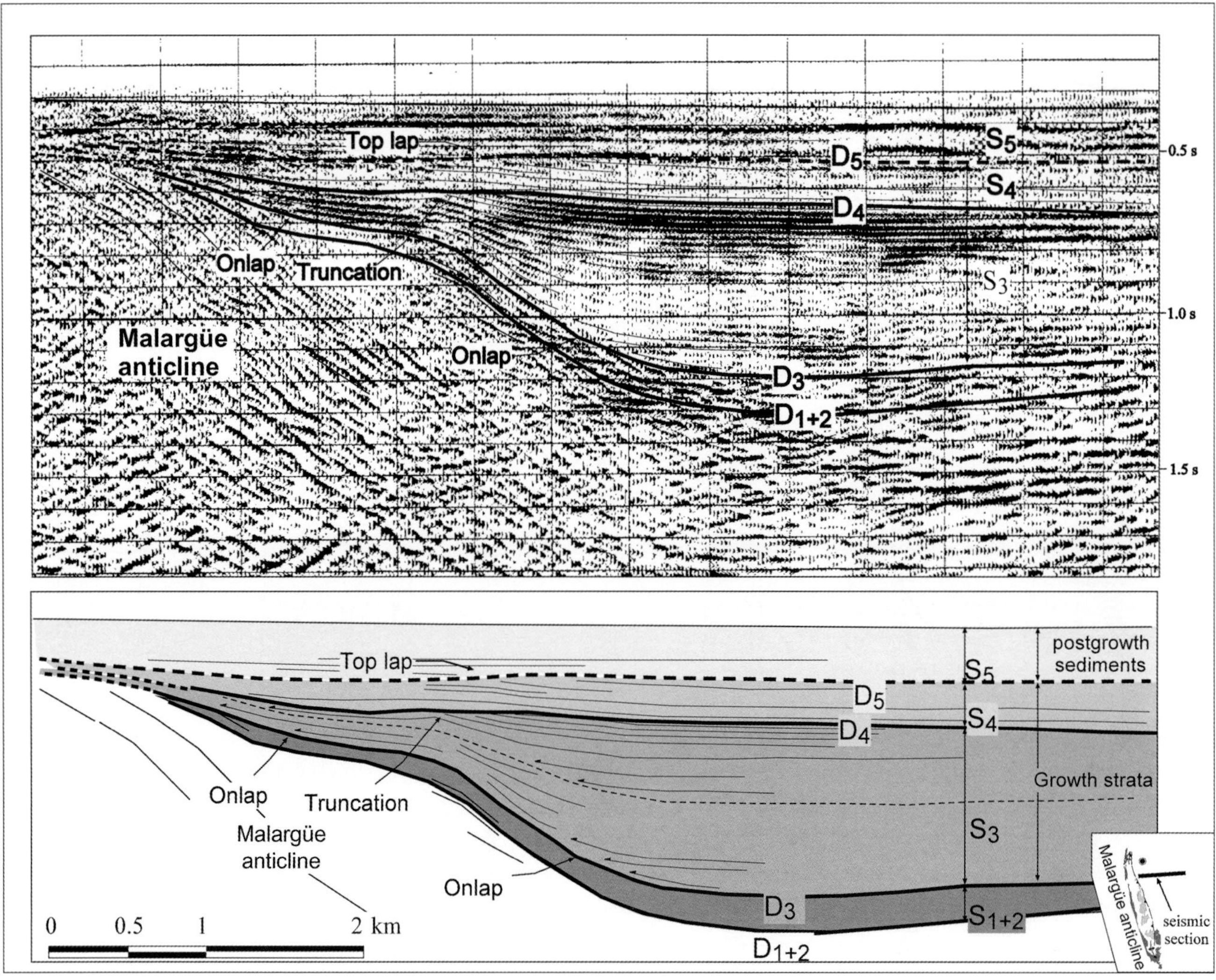

Figure 9. Seismic stratigraphic interpretation of the Malargüe megasequence composed of S3 and S4 growth strata sequences deposited along the eastern flank of the Malargüe anticline. S1–S5 = sequences; D1–D5 = unconformities.

and Torrecillas anticlines and La Brea-Dona Juana triangular zone was synchronous with deposition of the middle–upper Miocene Pincheira megasequence, the activity continued during the Pliocene.

The presence of Triassic sediments of the Tronquimalal Group associated to normal faults in the core of the Malargüe anticline (Figure 2B, C) suggests that this structure was produced by the inversion of a Triassic half graben. The integration of surface and seismic data allows interpreting the Malargüe anticline as formed by an anticline and an out-of-sequence thrust. The basement fold is well exposed in the surface toward the southern plunge of the anticline (Figure 1), whereas in the central and northern zones, the Malargüe thrust sheet is well developed and the anticline is buried under Tertiary deposits and only recognized in seismic data. Growth strata age and geometry show that the Malargüe anticline was active during the Pliocene, whereas the Malargüe thrust was emplaced during the late Pliocene–early Pleistocene and then fossilized under Pleistocene deposits. The Malargüe anticline is interpreted as a rotated anticline forelimb produced by the inversion of preexisting half grabens that balance the shortening of the entire structure (Mitra and Mount, 1998) that is finally transported and folded by fault-bend folding processes (Figures 8, 11). The shortening of the late episode is transferred to a back thrust that duplicates the Neuquén Group as evidenced by well data along the forelinb of the anticline. The Pampa Amarilla anticline is interpreted as a satellite backthrust structure developed at the southern plunge of the Malargüe anticline. It is interpreted as originated by inversion of an east-dipping normal fault that transfers shortening to the cover producing the Tronquimalal thrust identified by exploratory wells

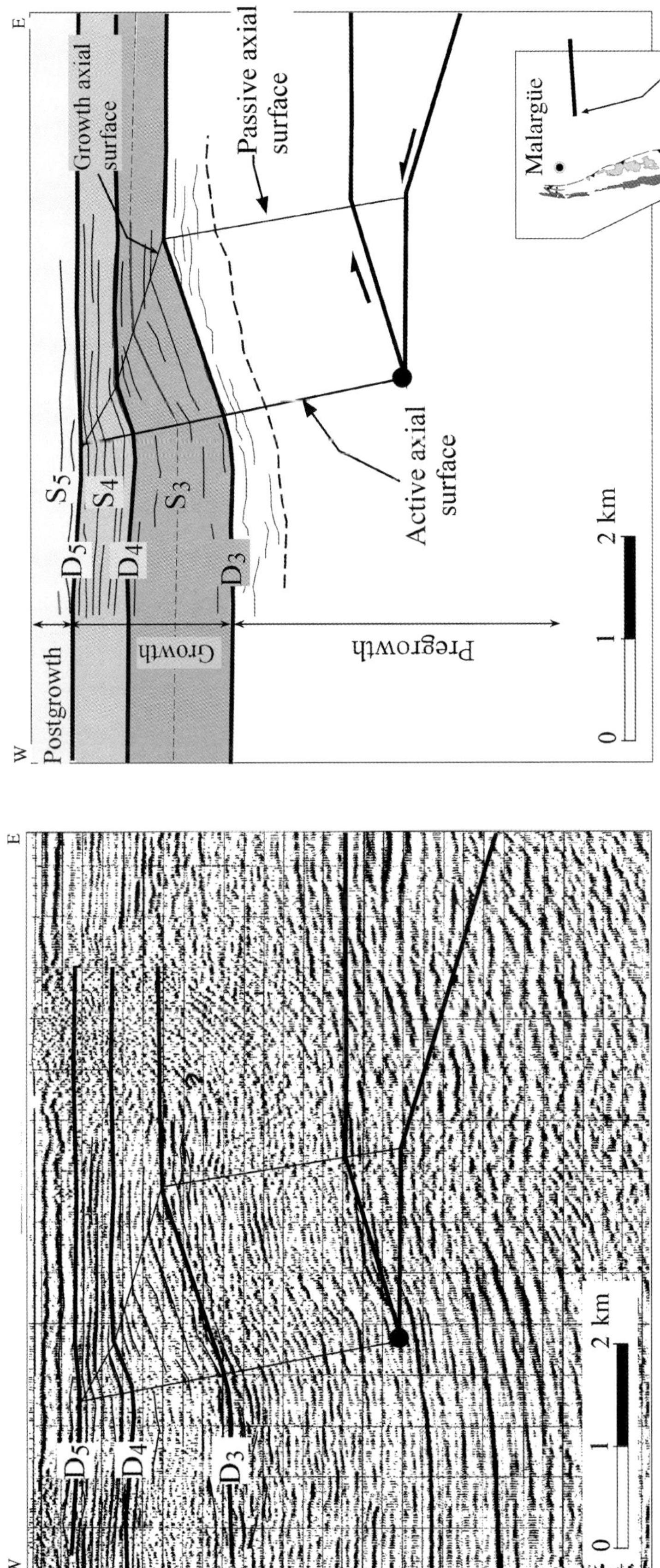

Figure 10. Seismic stratigraphic interpretation of the Malargüe megasequence composed of S3 and S4 growth strata sequences deposited along the eastern flank of the Malargüe anticline. D3–D5 = unconformities; S3–S5 = sequences.

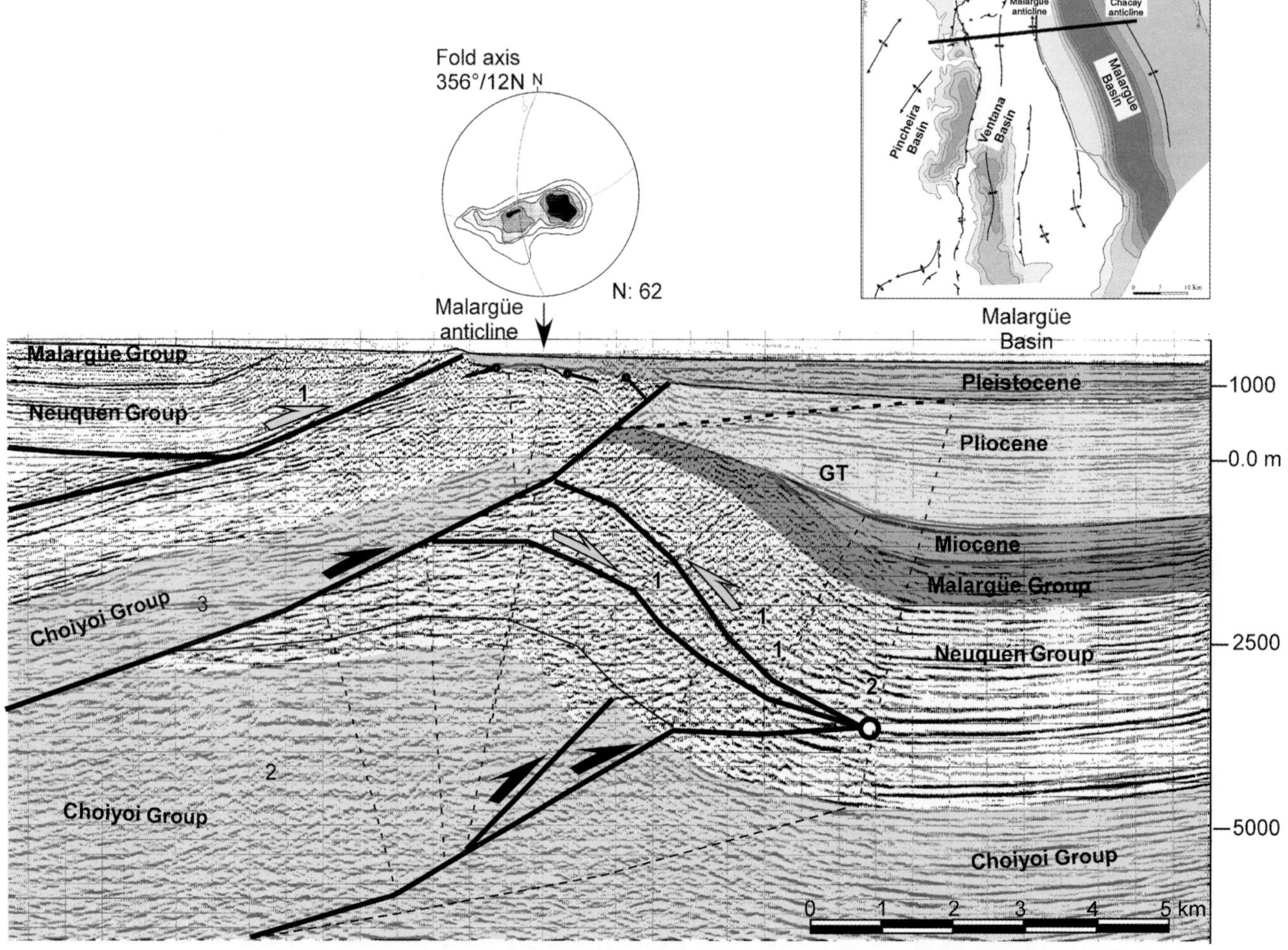

Figure 11. Seismic section showing the main structural features of the Malargüe anticline at the northern plunge of the structure. The stereonet shows the north-plunging axis of the folded Malargüe thrust sheet. The numbers indicate the sequence of structural emplacement that best explains the features observed in the seismic section. (1) Cover backthrust emplacement and folding of the Miocene megasequence; (2) emplacement of the Malargüe anticline basement thrust, backthrust reactivation, and formation of a Pliocene growth triangle (GT); (3) emplacement of the Malargüe out-of-sequence thrust. The unconformable deposition of Pleistocene–Holocene deposits over the Malargüe thrust establishes as late Pliocene–Pleistocene the last pulse of deformation of the fold belt.

(Figure 8). The Pampa Amarilla back thrust is partially consuming the displacement that, at the central and northern Malargüe anticline, is consumed by the Malargüe thrust.

Although the main period of structural activity of the Malargüe anticline is recorded during the Pliocene, an early period of Miocene activity is evidenced by the thinning of the Pincheira-Ventana megasequence against the flanks of the anticline (Figures 9, 11).

The Chacay anticline is interpreted as a half graben bounded by an east-dipping inverted normal fault. The growth triangle developed in Tertiary sediments (Figure 10) is compatible with a basement wedge and a back thrust detached in the upper levels of the Rayoso Group.

The Bardas Blancas anticline is somehow anomalous because it shows a decrease in thickness of the Jurassic sediments instead of an increase like in other anticlines of the studied area (Figure 2). This suggests that this anticline was probably a structural high or accommodation zone during the Jurassic rifting that was reactivated during the Tertiary compression (Manceda and Figueroa, 1995). The total shortening of the structure was distributed into the internal deformation of the basement wedge, and the remaining was transferred to the sedimentary cover that formed the Cerro Dona Juana triangular zone in agreement with the model proposed by Dimieri (1997).

Well data show the development of folds and duplex structures detached in the shales of the Vaca Muerta

Formation and the Rayoso Group developed below 500 m (1640 ft) of deformed evaporates, whereas the Neuquén and Malargüe groups are incorporated passively in the deformation.

When the foreland displacement of the wedge stopped because of facies change at the decollement levels, the shortening is compensated inside the wedge producing duplex structures along the frontal limb of the structure with the consequent steepening of the fold flank and wedge thickening (Figure 8).

The two balanced cross sections allowed palinspastically restoring the paleoposition of the Pincheira-Ventana and Malargüe basins during the Miocene and Pliocene and estimate a total shortening of nearly 30 km (19 mi) for the studied area (Figure 8A, B).

PALEOGEOGRAPHIC EVOLUTION

The Malargüe fold belt shows two main episodes of structural activity recorded in two wedge-top basins of Miocene and Pliocene age that accommodated a sedimentary thickness of more than 2000 m (6562 ft) each.

The palinspastic restoration of two balanced cross sections (Figure 8) allowed estimation of the spatial position of both wedge-top basins during the evolution of the fold belt and consequently allowed performing the repositioning of the stratigraphic sections to create transverse and longitudinal correlation panels (Figures 12, 13) across the basins.

This information was used to construct the paleogeographic maps of each of the sequences identified in the Pincheira-Ventana and Malargüe megasequences (Figure 14).

First Stage: Pincheira-Ventana Megasequence (Middle–upper Miocene)

The main active structures during this period were the Malargüe anticline to the east and the Torrecillas, La Valenciana, and Bardas Blancas anticlines to the west. The La Brea-Dona Juana back thrust acted as an internal high that separated the Pincheira-Ventana Basin into a western and an eastern zone. The minimum measured thickness of the Pincheira-Ventana megasequence is about 1600 m (5249 ft).

During deposition of sequence S1 (Figure 13), the western zone shows a prevalence of volcanic facies of the Molle formation explained by the proximity of volcanic effusive centers and the possibility that these volcanic centers acted as a barrier for the entry of clastic deposits of the Agua de la Piedra formation coming from the west. The fact that the La Brea-Dona Juana internal high shows a decreasing structural relief toward the south implies that it may have acted as an effective barrier for the volcanic facies along its northern part (Figure 12B) but had much less impact on the eastward distribution of clastic deposits of the Agua de la Piedra formation that becomes the dominant facies replacing the Molle formation at the southeastern part of the basin (Figures 12D, 13).

The Malargüe anticline acted as a topographic high separating the Pincheira-Ventana and Malargüe basins and produced a thickness variation of nearly 800 m (2625 ft) between these two basins that can be explained as independent base levels, very steep gradients, or differential subsidence (Figure 12B, D). The observed thinning of this megasequence along the eastern flank of the Malargüe anticline (Figures 9, 11) suggests that this anticline may have acted as a local source of detritus.

Deposition of S2 begins with fine lacustrine facies of the Butaló formation, evidencing a period of restricted volcanic and structural activity (Figure 14). When the volcanic activity restarts, the volcanic facies of the Molle formation are limited to the southwestern corner of the basin. The structural activity resumes as evidenced by the arrival of coarse conglomerates of the Pincheira formation coming from the erosion of the volcanic relief situated to the west (Figure 14). The La Brea-Dona Juana back thrusts are reactivated at the northeastern corner of the basin, and thrust sheets of Neuquén Group sandstones are eroded, producing local clastic wedges intercalated in the upper levels of the Pincheira formation (Figure 6A), whereas toward the south, the structural expression of the back thrust diminishes progressively (Figure 12D). The Butaló formation is represented by two isolated depocenters to the north and south of the La Brea-Dona Juana back thrusts (Figure 14), but the possibility of a single lake deposit eroded by the overlying high-energy deposits of the Pincheira formation cannot be ruled out. The differential thickness of nearly 1200 m (3937 ft) between the Malargüe and Pincheira-Ventana basins suggests that the combined relief of the La Brea-Dona Juana back thrusts to the north and the Malargüe anticline to the south trapped most of the detritic material of the Pincheira-Ventana megasequence along the western depocenters, suggesting the existence of independent base levels during the Miocene (Figure 12B, D). The basin architecture, infill, and relationships with surrounding structures are typical of a wedge-top basin in a subaerial setting as defined by DeCelles and Giles (1996).

Second Stage: Malargüe Megasequence (Pliocene)

During deposition of the Malargüe megasequence, the Malargüe anticline continues its structural uplift, whereas

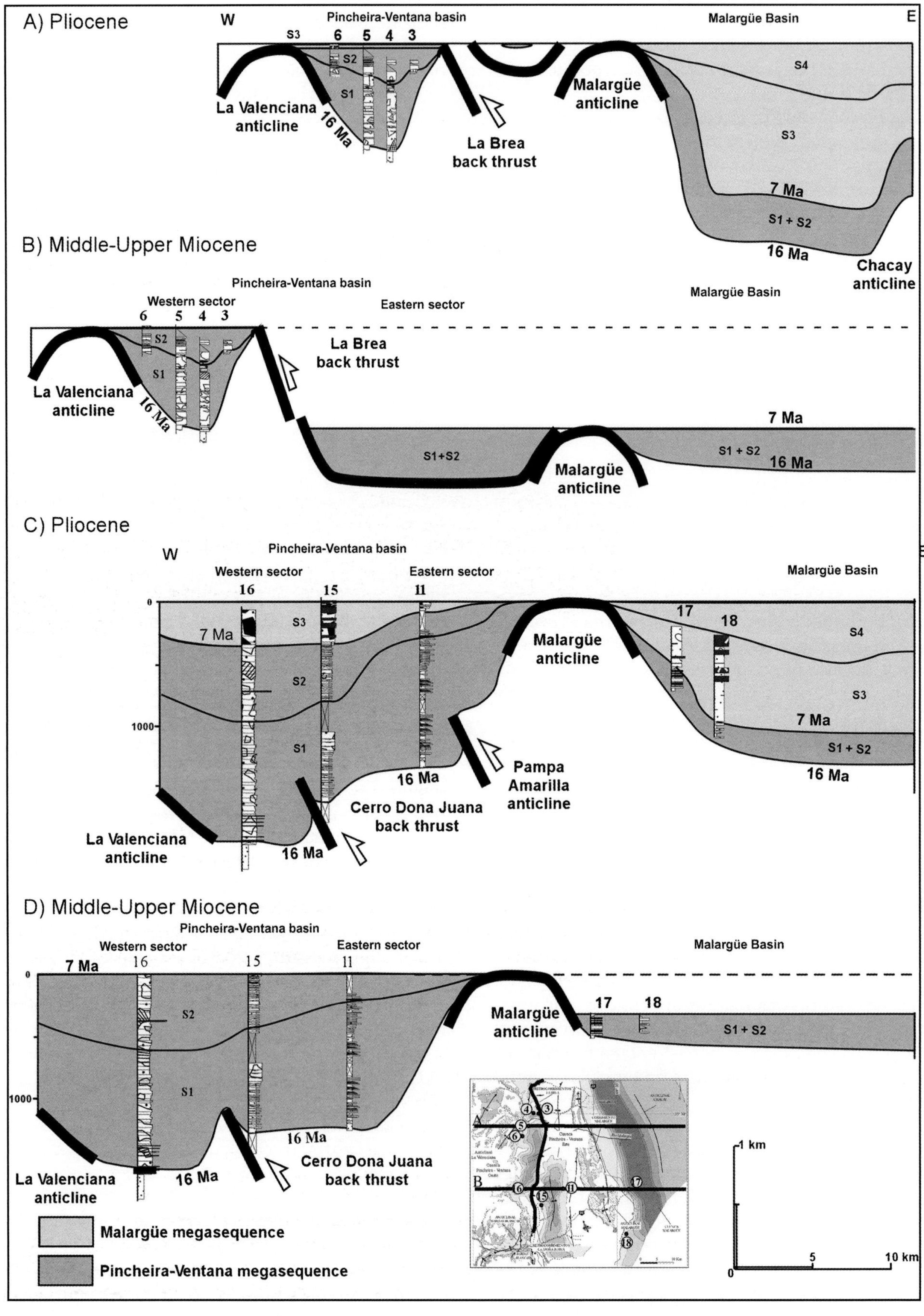
A) Pliocene
W
Pincheira-Ventana basin
Malargüe Basin
E
S3 6 5 4 3
S2
S1
S4
S3
La Valenciana anticline
16 Ma
La Brea back thrust
Malargüe anticline
7 Ma
S1 + S2
16 Ma
Chacay anticline
B) Middle-Upper Miocene
Pincheira-Ventana basin
Western sector
Eastern sector
Malargüe Basin
6 5 4 3
S2
S1
La Valenciana anticline
16 Ma
La Brea back thrust
7 Ma
S1+S2
Malargüe anticline
S1 + S2
16 Ma
C) Pliocene
W
Pincheira-Ventana basin
Western sector
Eastern sector
Malargüe Basin
E
16
15
11
17
18
0
1000
7 Ma
S3
S2
S1
Malargüe anticline
S4
S3
7 Ma
S1 + S2
16 Ma
16 Ma
Pampa Amarilla anticline
Cerro Dona Juana back thrust
La Valenciana anticline
16 Ma
D) Middle-Upper Miocene
Pincheira-Ventana basin
Western sector
Eastern sector
Malargüe Basin
7 Ma
16
15
11
0
S2
S1
1000
Malargüe anticline
17
18
S1 + S2
16 Ma
Cerro Dona Juana back thrust
La Valenciana anticline
16 Ma
Malargüe megasequence
Pincheira-Ventana megasequence
1 km
0
5
10 km

the Chacay anticline to the east starts to be active, these two structures being the main controls on the sedimentary deposition during the Pliocene. To the west of the Malargüe anticline, the whole region is uplifted and a limited accommodation space is available as is evidenced by the scarce 350 m (1148 ft) of S3 deposited in this area (Figure 13). By contrast, the Malargüe Basin located to the east of the Malargüe anticline records more than 2000 m (6562 ft) of Pliocene deposits demonstrating the migration of the Pliocene depocenter to the east (Figure 12A, C). Sequence S3 shows a dominance of basaltic composition in volcanic facies deposited mainly to the west and south of the Malargüe anticline (Figures 7, 14). Radiometric dating of the intrusive body of Cerro La Batra located south of the Malargüe anticline yielded ages of 7 Ma, suggesting that this was one of the effusive centers that sourced some of the volcanic facies of S3. At seismic sequence scale, the combination of a basal onlap followed by a prograding arrangement of reflections plus the increased steepness of the growth axial surface in the Chacay anticline (Figures 9, 10) supports the interpretation of a general decrease in the relative rate of structural uplift during deposition of S3.

The subsurface deposition of S4 is restricted to the east of the Malargüe anticline, suggesting that the whole western area was uplifted, becoming a sedimentary source for this interval (Figure 14). The prograding arrangement of reflectors and continued steepening of the growth axial surface observed in S4 (Figures 9, 10) suggest a continued decrease in the relative uplift rates of the Malargüe and Chacay anticlines, a tendency already observed in sequence S3. The last structural activity of the fold belt corresponds to the emplacement of the out-of-sequence Malargüe thrust (Figure 11) and the reactivation of the eastern flank of the Malargüe anticline evidenced by the folding of Pliocene deposits (Figures 8, 11). The structures become fossilized under the Pleistocene and Holocene deposits of sequence S5 (Figures 9–11).

The general external geometry of the Malargüe megasequence suggests that the Pliocene deposits were controlled not only by local structures such as the Malargüe and Chacay anticlines but also by a long-wavelength tilting responsible for the wedgelike geometry of this megasequence. This tilting can be originated either by a long-wavelength flexure produced by the structural loading of the fold belt or by a major block rotation beyond the studied area, or a combination of both. This suggests that although the Malargüe Basin has been transported by deep basement thrusts, as the Chacay and LPis x-1 anticlines evidence, and shares some attributes typical of piggyback basins, it also displays features typical of foredeep basins (DeCelles and Giles, 1996).

DISCUSSION

Kinematics of the Fold Belt and Application of the Critical Wedge Taper Model

The dynamics and geometric features of the syntectonic basins of the studied area show a foreland propagation of the depocenters and related structures (Figure 15).

Following the critical taper model applied to fold and thrust belts (Dahlen and Suppe, 1988; Dahlen, 1990), the foreland propagation of the wedge starts when the angle of the topographic slope (α_c) together with the angle of detachment (β_c) reach a critical taper (θ_c). The stratified nature of the basement in the studied area is compatible with the assumed existence of a deep westward-dipping low-angle planar basal decollement and the use of the critical taper model as a tool to explain the general evolution of the fold belt. This model may be applied assuming that the taper is in a critical equilibrium during its whole evolution, such as the case of the Aconcagua fold belt located to the north of the studied area where this model was applied to the Cenozoic evolution (Hilley et al., 2004).

The analysis of the kinematics evolution of the Malargüe fold and thrust belt, instead of showing a continuous evolution of the belt, shows at least two main stages of folding, erosion and sedimentation, the first during the Miocene and the second during the Pliocene. This phenomenon is frequently observed in many fold belts around the world (DeCelles and Mitra, 1995) and in scale models where the mechanical behavior of the crust is reproduced in a realistic way (Lohrmann et al., 2003). Each stage is developed when the taper is in a subcritical state ($\theta < \theta_c$) or supracritical state ($\theta > \theta_c$), and it needs to recover the steady state increasing or decreasing the topographic slope (α_c) and the angle of detachment (β_c). The steady state of the taper in nature is recovered during a period of transition constrained in time. The geometry of the wedge depends on the erosion rate, the sediments added to the wedge, the internal

Figure 12. Palinspastically corrected east–west stratigraphic correlation panel of the Pincheira-Ventana and Malargüe basins of the (A, B) northern and (C, D) southern parts of the studied area. The strong thickness change of the Pincheira-Ventana and Malargüe megasequences related to the La Brea-Dona Juana back thrust and the Malargüe anticline suggests the existence of subaerial perched basins with independent base levels and an eastward foreland migration of depocenters. S1–S4 = sequences.

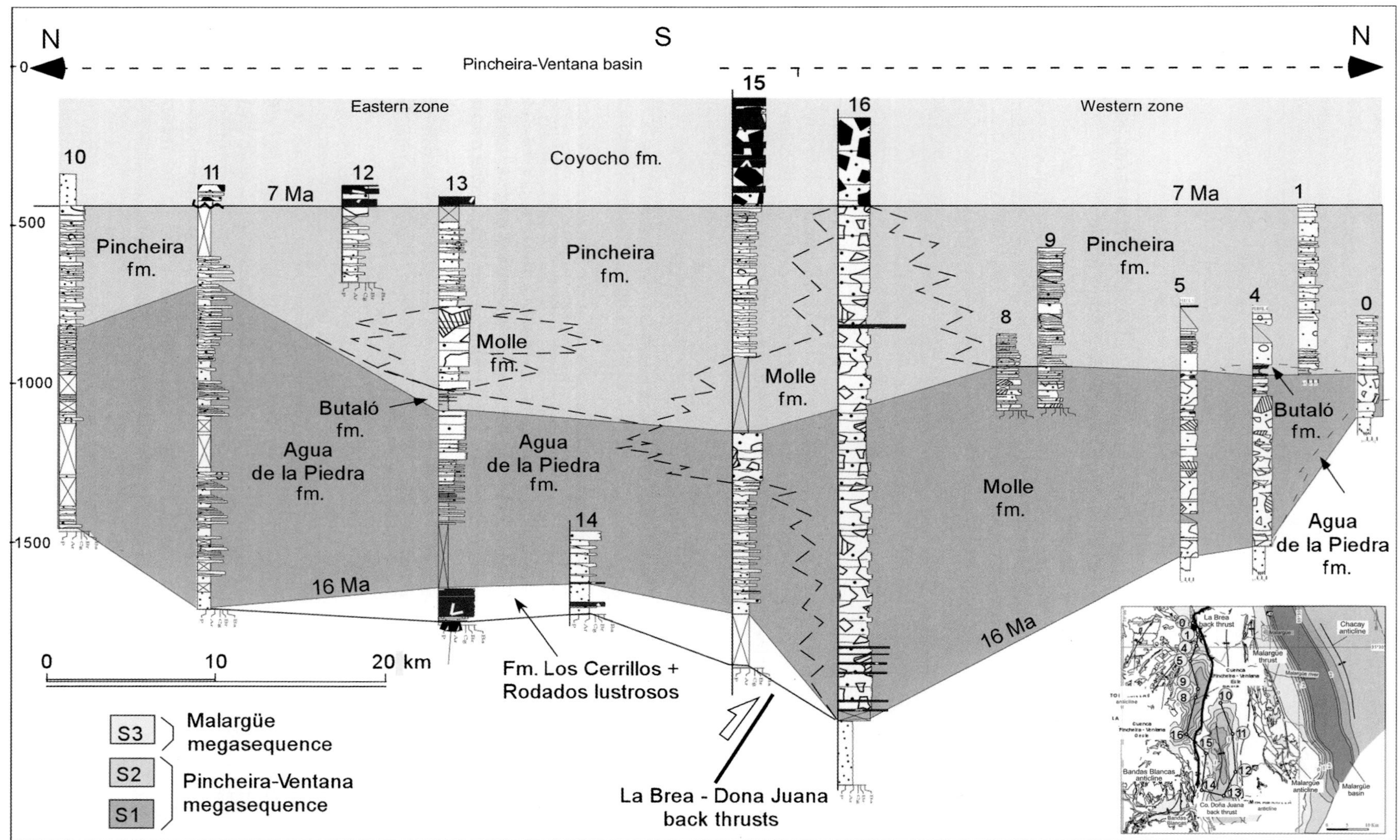

Figure 13. North–south stratigraphic correlation panel linking the western and eastern Pincheira-Ventana basin zones. Sequence S1 shows a dominant volcanic facies composition of the Molle formation at the western zone, whereas the clastic facies of the Agua de la Piedra formation dominates the eastern zone. Sequence S2 is dominated by epiclastic and pyroclastic facies of the Pincheira formation, and the volcanic facies of Molle formation are less developed, suggesting a decrease in the volcanic activity during this time. Fm. = Formation; Co. = Cerro (hill).

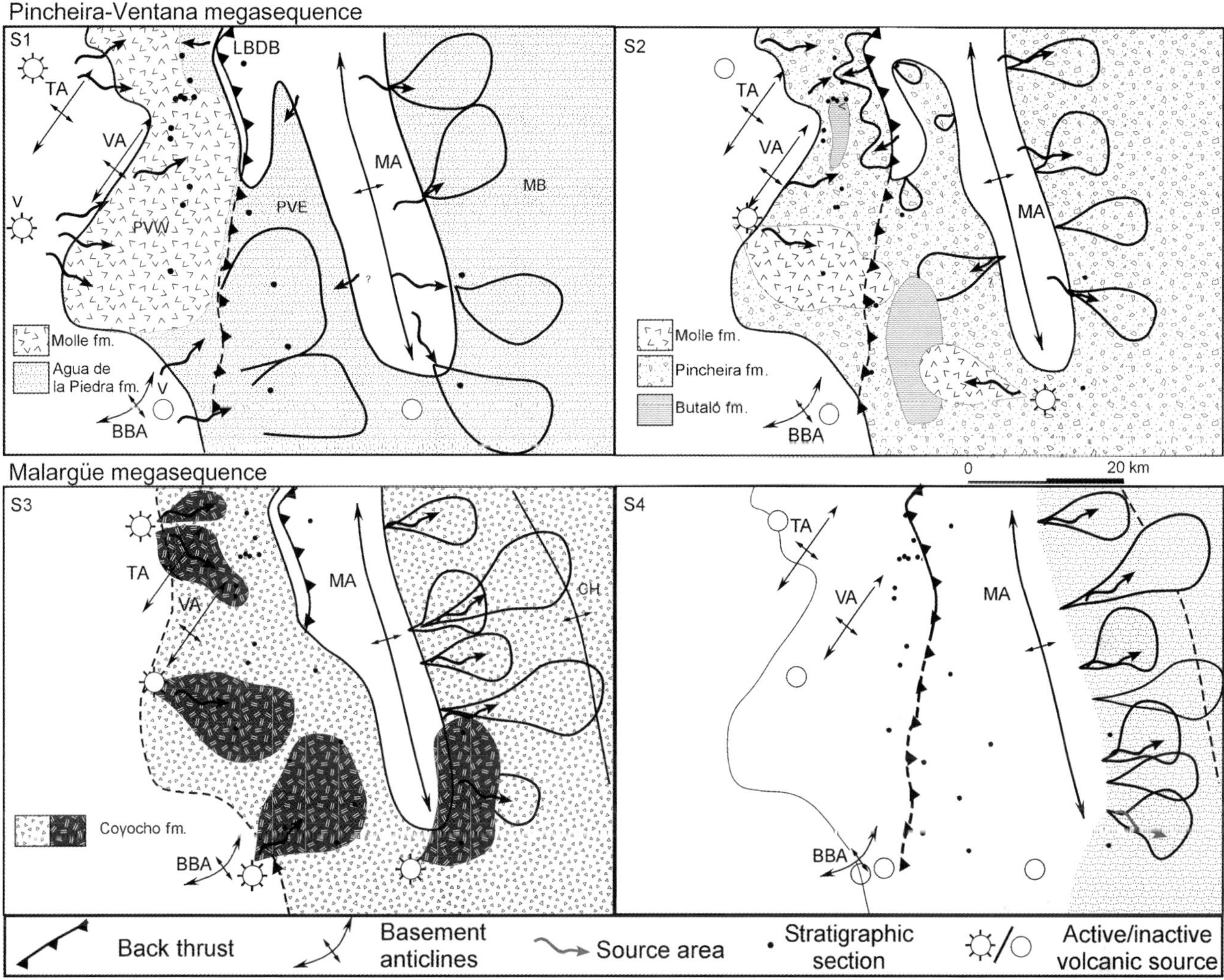

Figure 14. Paleogeographic evolution of the studied area. During deposition of the Pincheira-Ventana megasequence, the main active structures were the Malargue anticline (MA) to the east and the Torrecillas (TA), La Valenciana (VA), and Bardas Blancas (BBA) to the west. The volcanic effusive centers to the west were the main source area for this interval. S1. The volcanic deposits of the Molle formation are the dominant facies at the western zone (PVW), and the La Brea-Dona Juana back thrust (LBDB) limited the expansion of these deposits toward the eastern zone (PVW) where the clastic deposit of the Agua de la Piedra formation becomes dominant. S2: The volcanic activity decreases and low-energy lacustrine deposits of the Butaló formation are deposited at the depocenters covered by the coarse deposits of the Pincheira formation. The structural activity of the northern La Brea back thrusts produced local clastic wedges intercalated in the Pincheira formation (Figure 6). S3: The volcanic activity restarts but with a dominant basaltic composition to the west, and the active Malargüe anticline (MA) becomes a local detritic source of sediments prograding into the basin. The Chacay anticline (CH) becomes active during this stage. S4: The area west of the Malargüe anticline is uplifted and becomes a major detritic source. The Malargüe and Chacay anticlines are still active during this time but with a decreasing relative uplift rate. After deposition of S4, the Malargüe anticline is reactivated, folding the Pliocene megasequence followed by the emplacement of the Malargüe thrust, both structures being unconformably covered by the posttectonic sequence S5 of Pleistocene–Holocene age. MB = Malargüe Basins; fm. = formation; PVE = eastern zone; S1–S4 = sequences.

and basal frictional state, the development of out-of-sequence thrusts, the pore pressure, etc.

Therefore, the explanation of the evolution of the Malargüe area begins with a Miocene stage where the deformation front moves to the foreland producing the uplift of the Torrecillas, La Valenciana, Bardas Blancas, and Malargüe anticlines and the La Brea-Doña Juana back thrusts, which controlled the sedimentary evolution of the Pincheira-Ventana Basin. The progressive decrease in the rate of uplift and the filling of the

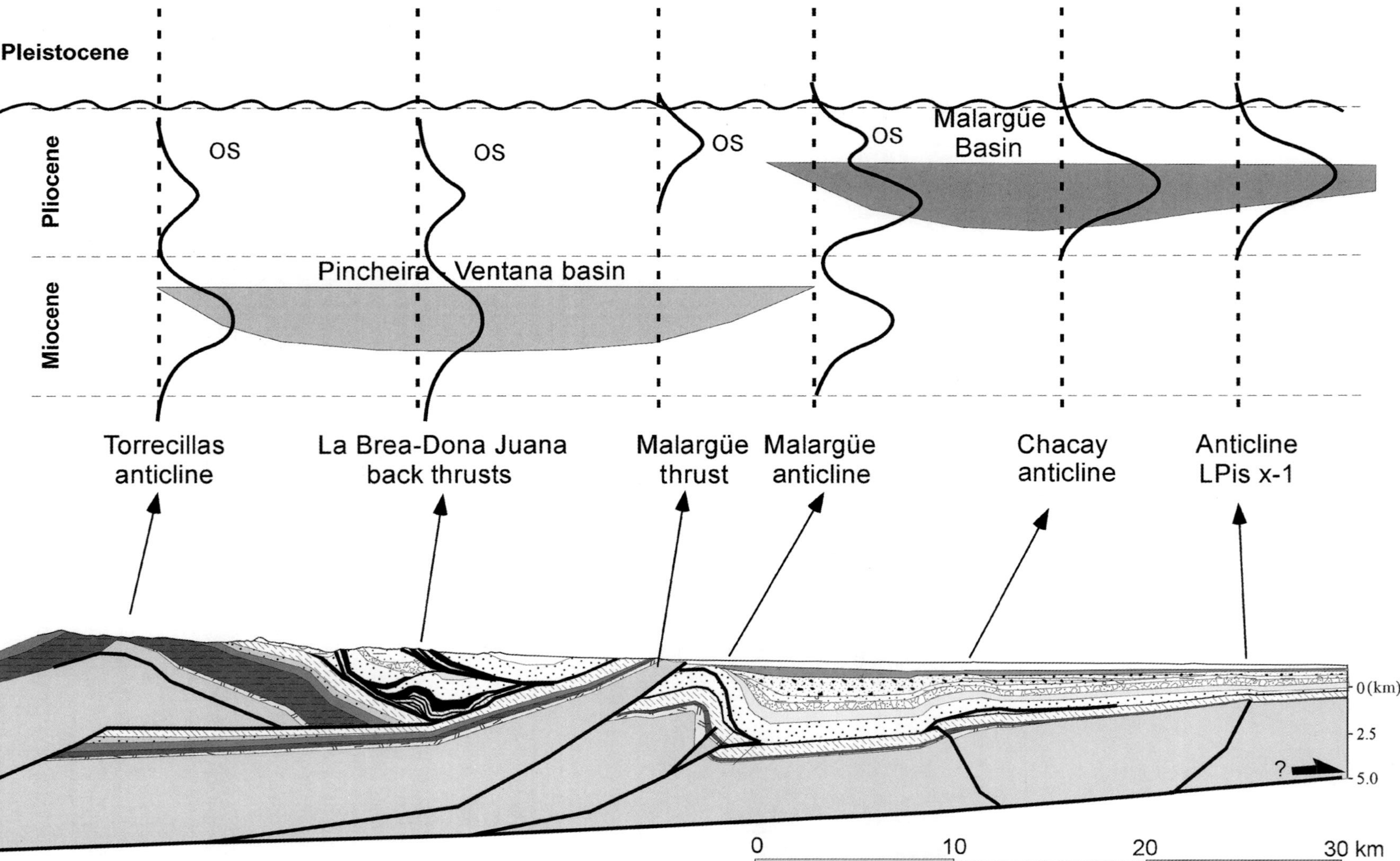

Figure 15. The first-order kinematics of the fold belt is evidenced by the eastward jump of the Miocene and Pliocene depocenters toward the foreland. During each major tectonic event, several structures are active at the same time. During the major displacement of structural activity toward the foreland, the hinterland structures are reactivated or emplaced as out-of-sequence folds and thrust (OS).

basin at the end of this time shows that the deformation front starts to become inactive. This could be explained by considering a subcritical wedge ($\theta < \theta_c$) where the topographic slope decreases because of the sediments added to the orogenic front (Storti and McClay,1995) as well as the propagation of the deformation to the foreland (Dahlen, 1990).

The beginning of the second stage during the Pliocene is shown by a new foreland propagation of the deformation where the Chacay anticline becomes active. The reactivation of the Malargüe and Torrecillas anticlines as out-of-sequence structures shows that this mechanism is relevant to thicken the taper (Morley, 1988), which becomes critical or supercritical ($\theta > \theta_c$) to start to move to the foreland (DeCelles and Mitra, 1995). Toward the end of the Pliocene, the rate of uplift decreases again and the Malargüe Basin is fulfilled, showing that the deformation front becomes inactive after reaching the subcritical state ($\theta < \theta_c$) due to deposition of sediments at the front and the migration to the foreland. The last reactivation of the Malargüe anticline and the out-of-sequence emplacement of the Malargüe thrust could be interpreted as an aborted intent to thicken the wedge, which is not followed by propagation to the foreland as shown by the fossilized deformation front under Pleistocene sediments.

The development of the large back-thrust system of La Brea-Doña Juana could be explained following the weak basal-wedge model of Davis and Engelder (1985) and Smit et al. (2003) where the structures are developed synchronously and with variable vergence. In the Malargüe fold belt, the development of back thrusts is almost always related to the presence of evaporites as extremely weak horizons of detachment (Figure 8). In the same way, basement structures tend to reactivate synchronously (Figure 15), suggesting a weak detachment level related to basement fabrics or preexistent faults reactivated as inversion structures (Figure 2).

Evolution of the Deformation Front Between 33° and 37°

The analysis of the timing of deformation in a regional framework shows a renewal of the tectonic activity from south to north. In this sense, the timing at the studied latitude (35–36°) is similar to the evolution of the Atuel River area (34–35°) where the sequence of folding defined by Kozlowski (1984) starts in the west with the uplift of basement structures at the same time that the Agua de la Piedra formation is deposited. The propagation to the east and the development of shallow detachments occur at the time of deposition of the Loma Fiera Formation, dated by Combina and Nullo (2000) in 10.7 Ma. The tectonic activity at this latitude concludes with the uplift of basement structures at the deformation front after 5.3 Ma (Combina et al., 1993). To the north of 34°, the deformation front keeps active up to the present, showing an increment in the involvement of basement structures to the north of 33° (Giambiagi et al., 2003). To the south at 37°30′, Ramos and Barbieri (1989) constrained the age of the deformation front between 18 and 9 Ma based on radiometric ages obtained from basalts and andesitic intrusives of the Huantraico syncline. New data published by Kay (2005) show the propagation of the deformation to the east between 11 and 4 Ma. Finally, to the south of 37°30′, the Agrio fold belt becomes inactive during the upper Miocene before about 9 Ma (Zapata et al., 2002) (Figure 16).

To summarize, these data show that the Neogene deformation front of the Andes is active between 33° and 34° with neotectonic and seismic activity; to the south between 34° and 37°, it was fossilized during the Pliocene; and finally to the south of 37°30′, it becomes inactive during the Miocene. This phenomenon is explained by Folguera et al. (2004) by a progressive stepping of the subducted slab and the migration of the deformation to the internal part of the fold and thrust belt.

CONCLUSIONS

The analysis of the structures and Tertiary growth strata from surface and subsurface allows us to define the geometry and timing of deformation of the Malargüe fold and thrust belt located at the Andean front between 35° and 36°. The hybrid structural style of the fold and thrust belt consists of a mixture of thick- and thin-skinned structures. The basement structures are interpreted as half grabens and transfer extensional zones, reactivated as inversion structures. The shortening of the basement is transferred to the cover following detachment levels of evaporites and shales, where thrusts, back thrusts, duplexes, and salt structures are developed. The whole estimated shortening in the cross sections is in the order of 30 km (19 mi).

The stratigraphic information from surface, radiometric dating, and seismic stratigraphic interpretation allows us to define two megasequences: the Miocene (16–7 Ma) Pincheira-Ventana megasequence composed of two syntectonic sequences (S1 and S2) and the Pliocene (5–1 Ma) Malargüe megasequence composed of the syntectonic sequences (S3 and S4). The Pleistocene posttectonic sequence (S5) defines the end of the structural activity of the fold and thrust belt at this latitude.

These continental basins are dominated by pyroclastic and vulcaniclastic facies with participation of fine

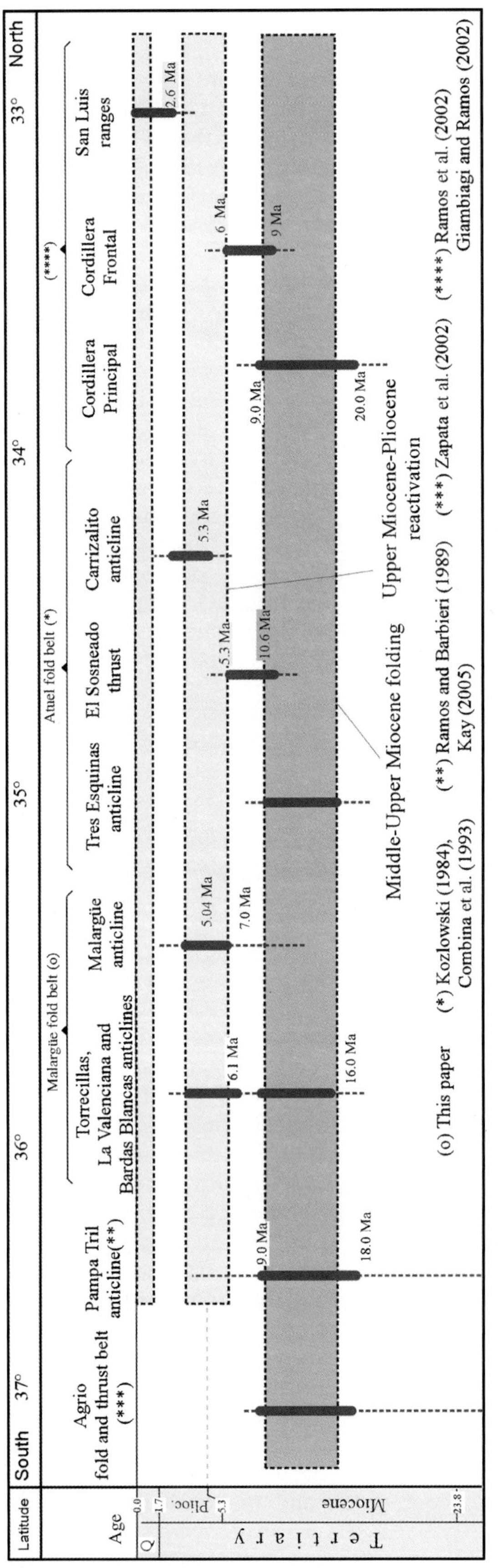

Figure 16. Diagram showing the progressive deactivation of the Neogene deformation front toward the south. North of 34°, the deformation front is still active; between 34° and 36°, it was fossilized during the Pliocene, whereas south of 37°, it was fossilized during the Miocene. Plioc. = Pliocene.

deposits restricted to the central position of the basin, constrained only to the periods of structural and volcanic quiescence. The uplift of internal structures acted as boundaries given place to partially isolated and perched depocenters as suggested by the palinspastic and paleogeographic restitution of the basins.

The tectonic evolution of the fold and thrust belt shows two main steps: the first during the Miocene and the second during the Pliocene, showing a foreland propagation of the system together with the migration of the syntectonic basins. During the Miocene, the Pincheira-Ventana Basin is developed with the center of subsidence to the west of the Malargüe anticline. During the Pliocene, the depocenter moves to the east developing the Malargüe Basin.

Following the critical taper model, the supercritical state is achieved by the thickening of the wedge through emplacements of out-of-sequence thrusts. This supercritical state allows the foreland propagation of the system.

The decrease in the rate of uplift and the fulfilling of the basins at the end of each stage shows that the deformation front becomes inactive after reaching the subcritical state. The beginning of the next stage occurs when the old deformation front is reactivated and involved in the wedge as out-of-sequence structures reaching a critical or supercritical state, allowing the propagation of the deformation front to the foreland.

The analysis of the published data about the Neogene evolution of the Andean deformation front between 33° and 37° shows a renewal of the tectonic activity from south to north, explained by a progressive stepping of the subducted slab.

ACKNOWLEDGMENTS

The authors thank the Argentine National Research Council CONICET for the financial support and TRITON for providing subsurface information.

REFERENCES CITED

Artabe, A., E. Morel, L. Spalletti, and M. Brea, 1998, Paleoambientes sedimentarios y paleoflora asociada en el Triásico tardío de Malargüe, Mendoza: Revista de la Asociación Geológica Argentina, v. 53, no. 4, p. 526–548.

Baranzaghi, M., and B. I. Isacks, 1976, Spatial distribution of earthquakes and subduction of the Nazca plate beneath South America: Geology, v. 4, p. 686–692, doi:10.1130/0091-7613(1976)4<686:SDOEAS>2.0.CO;2.

Cahill, T., and B. I. Isacks, 1992, Seismicity and the shape of the subducted Nazca plate: Journal of Geophysical Research, v. 97, p. 17,503–17,529, doi:10.1029/92JB00493.

Combina, A., and F. Nullo, 2000, La Formación Loma Fiera (Mioceno superior) y su relación con el volcanismo y el tectonismo neógeno, Mendoza: Revista de la Asociación Geologica Argentina, v. 55, no. 3, p. 201–210.

Combina, A., F. Nullo, and G. Stephens, 1993, Depósitos Terciarios en el Pie de Sierra del area de las Aucas, Sur de Mendoza: XII Congreso Geológico Argentino, v. Actas II, p. 180–186.

Criado Roque, P., 1950, El Terciario del Sur de la Provincia de Mendoza: Revista de la Asociación Geológica Argentina, v. 5, no. 4, p. 223–255.

Dahlen, F. A., 1990, Critical taper model of fold and thrust belts and accretionary wedges: Annual Reviews of Earth and Planetary Sciences, v. 18, p. 55–99, doi:10.1146/annurev.ea.18.050190.000415.

Dahlen, F. A., and J. Suppe, 1988, Mechanics, growth, and erosion of mountain belts, *in* S. P. Clark, B. C. Burchfield, and J. Suppe, eds., Processes in continental lithospheric deformation: Geological Society of America Special Paper, p. 161–178.

Davis, G. A., and T. Engelder, 1985, The role of salt in fold-and-thrust belts: Tectonophysics, v. 119, p. 67–88, doi:10.1016/0040-1951(85)90033-2.

DeCelles, P. G., and K. A. Giles, 1996, Foreland basin systems: Basin Research, v. 8, p. 105–375, doi:10.1046/j.1365-2117.1996.01491.x.

DeCelles, P. G., and G. Mitra, 1995, History of the Sevier orogenic wedge in terms of critical taper models, northeast Utah and southwest Wyoming: Geological Society of America Bulletin, v. 107, p. 454–462, doi:10.1130/0016-7606(1995)107<0454:HOTSOW>2.3.CO;2.

Dimieri, L., 1997, Tectonic wedge geometry at Bardas Blancas, southern Andes (36°S), Argentina: Journal of Structural Geology, v. 19, no. 11, p. 1419–1422.

Folguera, A., V. A. Ramos, R. L. Hermanns, and J. Naranjo, 2004, Neotectonics in the foothills of the southernmost central Andes (37°–38°S): Evidence of strike-slip displacement along the Antiñir-Copahue fault zone: Tectonics, v. 23, p. TC5008, doi:10.1029/2003TC001533.

Giambiagi, L. B., and V. A. Ramos, 2002, Structural evolution of the Andes in a transitional zone between flat and normal subduction (33°30′–33°45′S), Argentina and Chile: Journal of South American Earth Sciences, v. 15, p. 101–116.

Giambiagi, L. B., E. Godoy, P. Alvarez, S. Orts, and V. A. Ramos, 2003, Cenozoic deformation and tectonic style of the Andes, between 33° and 34° south latitude: Tectonics, v. 22, no. 4, p. 1–23.

Groeber, P., 1946, Observaciones geológicas a lo Largo del Meridiano 70: Buenos Aires, Argentina, Asociación Geológica Argentina, Serie C Reimpresiones (1980), 174 p.

Gulisano, C., and A. Gutierrez Pleimling, 1995, Field guide: The Jurassic of the Neuquén Basin, b) Mendoza Province: Buenos Aires, Asociación Geológica Argentina, Serie E, v. 3, 103 p.

Hilley, G. E., M. R. Strecker, and V. A. Ramos, 2004, Growth and erosion of fold-and-thrust belts, with an application to the Aconcagua fold-and-thrust belt, Argentina: Journal of Geophysical Research, Solid Earth, v. 109, doi:10.1029/2002JB002282.

Jordan, T. E., B. L. Issacks., R. W. Allmendinger, J. A. Brewer, V. A. Ramos, and C. J. Ando, 1983, Andean tectonics related to geometry of subducted Nazca plate: Geological Society of America Bulletin, v. 94, p. 341–361, doi:10.1130/0016-7606(1983)94<341:ATRTGO>2.0.CO;2.

Kay, S. M., 2005, Tertiary to recent evolution of the Andean arc and backarc magmas between 36°s and 38°s and evidence for shallowing of the Nazca plate under the Neuquén Basin: 6th International Symposium on Andean Geodynamics: Montpellier, France, IRD Editions, p. 420–423.

Kozlowski, E., 1984, Interpretación Estructural de la Cuchilla de la Tristeza, Provincia de Mendoza: IX Congreso Geológico Argentino, v. Actas II, p. 381–395.

Kozlowski, E., C. E. Cruz, and G. A. Rebay, 1987, El Terciario Vulcanoclástico de la zona de Puntilla de Huincán, Provincia de Mendoza Argentina: Simposio de Vulcanismo Andino, X Congreso Geológico Argentino, v. Actas IV, p. 229–232.

Kozlowski, E., C. Cruz, P. Condat, and R. Manceda, 1989, Informe geológico zona Malargüe Occidental: Buenos Aires, Argentina, YPF, Informe Inédito, p. 47.

Kozlowski, E., R. Manceda, and V. Ramos, 1993, Estructura, *in* V. Ramos, ed., Geología y Recursos Naturales de Mendoza: Relatorio del XII Congreso Geológico Argentino y II Congreso de Exploración de Hidrocarburos, p. 235–256.

Kozlowski, E., C. Cruz, and C. Sylwan, 1997, Modelo Exploratorio de la Faja Corrida de la Cuenca Neuquina, Argentina: VI Simposio Bolivariano de Exploración Petrolera en Cuencas Subandinas, p. 15–31.

Kraemer, P., and G. Zulliger, 1994, Sedimentación cenozoica sinorogénica en la faja plegada andina a los 35°S, Malargüe-Mendoza, Argentina: VII Congreso Geológico Chileno, v. 1, p. 460–464.

Kraemer, P. E., 2000, Kinematic of the Andean fold-belt inferred from the geometry and age of syntectonic sediments, Malargüe (35°30′S), Mendoza, Argentina: Geological Society of America, Abstracts with Programs A, p. 506.

Legarreta, L., D. A. Kokogian, and D. A. Boggetti, 1989, Depositional sequences of the Malargüe Group (Upper Cretaceous–lower Tertiary), Neuquén Basin, Argentina: Cretaceous Research, v. 10, p. 337–356, doi:10.1016/0195-6671(89)90009-8.

Linares, E., 2001, Catálogo de edades radimétricas de la República Argentina: Parte II. Años 1988–2000: Asociación Geológica Argentina, Serie F, v. 1, Publicaciones en CD.

Lohrmann, J., N. Kukowski, J. Adam, and O. Oncken, 2003, The impact of analog material properties on the geometry, kinematics, and dynamics of convergent sand wedges: Journal of Structural Geology, v. 25, p. 1691–1711, doi:10.1016/S0191-8141(03)00005-1.

Manceda, R., and D. Figueroa, 1995, Inversion of the Mesozoic Neuquén rift in the Malargüe fold and thrust belt, Mendoza, Argentina, *in* A. Tankard, R. Suárez, and H. Welsink, eds., Petroleum basins of South America: AAPG Memoir 62, p. 369–382.

Mitra, S., and V. Mount, 1998, Foreland basement-involved structures: AAPG Bulletin, v. 82, no. 1, p. 70–109.

Morley, C. K., 1988, Out-of-sequence thrusts: Tectonics, v. 7, p. 539–561, doi:10.1029/TC007i003p00539.

Nocioni, A. D., 1996, Estudio estructural de la faja plegada y corrida de la Cuenca Neuquina surmendocina: XIII Congreso Geológico Argentino y III Congreso de Exploración de Hidrocarburos, v. 2, p. 353–372.

Ploszkiewicz, J., and R. Gorroño, 1988, Tectónica de inyección salina en la faja fallada y plegada del sur de Mendoza: BIP, v. 30, p. 29–34.

Ramos, V. A., 1999, Plate tectonic setting of the Andean Cordillera: Episodes, v. 22, no. 3, p. 183–190.

Ramos, V. A., and M. Barbieri, 1989, El volcanismo cenozoico de la zona de Huantraico: Edad y relaciones isotopicas iniciales, provincia de Neuquen: Revista Asociacion Geologica Argentina, v. 43, no. 2, p. 210–223.

Ramos, V. A., E. O. Cristallini, and D. J. Perez, 2002, The Pampean flatslab of the central Andes: Journal of South American Earth Sciences, v. 15, p. 59–78.

Smit, J. H. W., J. P. Brun, and D. Sokoutis, 2003, Deformation of brittle-ductile thrust wedges in experiments and nature: Journal of Geophysical Research, v. 108, B10, doi:10.1029/2002JB002190.

Storti, F., and K. McClay, 1995, Influence of syntectonic sedimentation on thrust wedges in analog models: Geology, v. 23, p. 999–1002, doi:10.1130/0091-7613(1995)023<0999:IOSSOT>2.3.CO;2.

Suppe, J., G. T. Chou, and S. C. Hook, 1992, Rates of folding and faulting determined from growth strata, *in* K. R. McClay, ed., Thrust tectonics: London, Chapman & Hall, p. 105–121.

Yrigoyen, M., 1972, Cordillera principal, *in* A. Leanza, ed., Geología regional Argentina: Cordobaba, Academia Nacional de Ciencias, p. 651–693.

Zapata, T., S. Corsico, F. Dzelajica, and G. Zamora., 2002, La faja plegada y corrida del Agrio: Análisis estructural y su relación con los estratos terciarios de la Cuenca Neuquina Argentina: V Congreso de Exploración de y desarrollo de Hidrocarburos, Mar del Plata, Argentina, publicaciones en CD.

12

Mount, Van S., Kevin W. Martindale, Thomas W. Griffith, and John O. D. Byrd, 2011, Basement-involved contractional wedge structural styles: Examples from the Hanna Basin, Wyoming, *in* K. McClay, J. H. Shaw, and J. Suppe, eds., Thrust fault-related folding: AAPG Memoir 94, p. 271 – 281.

Basement-involved Contractional Wedge Structural Styles: Examples from the Hanna Basin, Wyoming

Van S. Mount
Anadarko Petroleum Corporation, Houston, Texas, U.S.A.

Kevin W. Martindale
Anadarko Petroleum Corporation, Houston, Texas, U.S.A.

Thomas W. Griffith
Anadarko Petroleum Corporation, Houston, Texas, U.S.A.

John O. D. Byrd
Hess Corporation, Houston, Texas, U.S.A.

ABSTRACT

In a typical wedge structure developed in layered sedimentary rocks, displacement is transferred from a deep fault to a shallow fault. The deep and shallow faults dip in opposite directions and merge with one another to form a wedge-shaped indentor. Displacement within the wedge fault system is transferred vertically into structures overlying the faults instead of being transferred laterally toward the foreland as in a typical fault-bend fold. Wedge structures associated with basement-involved contractional uplifts are commonly observed where faults that uplift a basin margin emanate from competent basement rocks into layered sedimentary rocks within the basin. Two categories of basement-involved wedge structures have been identified based on the level of the wedge point with respect to the basement surface. In the first category, the wedge point occurs within the sedimentary section overlying the basement, and in the second category, the wedge point occurs within the basement material. Interpretations of both categories of basement-involved wedge structures imaged in long-offset seismic data from the Hanna Basin in south-central Wyoming are presented.

INTRODUCTION

Wedge structures are a class of fault-related folds developed in the basement and layered sedimentary rocks in which displacement is transferred from a deep fault to an overlying fault with an opposite dip direction. Previous studies (Medwedeff, 1990; Bellotti et al., 1995; Erslev et al., 2001; Shaw et al., 2004) have recognized and

DOI:10.1306/13251341M941003

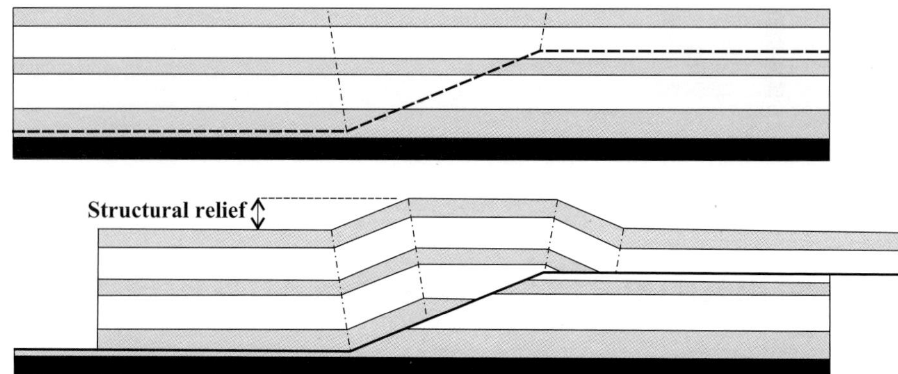

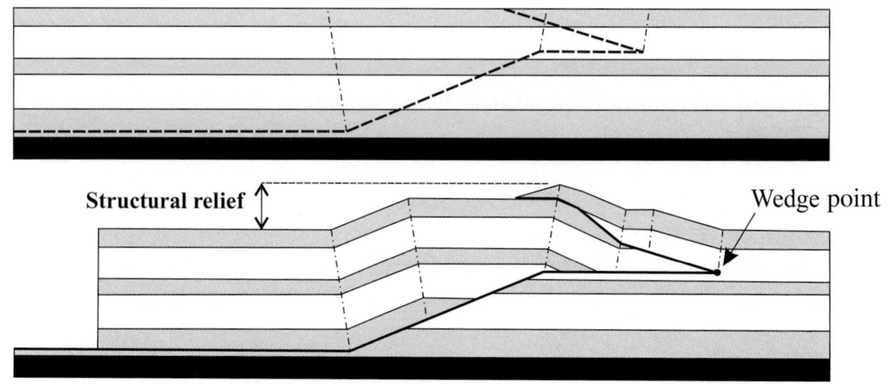

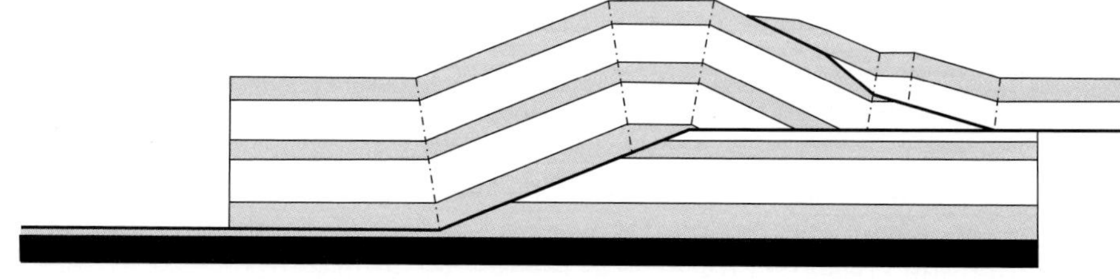

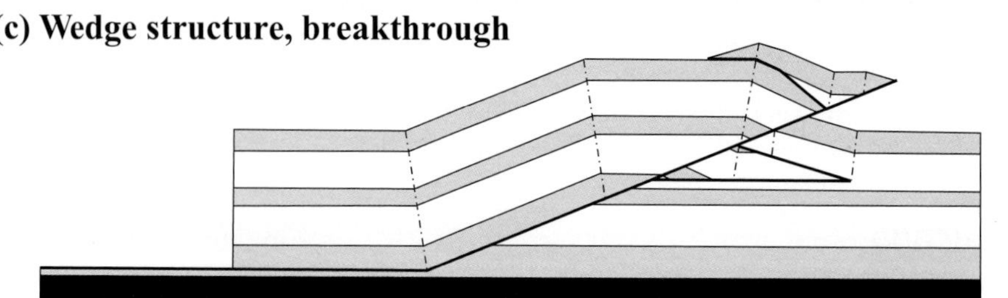

Figure 1. (a) Typical fault-bend fold in which displacement is transferred from the left side of the model over the ramp, creating a fault-bend fold and transferring slip off the right side of the model. (b) Typical wedge structure in which displacement on the lower fault is wedged back to the left at the wedge point. Notice that no displacement is transferred off the right side of the model. (c) Model illustrating partitioning of slip between the fault-bend fold and wedge structure. (d) Breakthrough wedge structure. The original wedge structure and wedge point are left in the footwall block of the later broken-through reverse fault.

described contractional wedge structures in which slip is transferred vertically up section, by means of wedging, to overlying, shallower structural levels (Figure 1b) instead of horizontally to laterally adjacent structures (Figure 1a). In a simple wedge structure, the intersection of the deep and shallow fault segments creates a wedge point in cross section view (Figure 1b) and a wedge line in three dimensions. Because slip is transferred from the deep fault to the shallow fault, where displacement on the deep fault decreases to zero laterally along the wedge line, displacement on the shallow fault also decreases to zero because the faults are linked. Consistency in magnitude of displacement on the deep and shallow faults can help discriminate true wedge structures from other structural styles such as those generated by strike-slip deformation.

Figure 1 illustrates the difference between a fault-bend fold and a wedge structure developed over an identical fault ramp. In the fault-bend fold model (Figure 1a), slip over the fault ramp generates a hanging-wall anticline and most of the slip is transferred off the right side of the model (with some of the slip being consumed in folding). In the wedge model (Figure 1b), the slip that generates the hanging-wall anticline is not transferred off the right side of the model but is wedged back to the left at the wedge point. The location of the wedge point in this model is not fixed but migrates to the right with increasing slip on the deep fault. The slip generating the wedge structure is transferred upward to overlying, shallower structural levels, generating a structure that is a composite of two fault-related folds overlying one another. The upward transfer of slip in the wedge structure results in greater structural relief than in the fault-bend fold model, and the magnitude of structural relief increases within the wedge structure at shallower structural levels. In contrast, slip in the fault-bend fold model (Figure 1a) is transferred off the right side of the model in a lateral sense, to adjacent structures, or the ground surface, and the maximum structural relief associated with the structure is equal to the height of the underlying fault ramp. Figure 1a and b illustrate end-member models. Intermediate cases in which slip along the upper flat in the model is partitioned, with a component being wedged back to the left and the remainder being sent off the right side of the model, are to be expected (Figure 1c). In addition, breakthrough structures (Suppe and Medwedeff, 1990; Martindale et al., 2004) initiate as wedge structures and, with additional slip, are subsequently bypassed. In this model, the lower part of the early formed wedge structure is stranded in the footwall block, and the upper level parts of the early formed wedge structure are carried in the hanging-wall block (Figure 1d).

Possible factors that may contribute to the formation of wedge structures, as opposed to classic fault-bend folds, include changes in mechanical strength associated with stratigraphic changes within the interval in which the detachment is located, variation in geopressure along the detachment altering the frictional properties of the detachment surface, and intersection of the deep detachment with a preexisting fault or buttress.

BASEMENT-INVOLVED WEDGE STRUCTURAL STYLES

The models in Figure 2 illustrate differences between basement-detached and basement-involved wedge structures. For purposes of this discussion, a mechanical basement is envisioned to be mechanically competent rocks (i.e., granitic, metamorphic rocks, or even thick carbonate sequences) in comparison to the cover sequence, which is envisioned to consist of layered sedimentary rocks. In the basement-detached structure (Figure 2a), all deformation is occurring above the basement level, whereas in Figure 2b and c, the basement and the overlying cover sequence are involved in the deformation. Two categories of basement-involved wedge structures are recognized based on the location of the wedge point with respect to the basement surface. In the first category (Figure 2b), the wedge point occurs within the overlying sedimentary cover section, and in the second category (Figure 2c), the wedge point occurs within the basement material.

Slip on the deeper fault in the basement-involved models shown in Figure 2a and b (numbers 1–3) is transferred from left to right, wedged back to the left on the upper fault, and sent off the left edge of the model. Figure 2a and b (numbers 1–3) illustrate the amount of slip that is transferred to the upper system that needs to be consumed in folding, or intersect the ground surface, to eliminate the void at the left side of the model. Figure 2b (number 4) illustrates a scenario in which the shallow-level slip is consumed in a fault-propagation fold and does not exit the upper left side of the model. In the second category of basement-involved wedge structures, in which the wedge point is within the basement (Figure 2c), the small amount of slip on the right-dipping back thrust is consumed in folding above the fault tip and no displacement is sent off the left side of the model. In this model, the early formed wedge forms a basement pop-up structure, which is translated with the dominant reverse fault in subsequent deformation.

One of the diagnostic features of basement-involved structures in general, including basement-involved wedge structures, is that a step in structural relief across the structure exists. Notice that, in the basement-involved models in Figure 2b and c, the elevation of the left side of the model is higher than the elevation of the same unit on the right side of the model because of slip on the underlying basement fault, which dips to the left. In the basement-detached model in Figure 2a, the elevation of each unit is the same on the left side of the model as on the right side of the model. In both basement-involved models, the fault dip decreases at the point where the fault enters the sedimentary cover, creating a monoclinal fold geometry, similar to geometries predicted and observed in typical Rocky Mountain foreland structures by Mitra and Mount (1998).

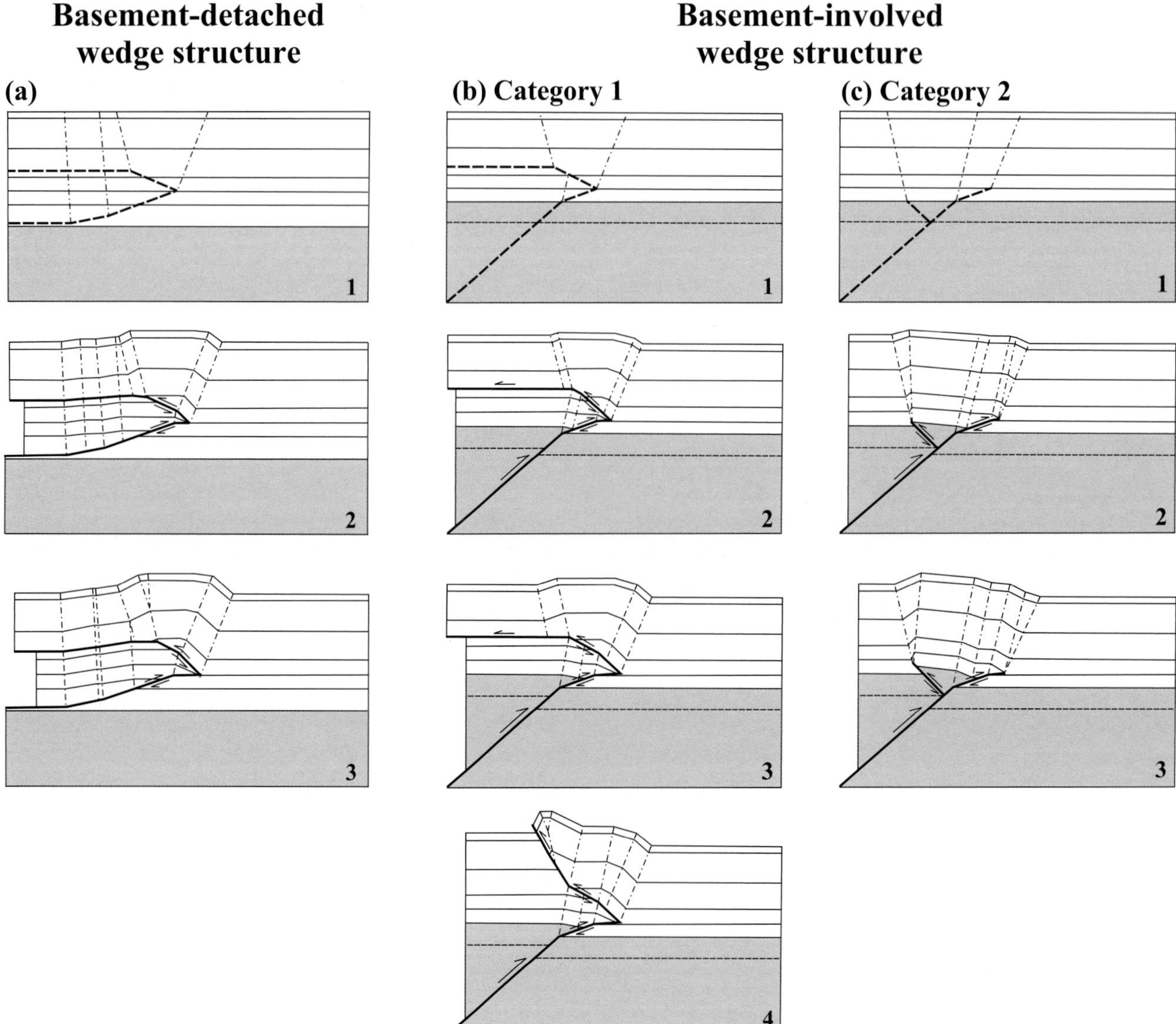

Figure 2. Basement-detached and basement-involved structural models. (a) Basement-detached model in which all deformation occurs above the basement; the basement is not involved in the structure (modified from Medwedeff, 1990). (b) Model of a category 1 basement-involved wedge structure in which the wedge point is within the sedimentary cover. (c) Model of a category 2 basement-involved structure in which the wedge point is within the basement.

The fault-propagation fold structures, which ultimately consume slip at the shallow fault tip in the basement-involved wedge structure models in Figure 2b and c, are genetically similar to the back-thrust-tip anticlines described by Erslev et al. (2001).

HANNA BASIN CONTRACTIONAL WEDGE EXAMPLES

Excellent examples of basement-involved wedge structures are imaged in seismic data acquired by the Anadarko Petroleum Corporation along the margins of the Hanna Basin in the Rocky Mountain foreland province of the western United States.

Generalized Hanna Basin Structure and Stratigraphy

The Hanna Basin is a Laramide age (Late Cretaceous–Eocene) intermontane basin located in south-central Wyoming (Figure 3). With a depth to the basement surface of approximately 15 km (9 mi) in its deepest part, the Hanna Basin has the distinction of being the deepest basin in the entire Rocky Mountain region. The basin is strongly asymmetric, the western flank dips

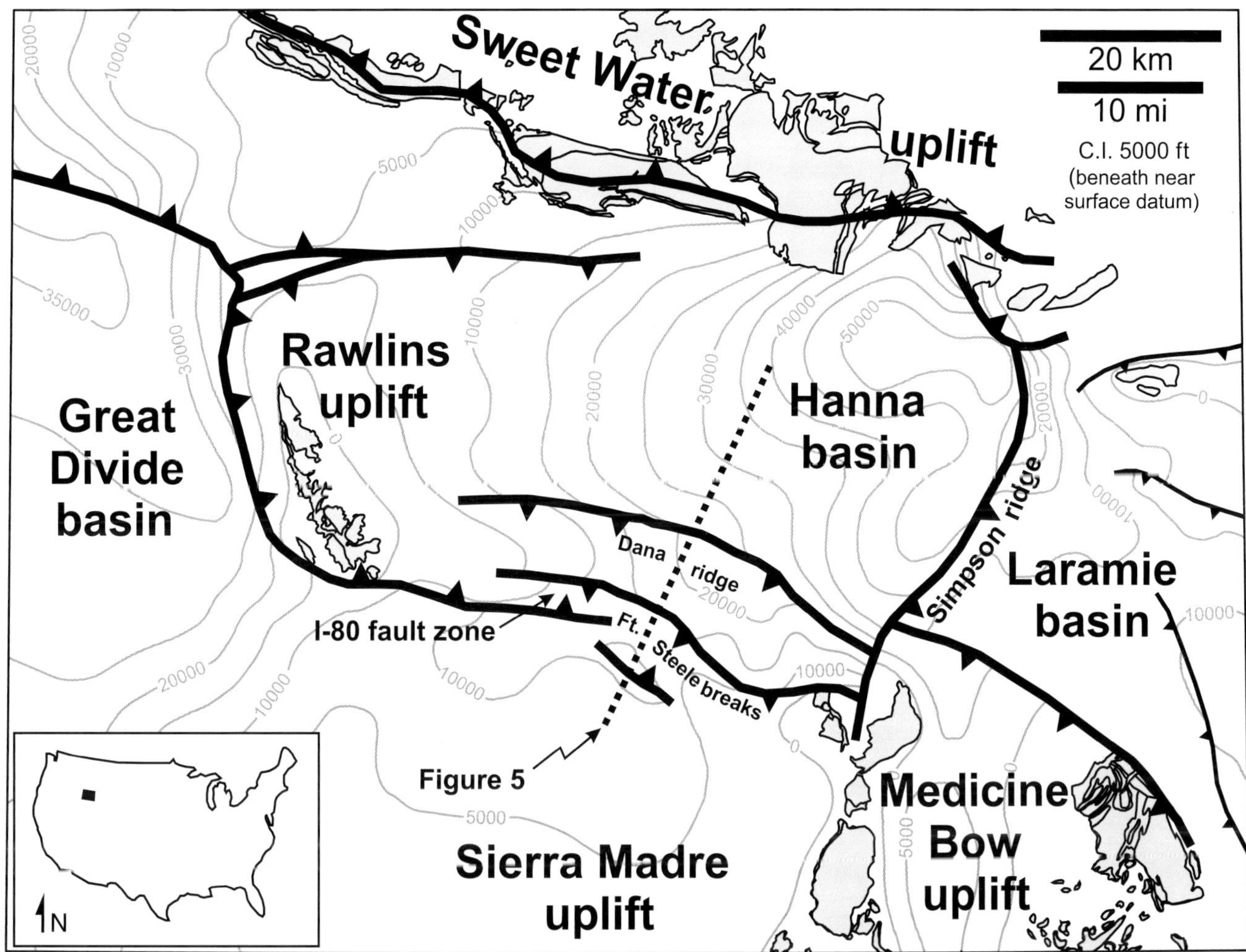

Figure 3. Simplified location map of the Hanna basin in south-central Wyoming and surrounding basement uplifts. Contours (5000 ft [16,404 ft]) are on the surface near the top Precambrian basement below a 2400-m (7874-ft) datum elevation. Gray shaded units represent a surface outcrop of the Pennsylvanian Tensleep Formation and older rocks. Dana ridge and Fort Steele breaks are southwest-vergent surface anticlines associated with underlying basement-involved wedge structures. Contour interval (C. I.) is 5000 ft (1524 m). Ft. = Fort.

gently to the east from the Rawlins uplift, and the northern and eastern boundaries exhibit steeply dipping to overturned bedding associated with reverse faults underlying the Sweetwater uplift on the north and Simpson Ridge on the east. The southern boundary of the Hanna Basin comprises a complex system of wedge structures, which are illustrated and discussed in this chapter.

Paleozoic through Triassic rocks in the Hanna Basin were deposited on a gradually emerging west-facing passive margin and are approximately 1500 m (4921 ft) thick (Figure 4). Major units include the Flathead, Buck Spring, Madison, Amsden, Tensleep, Goose Egg, Chugwater, Sundance, and Morrison formations. During the Cretaceous, all of Wyoming began to experience a dramatic reorientation of its depositional systems. Thrust deformation in Utah and far western Wyoming (Sevier orogeny) led to the formation of a large foreland basin along the western margin of the Interior Cretaceous seaway. The Lower Cretaceous Cloverly and Muddy sandstones record this transition. In the Upper Cretaceous, a thick sequence of clastic units was deposited, including the Mowry, Frontier, Steele, Mesaverde, and Lewis formations. The final withdrawal of the seaway occurred prior to the deposition of the Upper Cretaceous Medicine Bow Formation. The total thickness of the Cretaceous section in the center of the Hanna Basin is approximately 6000 m (19,685 ft).

From the latest Cretaceous through the Eocene, the Laramide orogeny generated large uplifted basement blocks and associated intermontane basins. In the Hanna Basin, the Paleocene Ferris and Hanna formations record a dramatic uplift-subsidence couple between

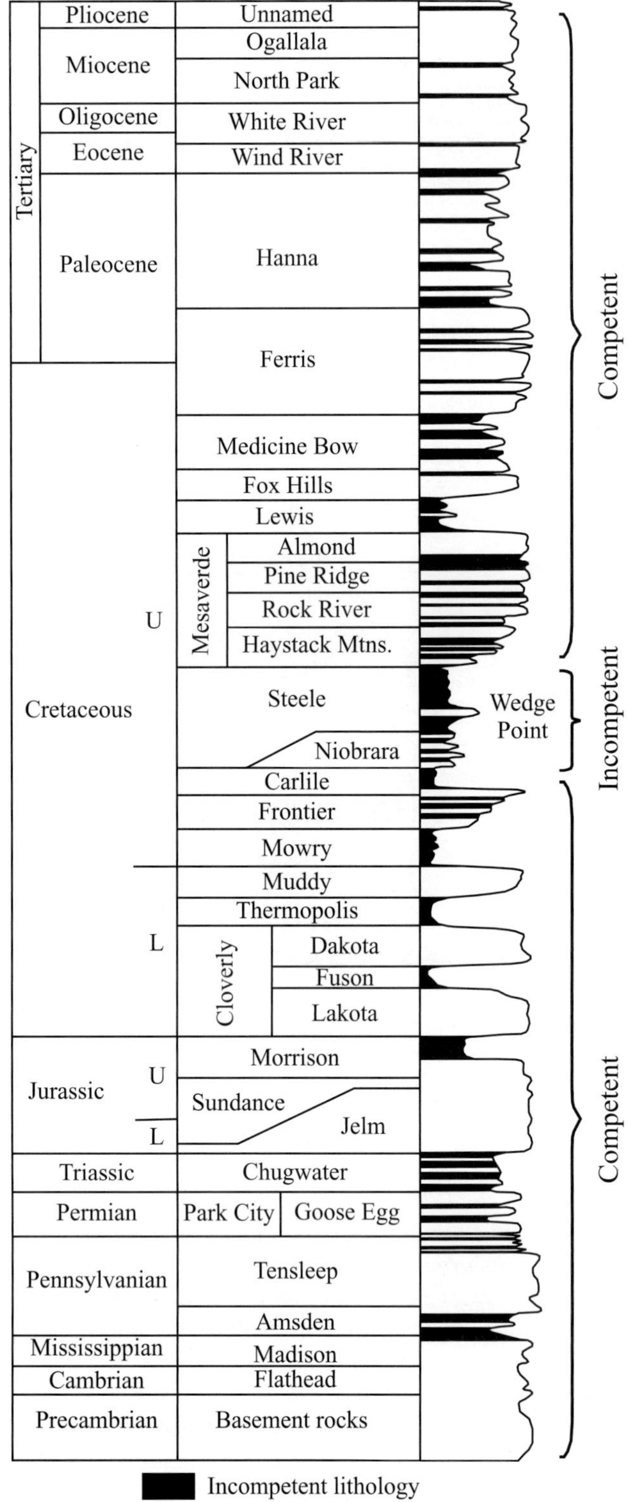

Figure 4. Hanna Basin stratigraphic column. The column on the right summarizes mechanical stratigraphy. Black units represent shale lithologies; white units represent granitic, sandstone, siltstone, and carbonate lithologies. The wedge point in the Hanna Basin occurs within the Upper Cretaceous Steele Shale/Niobrara formations. Mtns. = Mountains; U = Upper; L = Lower.

the surrounding mountain ranges and the basin proper (Figure 4). In the center of the Hanna Basin, Paleocene sediments are greater than 5500 m (18,045 ft) thick. Wireline sonic logs, vitrinite reflectance, and two-dimensional (2-D) seismic data indicate that up to 2000 m (6562 ft) of sediment has been removed in the Hanna Basin region as a result of post-Eocene regional-scale exhumation and erosion.

Hanna Basin Mechanical Stratigraphy

The mechanical stratigraphy of the sedimentary section overlying a granitic basement within the Hanna Basin has an important effect on the structural styles observed along the margins of the basin. With minor exceptions, the Hanna Basin sedimentary section above the basement can be divided into three mechanical units: two competent units separated by a thick regional incompetent unit (Figure 4). The deep competent unit directly overlying the basement comprises Paleozoic through Lower Cretaceous strata, predominantly consisting of layered sandstone and carbonate rocks. The middle incompetent unit is the lower Steele Shale/upper Niobrara Formation, which is overlain by mechanically competent sandstone and siltstone of the Upper Cretaceous Mesaverde Group through the Eocene Hanna Formation, dominated by layered lithologies. Both the upper and lower competent units have maintained approximately constant bed thickness and deformed predominantly by means of flexural slip and faulting deformation mechanisms during the Laramide orogeny, with minor penetrative deformation. In contrast, the intervening, thick, incompetent Steele Shale/Niobrara unit is characterized by significant penetrative deformation and displays bed-thickness variations in response to Laramide age deformation.

Hanna Basin Seismic Program

During 2001 and 2002, Anadarko Petroleum acquired 17 regional, high-quality 2-D seismic lines of 983 km (611 mi) spanning the Hanna Basin. This proprietary program was acquired because the short line length and limited far offsets of the legacy seismic were considered inadequate for interpreting the structural geometry and complexities (including steeply dipping strata) associated with the major fault systems known to exist in the basin. These data were acquired using vibrator vehicles as the primary source, supplemented with shothole dynamite sources in rugged terrain. Typical lines, including the example in Figure 5, are a nominal 50–75-fold with an approximately 5000-m (16,404-ft) maximum receiver offset in a symmetric split-spread configuration. All of these data were prestack time

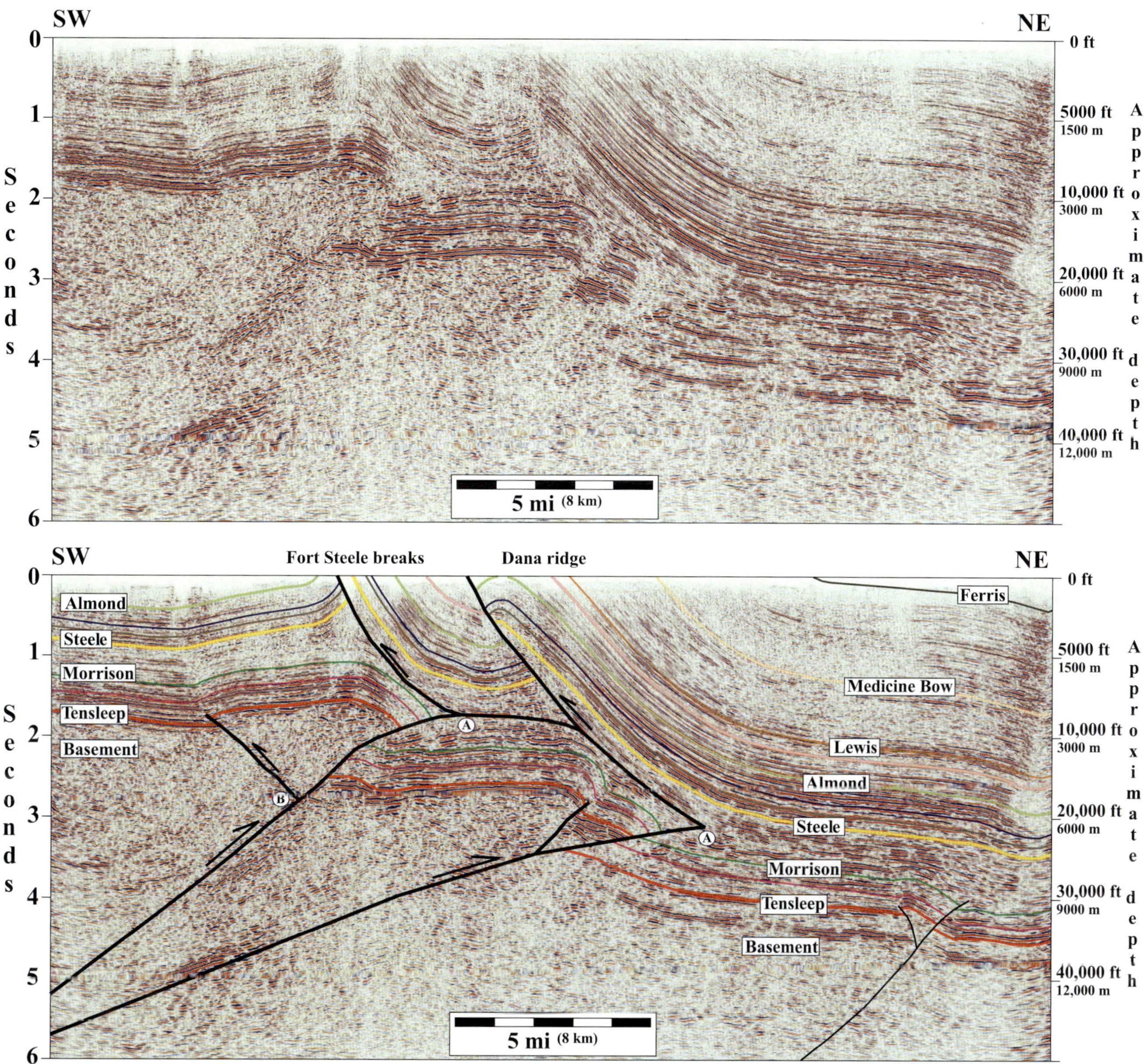

Figure 5. Hanna Basin regional 2-D seismic line. The vertical scale on the southwest end of the section is the time in seconds with a near-surface datum. The approximate depth scale is shown on the northeast end of the section. Stratigraphic picks are based on well data tied to 2-D seismic grid interpretation. Points labeled A signify wedge points of category 1 wedges in the Steel Shale/Niobrara unit. Point B signifies the wedge point of a category 2 basement-involved wedge in which the wedge point is within the basement.

migrated with refraction tomographic statics used in the velocity model building.

Hanna Basin Example 1: Wedge Point in Sedimentary Cover

The regional, long-offset, 2-D seismic data across the southern margin of the Hanna Basin between the Rawlins uplift and Simpson Ridge (referred to informally in the past as the I-80 fault zone) provide a clear image of the basement-involved wedge structures, which define this margin (Figures 3, 5). In this area, the deep faults associated with wedge structures are southwest-dipping, basement-involved reverse faults, which uplift the northern part of the Sierra Madre and Medicine Bow Mountains. These faults cut up section through competent basement, Paleozoic, Triassic, Jurassic, and Lower

Cretaceous strata, ultimately merging with shallow northeast-dipping reverse faults, which detach within the incompetent Upper Cretaceous Steele Shale/ Niobrara Formation (wedge points in Steele Shale/ Niobrara labeled A in Figure 5). Deformation associated with displacement over the shallow northeast-dipping faults generates the southwest-vergent Dana ridge and Fort Steele breaks fault-propagation fold structures observed at the surface (Figures 3, 5). The seismic interpretation in Figure 5 features two basement-involved wedge structures, both similar to the model proposed in Figure 2b (number 4), in which displacement is ultimately consumed up section in a fault-propagation fold structure. Notice that very little penetrative deformation is imaged within either the lower competent sedimentary section (Paleozoic through Jurassic age strata) or the upper competent sedimentary section (lower Tertiary age strata). The lower competent units tend to maintain constant bed thickness over the entire transect. Forelimb folding of the lower competent section is related to a decrease in fault dip at the basement-cover interface, an observed association in other foreland basement-involved structures (Mitra and Mount, 1998). The wedge point in this interpretation is within the Steele Shale/Niobrara Formation (Figures 4, 5). Notice that the Steele Shale/Niobrara stratigraphic unit does not maintain constant thickness but tends to increase in thickness within the anticlinal cores of the Dana ridge and Fort Steele breaks structures and to decrease in thickness in the forelimb areas of the deeper folds associated with the southwest-dipping basement-involved faults. This implies significant penetrative deformation within the Steel Shale/Niobrara interval. The upper competent section (lower Tertiary) tends to maintain a constant bed thickness, suggesting that the dominant deformation mechanism is a nonpenetrative, flexural slip folding. The contrast in mechanical stratigraphy between the thick, incompetent Steele Shale/ Niobrara Formation and the surrounding more competent sedimentary units appears to be an important factor in controlling the stratigraphic position of the wedge point along the southern margin of the Hanna Basin (Figure 4).

Figure 6 is a restoration of the basement-involved wedge interpretation in Figure 5. The restoration assumed a constant line length, a flexural slip deformation mechanism to restore the competent upper and lower layered sedimentary sections, and an area balance of the incompetent Steele Shale/Niobrara interval (Mitra and Namson, 1989). The area of the Steele Shale/Niobrara unit was divided by the length of the top Morrison Formation to calculate the undeformed thickness of the Steele Shale/Niobrara unit. The calculated undeformed thickness for the Steele Shale/ Niobrara unit is approximately 1500 m (4921 ft), which is consistent with the observed thickness of the unit in undeformed areas at the northeast and southwest ends of the section (Figure 5). The basement was restored using an inclined shear deformation mechanism, which varied from vertical to approximately 45° (dipping southwest).

Displacements on the deep and shallow fault systems track each other along strike such that displacement on the deep southwest-dipping fault system decreases to zero directly beneath the lateral termination of the fault-propagation fold structure associated with the shallow, northeast-dipping fault system, indicating that the deep and shallow systems are linked.

Note that Kraatz (2002) presented a very similar wedge geometry structural interpretation for the eastern boundary of the Hanna Basin based on surface, well, and older vintage seismic data (Figure 3). Kraatz (2002) interpreted the Simpson Ridge structure as an east-dipping, west-vergent basement-involved wedge structure with a wedge point within the Steele Shale, which generates a shallow-level, east-vergent fault-propagation fold over a west-dipping reverse fault. The interpretation of the southern margin of the Hanna Basin presented in this chapter and Kraatz's (2002) interpretation of the eastern margin of the basin suggest that the deep and shallow Laramide deformation within the Hanna Basin was mostly decoupled within the Steele Shale/Niobrara Formation.

Hanna Basin Example 2: Wedge Point Within the Basement

The regional, long-offset seismic data across the southern margin of the Hanna Basin also image an example of the second category of basement-involved wedge structure, in which the wedge point is within the basement unit (Figure 2c). At the southwestern end of Figure 5, a small displacement wedge structure where the wedge point is within the basement exists (labeled B in Figure 5), which generates a pop-up structure. In the fault system that generated this pop-up structure, displacement on the deep southwest-dipping reverse fault was partitioned between (1) slip sent to the northeast into the basin (and ultimately wedged back to the southwest at the Steele/Niobrara level) and (2) southwest-vergent slip on the northeast-dipping shallow fault or the upper fault of the basement wedge structure. This interpretation is similar to the basement-detached model in Figure 1c, in which displacement is partitioned between the wedge structure and slip off the right side of the model.

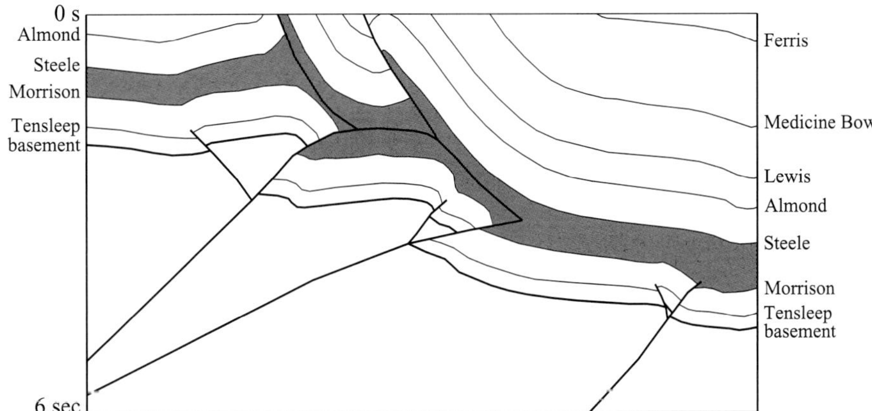

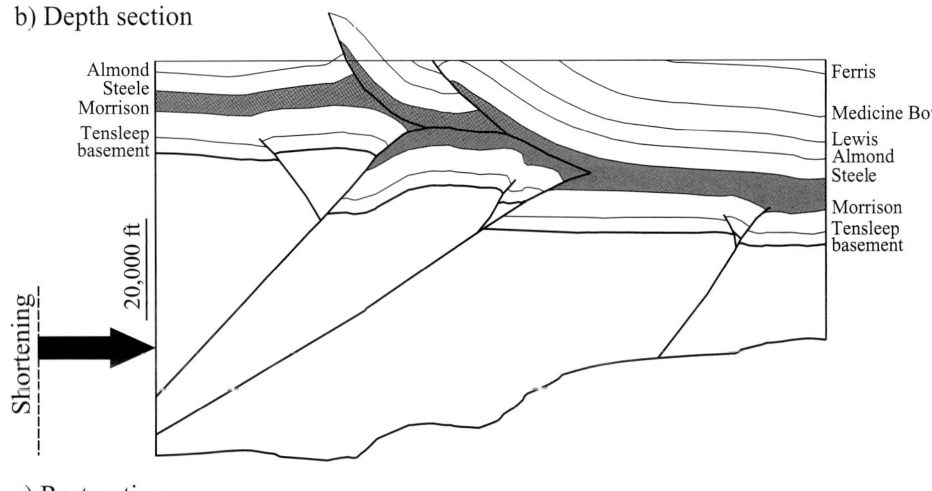

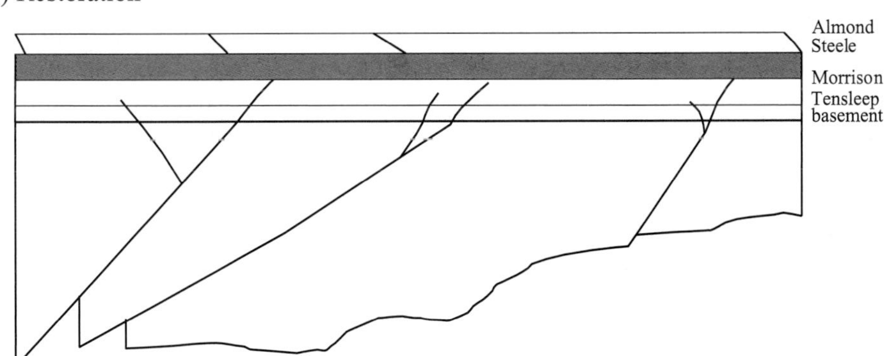

Figure 6. (a) Interpreted time section shown in Figure 5. The incompetent Steel Shale/Niobrara unit is shaded in gray. (b) Depth-converted section; vertical scale is equal to horizontal scale. Velocities used in depth conversion were based on well data and seismic stacking velocities. (c) Restoration of depth section. The incompetent Steel Shale/Niobrara unit shaded gray is restored by area balancing, whereas competent units are restored using constant line lengths assuming a flexural slip deformation mechanism. Basement is restored using a combination of inclined and vertical shear. The restoration indicates approximately 6 km (3.7 mi) of shortening required in the wedge system to produce the observed uplift and shallow-level deformation. Minor overlaps and gaps between blocks in the restored section have been eliminated. The depth conversion and restoration were done using GeoLogic System's LithoTect program.

DISCUSSION

An understanding of the geometry of basement-involved wedge structures developed where they are well imaged, such as along the southern margin of the Hanna Basin, can assist in the interpretation of these structures where they are not well imaged or are partially eroded.

Rabbit Ear or Out-of-syncline Structures

Rabbit ear structures (Brown, 1984) or out-of-syncline structures are secondary structures that are commonly observed along synclines in front of large first-order basement-involved uplifts. These structures have been attributed to crowding in the synclinal core of such structures, resulting in small displacement faulting and

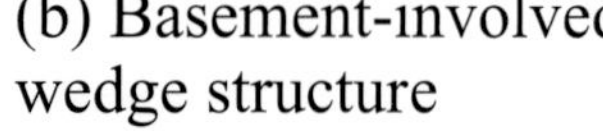

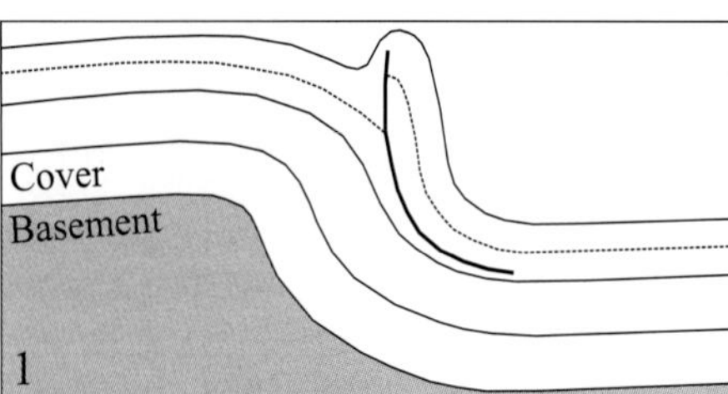

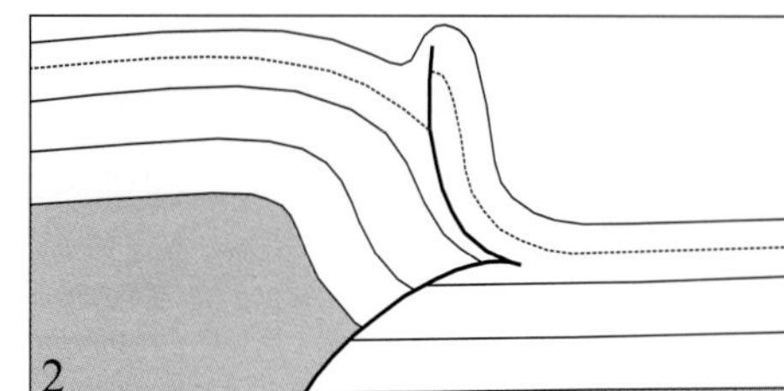

Figure 7. (a) Rabbit ear structure formed by synclinal crowding. (b) Basement-involved wedge structure forming a second-order anticline on the steep limb of the basement-involved structure. (c) Effect of the erosion level of a basement-involved wedge structure on potential interpretations. See the text for details.

associated folding (Figure 7a). If such structures are located on the steep limb of a large first-order, basement-involved structure and have significant displacement, then the possibility exists that these structures are, in fact, basement-involved wedge structures (Figure 7b). In the case where these second-order structures are in fact wedge structures, the reverse fault underlying the rabbit ear fold or out-of-syncline fault actually merges at depth with the large basement-involved fault, which dips in the opposite direction (displacement on the deep fault is responsible for the step in structural relief across the structure and formation of the frontal syncline from which the out-of-syncline fault emanates).

Lost Displacement in an Eroded Wedge Structure Interpretation

The structural style and slip magnitude associated with basement-involved wedge structures can be misinterpreted if the upper part of the structure is eroded. Figure 7c illustrates an example where the upper part of a basement-involved wedge structure has been eroded. If the upper fault of the wedge structure is not recognized, the interpretation would be that displacement on the underlying basement fault decreases to zero where it intersects the synclinal axial surface (or slip is sent off the section laterally). However, in reality, the slip has been wedged back on the upper fault. Given that the wedge point associated with basement-involved wedge structures commonly occurs in thick, incompetent shale sections (such as the Steele Shale/Niobrara interval in the Hanna Basin examples), detecting a nearly bedding-parallel fault at the surface within this interval would be difficult. This situation can make recognition of the wedge structural style history of the structure difficult.

Confusion with Strike-slip-related Interpretations

Because the faults underlying the structural culmination of basement-involved wedge structures dip in opposite directions and merge at depth, wedge structures can easily be confused with, or misinterpreted as, strike-slip flower structures. However, inspection of displacement on the deep and shallow faults (in both dip and strike direction) and the associated structural relief over the deep and shallow parts of the structure can help discriminate between basement-involved wedge and strike-slip-related structural interpretations. The southern margin of the Hanna Basin has been interpreted in the past as a positive flower structure associated with deep strike-slip deformation (the I-80 fault zone). The strike-slip interpretation is in part based on the observation that reverse faults that dip in opposite directions are interpreted to merge into a near-vertical throughgoing strike-slip fault at depth, one of the criteria used to identify flower structures. However, the new seismic data and interpretation presented in this study indicate that the faults, which dip in opposite directions, actually merge at depth at a wedge line and the observed deformation was generated by a predominantly dip-slip deformation associated with a structural wedge. Although the predominant deformation in the structures discussed along the southern margin of the Hanna Basin is interpreted to be dip slip, a component of oblique slip is interpreted to have been involved. The I-80 fault zone occurs at the intersection of the northeast-vergent Sierra Madre and Medicine Bow uplifts to the south with the southwest-vergent Rawlins uplift to the north (Figure 3), which would likely have generated a left-lateral shear couple and oblique left-lateral slip along the faults. Based on regional seismic mapping, the magnitude of strike slip is interpreted to be small on the order of the dip-slip magnitude.

ACKNOWLEDGMENTS

The authors thank Anadarko Petroleum Corporation for the permission to show and discuss the Hanna Basin seismic data presented in this chapter.

REFERENCES CITED

Bellotti, H. J., L. L. Saccavino, and G. A. Schachner, 1995, Structural styles and petroleum occurrence in the sub-Andean fold and thrust belt of northern Argentina, *in* A. J. Tankard, R. Suarez Soruco, and H. J. Welsink, eds., Petroleum basins of South America: AAPG Memoir 62, p. 545–555.

Brown, W. G., 1984, Basement-involved tectonics—Foreland areas: AAPG Continuing Education Course Note Series 26, 92 p.

Erslev, E., P. Hennings, and C. Zahm, 2001, Kinematics and structural closure of basement-involved anticlines in the central Rocky Mountains Petroleum Province (abs.): AAPG Annual Meeting, v. 85, p. A58.

Kraatz, B. P., 2002, Structural and seismic-reflection evidence for development of the Simpson Ridge anticline and separation of the Hanna and Carbon basins, Carbon County, Wyoming: Rocky Mountain Geology, v. 37, no. 1, p. 75–96, doi:10.2113/gsrocky.37.1.75.

Martindale, K., J. Byrd, V. S. Mount, T. Bergstresser, and S. Young, 2004, Along-strike variations in termination of the Darby (Hogsback) thrust, Uinta County, Wyoming: Insights from new seismic data (abs.): AAPG Bulletin, v. 88, no. 13, 1 p.

Medwedeff, D. A., 1990, Geometry and kinematics of an active, laterally propagating wedge-thrust, Wheeler Ridge, California (abs.): AAPG Bulletin, v. 74, p. 717.

Mitra, S., and V. S. Mount, 1998, Foreland basement-involved structures: AAPG Bulletin, v. 82, no. 1, p. 70–109.

Mitra, S., and J. S. Namson, 1989, Equal-area balancing: American Journal of Science, v. 289, p. 563–599.

Shaw, J. H., C. Connors, and J. Suppe, 2004, Seismic interpretation of contractional fault-related folds: AAPG Studies in Geology 53, 156 p.

Suppe, J., and D. A. Medwedeff, 1990, Geometry and kinematics of fault-propagation folding: Eclogae Geologicae Helvetiae, v. 83, no. 3, p. 409–454.

13

Mencos, J., J. A. Muñoz, and S. Hardy, 2011, Three-dimensional geometry and forward numerical modeling of the Sant Corneli anticline (southern Pyrenees, Spain), *in* K. McClay, J. H. Shaw, and J. Suppe, eds., Thrust fault-related folding: AAPG Memoir 94, p. 283–300.

Three-dimensional Geometry and Forward Numerical Modeling of the Sant Corneli Anticline (Southern Pyrenees, Spain)

J. Mencos

Institut de Recerca GEOMODELS and GGAC (Grup de Geodinàmica i Anàlisi de Conques), Departament Geodinàmica i Geofísica, Universitat de Barcelona, Barcelona, Spain

J. A. Muñoz

Institut de Recerca GEOMODELS and GGAC (Grup de Geodinàmica i Anàlisi de Conques), Departament Geodinàmica i Geofísica, Universitat de Barcelona, Barcelona, Spain

S. Hardy

ICREA (Institució Catalana de Recerca i estudis Avaneats) and GGAC (Grup de Geodinàmica i Anàlisi de Conques), Departament Geodinàmica i Geofísica, Facultat de Geologia, Universitat de Barcelona, Barcelona, Spain

ABSTRACT

This work is based on a three-dimensional (3-D) reconstruction methodology for geologic structures from field and subsurface data. The methodology consists of several steps: (1) collection and georeferencing of data, i.e., 3-D digitalization; (2) analysis of data and definition of a 3-D geometric and stratigraphic model, i.e., structural and stratigraphic analysis; and finally, (3) the reconstruction of key surfaces that form the structure, i.e., reconstruction of the reference surface and reconstruction of additional surfaces, both honoring the defined geometric model. The methodology has been applied to a natural example, the Sant Corneli anticline, a thrust-related fold located in the southern Pyrenees. This fold, oriented approximately east–west, has a complex 3-D geometry, with stratigraphic and structural variations, both laterally and vertically. This chapter focuses on the study of the Late Cretaceous postrift series. Outcrop information has been collected to reconstruct the superficial geometry of the Sant Corneli anticline. Seismic profiles that cross the area have been interpreted to reconstruct the geometry of the thrust fault at depth. The reconstruction methodology at surface and at depth is made following the same workflow, adapting it to the different nature of the original data sets. At the same time, the use of 3-D forward numerical modeling allows us to explore the relationship between the fold and the thrust, the main factors that influence fold development through space and time, and to find the kinematic model(s) that best fit the proposed 3-D reconstruction.

The main contribution of this work is the incorporation of forward modeling techniques in the 3-D reconstruction workflow to independently establish the relationship between the

DOI:10.1306/13251342M943434

various reconstructed geologic surfaces. Once this relationship is known, it is possible to refine the reconstructed surfaces and integrate them in a single and comprehensive 3-D model that honors both the available data and the established kinematic model. This technique is also used to analyze the possible kinematic evolution of a fault-related structure that best reproduces the deformed geometry obtained through the 3-D reconstruction process. Moreover, this study also improves the developed 3-D reconstruction methodology as it incorporates the use of isopach maps.

INTRODUCTION

Much of the work concerning the characterization of fold and thrust structures and the relationship between them has classically focused on the search for, and definition of, two-dimensional (2-D) geometric and kinematic models, typically parallel to the thrust transport direction (Suppe, 1983; Suppe and Medwedeff, 1990; Mitra, 2002). Recently, the development of new techniques for data acquisition, management, analysis, and modeling and visualization has opened up the possibility to analyze the relationships between folds and faults directly in three dimensions (i.e., acquisition of three-dimensional [3-D] seismic data sets and recent 3-D interpretation and modeling platforms such as The Kingdom Suite© manufactured by Seismic Micro-Technology, GOcad© by Paradigm, etc.). As a consequence, nowadays, the 3-D reconstruction and modeling of structures constitute a main goal for both industry and the academic community. Actually, most of these techniques are strongly dependent on the amount and quality of the data, such as 3-D seismic survey data sets. In the cases where data rely on 2-D seismic lines or surface field data, 3-D geometric models of thrust-related folds have mostly been based on the linear interpolation of serial 2-D cross sections (Husson and Mugnier, 2003; Banerjee and Mitra, 2004, 2005). With such a procedure, the reliability of the 3-D model will depend on the quality of the constructed cross sections, their spacing, and the match between the interpolation vectors and the plunge lines characterizing the folds involved (Carrera et al., 2009). The 3-D reconstruction of a fault-related fold from cross sections will also be biased by the fault-fold deformation model applied to interpret the structure during the process of cross section construction or even to interpolate data at an earlier stage.

This chapter aims to describe a new methodology for the 3-D reconstruction of fault-related folds and their interpretation, and its application to the Sant Corneli anticline (southern Pyrenees, Spain). This study deals with a purely 3-D approach for both structural reconstruction and forward numerical modeling. The cornerstone of the workflow relies on the reconstruction of the folded surfaces independently of the fault surface at depth. This approach has been chosen with consideration for the differences among available data sets. The source data for the geologic horizons come mostly from field data and well information, whereas the source data for the thrust correspond only to subsurface information. The relationship between fold and fault cannot be observed directly in the field for all of the studied area. We subsequently analyze these relationships by 3-D forward numerical models, avoiding the need to establish a fault-fold deformation model prior to the 3-D reconstruction to reproduce both structures. In practice, this consideration gives priority to the original source data over a predefined theoretical model. The methodology of 3-D reconstruction deals with data in their present 3-D coordinates without projection into cross-sectional planes and is made based on geometric construction rules and a dip-domain assumption. Thus, this approach minimizes errors related to data projection and favors the detection of along-strike irregularities. It can be applied both to surface and subsurface data and takes advantage of 3-D mapping techniques in a Geographic Information System (GIS) environment. This methodology has been applied to the Sant Corneli anticline in the southern Pyrenees. The excellent outcrops of this major thrust-related fold, the structural level of exposure, the unique preservation of the growth sequence, and the availability of subsurface data (seismic sections and exploration wells) enable the development and application of the methodology explained in this chapter. Moreover, the availability of detailed digital topography and color orthophotomaps allows us to test 3-D mapping techniques, which are an important part of the 3-D reconstruction methodology.

GEOLOGIC SETTING OF THE SANT CORNELI ANTICLINE

Regional Geology and Stratigraphy

The Sant Corneli anticline is located at the thrust front of the Bóixols thrust sheet, in the southern Pyrenees

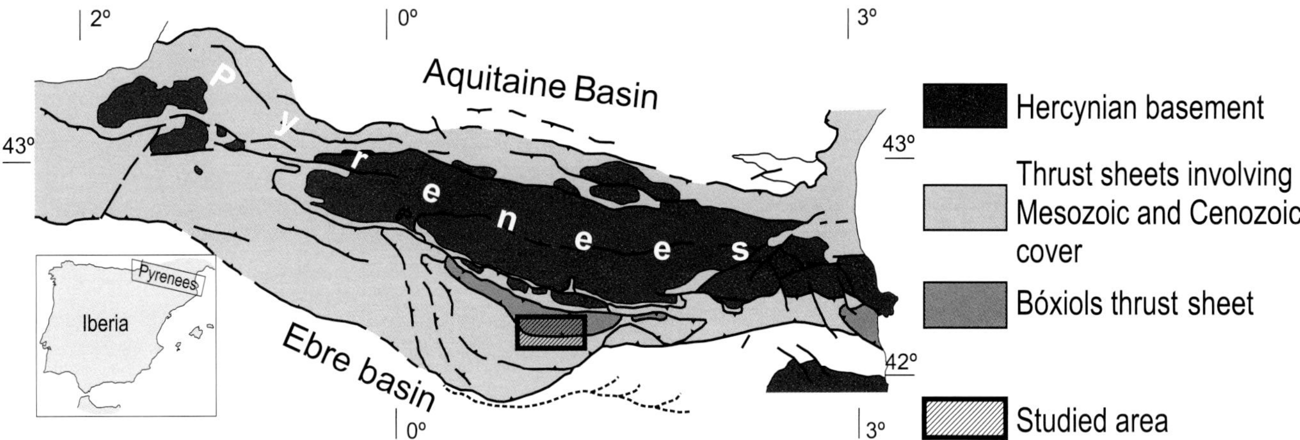

Figure 1. Simplified geologic map of the central Pyrenees. The hatched box indicates the study area.

(Figure 1). The Bóixols thrust sheet is a major structural unit of the central Pyrenees that incorporates Mesozoic rocks. It is bounded to the north by the Nogueres thrust sheet, constituted mainly by basement rocks, and to the south by the Graus-Tremp Basin, infilled with Mesozoic and Cenozoic rocks and being the foreland basin of the Bóixols thrust sheet during its emplacement.

The Bóixols thrust sheet resulted from the inversion in the Late Cretaceous of an earlier system of extensional basins (Bond and McClay, 1995; Muñoz, 2002). These basins are part of a rift system developed during the Early Cretaceous caused by the separation of Iberian and European plates. During this extensional period, a thick series (up to 4 km [2.5 mi]) of synrift marine sediments accumulated in the hanging walls of extensional faults, which limited these basins to the south (García-Senz, 2002).

Thereafter, during the Alpine orogeny, a change to convergence between the two plates first caused the inversion of these basins and their later incorporation into the Pyrenean thrust sheets. This is the case for the Organyà extensional basin, which was inverted and evolved into the subsequent Bóixols thrust sheet (Garrido-Mejías, 1973; Cámara and Klimowitz, 1985; Berástegui et al., 1990; Bond and McClay, 1995; García-Senz, 2002; Muñoz, 2002).

The Sant Corneli anticline involves a thick Mesozoic succession detached on top of Triassic evaporites. The prefolding sequence consists of up to 5 km (3.1 mi) of pre-, syn-, and postrift carbonates ranging in age from Jurassic to Upper Cretaceous. Prerift Jurassic materials are limestones and dolomites. Synrift Lower Cretaceous sediments evolve from a lower unit of limestones to upper marly formations (Font Bordonera, Lluçà formations, Figure 2) ending up with the development of a carbonate platform corresponding to the Coll d'Abella Formation, Figure 2. A section is presented (Figure 3) illustrating how this synrift sequence displays considerable thickness variations in a north–south direction. These thickness variations may be due to the paleogeographic control of the rift system location as it will be discussed in this chapter. The postrift is represented by a 1-km (0.6-mi)-thick sequence of upper Cenomanian–lower Santonian limestones (Simó, 1986), which corresponds to the Santa Fe Formation, Reguard and Congost formations, Collada-Gassó formation, and Montagut and Aramunt members of the Sant Corneli formation (Figure 2). These various postrift formations best delineate the geometry of the Sant Corneli anticline at surface (Figure 3). They have been the focus for detailed mapping of the structure and the 3-D reconstruction of the surfaces used to decipher the fold geometry. The synfolding sequence (upper Santonian to Maastrichtian) grades laterally and vertically from slope sediments (Collades member, Podega member, and Herbasavina formation, Figure 2) into the shallow-marine and fluvial formations (Mutti and Sgavetti, 1987; Arbués et al., 1996; Ardèvol et al., 2000) of the Areny group (Figure 2). They seal the main thrust and expand over all of the structure in the western part of the studied area (Figure 3).

Structure

The Sant Corneli anticline is a major east–west regionally trending fold, which crops out for more than 50 km (31 mi). In the eastern part of the studied area, it is cored at surface by the Lower Cretaceous synrift sediments. There, it is referred to as the Bóixols anticline (García-Senz, 2002). Younger sediments crop out westward in the core of the anticline as a result of its general plunge toward the west.

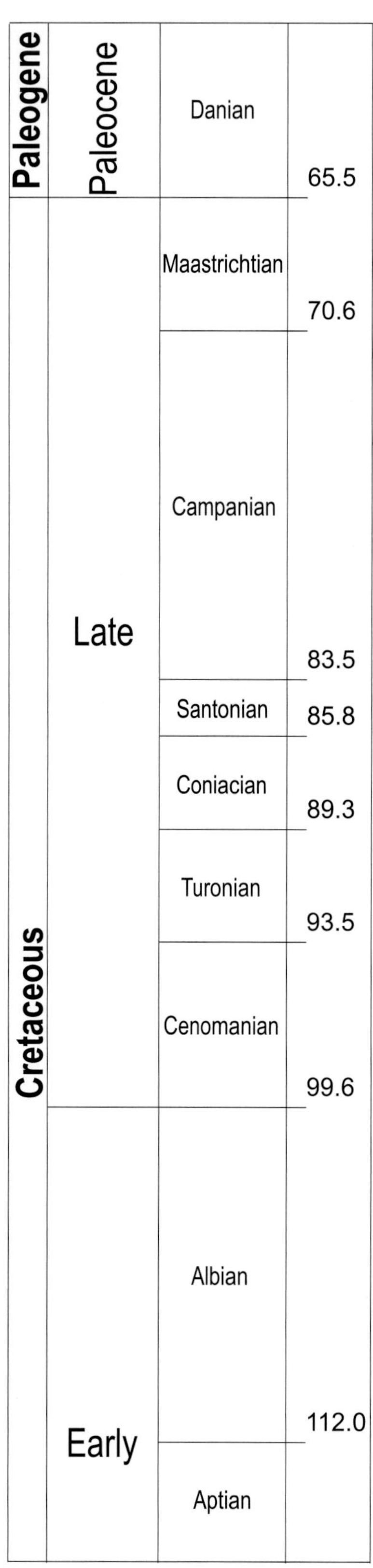

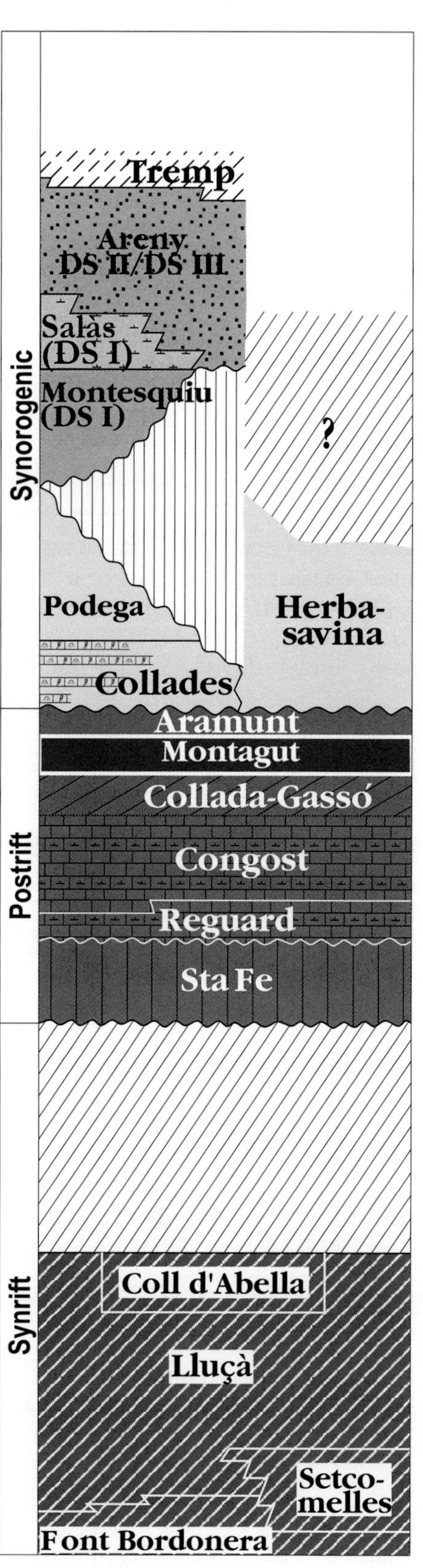

Figure 2. Synthetic stratigraphic diagram showing the differentiated units in the study area. The Montagut unit (highlighted) corresponds to the reference horizon for 3-D reconstruction. Sta Fe = Santa Fe; DS = depositional sequence.

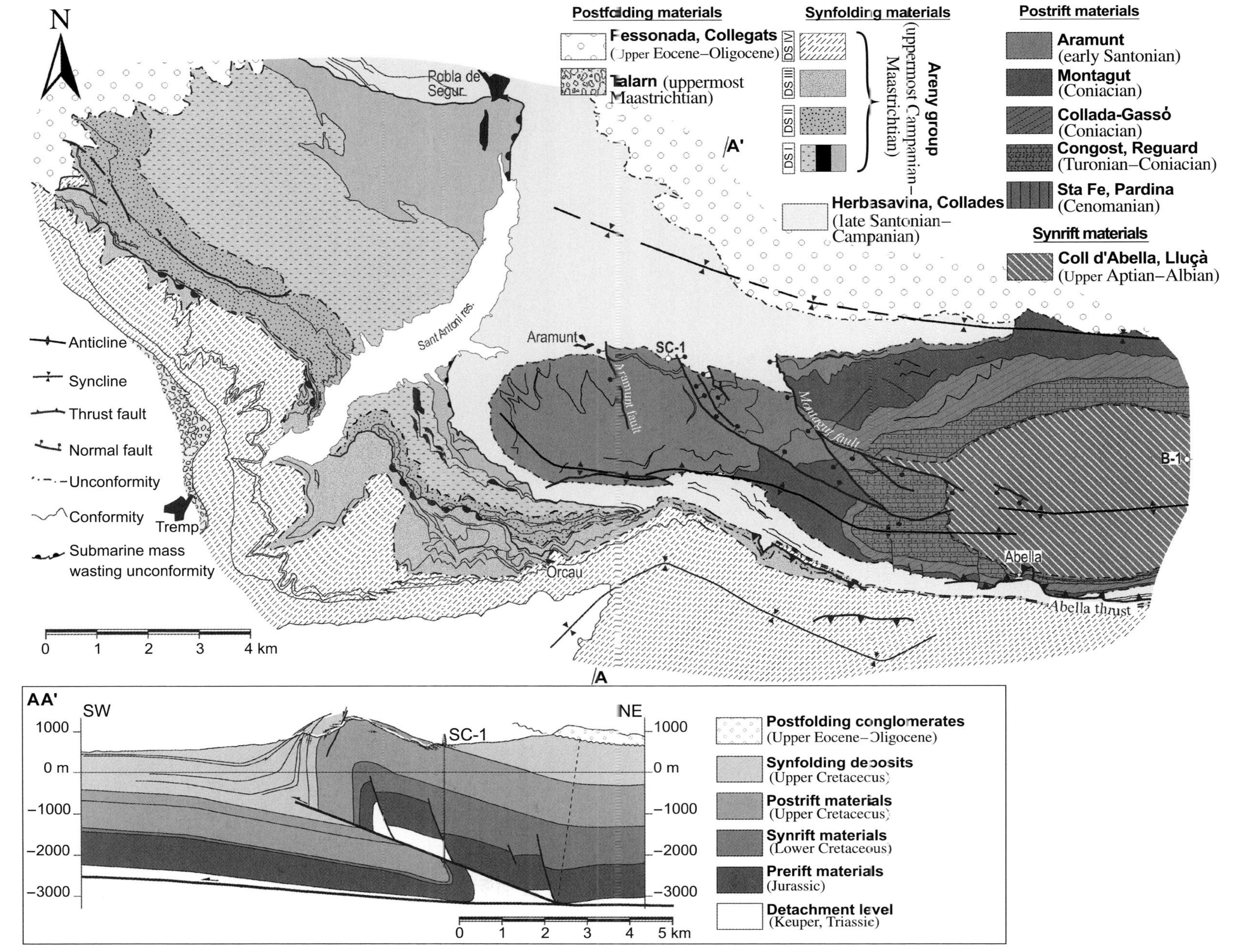

Figure 3. Geologic map of the Sant Corneli anticline and regional cross section across the structure (AA′). Mapping of synfolding sediments is based on the work of P. Arbués (P. Arbués, 1987, personal communication). SC = Sant Corneli; Sta Fe = Santa Fe; res. = reservoir.

Structure at Surface

In the studied area, the Sant Corneli anticline is characterized by a gently and constantly dipping backlimb and a vertical to overturned frontal limb. Because of this geometry, it has been interpreted as a fault-propagation fold (Bond and McClay, 1995). The thrust related to the Sant Corneli anticline (Bóixols thrust) does not crop out in the studied area. It does, however, crop out farther east (García-Senz, 2002). However, in the study area, some minor thrusts with hundreds of meters of displacement deform the frontal limb. One of them spectacularly crops out in the Abella de la Conca village, where a 2.5-m (8.2-ft)-wide shear band deformation zone with an S-C (S, from the French term Schistosite, the main foliation oblique to shear zone boundary; C, from the French term Cisaillement, shear zone boundary) fabric delineates the thrust at surface (Figure 3).

In the western part of the studied area, close to the Sant Antoni reservoir (Figure 3), the topographic termination of the fold is controlled by the periclinal closure of the postrift carbonates below the marly Herbasavina formation. No agreement exists on the significance of this termination, which coincides with an abrupt increase of the fold plunge. The classic interpretation is that this periclinal feature corresponds with the western termination of the structure, but it is not clear if the fold continues farther west below the synfolding sediments. This is one of the questions that have been addressed with the 3-D reconstruction and forward modeling discussed later.

Oblique extensional faults occur in the backlimb of the anticline. They show a concave geometry in map view, trending northwest–southeast in the north and becoming parallel to the east–west trend of the fold southward (Figure 3). These extensional faults are restricted to an 8-km (5-mi)-wide strip in the anticline. They have dips ranging from 60° to almost vertical and are observed at different scales from a few meters to hundreds of meters. The most important ones (i.e., Montagut fault) dip to the southwest favoring the outcrop of younger sediments to the west as the plunge of the fold does (Figure 3). Conjugate northwest-dipping normal faults also occur leading to a significant compartmentalization of the backlimb into fault blocks.

Subsurface Information and Structure at Depth

The available subsurface information in the area consists of five seismic profiles, three of them perpendicular to the structure (north–south) and the other two longitudinal (west–east) (Figure 4). These profiles belong to a seismic survey undertaken together with the drilling of several exploration wells during the sixties, two of them in the studied area (Sant Corneli 1 well [SC-1] and Bóixols 1 well [B-1]; Lanaja, 1987) (Figure 4). The quality of the seismic profiles is relatively poor but good enough to reasonably constrain the geometry of the thrust and the backlimb of the anticline. Moreover, they provide the opportunity to observe a first-order relationship between the fold and the thrust, which is a parallelism between the fold backlimb and the thrust ramp. The wells were drilled into the northern limb of the fold and have been used to validate the velocity model necessary to convert seismic data from time to depth and to facilitate the integration of subsurface and surface data.

From the cross sections (Figures 3, 4), we can observe that most of the synrift sequence terminates against high-angle extensional faults in the hanging wall of the Bóixols thrust. These faults are well preserved and exposed farther east, outside the studied area, where lower structural levels are exposed (Berástegui et al., 1990; García-Senz, 2002).

The SC-1 well (Lanaja, 1987) reached the Bóixols thrust and continued into its footwall, which is characterized by an overturned succession of Upper Cretaceous, Lower Cretaceous, and Jurassic sediments (Figures 3, 4). The Lower Cretaceous of the footwall is much reduced (no more than 30 m [98 ft] in thickness) and strongly contrasts with the thick succession that was drilled in the hanging wall, which reaches 750 m (2461 ft).

The B-1 well has a total depth of 1750 m (5741 ft) and was drilled into the Lower Cretaceous package in the hanging wall of the thrust, without reaching either the top or the base of the synrift sediments, demonstrating the thickening and expansion of the synrift sequence not only northward but also eastward (Figure 3).

Taken together, these relationships are consistent with the idea that the Bóixols thrust resulted from the inversion of an Early Cretaceous extensional fault system, similar to that observed farther east (García-Senz, 2002). The SC-1 well is interpreted to have reached one of these extensional faults because here, the upper part of the Jurassic succession lies directly on top of the Triassic evaporites (Figures 3, 4) (García-Senz, 2002).

Thus, it appears that the forelimb of the Sant Corneli anticline developed in the footwall of the main extensional fault bounding the Lower Cretaceous basin to the south, with this main fault at present remaining in the fold core. Thus, the Bóixols thrust represents a shortcut of the previous extensional Organyà fault that was developed during the reactivation of the extensional fault system. The geometry of the extensional faults and the thickness of the synrift sequence suggest that these faults probably involved the Variscan basement, with a downdip displacement able to accommodate almost 4000 m (13,123 ft) of synrift sediments (García-Senz, 2002).

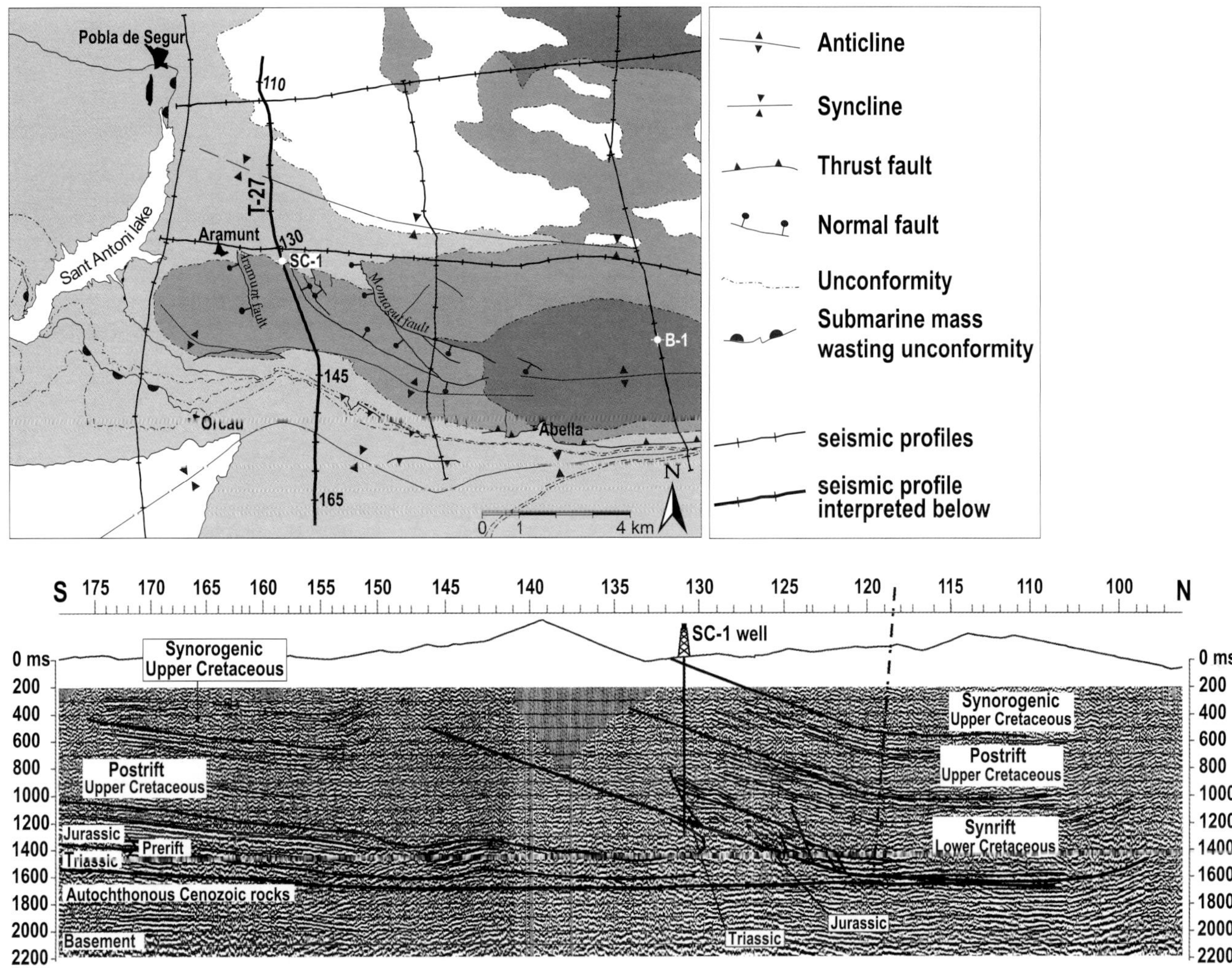

Figure 4. Location (above) and example (below) of the seismic profiles used in this study.

However, we can observe that rocks no older than the upper Triassic evaporites are involved in the Bóixols thrust sheet (Figure 3). Its basement would be part of the Nogueres thrust sheet, at present cropping out farther north (Beaumont et al., 2000). This is a key problem to explain in the evolution of the Bóixols thrust sheet and the Sant Corneli anticline and one of the main issues to be addressed with 3-D reconstruction and forward modeling.

The present configuration suggests two distinct phases in the evolution of the Sant Corneli anticline. The first phase is a stage of inversion and structural uplift of the Early Cretaceous basin, characterized by the reactivation of the faults controlling the rift system in an inversion tectonic regime. The second is a phase in which the whole basin was detached on top of the Triassic evaporates and thrust southward, being incorporated into the Bóixols thrust sheet, characterized by the development of the Bóixols thrust in a thin-skinned tectonic system.

THREE-DIMENSIONAL RECONSTRUCTION METHODOLOGY

The methodology used in this chapter aims to reconstruct the geologic surfaces that best represent the 3-D geometry of the fold (Figure 5). The term "reconstruction" defines the process of reproducing in a 3-D georeferenced environment the geometry of folded surfaces from original scattered and incomplete data. This process occurs by means of the interpolation and extrapolation of data following some methodological and geometric assumptions explained in detail in this section. The 3-D reconstruction methodology comprises all steps from data acquisition to the creation of the final 3-D image of the structure under study and is an interactive and iterative process in which each step is used to validate the previous one and is used as a starting point for the following one.

In this study, the reconstruction methodology deals in particular with the geometry of the resistant beds of

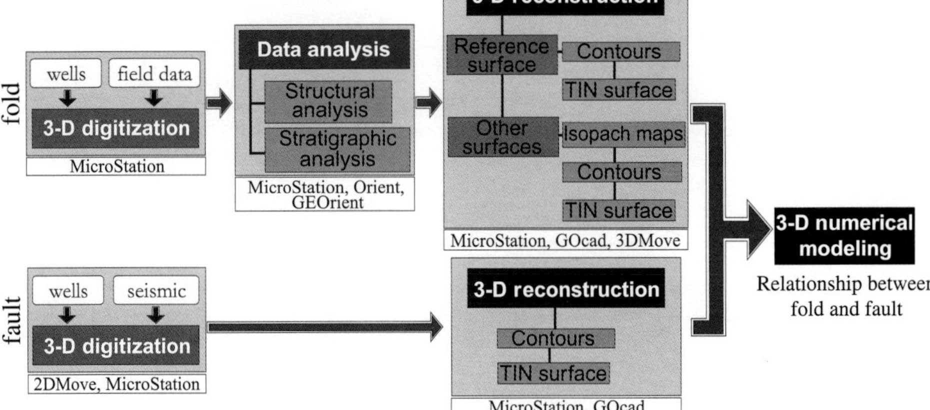

Figure 5. Workflow for the methodology reported here. The reconstruction of the fold at surface and of the fault at depth has been done independently and thereafter related using a 3-D numerical modeling technique. The final 3-D reconstruction must honor all data according to the degree of certainty of the original data and the geometric and stratigraphic model defined. TIN = triangulated irregular network.

the postrift sequence. In this way, we avoid structural complexities observed in the older synrift sediments as a result of the inversion of the previous extensional features. Moreover, the postrift beds crop out extensively in the studied area and control the present topography, which facilitates the use of 3-D mapping techniques (Figure 6). The 3-D methodology applied is based on the concepts and the workflow previously explained by Fernández (2004). Here, we go further in that we incorporate 3-D forward numerical modeling in the workflow to investigate the relationships between the fold and the thrust to which it is related (Figure 5). Furthermore, we also introduce a semiautomated technique based on isopach maps to reconstruct additional surfaces once an initial reference geologic horizon is generated.

The methodology proposed employs any kind of available data; in the studied case, the data are derived from 3-D mapping techniques, field data, and subsurface data. The process of reconstruction involves distinct steps, each one of them undertaken using dedicated software (Fernández, 2004; Fernández et al., 2004).

Three-dimensional Digitization

The initial part of this process consists of the compilation of field data and their positioning in 3-D space. This occurs using MicroStation© (manufactured by Bentley Systems, Incorporated). When working with seismic data, The Kingdom Suite, and 2DMove™ (by Midland Valley), are used.

Field data have been positioned and georeferenced using digital terrain models (DTMs) constructed from digital topography and draped by orthophotographs at a scale of 1:5000 (Figure 6). This process has a degree of uncertainty of ±5 m (16 ft) with respect to the original information, which is the size of the DTM mesh.

Map traces of stratigraphic horizons and formation contacts as well as faults have been traced on the 3-D terrain model (Figure 6). Several steps can be performed to analyze, validate, or densify field data. This is possible because of the use of an in-house software designed specifically for this purpose (Fernández, 2004, 2005; Fernández et al., 2004), implemented within MicroStation software. Some of these operations are explained as follows.

1) Three-dimensional digitization enables us to extract dip data by calculating best-fit planes through the planar regression of digitized map traces (Fernández, 2005). Thus, we can maximize the information derived from geologic mapping because it is possible to extract information from the DTM. Orientation values can be compared to the dip data directly obtained in the field or even to validate them by intersecting a plane with a known orientation with the DTM. In this way, map trace analysis can provide values of orientation for areas without direct field measurements and help densify the field structural data.
2) Knowing the 3-D location and orientation of two single points (i.e., two different horizons at two different points), calculating the stratigraphic separation between them is possible.
3) Given the location and orientation of a map trace or any type of georeferenced element, this can be projected perpendicular to bedding to validate stratigraphic measurements done in the field or even to densify data.

In the same 3-D space, we can georeference all the information from subsurface media.

Structural and Stratigraphic Analysis

The objective of this step is to define a geometric model that will be the framework for the generation of the multiple surfaces that form the structure. This geometric

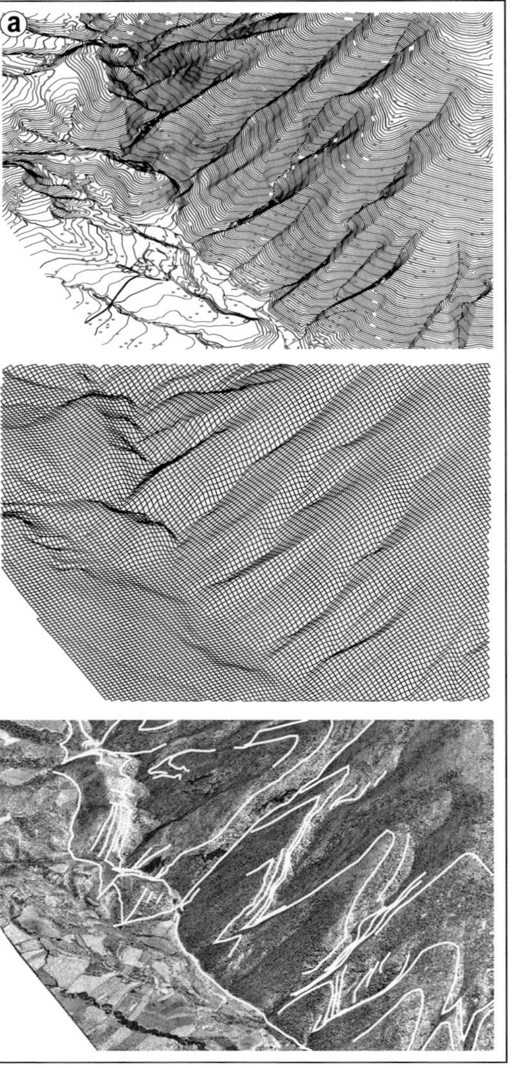

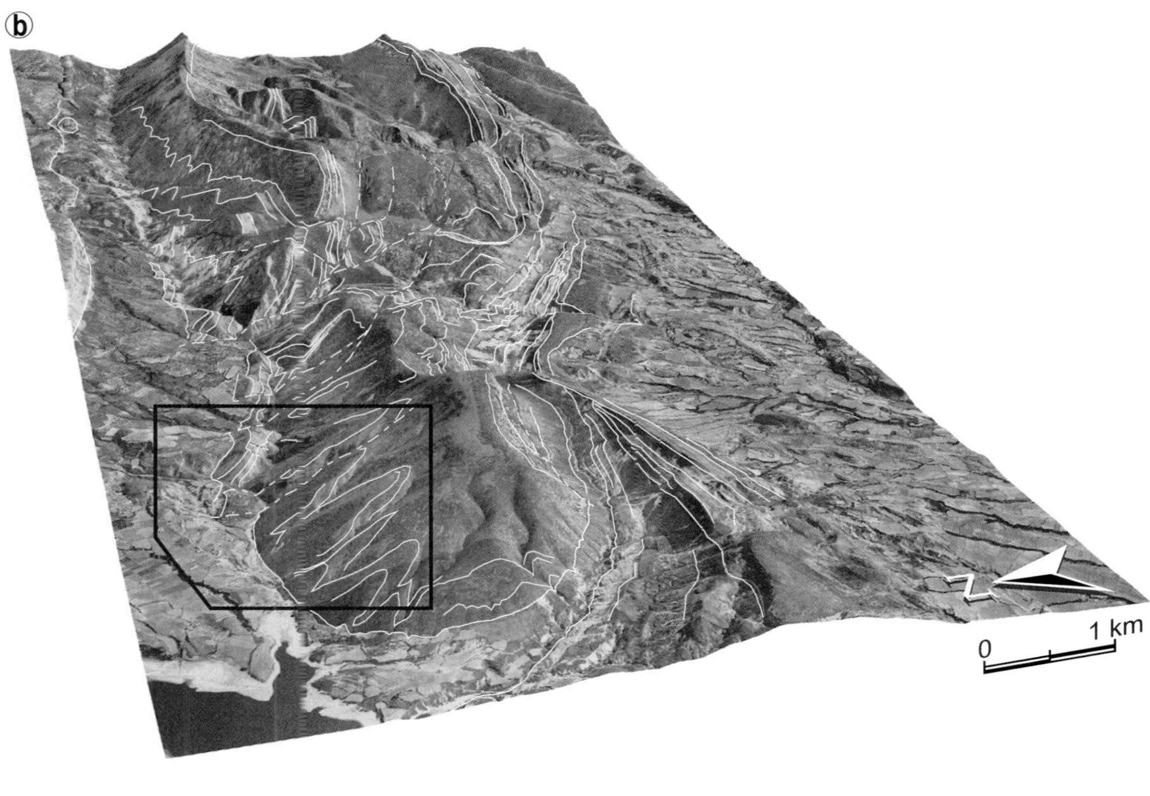

Figure 6. (a) Process of construction of a digital terrain model (DTM) from digital topography and orthophotomaps. (b) View of the 3-D digitization performed in the Sant Corneli anticline. White lines are digitized stratigraphic contacts and map traces; dotted lines correspond to faults.

model has to incorporate the maximum amount of data possible, considering the main structural and stratigraphic features of the structure. This step occurs through the structural analysis of the available data together with the determination of stratigraphic relationships between georeferenced elements.

This process is performed with the aid of in-house applications developed within MicroStation software to assist in the structural and stratigraphic analysis of data (Fernández, 2004), in combination with classical structural analysis programs such as Orient© (Charlesworth, H.) or GEOrient© (Holcombe, R.).

In our example, the geometric model created is based on the dip-domain method, which assumes that structures can be subdivided in volumes within which the attitudes of stratigraphic horizons remain constant (Figure 7). This is equivalent to the dip-domain method used in the construction of sections in two dimensions. The details of this method can be found in Fernández et al. (2004). The result of the structural analysis will be a 3-D representation of all available structural data (even field data or secondary data obtained through 3-D digitization) subdivided into volumes of constant orientation, together with the location and orientation of the dip-domain boundaries. Moreover, main structural vectors and secondary structures will also be identified and incorporated into the geometric model (i.e., fold axis orientations, existence of minor folds, etc.).

The geometric model is completed with the analysis of stratigraphic separations, which consists of the determination of stratigraphic thicknesses between geologic horizons and their variation all over the studied area. This is done using primary field measurements or information obtained during the 3-D digitization process. This is necessary to optimize the structural analysis because it enables the stratigraphic projection of data to constrain areas of a particular surface with incomplete information. It can be easily incorporated into the geometric model as annotations in the 3-D space.

Using this methodology, a dip-domain model has been generated and combined with a 3-D model of stratigraphic separations to project dip data into the surfaces that will be reconstructed.

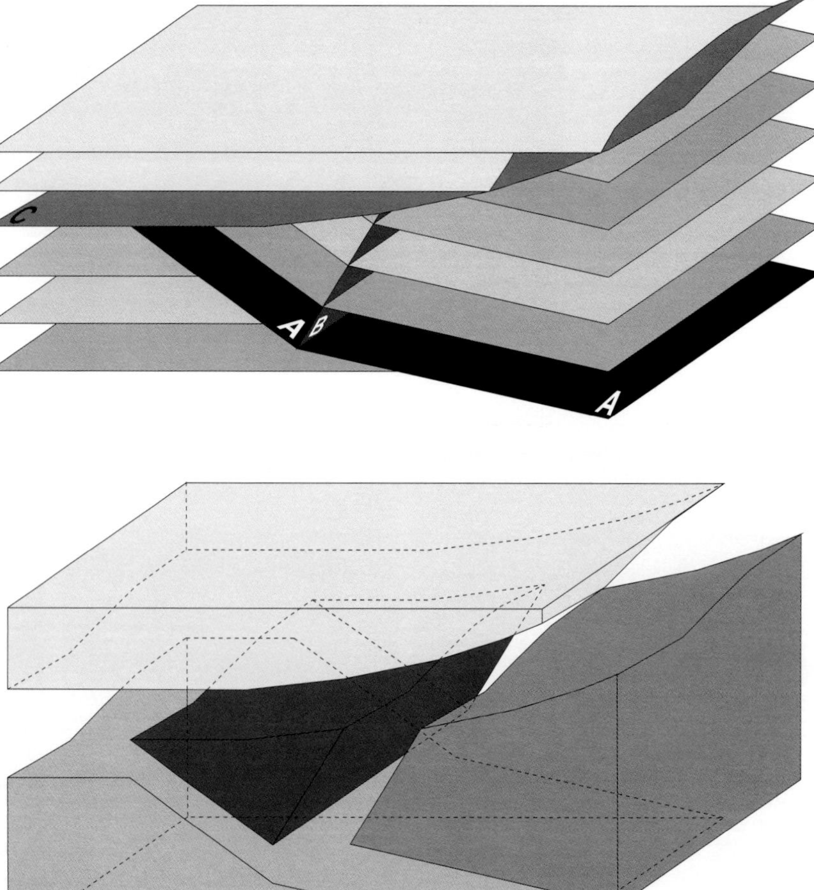

Figure 7. Schematic diagram of the dip-domain method used for the geometric analysis of dip data. (top) Stratigraphic horizons distributed in dip domains limited by different types of boundary surfaces. A = thrust surface; B = axial surface; C = unconformity. (below) Volumes within which the attitude of stratigraphic horizons remains constant. Modified from Fernández (2004).

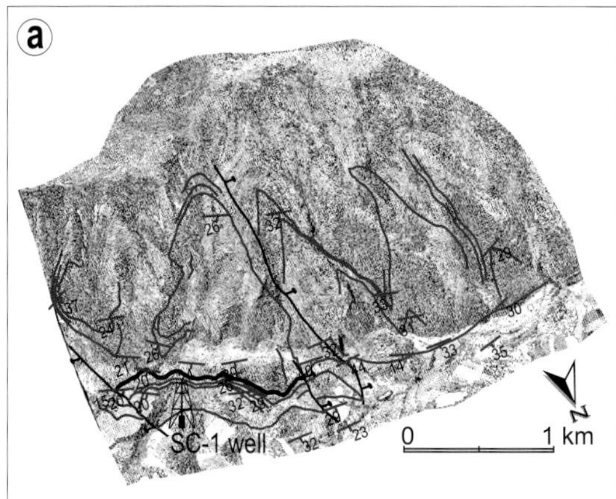

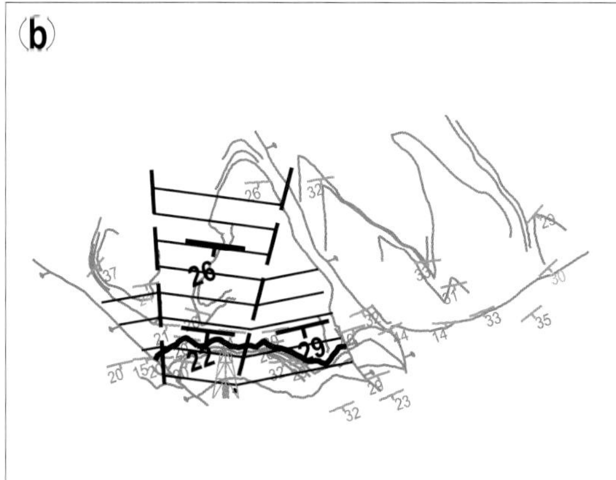

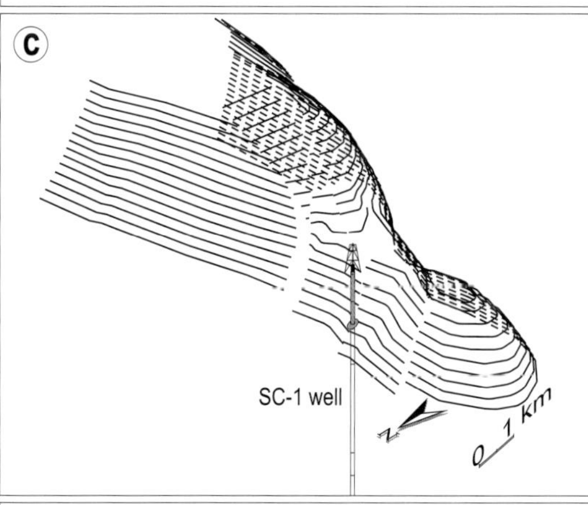

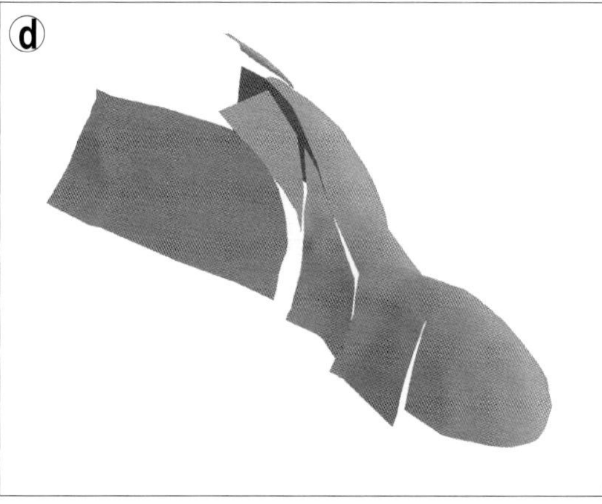

For 3-D reconstruction, the best practice involves working with a single surface at a time. Once a satisfactory geometric model is obtained, the process of horizon reconstruction begins with the generation of a particular surface called the reference surface (Figure 8).

Reconstruction of the Reference Surface

The reference surface represents a geologic surface (commonly a stratigraphic horizon) for which we have a great deal of information and/or which can be identified all along the structure. The importance of defining such a reference surface is that accumulated uncertainty during surface generation will be minimized because the reconstruction will have less degree of freedom. In this study, this surface corresponds to the base of the Montagut unit (Coniacian).

Having defined the geometric model and knowing the stratigraphic position of this reference horizon, we proceed with the generation of the structural contours for that surface (Figure 8). This is done within MicroStation software and taking advantage of in-house applications. Remember that the generation of structural contours is constrained by the position and boundaries of the defined dip-domain volumes and the model of stratigraphic variations. This means that once the stratigraphic position of a surface within a dip domain is known, the generated contours for that surface part will be extended until reaching the dip-domain boundary. This has to be done consecutively through all the dip-domain volumes, honoring both the stratigraphic position and the mean orientation calculated for that surface part in the corresponding dip domain.

Thereafter, using GOcad (manufactured by Paradigm), or alternatively 3DMove© (manufactured by Midland Valley), a triangulated irregular network (TIN) surface is generated (Figure 8). The algorithms used to generate the TIN surface are based on the discrete smooth interpolation method (DSI of Mallet, 1989) when working in GOcad, or alternatively based on the Delaunay triangulation method (Delaunay, 1934) when using 3DMove.

During this stage, defining and reconstructing key structures that can control the position and geometry of the dip domains is necessary (i.e., A in Figure 7). In

Figure 8. Sequence of images illustrating the process of reconstruction of the reference surface. (a) Available data georeferenced. (b) Construction of the geometric and stratigraphic model following the dip-domain method. (c) Construction of the structural contours of the reference surface. Note the presence of the SC-1 well used to constrain the stratigraphic position of the reference horizon. (d) Triangulated irregular network (TIN) surface generated. SC = Sant Corneli.

this study, these are several oblique normal faults that compartmentalize the structure as well as minor thrusts affecting the frontal limb. They correspond to a special type of dip-domain boundary, the so-called discontinuity boundary (Fernández, 2004).

The reconstruction of these fault surfaces is similar to the reconstruction of horizons and is done from field data, when available; otherwise, it is inferred from the geometry of the displaced horizons and their map attitude.

With the main faults and the reference horizon reconstructed, the other significant horizons of the structure have to be built. The selection of these horizons is made following geologic criteria with the final objective being the creation of an overall view of the structure, which is as comprehensive as possible. Using the reference surface and the 3-D model of stratigraphic separations, other surfaces can then be reconstructed automatically in an efficient manner.

Reconstruction of Additional Surfaces

We propose an automated method to complete this step based on isopach maps, which represent the stratigraphic thickness between the reference horizon and the new horizon that has to be reconstructed. These isopach maps have been created from field data, wells, and the 3-D digitization, following the model of stratigraphic separations defined during the structural analysis.

The reconstruction of the geometry of additional surfaces has been attempted using 3DMove software (Midland Valley). The surface construction tool can deal with either constant or variable thickness isopachs. The information required consists of a 3-D surface (in this case, the reference surface) and a 2-D isopach map that reflects the thickness variations. The limitations of the method depend on the dip of the 3-D surface and the detail of the isopach map.

Two additional stratigraphic horizons have been reconstructed, which correspond to, from the youngest to oldest, respectively, the base of the Santa Fe unit, upper Cenomanian in age, and the top of the Aramunt unit, lower Santonian in age.

Structure at Depth

Concurrently with the 3-D reconstruction at surface, we proceed to reconstruct the geometry of the Bóixols thrust from subsurface data. In the same manner as the horizon reconstruction, we have first generated the structural contours of the thrust surface, and afterward, a TIN surface has been created (Figure 9).

The degree of uncertainty in the reconstruction of this structure is higher than that in the reconstruction of the anticline at surface, as a result of the irregular spacing, quality, and scale of original data (Figure 4).

Forward Numerical Modeling

Once the 3-D models of the fold at surface and the thrust at depth have been created, a forward modeling technique is applied to explore the potential relationships between both structures (Figure 10). This methodology consists in defining initial conditions and applying a kinematic model to obtain a 3-D model of the structure. Through this process, the influence of different parameters and processes in the structure's evolution can be explored. The 3-D character of this approach also makes it possible to study the influence of their variation along the structure. With the application of different kinematic models, different hypotheses about the structure can be studied and compared to the 3-D reconstruction. Thus, the 3-D reconstruction can be assessed (Figure 9).

We use an in-house 3-D numerical forward modeling program developed specifically to address the geometry of pre- and syntectonic sediments and their relationship to deeper structures. The model is a combined kinematic and sedimentary model in which all beds and structures are modeled in a Lagrangian framework to allow overturned structures to develop (Waltham and Hardy, 1995). A variety of different kinematic models can be applied or tested with respect to a particular structure (Hardy and Poblet, 1995; Hardy and Ford, 1997). The resolution of the model is 25 m (82 ft), and studied areas can be up to 50 × 50 km (31 × 31 mi) in surficial extent.

The objective behind the forward modeling is not to obtain an exact representation of the structure but to obtain a 3-D model consistent with some of the general parameters defined. The aim of this modeling technique is not, in this study, to obtain quantitative information about the structure geometry and evolution but instead to have qualitative models with which to interpret what we see in reality.

The first step in the numerical modeling is to establish the initial geometric conditions, constrained from the 3-D model and from knowledge of the structure, which represent the most important features that the numerical model has to honor. In this case, some geometric constraints are the following (see the Results and Discussion section for details).

- A footwall ramp with a cutoff angle between 22 and 30° to the north, near-horizontal flat
- A backlimb length approximately twice the displacement value
- A fold with a long backlimb, a short crest, and a near-vertical to slightly overturned frontal limb

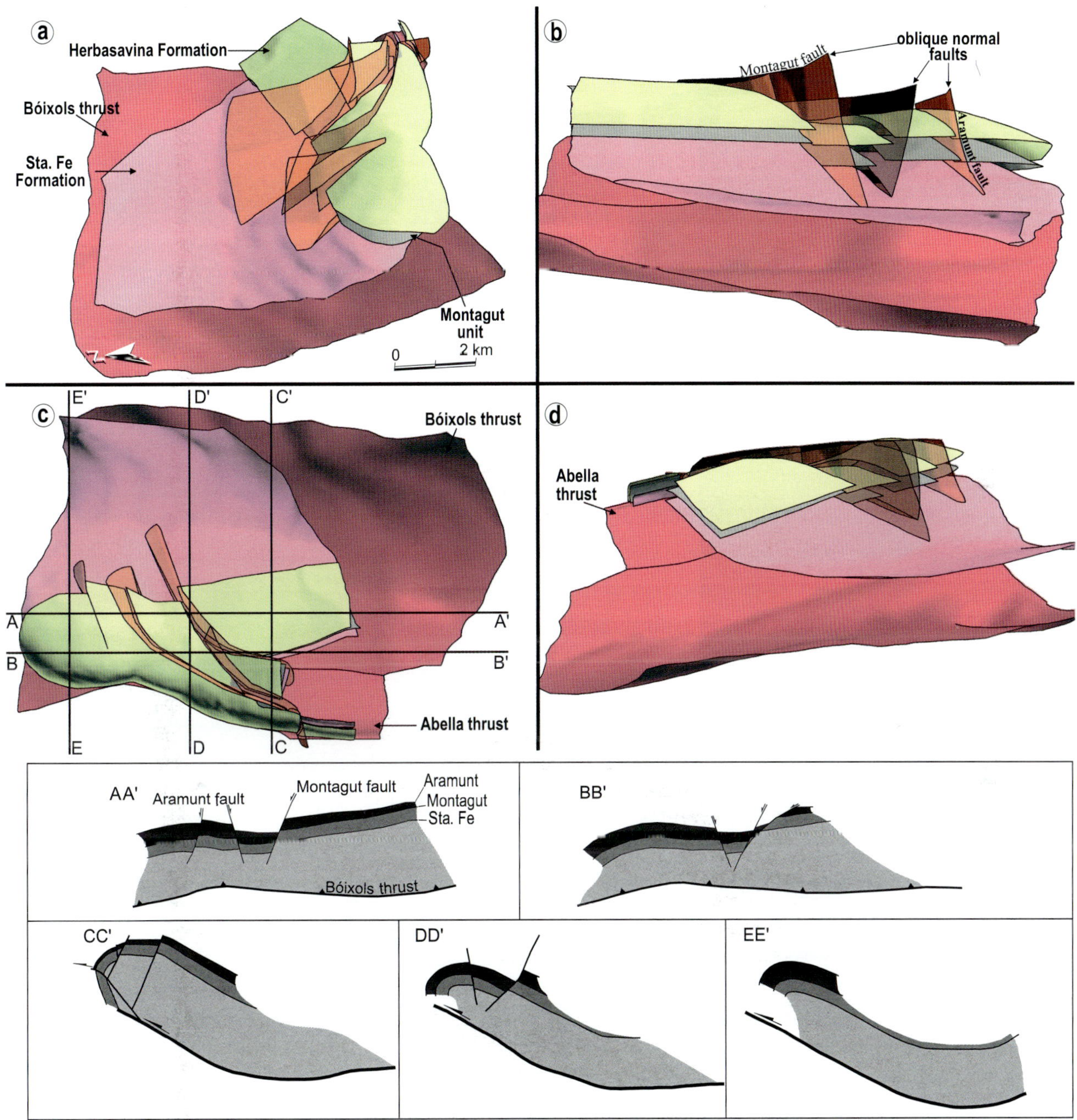

Figure 9. (top) Different views of the 3-D reconstruction achieved. (a) View from the northwest. (b) View from the north. The dark folded surface corresponds to the reference horizon or base of the Montagut unit. (c) Map view (with the location of the sections shown below). (d) View from the northeast. (below) Sections obtained from 3-D reconstruction. AA′, BB′ = longitudinal to structure; CC′, DD′, EE′ = perpendicular to structure. Sta. Fe = Santa Fe.

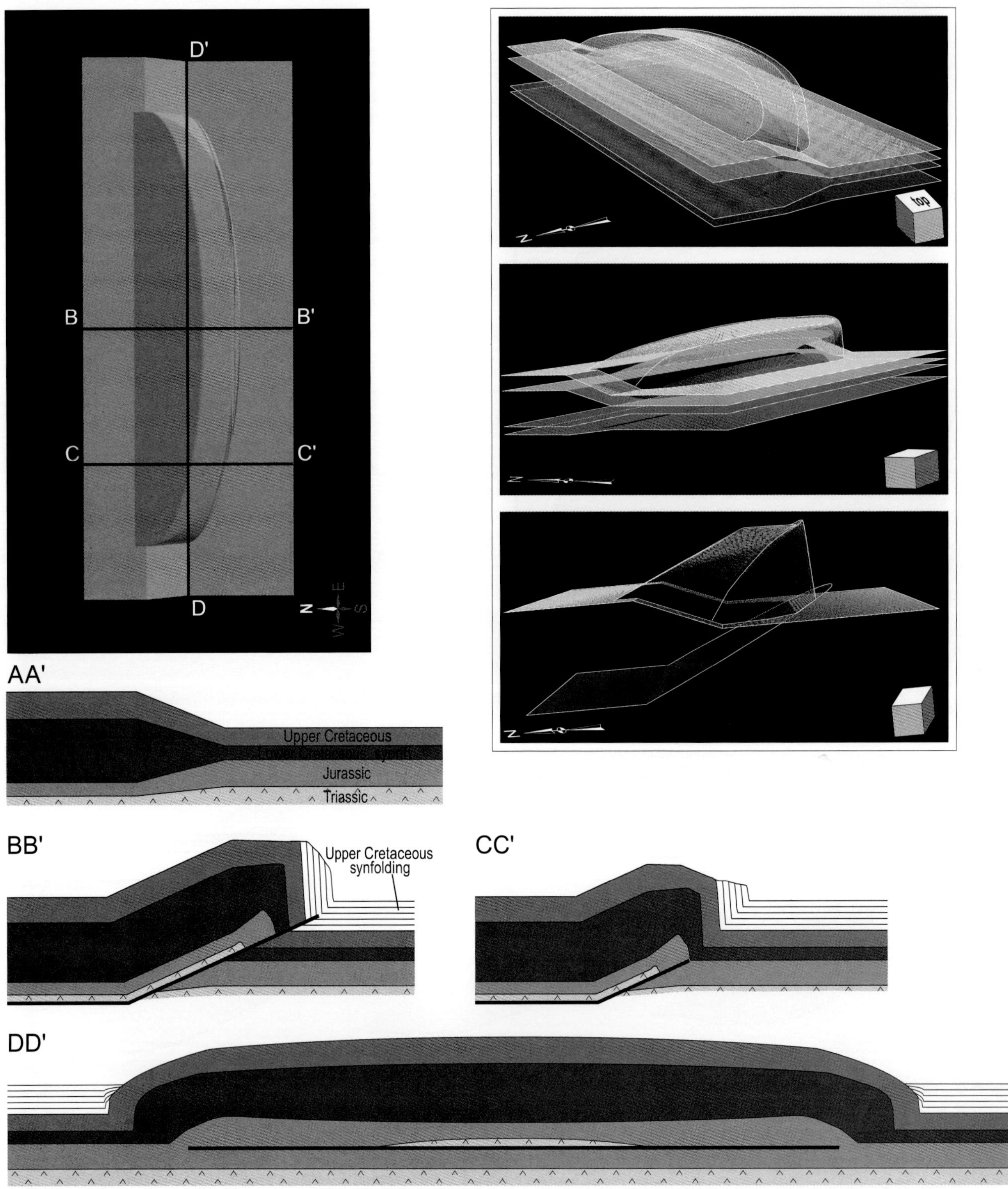

Figure 10. Three-dimensional (3-D) forward modeling performed. (top right) Different views of the generated 3-D image of the structure. (top left) Map view with locations of the cross sections presented below, derived from the 3-D forward model. AA′ corresponds to the initial geometry conditions for the simulation. As well as BB′ and CC′ cross sections, it is parallel to the displacement direction, whereas DD′ is perpendicular. Note the differences in thickness in the synrift materials.

- The existence of dramatic thickness variations between footwall and hanging-wall series, especially in the synrift sequence
- Different altitude in the regional level of the base of the postrift sequence (between the hanging wall and footwall of the thrust)
- A tip line sealed by the synfolding materials (Areny group)

Using these features as a guideline, we have iteratively applied several different kinematic models (fold-bend folding [Suppe, 1983], fixed-axis, constant thickness fault-propagation folding [Suppe and Medwedeff, 1990], and trishear fold-propagation folding [Erslev, 1991]) in an attempt to reproduce the characteristics of the Bóixols thrust and the Sant Corneli anticline. The numerical modeling performed suggests that fault-propagation fold models with constant bed thickness (Suppe and Medwedeff, 1990) account best for most of the features of the Sant Corneli anticline. In Figure 10, we show a best-fit model with the dimensions of the Sant Corneli anticline, in which the displacement is almost constant all along the thrust fault with a slight decrease close to the tips.

To match the observed geometries of the fault, the fold, and the relationship between them, it is necessary to propose the existence of an inversion of the extensional Early Cretaceous system and generation of structural uplift prior to the development of the Bóixols thrust. Otherwise, the present-day geometry cannot be satisfactorily reproduced (Figure 3). The main features illustrating this preexisting configuration are thickness variation in the synrift materials together with different altitude of the regional level of the postrift sequence base. This idea is supported by knowledge of the regional geology and discussed in the next section.

However, although the numerical modeling performed reproduces most of the geometric features of the folded postrift sequence, it does not reproduce as well the geometries displayed by the synfolding materials.

RESULTS AND DISCUSSION

Reconstruction at Surface

Three stratigraphic horizons have been reconstructed (as explained previously): the first postrift horizon (base of the Santa Fe unit, upper Cenomanian); base of the Montagut unit, Coniacian (reference surface); and the ultimate surface from the postrift sequence, lower Santonian in age (Aramunt unit). In addition, four oblique normal faults and one minor thrust in the frontal limb have also been reconstructed (Figure 9). The degree of uncertainty of the 3-D reconstruction ranges from ±5 m (16 ft) in well-constrained areas (which is the inherent error associated with the scale of the DTM) to ±20 m (66 ft) in zones with poor surface and subsurface control.

The 3-D reconstruction shows an asymmetric fold with a backlimb dipping 24–33° to the north, a forelimb dipping 60° south to 75° north (overturned), and a wide crest. It also shows along-strike geometric variations that were unknown before the 3-D reconstruction being made (Figure 9). The most remarkable ones are changes in the plunge of the fold axis overprinting the general plunge of the fold toward the west. These changes are restricted to two different areas of the fold: the area where most of the oblique extensional faults are located and the area near the Sant Antoni reservoir where an abrupt increase of the fold plunge results in the disappearance of the surface expression of the fold westward.

From seismic interpretation and the regional cross sections, we can observe that the length of the fold backlimb is approximately twice the displacement along the thrust fault, so it is consistent with a fault-propagation fold (Suppe and Medwedeff, 1990).

Reconstruction at Depth

The Bóixols thrust consists of a near-horizontal flat and a frontal ramp, which dips 22–30° to the north in the study area (as does the backlimb of the Sant Corneli anticline). The intersection of the ramp and flat is a west-dipping line with a sudden dip increase near the Sant Antoni reservoir.

Discussion

From a regional point of view, the general plunge of the fold toward the west is clearly consistent with a deepening of the thrust (Figure 9). Although, comparing the geometry of the fold and the thrust, we can observe that whereas the thrust geometry at depth remains overall constant, the fold geometry at surface shows along-strike variations, particularly in the plunge of the fold axis (as it has been explained above).

One of the areas that demonstrates the most significant changes in the attitude of the fold axis, and changes in the fold geometry itself, coincides with the area where synrift deposits undergo considerable thinning and also with the occurrence of the oblique normal faults that compartmentalize the fold (Figure 9). Previous studies (García-Senz, 2002) pointed out the geometry of the Lower Cretaceous rift system, and our study area is located in the western termination of one of these extensional basins (Organyà Basin). This fact is crucial in explaining the longitudinal variation of the fold geometry and can be related to the position and geometry of the extensional basin margin. Moreover, the position of the oblique normal faults is consistent with the position and orientation of that western boundary of the Organyà Basin.

The other area with remarkable changes in the fold plunge is near the Sant Antoni reservoir. There, the 3-D reconstruction performed shows that the western termination of the anticline at surface does not represent the termination of the structure as demonstrated by the parallelism between the plunge of the fold and the dip of the thrust ramp at depth (Figure 9). Furthermore, the forward modeling has demonstrated that a decrease in slip can also increase the plunge of the fold axis (Figure 10, DD′). This hypothesis has been tested by comparing seismic sections at both sides of the Sant Antoni reservoir, and we found out that together with a deepening of the thrust, a minor structural relief (as imposed by the thickness decrease of the folded succession) and a decrease in fault displacement can also be observed. We prove that the Sant Corneli anticline continues at depth below the synfolding strata.

The Sant Corneli anticline was developed above the frontal ramp of a blind thrust, which occurred in the footwall of a previous extensional fault (Figure 11). At least part of the inversion and creation of structural relief occurred before the beginning of thrusting, as is shown by the position and characteristics of most of the synfolding deposits (Areny group), at present cropping out only in the southern limb of the structure (Figure 11). This result is supported by the forward model, in which the existence of a structural uplift previously to the thrust development was necessary. This preexisting structural relief was solved by defining an initial situation for the forward modeling in which the synrift sequence demonstrated dramatic thickness changes parallel to the thrust transport direction (Figure 10, AA′) as well as being already uplifted above the regional level. Then we ran the forward modeling algorithm and created the Bóixols thrust over this preexisting configuration.

CONCLUSIONS

The objective of this study was to develop and apply a new methodology for the 3-D reconstruction and interpretation of fault-related structures based on field and subsurface data.

This methodology has been tested in the Sant Corneli anticline, located in the south-central Pyrenees. This anticline and the thrust that it is related to (Bóixols thrust) are the result of the complex evolution of a previous rift system. Thus, the present-day geometry and stratigraphy of this fold are clearly three-dimensional, reflecting the structural and paleogeographic evolution of the region.

This study has demonstrated that a full 3-D approach, like the methodology described in this chapter, is the only way to constrain and define the 3-D geometry and evolution of this kind of structure in a reasonable manner. Along-strike geometric variations of the structure that otherwise would remain unknown can become apparent following this 3-D technique of reconstruction.

A 3-D geometric model of the folded postrift materials (Upper Cretaceous) has been undertaken from field data together with a 3-D simplified geometric model of the thrust surface at depth (in that case, obtained from subsurface data). The application of 3-D forward numerical modeling techniques has permitted the exploration and characterization of the relationship between both related structures and has given a kinematic interpretation of the structure evolution. However, we have to go farther in to better define the function of the synfolding sequence.

Finally, this study has suggested that the Sant Corneli anticline is a fault-propagation fold related to a thrust, which was developed as a shortcut of a previous extensional system. The location and geometry of this system and its inversion, confirmed in this study, controlled most of the geometric features shown by the fold at surface.

ACKNOWLEDGMENTS

This work was conducted under the financial support of the Modelización Estructural 4-D project (CGL2007-66431-C02-01). Research by J. Mencos is funded by a predoctoral grant from the Ministerio de Ciencia e Innovación (2004-1824). We wish to acknowledge Grup de Geodinàmica i Anàlisi de Conques (2001SGR-000074). Part of the seismic interpretation used The Kingdom Suite software, which was generously provided by Seismic Micro-Technology via the University Gift Program. Midland Valley is acknowledged for providing 2DMove and 3DMove, which were used for part of the seismic interpretation and 3-D reconstruction and visualization. We also acknowledge Bentley Systems, Inc. for providing MicroStation, which was used for 3-D digitization and 3-D reconstruction. Part of the 3-D reconstruction process was performed using GOcad by Paradigm. P. Arbués and Servei Geològic de Catalunya are acknowledged for providing mapping information. The manuscript has been improved by the comments of Jürgen Adam, Massimo Bonora, and two anonymous reviewers.

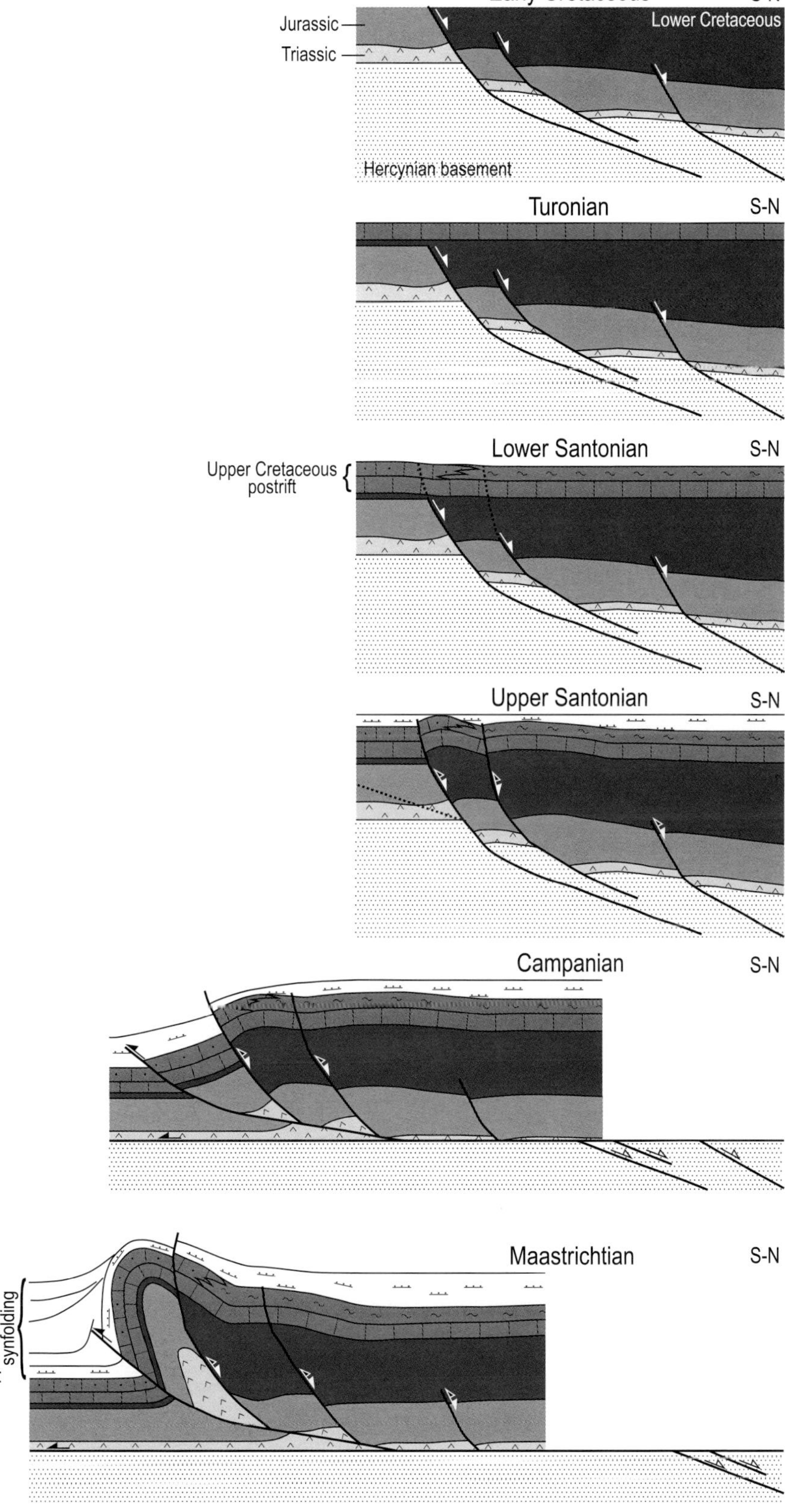

Figure 11. Schematic diagram showing the evolution of the Sant Corneli anticline (not to scale).

REFERENCES CITED

Arbués, P., E. Pi, and X. Berástegui, 1996, Relaciones entre la evolución sedimentaria del Grupo de Arén y el cabalgamiento de Bóixols (Campaniense terminal-Maastrichtiense del Pirineo meridional-central): Geogaceta, v. 20, no. 2, p. 446–449.

Ardèvol, L., J. Klimowitz, J. Malagón, and P. Nagtegaal, 2000, Depositional sequence response to foreland deformation in the upper Cretaceous of the southern Pyrenees, Spain: AAPG Bulletin, v. 84, no. 4, p. 566–587.

Banerjee, S., and S. Mitra, 2004, Remote surface mapping using orthophotos and geologic maps draped over digital elevation models: Application to the Sheep Mountain anticline, Wyoming: AAPG Bulletin, v. 88, no. 9, p. 1227–1237, doi:10.1306/02170403091.

Banerjee, S., and S. Mitra, 2005, Fold-thrust styles in the Abaroka thrust sheet, Caribou National Forest area, Idaho-Wyoming thrust belt: Journal of Structural Geology, v. 27, p. 51–56, doi:10.1016/j.jsg.2004.07.004.

Beaumont, C., J. A. Muñoz, J. Hamilton, and P. Fullsack, 2000, Factors controlling the Alpine evolution of the central Pyrenees inferred from a comparison of observations and geodynamical models: Journal of Geophysical Research, v. 105, no. B4, p. 8121–8145, doi:10.1029/1999JB900390.

Berástegui, X., J. García-Senz, and M. Losantos, 1990, Tecto-sedimentary evolution of the Organyà extensional basin (central south Pyrenean unit, Spain) during the Lower Cretaceous: Bulletin de la Société Géologique de France (8), v. 6, no. 2, p. 251–264.

Bond, R. M. G., and K. R. McClay, 1995, Inversion of a lower Cretaceous extensional basin, south central Pyrenees, Spain, *in* J. G. Buchanan and P. G. Buchanan, eds., Basin inversion: Geological Society (London) Special Publication 88, p. 415–431.

Cámara, P., and J. Klimowitz, 1985, Interpretación geodinámica de la vertiente centro-occidental surpirenaica: Estudios Geológicos, v. 41, p. 391–404.

Carrera, N., J. A. Muñoz, and E. Roca, 2009, 3D reconstruction of geological surfaces by the equivalent dip-domain method: An example from field data of the Cerro Bayo Anticline (Cordillera Oriental, NE Argentine Andes): Journal of Structural Geology, v. 31, no. 12, p. 1573–1585.

Delaunay, B., 1934, Sur la sphère vide: Izvestia Akademii Nauk SSSR, Otdelenie Matematicheskikh i Estestvennykh Nauk, v. 7, p. 793–800.

Erslev, E. A., 1991, Trishear fault-propagation folding: Geology, v. 19, p. 617–620, doi:10.1130/0091-7613(1991)019 <0617:TFPF>2.3.CO;2.

Fernández, O., 2004, Reconstruction of geological surfaces in 3D: An example from the southern Pyrenees: Ph.D. thesis, Department of Geodynamics and Geophysics, University of Barcelona, 321 p.

Fernández, O., 2005, Obtaining a best fitting plane through 3D georeferenced data: Journal of Structural Geology, v. 27, p. 855–858, doi:10.1016/j.jsg.2004.12.004.

Fernández, O., J. A. Muñoz, P. Arbués, O. Falivene, and M. Marzo, 2004, Three-dimensional reconstruction of geological surfaces: An example of growth strata and turbidite systems from the Ainsa Basin (Pyrenees, Spain): AAPG Bulletin, v. 88, p. 1049–1068, doi:10.1306/02260403062.

García-Senz, J., 2002, Cuencas extensivas del Cretacico Inferior en los Pirineos Centrales, formación y subsecuente inversión: Ph.D. thesis, Department of Geodynamics and Geophysics, University of Barcelona, 310 p.

Garrido-Mejías, A., 1973, Estudio geológico y relación entre tectónica y sedimentación del Secundario y Terciario de la vertiente meridional pirenaica en su zona central (provincias de Huesca y Lérica): Ph.D. thesis, University of Granada, 335 p.

Hardy, S., and M. Ford, 1997, Numerical modeling of trishear fault propagation folding: Tectonics, v. 16, no. 5, p. 841–854, doi:10.1029/97TC01171.

Hardy, S., and J. Poblet, 1995, The velocity description of deformation: Paper 2. Sediment geometries associated with fault-bend and fault-propagation folds: Marine and Petroleum Geology, v. 12, no. 22, p. 165–176, doi:10.1016/0264-8172(95)92837-M.

Husson, L., and J. L. Mugnier, 2003, Three-dimensional reconstruction from outcrop structural data, restoration, and strain field of the Baishai anticline, western Nepal: Journal of Structural Geology, v. 25, p. 79–90, doi:10.1016/S0191-8141(02)00044-5.

Lanaja, J. M., 1987, Contribución de la exploración petrolífera al conocimiento de la geología de España: Madrid, Instituto Geológico y Minero de España, 465 p.

Mallet, J. L., 1989, Discrete smooth interpolation: ACM Transactions on Graphics, v. 8, no. 2, p. 121–144, doi:10.1145/62054.62057.

Mitra, S., 2002, Structural models of faulted detachment folds: AAPG Bulletin, v. 86, no. 9, p. 1673–1694.

Muñoz, J. A., 2002, Alpine tectonics: I. The Pyrenees, *in* W. Gibbons and T. Moreno, eds., Geology of Spain: Bath, United Kingdom, The Geological Society, p. 367–401.

Mutti, E., and M. Sgavetti, 1987, Sequence stratigraphy of the Maastrichtian Arén region, south-central Pyrenees, Spain: Distinction between eustatically and tectonically controlled depositional sequences: Annali dell Università degli Studi di Ferrara, Sezione Scienze della Terra, v. 1, no. 1.

Simó, A., 1986, Carbonate platform depositional sequences, Upper Cretaceous, south-central Pyrenees (Spain): Tectonophysics, v. 129, p. 205–231, doi:10.1016/0040-1951(86)90252-0.

Suppe, J., 1983, Geometry and kinematics of fault-bend folding: American Journal of Science, v. 283, p. 684–721.

Suppe, J., and A. Medwedeff, 1990, Geometry and kinematics of fault-propagation folding: Eclogae Geologicae Helveticae, v. 83, no. 3, p. 409–454.

Waltham, D., and S. Hardy, 1995, The velocity description of deformation: Paper 1. Theory: Marine and Petroleum Geology, v. 12, no. 2, p. 153–163, doi:10.1016/0264-8172(95)92836-L.

14

Wu, Jonathan E., and Ken R. McClay, 2011, Two-dimensional analog modeling of fold and thrust belts: Dynamic interactions with syncontractional sedimentation and erosion, *in* K. McClay, J. H. Shaw, and J. Suppe, eds., Thrust fault-related folding: AAPG Memoir 94, p. 301 – 333.

Two-dimensional Analog Modeling of Fold and Thrust Belts: Dynamic Interactions with Syncontractional Sedimentation and Erosion

Jonathan E. Wu

Fault Dynamics Research Group, Department of Earth Sciences, Royal Holloway University of London, Egham, Surrey, United Kingdom

Ken R. McClay

Fault Dynamics Research Group, Department of Earth Sciences, Royal Holloway University of London, Egham, Surrey, United Kingdom

ABSTRACT

The effects of syncontractional sedimentation and erosion on simple, critically tapered Coulomb wedges were evaluated by conducting twelve two-dimensional analog model sandbox experiments. All 12 models produced critically tapered Coulomb wedges with topographic slopes of 6–10° above horizontal basal detachments. The model without syncontractional sedimentation or erosion exhibited a general forward-breaking sequence with synchronous thrust activity. Syncontractional sedimentation produced longer wedges composed of fewer major forward-vergent thrusts and lowered thrust activities in the foreland. Syncontractional erosion inhibited forward propagation of the deformation front, decreased the number of major thrusts, and increased thrust activities in the hinterland. Where combined, the effects of syncontractional sedimentation and erosion were complementary.

At the scale of individual folds, syncontractional sedimentation altered fold evolution by producing limb rotation and a front-limb trishear zone formed by tip-line thrust splays. At this scale, syncontractional erosion did not cause significant changes to the fold geometries as they developed.

Comparisons of the model thrust wedges with natural fold and thrust wedges indicate that the Nankai accretionary prism, with its well-ordered array of closely spaced thrusts, would be typical of fold and thrust belts with low rates of surface processes. In contrast, the fold and thrust belts of offshore Niger Delta, the central Apennines, and the sub-Andes are characterized by buried, widely spaced, low-activity thrusts in the foreland that would be typical of

DOI:10.1306/13251343M9450

high syncontractional sedimentation. High syncontractional erosion would produce very active hinterland thrusts resembling the present-day Taiwan fold and thrust belt. Changes in thrust-wedge dynamics caused by increased syncontractional erosion in the model wedges imply that subaerial fold and thrust belts, with higher erosion, would evolve differently from their submarine counterparts.

INTRODUCTION

Fold and thrust belts in both subaerial and submarine settings are typically wedge shaped and defined by an upper topographic surface that dips toward the foreland and by a basal detachment surface that climbs up section from the hinterland to the foreland (Figures 1, 2). The frontal parts of these thrust systems have commonly been interpreted to be critically tapered Coulomb wedges (Figure 1a) following the models proposed by Chapple (1978) and quantified by Davis et al. (1983) and Dahlen (1990) and others. The geometries and shapes of such critically tapered Coulomb wedges may be modified by surface processes such as syntectonic erosion and syntectonic sedimentation, both of which will change the surface topography and hence the wedge dynamics as well as the internal thrust fault and fold geometries (Figures 1b, 3) (e.g., Koons, 1990; Beaumont et al., 1992; Willett, 1992).

This chapter describes the effects of syntectonic sedimentation and syntectonic erosion on simple critically tapered Coulomb wedges formed in two-dimensional analog model experiments. These experiments tested the sensitivity of the wedge systems to surface processes and also their effects on thrust-related fold geometries. The models build on previous experimental studies of critically tapered Coulomb wedge-thrust systems (e.g., Liu et al., 1992; Storti and McClay, 1995; Mugnier et al., 1997; McClay and Whitehouse, 2004). The results of the scaled analog models are compared to natural examples of fold-thrust belt systems in both subaerial and submarine environments (e.g., Figure 2).

Coulomb Wedge Dynamics

The critically tapered wedge model (Chapple, 1978; Davis et al., 1983; Dahlen, 1984, 1990; Dahlen et al., 1984) has been successfully applied to explain the mechanical and kinematic evolution of fold and thrust belts (e.g., Koons, 1990; Horton, 1999; Bilotti and Shaw, 2005; Morley, 2007). This model predicts that a horizontal compressive force applied to a layer of Coulomb material causes it to deform internally to form a forward-tapering wedge until the combined angles of the basal detachment dip (β) and the upper surface slope (α) reach a critical taper angle ($\alpha + \beta$) (Figure 4a). Continued compression and shortening would result in the stable sliding of this critically tapered wedge on the basal detachment with new thrusts initiating in undeformed material at the front of the wedge. The critically tapered wedge would grow by frontal accretion of new material by shortening on the most frontal thrusts. The wedge would grow in size, but the critical taper (basal detachment dip β + surface slope α) would remain unchanged. Wedge growth would continue as long as new material could be accreted at the front of the wedge.

Davis et al. (1983), Dahlen (1990), and others analyzed the mechanics of these wedges for homogeneous and anisotropic materials and have included pore-fluid pressures within the wedge system. The principal controls on the angle of the wedge taper ($\alpha + \beta$) are the coefficient of friction along the basal detachment and the angle of internal friction of the material forming the wedge. In addition, Davis et al. (1983) recognized that changes in the basal decollement dip and changes in the surface slopes of the wedge would generate internal readjustments within the wedge until an equilibrium critical taper was re-established (Figure 1b).

Surface Processes in Fold and Thrust Belts

Syntectonic sedimentation and syntectonic erosion will change both the loading in the thrust wedge and more importantly the geometry of the upper surface of the thrust wedge and its slope angle (α). The removal of material from the hinterland of the thrust wedge by erosion and its redeposition in the foreland as synkinematic sediments (Figure 4b) generally reduce the dip of the upper surface of the wedge (α), causing the wedge taper angle ($\alpha + \beta$) to fall below the critical taper, a state termed "subcritical" (Figure 1b) (Davis et al., 1983). In some cases, foreland sedimentation can cause local parts of the wedge to increase above the critical taper (e.g., Ghiglione and Ramos, 2005; Stockmal et al., 2007), a state termed "supercritical" (Figure 1b). The critical wedge model predicts that shortening of a subcritical wedge (Figure 4b) would cause internal deformation that would raise the surface slope such that the wedge would regain the critical taper. This internal deformation within the thrust wedge would increase the wedge height in the hinterland and could include back thrusting, thrust reactivation, out-of-sequence thrusting, or

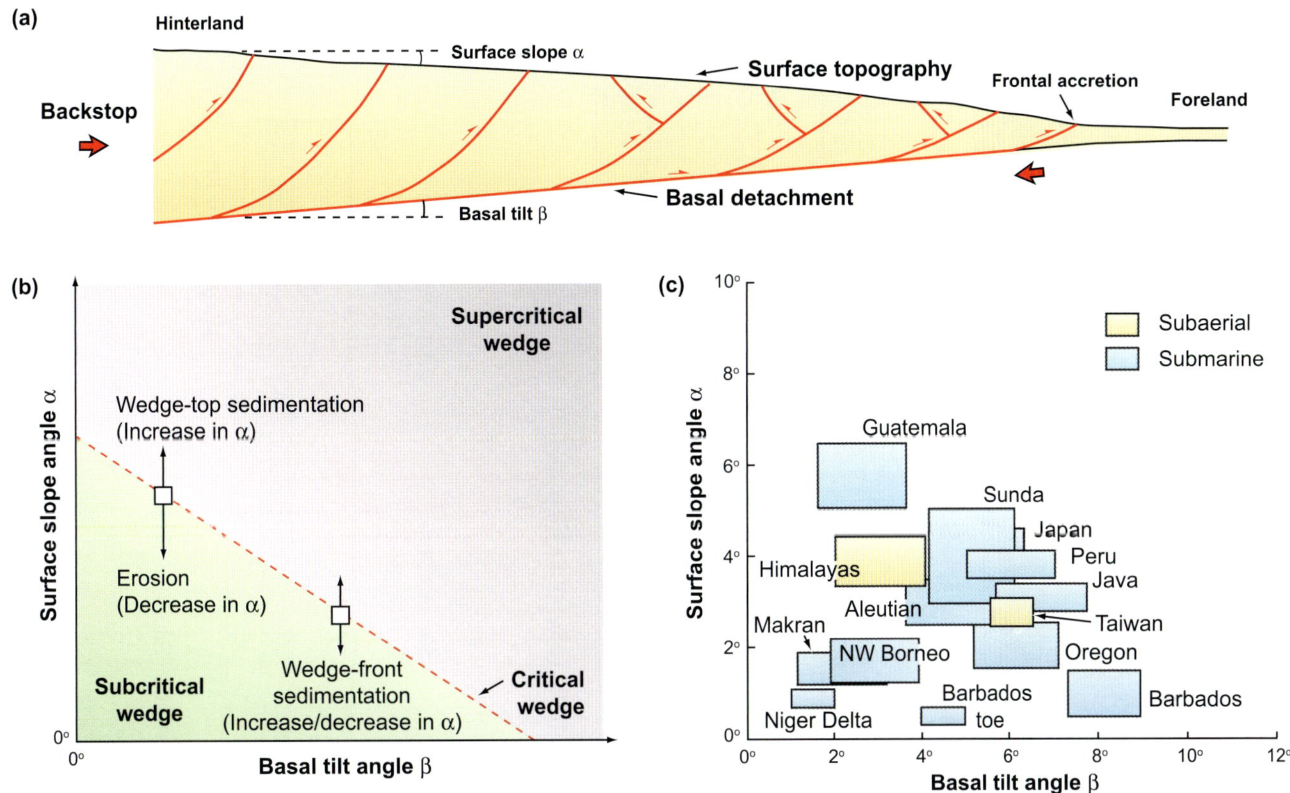

Figure 1. (a) Two-dimensional cross-sectional profile of a critically tapered Coulomb wedge. (b) Effects of synkinematic sedimentation and erosion on the state of a critically tapered wedge (modified from Davis et al., 1983). (c) Surface slope and basal tilt wedge geometries of natural fold and thrust belts (modified from Davis et al., 1983; Bilotti and Shaw, 2005; Morley, 2007).

basal accretion by duplexing at the base of the wedge. In contrast, shortening of a supercritical wedge could cause either a rapid advance of the deformation front or extensional collapse at the rear of the wedge, both of which would lower the surface slope (Davis et al., 1983; Platt, 1986).

Rates of Syntectonic Erosion, Syntectonic Sedimentation, and Fault Slip in Fold and Thrust Belts

The rate of thrusting, together with variations in fold kinematics and the rates of syntectonic sedimentation and erosion, influences the final growth stratal geometries above individual fault-related fold structures (e.g., Suppe et al., 1992; Hardy and Poblet, 1994). The medium- to long-term rates (10^4–10^6 yr) of syntectonic sedimentation reported in literature lie between 0.1 and 1 mm/yr (0.004 and 0.04 in./yr) (Figure 5a), and the medium- to long-term rates (10^4–10^6 yr) of erosion in fold and thrust belts fall in the range of 0.1 to 6 mm/yr (0.004 to 0.24 in./yr) (Figure 5b). These ranges indicate variations of at least an order of magnitude in the rates of surface processes in natural fold and thrust belt wedges. Similarly, long-term rates of thrusting in fold and thrust belts also vary by an order of magnitude over typical ranges of 0.30 to 2.5 mm/yr (0.01 to 0.10 in./yr) (e.g., Hardy et al., 1996). These wide-ranging variations provide the basis for this study, which was designed to systematically test a wide range of synkinematic sedimentation and/or erosion rates to determine their effect on model thrust wedges and hence give insights as to how natural thrust-wedge systems might behave under equivalent conditions.

Analog Models of Coulomb Wedge-thrust Systems

Previous analog model studies of thrust-wedge systems have demonstrated a dynamic interaction between synkinematic sedimentation and erosion and the evolution of Coulomb wedge-thrust systems (e.g., Storti and McClay, 1995; Mugnier et al., 1997; Nieuwland et al., 2000; Storti et al., 2000; Barrier et al., 2002; Persson and Sokoutis, 2002; Nalpas et al., 2003; McClay and Whitehouse, 2004; Konstantinovskaia and Malavieille, 2005; Hoth et al., 2006; Bonnet et al., 2007; Cruz et al., 2008; Wu et al., 2008; Malavieille and Trullenque, 2009).

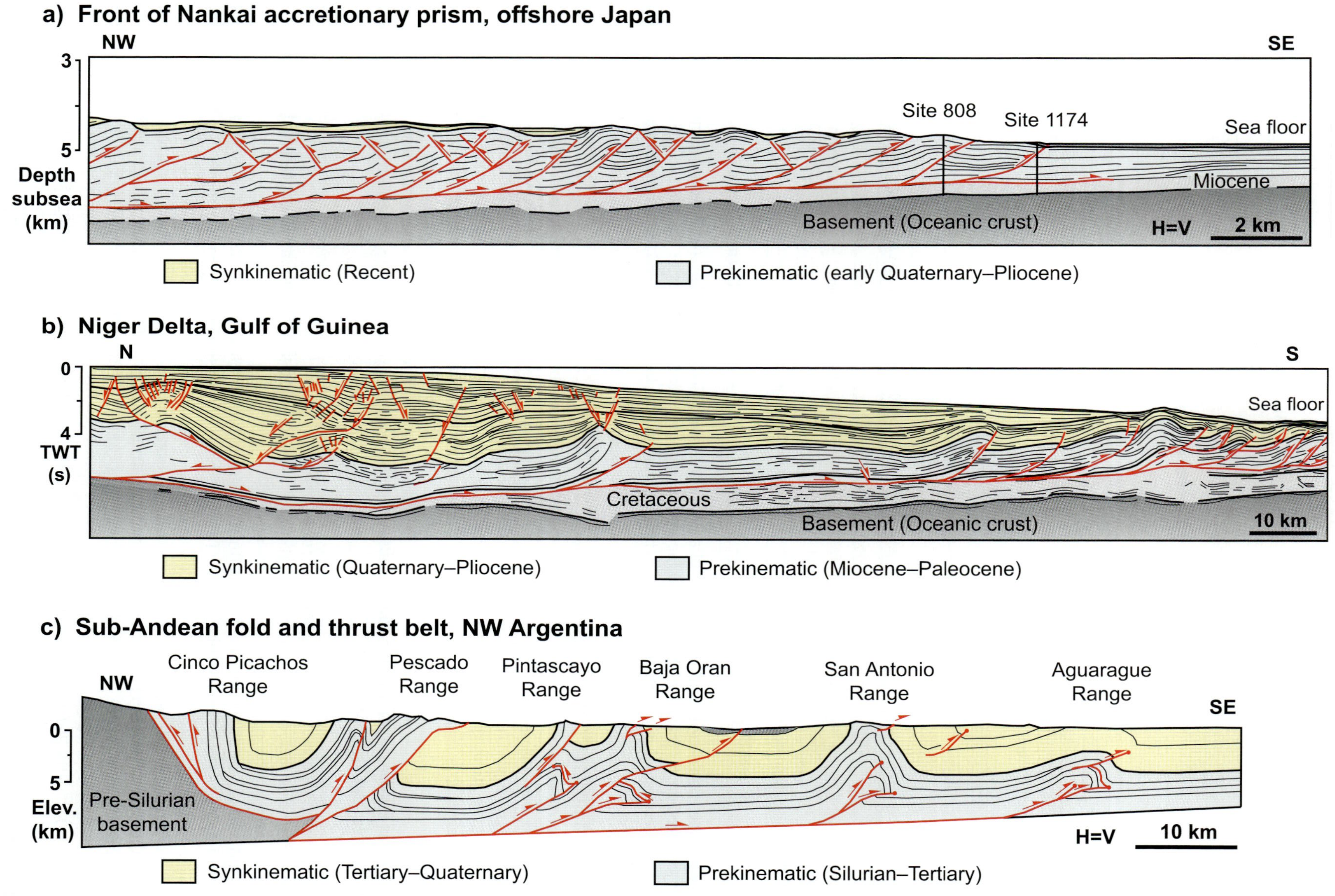

Figure 2. Structural styles of three fold and thrust belts showing variations in erosion and sedimentation. (a) Frontal zone of the Nankai accretionary complex, offshore Japan, reinterpreted from data by Hills et al. (2001). (b) Eastern lobe of the offshore Niger Delta, Gulf of Guinea (modified from Ajakaiye and Bally, 2002). (c) Sub-Andean fold and thrust belt, northwest Argentina (modified from Echavarria et al., 2003). TWT = two-way time; H = V = no vertical exaggeration (horizontal and vertical scales equal); Elev. = elevation.

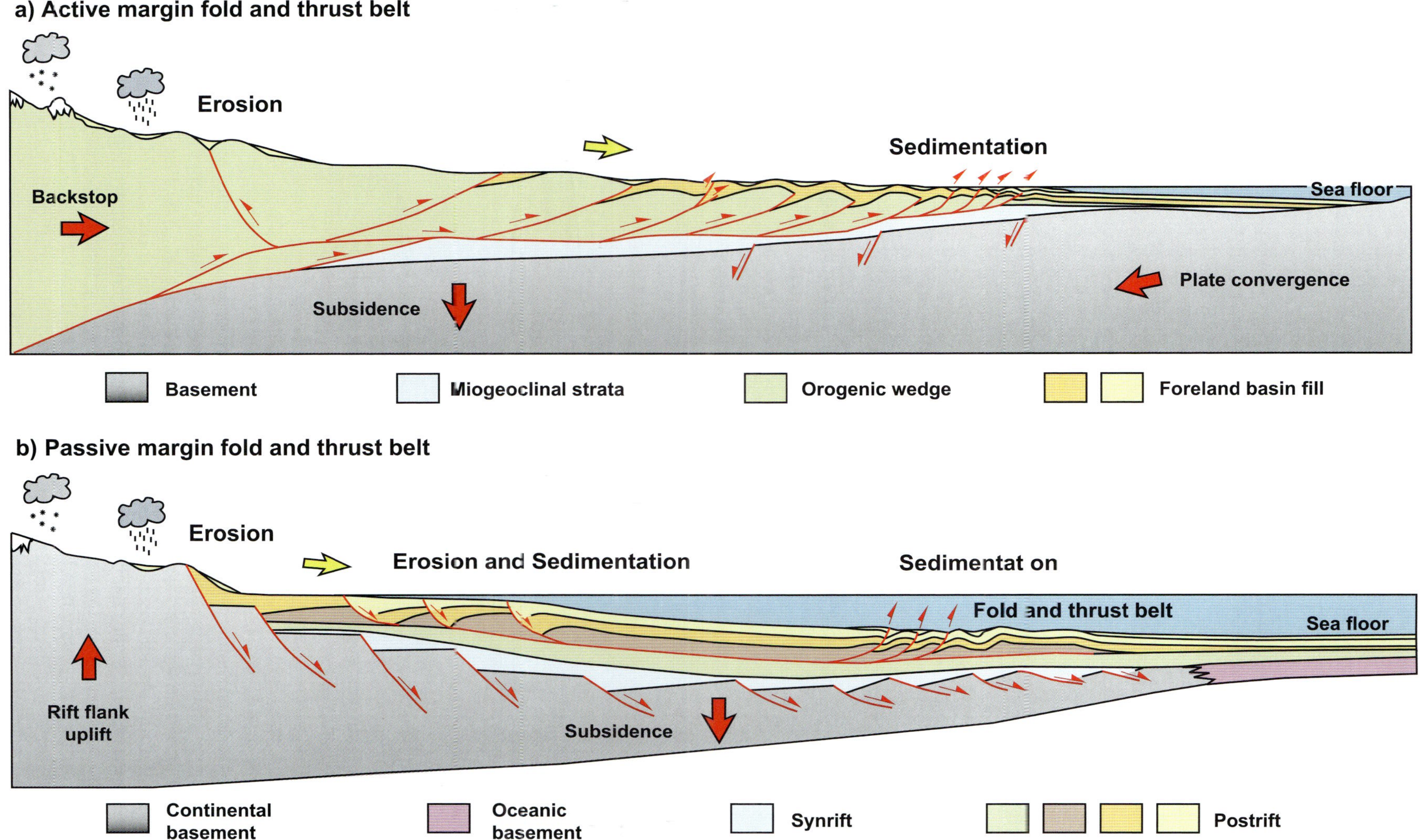

Figure 3. Conceptual cartoon showing the interaction of synkinematic sedimentation and synkinematic erosion with fold and thrust belts in (a) an active margin and (b) a passive margin.

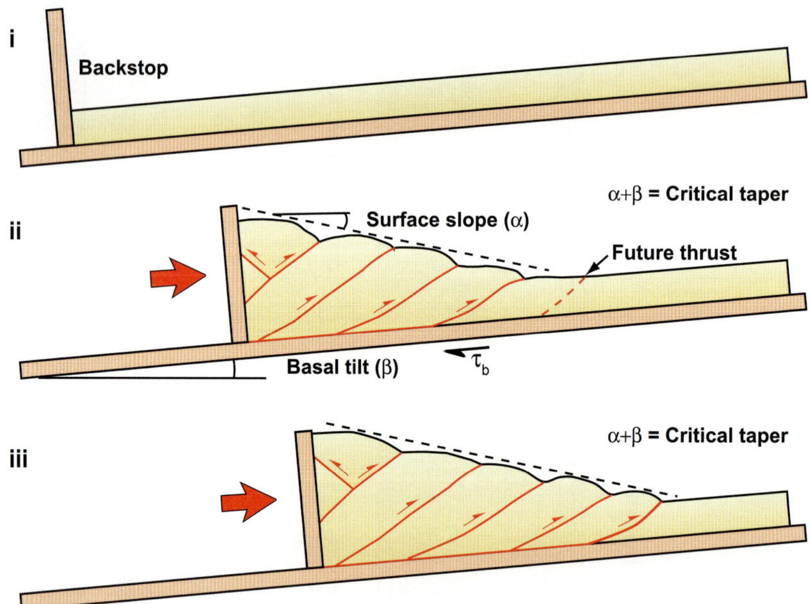

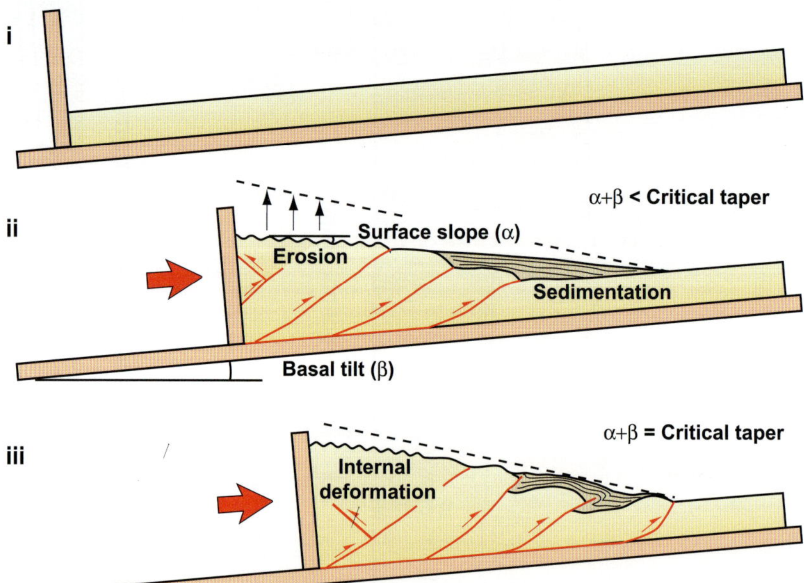

Figure 4. Progressive evolutionary sequence of a simple critically tapered wedge (a) without and (b) with synkinematic sedimentation and synkinematic erosion (modified from Davis et al., 1983).

In these models, synkinematic erosion and sedimentation controlled the interplay between frontal imbrication and the basal accretion of hinterland passive roof duplexes and antiformal thrust stacks (Mugnier et al., 1997; Konstantinovskaia and Malavieille, 2005; Bonnet et al., 2007; Malavieille and Trullenque, 2009). Deformation and material transfer paths responded immediately to the rate and spatial distribution of erosion (Hoth et al., 2006; Cruz et al., 2008). Erosion and sedimentation increased the activities of faults in the hinterland and triggered out-of-sequence thrusting (Nieuwland et al., 2000; Storti et al., 2000; Persson and Sokoutis, 2002; McClay and Whitehouse, 2004). The rate of synkinematic sedimentation controlled the total number of foreland-verging thrusts (Storti and McClay, 1995) and back thrusts (Bonnet et al., 2008). Sedimentation patterns altered the spatial distribution of thrust activities (Wu et al., 2008).

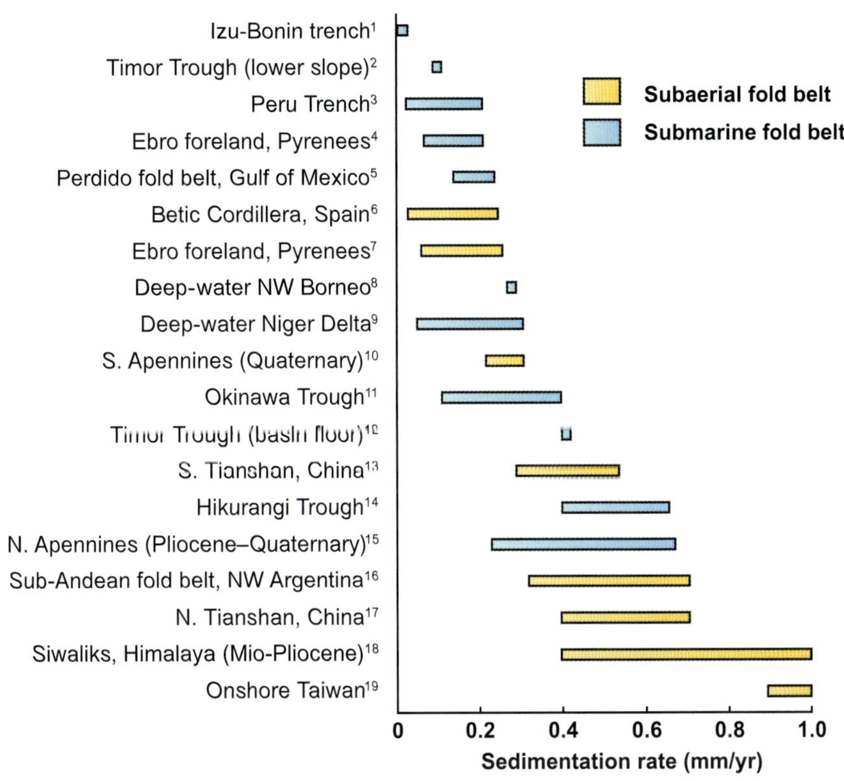

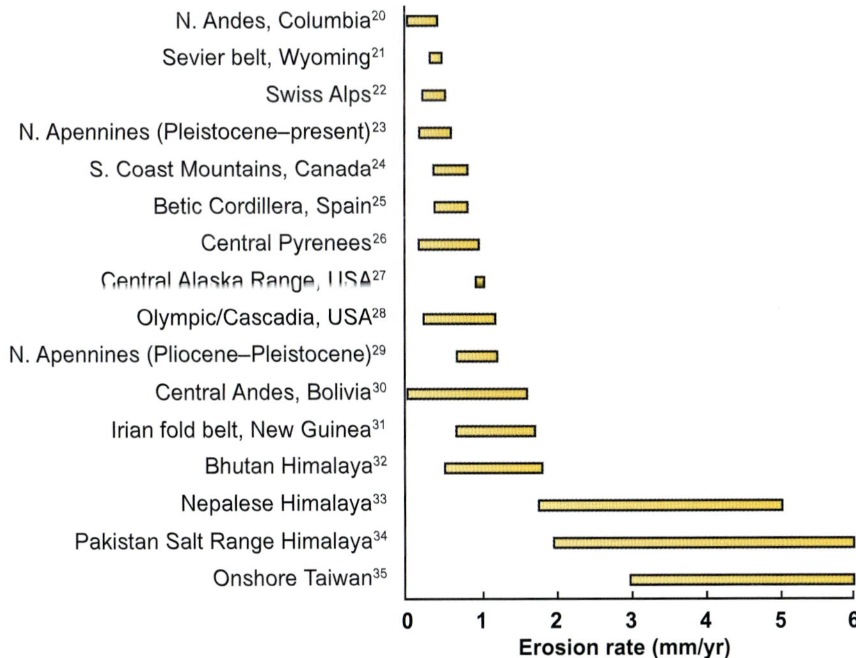

Figure 5. Rates of (a) syntectonic sedimentation and (b) syntectonic erosion in fold and thrust belts over medium- to long-term time scales (10^4-10^6 yr). ([1]ODP Shipboard Scientific Party, 2000; [2]Charlton, 1988; [3]Skilbeck and Fink, 2006; [4]Verges et al., 2002; Holl and Anastasio, 1993; [5]Mount et al., 1990; [6]Iribarren et al., 2009; [7]Jones et al., 2004; Burbank and Verges, 1994; [8]Morley and Leong, 2008; [9]Ngala, 2007; [10]Amato et al., 2003; [11]Xiong et al., 2005; [12]Charlton, 1988; [13]Charreau et al., 2006; Hubert-Ferrari et al., 2007; [14]Barnes and Mercier de Lepinay, 1997; Lewis et al., 1998; [15]Bartolini, 1999; [16]Echavarria et al., 2003; [17]Charreau et al., 2008; [18]Kumar et al., 2003; [19]Mouthereau et al., 2001; [20]Restrepo-Moreno et al., 2009; [21]Warner and Royse, 1987; [22]Schlunegger and Willett, 1999; [23]Cyr and Granger, 2008; Bartolini, 1999; [24]Ehlers et al., 2006; [25]Johnson, 1997; [26]Rahl et al., 2007; [27]Fitzgerald et al., 1995; [28]Brandon et al., 1998; [29]Zattin et al., 2002; Balestrieri et al., 2003; [30]Barnes and Pelletier, 2006; Uba et al., 2007; [31]Weiland and Cloos, 1996; [32]Grujic et al., 2006; [33]Whipp et al., 2007; Copeland and Harrison, 1990; Husson et al., 2004; [34]Burbank and Beck, 1991; Burbank and Anderson, 2000; [35]Dadson et al., 2003; Willett et al., 2003).

Synkinematic sedimentation also dynamically influenced the evolution of individual folds in analog models. Increases in synkinematic sedimentation caused thrust fault ramp angles to steepen through the synkinematic strata (Storti and McClay, 1995; Nieuwland et al., 2000) and develop arrays of steeper or shallower dipping tip-line thrust splays in response to the sedimentation versus uplift rate (Barrier et al., 2002; Nalpas et al., 2003).

Numerical Models of Coulomb Wedge-thrust Systems

The dynamic interaction of syncontractional sedimentation and erosion has also been demonstrated in numerical models of thrust-wedge systems. Beaumont et al. (1992) showed that the evolution of an orogen could be dictated by spatial erosion patterns. Other models have demonstrated the control of sedimentation and erosion on the spatial and temporal distribution of fault activities (Hardy et al., 1998; Leturmy et al., 2000; Simpson, 2006; Stockmal et al., 2007). In these models, erosion generally promoted continued or out-of-sequence fault activities, whereas sedimentation inhibited fault activity. Sedimentation increased the spacing between thrusts and caused the deformation front to step farther out into the foreland (e.g., Stockmal et al., 2007).

At the scale of individual fold-fault structures, distinct-element models by Strayer et al. (2004) showed a dynamic link that existed between syntectonic sedimentation and the final geometry of fault-related folds. Folds with little or no sedimentation developed open, gentle fault-bend-style folds, whereas thick sequences of syncontractional sedimentation caused fold tightening leading to the development of fault-propagation folds with strongly overturned front limbs.

In this study, synkinematic sedimentation and/or erosion were systematically varied in simple scaled analog models of thrust wedges to investigate their effects on fold and thrust belt evolution. The progressive evolution of the analog models was analyzed in detail as well as that of individual fold and fault structures. The results of these experiments were critically compared to theoretical and numerical models as well as to natural examples of fold-thrust belt systems.

EXPERIMENTAL METHODOLOGY

Modeling Materials

The results of 12 analog experiments of simple thrust wedges are described in this chapter (Table 1). The physical models were constructed in a metal rig with a static, horizontal, rigid metal base with two glass sidewalls (Figure 6). The glass was treated with a polymer to reduce sidewall friction, and the base of the rig was lined with a low-friction material. Horizontal shortening was achieved by moving the left-hand, 20-cm (7.9-in.)-high vertical end wall at a rate of 30 cm/hr (11.8 in./hr) by a stepper motor. At the end of each experiment, the models were impregnated with a gelling agent, serially sectioned at 1-cm (0.4-in.) thickness and digitally photographed.

Table 1. List of physical models.

Model number	Sedimentation rate*	Erosion rate**	Total shortening (cm)
1	None	None	50.2
2	Low	None	49.2
3	High	None	41.2
4	None	Low	40
5	Low	Low	39
6	High	Low	39
7	None	Moderate	40
8	Low	Moderate	40
9	High	Moderate	35
10	None	High	40
11	Low	High	38.8
12	High	High	38.8

*Sedimentation rates: high: 1–2 mm/1 cm shortening; low: 0.5 mm/1 cm shortening.
**Erosion rates: high: 1 mm/1 cm shortening; moderate: 1 mm/4 cm shortening; low: 1 mm/8 cm shortening.

In each experiment, the initial prekinematic sand pack consisted of a 2.5-mm (0.10-in.)-thick basal detachment layer of glass beads above which nine 2.5-mm (0.10-in.)-thick horizontal layers existed, consisting of alternating blue, white, and black dry quartz sand layers (Figure 6). The initial dimensions of the sand pack were 30 cm (11.8 in.) wide, 110 cm (43 in.) long, and 2.5 cm (0.10 in.) thick (Figure 6a).

The dry quartz sand had a grain size of 90–190 μm, a density of 1.42 g/cm^3, a peak internal friction coefficient of 0.65, a dynamic internal friction coefficient of 0.61, and a negligible cohesion (85–105 Pa). The glass beads had an average grain size of 100–200 μm, a density of 1.45 g/cm^3, a peak internal friction coefficient of 0.51, a dynamic internal coefficient of 0.47 and a negligible cohesion (46–51 Pa). The high degree of sphericity of the glass beads and its lower coefficient of internal friction relative to the sand made it an ideal basal detachment layer in combination with the low-friction base of the deformation rig. Dry quartz sand and glass beads, both of which have a Navier-Coulomb rheology, have been used extensively to simulate the brittle deformation of sedimentary rocks in the shallow upper 10 km (6.2 mi) (e.g., Davis et al., 1983; Malavieille, 1984; McClay, 1990). The scaling factor between model and prototype is 10^{-5} such that 1 cm (0.4 in.) in the models

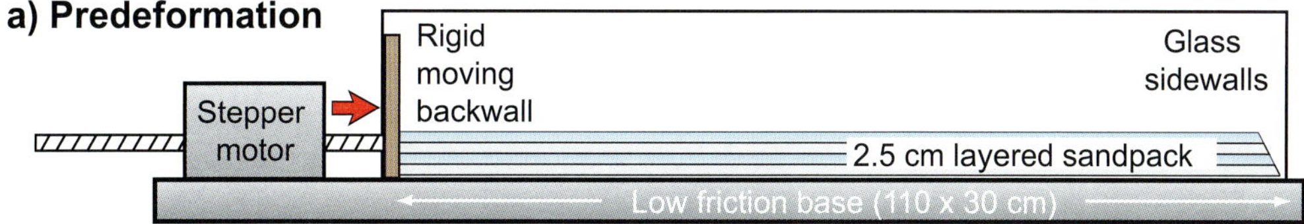

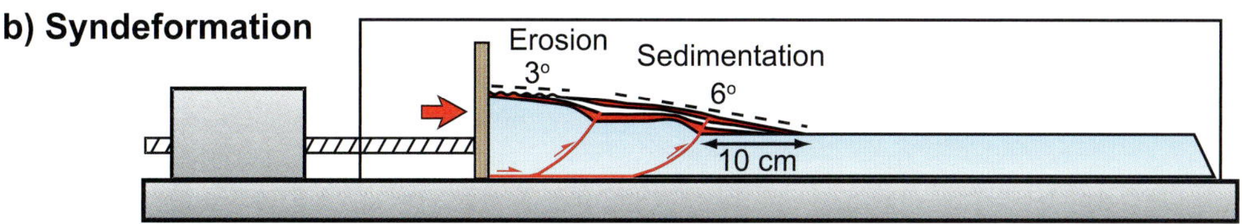

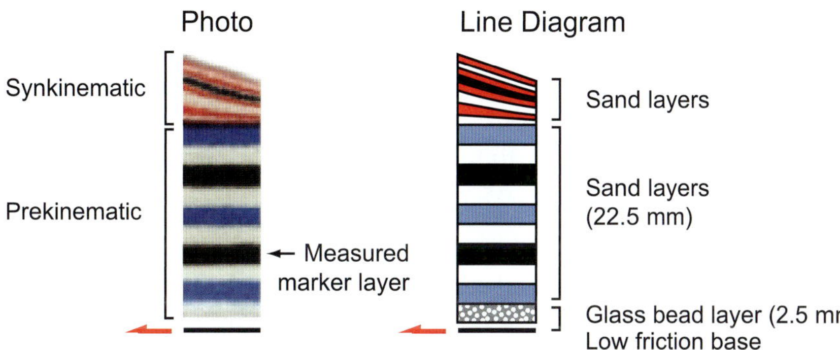

Figure 6. (a) Diagram of the experimental apparatus used to simulate thrust wedges. (b) Geometry of material removed and/or added to simulate synkinematic erosion and synkinematic sedimentation, respectively. (c) Stratigraphy used in the analog models.

approximates to 1 km (0.6 mi) of brittle, sedimentary rock in the upper crust (cf. McClay, 1990). We noted, however, that, in the brittle deformation field, many structures, faults and folds, are mostly scale independent with similar structures formed over a range of scales from tens of centimeters to kilometers in scale.

Experimental Procedure

Three series of experiments were conducted.

Series 1: No synkinematic sedimentation and increased synkinematic erosion (models 1, 4, 7, and 10, Table 1)
Series 2: Low synkinematic sedimentation and increased synkinematic erosion (models 2, 5, 8, and 11, Table 1)
Series 3: High synkinematic sedimentation and increased synkinematic erosion (models 3, 6, 9, and 12, Table 1)

Horizontal shortening in all models was achieved by moving the left end-wall of the deformation rig in 2-mm (0.08-in.) increments. After each 2 mm (0.08 in.) of shortening, a high-resolution digital photograph of the model was taken though the glass sidewall of the apparatus.

In all experiments, an initial 10 cm (3.9 in.) of shortening was applied such that a stable thrust wedge began to form (Table 1). After this initial 10 cm (3.9 in.) of shortening, synkinematic erosion and synkinematic sedimentation were applied in the following increments

1) In Series 1 experiments (no synkinematic sedimentation and varying rates of synkinematic erosion), model 1 was the baseline model and had no erosion; erosion was conducted after every 8 cm (3.1 in.) of shortening (low erosion, model 4), after every 4 cm (1.6 in.) of shortening (intermediate erosion, model 7), and after every 1 cm (0.4 in.) of shortening (high erosion, model 10).
2) In Series 2 experiments (low synkinematic sedimentation and varying rates of synkinematic erosion), model 2 had no erosion and models 5, 8, and 11 underwent successive erosional episodes after every 8, 4, and 1 cm (3.1, 1.6, and 0.4 in.) of shortening, respectively, as for the Series 1 experiments.
3) In Series 3 experiments (high synkinematic sedimentation and varying rates of synkinematic erosion),

model 3 had no erosion and models 6, 9, and 12 underwent successive erosional episodes after every 8, 4, and 1 cm (3.1, 1.6, and 0.4 in.) of shortening, respectively, as for the Series 1 experiments.

Synkinematic sedimentation was incrementally applied by hand sieving a layer of colored sand (alternating red, white, or black) on to the thrust wedge every 1 cm (0.4 in.) of shortening (after the initial 10 cm [3.9 in.] of shortening) to form an approximate 6° taper on the frontal slope of the model thrust wedge terminating 10 cm (3.9 in.) in front of the active frontal thrust (Figure 6b). For low synkinematic sedimentation, the layers were approximately 0.5–1 mm (0.02–0.04 in./yr) thick, whereas for high synkinematic sedimentation, the layers were about 1–2 mm (0.04–0.08 in.) thick.

Synkinematic erosion was incrementally achieved at 8-, 4-, and 1-cm (3.1-, 1.6-, and 0.4-in.) intervals (as described above) by removing approximately 1 mm (0.04 in.) of sand from the top surface of the wedge (but preserving synkinematic deposition in the synclines). Critically tapered wedges in analog models (and in natural examples) typically have two domains: an uplifted hinterland domain with a low slope forming part of the backstop where the thrust faults have been rotated to very steep dips and have become inactive, and a frontal domain with a low taper where the wedge is actively deforming internally and by frontal accretion (e.g., Mulugeta and Koyi, 1987; Koyi, 1995; Lohrmann et al., 2003; this study). In the models described in this chapter, erosion was focused on the uplifted hinterland domain (Figure 6b). The volumes of material removed by syntectonic erosion and resedimented in the models were not balanced because in natural fold and thrust belt systems, much of the eroded material is typically transported outboard of the thrust wedge in both subaerial, e.g., the Himalayas (Curray et al., 2002), and in submarine settings, e.g., offshore northwest Borneo (Morley and Leong, 2008).

RESULTS

In this study, 12 representative experiments (Table 1) were analyzed in detail. The baseline model (model 1) had no synkinematic erosion or sedimentation. Models 2 and 3 had no syntectonic erosion and low to high synkinematic sedimentation, respectively (Table 1). Models 5, 8, and 11 had low synkinematic sedimentation but were subjected to increased amounts of synkinematic erosion, respectively, whereas models 6, 9, and 12 had high synkinematic sedimentation and were also subjected to increased amounts of synkinematic erosion (Table 1). Four end-member experiments (models 1, 3, 10, and 12) are described in detail and shown in Figures 7–12 inclusively. Figure 13 displays a final vertical cross section for all 12 experiments considered in this chapter.

Model 1 Baseline Model: No Synkinematic Sedimentation or Erosion

Figure 7 shows the progressive evolution of the baseline model (no synkinematic sedimentation and no synkinematic erosion) in seven stages up to a maximum of 50.2 cm (19.8 in.) of shortening. In this model, the initial 5 cm (2 in.) of shortening (Figure 7a) formed a forward-breaking sequence of closely spaced foreland-vergent imbricate thrusts together with a back thrust and its associated flat-topped, hanging-wall anticline. A low-relief detachment fold forming a gentle bulge developed in front of the foreland-vergent thrust system, marking the position of the incipient T1 thrust (Figure 7a). By 10 cm (3.9 in.) of contraction thrust, T1 had propagated to the surface forming a frontal ramp fold with an associated small displacement back thrust (Figure 7b).

Subsequent shortening shown in Figure 7c–g produced an imbricate fan of six thrusts, T1–T6, that nucleated in a forward-breaking sequence. At each stage, a low-amplitude detachment fold formed in front of the thrust wedge (e.g., T3 in Figure 7c) as the precursor to a thrust fault breaking through the layers and propagating to the surface (e.g., T3 in Figure 7d). The main foreland-vergent thrusts formed at approximately regular intervals (average of 9.4 cm [3.7 in.]) after each approximately 7 to 8 cm (3.1 in.) of shortening as the imbricate fan of the thrust wedge developed. In addition to the hanging-wall ramp fold formed above each thrust, small high-angle, low-displacement back thrusts formed in each thrust sheet. These back thrusts nucleated in the hanging wall from the region where the main foreland-vergent thrust climbed up section from the main basal detachment surface (Figure 7b–g).

After a total of 50.2 cm (19.8 in.) of shortening, the final model consisted of a thrust wedge with a forward-sloping surface topography formed by six foreland-vergent thrusts each with their associated hanging-wall ramp folds (Figure 7g–h). These thrusts had nucleated in a forward-breaking sequence. In front of thrust T6, a small displacement detachment fold and the incipient position of the next thrust fault (T7) were observed. The frontal part of the thrust wedge had a critical taper angle of 10° (frontal surface slope). The hinterland section of the wedge had a surface slope of 4°. Here the earliest formed thrusts had been rotated to very steep ramp angles (40–50°) relative to the frontal thrust T6 (32°) and

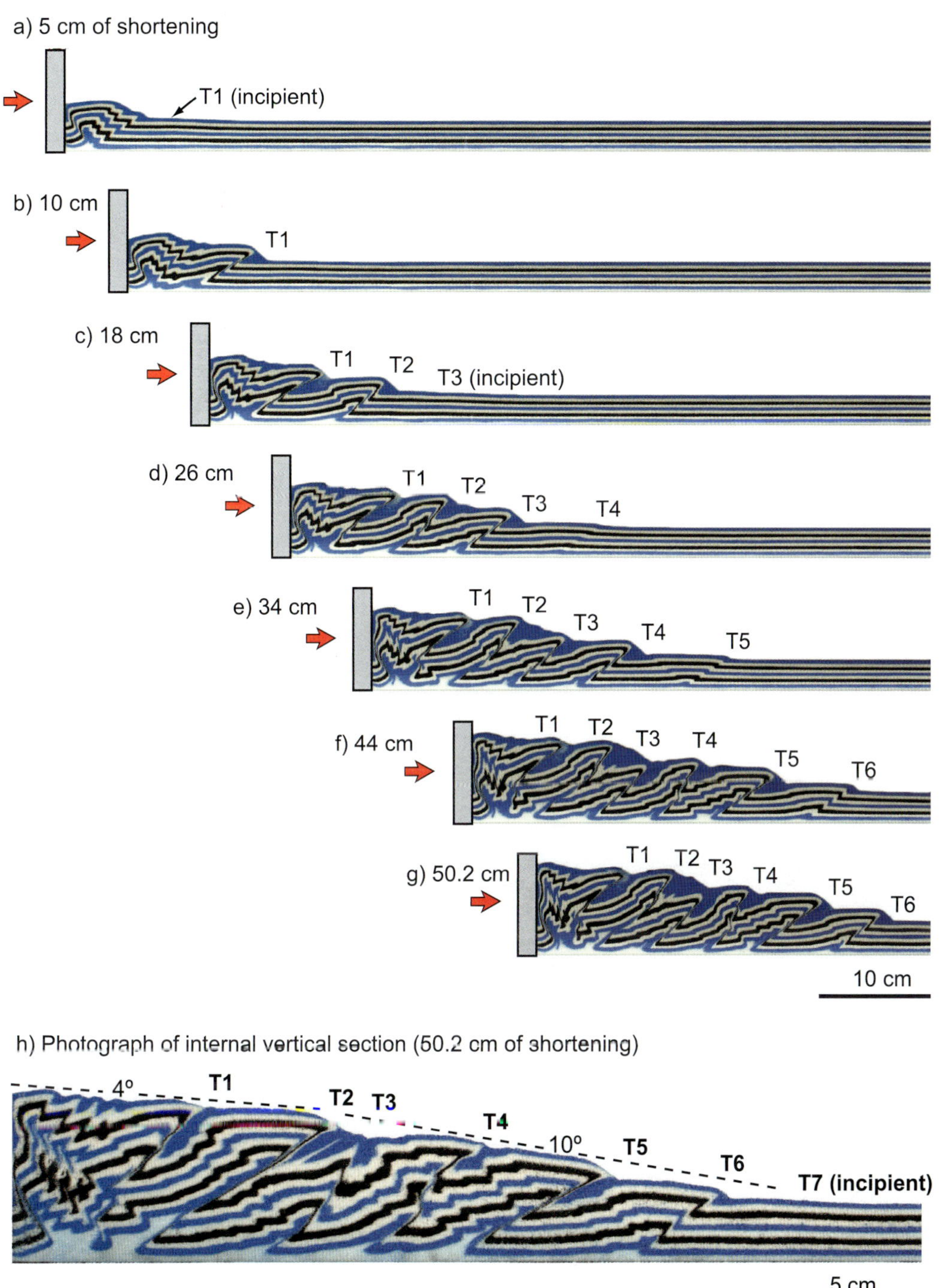

Figure 7. Model 1 with no synkinematic sedimentation or erosion. Progressive evolution illustrated by sequential photographs after (a) 5, (b) 10, (c) 18, (d) 26, (e) 34, (f) 44, and (g) 50.2 cm (2, 3.9, 7.1, 10.2, 13.4, 17.3, and 19.8 in.) of shortening. (h) Photograph of the internal vertical section after model preservation. T1 to T*n* = sequence of thrusts ordered by nucleation.

had become inactive, effectively becoming part of the backstop (Figure 7h).

Model 3: High Synkinematic Sedimentation with No Erosion

In model 3 (Figure 8), the initial geometries and properties of the sand pack were identical with those in model 1 described above (Table 1). After the initial 10 cm (3.9 in.) of shortening, synkinematic sand layers were carefully sieved on to the frontal slope of the growing thrust wedge at a rate of 1–2 mm (0.04–0.08 in.) after each additional 1-cm (0.4-in.) increment of shortening (Table 1).

At 10 cm (3.9 in.) of shortening, the model consisted of a short thrust wedge with a narrow zone of closely spaced thrusts near the rigid backstop and one major foreland-vergent thrust and associated hanging-wall

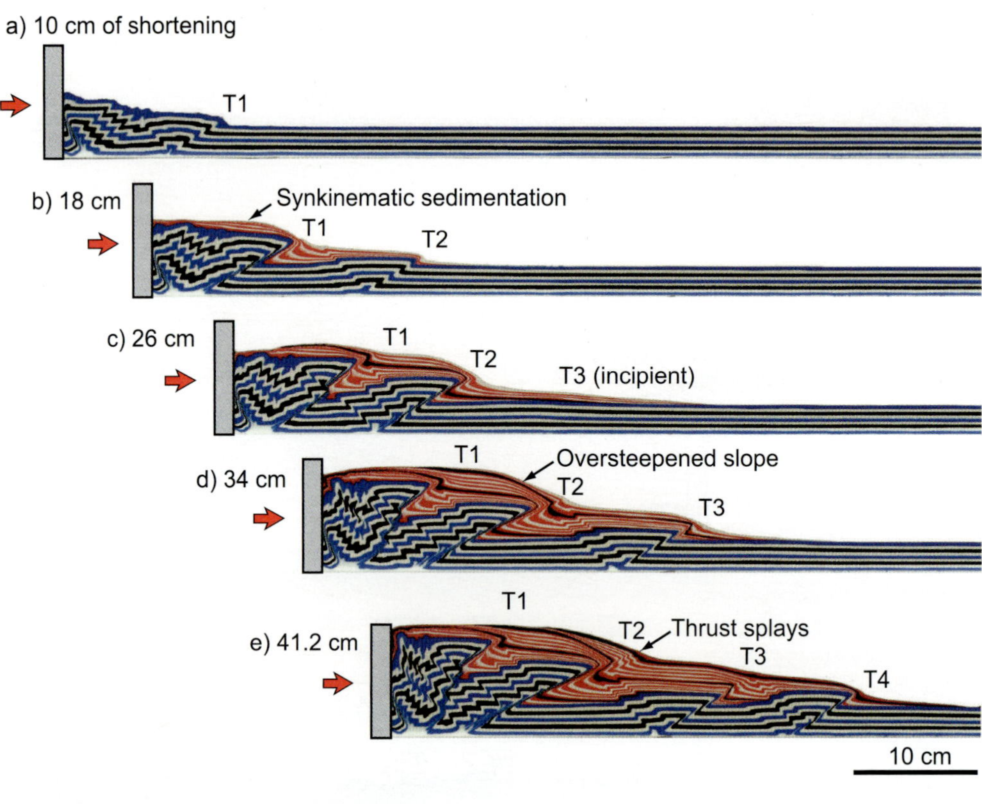

Figure 8. Model 3 with synkinematic sedimentation and no erosion. Progressive evolution illustrated by sequential photographs after (a) 10, (b) 18, (c) 26, (d) 34, and (e) 41.2 cm (3.9, 7.1, 10.2, 13.4, and 16.2 in.) of shortening. (f) Photograph of the internal vertical section after model preservation. T1 to T*n* = sequence of thrusts ordered by nucleation.

anticline at the front of the model (Figure 8a) similar to model 1 with equivalent shortening (Figure 7b). Continued shortening with the addition of synkinematic sedimentation at a high rate resulted in the formation of three additional widely spaced (average of 13.9 cm [5.5 in.]) foreland-vergent thrusts that nucleated at intervals of between 8 and 11 cm (3.1 and 4.3 in.) of shortening (Figure 8b–e). Large fault displacements of thrust T2 caused an oversteepened surface to develop (Figure 8d). At 51.2 cm (20.1 in.) of shortening, the wedge ceased to deform internally and simply slid over the basal detachment. Individual thrusts were characterized by hanging-wall ramp anticlines and associated small-displacement back thrusts. The major foreland-vergent thrusts, T1–T4, formed with ramp angles of 30–33° and had planar trajectories that cut through the synkinematic layers (Figure 8e). Small fault splays formed at the tip of thrust T2 (Figure 8e–f).

Model 10: High Synkinematic Erosion and No Sedimentation

In model 10 (Figure 9), the initial geometries and properties of the sand pack were identical with those in model 1 described above (Table 1). After 10 cm (3.9 in.) of shortening, the model consisted of a narrow thrust wedge with closely spaced thrusts near the rigid backstop and one major foreland-vergent thrust and its associated hanging-wall anticline at the front of the model (Figure 9a).

Subsequent shortening increments, beyond the initial 10 cm (3.9 in.) of contraction, were accompanied by the introduction of syntectonic erosion by removing 1 mm (0.04 in.) from the top surface of the model per 1 cm (0.4 in.) of shortening. The applied erosion reduced the height of the rear of the wedge such that it became subcritical (e.g., Figure 1b).

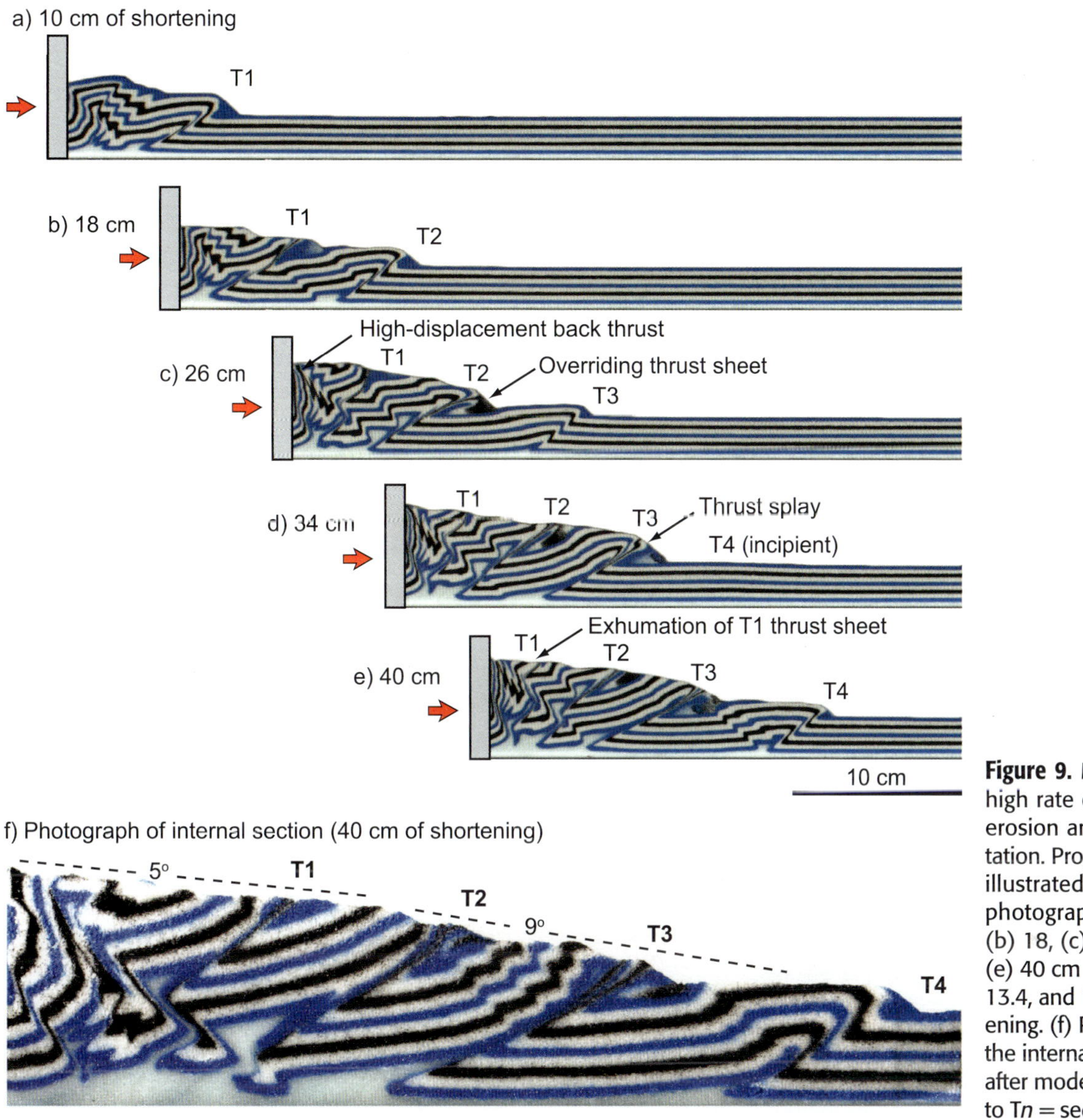

Figure 9. Model 10 with high rate of synkinematic erosion and no sedimentation. Progressive evolution illustrated by sequential photographs after (a) 10, (b) 18, (c) 26, (d) 34, and (e) 40 cm (3.9, 7.1, 10.2, 13.4, and 15.7 in.) of shortening. (f) Photograph of the internal vertical section after model preservation. T1 to T*n* = sequence of thrusts ordered by nucleation.

With continued shortening under these conditions, high-displacement back thrusts formed at the rear of the model, and new foreland-vergent thrusts, T2–T4, nucleated in sequence at the front of the model. Thrust T2 initiated after 14 cm (5.5 in.) of shortening, whereas thrusts T3 and T4 initiated at about 24 and 34 cm (9.4 and 13.4 in.) of shortening, respectively, generating a spacing between major thrusts that averaged 11.8 cm (4.6 in.). As in other models described above, thrust-fault initiation was preceded by the formation of a low-amplitude detachment fold (e.g., Figure 9d). After 40 cm (15.7 in.) of contraction, no further internal deformation was observed. By this stage, hinterland erosion had removed significant parts of the hanging walls of both T1 and T2 thrust sheets producing a pronounced erosional unconformity in this part of the model (Figure 9f). The principal foreland-vergent thrusts, T1–T4, formed with ramp angles of 30–33° and had planar trajectories (Figure 9f).

Model 12: High Synkinematic Sedimentation and High Erosion

In model 12 (Figure 10), the initial geometries and properties of the sand pack were identical with those in model 1 described above (Table 1). After an initial shortening of 10 cm (3.9 in.), the hinterland of the wedge was eroded at a rate of 1 mm (0.04 in.) per 1 cm (0.4 in.) of shortening. Synkinematic sand layers were also carefully

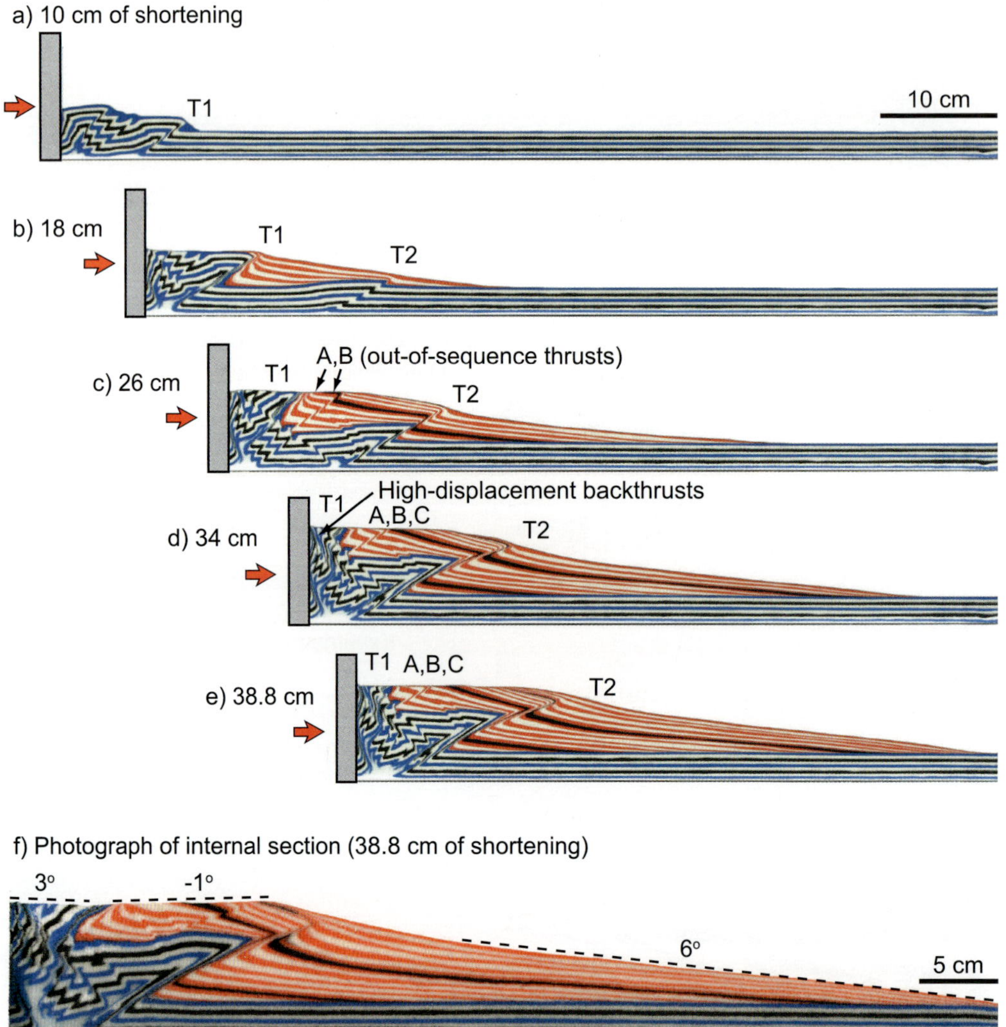

Figure 10. Model 12 with high rates of both synkinematic sedimentation and erosion. Progressive evolution illustrated by sequential photographs after (a) 10, (b) 18, (c) 26, (d) 34, and (e) 38.8 cm (3.9, 7.1, 10.2, 13.4, and 15.3 in.) of shortening. (f) Photograph of the internal vertical section after model preservation. T1 to T*n* = sequence of thrusts ordered by nucleation; A, B, C = out-of-sequence thrusts.

sieved onto the frontal slope of the growing thrust wedge at a rate of 1–2 mm (0.04–0.08 in.) after each additional 1-cm (0.4-in.) increment of shortening (Table 1).

After 10 cm (3.9 in.) of shortening, the model consisted of a single major foreland-vergent thrust, T1, with a dip of 32°. The hinterland was formed by a single back thrust together with several closely spaced foreland-vergent imbricate thrusts (Figure 10a).

With continued deformation, and with continued syntectonic erosion and sedimentation, only one more large thrust formed, T2, after 17 cm (6.7 in.) of contraction, and the model ceased to deform internally after a total of 38.8 cm (15.3 in.) of shortening (Figure 10f).

At 26 cm (10.2 in.) of shortening, two foreland-vergent, out-of-sequence thrusts, A and B, initiated behind thrust T2 and in front of thrust T1 (Figure 10c). Growth stratal geometries indicate that T2 continued to deform during this period. In addition, two small-displacement back thrusts formed in the hanging wall of thrust T2. A third high-displacement back thrust propagated from the base of T2, offsetting thrusts A and B such that they became inactive. In the hinterland of the wedge, thrust T1 was exhumed and eroded with removal of most of its front limb (Figure 10e). The front of the model was buried by sedimentation creating a major downlapping unconformity between the pre- and synkinematic strata (Figure 10e–f).

ANALYSIS OF DEFORMATION PATTERNS AND THRUST-FOLD GEOMETRIES

Thrust Activities

Thrust activities and relative displacements along the fault plane of the key marker layer (Figure 6c) in the four key models described above are plotted against

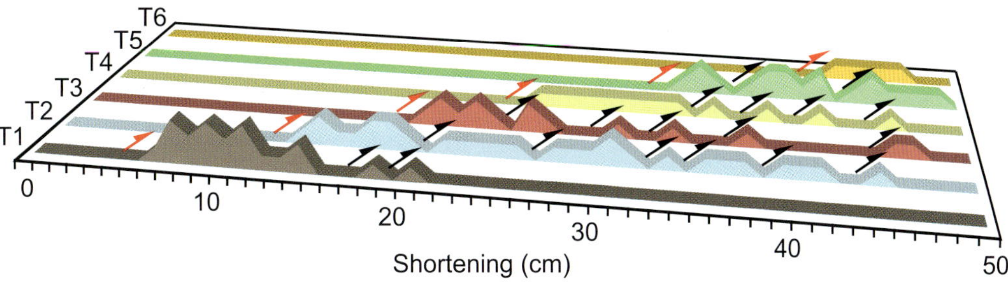

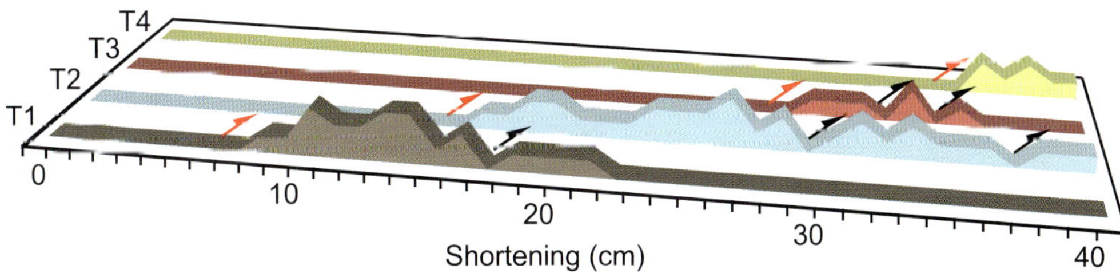

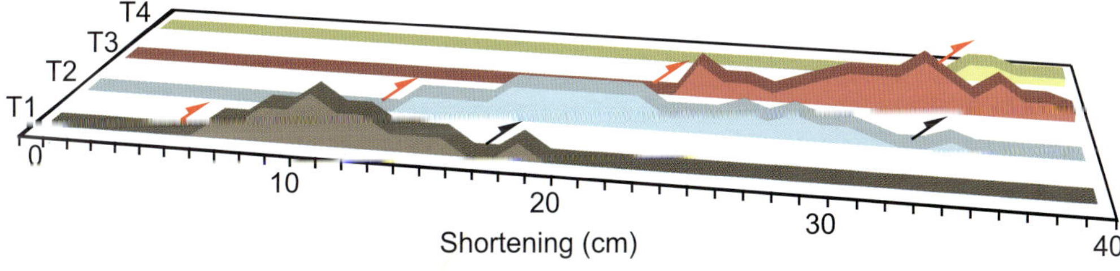

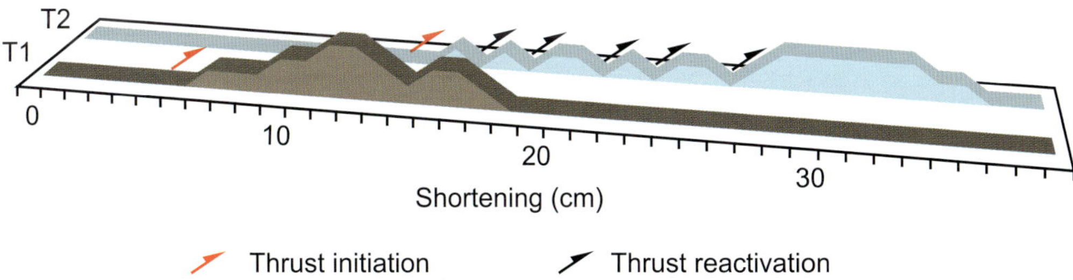

Figure 11. Three-dimensional plot of relative thrust activities against incremental shortening plotted from displacement of the key marker layer in Figure 6c. See Figure 14 for the location of thrust faults.

incremental shortening in Figure 11. These three-dimensional plots show the initiation of the main foreland-vergent thrusts, the duration of displacement on each thrust as well as episodes of thrust-fault reactivation. In the four key models, the first major foreland-vergent thrust, T1, formed between 7 and 8 cm (2.7 and 3.1 in.) of shortening with variable displacement continuing to around 19–22 cm (7.5–8.7 in.) of shortening (Figure 11a–d). The second major foreland-vergent thrust, T2, in each model initiated at around 14 cm (5.5 in.) of shortening for models without synkinematic sedimentation and at around 16 cm (6.3 in.) of shortening in models with high synkinematic sedimentation (Figure 11).

In all of these four models, synchronous and episodic thrust activities were recorded. This is particularly evident in models 1, 3, and 10 where two or more thrusts were active throughout the deformation history

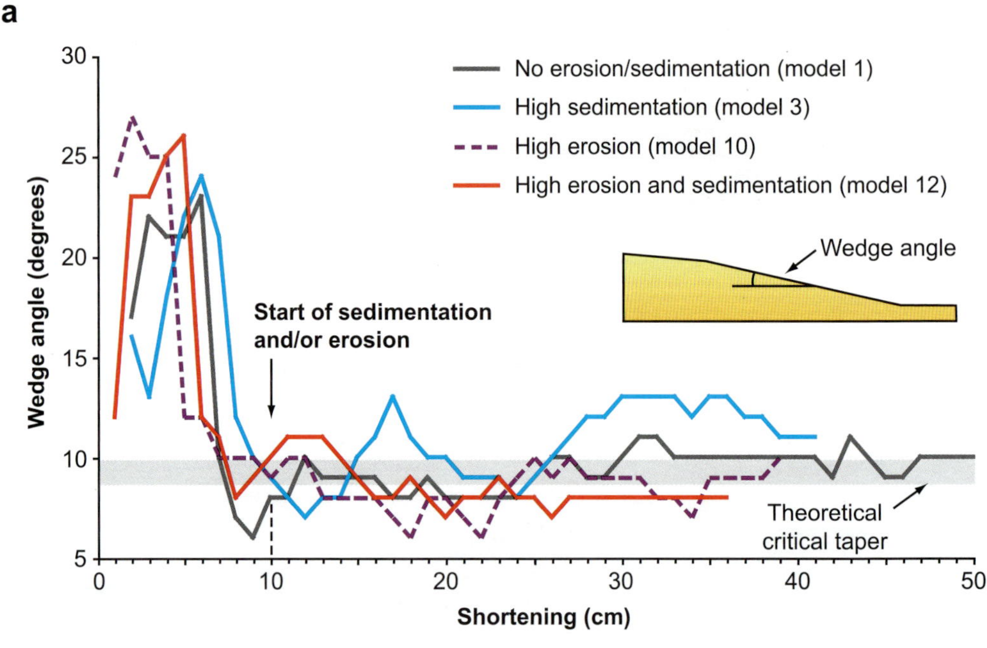

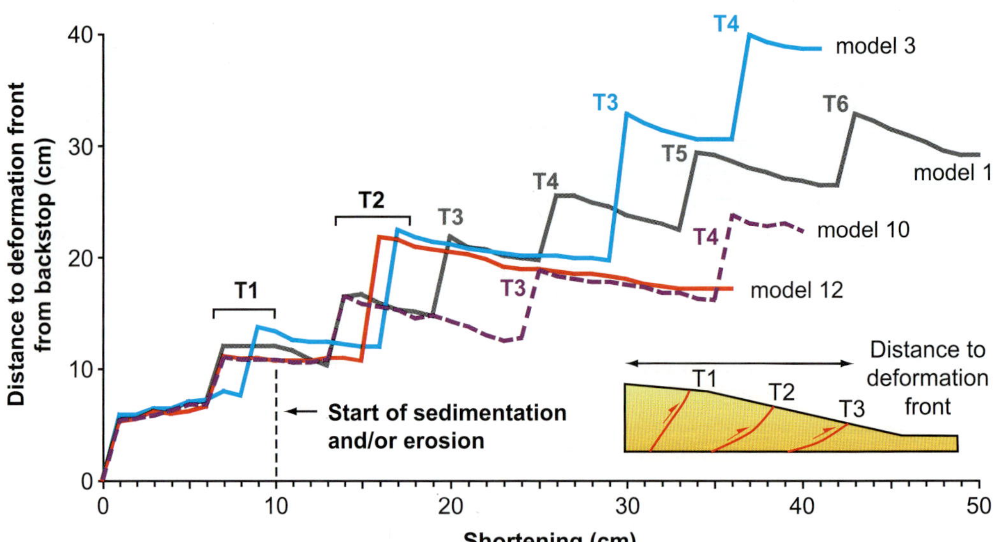

Figure 12. Graphs showing a comparison of (a) the frontal wedge taper angle versus shortening for models 1, 3, 10, and 12, and (b) the distance to the deformation front measured from the backstop. Measurements were made every 1 cm (0.4 in.) of shortening at the glass sidewalls.

(Figure 11a–c). Thrust reactivation (indicated by the black arrows in Figure 11) occurred in all models and was strongest in models 1 and 12.

Critical Wedge Geometries

Figure 12a shows the incremental wedge geometries as indicated by their taper angle, $\alpha + \beta$, plotted against horizontal shortening. In these models, the taper angle, $\alpha + \beta$, was measured as the frontal topographic slope, α, because the dip of the basal detachment, β, was 0°. The theoretical critical taper was calculated to be 9–10° from the known basal and internal angles of friction within the wedge using the equation from Davis et al. (1983) and an estimate of sidewall friction from the glass sidewalls.

In all models, the wedge geometries had very high slopes (greater than 20°) for the first 9 cm (3.5 in.) of shortening as the backstop geometries were formed (Figure 12a). Beyond the first 10 cm (3.9 in.) of shortening, model 1 with no synkinematic sedimentation or erosion and model 3 with high synkinematic erosion displayed a cyclic behavior with taper angles oscillating above and below the theoretical critical taper angle. The changes in taper angle from as low as 6° to as high as 12° for model 3 with high synkinematic sedimentation reflect the localization of deformation on to

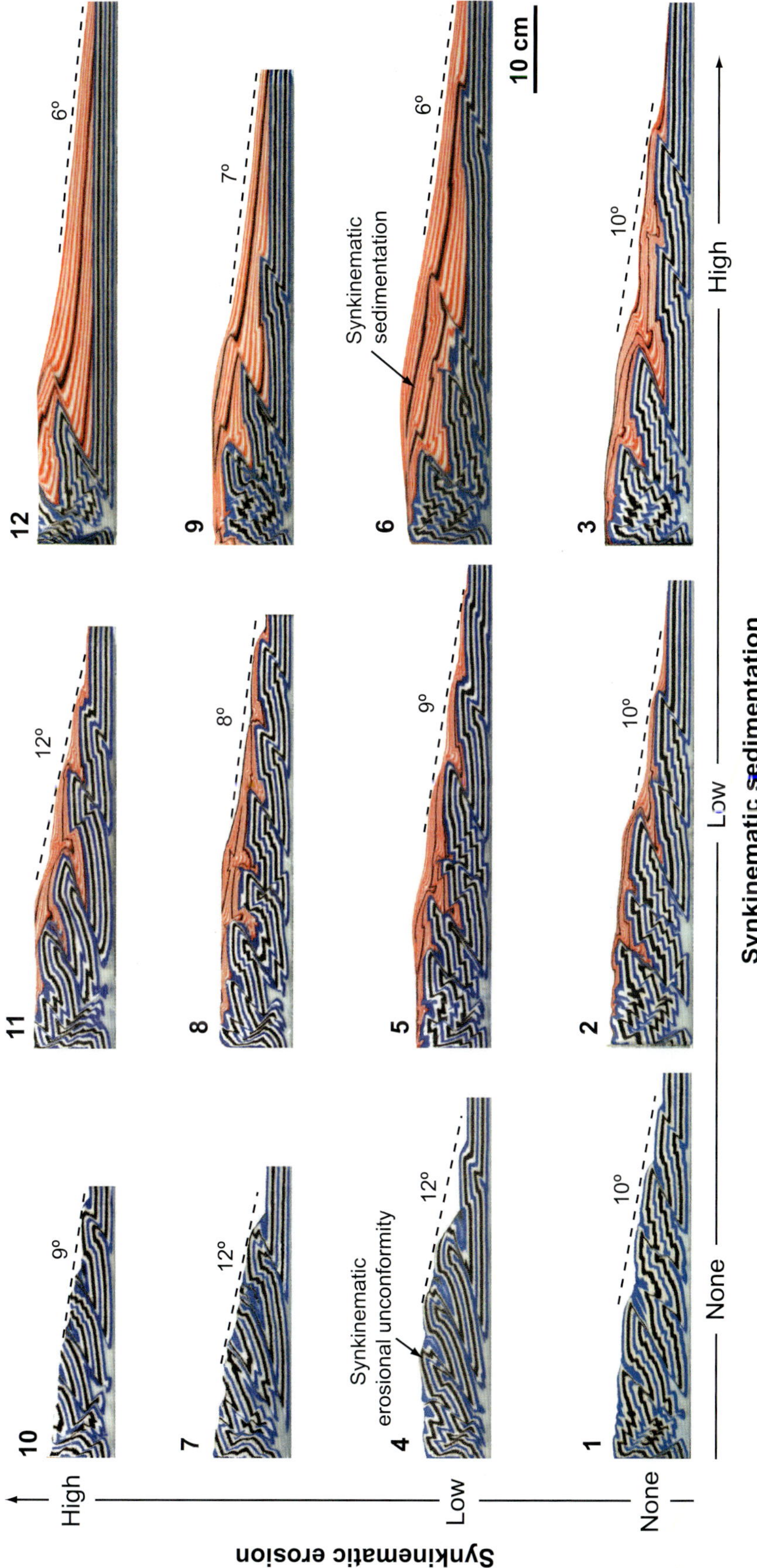

Figure 13. Summary table showing internal vertical sections and frontal taper angles of analog models with systematic variations in the rates of synkinematic sedimentation and/or synkinematic erosion. All models are shown at the same scale.

individual thrust surfaces within the wedge such that the taper angle is modified by episodic motion on particular high-displacement thrusts. Both model 10 with high synkinematic erosion and model 12 with high synkinematic sedimentation and erosion had taper angles lower than the theoretical critical taper. In model 10, this was caused by constant removal of material from the upper surface of the model, whereas in model 12, wedge taper was passively built up by the foreland sedimentation and not by fault-related deformation, allowing the taper angle to decrease below the theoretical angle.

The cyclic nature of the thrust-wedge geometries and of the distribution of the internal deformation within the overall thrust wedge is also reflected in the plot of wedge length as indicated by the distance between the deformation front in the foreland and the backstop (Figure 12b). Significant increases in the lengths of the imbricate thrust wedges occurred each time a new frontal foreland-vergent thrust developed. Following this step change in the dimensions of the thrust wedge, subsequent finite displacement on the newly formed frontal thrust due to shortening from the moving backwall of the experimental apparatus resulted in a measurable decrease in the wedge length until the next major foreland-vergent thrust was initiated (Figure 12b). Propagation of the deformation front into the foreland was inhibited by syntectonic erosion and is reflected by a decreased slope in the curves for models 10 and 12, both of which had a component of syntectonic erosion, relative to the other models (Figure 12b). The graphs of Figure 12a and b clearly show the cyclic nature of thrust activities with all the model wedges and highlight the effect of foreland sedimentation in building wedge taper as well as the effect of syntectonic erosion in delaying the propagation of the deformation front into the foreland.

DISCUSSION

Analog Model Geometries and Evolution

The final geometries of all 12 models (Figure 13) and the interpreted principal structural geometries of key models (Figure 14) demonstrate the wide range of possible thrust-wedge geometries that were generated by variations in synkinematic sedimentation and/or erosion. All 12 models generated critically tapered Coulomb wedges with topographic slopes of 6 to 10° above horizontal basal detachments (Figure 13). The thrust wedges were formed of imbricate fans composed of two to six forward-vergent thrusts with associated low-displacement back thrusts. At the rear of the wedges, high-displacement back thrusts were produced that were associated with thrust T1 and in some cases thrust T2.

In models 1, 2, and 3 with synkinematic sedimentation and without erosion (Figure 14a–c), longer thrust wedges composed of progressively fewer major forward-vergent thrusts (six in model 1 to four in model 3) were formed as the rate of synkinematic sedimentation rate was increased. The lengthening of the thrust wedges combined with the reduction in the number of major forward-vergent thrusts effectively widened the spacing between the thrusts producing a distinct wedge geometry composed of thrusts separated by broad thrust top basins. Similar results were observed in the models regardless of whether synkinematic erosion was applied (Figure 13). This geometry highlights the function of synkinematic sedimentation in achieving critical taper in a thrust wedge; instead of relying only on thrust imbrication to internally deform the wedge to attain the critical taper such as in the case of models 1, 4, 7, and 10 (Figures 13; 14a, d–f), the critical taper was achieved through a combination of thrust imbrication and passive sedimentary infill in the models with synkinematic sedimentation (Figures 13; 14b, c) thereby allowing the wedges to rapidly propagate forward into the foreland.

The building of wedge taper by synkinematic sedimentation lowered thrust activities in the foreland relative to the hinterland and was reflected in the conspicuous difference in displacement of thrusts T3 and T4 in the foreland relative to thrusts T1 and T2 in the hinterland of model 3 (Figure 14c). In more extreme cases, accumulation of synkinematic sedimentation in the foreland decreased the frontal wedge taper below the critical taper angle rendering the wedge subcritical. This case was observed in models 6, 9, and 12, all of which had high rates of synkinematic sedimentation (Figure 13). These models produced wedges with taper angles of 6 to 7° (Figure 13), all of which were lower than the theoretical critical taper of 9–10° (e.g., Figure 12a). Similar results were obtained in other physical models (Storti and McClay, 1995) and numerical models (Simpson, 2006) in which the highest surface process rates also produced the lowest wedge tapers.

In models 1, 4, 7, and 10 with variable synkinematic erosion and no sedimentation (Figure 14a, d–f), shorter thrust wedges composed of fewer major forward-vergent imbricate thrusts (from six in model 1 to four in models 7 and 10) were produced when the rate of synkinematic erosion was progressively increased. The higher erosion rate caused the thrust wedge to become progressively more subcritical (e.g., Figure 1b), which was demonstrated by the progressive decrease in wedge height at the rear of the models (Figure 14d–f). Because of the subcritical wedge state, the rear of the model was exhumed through back thrusting and thrust reactivation

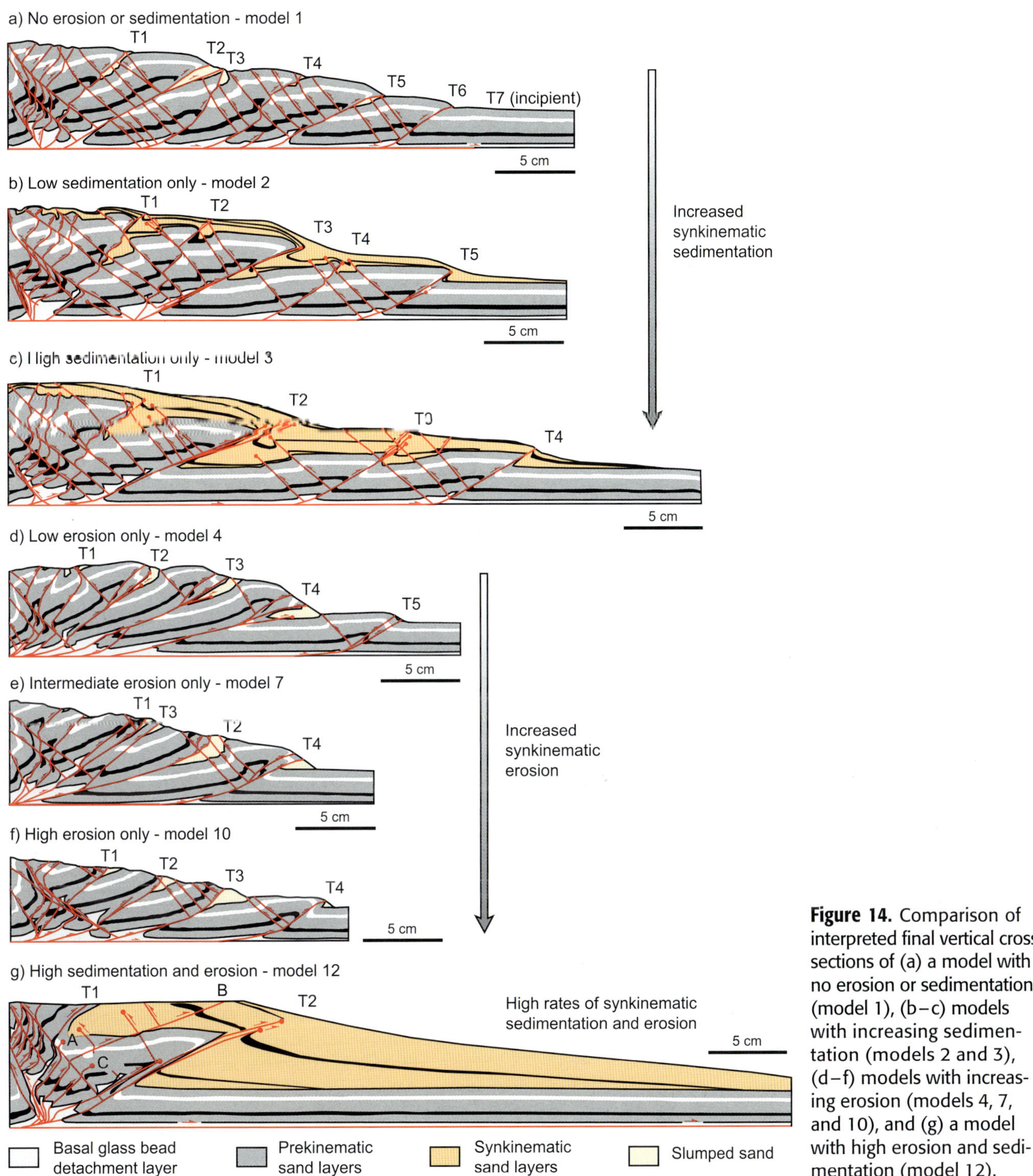

Figure 14. Comparison of interpreted final vertical cross sections of (a) a model with no erosion or sedimentation (model 1), (b–c) models with increasing sedimentation (models 2 and 3), (d–f) models with increasing erosion (models 4, 7, and 10), and (g) a model with high erosion and sedimentation (model 12).

to reestablish the height of the wedge according to the critically tapered Coulomb wedge theory of Davis et al. (1983), Dahlen (1990), and others. These internal deformations inhibited the propagation of the deformation front into the foreland. Because no synkinematic sedimentation existed to passively lengthen the wedge into the foreland, the net result was the formation of shorter thrust wedges with fewer major forward-vergent thrusts.

In the models that combined synkinematic sedimentation together with synkinematic erosion, the final geometries had characteristics of both the models with synkinematic sedimentation only (Figure 14b, c) and those with synkinematic erosion only (Figure 14d–f). This is best illustrated by a comparison of model 12, which had high syntectonic sedimentation and erosion, to the other 11 models in the study (Figure 13). In model 12, the lengthened thrust wedge and the lowered frontal taper angle were similar to models with syntectonic sedimentation only, whereas the highly exhumed hinterland and the shortened distance between the backstop and the deformation front were similar to the models with syntectonic erosion only. Model 12 also developed the fewest number of major forward-vergent thrusts relative to the other eleven models (Figure 13), an effect similar to that observed when either synkinematic sedimentation or synkinematic erosion was applied (Figure 14). These results indicate that effects of synkinematic sedimentation and erosion in the models, when combined, were complementary and noncompetitive. The reason for this was because at the scale of an entire thrust wedge, both syntectonic sedimentation and syntectonic erosion caused the model wedges to become subcritical (e.g., Figure 1b), but they acted on opposing ends of the wedge.

Analog Model Deformation Patterns

Although all material in critically tapered wedges is considered to be in a state of yield, in reality, wedges are mechanically partitioned requiring failure to occur at a finite number of preexisting zones of weakness or by forming new faults (e.g., Nieuwland and Saher, 2002; Lohrmann et al., 2003). Consequently, when one segment of a wedge is brought into equilibrium through internal deformation, the equilibrium of a different segment may be concurrently destroyed. This results in cyclic deformation resulting in the oscillation of taper angles and the synchronous and/or out-of-sequence thrust activities that were observed in the models (e.g., Figures 11, 12) particularly in model 1, which had no synkinematic sedimentation or erosion.

This cyclic deformation was dynamically altered by synkinematic sedimentation and synkinematic erosion, which maintained or destroyed equilibrium by building or removing material from the upper surface of the wedge. When synkinematic sedimentation occurred in the foreland only, such as in models 2 and 3 (Figure 14b, c), wedge equilibrium was enhanced in the foreland as shown by the lower thrust displacements in the foreland relative to the hinterland (Figure 14b, c). In contrast, synkinematic sedimentation on both the foreland and hinterland of a wedge stabilized the entire wedge and promoted frontal imbrication (Wu et al., 2008). Synkinematic sedimentation in thrust top basins created local stable supercritical wedge segments that did not internally deform (e.g., Mugnier et al., 1997; Stockmal et al., 2007). Synkinematic erosion in the hinterland, such as in models 4, 7, and 10 (Figure 14d–f), destroyed wedge equilibrium in the hinterland and increased deformation at the rear of the wedge in the form of high-displacement back thrusting.

Thrust Fault-related Fold Geometries

Figure 15 shows examples of the geometries of thrust-related folds formed in the models. In model 1 without synkinematic sedimentation or synkinematic erosion, fold evolution began with layer-parallel shortening producing a broad, unfaulted, low-relief detachment fold. Folding was succeeded by the breakthrough of a forward-vergent thrust and associated low-displacement back thrust producing a flat-topped hanging-wall ramp fold (Figure 15a). Continued shortening caused the passive transport of the fold up the ramp, and in some cases, a lower-angle shortcut thrust developed as the fold traversed the ramp-flat segment of the fault at the upper surface (Figure 15b). The dip of the backlimb was lower than the ramp angle, resulting in a geometry similar to that described for pure-shear fault-bend folds (Suppe et al., 2004). This general pattern of low-amplitude detachment folding followed by limb breakthrough was consistently observed in all experiments.

In experiments with synkinematic sedimentation, folds also initiated as low-relief detachment folds (e.g., T3 in Figure 8c or T4 in Figure 9d). As the folds amplified growth, stratal architectures with fanning dips were developed on both the backlimbs and front limbs, indicating limb rotation (Figure 15c). The frontal sections of these growth folds were characterized by zones of trishear and showed rounded geometries cut in places by tip-line thrust splays (Figure 15c) similar to features seen in field and subsurface examples of natural fold and thrust belts (e.g., Erslev, 1991; Schelling, 1999; Butler and McCaffrey, 2004). The thickness of the pregrowth strata was effectively increased by synkinematic sedimentation ahead of the deformation front (Figure 15d), resulting in additional loading that increased shear stresses at the basal detachment. This caused the foreland-vergent thrust that broke through the front limb to develop with a lower ramp angle (e.g., 27° in Figure 15d) compared to the model without synkinematic sedimentation (e.g., 34° in Figure 15a). In addition, the number of associated low-displacement back thrusts displacing the backlimb of the fold was reduced (compare model 3 in Figure 14c with model 1 in Figure 14a). In contrast, fold geometries in the models with synkinematic erosion (Figure 15b) did not differ greatly from folds in model 1 without synkinematic erosion or sedimentation (Figure 15a), indicating that synkinematic erosion did not appear to

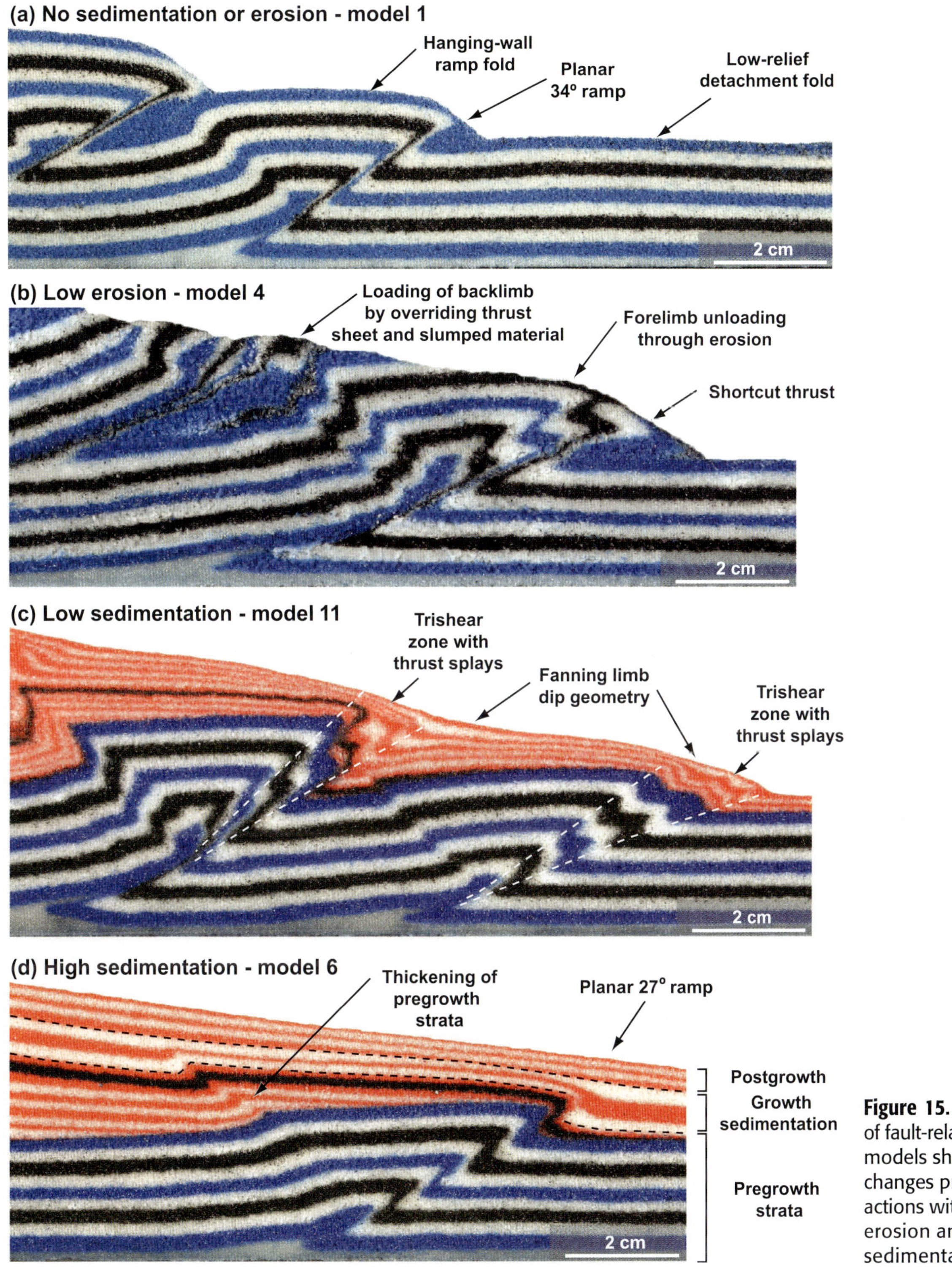

Figure 15. (a–d) Examples of fault-related folds in the models showing geometric changes produced by interactions with synkinematic erosion and synkinematic sedimentation.

have a major influence on individual fold geometries during fold growth.

Geometric and kinematic models of fault-related folds that amplify in a self-similar evolutionary manner (e.g., Suppe, 1983; Suppe and Medwedeff, 1990) have been extensively used to understand fault or fold evolutionary histories and their resultant growth stratal architectures (e.g., Suppe et al., 1992; Hardy et al., 1996; Shaw et al., 2005). The fault-related fold geometries in Figure 15 demonstrate that particularly in cases with synkinematic sedimentation, fold growth will respond to the sedimentation, thus posing a challenge for these simplified models of growth stratal architectures generated using self-similar kinematics. These simplified models contain the inherent assumption that fold evolution would be independent from syncontractional sedimentation and syncontractional erosion and do not allow for any geometric or kinematic divergences in evolution caused by the dynamic interactions as illustrated in the analog models in this study.

Model Limitations

The experiments in this study were designed to demonstrate changes to thrust-wedge geometries and activities caused by synkinematic sedimentation and synkinematic erosion. However, as in all models, both physical and numerical, limitations due to assumptions about the mechanics and kinematics of the system being studied as well as limitations imposed by the boundary conditions of the models exist. The rigid horizontal base used in the models in this study did not allow a flexural isostatic response. Other physical models (e.g., Hoth et al., 2006) and numerical models (e.g., Simpson, 2006; Stockmal et al., 2007) in which a flexible base plate was used allowed a progressive basal tilt of 2–3° to develop toward the hinterland due to isostatic loading. The results of these studies were comparable, showing that the incorporation of flexural isostasy would not materially change the experimental conclusions. Differences in mechanical stratigraphy were also not included in the experimental configuration. In particular, analog models incorporating synkinematic sedimentation and/or erosion with an intermediate detachment level (e.g., Gutscher et al., 1996; Mugnier et al., 1997; Bonnet et al., 2007; Malavieille and Trullenque, 2009) showed that the thrust wedge responded to erosion of the hinterland by underthrusting as an alternative to high-displacement back thrusts at the rear of the thrust wedge. In the models, pore-fluid effects, thermal effects, competency contrasts, and anisotropies were not simulated. Pore-fluid effects are important in weakening faults and reducing the basal friction acting on the thrust wedge. Its incorporation in the experimental configuration (Cobbold et al., 2001; Mourgues and Cobbold, 2006) would likely have reduced the critical-taper angle of the wedges closer to natural prototypes (e.g., Figure 1c). Synkinematic sedimentation and erosion were applied in this study without considering feedback from tectonic or climatic changes, surface topography, or sedimentary processes. In natural orogenic systems, complex feedbacks exist between these factors that appear to be primarily driven by climate (e.g., Whipple and Meade, 2004, 2006; Whipple, 2009). The methodology used for synkinematic sedimentation and erosion in this study was designed to create a first-order approximation of surface processes in natural systems but was not designed to address their intricacies.

Comparisons with Natural Prototypes

Submarine Fold and Thrust Belts

In the Nankai accretionary prism, offshore Japan, a cross section across the frontal section of the prism indicates both low synkinematic sedimentation and erosion rates based on growth stratal thicknesses and the absence of major erosional unconformities (Figure 2a). The Nankai fold and thrust belt is formed of a well-ordered array of closely spaced (1–2 km [0.6–1.2 mi]), forward-vergent thrusts with associated minor back thrusts similar to the structural styles shown in model 1 (Figure 7). The frontal folds in the Nankai section have long, gentle backlimbs that dip less than the fault ramp and have been successfully geometrically modeled as pure-shear fault-bend folds by Suppe et al. (2004). The architecture of the frontal fold is a broad, low-relief anticline with a thrust that has broken through its front limb similar to the frontal folds in the models (Figure 15a).

In the deep-water fold belt of eastern offshore Niger Delta, Gulf of Guinea (Figure 2b), rates of synkinematic sedimentation appear to be higher in comparison to the frontal Nankai accretionary prism such that the fold belt is completely buried. The lack of major erosional unconformities within the stratigraphic section indicates a low rate of synkinematic erosion. The wedge geometries formed by the sea floor and basal detachment appear to fit with those predicted by critically tapered wedge mechanics (Bilotti and Shaw, 2005). In Figure 2b, the imbricate thrusts that form the fold and thrust belt are widely spaced, have small displacements, and are completely buried (e.g., shut off by the synkinematic sedimentation in a manner similar to model 3 of this study shown in Figure 8). Within the Niger Delta deep-water fold belt, evidence exists on seismic profiles for the development of a trishear zone in the front limb of the growth folds (Briggs et al., 2006) or tip-line thrust splays with a folded growth section at the front limb of the fold (Kostenko et al., 2008) resembling fold geometries in the experimental models with synkinematic sedimentation (Figure 15c, d).

Subaerial Fold and Thrust Belts

A cross section through the sub-Andean fold and thrust belt in northwestern Argentina (Figure 2c) shows an example of a subaerial fold and thrust belt where a high rate of synkinematic sedimentation has produced thrust top basins filled with more than 7 km (4.3 mi) of Neogene sediments (Echavarria et al., 2003). A surface unconformity gives evidence for synkinematic erosion, and the climate is arid (Strecker et al., 2007), indicating low to moderate erosion rates. In the foreland, the thrust belt is characterized in cross section by broad synclines approximately 30 km (19 mi) wide divided by fault-propagation folds with steep flanks and narrow crests, whereas in the hinterland, a high-displacement forward-vergent thrust associated with a broad synform has developed ahead of a backstop formed by the pre-Silurian basement. At the fold and thrust belt scale, the sub-Andes cross section in Figure 2c shows similarities to the geometries of models 3, 6, and 9 (Figure 13) in which fault-related folds are separated by wide expanses of synkinematic sedimentation. Reconstructions of thrust sequences by Echavarria et al. (2003) show a similar kinematic history to these models (models 3, 6, and 9) in which thrusts in the hinterland were highly active in relation to the buried thrusts in the foreland.

A cross section through the central Apennines, Italy, shows a fold and thrust belt that is partially submerged under the Adriatic Sea at present (Figure 16). The rear of the wedge is subaerial and has undergone surface erosion. High synkinematic sedimentation at the front of the wedge is evident from the progradational geometries of the sedimentary fill that have completely buried the frontal thrusts (Figure 16a) and is revealed in the isopach maps of synkinematic sediment thickness (Figure 16b). The wedge geometries of Figure 16a, with multiple high-displacement thrusts in the rear and widely spaced (up to 30 km [19 mi] apart), small-displacement thrusts at the front, resemble geometries in model 6 with high synkinematic sedimentation and low synkinematic erosion (Figure 16c). Imbricate thrusts at the front of the thrust wedge in the central Apennines have formed fault-propagation folds (Figure 16a). In one of these folds, tip-line thrust splays have formed in the front limb similar to models in this study with synkinematic sedimentation (e.g., Figure 15c, d).

The Taiwan orogen provides an example of a fold and thrust belt that has been subjected to one of the highest persistent (e.g., from short- to long-term time scales) erosion rates in the world at 3–6 mm/yr (0.12–0.24 in./yr) (e.g., Figure 5b). Global Positioning System (GPS) velocity vectors (Yu et al., 1997) show that the interior of the orogen is being transported westward faster than the foreland (Figure 17a). Measurements of erosion rates show that the highest long-term rates of synkinematic erosion also occur in the interior of the orogen (Figure 17b). An interpreted regional cross section through southwestern Taiwan (Figure 17c) shows an imbricate fan with forward-vergent thrusts with high displacements in the hinterland and low thrust displacements in the foreland, matching the present-day pattern of GPS surface velocities. This deformation pattern compares well to the analog models with synkinematic erosion (models 4, 7, and 10 in Figure 13) that had short wedge lengths and highly active thrusts in the hinterland.

The central sub-Andes fold and thrust belt, Bolivia (Figure 18), is an example of a fold and thrust belt with long-term along-strike climatic variations that are inferred to have produced vastly different synkinematic sedimentation and erosion rates across the same orogen (Horton, 1999; Barnes and Pelletier, 2006; McQuarrie et al., 2008a). Precipitation is much higher in the north compared to the semiarid south, and long-term denudation rates are more than twice as high in the north compared to the south (Barnes and Pelletier, 2006). Assuming lithologies and mechanical stratigraphy are relatively similar along strike (Sempere, 1995; McQuarrie, 2002), we infer that the higher rates of synkinematic sedimentation and erosion in the north have caused the fold and thrust belt in the north to be narrower (McQuarrie et al., 2008b). In this study, high syntectonic sedimentation and erosion inhibited the forward propagation of the deformation front into the foreland of the model thrust wedges and reduced the number of imbricate thrusts (e.g., model 12 in Figure 13), providing a possible explanation for changes in the width of the central sub-Andes fold and thrust belt.

Comparisons of Subaerial and Submarine Fold and Thrust Belts

Syntectonic sedimentation rates in both subaerial and submarine settings are roughly comparable (Figure 5a), whereas syntectonic erosion rates, although poorly quantified for submarine fold and thrust belt systems, are likely in the order of a magnitude lower in submarine compared to subaerial settings (Mitchell et al., 2003). Considering the dramatic differences in the results of the 12 models in this study (Figure 13) and in particular the contrasts produced in the models by increased rates of synkinematic erosion, this distinction in syntectonic erosion rates between the two environments indicates that wedge geometry and kinematics may differ between subaerial and submarine fold and thrust belts.

Implications for Hydrocarbon Exploration

In fold and thrust belts, hydrocarbons are commonly trapped in structural culminations formed by hanging-wall fault-related folds. Correspondingly, the timing

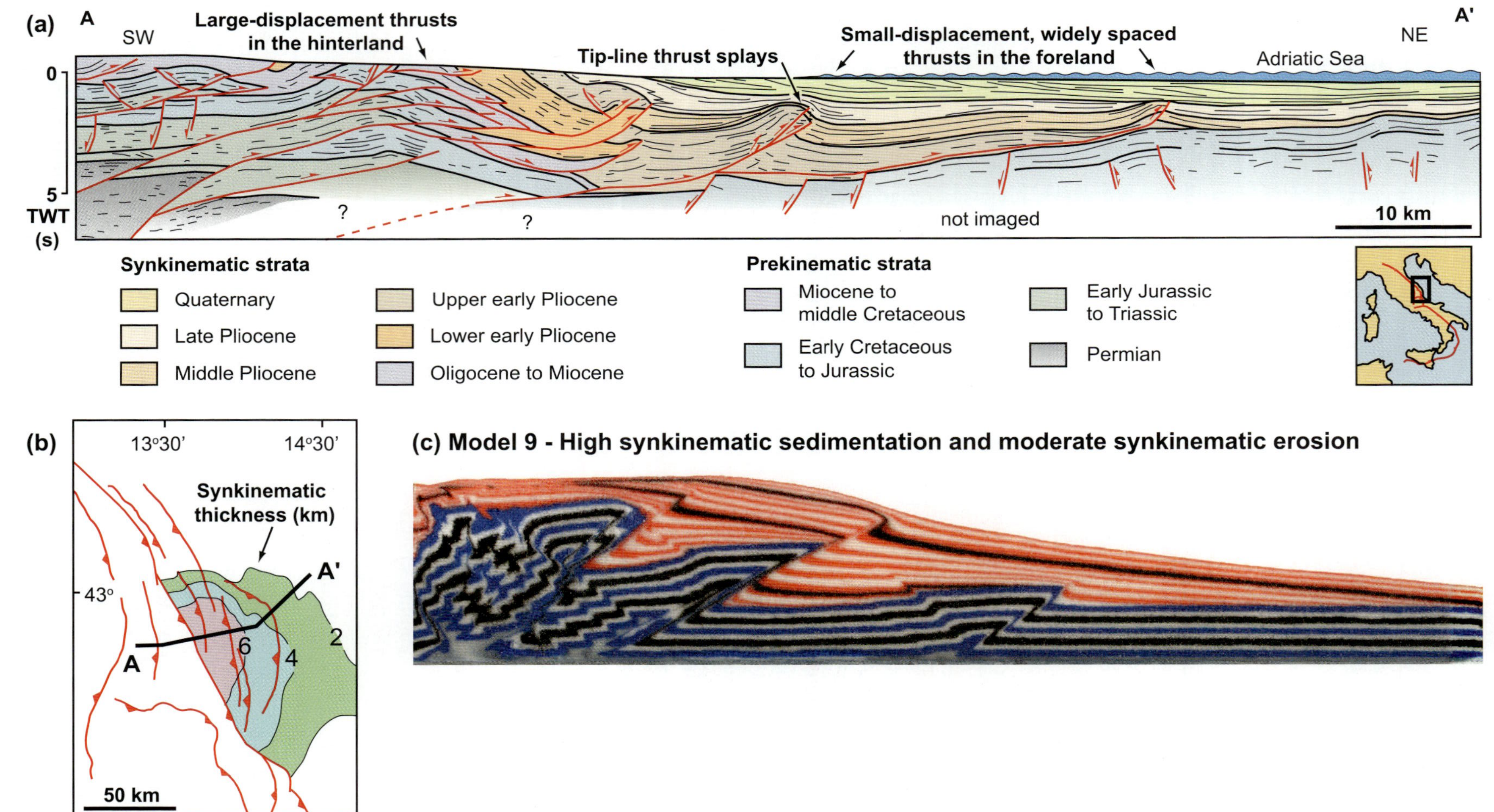

Figure 16. (a) Geological cross section through the central Apennines (modified from Scisciani and Montefalcone, 2006). (b) Map showing the cross section location, major thrusts, and isopach of synkinematic fill thickness modified from De Alteriis (1995). (c) Comparison to a final internal vertical section from model 9 with high synkinematic sedimentation and moderate synkinematic erosion showing high-displacement thrusts in the hinterland and buried, low-displacement thrusts in the foreland. TWT = two-way time.

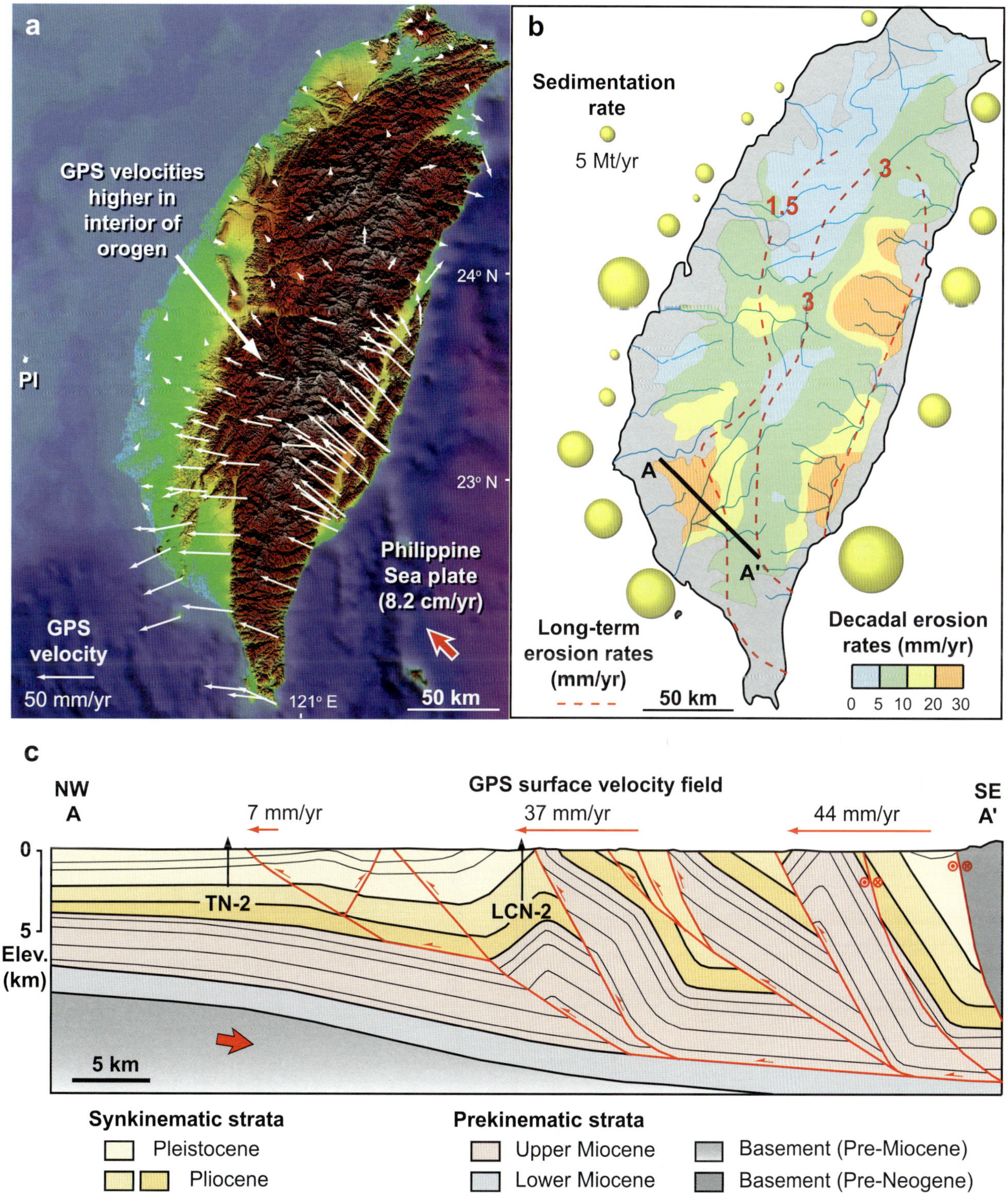

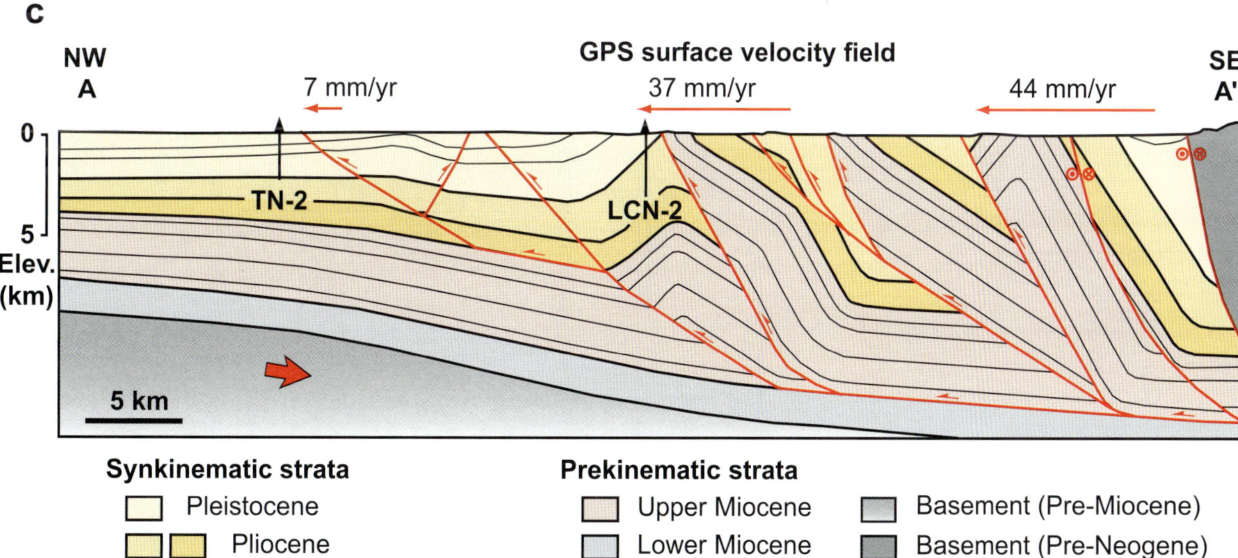

Figure 17. (a) SRTM DEM of Taiwan with Global Positioning System (GPS) surface velocity vectors from Yu et al. (1997) plotted relative to Penghu Island (PI). (b) Distribution of long-term and decadal erosion and sedimentation rates in Taiwan (Dadson et al., 2003) and location of cross section AA′; (c) Interpreted cross section AA′ through southwest Taiwan (modified from Hickman et al., 2002; Huang et al., 2006, 2009) with along-section GPS velocities of Yu et al. (1997). These GPS velocities show relatively higher thrust activities in the hinterland relative to the foreland and compare well to models 4, 7, and 10 with synkinematic erosion. SRTM = shuttle radar topography mission; DEM = digital elevation model; Elev. = elevation.

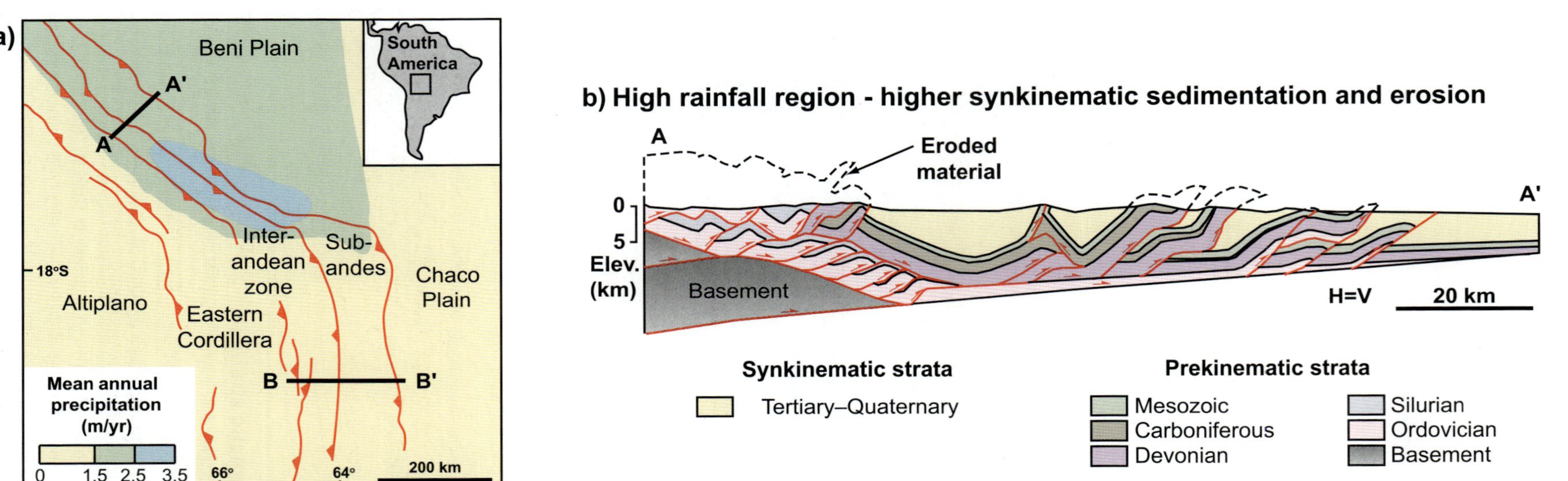

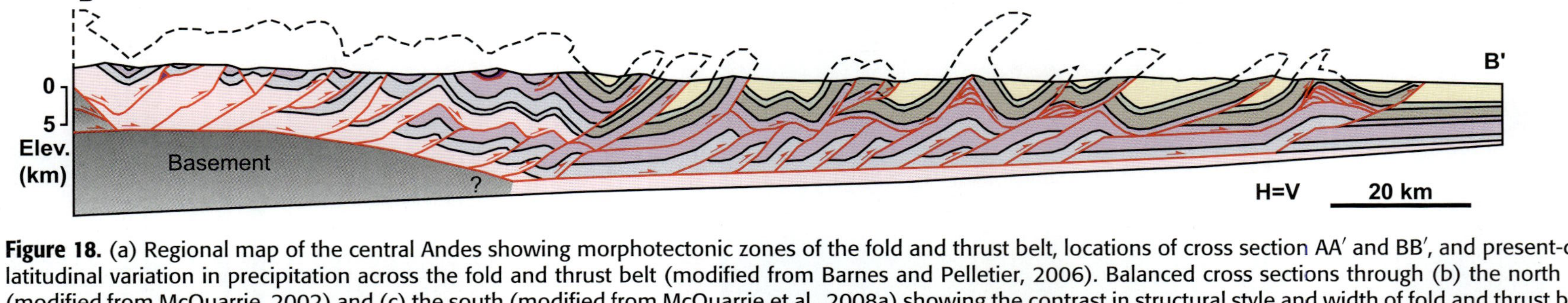

Figure 18. (a) Regional map of the central Andes showing morphotectonic zones of the fold and thrust belt, locations of cross section AA′ and BB′, and present-day latitudinal variation in precipitation across the fold and thrust belt (modified from Barnes and Pelletier, 2006). Balanced cross sections through (b) the north (modified from McQuarrie, 2002) and (c) the south (modified from McQuarrie et al., 2008a) showing the contrast in structural style and width of fold and thrust belt, which is inferred to be a result of long-term climatic differences (modified from McQuarrie et al., 2008b). H = V = no vertical exaggeration (horizontal and vertical scales equal); Elev. = elevation.

of trap formation relative to the generation and expulsion of hydrocarbons is a critical risk in fold and thrust belt exploration (Meneses-Rocha and Yurewicz, 1999; Cooper, 2007). In addition, the leakage of hydrocarbons in a structural culmination caused by thrust reactivation or unroofing by erosion is an additional risk for fault or top-seal breaching (e.g., northwest Borneo, Ingram et al., 2004).

To mitigate the risk associated with trap formation or fault reactivation, structural restorations are commonly constructed at the scale of the fold and thrust belt (e.g., Echavarria et al., 2003) or at the scale of an individual structure (e.g., Kostenko et al., 2008). Such restorations should consider that, in fold and thrust belts, cyclic thrust deformation characterized by synchronous or out-of-sequence thrusting may be very common. The models in this study demonstrate that additional geometric and kinematic possibilities at both the scale of a fold and thrust belt (Figure 19a) and at the scale of individual folds (Figure 19b) should be considered because of the effects of synkinematic sedimentation and erosion. In particular, hydrocarbon traps located in areas with higher synkinematic sedimentation may have a lower risk of thrust reactivation because of the positive effect of sedimentation in building and maintaining critical taper within the fold and thrust belt.

CONCLUSIONS

The twelve thrust-wedge experiments discussed in this chapter demonstrated that syntectonic sedimentation and syntectonic erosion exert a first-order control on geometries and activities of thrust-fault systems in analog models of critically tapered Coulomb thrust wedges. Systematic variations in the rates of syntectonic sedimentation in the analog models showed that syntectonic sedimentation increased the length of the thrust wedges, reduced the number of major forward-vergent thrusts, reduced thrust activities in the foreland, and allowed abnormally low frontal wedge taper angles to develop. In contrast, syntectonic erosion reduced the number of major forward-vergent thrusts, increased exhumation and thrust activities at the rear of the thrust wedge, and inhibited the forward propagation of the deformation front into the foreland (Figure 19a). When syntectonic sedimentation and syntectonic erosion were combined, their respective effects were complementary because they caused opposing parts of the wedge to become subcritical (erosion primarily in the hinterland and sedimentation in the foreland), thus lowering the surface slope and the wedge taper angle.

At the scale of individual faults or folds, syntectonic sedimentation produced rotation of the front limbs and backlimbs of individual thrust-related folds and formed a trishear zone with tip-line thrust splays on the front limb (Figure 19b). In contrast, syntectonic erosion did not produce significant changes. The results of these analog models challenge the popular self-similar kinematic models of thrust-related growth folds and indicate that more complex kinematic and geometric models are necessary in considering the dynamic interactions of thrust-related fold systems with synkinematic sedimentation and erosion.

The experimental results shown in this chapter suggest that fold and thrust belts with low syntectonic sedimentation and erosion would evolve in a self-similar fashion to produce well-ordered arrays of closely spaced thrusts similar to the frontal Nankai accretionary prism, offshore Japan. Fold and thrust belts with high syntectonic sedimentation would evolve with buried, widely spaced thrusts in the foreland similar to fold and thrust belts in the offshore Niger Delta, the central Apennines, and the sub-Andes. High syntectonic erosion would produce fold and thrust belts with very active hinterland thrusts resembling the present-day thrust activities of the onshore Taiwan fold and thrust belt. Reduced forward propagation of the deformation front due to high syntectonic erosion in the models provides a possible explanation for changes in the width of the central sub-Andean fold and thrust belt.

The model results indicate that subaerial fold and thrust belts, with high syntectonic erosion, should have wedges with high thrust activities and high exhumation at the rear of the wedge, low numbers of major thrusts, and a slow forward propagation of the deformation front. In contrast, submarine fold and thrust belts, with low syntectonic erosion, should be formed of longer wedges with lower exhumation rates that propagate forward quickly especially if syntectonic sedimentation rates are high. The model results also suggest that, for hydrocarbon exploration, traps located in areas with higher synkinematic sedimentation have a lower risk of thrust reactivation because of the effect of syntectonic sedimentation in building and maintaining critical taper within a fold and thrust belt.

ACKNOWLEDGMENTS

Saifal Janai, Matt Woollard, Lela Isa, and Rachel Evans are gratefully acknowledged for conducting several of the analog model experiments as part of the M.Sc. tectonics course at Royal Holloway, University of London, United Kingdom. Frank Despinois of the Fault Dynamics Research Group, Royal Holloway, is thanked for his expertise and assistance in preparing animations that enabled the analysis of the models. Funding for

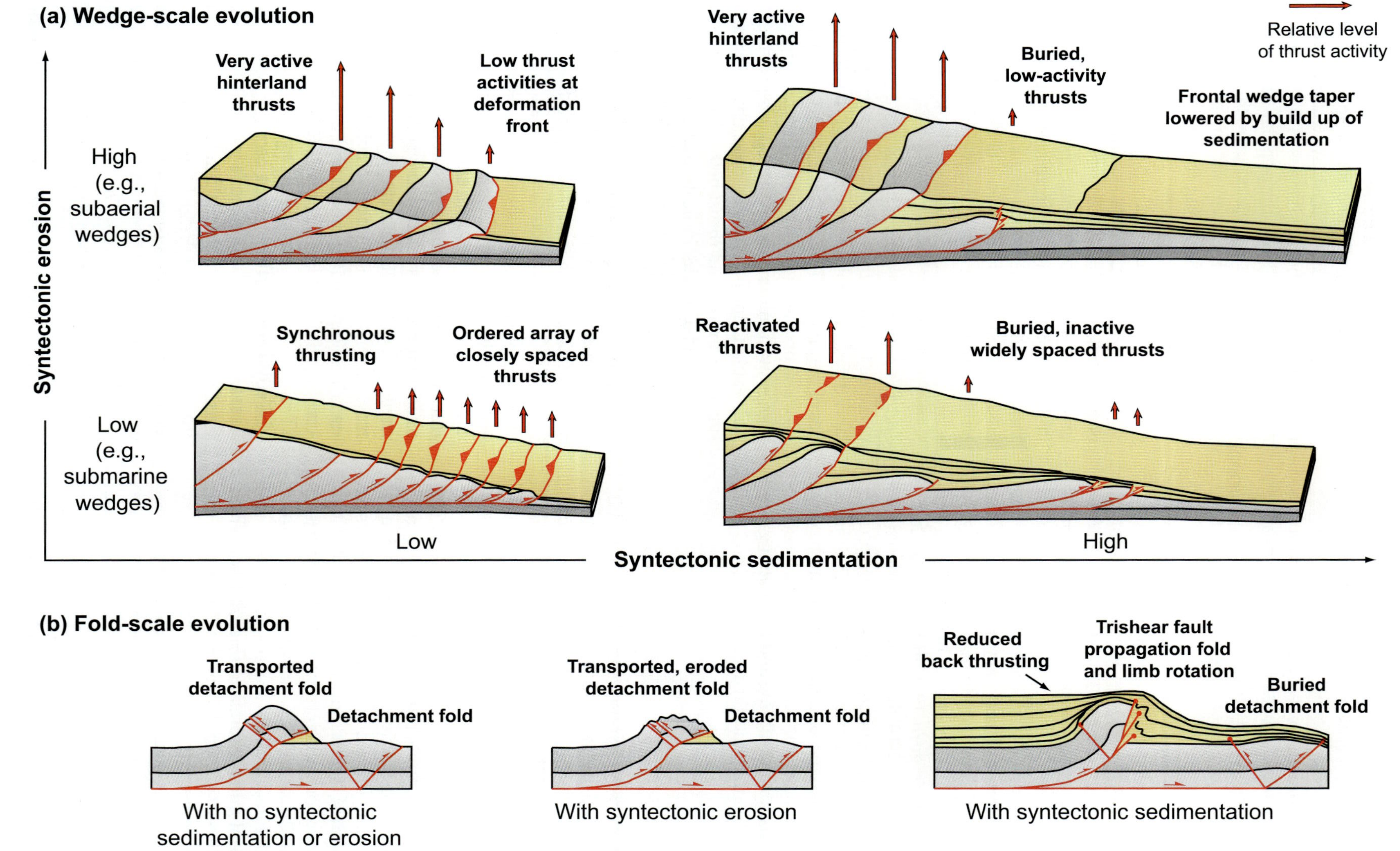

Figure 19. Summary of results of this study showing kinematic and geometric changes in the analog models caused by interactions with syntectonic sedimentation and syntectonic erosion at the scale of (a) the entire thrust wedge and (b) individual folds and thrusts.

Jonathan Wu's Ph.D. research at Royal Holloway is from Shell International E&P. We are grateful to Cesar Witt and Jose De Vera for many helpful comments and suggestions. Ben Clements and Rob Barnes are thanked for improving the final manuscript. Juergen Adam is thanked for advice and discussions on ring shear testing of analog model materials.

REFERENCES CITED

Ajakaiye, D. E., and A. W. Bally, 2002, Some structural styles on reflection profiles from offshore Niger Delta: Course manual and atlas of structural styles from the Niger Delta: AAPG Continuing Education Course Note 41, 107 p.

Amato, A., P. P. C. Aucelli, and A. Cinque, 2003, The long-term denudation rate in the Southern Apennines Chain (Italy): A GIS-aided estimation of the rock volumes eroded since middle Pleistocene time: Quaternary International, v. 101–102, p. 3–11, doi:10.1016/S1040-6182(02)00087-3.

Balestrieri, M. L., M. Bernet, M. T. Brandon, V. Picotti, P. Reiners, and M. Zattin, 2003, Pliocene and Pleistocene exhumation and uplift of two key areas of the Northern Apennines: Quaternary International, v. 101–102, no. 1, p. 67–73, doi:10.1016/S1040-6182(02)00089-7.

Barnes, J. B., and J. D. Pelletier, 2006, Latitudinal variation of denudation in the evolution of the Bolivian Andes: American Journal of Science, v. 306, no. 1, p. 1–31, doi:10.2475/ajs.306.1.1.

Barnes, P. M., and B. Mercier de Lepinay, 1997, Rates and mechanics of rapid frontal accretion along the very obliquely convergent southern Hikurangi margin, New Zealand: Journal of Geophysical Research B, Solid Earth, v. 102, no. 11, p. 24,931–24,952.

Barrier, L., T. Nalpas, D. Gapais, J. N. Proust, A. Casas, and S. Bourquin, 2002, Influence of syntectonic sedimentation on thrust geometry. Field examples from the Iberian Chain (Spain) and analog modeling: Sedimentary Geology, v. 146, no. 1–2, p. 91–104, doi:10.1016/S0037-0738(01)00168-3.

Bartolini, C., 1999, An overview of Pliocene to present-day uplift and denudation rates in the Northern Apennine: Geological Society (London) Special Publication 162, p. 119–125.

Beaumont, C., P. Fullsack, and J. Hamilton, 1992, Erosional control of active compressional orogens, *in* K. R. McClay, ed., Thrust tectonics: London, Chapman & Hall, p. 1–18.

Bilotti, F., and J. H. Shaw, 2005, Deep-water Niger Delta fold and thrust belt modeled as a critical-taper wedge: The influence of elevated basal fluid pressure on structural styles: AAPG Bulletin, v. 89, no. 11, p. 1475–1491, doi:10.1306/06130505002.

Bonnet, C., J. Malavieille, and J. Mosar, 2007, Interactions between tectonics, erosion, and sedimentation during the recent evolution of the Alpine orogen: Analog modeling insights: Tectonics, v. 26, no. 6, p. TC6016, doi:10.1029/2006TC002048.

Bonnet, C., J. Malavieille, and J. Mosar, 2008, Surface processes versus kinematics of thrust belts: Impact on rate of erosion, sedimentation, and exhumation— Insights from analog models: Bulletin de la Societe Geologique de France, v. 179, no. 3, p. 297–314, doi:10.2113/gssgfbull.179.3.297.

Brandon, M. T., M. K. Roden-Tice, and J. I. Carver, 1998, Late Cenozoic exhumation of the Cascadia accretionary wedge in the Olympic Mountains, northwest Washington State: Geological Society of America Bulletin, v. 110, no. 8, p. 985–1009, doi:10.1130/0016-7606(1998)110<0985:LCEOTC>2.3.CO;2.

Briggs, S. E., R. J. Davies, J. A. Cartwright, and R. Morgan, 2006, Multiple detachment levels and their control on fold styles in the compressional domain of the deep water west Niger Delta: Basin Research, v. 18, no. 4, p. 435–450, doi:10.1111/j.1365-2117.2006.00300.x.

Burbank, D. W., and R. S. Anderson, 2000, Tectonic geomorphology: Ann Arbor, Wiley-Blackwell, 288 p.

Burbank, D. W., and R. A. Beck, 1991, Rapid, long-term rates of denudation: Geology, v. 19, no. 12, p. 1169–1172, doi:10.1130/0091-7613(1991)019<1169:RLTROD>2.3.CO;2.

Burbank, D. W., and J. Verges, 1994, Reconstruction of topography and related depositional systems during active thrusting: Journal of Geophysical Research, v. 99, no. B10, p. 20,281–20,297.

Butler, R. W. H., and W. D. McCaffrey, 2004, Nature of thrust zones in deep water sand-shale sequences: outcrop examples from the Champsaur sandstones of SE France: Marine and Petroleum Geology, v. 21, no. 7, p. 911–921, doi:10.1016/j.marpetgeo.2003.07.005.

Chapple, W. M., 1978, Mechanics of thin-skinned fold and thrust belts: Geological Society of America Bulletin, v. 89, no. 8, p. 1189–1198, doi:10.1130/0016-7606(1978)89<1189:MOTFB>2.0.CO;2.

Charlton, T. R., 1988, Tectonic erosion and accretion in steady-state trenches: Tectonophysics, v. 149, no. 3–4, p. 233–243, doi:10.1016/0040-1951(88)90175-8.

Charreau, J., S. Gilder, Y. Chen, S. Dominguez, J.-P. Avouac, S. Sen, M. Jolivet, Y. Li, and W. Wang, 2006, Magnetostratigraphy of the Yaha section, Tarim Basin (China): 11 Ma acceleration in erosion and uplift of the Tian Shan mountains: Geology, v. 34, no. 3, p. 181–184, doi:10.1130/G22106.1.

Charreau, J., J.-P. Avouac, Y. Chen, S. Dominguez, and S. Gilder, 2008, Miocene to present kinematics of fault-bend folding across the Huerguosi anticline, northern Tianshan (China), derived from structural, seismic, and magnetostratigraphic data: Geology, v. 36, no. 11, p. 871–874, doi:10.1130/G25073A.1.

Cobbold, P. R., S. Durand, and R. Mourgues, 2001, Sandbox modeling of thrust wedges with fluid-assisted detachments: Tectonophysics, v. 334, no. 3–4, p. 245–258, doi:10.1016/S0040-1951(01)00070-1.

Cooper, M., 2007, Structural style and hydrocarbon prospectivity in fold and thrust belts: a global review: Geological Society (London) Special Publication 272, p. 447–472.

Copeland, P., and T. M. Harrison, 1990, Episodic rapid uplift in the Himalaya revealed by $^{40}Ar/^{39}Ar$ analysis of detrital K-feldspar and muscovite, Bengal fan: Geology, v. 18, no. 4, p. 354–357, doi:10.1130/0091-7613(1990)018<0354:ERUITH>2.3.CO;2.

Cruz, L., C. Teyssier, L. Perg, A. Take, and A. Fayon, 2008, Deformation, exhumation, and topography of experimental doubly-vergent orogenic wedges subjected to asymmetric erosion: Journal of Structural Geology, v. 30, no. 1, p. 98–115, doi:10.1016/j.jsg.2007.10.003.

Curray, J. R., F. J. Emmel, and D. G. Moore, 2002, The Bengal Fan: Morphology, geometry, stratigraphy, history and processes: Marine and Petroleum Geology, v. 19, no. 10, p. 1191–1223, doi:10.1016/S0264-8172(03)00035-7.

Cyr, A. J., and D. E. Granger, 2008, Dynamic equilibrium among erosion, river incision, and coastal uplift in the Northern and central Apennines, Italy: Geology, v. 36, no. 2, p. 103–106, doi:10.1130/G24003A.1.

Dadson, S. J., et al., 2003, Links between erosion, runoff variability and seismicity in the Taiwan orogen: Nature, v. 426, no. 6967, p. 648–651, doi:10.1038/nature02150.

Dahlen, F. A., 1984, Noncohesive critical Coulomb wedges: An exact solution: Journal of Geophysical Research, v. 89, no. B12, p. 10,125–10,133.

Dahlen, F. A., 1990, Critical taper model of fold-and-thrust belts and accretionary wedges: Annual Review of Earth and Planetary Sciences, v. 18, no. 1, p. 55–99.

Dahlen, F. A., J. Suppe, and D. Davis, 1984, Mechanics of fold-and-thrust belts and accretionary wedges: Cohesive Coulomb theory: Journal of Geophysical Research, v. 89, no. B12, p. 10,087–10,101.

Davis, D., J. Suppe, and F. A. Dahlen, 1983, Mechanics of fold and thrust belts and accretionary wedges: Journal of Geophysical Research, v. 88, no. B2, p. 1153–1172, doi:10.1029/JB088iB02p01153.

De Alteriis, G., 1995, Different foreland basins in Italy: Examples from the central and southern Adriatic Sea: Tectonophysics, v. 252, no. 1–4, p. 349–373, doi:10.1016/0040-1951(95)00155-7.

Echavarria, L., R. Hernandez, R. Allmendinger, and J. Reynolds, 2003, Subandean thrust and fold belt of northwestern Argentina: Geometry and timing of the Andean evolution: AAPG Bulletin, v. 87, no. 6, p. 965–985, doi:10.1306/01200300196.

Ehlers, T. A., K. A. Farley, M. E. Rusmore, and G. J. Woodsworth, 2006, Apatite (U-Th)/He signal of large-magnitude accelerated glacial erosion, southwest British Columbia: Geology, v. 34, no. 9, p. 765–768, doi:10.1130/G22507.1.

Erslev, E. A., 1991, Trishear fault-propagation folding: Geology, v. 19, no. 6, p. 617–620, doi:10.1130/0091-7613(1991)019<0617:TFPF>2.3.CO;2.

Fitzgerald, P. B., R. B. Sorkhabi, T. F. Redfield, and E. Stump, 1995, Uplift and denudation of the central Alaska Range: A case study in the use of apatite fission track thermochronology to determine absolute uplift parameters: Journal of Geophysical Research, v. 100, no. B10, p. 20,175–20,191.

Ghiglione, M. C., and V. A. Ramos, 2005, Progression of deformation and sedimentation in the southernmost Andes: Tectonophysics, v. 405, no. 1–4, p. 25–46, doi:10.1016/j.tecto.2005.05.004.

Grujic, D., I. Coutand, B. Bookhagen, S. Bonnet, A. Blythe, and C. Duncan, 2006, Climatic forcing of erosion, landscape, and tectonics in the Bhutan Himalayas: Geology, v. 34, no. 10, p. 801–804, doi:10.1130/G22648.1.

Gutscher, M. A., N. Kukowski, J. Malavieille, and S. Lallemand, 1996, Cyclical behavior of thrust wedges: Insights from high basal friction sandbox experiments: Geology, v. 24, no. 2, p. 135–138, doi:10.1130/0091-7613(1996)024<0135:CBOTWI>2.3.CO;2.

Hardy, S., and J. Poblet, 1994, Geometric and numerical model of progressive limb rotation in detachment folds: Geology, v. 22, no. 4, p. 371–374, doi:10.1130/0091-7613(1994)022<0371:GANMOP>2.3.CO;2.

Hardy, S., J. Poblet, K. R. McClay, and D. Waltham, 1996, Mathematical modeling of growth strata associated with fault-related fold structures: Geological Society (London) Special Publication 99, p. 265–282.

Hardy, S., C. Duncan, J. Masek, and D. Brown, 1998, Minimum work, fault activity and the growth of critical wedges in fold and thrust belts: Basin Research, v. 10, no. 3, p. 365–373, doi:10.1046/j.1365-2117.1998.00073.x.

Hickman, J. B., D. V. Wiltschko, J.-H. Hung, P. Fang, and Y. Bock, 2002, Structure and evolution of the active fold-and-thrust belt of southwestern Taiwan from Global Positioning System analysis, *in* T. B. Byrne and C.-S. Liu, eds., Geology and geophysics of an arc-continent collision, Taiwan: Geological Society of America Special Paper 358, p. 75–92.

Hills, D. J., G. F. Moore, N. L. Bangs, S. S. Gulick, and Leg 196 Shipboard Scientific Party, 2001, Preliminary results from integration of 2D PSDM and ODP Leg 196: LWD velocity data in the Nankai accretionary prism: Eos, Transactions of the American Geophysical Union, v. 82, p. F1221.

Holl, J. E., and D. J. Anastasio, 1993, Paleomagnetically derived folding rates, southern Pyrenees, Spain: Geology, v. 21, no. 3, p. 271–274, doi:10.1130/0091-7613(1993)021<0271:PDFRSP>2.3.CO;2.

Horton, B. K., 1999, Erosional control on the geometry and kinematics of thrust belt development in the central Andes: Tectonics, v. 18, no. 6, p. 1292–1304, doi:10.1029/1999TC900051.

Hoth, S., J. Adam, N. Kukowski, and O. Oncken, 2006, Influence of erosion on the kinematics of bivergent orogens: Results from scaled sandbox simulations, *in* S. D. Willett, N. Hovius, M. T. Brandon, and D. M. Fisher, eds., Tectonics, climate, and landscape evolution: Geological Society of America Special Paper 398, p. 201–225.

Huang, M.-H., J.-C. Hu, C.-S. Hsieh, K.-E. Ching, R.-J. Rau, E. Pathier, B. Fruneau, and B. Deffontaines, 2006, A growing structure near the deformation front in SW Taiwan as deduced from SAR interferometry and geodetic observation: Geophysical Research Letters, v. 33, no. 12, p. L12305.

Huang, M.-H., J.-C. Hu, K.-E. Ching, R.-J. Rau, C.-S. Hsieh, E. Pathier, B. Fruneau, and B. Deffontaines, 2009, Active deformation of Tainan tableland of southwestern Taiwan based on geodetic measurements and SAR interferometry: Tectonophysics, v. 466, no. 3–4, p. 322–334.

Hubert-Ferrari, A., J. Suppe, R. Gonzalez-Mieres, and X. Wang, 2007, Mechanisms of active folding of the landscape (southern Tian Shan, China): Journal of Geophysical Research B: Solid Earth, v. 112, no. 3, p. B03S09.

Husson, L., J.-L. Mugnier, P. Leturmy, and G. Vidal, 2004, Kinematics and sedimentary balance of the sub-Himalayan zone, western Nepal, *in* K. R. McClay, ed., Thrust tectonics and hydrocarbon systems: AAPG Memoir 82, p. 115–130.

Ingram, G. M., T. J. Chisholm, C. J. Grant, C. A. Hedlund, P. Stuart-Smith, and J. Teasdale, 2004, Deep water north west Borneo: Hydrocarbon accumulation in an active fold and thrust belt: Marine and Petroleum Geology, v. 21, no. 7, p. 879–887, doi:10.1016/j.marpetgeo.2003.12.007.

Iribarren, L., J. Verges, and M. Fernandez, 2009, Sediment supply from the Betic-Rif orogen to basins through Neogene: Tectonophysics, v. 475, no. 1–2, p. 68–84, doi:10.1016/j.tecto.2008.11.029.

Johnson, C., 1997, Resolving denudational histories in orogenic belts with apatite fission-track thermochronology and structural data: An example from southern Spain: Geology, v. 25, no. 7, p. 623–626, doi:10.1130/0091-7613(1997)025<0623:RDHIOB>2.3.CO;2.

Jones, M. A., P. L. Heller, E. Roca, M. Garces, and L. Cabrera, 2004, Time lag of syntectonic sedimentation across an alluvial basin: Theory and example from the Ebro Basin, Spain: Basin Research, v. 16, no. 4, p. 467–488, doi:10.1111/j.1365-2117.2004.00241.x.

Konstantinovskaia, E., and J. Malavieille, 2005, Erosion and exhumation in accretionary orogens: Experimental and geological approaches: Geochemistry Geophysics Geosystems (G3), v. 6, no. 2, p. Q02006.

Koons, P. O., 1990, Two-sided orogen: Collision and erosion from the sandbox to the Southern Alps, New Zealand: Geology, v. 18, no. 8, p. 679–682, doi:10.1130/0091-7613(1990)018<0679:TSOCAE>2.3.CO;2.

Kostenko, O. V., S. J. Naruk, W. Hack, M. Poupon, H.-J. Meyer, M. Mora Glukstad, C. Anowai, and M. Mordi, 2008, Structural evaluation of column-height controls at a toe-thrust discovery, deep-water Niger Delta: AAPG Bulletin, v. 92, no. 12, p. 1615–1638, doi:10.1306/08040808056.

Koyi, H., 1995, Mode of internal deformation in sand wedges: Journal of Structural Geology, v. 17, no. 2, p. 293–300, doi:10.1016/0191-8141(94)00050-A.

Kumar, R., S. K. Ghosh, and S. J. Sangode, 2003, Mio-Pliocene sedimentation history in the northwestern part of the Himalayan Foreland Basin, India: Current Science, v. 84, no. 8, p. 1006–1013.

Leturmy, P., J. L. Mugnier, P. Vinour, P. Baby, B. Colletta, and E. Chabron, 2000, Piggyback basin development above a thin-skinned thrust belt with two detachment levels as a function of interactions between tectonic and superficial mass transfer: The case of the Subandean Zone (Bolivia): Tectonophysics, v. 320, no. 1, p. 45–67, doi:10.1016/S0040-1951(00)00023-8.

Lewis, K. B., J.-Y. Collot, and S. E. Lallemand, 1998, The dammed Hikurangi Trough: A channel-fed trench blocked by subducting seamounts and their wake avalanches (New Zealand-France GeodyNZ Project): Basin Research, v. 10, no. 4, p. 441–468, doi:10.1046/j.1365-2117.1998.00080.x.

Liu, H., K. R. McClay, and D. Powell, 1992, Physical models of thrust wedges, *in* K. R. McClay, ed., Thrust tectonics: London, Chapman & Hall, p. 71–81.

Lohrmann, J., N. Kukowski, J. Adam, and O. Oncken, 2003, The impact of analog material properties on the geometry, kinematics, and dynamics of convergent sand wedges: Journal of Structural Geology, v. 25, no. 10, p. 1691–1711, doi:10.1016/S0191-8141(03)00005-1.

Malavieille, J., 1984, Modelisation experimentale des chevauchements imbriquees: Application aux chaines de montagnes: Bulletin de la Societe Geologique de France, v. 26, no. 1, p. 129–138.

Malavieille, J., and G. Trullenque, 2009, Consequences of continental subduction on forearc basin and accretionary wedge deformation in SE Taiwan: Insights from analog modeling: Tectonophysics, v. 3–4, p. 377–394, doi:10.1016/j.tecto.2007.11.016.

McClay, K. R., 1990, Deformation mechanics in analog models of extensional fault systems: Geological Society (London) Special Publication 54, p. 445–453.

McClay, K. R., and P. S. Whitehouse, 2004, Analog modeling of doubly vergent thrust wedges, *in* K. R. McClay, ed., Thrust tectonics and hydrocarbon systems: AAPG Memoir 82, p. 184–206.

McQuarrie, N., 2002, The kinematic history of the central Andean fold-thrust belt, Bolivia: Implications for building a high plateau: Geological Society of America Bulletin, v. 114, no. 8, p. 950–963, doi:10.1130/0016-7606(2002)114<0950:TKHOTC>2.0.CO;2.

McQuarrie, N., J. B. Barnes, and T. A. Ehlers, 2008a, Geometric, kinematic, and erosional history of the central Andean Plateau, Bolivia (15–17°S): Tectonics, v. 27, no. 3, p. TC002054, doi:10.1029/2006TC002054.

McQuarrie, N., T. A. Ehlers, J. B. Barnes, and B. Meade, 2008b, Temporal variation in climate and tectonic coupling in the central Andes: Geology, v. 36, no. 12, p. 999–1002, doi:10.1130/G25124A.1.

Meneses-Rocha, J., and D. A. Yurewicz, 1999, Petroleum exploration and production in fold and thrust belts; ideas from a Hedberg research symposium: AAPG Bulletin, v. 83, no. 6, p. 889–897.

Mitchell, N. C., W. B. Dade, and D. G. Masson, 2003, Erosion of the submarine flanks of the Canary Islands: Journal of Geophysical Research, v. 108, no. F1, p. 3.1–3.11.

Morley, C. K., 2007, Interaction between critical wedge-geometry and sediment supply in deep-water fold belt: Geology, v. 35, no. 2, p. 139–142, doi:10.1130/G22921A.1.

Morley, C. K., and L. C. Leong, 2008, Evolution of deep-water synkinematic sedimentation in a piggyback basin, determined from three-dimensional seismic reflection data: Geosphere, v. 4, no. 6, p. 939–962, doi:10.1130/GES00148.1.

Mount, V. S., J. Suppe, and S. C. Hook, 1990, A forward modeling strategy for balancing cross sections: AAPG Bulletin, v. 74, no. 5, p. 521–531.

Mourgues, R., and P. R. Cobbold, 2006, Thrust wedges and

fluid overpressures: Sandbox models involving pore fluids: Journal of Geophysical Research, v. 111, p. B05404, doi:10.1029/2004JB003441.

Mouthereau, F., O. Lacombe, B. Deffontaines, J. Angelier, and S. Brusset, 2001, Deformation history of the southwestern Taiwan foreland thrust belt: Insights from tectono-sedimentary analyses and balanced cross-sections: Tectonophysics, v. 333, no. 1–2, p. 293–322, doi:10.1016/S0040-1951 (00)00280-8.

Mugnier, J. L., P. Baby, B. Colletta, P. Vinour, P. Bale, and P. Leturmy, 1997, Thrust geometry controlled by erosion and sedimentation: A view from analog models: Geology, v. 25, no. 5, p. 427–430, doi:10.1130/0091-7613(1997) 025<0427:TGCBEA>2.3.CO;2.

Mulugeta, G., and H. Koyi, 1987, Three-dimensional geometry and kinematics of experimental piggyback thrusting: Geology, v. 15, no. 11, p. 1052–1056, doi:10.1130/0091 -7613(1987)15<1052:TGAKOE>2.0.CO;2.

Nalpas, T., D. Gapais, J. Verges, L. Barrier, V. Gestain, G. Leroux, D. Rouby, and J.-J. Kermarrec, 2003, Effects of rate and nature of synkinematic sedimentation on the growth of compressive structures constrained by analog models and field examples: Geological Society (London) Special Publication 208, p. 307–319.

Ngala, U., 2007, 4-D delta tectonics: Tectonic evolution of the deep water south central Niger Delta: Ph.D. thesis, Royal Holloway, University of London, 481 p.

Nieuwland, D. A., and M. Saher, 2002, The mechanics of a growing critical taper: New insights from high resolution video-laser scans and in-situ stress measurements in analog models: AAPG Hedberg Conference 2002.

Nieuwland, D. A., J. H. Leutscher, and J. Gast, 2000, Wedge equilibrium in fold and thrust belts: Prediction of out-of-sequence thrusting based on sandbox experiments and natural examples: Geologie en Mijnbouw, v. 79, no. 1, p. 81–91.

ODP Shipboard Scientific Party, 2000, Leg 185 Summary: inputs to the Izu-Mariana subduction system, *in* T. Plank, J. N. Ludden, and C. Escutia, eds., Proceedings of the Ocean Drilling Program, Initial Reports 185, p. 1–63.

Persson, K. S., and D. Sokoutis, 2002, Analog models of orogenic wedges controlled by erosion: Tectonophysics, v. 356, no. 4, p. 323–336, doi:10.1016/S0040-1951(02)00443-2.

Platt, J. P., 1986, Dynamics of orogenic wedges and the uplift of high-pressure metamorphic rocks: Geological Society of America Bulletin, v. 97, no. 9, p. 1037–1053, doi:10.1130 /0016-7606(1986)97<1037:DOOWAT>2.0.CO;2.

Rahl, J. M., T. A. Ehlers, and B. A. van der Pluijm, 2007, Quantifying transient erosion of orogens with detrital thermochronology from syntectonic basin deposits: insights from the central Pyrenees, Spain: Eos, Transactions of the American Geophysical Union, 2007 Fall Meeting, p. T23C–1541.

Restrepo-Moreno, S. A., D. A. Foster, D. F. Stockli, and L. N. Parra-Sanchez, 2009, Long-term erosion and exhumation of the "Altiplano Antioqueno", Northern Andes (Colombia) from apatite (U-Th)/He thermochronology: Earth and Planetary Science Letters, v. 278, no. 1–2, p. 1–12.

Schelling, D. D., 1999, Frontal structural geometries and detachment tectonics of the northeastern Karachi arc, southern Kirthar Range, Pakistan: Geological Society of America Special Paper 328, p. 287–302.

Schlunegger, F., and S. D. Willett, 1999, Spatial and temporal variations in exhumation of the central Swiss Alps and implications for exhumation mechanisms, *in* U. Ring, M. T. Brandon, G. S. Lister, and S. D. Willett, eds., Geological Society (London) Special Publication 154, p. 157–179.

Scisciani, V., and R. Montefalcone, 2006, Coexistence of thin- and thick-skinned tectonics: An example from the Central Apennines, Italy, *in* S. Mazzoli and R. W. H. Butler, eds., Styles of continental contraction: Geological Society of America Special Paper 414, p. 33–54.

Sempere, T., 1995, Phanerozoic evolution of Bolivia and adjacent regions, *in* A. J. Tankard, R. Suarez-Soruco, and H. J. Welsink, eds., Petroleum basins of South America: AAPG Memoir 62, p. 207–230.

Shaw, J. H., C. Connors, and J. Suppe, 2005, eds., Seismic interpretation of contractional fault-related folds: An AAPG seismic atlas: AAPG Studies in Geology 53, 156 p.

Simpson, G. D. H., 2006, Modeling interactions between fold-thrust belt deformation, foreland flexure and surface mass transport: Basin Research, v. 18, no. 2, p. 125–143, doi:10.1111/j.1365-2117.2006.00287.x.

Skilbeck, C. G., and D. Fink, 2006, Data report: Radiocarbon dating and sedimentation rates for Holocene–upper Pleistocene sediments, eastern equatorial Pacific and Peru continental margin, *in* B. B. Jorgensen, S. L. D'Hondt, and D. J. Miller, eds., Proceedings of the Ocean Drilling Program, Scientific Results 201, p. 1–15.

Stockmal, G. S., C. Beaumont, M. Nguyen, and B. Lee, 2007, Mechanics of thin-skinned fold and thrust belts: Insights from numerical models, *in* J.W. Sears, T. A. Harms, and C. A. Evenchick, eds., Whence the mountains? Inquiries in the evolution of orogenic systems: A volume in honor of Raymond A. Price: Geological Society of America Special Paper 433, p. 63–98.

Storti, F., and K. McClay, 1995, Influence of syntectonic sedimentation on thrust wedges in analog models: Geology, v. 23, no. 11, p. 999–1002, doi:10.1130/0091-7613(1995)023 <0999:IOSSOT>2.3.CO;2.

Storti, F., F. Salvini, and K. McClay, 2000, Synchronous and velocity-partitioned thrusting and thrust polarity reversal in experimentally produced, doubly-vergent thrust wedges: Implications for natural orogens: Tectonics, v. 19, no. 2, p. 378–396, doi:10.1029/1998TC001079.

Strayer, L. M., S. G. Erickson, and J. Suppe, 2004, Influence of growth strata on the evolution of fault-related folds-distinct-element models, *in* K. R. McClay, ed., Thrust tectonics and hydrocarbon systems: AAPG Memoir 82, p. 413–437.

Strecker, M. R., R.N. Alonso, B. Bookhagen, B. Carrapa, G. E. Hilley, E. R. Sobel, and M. H. Trauth, 2007, Tectonics and climate of the southern Central Andes: Annual Review of Earth and Planetary Sciences, v. 35, no. 1, p. 747–787.

Suppe, J., 1983, Geometry and kinematics of fault-bend folding: American Journal of Science, v. 283, no. 7, p. 684–721.

Suppe, J., and D. Medwedeff, 1990, Geometry and kinematics of fault-propagation folding: Eclogae Geologicae Helvetiae, v. 83, p. 409–454.

Suppe, J., G. T. Chou, and S. C. Hook, 1992, Rates of folding and faulting determined from growth strata, *in* K. R. McClay, ed., Thrust tectonics: London, Chapman & Hall, p. 105–121.

Suppe, J., C. D. Connors, and Y. Zhang, 2004, Shear fault-bend folding, *in* Thrust tectonics and hydrocarbon systems: AAPG Memoir 82, p. 303–323.

Uba, C. E., M. R. Strecker, and A. K. Schmitt, 2007, Increased sediment accumulation rates and climatic forcing in the central Andes during the late Miocene: Geology, v. 35, no. 11, p. 979–982, doi:10.1130/G224025A.1.

Verges, J., M. Marzo, and J. A. Munoz, 2002, Growth strata in foreland settings: Sedimentary Geology, v. 146, no. 1–2, p. 1–9, doi:10.1016/S0037-0738(01)00162-2.

Warner, M. A., and F. Royse, 1987, Thrust faulting and hydrocarbon generation: Discussion: AAPG Bulletin, v. 71, p. 882–889.

Weiland, R. J., and M. Cloos, 1996, Pliocene–Pleistocene asymmetric unroofing of the Irian fold belt, Irian Jaya, Indonesia: Apatite fission-track thermochronology: Geological Society of America Bulletin, v. 108, no. 11, p. 1438–1449, doi:10.1130/0016-7606(1996)108<1438:PPAUOT>2.3.CO;2.

Whipp, J. D. M., T. A. Ehlers, A. E. Blythe, K. W. Huntington, K. V. Hodges, and D. W. Burbank, 2007, Plio-Quaternary exhumation history of the central Nepalese Himalaya: 2. Thermokinematic and thermochronometer age prediction model: Tectonics, v. 26, no. 3, p. TC3003, doi:10.1029/2006TC001991.

Whipple, K. X., 2009, The influence of climate on the tectonic evolution of mountain belts: Nature Geoscience, v. 2, no. 2, p. 97–104, doi:10.1038/ngeo413.

Whipple, K. X., and B. J. Meade, 2004, Controls on the strength of coupling among climate, erosion, and deformation in two-sided, frictional orogenic wedges at steady state: Journal of Geophysical Research, v. 109, p. F01011, doi:10.1029/2003JF000019.

Whipple, K. X., and B. J. Meade, 2006, Orogen response to changes in climatic and tectonic forcing: Earth and Planetary Science Letters, v. 243, no. 1–2, p. 218–228.

Willett, S. D., 1992, Dynamic and kinematic growth and change of a Coulomb wedge, *in* K. R. McClay, ed., Thrust tectonics: London, Chapman & Hall, p. 19–31.

Willett, S. D., D. Fisher, C. Fuller, E.-C. Yeh, and C.-Y. Lu, 2003, Erosion rates and orogenic-wedge kinematics in Taiwan inferred from fission-track thermochronometry: Geology, v. 31, no. 11, p. 945–948, doi:10.1130/G19702.1.

Wu, J. E., K. R. McClay, F. Despinois, M. Woollard, R. Evans, L. Isa, and S. Janai, 2008, Analog modeling of deep water fold and thrust belts: dynamic interaction with syntectonic sedimentation: International Meeting of Young Researchers in Structural Geology and Tectonics (YORSGET) Abstracts Volume, Oviedo, Spain.

Xiong, Y., Z. Liu, T. Li, Y. Liu, and H. Yu, 2005, The sedimentation rates in the Okinawa Trough during the late Quaternary: Acta Oceanologica Sinica, v. 24, no. 4, p. 146–154.

Yu, S. B., H. Y. Chen, and L. C. Kuo, 1997, Velocity field of GPS stations in the Taiwan area: Tectonophysics, v. 274, no. 1–3, p. 41–59.

Zattin, M., V. Picotti, and G. G. Zuffa, 2002, Fission-track reconstruction of the front of the northern Apennine thrust wedge and overlying Ligurian unit: American Journal of Science, v. 302, no. 4, p. 346–379, doi:10.2475/ajs.302.4.346.

15

Zalán, Pedro Victor, 2011, Fault-related folding in the deep waters of the equatorial margin of Brazil, *in* K. McClay, J. H. Shaw, and J. Suppe, eds., Thrust fault-related folding: AAPG Memoir 94, p. 335–355.

Fault-related Folding in the Deep Waters of the Equatorial Margin of Brazil

Pedro Victor Zalán

Petrobras/E&P/E&P-EXP/GPE/NNE, Rio de Janeiro, RJ, Brazil

ABSTRACT

Two gravitational fold and thrust belts (GFTBs) from the deep waters of the Pará-Maranhão and Barreirinhas basins were interpreted in two-dimensional seismic sections and analyzed using the concepts of fault-related folding and taper-wedge mechanics. These basins lie in the equatorial Atlantic Ocean margin of Brazil, a classic example of transform to oblique (transtensional) continental margin. There, deep-water anoxic shales of probably Turonian age served as decollement zones to Late Cretaceous and Paleogene predominantly siliciclastic wedges to slide down from upper slope (extension) to lower slope (contraction) realms. The structural style of both fold belts is typical thin-skinned tectonics with imbricate thrust faults branching upward from the detachment level with associated fault-related folding. The direction of tectonic transport is from the coast to offshore (southwest to northeast). The similarities and differences between the GFTBs were highlighted. The most striking difference regards the distribution of contractional strain throughout the fold belt. In the Pará-Maranhão GFTB, a single major contraction event was achieved via eight regularly spaced imbricate thrust faults. Intervening slices present all types of classic fault-related folding (fault-bend, fault-propagation, and detachment folds). In the Barreirinhas GFTB, an earlier minor contraction was also spread over regularly spaced faults; however, a major late contraction was achieved by one main thrust fault with one major fault-related fold associated to it. This fold is predominantly a shear fault-bend fold whose geometry varies slightly along strike. Their main similarities regard the discrete nature of the detachment level, their structural coherence and narrowness, their low to moderate wedge taper, and their noncritical nature of the taper wedge. Both present syntectonic growth strata that record variations in the balance between the rates of sedimentation and structural uplift of the fold, and folding by limb rotation. The fault-related folding thus determined and the parameters established for wedge taper are slightly different from those presented by active submarine fold and thrust belts at convergent margins and passive margins throughout the world. The decollement dip of the GFTBs is significantly lower, and the bathymetric slope is somewhat higher than elsewhere around the globe.

DOI:10.1306/13251344M943089

INTRODUCTION

The continental margin of northeastern Brazil was developed in a plate tectonic environment of transform to oblique (transtensional) breakup and separation (Mascle and Blarez, 1987; Azevedo, 1991; Matos, 2000). Its present-day overall trend is strongly influenced by three major oceanic fracture zones (from south to north, Chain, Romanche, and Saint Paul, Figure 1) (Zalán, 1984) that played fundamental roles in the evolution of the equatorial Atlantic Ocean and also on the geology of the counterpart South American and African continental passive margins (Matos, 2000; Basile et al., 2005, their figure 1). Where these fracture zones intercept the South American margin, they imprint their east–west trend upon the coastline, the shelf break, and the continental-oceanic crust boundary (C.O.B., Figure 1). Strong transpressional deformation confined to narrow east-west–trending belts characterizes the structural styles of the sedimentary packages along these zones (Zalán et al., 1985; Attoh et al., 2004; Tari et al., 2004). Between these intersections, the continental margin segments trend northwest–southeast, and the sedimentary filling resembles more normal passive margin depositional sequences; that is, the postrift strata seem to be predominantly deformed by gravitational processes. In the equatorial margin, rifting started in the Aptian and ended in the latest Albian (Souza-Lima and Hamsi, 2003; Zalán et al., 2003). Seafloor spreading is believed to have started ca. 100 Ma (Attoh et al., 2004, their figure 11).

The continental margin of Brazil situated between the Romanche and Saint Paul fracture zones encompasses two large sedimentary basins: the Barreirinhas (onshore and offshore) and Pará-Maranhão (offshore) basins (Figure 1). These basins present very characteristic deformation patterns in their deep and ultradeep waters. Several fold and thrust belts with areas between 1000 and 2000 km^2 (386 and 772 mi^2) are present at the base of the continental slope (see example in Figure 2 and its location in Figure 1) and affect a Coniacian to Holocene siliciclastic sedimentary section, which can attain thicknesses of up to 10 km (6 mi) deposited upon oceanic crust. Gliding and contraction are repetitive in both time and space. Seismic sections typically show evidence for superposition of deformation events: an older event involving Late Cretaceous to Paleogene sediments and a younger event involving Neogene to Holocene sediments. Commonly, the younger phase of contraction appears to reactivate the preexisting detachment level, also reactivating older thrust ramps.

Precise dating of such events is difficult to constrain because there are practically no petroleum exploration wells drilled in the deep waters of these basins. The tying of seismic sections between the shelf area (where plenty of wells exist) and the deeper areas beyond the shelf break is almost impossible because of a significant decrease in seismic resolution exactly below this geological feature. The deeper detachment levels are tentatively attributed to the Cenomanian–Turonian shales whereas the more rare shallower detachment zones to the Paleocene shales. Thick layers of anoxic shales of such ages are known to occur on both sides of the equatorial Atlantic. Is this large amount of fold belts in any way related to the proximity of the Romanche and Saint Paul fracture zones? Are they common in such basins because of the seismic activity of the fracture zones? This subject is still open to debate.

Fold and thrust belts dominate the deep and ultradeep waters of most basins of divergent continental margins (Zalán, 2005) and are part of larger linked extensional-compressional systems (Figure 2 and see location of line A in Figure 1) that develop as a response to gravitational failure of sedimentary successions in the outer parts of continental shelves and upper parts of continental slopes (extensional domain). These allochthonous masses of sediments slide down the slope (translational domain) over plastic lithology that detaches the moving rocks above from the autochthonous rocks below. When the frontal parts of the allochthon diminish their pace due either to a decrease in the gradient of the detachment zone, to a change in friction along decollement, or to the buttressing effect of any physical barrier (e.g., a ramp, a volcanic edifice, etc.), the incoming volume of allochthonous rocks collides and contraction occurs (compressional domain). A wedge of tectonically piled-up rocks strongly deformed by compression develops as syntectonic and posttectonic strata cover them in the deep submarine realm. These contractional entities have been studied in the last decade all over the world because of their potential for developing their own petroleum systems. Significant reserves of hydrocarbons have been discovered in such environments in the Gulf of Mexico (e.g., Trident, Great White, Tiger, and Tobago; Meyer et al., 2005) (e.g., Mad Dog and K2/Timon; Rowan et al., 2004), as well as the Atlantis, Neptune, and Champlain fields, Nigeria (e.g., Agbami, Bonga, Chota, Ngolo, and Nnwa fields; Bilotti and Shaw, 2005), as well as the Akpo and Bobo fields, and Equatorial Guinea (e.g., La Ceiba field; Dailly et al., 2002). Different names have been coined to designate them: gravity-driven fold belts (Rowan et al., 2004), gravitational fold and thrust belts (GFTBs, Zalán, 2005), or simply deep-water fold and thrust belts (Corredor et al., 2005).

The GFTBs from the Pará-Maranhão and Barreirinhas basins (Figure 1) were studied in detail, focusing on the geometries and nature of their fault-related folds as well as the kinematic evolution. The GFTB of the Pará-Maranhão Basin (see the location of line B in Figure 1)

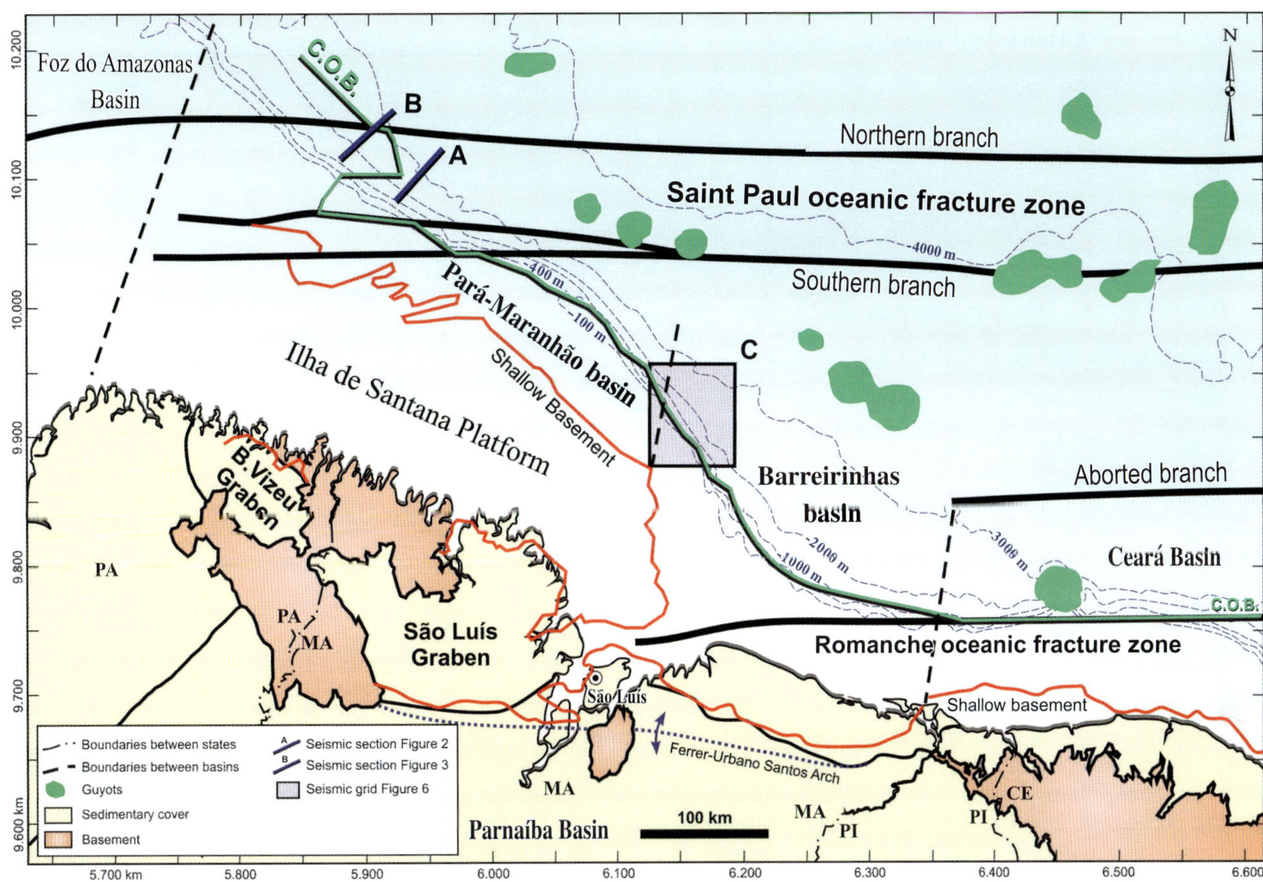

Figure 1. Map of the regional structural framework of the Pará-Maranhão and Barreirinhas basins (between dashed lines, down to the 3000-m (9842-ft) bathymetric contour), showing the Saint Paul and Romanche fracture zones, the continental-oceanic crust boundary (C.O.B.), surrounding basins, shallow basement platforms, and bathymetry. Line A corresponds to the seismic section shown in Figure 2, line B to the seismic section shown in Figure 3, and area C to the seismic grid shown in Figure 6. Green circular areas correspond to guyots. CE = Ceará; MA = Maranhão; PI = Piauí; PA = Pará; B. Vizeu = Bragança Vizeu graben.

presents a single tectonic event where the compressional deformation is evenly distributed among a discrete number (eight) of well-defined thrust faults, each with its own fault-related folds. In opposition to this, the GFTB from the Barreirinhas basin (see the location of area C in Figure 1) displays most of the contraction being consumed by one dominant thrust fault and one shear fault-bend fold.

All maps shown in this chapter are displayed in polyconic projection, datum SAD69 (South American Datum defined in 1969), central meridian 54°, false north of 10,000,000, and false east of 5,000,000. All seismic lines studied and presented in this work are from two-dimensional (2-D) seismic surveys.

GRAVITATIONAL FOLD AND THRUST BELT FROM THE PARÁ-MARANHÃO BASIN

This GFTB occurs in the deep waters (bathymetry between 1500 and 3000 m [4921 and 9842 ft]) of the Pará-Maranhão Basin. It presents all three components of a well-developed linked extensional-compressional system (Figure 3). The extensional domain is at least 20 km (12 mi) wide (diffuse extension is widespread in the upper slope of the basin) and is characterized by thick wedges of growth strata in rollovers associated to listric normal faults. The translational domain presents subhorizontal sediments for about 10 km (6 mi). The compressional domain is 20 km (12 mi) wide and displays strong imbricate thrusting and folding affecting pretectonic (parallel layering) strata covered by a thin veneer of growth strata that thickens subtly outward of the tectonic culmination. The timing of the activity of the faults, in both extensional and compressional domains, and related deformation and their influence upon sedimentation of the growth strata point to a single long-lasting (Eocene to Oligocene) event of sliding and contraction.

Figure 3 shows a prestack depth-migrated seismic section with no vertical exaggeration, illustrating the true appearance of a linked extensional-compressional

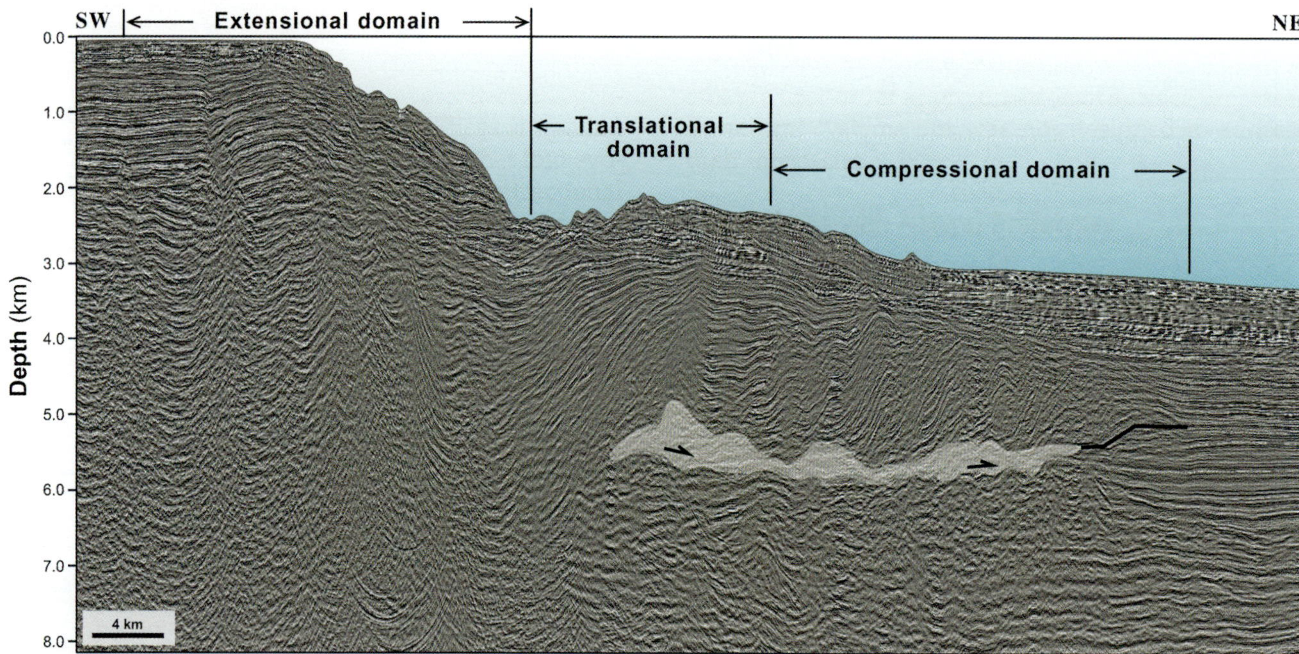

Figure 2. Line A in Figure 1. Time-migrated seismic section converted to depth (V.E. = 4×) from the deep-water Pará-Maranhão Basin showing the typical expression of a linked extensional-compressional system of gravitational nature and illustrating the three major tectonic domains. The width of the GFTB is around 25 km (15 mi), and its relationship with the underlying detachment zone (shaded area) is complex because it seems to involve some plastic flow. The section underwent high-resolution processing aimed at recovering high frequencies. V. E. = vertical/exaggeration.

system of gravitational nature. Zalán (2005) presented the same seismic line to show a regular zonation in the distribution of the different fault-related folds within the contractional wedge. Fault-bend folds are concentrated in the frontal part (between faults 1 and 5, Figure 3), fault-propagation folds in the middle part (between faults 5 and 8, Figure 3), and detachment folds that occur in the innermost part of the fold belt (immediately to the left of fault 8). The author also tried to correlate syntectonic (growth) strata from the forelimb (nondeformed strata onlapping unconformities) to the backlimb (deformed strata below unconformities) of the contractional wedge through a series of time-transgressive unconformities developed atop it. The correlation accomplished seemed to indicate that the same mechanism proposed by Medwedeff (1989) for a fault-bend fold in the San Joaquin Valley in California worked in the deep waters of northeastern Brazil. Coeval syntectonic sediments were deposited onlapping the passive forelimb (because, as discussed later, the style of deformation is break-backward) of the GFTB and in stratigraphic concordance with pretectonic sediments in the active backlimb of the GFTB, physically separated by time-transgressive unconformities developed atop the locus of uplift and deformation (Figure 3). This tectonic-sedimentary relationship also occurs in folds in the deep waters of the Gulf of Mexico (M. P. Rowan, 2005, personal communication).

Palinspastic restoration of the interpretation presented in Figure 3 indicated a shortening of 17 km (10 mi) (36% for the contractional domain) (Figure 4), 90% of which was achieved in a relatively short time (between stages 1 and 2 of Figure 4 and during the deposition of the yellow sequence, possibly of Eocene age, in Figure 3). Employing the concepts of wedge taper (Davis et al., 1983), we used Figure 3 to determine the surface slope of the top of the deformed wedge (measured between points 1 and 2) of the contractional wedge. A value of 5° was obtained. The detachment zone is practically horizontal (0.7°, measured through the entire length of the section). The shape of the contractional wedge is thus constrained by a wedge taper of 5.7°. Such a value is classified as a moderate taper wedge (Bilotti and Shaw, 2005). This figure together with the palinspastic sequential restoration (Figure 5) of the interpretation presented in Figure 3, where break-backward propagation dominated the evolution of the thrust faults with no out-of-sequence thrusting, indicated that this GFTB was a noncritical taper wedge. Its overall coherence and narrowness (20 km [12 mi]) with respect to the entire linked extensional-compressional system (50 km [31 mi]) corroborate this statement.

Break-backward propagation of thrust faults in GFTBs seems to be the rule in the equatorial Atlantic margin of Brazil. The same mechanism has been consistently

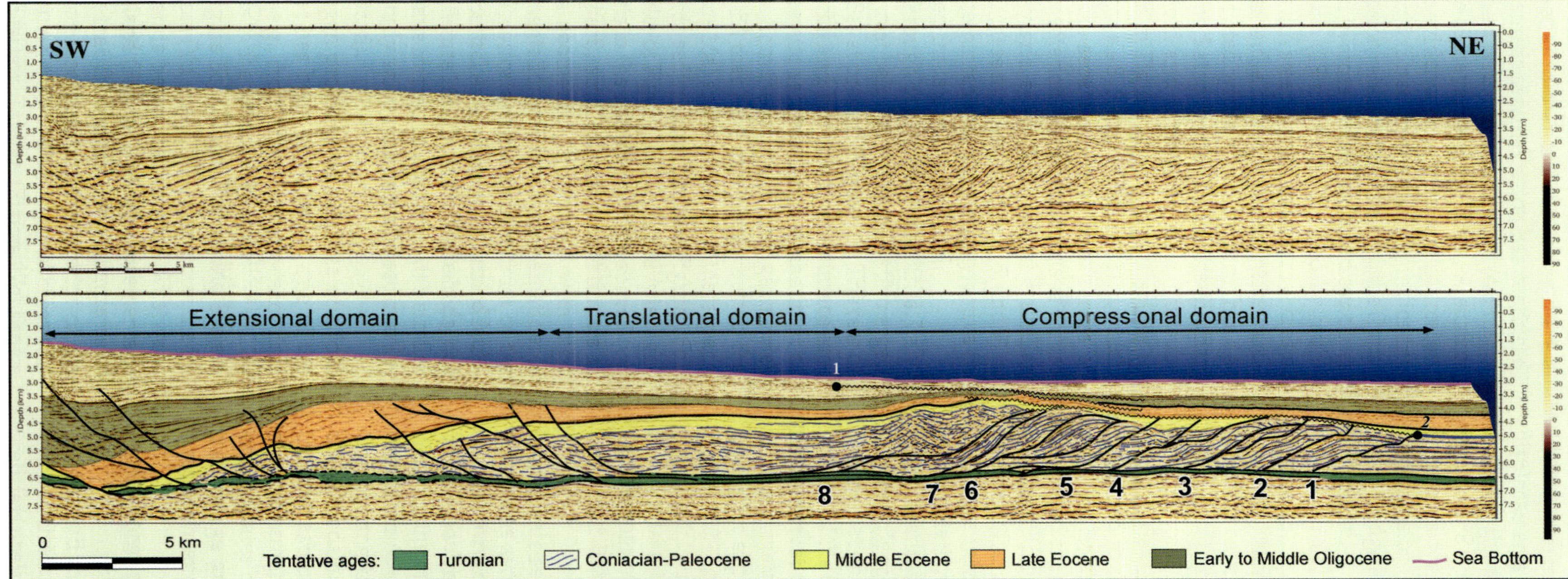

Figure 3. Line B in Figure 1. Prestack depth-migrated seismic section (without vertical exaggeration) from the deep-water Pará-Maranhão Basin showing the true expression of a linked extensional-compressional system of gravitational nature, with emphasis on the internal geometry of the fold and thrust belt. Fault-bend folds dominate most external thrust slices, fault-propagation folds occur in the central part, and a detachment fold is visible at the innermost part of the fold belt. Three syntectonic sequences are highlighted with colors. The shape of the contractional wedge is constrained by a top surface slope of 5°, measured between points 1 and 2, and by a practically horizontal basal detachment (0.7° dip), resulting in a wedge taper of 5.7°. Step-up angles for eight thrust ramps range between 19 and 35° and averaging 27°. Top of the brown sequence is tentatively assigned to the middle Oligocene, whereas the detachment zone (green) is probably nucleated in the Turonian shales.

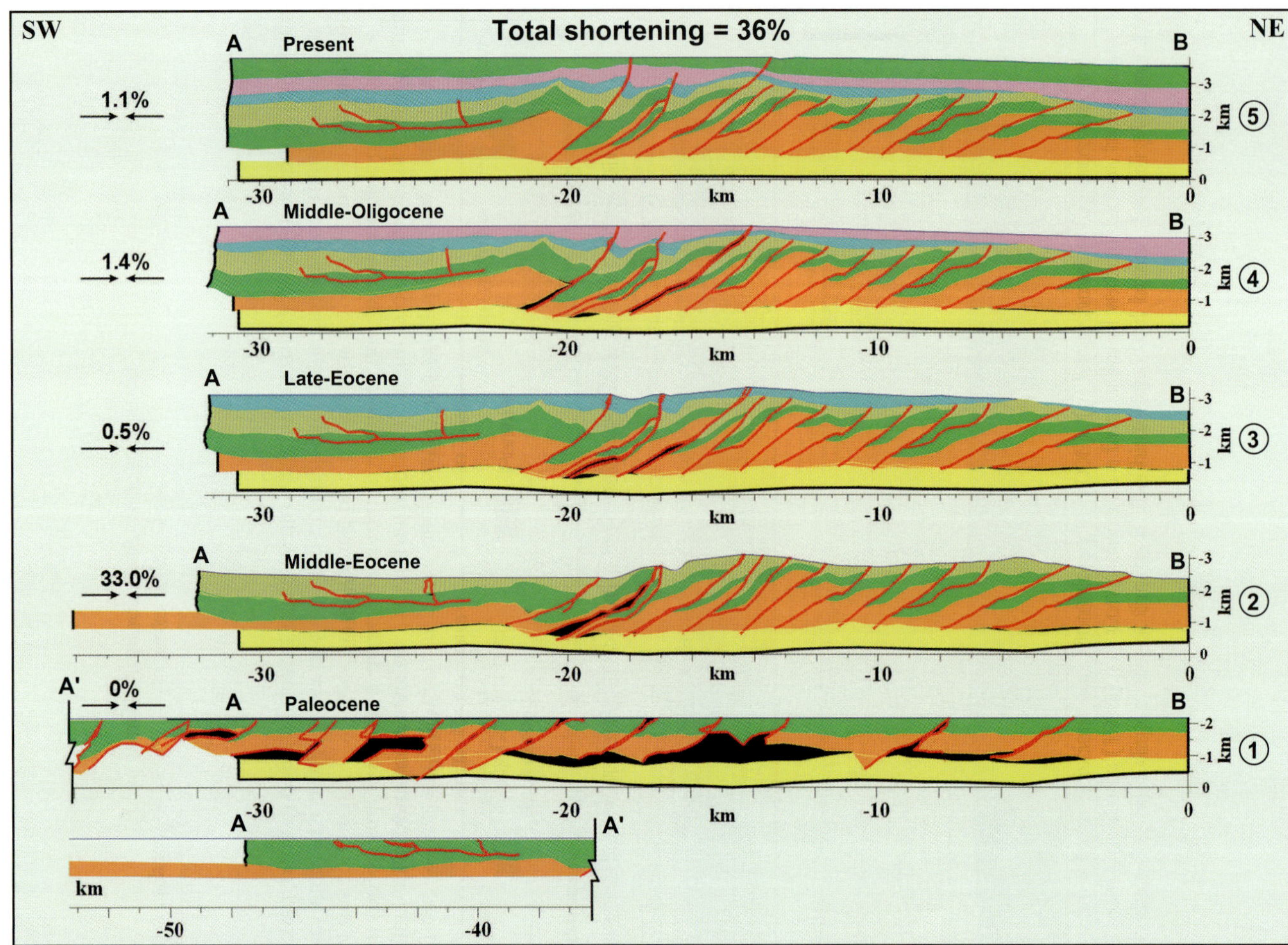

Figure 4. Sequential restoration of the interpretation of the seismic section shown in Figure 3. Orange and dark green sequences are precontraction sequences. The next four sequences represent growth strata. Overall shortening achieved 36%, most of it during deposition of the light green sequence (from stage 1 to 2). To allow comparisons, this sequence is equivalent to the yellow sequence of Figure 3. Voids (black areas) in the undeformed stage (1) are probably caused by some amount of plastic flow of the detachment zone (yellow).

deduced in several GFTBs in the Foz do Amazonas, Barreirinhas, and Pará-Maranhão basins (Zalán, 2005). The interpreted explanation invokes the frontal part of a sliding allochthonous mass of rock coming to a halt or diminishing its velocity relative to the rest of the system (for various reasons: obstacles, change in gradient of slope, etc.). Such difference in relative velocities develops compressional stresses that are gradually and continuously liberated by break-backward piling up of incoming rocks. Topography rises backward as well, and the onlapping of increasingly younger growth strata toward the higher elevations, as well as the accompanying deformation, indicates the sense of thrust propagation.

The regular distribution of contractional strain within this GFTB points to a wedge with a relatively high strength composed of fairly uniform material. Eight major thrust faults, evenly spaced (at about 2 km [1.2 mi]), ramp up from the detachment zone and form equally wide folded thrust slices (Figure 3). The basal step-up angles of the eight thrust ramps were measured, ranging between 19 and 35°, with an average of 27°. The thrusts become steeper in a backward direction.

GRAVITATIONAL FOLD AND THRUST BELT FROM THE BARREIRINHAS BASIN

This GFTB occurs in the deep waters (bathymetry between 1900 and 2600 m [6233 and 8530 ft]) of the Barreirinhas Basin. It presents only the extensional and the compressional domains of a linked extensional-compressional system (Figures 6–9). The extensional domain is 30 km (19 mi) wide and is characterized by listric normal faults

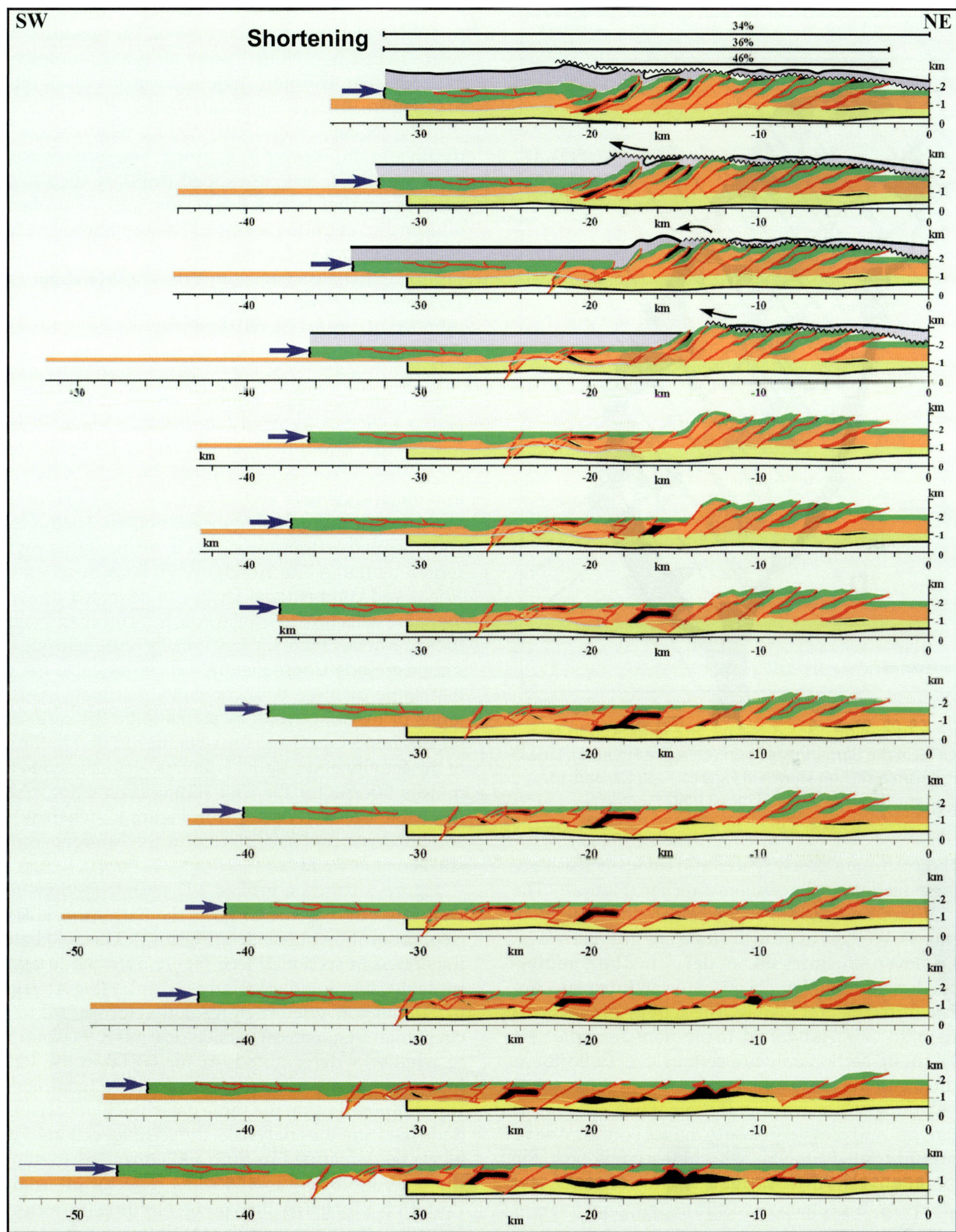

Figure 5. Sequential structural restoration of only the pretectonic sequences (orange and yellow) mapped fault by fault. In the last four stages, a syntectonic sequence (purple) was added to determine the structural evolution of such sequence. We can clearly see that by the end of the process, this sequence is cut by a time-transgressive unconformity separating nondeformed rocks above, in the forelimb, from deformed rocks below, in the backlimb.

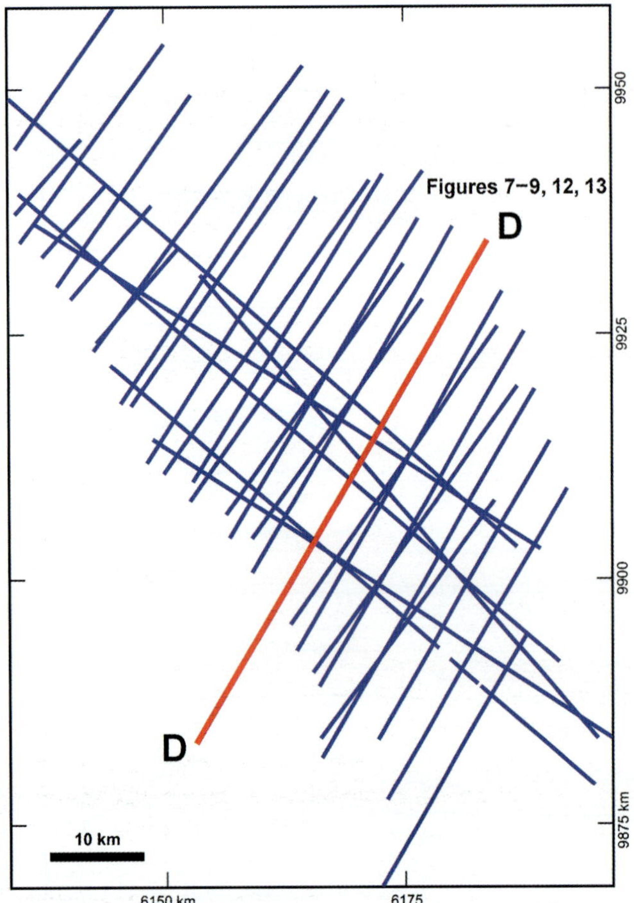

Figure 6. Two-dimensional seismic grid employed to map the GFTB in the Barreirinhas basin (area C in Figure 1). Line D is the seismic section shown in Figures 7–9, 12, and 13. GFTB = gravitational fold and thrust belt.

that become more closely spaced and cut up increasingly younger sections in the offshore direction (Figure 9). The last one crops out at sea bottom, close to the shelf break (Figure 9). Wedges of growth strata are thus narrower, and rollover structures do not develop. The compressional domain is 40 km (25 mi) wide and displays imbricate thrusting and folding. The activity of the thrust faults in the GFTB is twofold. In the more distal part, six faults cut up section only to a certain level (reflector X, Figure 9). In the more internal part, four faults cut up much younger sections, one of them (fault 7, Figure 9) flattens out and dies very close to the sea bottom. A very large fault-related anticline, whose crest crops out at the sea bottom, occurs in the hanging wall of such fault, and the study of its geometry constitutes the main focus of this chapter (Figures 9–13). A secondary detachment level develops in the more internal zone of the GFTB (Figure 9). Very young growth strata develop associated to the rising and deformation of the cited large anticline (Figures 9, 13).

The translational domain is absent. The reason for that seems to be a two-phase history of sliding/contraction in this system. The more internal normal faults (faults C to G, Figure 9) and thrust faults (faults 7 to 10, Figure 9) display geometries and crosscutting relationships indicative of an imprint of a strong, late tectonic event upon an older, less developed event. If such is the case, the earlier translational domain was probably faulted and engulfed in this last deformation. The first event affected a restricted package of rocks (probably Coniacian to Paleocene) with a thickness of about 1.0–1.5 km (0.6–0.9 mi) and promoted a grossly even distribution of compressional deformation among 10 faults (Figure 9). Two coeval extensional faults can be seen at the southwestern end of the detachment level affecting a progradational sequence characterized by offlaping reflectors situated below the continental platform (faults A and B, Figure 9). The second episode is constituted by five large listric normal faults that cut up into much shallower sections (faults C to G, Figure 9), the most external affecting the sea bottom. Their coeval and correlative compressional deformation is represented by faults 7 to 10 (Figure 9). Because all extensional and contractional faults can be traced down to the same detachment level (probably Turonian shales), the same detachment layer was likely reused during the second episode. Consequently, it is also logical to assume that some of these younger faults reactivated earlier faults of the first event, as seems to be the case with normal faults C to G and thrust faults 7 to 10. The timing of the activity of the faults is tentatively assigned to the middle Eocene for the first sliding-contraction phase (probable age of reflector X in Figure 9), whereas the second event probably started in the Miocene and is still active nowadays.

Figure 6 shows a grid of 2-D seismic sections that were used to map the GFTB of the Barreirinhas Basin (see the location of area C in Figure 1). This fold belt is illustrated in section D (see Figure 6 for the location) presented here in a time-migrated display (Figure 7) and in a depth-converted high-resolution (processed to recover high frequencies) version (Figures 8, 9). Contrary to what was determined for the GFTB in the Pará-Maranhão Basin, two episodes of sliding/contraction superimposed upon the same detachment level can be deduced from the analysis of the seismic sections. Four levels were mapped in time and converted to depth: the sea bottom (Figures 9, 10), a reflector internal to the GFTB affected by the younger folding (reflector X, Figures 9, 11), the main thrust fault (Figures 9, 14), and the detachment zone (Figures 9, 15). A detailed view (without vertical exaggeration) of the dominant thrust fault and of the fault-related fold is presented in Figures 12 and 13. The fault ramp is clearly distinguishable by a

Figure 7. Time-migrated seismic section from the deep waters of the Barreirinhas Basin showing the shelf break and the upper part of the continental slope dominated by normal faults, whereas the lower part of the slope is dominated by one large contractional fold related to a thrust ramp. The core of the fold outcrops at the sea bottom, with its backlimb undergoing active deformation and being presently eroded while the forelimb is buried under onlapping growth strata. Compare to Figure 8.

fault-plane reflection that steps up from the detachment zone at 28°, increasing to 45°, and diminishing up section to subhorizontal attitudes through a series of minor ramps and flats.

The shape of the fold can be characterized as follows (Figure 13): the younger, larger, outer anticline is asymmetric; it presents a wide backlimb that dips (16°30′) less than the fault ramp (average of 33°); and a narrower front limb. A major synclinal axial surface emanates from the foot of the thrust ramp and defines two major kink bands from the detachment zone up to the sea bottom defining the shape of this major young fold. According to Shaw et al. (2005), this is indicative of a simple-shear end-member fault-bend fold. No evidence of thickening of sheared decollement layers above the ramp exists (pure shear end-member) (Figure 13). Plotting the values of the ramp dip (average of 33°) and of the backlimb dip (16°30′) on the graph presented by the cited authors (their page 41), a shear value of 60° is obtained. The forelimb behaves as a normal fault-bend fold from point P on the fault (Figure 13) toward its termination. This overall geometry is similar to deep-water shear fault-bend folds described by Corredor et al. (2005) in the Niger Delta and by Shaw et al. (2005) in the Cascadia accretionary prism, Canada.

The deeper part of the anticline presents complex deformation because of the refolding of the older thrusts from the first contraction episode. Multipanel kink bands are defined by several anticlinal and synclinal axial surfaces emanating from bends in the older thrusts that do not extend above reflector X (Figure 13). The deeper core of the anticline is structurally thickened by the refolding of the older thrusts and the associated repeated sections. We do not favor the interpretation of shortening and thickening of the detachment level by pure shear as the cause for the thickening of the fold core.

Growth strata (sequences 3 to 5, Figure 13) cover the shear fault-bend fold, atop sequences 1 and 2 (Figure 13), which are pretectonic to the second episode of contraction. Sequences 3 and 4 (Figure 13) display thickening-down-to-basin and thinning-up-the-fold profiles and fanning of the dips in the forelimb, suggesting an equal to higher sedimentation rate compared to structural uplift rate, and folding with a component of limb rotation. Sequence 5 (Figure 13) presents growth onlapping in the forelimb and fanning of dips in the backlimb, suggesting

Figure 8. Same section as in Figure 7 converted to depth (V.E. = 5.5×). The section underwent high-resolution processing aimed at recovering high frequencies. See the interpretation in Figure 9. V.E. = vertical exaggeration.

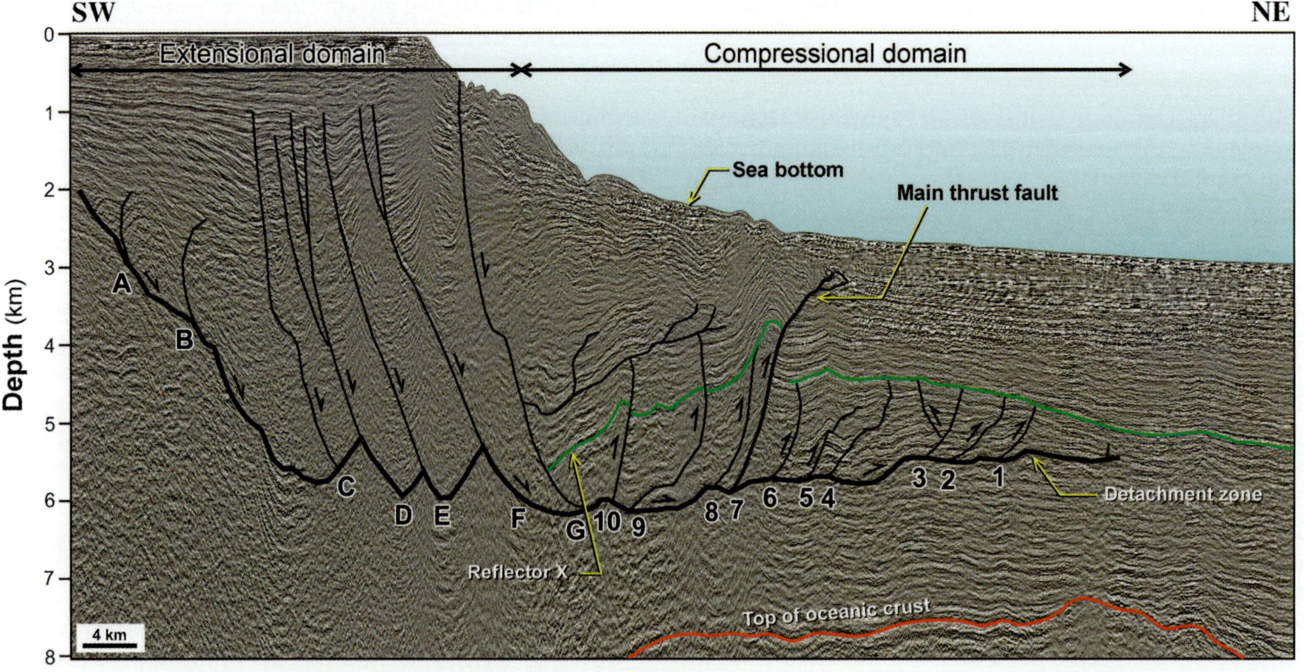

Figure 9. Interpreted section of Figure 8. A complete view of the linked extensional-compressional system of gravitational nature is presented. An older event of contraction occurring between the detachment zone and the green level (reflector X) is clearly seen. In this first event, deformation was more or less regularly distributed among several thrust faults. Modern GFTB is characterized by a dominant thrust fault and a major fault-related fold that crops out at sea bottom. The section underwent high-resolution processing aimed at recovering high frequencies.

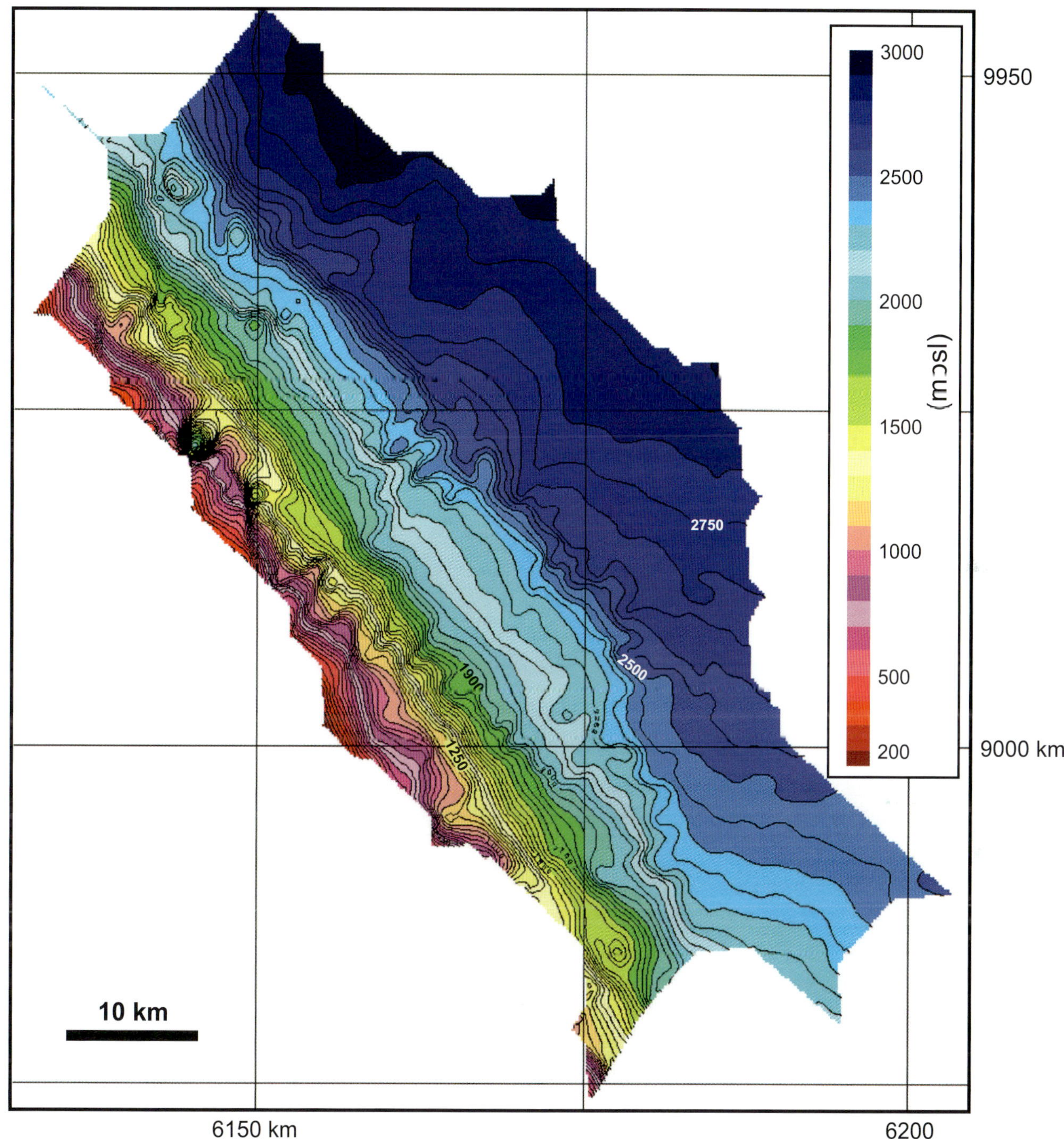

Figure 10. Sea-bottom topography (meters below sea level) underneath the studied area in the deep waters of the Barreirinhas Basin (area C in Figures 1, 6). A pronounced bench occurs between the bathymetric contours of 1900 and 2500 m (6233 and 8202 ft), and it seems to correspond to the outcrop of the modern fold belt shown in Figures 8–10 (the major thrust fault is blind). C.I. = 50 m (164 ft). mbsl = meters below sea level; C.I. = contour interval.

a change to a sedimentation rate lower than the uplift rate in the structure. A time-transgressive unconformity physically separates the coeval sections of sequence 5 in the forelimb and in the backlimb (Figure 13).

The shape of the GFTB is constrained by a surface slope of 6° and by a basal detachment angle of 2°, resulting in a relatively high wedge taper of 8° (Figure 13). The overall coherence and narrowness (20 km [12 mi]) with

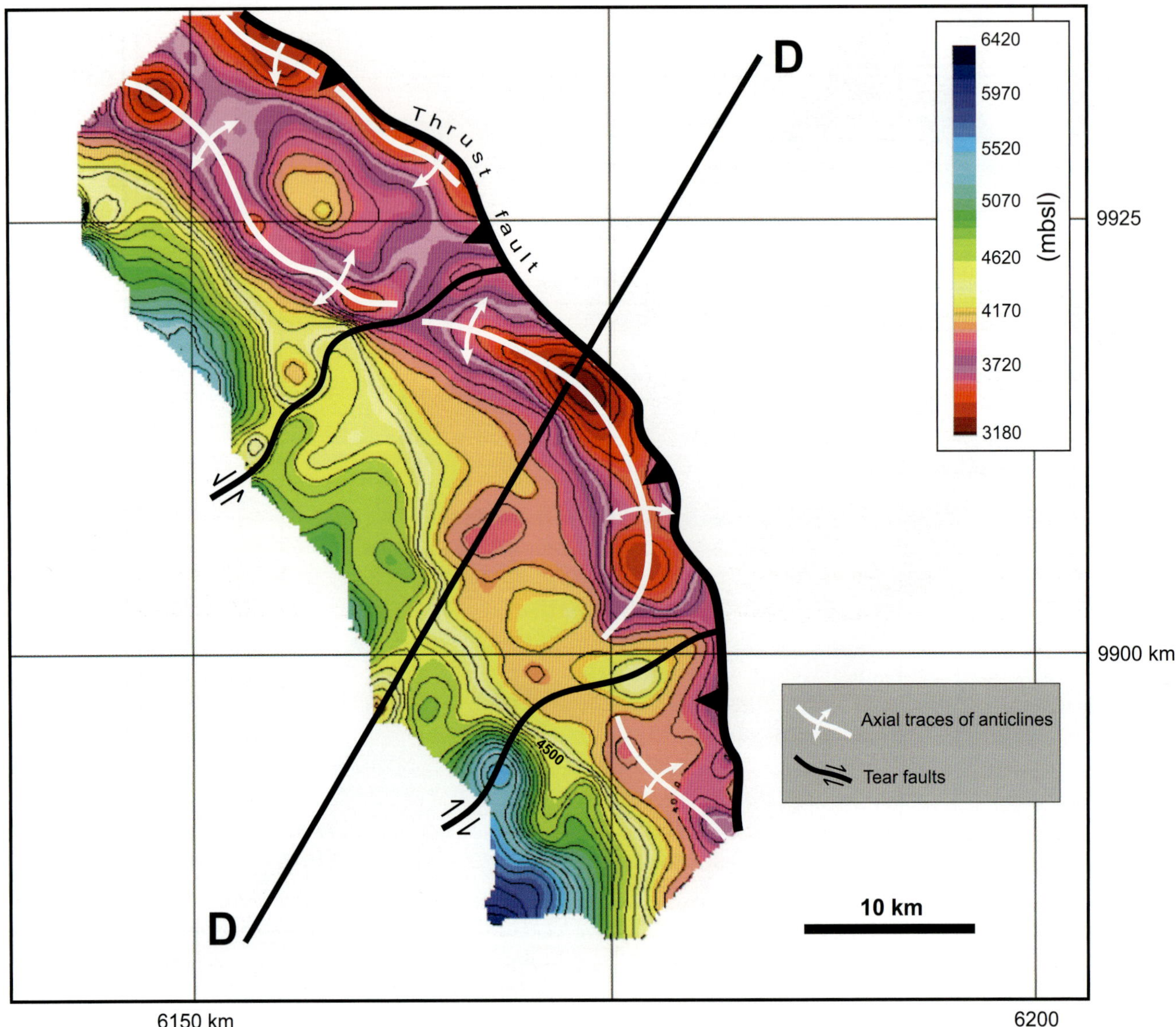

Figure 11. Structural contour map (meters below sea level) of the modern GFTB at the reflector X level (see Figure 9) in the deep waters of the Barreirinhas Basin. The fold belt is dominated by one single thrust fault and a related major fold (see axial traces on the figure). Seismic section D illustrates this fault-related fold in Figures 7–9, 12, and 13. Two zones of tear faulting offset it and change the expression of such fold laterally. To the northwest, the fold is split into two subparallel fold trends, whereas to the southeast, it is highly subdued. C.I. = 100 m (328 ft). mbsl = meters below sea level; C.I. = contour interval.

respect to the entire linked extensional-compressional system (40 km [25 mi]) indicate that this GFTB was a noncritical taper wedge. A critical taper wedge is a wedge that grows self-similarly and maintains the taper by internal deformation of the wedge. A noncritical wedge first builds taper by internal deformation and only propagates basinward (widening) if it reaches critical taper. All stratigraphic evidence indicates only break-backward propagation of the thrusts. The stronger the basal detachment and the weaker the wedge material, the larger the wedge taper and the narrower the GFTB (Bilotti and Shaw, 2005).

The three-dimensional (3-D) visualization of the GFTB (Figures 16–19) highlights the variable geometries that different profiles of the folded strata can present. Figures 11 and 16 show that the central part of the fold belt is dominated by a major single northwest-southeast–trending fold that changes expression across two northeast-southwest–trending tear faults (Figure 11). To the northwest of the northern transverse zone, the

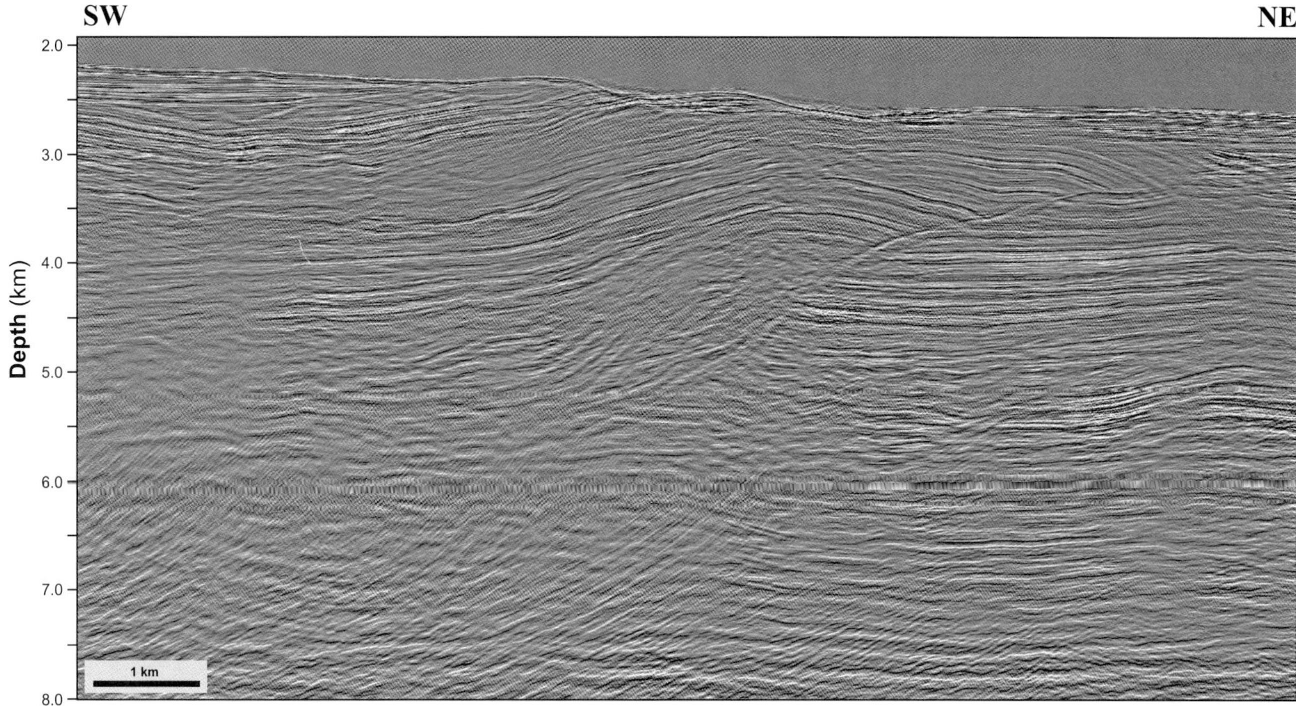

Figure 12. Time-migrated seismic section converted to depth (with no vertical exaggeration). The section underwent high-resolution processing aimed at recovering high frequencies. Notice a fault-plane reflection of the dominant thrust fault, an older thrusting to the right of it, a complex deformation pattern in the central lowermost part of the modern fold, and the outcropping and consequent erosion of the core and backlimb of the fold.

fold is split into two subparallel fold trends, whereas to the southeast, it is abruptly terminated against the southern tear fault (see axial traces of anticlines in Figure 11). The two tear faults do not cross the thrust fault but instead seem to emanate from it, their location being controlled by a marked northeast-southwest–trending structural grain present in the detachment zone (Figure 15).

Figures 17 and 18 show several profiles of the fault-related fold sliced parallel to the direction of the tectonic transport. From southeast to northwest, the geometry of the fold, always asymmetric, changes as follows: Figures 17A and 18A show a fold profile (see fold traces on cited figures) typical of a classic fault-bend fold with a total offset smaller than the ramp length. The separation between a tilted kink band (parallel to the fault trace) and a horizontal band of nondeformed strata projects in the midpoint of the fault ramp. Figure 17B displays a fold profile of fault-bend fold passing upward to a shear fault-bend fold. The lower part of the fold trace runs parallel to the fault trace, but the upper termination of the fold trace is already dipping less than the fault trace. Figure 17C–E presents wide backlimbs with fold traces that dip much less than the fault trace, and very narrow forelimbs. The lateral passage (from southeast to northwest) from a typical fault-bend fold (Figures 17A, 18A) to a fully developed shear fault-bend fold (Figures 17E, 18B) seems to be gradual via several intermediate stages (Figure 17B–D). This suggests that, for the same train of folds, their geometric classification in the scheme of fault-related folding may change along strike.

Figure 19 is a 3-D view of a strike direction of the fold train showing at least five different structural culminations of the fold. From this point of view, no conclusions regarding the geometry of the fold can be drawn. The spatial geometries of the thrust fault and of the detachment zone are fully displayed.

DISCUSSION OF THE RESULTS

Two GFTBs studied in the deep waters of the Pará-Maranhão and Barreirinhas basins present strikingly different characteristics. Both are moderate to high (5.7 to 8°) noncritical taper wedges. These values are comparable to data gained from several contractional wedges from different parts of the world (Figure 20). Maximum values for wedge taper range between 10 and 11°, whereas the minimum values are in the range of 2–6° (Bilotti and Shaw, 2005, their figure 7). Both GFTBs were developed above single and discrete low-dipping (0.7 and 2°,

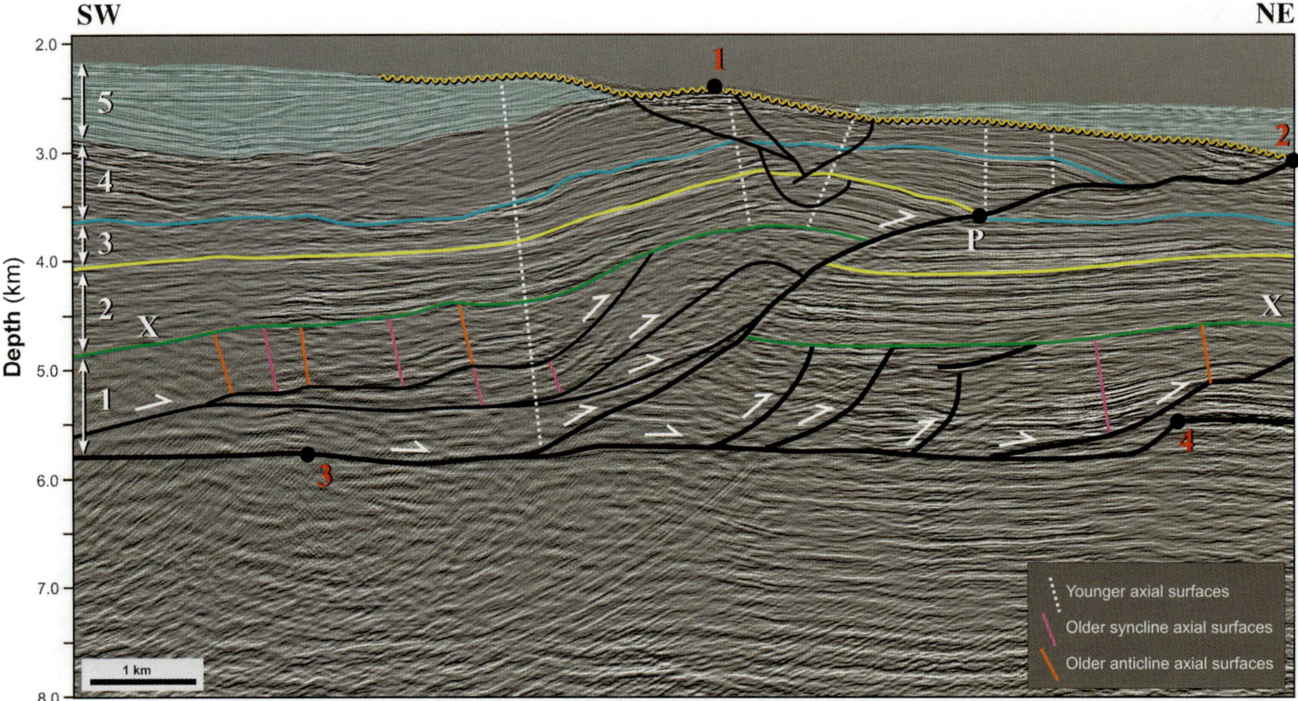

Figure 13. Interpreted time-migrated seismic section converted to depth (with no vertical exaggeration). The section underwent high-resolution processing aimed at recovering high frequencies. Two contractional events can be observed: an older event constituted by several minor thrust faults that stop at reflector X and a younger event dominated by a single thrust fault that was imprinted upon an older fault of the first event. Several axial surfaces separate kink bands in the deeper section and do not project into the overlying sequences. However, a major synclinal axial surface (white dotted line) defines kink bands from the detachment zone up to the sea bottom defining the shape of the major young fold. Overall shape of the fold suggests a simple-shear fault-bend fold (discussion in text). The shape of the contractional wedge is constrained by a top surface slope of 6°, measured between points 1 and 2 (red), and by a basal detachment that dips 2°, measured between points 3 and 4 (red), resulting in a wedge taper of 8°. Step-up angles for the thrust ramp start at 28° and increase to 45°, with an average of 33°. Sedimentary sequences 1 to 5 are discussed in the text. Ages of sequences are hard to establish in a basin with no wells in the deep waters. The detachment zone is probably within the Turonian shales.

respectively) detachment zones. The age and nature of this lithology are tentatively assigned to Cenomanian–Turonian (most probably Turonian) shales as far as scant regional seismic tying and mapping are allowed to deduce. Such lithology did not develop significant ductile flow. Neither evident signs of the formation of mud pillows and diapirs nor of the involvement with the deformation in the allochthon are observed. Back thrusts are noticeably absent in both GFTBs.

The GFTB from the Pará-Maranhão Basin is the result of a single sliding/contraction process that possibly occurred between the Eocene and middle Oligocene. The three tectonic domains (extensional, translational, and compressional) are well developed (Figure 3). Compressional strain (36% of shortening) was distributed on eight major imbricate thrust faults that step up from the detachment zone at angles ranging between 19 and 35°. The thrusts become steeper in a backward direction. Their overall geometry resembles that of duplexes; however, the faults and the slivers between them were never enveloped by a roof thrust. Their upper surface is always an unconformity. All types of fault-related folding are developed in the tectonic slices: fault-bend folds are dominant in the outer slivers, fault-propagation folds dominate the center wedges, and detachment folding is present in the innermost region (Figure 3). Growth strata present thickening-down-to-basin and thinning-up-the-fold sequences and fanning of dips, suggesting folding with a component of limb rotation, as suggested by the decreasing dips toward the top (Figure 3). The oldest growth section was in fact involved in the thrusting and folding (see the yellow sequence in Figure 3).

The GFTB from the Barreirinhas Basin is the result of two sliding/contraction events, with the younger event reactivating or using the detachment zone and the thrust faults of the older. The younger fold belt presents a narrower expression (24 km [15 mi]) than the older (40 km [25 mi]) but a much larger and spectacular amplitude

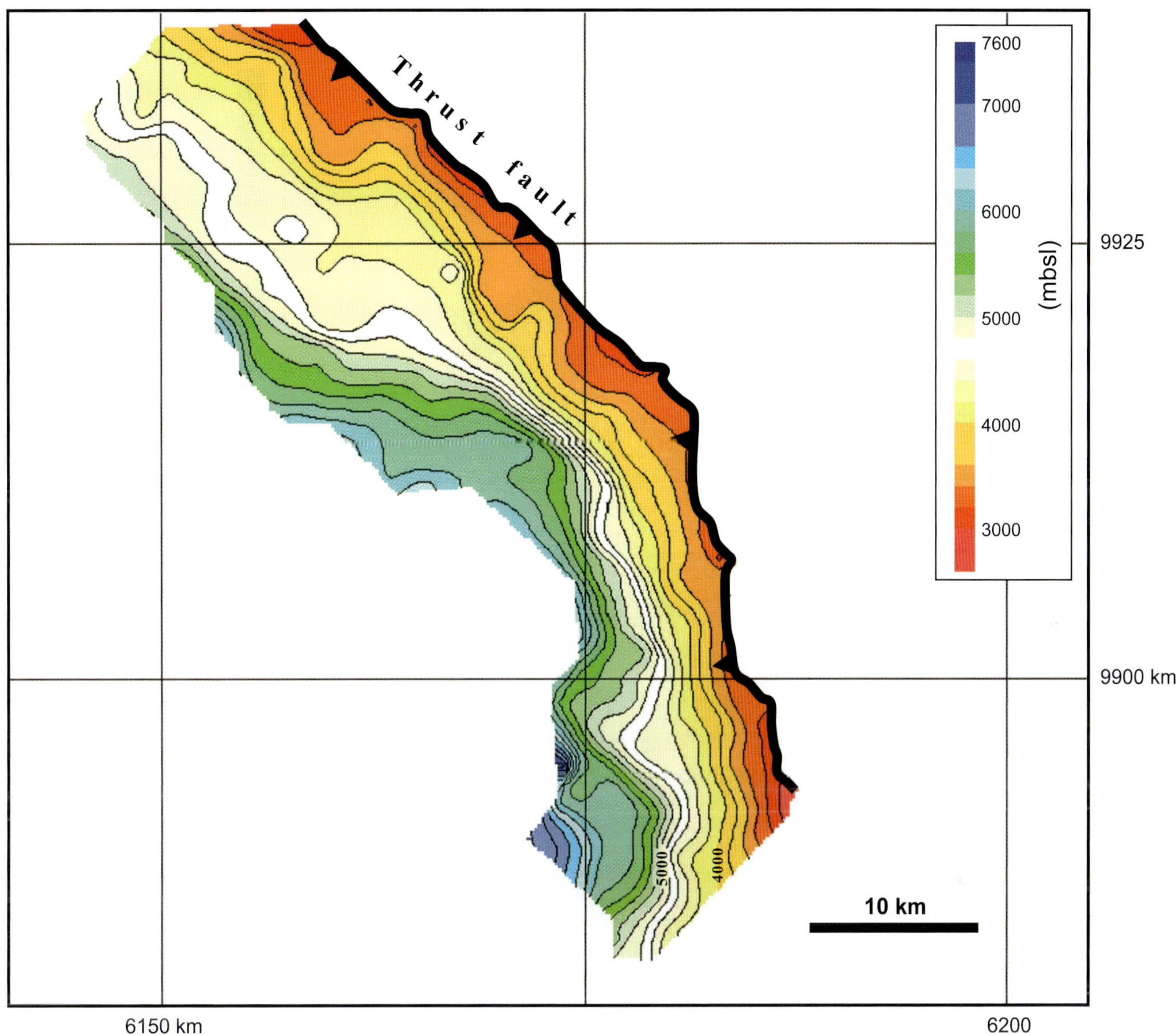

Figure 14. Structural contour map (mbsl = meters below sea level) of the dominant thrust fault in the GFTB of the Barreirinhas Basin. C.I. = 200 m (656 ft). C.I. = contour interval.

(3.5 km [2.2 mi] of vertical section involved compared to 1.5 km [0.9 mi] of vertical section affected by the first event, Figure 9). The extensional and compressional domains are roughly equal in width even when considering each sliding-contraction event. The translational domain is not identifiable (Figure 9) probably because of the imprint of a strong, late tectonic event upon an older, less developed event. Contractional strain is concentrated in one dominant thrust fault and one major associated asymmetric thrust-fault-related fold. The nature of the fault-related folding is predominantly simple-shear fault-bend folding, but it varies along strike to classic fault-bend fold (Figures 17, 18). One shear measurement indicated 60° of shearing. Growth strata record a change from equal or higher to lower sedimentation rates compared to structural uplift rates. Folding of growth strata was possibly accomplished with a component of limb rotation as suggested by their decreasing dips toward the top (Figure 13).

A comparison of some critical parameters between the Brazilian and Nigerian GFTBs (all developed in passive margins) is shown in Table 1. The Pará-Maranhão and Barreirinhas basins' GFTBs present higher wedge tapers and step-up angles of thrusts than the Niger Delta GFTB. They are also narrower and display a uniform compressional domain instead of the tripartite division of the Nigerian GFTB (Bilotti and Shaw, 2005; Corredor et al., 2005). The detachment zone shows strong plastic

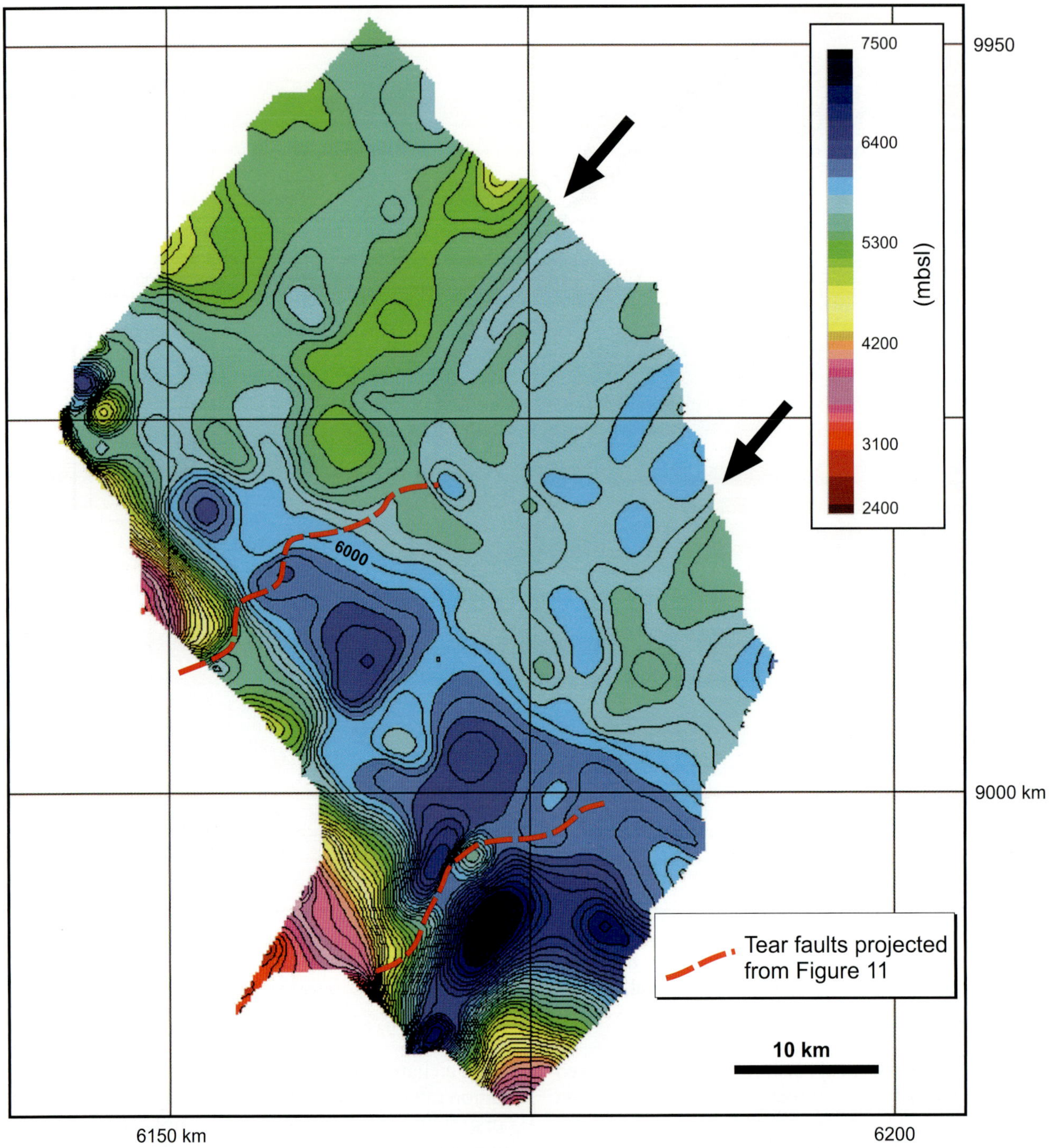

Figure 15. Structural contour map (mbsl = meters below sea level) of the detachment zone in the GFTB of the Barreirinhas Basin. Tear faults at the fold level seem to have been nucleated in the strong linear structural grain present in the detachment zones (arrows). C.I. = 100 m (328 ft). C.I. = contour interval.

behavior in Nigeria, where the Akata shale is involved in the deformation of the thrusts and folds (Corredor et al., 2005). In Brazil, no marked signs of ductile behavior of the Turonian shales exist. A notable absence of back thrusts in the Brazilian GFTBs in contrast with the Niger Delta GFTB is observed (Corredor et al., 2005). This is probably due to a high cohesive strength of the wedge material and to a weak basal detachment. The frictional traction along the basal decollement was never high enough to impede forward propagation of the thrust

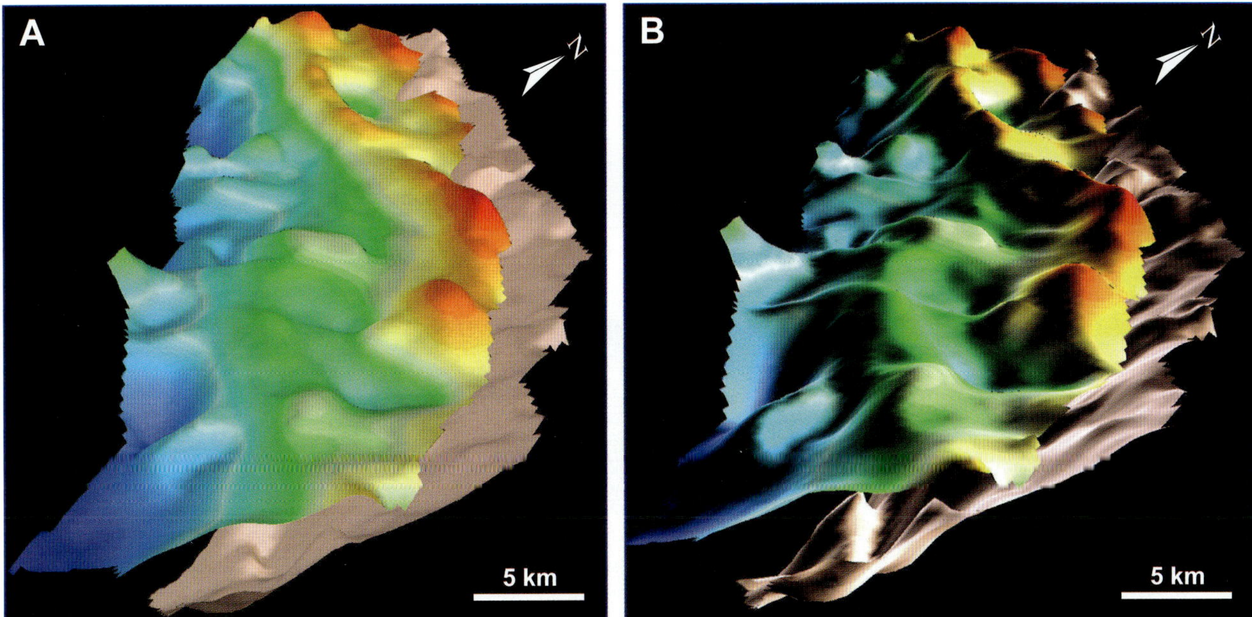

Figure 16. Three-dimensional visualization of the major young fold and the dominant thrust fault (A) without and (B) with shadows. View is from the southeast to northwest. Notice the abrupt termination of the fold to the southeast (caused by a northeast-southwest–trending tear fault) and the splitting of it into two parallel folds to the northwest (after a cut by another northeast-southwest–trending tear fault). Both tear faults are outlined in Figure 11. Images were constructed using VIS3D/SIGEO (proprietary software of Petrobras).

wedges. The wedges never stalled so the rocks under compression were not forced to rupture and fault in a backward direction (back thrusts).

Both GFTBs probably never achieved critical taper-wedge conditions (Table 1). All stratigraphic evidence of onlapping growth strata (see Figures 3, 13) upon the contractional folds and their deformation indicates that the accretion of material to the wedges was always via break-backward thrusting. According to Bilotti and Shaw (2005), a noncritical wedge first builds taper by internal deformation and then propagates basinward after reaching critical taper. Once the critical taper is reached, the wedge grows self-similarly and internal deformation of the wedge maintains the taper. According to Bilotti and Shaw (2005), a pronounced widening of a contractional wedge is an indication that it has reached critical taper and has continued to grow. The narrowness of the Pará-Maranhão (20 km [12 mi]) and Barreirinhas (40 km [25 mi]) basins' GFTBs is also an evidence of their subcritical stage, especially when compared to a fully developed critical taper wedge like the Niger Delta GFTB with a width of more than 100 km (62 mi) and a tripartite compartmentation of the compressional domain (Bilotti and Shaw, 2005).

The shapes of the fault-related folds thus determined in the Pará-Maranhão and Barreirinhas GFTBs are comparable to those found in other passive margin GFTBs throughout the world. Fault-bend folding and simple-shear fault-bend folding dominate the Brazilian GFTBs, whereas detachment folding and simple-shear fault-bend folding are prevalent in the Niger Delta GFTB (Table 1). In particular, the dominant major simple-shear fault-bend fold in the Barreirinhas Basin GFTB is very similar to the toe thrusts described by Corredor et al. (2005, their figure 8) for the Niger Delta. The growth of the fold and its relationship to syntectonic sedimentation are also similar to those seen in Nigeria. The difference lies in the fact that, for the African GFTB, the detachment zone was weakened by elevated basal fluid pressures, and this resulted in a much lower wedge taper (2.5°) and much wider and compartmented fold and thrust belts (40 to 100 km [25 to 62 mi]; Bilotti and Shaw, 2005). Because both GFTBs are detached within the local petroleum source rock, the question whether these differences have any significant implication regarding the petroleum potential of the detachment zones remains to be tested.

CONCLUSIONS

The GFTB from the Pará-Maranhão Basin involved relatively old material (Coniacian to Paleocene) gliding and

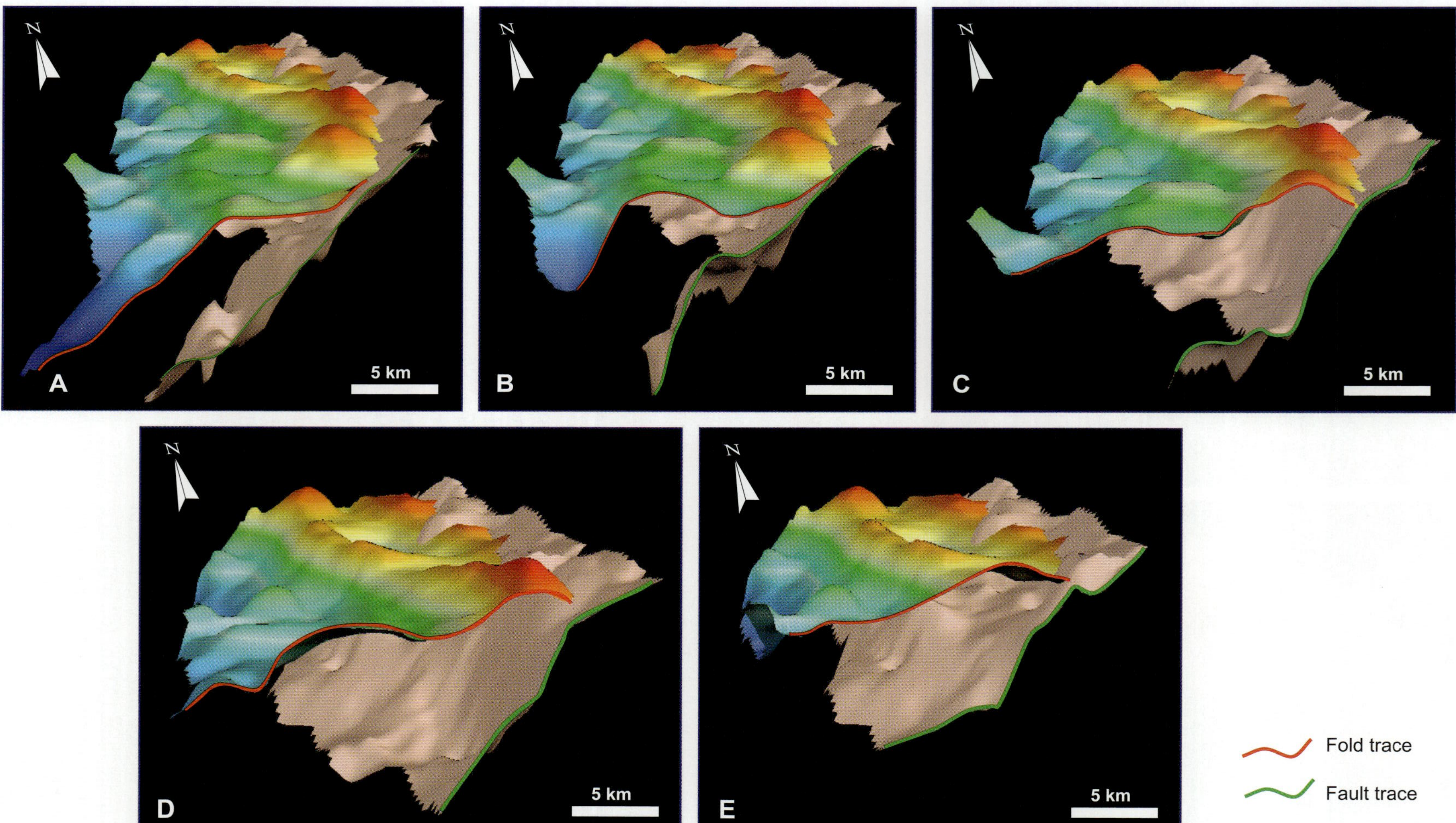

Figure 17. Five 3-D visualizations of profiles of the major young fold and the dominant thrust fault, sliced parallel to the tectonic transport, along their strike. View is from the southeast to northwest. Notice the different profiles of the fold. (A) Typical fault-bend fold, (B) fault-bend fold passing upward into a shear fault-bend fold, (C–E) typical shear fault-bend fold. Images were constructed using VIS3D/SIGEO (proprietary software of Petrobras).

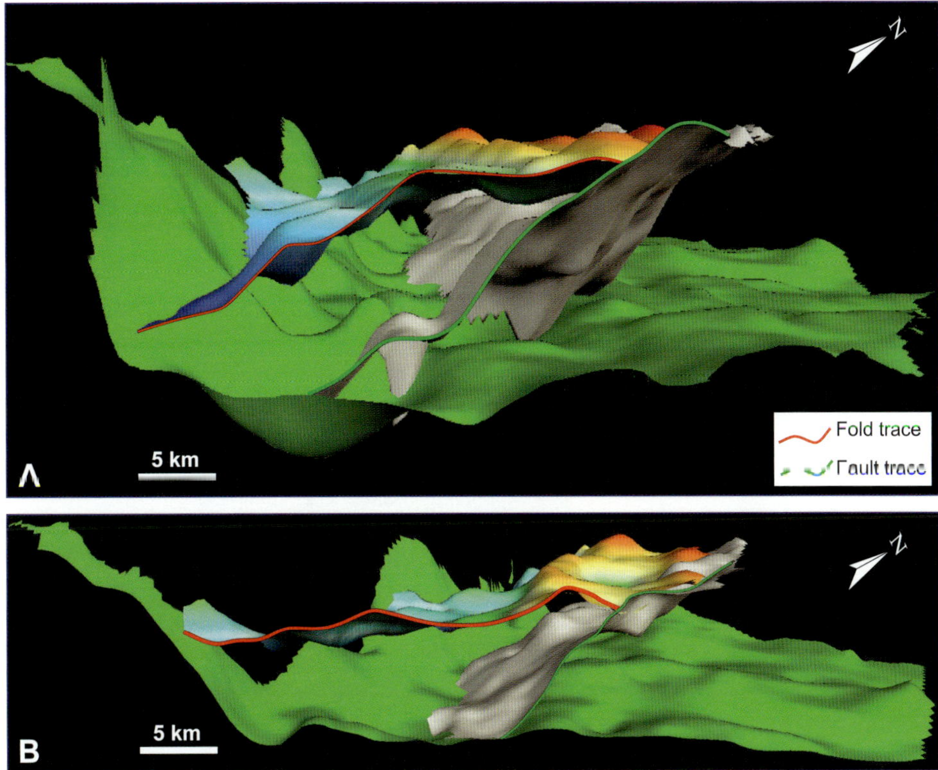

Figure 18. Three-dimensional visualizations of the major young fold, the dominant thrust fault (in gray), and the detachment zone (in green) in two sections parallel to the tectonic transport. View is from the southeast to northwest. Notice the different profiles of the fold: typical fault-bend fold in the upper frame and typical shear fault-bend fold in the lower frame. Images were constructed using VIS3D/SIGEO (proprietary software of Petrobras).

contracting during the Eocene/middle Oligocene. As a result, a narrow, noncritical contractional wedge with a moderate wedge taper of 5.7° developed. The high cohesive strength of the wedge of deforming material relative to the weaker basal detachment zone promoted fault-related folding in the form of fault-bend and fault-propagation folds, with a notable absence of back thrusting. Additionally, the detachment zone was not very weak because it did not allow the wide propagation of the compressional domain. In the GFTB of the Barreirinhas Basin, the same conditions regarding the material involved and timing probably prevailed during the first sliding/contraction event. However, in the second sliding/contraction event, the detachment zone became stronger than in the first event probably because of the larger age gap between the deformation (Neogene) and both the allochthonous rocks (Coniacian to present) and the detachment zone (Turonian). This led to the development of a contractional wedge of high taper (8°) dominated by a single thrust fault and a large simple-shear fault-bend anticline. Again, the absence of back thrusting is noticeable.

The noncritical taper-wedge condition of both GFTBs together with a lack of detachment zones with very weak cohesive strengths seems to indicate that the gravitational sliding that affected the Pará-Maranhão and Barreirinhas

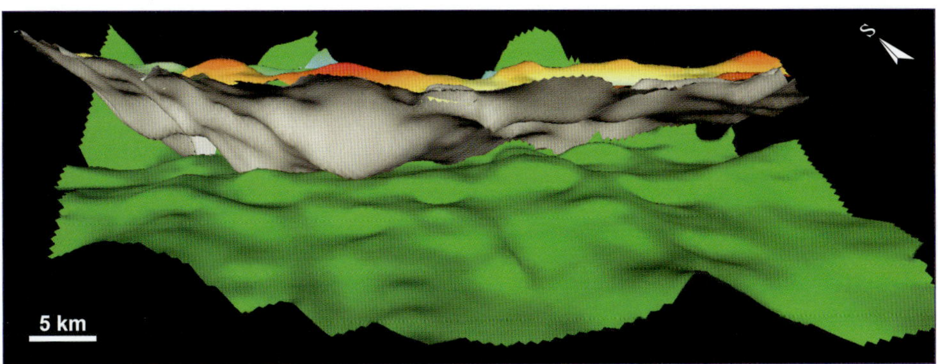

Figure 19. Frontal 3-D visualization of the major young fold, with only the tops of its anticlines shown above the dominant thrust fault (in gray) and the detachment zone (in green) parallel to their strike. View is from the northeast to southwest. Image was constructed using VIS3D/SIGEO (proprietary software of Petrobras).

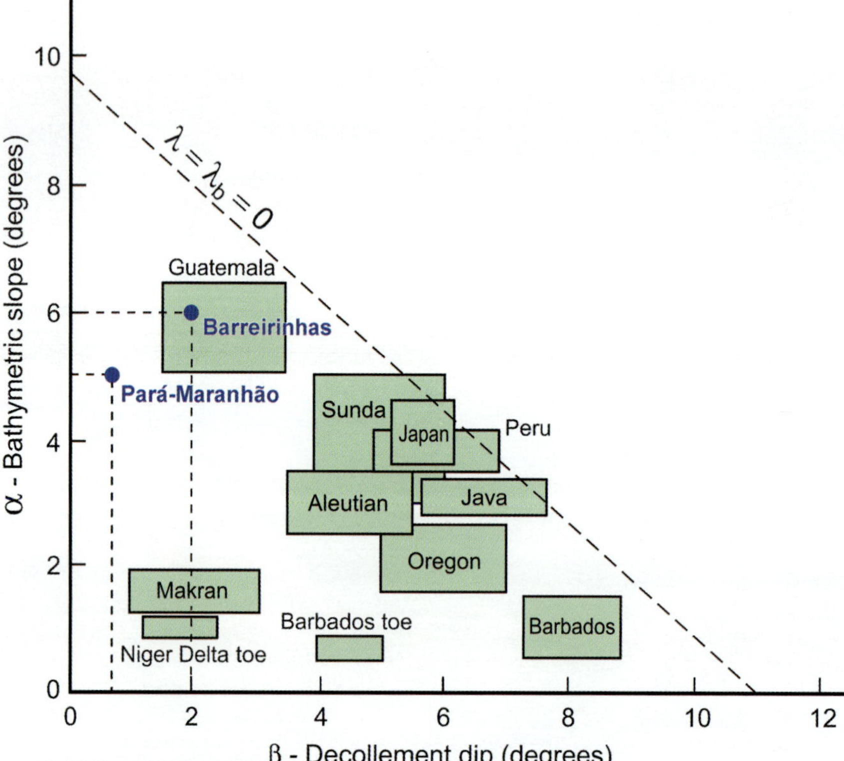

Figure 20. Plot of the decollement dip against the bathymetric slope of several contractional wedges from continental passive margins and accretionary prisms (modified from Bilotti and Shaw, 2005). Wedge shapes of the GFTBs studied in the deep waters of the Pará-Maranhão and Barreirinhas Basins are similar to the active submarine fold and thrust belt of the convergent margin of Guatemala.

basins was more the result of a tectonic origin than of the overpressure of petroleum source rocks (as described in the Niger Delta by Bilotti and Shaw, 2005). The proximity of these basins to very important oceanic fracture zones, namely, Romanche and São Paulo (Figure 1), suggests that during times of strong tectonic reactivation of such zones (especially the Eocene), the resulting earthquakes could have been as the triggering mechanisms for the gravitational processes that shaped the deep and ultradeep-water realms of these basins.

Table 1. Comparison of Pará-Maranhão and Barreirinhas GFTBs with the Niger Delta GFTB.

Parameters	Pará-Maranhão	Barreirinhas	Niger Delta*
Bathymetric slope (degrees)	5	6	0.7–1.3
Dip of detachment zone (degrees)	0.7	2	1.2–2.2
Wedge taper (degrees)	5.7	8	2.3–2.9
Step-up angle of thrusts (degrees)	27	28	22
Presence of detachment folding	Rare	Absent	Common
Presence of fault-propagation folding	Common	Absent	Rare
Presence of fault-bend folding	Common	Common	Rare
Presence of shear fault-bend folding	Absent	Dominant	Dominant
Presence of back thrusts	Absent	Absent	Abundant
Width of compressional domain	20 km	40 km	Up to 100 km
Compressional domain subdivisions	Unique	Unique	Threefold
Plastic flow in detachment zone	Some minor	No	Major
Presence of overpressure at detachment	No	No	Yes
Taper of the wedge	Subcritical	Subcritical	Critical

*Data are from Bilotti and Shaw (2005) and Corredor et al. (2005).

ACKNOWLEDGMENTS

The author thanks Petrobras for the permission to publish this manuscript and to geophysicists Walter Leitzke, Carlos Alves C. Filho, Tereza Cristina F. Ramos, and Adriana Perpetuo S. da Silva; to geologist Marcos Domingues; and to technician Haroldo M. Ramos, all from Petrobras, for their help in the processing and conversion to depth of the seismic sections, in the elaboration of maps, and in the drawing of the figures. Also, two reviewers, Ian Davison and Jose de Vera, presented thoughtful analyses on the first draft and stimulated a more profound discussion of the mechanisms involved in the folding, thus greatly improving the original manuscript.

REFERENCES CITED

Attoh, K., L. Brown, J. Guo, and J. Heanlein, 2004, Seismic stratigraphic record of transpression and uplift on the Romanche transform margin, offshore Ghana: Tectonophysics, v. 378, p. 1–16, doi:10.1016/j.tecto.2003.09.026.

Azevedo, R. P., 1991, Tectonic evolution of Brazilian equatorial continental margin basins: Ph.D. thesis, University of London, 455 p.

Basile, C., J. Mascle, and R. Guiraud, 2005, Phanerozoic geological evolution of the Equatorial Atlantic domain: Journal of African Earth Sciences, v. 43, p. 275–282, doi:10.1016/j.jafrearsci.2005.07.011.

Bilotti, F., and J. H. Shaw, 2005, Deep-water Niger Delta fold and thrust belt modeled as a critical-taper wedge: The influence of elevated basal fluid pressure on structural styles: AAPG Bulletin, v. 89, no. 11, p. 1475–1491, doi:10.1306/06130505002.

Corredor, F., J. H. Shaw, and F. Bilotti, 2005, Structural styles in the deep-water fold and thrust belts of the Niger Delta: AAPG Bulletin, v. 89, no. 6, p. 753–780, doi:10.1306/02170504074.

Dailly, P., P. Lowry, K. Goh, and G. Monson, 2002, Exploration and development of Ceiba field, Rio Muni Basin, southern Equatorial Guinea: The Leading Edge, v. 21, no. 11, p. 1140–1146, doi:10.1190/1.1523753.

Davis, D., J. Suppe, and F. A. Dahlen, 1983, Mechanics of fold and thrust belts and accretionary wedges: Cohesive Coulomb theory: Journal of Geophysical Research, v. 88, p. 1153–1172, doi:10.1029/JB088iB02p01153.

Mascle, J., and E. Blarez, 1987, Evidence for transform margin evolution from Ivory Coast-Ghana continental margin: Nature, v. 326, p. 378–381, doi:10.1038/326378a0.

Matos, R. M. D., 2000, Tectonic evolution of the Equatorial South Atlantic, *in* W. Mohriak and M. Talwani, eds., Atlantic rifts and continental margins: American Geophysical Union, Geophysical Monograph 115, p. 331–354.

Medwedeff, D. A., 1989, Growth fault-bend folding at southeast Lost Hills, San Joaquin Valley, California: AAPG Bulletin, v. 73, p. 54–67.

Meyer, D., L. Zarra, D. Rains, B. Meltz, and T. Hall, 2005, Emergence of the lower Wilcox Tertiary trend in the deepwater Gulf of Mexico: World Oil, v. 226, no. 5, p. 72–77.

Rowan, M. G., F. J. Peel, and B. J. Vendeville, 2004, Gravity-driven fold belts on passive margins, *in* K. R. McClay, ed., Thrust tectonics and hydrocarbon systems: AAPG Memoir 82, p. 157–182.

Shaw, J. H., C. Connors, and J. Suppe, 2005, Structural interpretation methods, *in* J. H. Shaw, C. Connors, and J. Suppe, eds., Seismic interpretation of contractional fault-related folds: AAPG Studies in Geology 53, p. 2–58.

Souza-Lima, W., and G. P. Hamsi Jr., 2003, Bacias sedimentares brasileiras-Bacias da margem continental: Phoenix, Fundação Paleontológica Phoenix, Aracaju, Brasil, Ano 5, no. 50, 4 p.

Tari, G., M. Kaminski, J. Molnar, and D. Valasek, 2004, New seismic data reveals unusual deepwater Ghana plays: Offshore, v. 64, no. 10, p. 37–39.

Zalán, P. V., 1984, Tectonics and sedimentation of the Piauí-Camocim sub-basin, Ceará Basin, offshore northeastern Brazil: Ph.D. thesis, Colorado School of Mines, 128 p.

Zalán, P. V., 2005, End members of gravitational fold and thrust belts (GFTBs) in the deep waters of Brazil, *in* J. H. Shaw, C. Connors, and J. Suppe, eds., Seismic interpretation of contractional fault-related folds: AAPG Studies in Geology 53, p. 147–153.

Zalán, P. V., E. P. Nelson, J. E. Warme, and T. L. Davis, 1985, The Piaui Basin: Rifting and wrenching in an Equatorial Atlantic transform basin, *in* K. Biddle and N. Christie-Blick, eds., Strike-slip deformation, basin formation and sedimentation: SEPM Special Publication 37, p. 177–192.

Zalán, P. V., P. R. Palagi, M. C. G. Severino, F. A. L. Martins, and E. P. Ferreira, 2003, Bacias sedimentares brasileiras—Bacia de Barreirinhas: Phoenix, Fundação Paleontológica Phoenix, Aracaju, Brasil, Ano 6, no. 64, 6 p.

16

Krueger, Scot W., and Neil T. Grant, 2011, The growth history of toe thrusts of the Niger Delta and the role of pore pressure, *in* K. McClay, J. H. Shaw, and J. Suppe, eds., Thrust fault-related folding: AAPG Memoir 94, p. 357–390.

The Growth History of Toe Thrusts of the Niger Delta and the Role of Pore Pressure

Scot W. Krueger
BP America Inc., Houston, Texas, U.S.A.

Neil T. Grant
ConocoPhillips (United Kingdom) Ltd., Aberdeen, United Kingdom

ABSTRACT

The evolution of thrust structures in the toe-thrust belt of the Niger Delta has been analyzed and reveals a common growth history. Once initiated, structures propagate rapidly along strike with minimal slip. Subsequently, the propagation slows or ceases, and slip accumulation begins to dominate. In the waning stage, deformation retreats toward structural culminations. The pattern of slip evolution across the deep-water thrust belt indicates that neighboring structures are active simultaneously. A structural restoration across the western Niger Delta suggests that a series of detachment advances in the late Miocene initiated when the Akata shale detachment horizon was buried to a depth of around 4 km (2.5 mi) below mud line. These produced a deep-water fold and thrust belt accommodating more than 12 km (7 mi) of shortening. Deformation combined with compaction disequilibrium may have helped produce elevated pore pressures approaching the fracture gradient in the detachment zone. Once the thrust structures began to attain structural height, pore-pressure transfer developed in interbedded sands and silts within the long backlimb of the fault bend folds. This process may have influenced the behavior of the fault zones. A simple model of pore-pressure evolution within the growing structures indicates that pressure transfer can help explain the localization of out-of-sequence thrust growth and might facilitate the focusing of slip toward the central high-relief parts of thrust structures.

INTRODUCTION

The Niger Delta is one of the largest and longest lived, wave-dominated deltas on earth. The delta is located at the mouth of the Benue Trough in the Gulf of Guinea. (Figure 1). The trough initiated in the Early Cretaceous as a failed rift triple junction and since then has been a focus for clastic sedimentation on the west African margin. Since the end of the Cretaceous, the Niger Delta has prograded out from the Benue Trough and onto the oceanic crust of the South Atlantic (Lehner and De Ruiter, 1977).

DOI:10.1306/13251345M943435

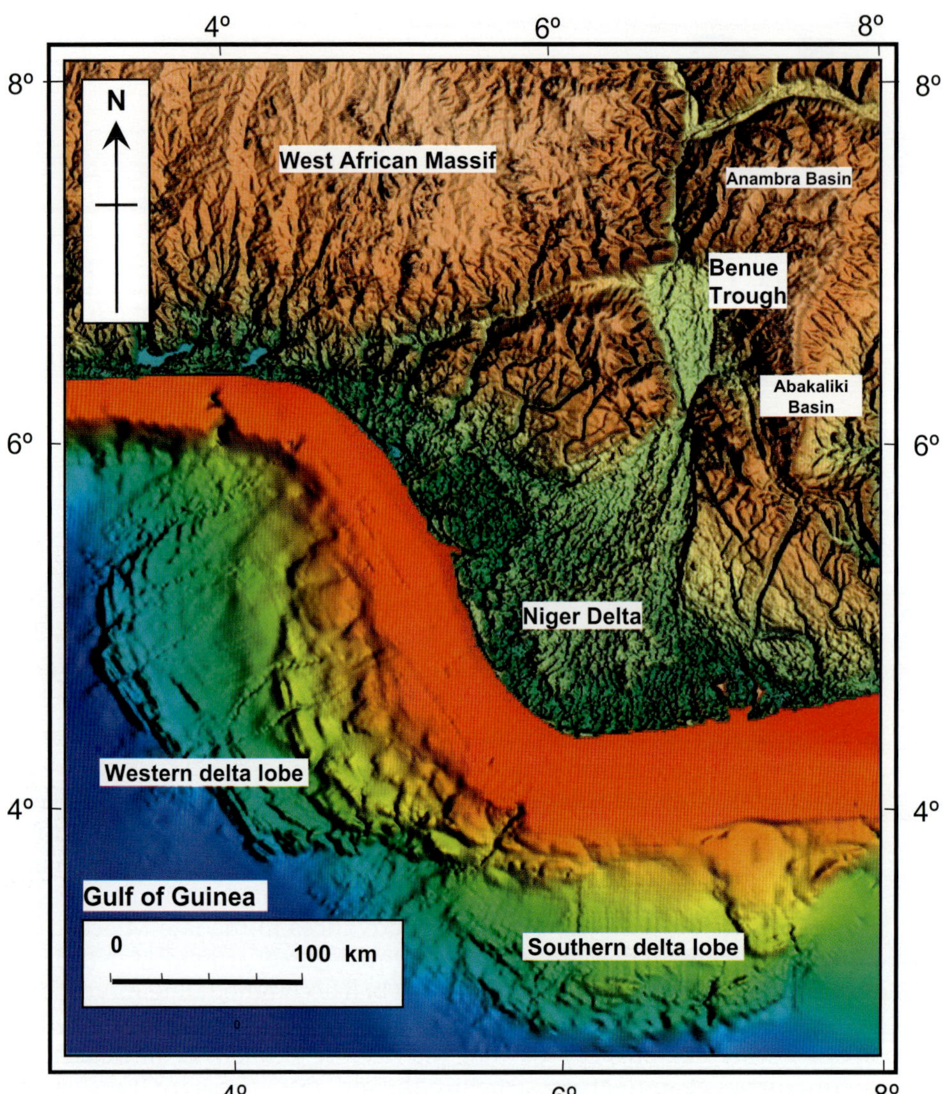

Figure 1. Location map of the Niger Delta showing onshore topography and seabed bathymetry. The main deep-water deformation belts are clearly seen as two separate submarine lobes.

The structure of the Niger Delta is dominated by the gravity spreading of the large clastic sedimentary wedge toward the ocean basin (Damuth, 1994). This has occurred on a substratum of mobile shale and has resulted in gravitationally driven extensional and compressional belts that have moved downslope on a deep-seated shale detachment zone (Figure 2). The delta top and offshore shelf comprise a series of arcuate extensional belts in which regional and counterregional faults form the dominant structures (Doust, 1990). Farther outboard, the structures are predominantly compressional (see Figures 2, 3). The deep-water part of the delta comprises two fold and thrust belts that accommodate the updip, gravity-driven extension on the continental shelf. These belts contain a wide variety of thin-skinned structural styles, including fault-bend, fault-propagation, and detachment folds as well as complex imbricate thrust systems and shale diapirs (Figure 3).

The aim of this chapter is to describe the overall pattern of fault activity in the outboard contractional belts in the deep-water Niger Delta and then to focus on the growth history of three example structures. These toe thrusts highlight a common three-stage growth history that may provide an analog for other thrust settings where the lack of preserved growth stratigraphy prevents a detailed interpretation of fault evolution. The Niger Delta is an ideal natural laboratory for studying the evolution of thrust faults because there is an extensive grid of regional two-dimensional (2-D) seismic lines, increasingly available three-dimensional (3-D) seismic volumes, and recent deep-water well penetrations that together provide an excellent data set. In addition, deposition rates near the toe of slope of the Niger Delta

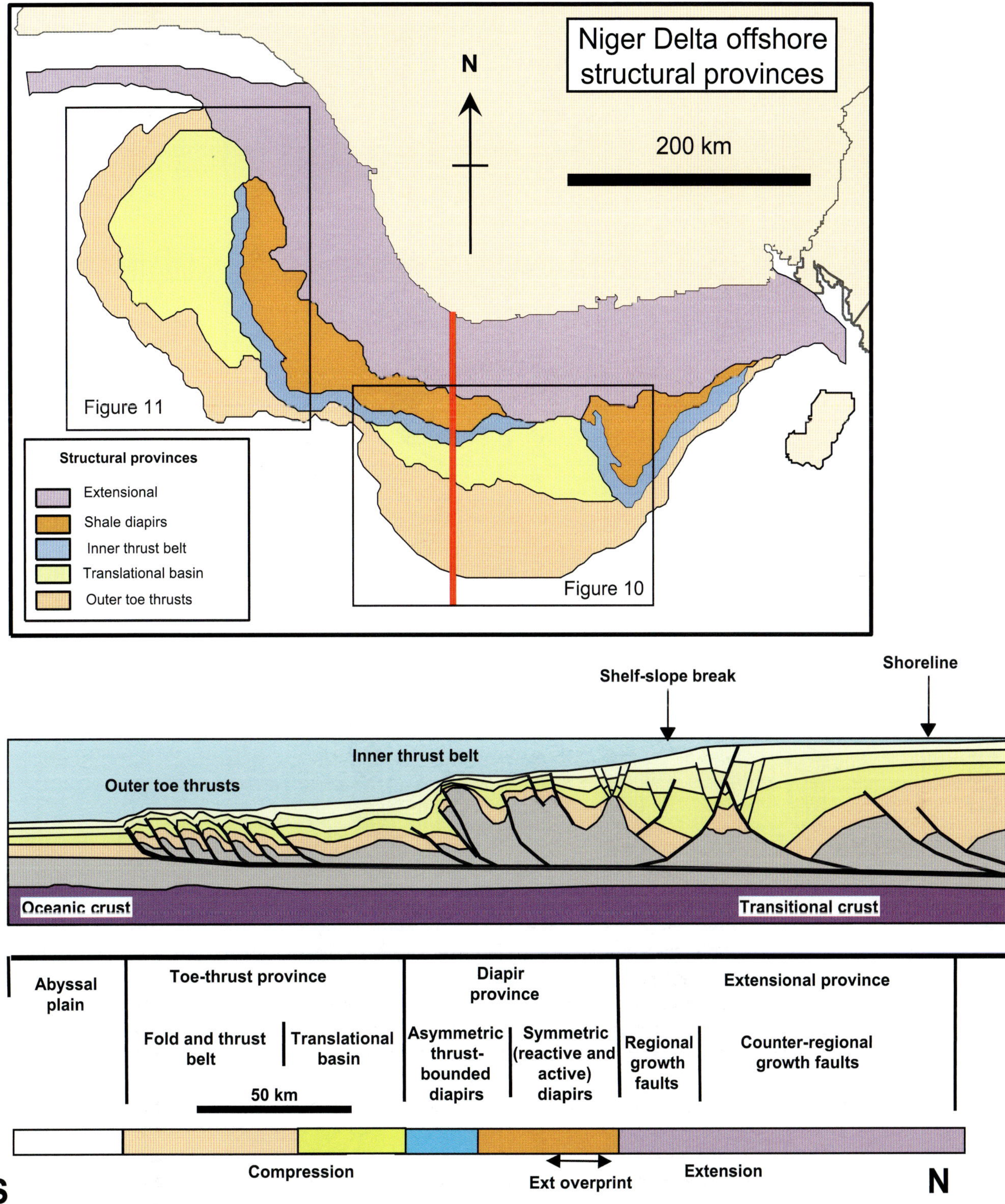

Figure 2. Regional map of the offshore Niger Delta showing the distribution of structural provinces. The accompanying schematic cross section shows the structural architecture of the Niger Delta from coastline to abyssal plain. The overall structure can be divided into five distinct structural provinces. From proximal to distal, these consist of a broad extensional belt, a zone of shale diapirs, a narrow belt of imbricate faulting called the inner thrust belt, a zone of broad translational basins, and an outer toe-thrust belt. The folds examined in this study come from the outer translational basin and the innermost part of the outer toe-thrust belt. Ext = extensional.

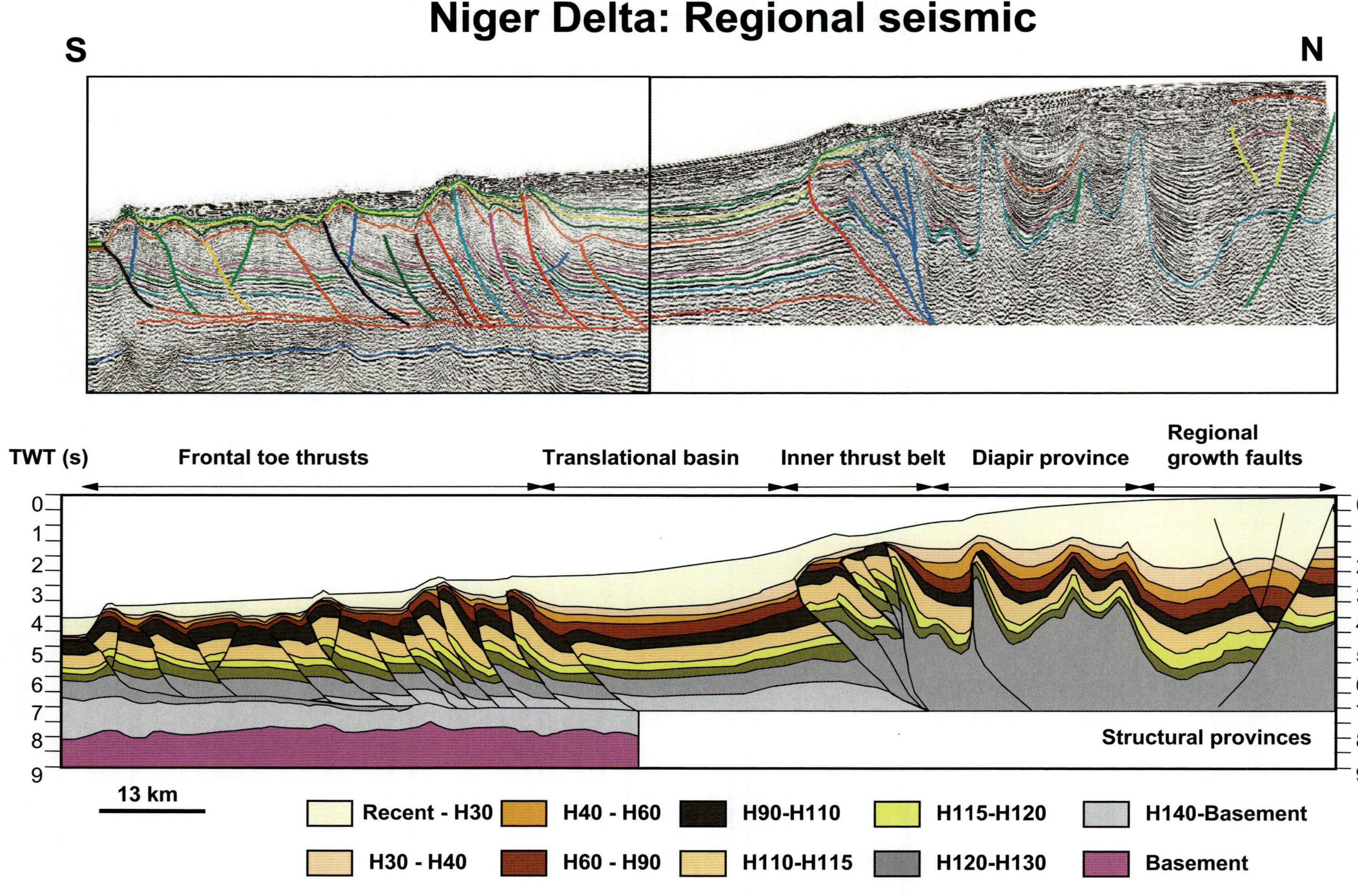

Figure 3. Regional seismic profile and cross section interpretation across the southern deep-water Niger Delta. The section highlights the linkage between the extensional province on the shelf and the contractional thrust systems at the toe of slope in the deep water. The stratigraphic units depicted in the cross section are defined in Figure 4. Seismic image courtesy of Mabon Geophysical. TWT = two-way time; H = horizon.

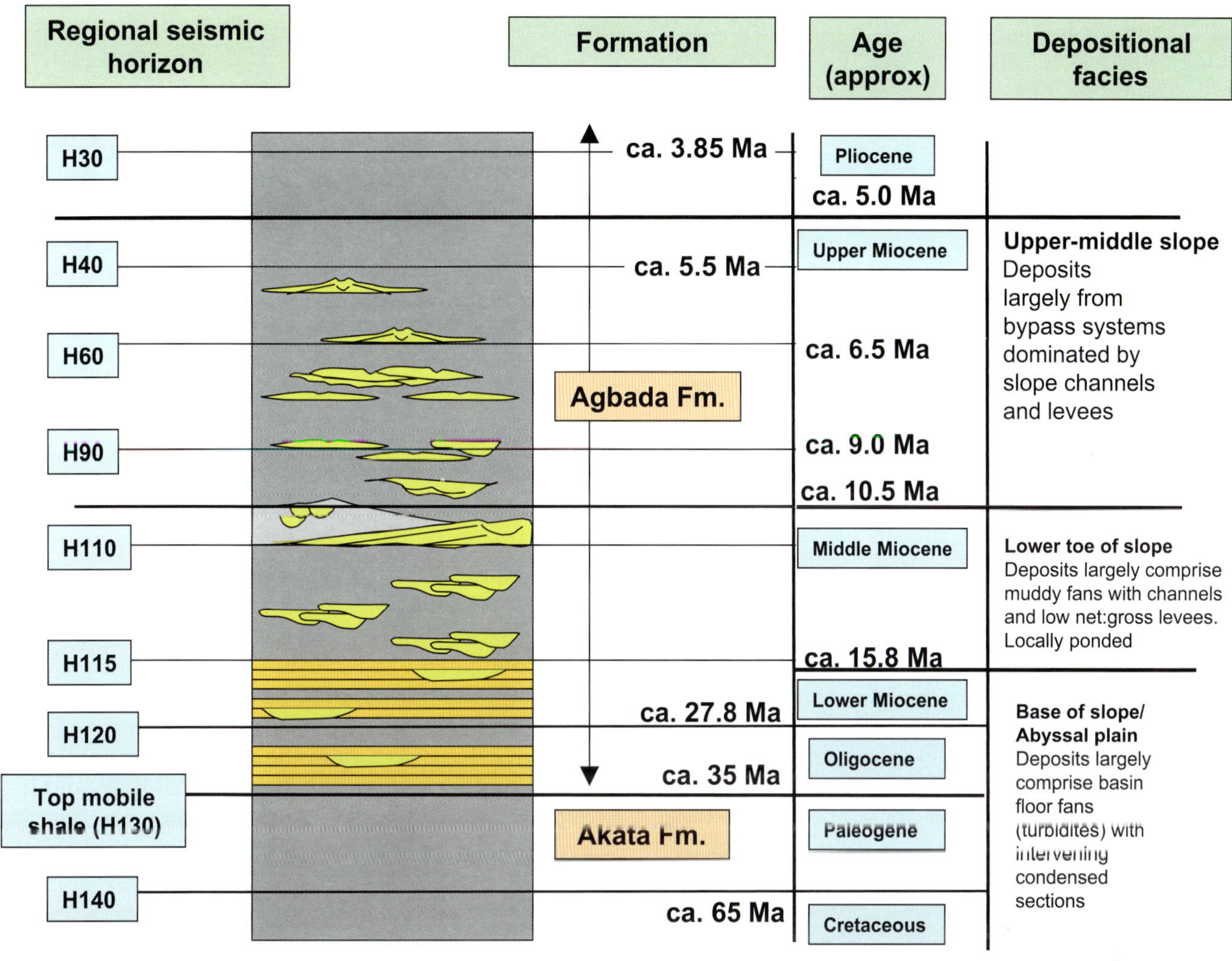

Figure 4. Summary stratigraphic chart for the deep-water depositional systems of the Niger Delta. The horizon nomenclature given is used internally by ConocoPhillips and is based on regional seismically tied events. Fm. = formation; H = horizon.

are sufficiently high to allow an extensive growth sedimentary record to be preserved, making detailed analysis of evolutionary history possible.

After studying the pattern of fault growth, some of the possible physical mechanisms involved in fault growth will be discussed, emphasizing in particular the role that fluid pressure may play in potentially influencing the fault behavior. The lateral transfer of fluid pressure toward the crest of structures is a common characteristic of deep-water toe-thrust systems and occurs along permeable strata that are tilted in the backlimbs of thrust-related folds. This excess pressure mechanism will be shown to be active in the Niger Delta. Its presence may influence fault activity by changing the stress state of the thrust fault, in particular lowering the shear stress required for fault reactivation. A simple model based on the 2-D centroid concept is presented that shows how fluid overpressures linked to ongoing compaction dynamically change as thrust structures grow. The role that this excess pressure mechanism plays in influencing thrust displacement patterns is then discussed.

Stratigraphy

The offshore Niger Delta comprises Cretaceous through Recent marine clastic deposits overlying a continent-ocean transition. The regional stratigraphy is summarized in Figure 4. In the deep-water delta, the two main stratigraphic units are the Akata and Agbada formations. These represent prodelta sediments deposited in a series of transgressive-regressive cycles (Doust and Omatsola, 1990). Turbidite sands, the main deep-water reservoirs, were deposited as slope-channel complexes and lowstand fans in ponded basins between the growing thrust structures (Damuth, 1994; Hooper et al.,

2002). These are interbedded with highstand hemipelagic and mass-transport shales that comprise the bulk of the prodelta strata. In the deep-water Niger Delta, the distinction between Akata and Agbada formations is blurred. The Akata shale is considered to be the main detachment horizon for the compressional toe structures (Cohen and McClay, 1996; Corredor et al., 2005). The overlying strata are grouped as the Agbada Formation and comprise the economically important deep-water mass-flow reservoirs. As such, the Agbada Formation in the deep-water province extends in age from the Oligocene to Holocene. The Akata Formation is then considered to be Late Cretaceous to Paleogene in age.

To aid the interpretation of the strata, several key regional seismic horizons (H) were interpreted across the delta. These have each been assigned an arbitrary horizon name (from the youngest mapped, the Pliocene H30, to the oldest, the Cretaceous H140) and have been dated using available well biostratigraphy. A summary scheme is also shown in Figure 4. Horizons above H120 have been intersected by current drilling and are dated from the Oligocene to Holocene (comprising the Agbada Formation). The deeper H120 to H140 horizons are assumed to be Paleogene or older in age and therefore belong to the Akata Formation.

Description of Thrust-belt Structures

The offshore Niger Delta can be divided into five main structural provinces based on structural styles (Figure 2).

1) The inboard extensional province is located beneath the continental shelf and is characterized by basinward dipping and counterregional normal growth faults.
2) The shale diapir belt, located beneath the upper continental slope, comprises passive, active, and reactive mud diapirs.
3) The inner fold and thrust belt consists of basinward verging imbricate thrusts attached to the front of the shale-diapir province.
4) The translational basin occurs beneath the lower continental slope and is populated by isolated broad detachment anticlines located possibly above ramps in the basal shale detachment.
5) The outer toe-thrust belt comprises both basinward and hinterland-verging thrust faults that combine to form complex imbricate systems. This forms the outer delta toe.

This structural division of the deep-water Niger Delta evolved through the Miocene as continued sedimentation on the delta top coupled with subsidence promoted the gravitational sliding of the sediment wedge toward the ocean. From the Oligocene to early Miocene, the deformation front was confined to the inner thrust belt. The outer toe-thrust belt began forming when the deformation front jumped outboard in the middle Miocene (around 10 Ma), and it has progressively advanced to its current location.

The basal shear inboard of the inner thrust belt is characterized by flow within a thick Cretaceous to Paleogene mobile shale sequence. The advance of the deformation front to the outer toe-thrust belt was initiated when some of the distributed shear within the mobile shale sequence began to transfer onto more discrete detachments in the Akata Formation shales beneath the abyssal fan instead of simply breaching to the surface. Broad regional anticlines developed where this slip transfer involved a climb in stratigraphic level, resulting in large shale-cored fault-bend folds above the ramp. The downslope transition from thick mobile shale flow to more discrete detachments has led to the complicated imbrication of the outer deep-water delta sediments.

Within the translational basins, the deformation is dominated by localized detachment folds and/or buckles above duplexes within the thin detachment layers. The boundary between the translational basin and the outer toe-thrust belt is loosely defined where the detached deformation once again breaks to the surface. The outer toe-thrust belt typically consists of an in-sequence set of 10 to 20 fault-fold structures. Although most of the structures are forward verging, local domains dominated by back thrusting and a frontal wedge exist. The structures are predominantly fault-bend folds, with frequent suggestions of initial break thrusting out of early low-relief buckles.

The sediments involved in the deformation of the outer toe-thrust belt are primarily unconsolidated to weakly consolidated mud, silt, and sand that behave in a predominantly ductile manner. Folds tend to develop very rounded forms, and the faults show only minor bending after ramping up from the detachment (Figure 5). Only minimal structural topography can develop because the surface sediment is unconsolidated. Downslope failure removes the crest of the growing structures. As a result, breaching thrusts and emergent hanging-wall fault-bend folds are quickly truncated at the sea floor.

The thrust faults are generally imaged as steep planar surfaces in seismic sections (Figure 5) occasionally marked by a prominent fault-plane reflection. These reflections generally have the reverse polarity to the seafloor reflection indicating a soft kick (i.e., the fault plane or zone represents a layer of decreased acoustic impedance contrast relative to the surrounding strata). This

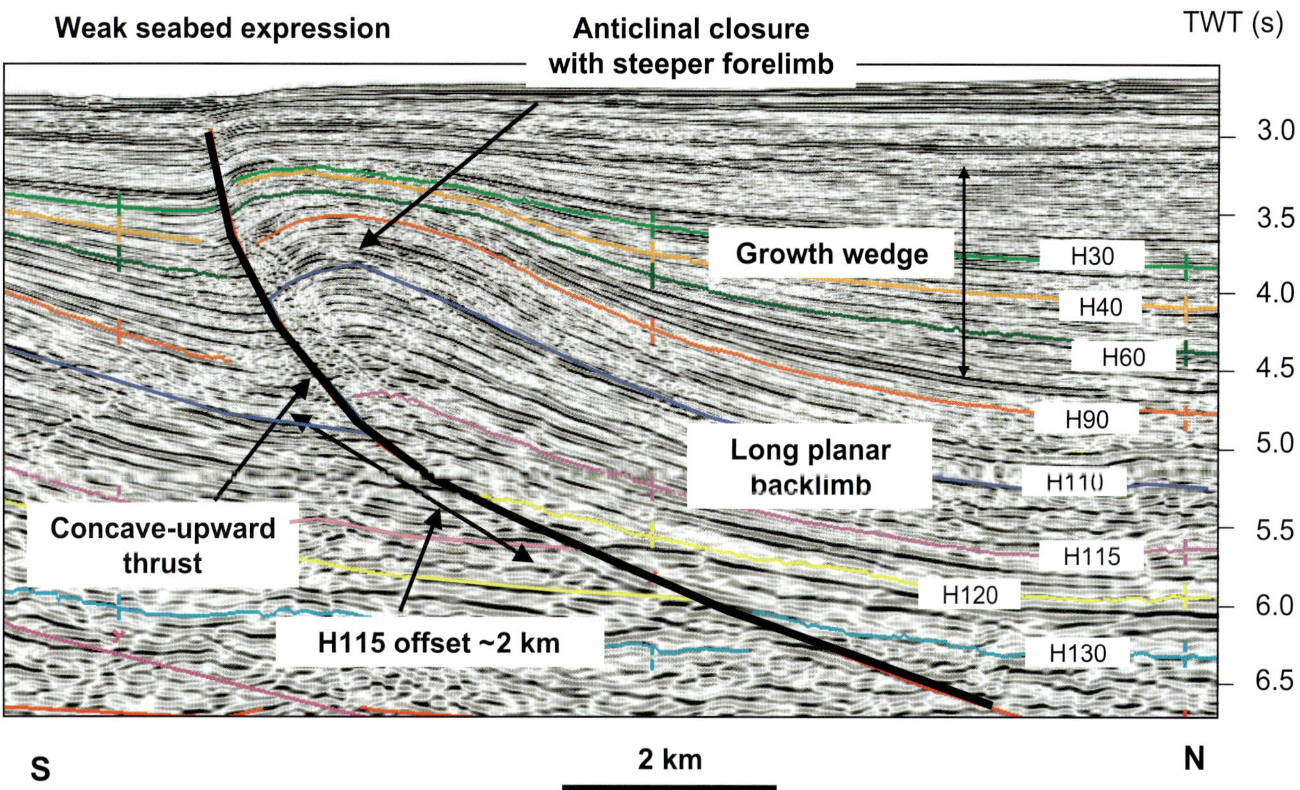

Figure 5. Seismic section showing the typical structural features associated with a deep-water toe thrust from the southern Niger Delta. The thrust is concave upward with a backlimb that dips as a long convex panel at a lower angle than the ramp. The forelimb is generally steeper dipping. The crest of the anticline is rounded and may or may not have seabed expression. Syntectonic strata form a growth wedge on the backlimb that either onlaps the structure or thins markedly across the anticline crest. The onset of thrust movement is dated from the first onlap onto the growing fold and postdates the H90 horizon (ca. 9 Ma). The thrust shown has an offset of around 2 km (1.2 mi), which typically forms an upper bound for displacements within the imbricate stack. The upward steepening of the thrust is consistent with growth during rapid sedimentation as demonstrated in sand box models by Storti and McClay (1995). Seismic data are courtesy of CGGVeritas. TWT = two-way time; H = horizon.

seismic response has also been noted from thrusts in accretionary prisms (Davis and Von Huene, 1987; Karig, 1990). The upward convexity of some thrusts toward leading fault tips suggests that they may originally have emerged onto the paleo–sea floor during their movement (Corredor et al., 2005). Syntectonic strata onlap the fold limbs where growth sedimentation rates are low relative to uplift rates. The fold crests may become emergent while the structure is active. As movement ceases, the thrusts are buried and only a weak expression of the underlying fold is seen at the seabed. This is the case with the thrust shown in Figure 5. Some toe thrusts remain buried throughout their movement history presumably because of lower uplift rates and/or early cessation of movement. In these structures, syntectonic strata thin instead of wedge across the resultant fold. The structures discussed in this chapter come predominantly from the outer toe-thrust belt.

A balanced and restored section from the southern deep-water Niger Delta is presented in Figures 6 and 7. This line is the same as that shown in Figure 3, but only the structures outboard of the diapir belt are shown. A palinspastic restoration (using GeoSec© software manufactured by Paradigm Geophysical) of the section to the H110 surface at ca. 13 Ma (the youngest prekinematic layer in the inner toe-thrust belt) suggests 19 km (12 mi) of total shortening, 12 km (7 mi) of which is partitioned in the outer toe-thrust belt (Figure 7). At the time of the onset of this shortening (around H90 ca. 9 Ma), the detachment surface within the Akata shale was buried to a depth of around 3.5–4 km (2.2–2.5 mi).

Timing of Fault Movement

The timing of latest thrust movement in the deep-water Niger Delta can be estimated using the age and pattern

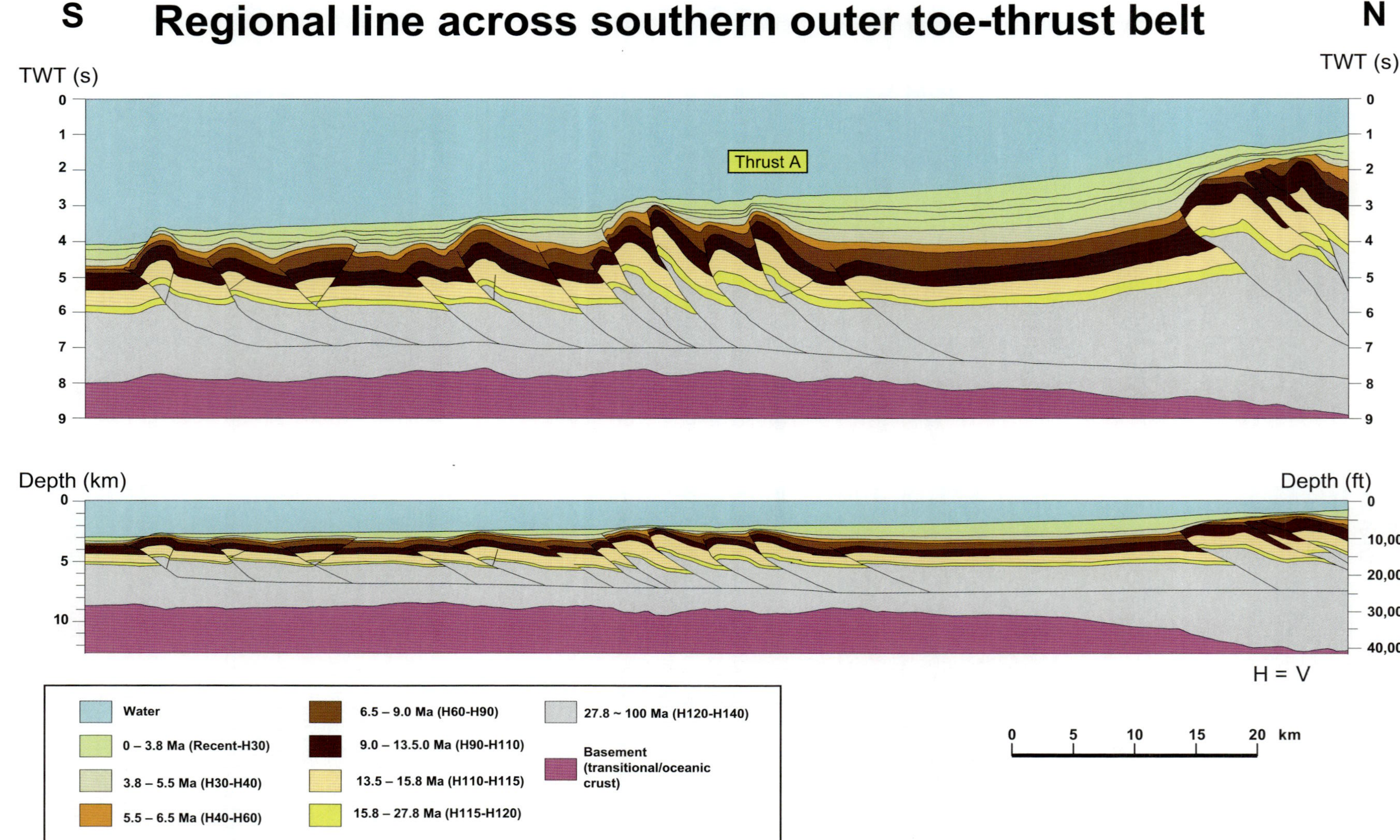

Figure 6. Time and depth cross sections of the outer part of the seismic line shown in Figure 3. The depth section has been adjusted to ensure structural balance, particularly the attitude of the basal decollement, which is depicted as a planar surface dipping gently toward the north. TWT = two-way time; H = V = horizontal = vertical scales; H = horizon.

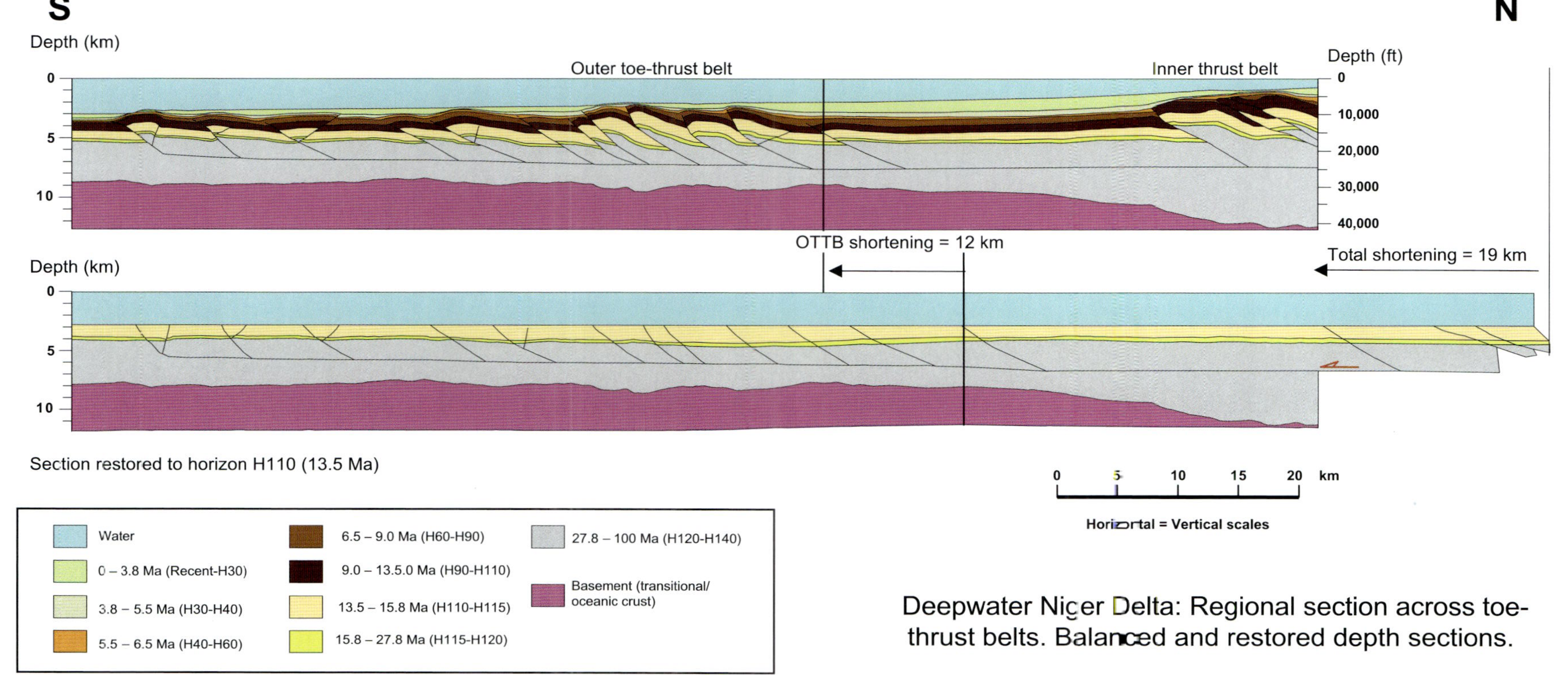

Figure 7. Restoration of the depth section shown in Figure 6. The section was restored using the Paradigm Geophysica GeoSec program. A combination of vertical simple shear, flexural slip, and area balancing was employed to retrodeform individual thrust horses. The section was then decompacted and isostatically adjusted assuming Airy isostasy. The resultant restoration suggests a minimum of 19 km (12 mi) of horizontal shortening is accommodated within the frontal and trailing imbricate systems. This does not consider possible distributed strain present as a precursor deformation field prior to fault slip. Individual thrusts ceased movement at different times. Within the plane of section, strictly identifying a fault propagation sequence is difficult. Out-of-sequence movements were common and can be seen in the variable elevations of individual anticlinal closures. OTTB = outer toe-thrust belt; H = horizon.

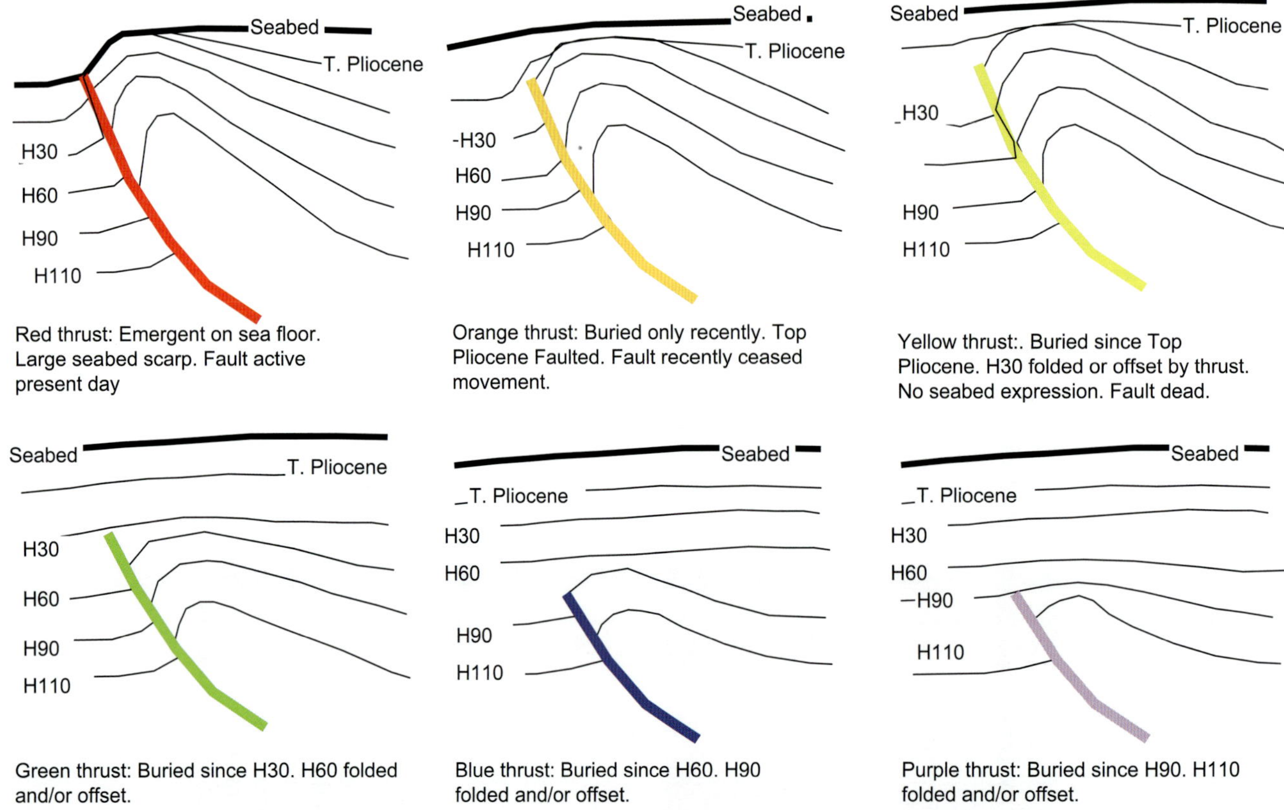

Figure 8. Schematic cross sections showing the classification scheme developed to describe the various stages of thrust activity and timing. The thrusts are classified and color coded based on the youngest stratigraphic surface deformed during movement. The color coding therefore reflects the relative timing of cessation of fault activity. See the text for discussion. T. = Top; H = horizon.

of growth strata preserved on the backlimbs of the structures. The concept is shown in Figure 8. We recognize that, to some extent, the timing can be subjective as strata can wedge against or drape across an extant structure in the absence of active growth. However, the principal that the geometry of strata can be employed in dating fault movement is considered a robust approximation provided the same parameters are applied consistently across the deep-water fold belts. To aid in visualization of the thrust activity pattern, a color scheme has been employed to identify the particular age of fault movement based on the series of regional horizons picked across the delta. The overall scheme is summarized in Figure 8.

A red thrust is considered to actively fold the sea floor or at least have had enough recent uplift to result in significant bathymetric change across the structure. Orange thrusts have only minor seabed expression and fold the Pliocene H30 surface. The oldest thrusts are in purple and only fold and/or offset the H90 seismic event. In effect, what is being color coded is the timing of last movement. When applied to the whole toe-thrust complex, this reveals the thrust complex death assemblage. Recent activity occurs on the red fault segments. Figure 9 shows two example seismic lines from the southern delta with the thrust faults color coded according to the schema of Figure 8. The color coding emphasizes that the patterns of displacement are complex with a common out-of-sequence movement. The overall results of this simple analysis are presented as maps of displacement activity for the western and southern delta lobes in Figures 10 and 11 using a fault pattern map at the lower Miocene H115 level for reference. Figure 10 shows the southern delta toe thrusts. Figure 11 shows the thrust activity pattern for the western thrust lobe. The patterns are instructive and show several perhaps surprising features.

1) The southern lobe is slightly older than the western lobe. The oldest fault displacement in the west generally postdates the H90 surface, whereas it predates the H90 surface in the south.

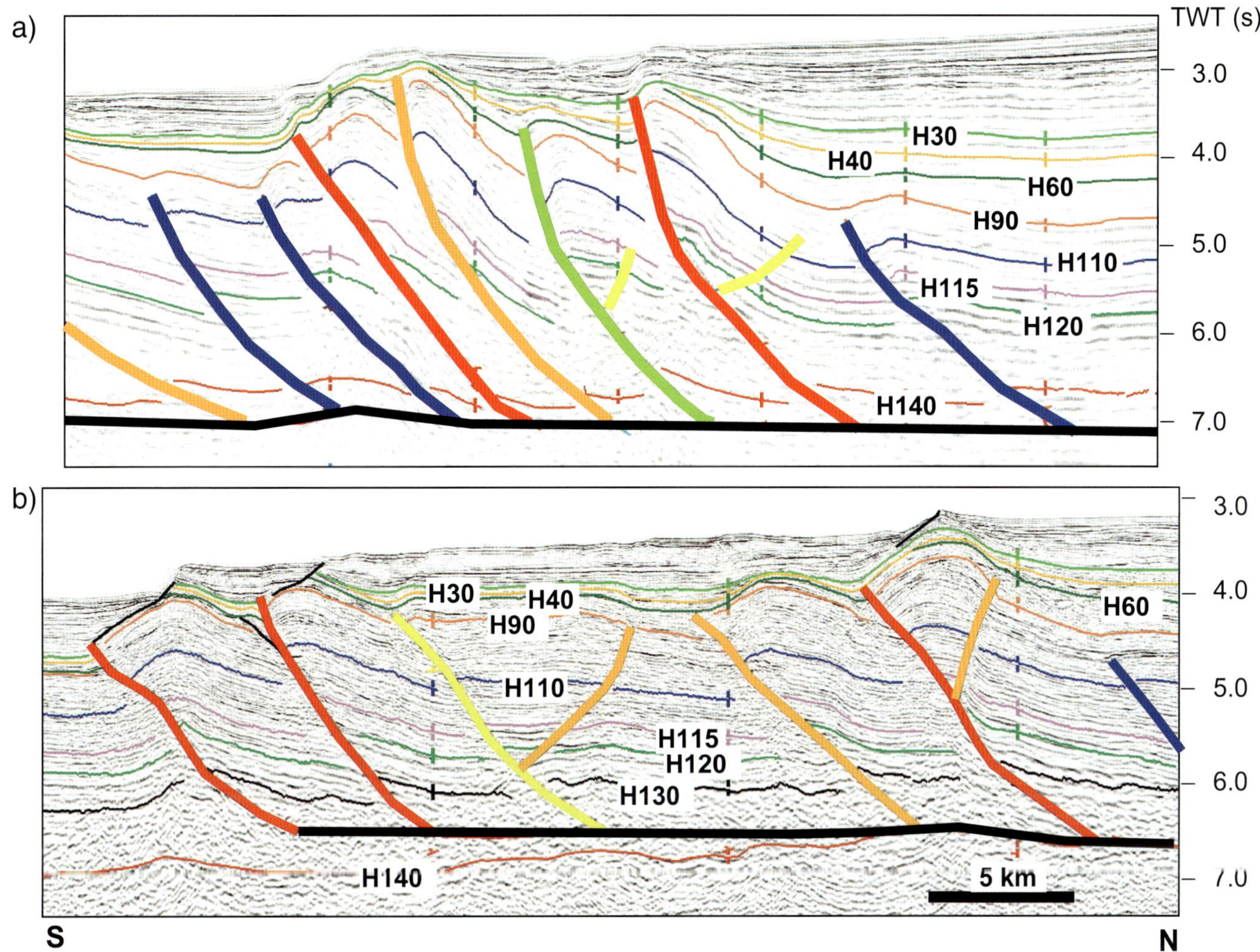

Figure 9. Example seismic lines showing buried closely spaced imbricate thrusts color coded according to the classification developed in Figure 8. (a) Larger scale view of thrust imbricates from the seismic section of Figure 3. The color coding of fault activity suggests that a complex sequence of movements has occurred. (b) Seismic line showing that the leading imbricate thrusts faults have had variable timings of activity. The frontal thrust strongly uplifts the sea floor associated with a slump scarp and is color coded red. Other trailing imbricate thrusts only weakly deform the seabed or cause mild folding of H30 horizon. These are color coded orange and yellow, respectively. Seismic data are courtesy of Mabon Geophysical and CGGVeritas. TWT = two-way traveltime; H = horizon.

2) The oldest parts of the thrust systems commonly occur at the lateral fault tips, and conversely, the youngest activity commonly occurs toward the fault centers where it forms distinct culminations. This pattern suggests faults propagated rapidly early in their history and then accumulated most of their displacement after lateral propagation had ceased. This growth history will be explored in more detail below.
3) Current activity (red thrusts) is not only concentrated as expected toward the frontal thrusts but is distributed throughout the imbricate thrust systems supporting the idea that more than one fault is moving simultaneously and that no simple sequence to the fault movements exists. In other words, although the thrusts may have propagated in a break-forward (or locally, break-backward) sequence, the ultimate displacement patterns do not show any systematic pattern that clearly reflects a systematic propagation sequence. For example, although an older thrust may have a younger moving thrust in its hanging wall at one location (suggesting break-back propagation), along strike, the pattern may be reversed (suggesting break-forward propagation).
4) A minority of thrusts display successive younger displacement activity toward one fault tip, indicating lateral propagation with time.

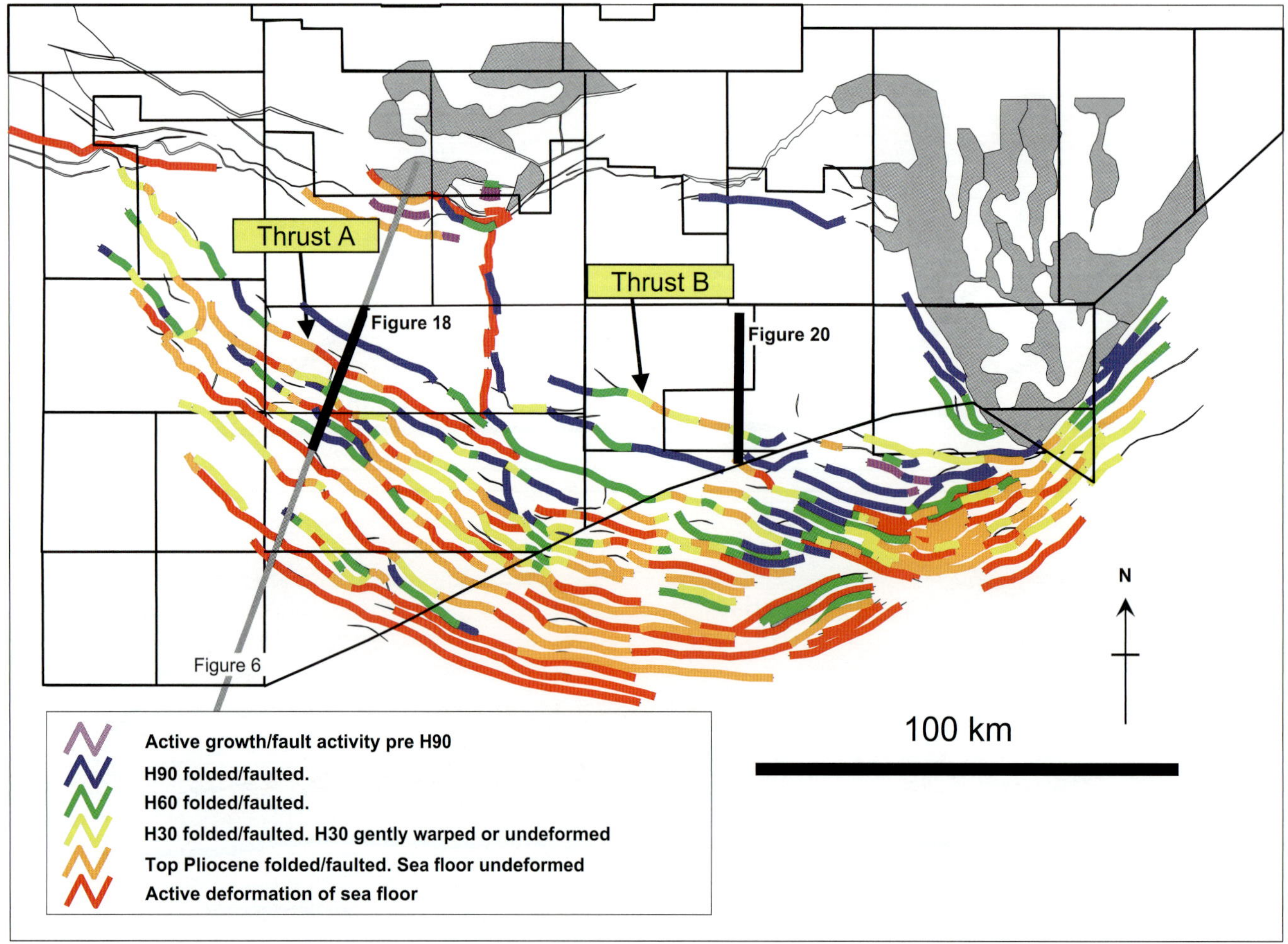

Figure 10. Color-coded thrust map from the southern Niger Delta recording the relative timing of the cessation of fault activity. Thrust-fault traces are based on the offset of the H115 seismic marker horizon. The color scheme used is that defined in Figure 8. Locations of structures and seismic lines described in the text are shown. Younger thrusts are found preferentially toward the front of the thrust belt, suggesting a piggyback propagation sequence. However, many of the imbricate thrusts show out-of-sequence movements indicative of a complex overall displacement history. H = horizon.

Detailed Growth Analysis of Thrust Structures

To shed further light on the thrust growth history, three structures have been analyzed in more detail to illustrate the pattern of displacement on the faults. This has been done by interpreting a closely spaced grid of 2-D seismic lines (dip lines) across the structures and mapping the fault displacement patterns using a series of seismic marker horizons. Thrust faults A and B come from the southern delta toe-thrust complex, whereas thrust C is located in the western delta complex. The locations of these structures are highlighted in Figures 10 and 11.

Thrust A is a large, forward-verging thrust with a prominent hanging-wall fault-bend fold located within the imbricate belt from the southern lobe of the Niger Delta (Figures 6, 12). The dominant thrust within the fold is now blind, with a fault tip imaged approximately a half second below the sea floor. A fairly prominent antithetic thrust is present in the hanging wall that persists over much of its strike length. A subtle thickening of strata within the core of the feature suggests an early phase of low-relief detachment folding before the thrust broke through. This fault does not link directly with any other toe thrusts, but a clear overlap zone at the eastern end with an en echelon thrust exists (see right end of Figure 12c). The extreme eastern tip zone of thrust A becomes difficult to image in the footwall of the other structure and could not be precisely defined. The central part of the fold appears to have grown in isolation of the neighboring structures.

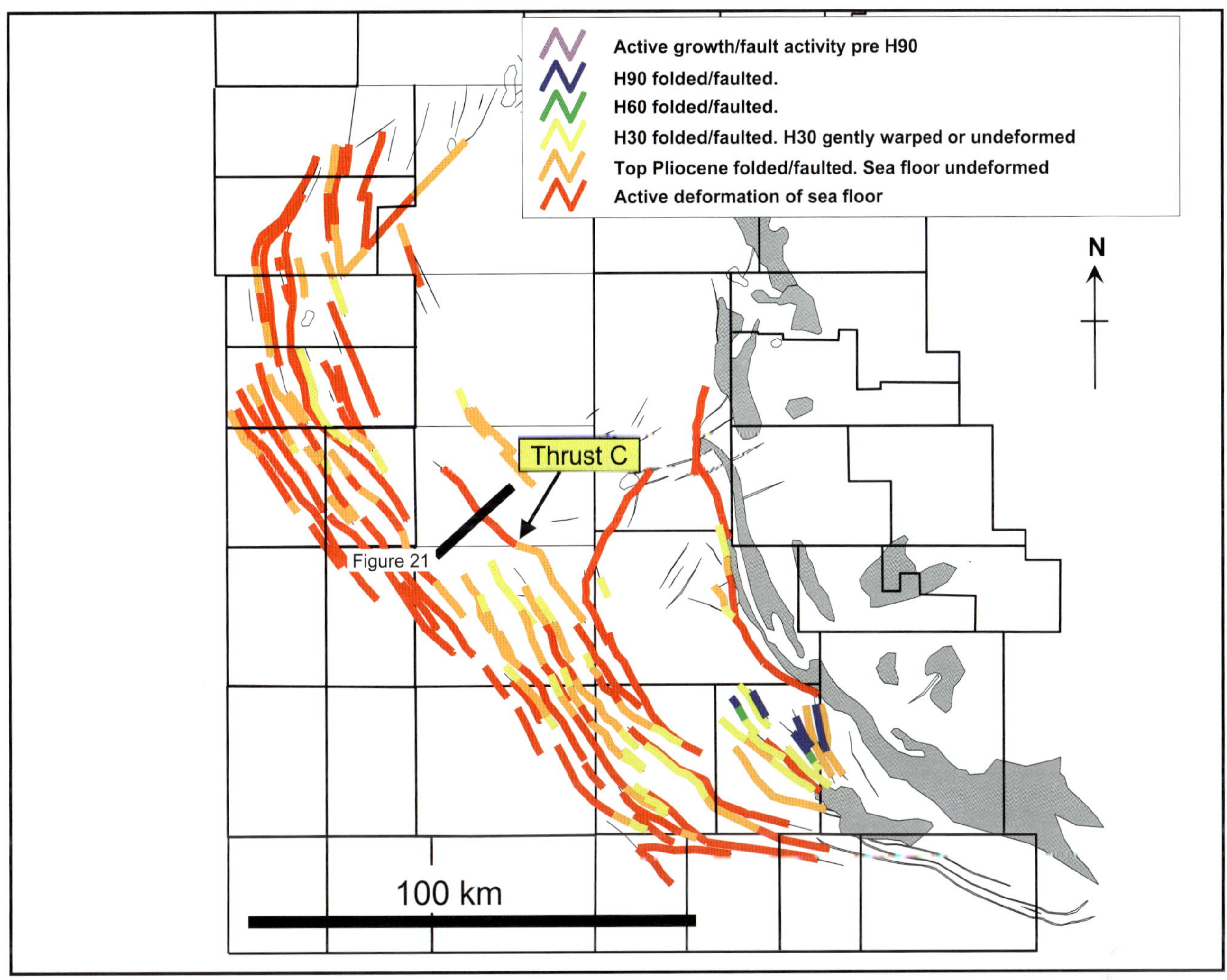

Figure 11. Color-coded thrust map from the western Niger Delta documenting the timing of fault activity cessation. See the text for discussion. H = horizon.

The timing of onset for thrust A is well constrained from the growth strata wedging seen in its hanging wall. A clear isopachous pregrowth unit up to the H90 surface is observed, overlain by a series of beds with clear growth on the backlimb (H90–H30). The timing of onset appears nearly synchronous across much of the central part of the structure, but a slight but discernable younging of age of onset can be seen along the flanks (Figure 12d).

The pattern of peak activity for thrust A (black thick bars in Figure 12d) shows that the maximum displacement activity started shortly after initiation. This period of rapid growth persisted for a distinctly longer period of time in the central part of the structure (until H30). The later stages of motion are more subdued. After the period of moderately rapid lateral propagation, activity on thrust A changed and retreated back toward the center. The onset of abandonment of the tips seems to have occurred while the central part of the structure was still undergoing rapid growth and uplift. The area of active deformation steadily shrank back toward the crestal region. Roughly 20% of the fault's length appears to have been recently active. This central active part of the fold is associated with a distinct step in the topography of the sea floor.

The second thrust structure analyzed is referred to as thrust B. This is a large, forward-verging thrust with a prominent fault-bend fold (Figure 13) and is the trailing imbricate structure in the central section of the southern outer toe-thrust belt (Figure 10). The dominant thrust within the fold is blind, with a fault tip imaged approximately a half second two-way time below the sea floor. Some thickening of strata within the core of the feature suggests an early phase of low-relief detachment

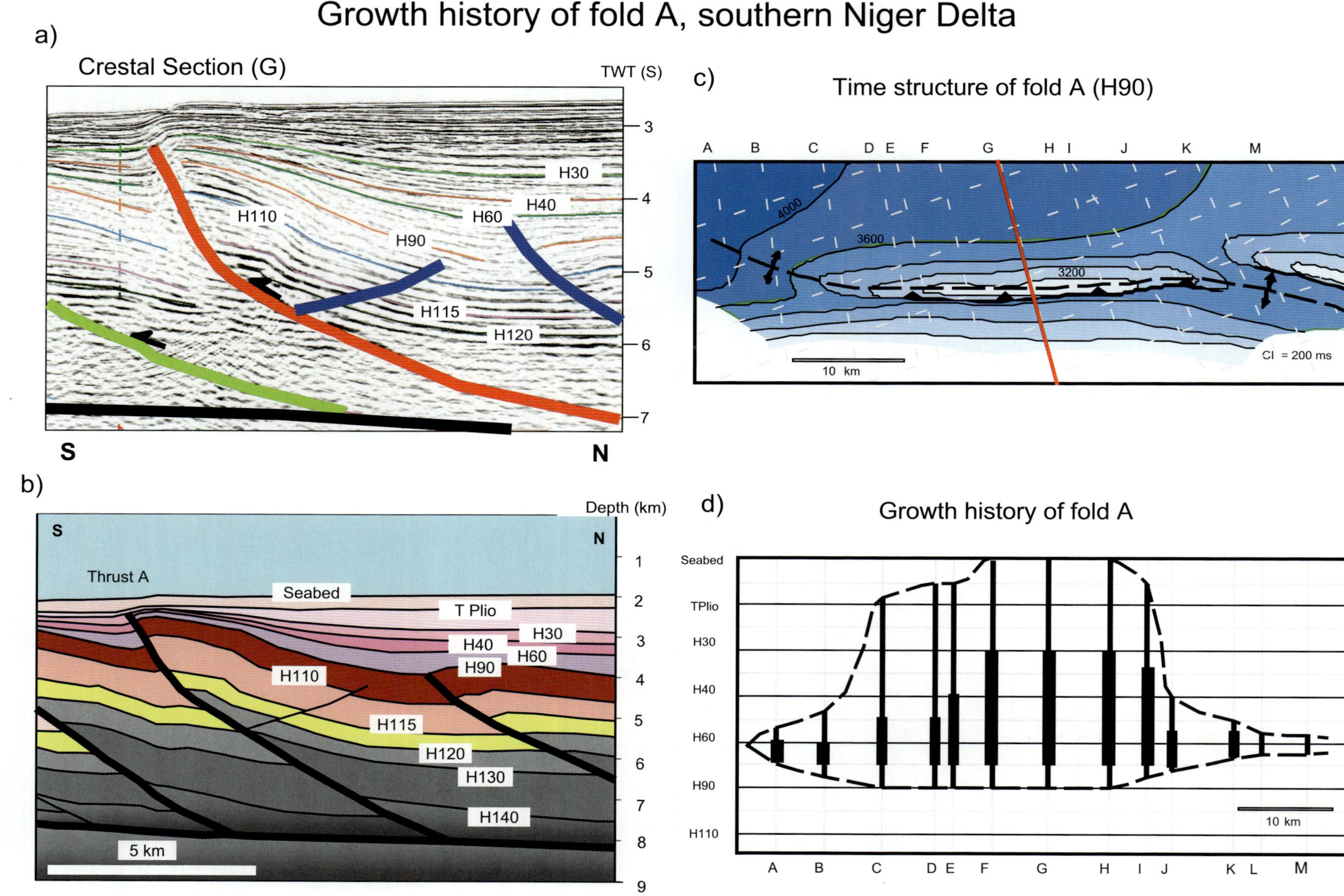

Figure 12. Growth history analysis for fold A, western Niger Delta. (a) Two-way time (TWT) dip section near the crest of the fold. (b) Depth-converted cross section based on the time section in panel a. (c) Time structure map showing the fold geometry at the H90 horizon. (d) Growth history plot showing the relative timing of deformation along strike. The lower and upper lines represent the timing of the first and last motion, respectively, based on the analysis of growth stratigraphy. The thick bars represent times of peak activity. Seismic data are courtesy of Mabon Geophysical. CI = contour interval in milliseconds; T Plio, TPlio = Top Pliocene; H = horizon.

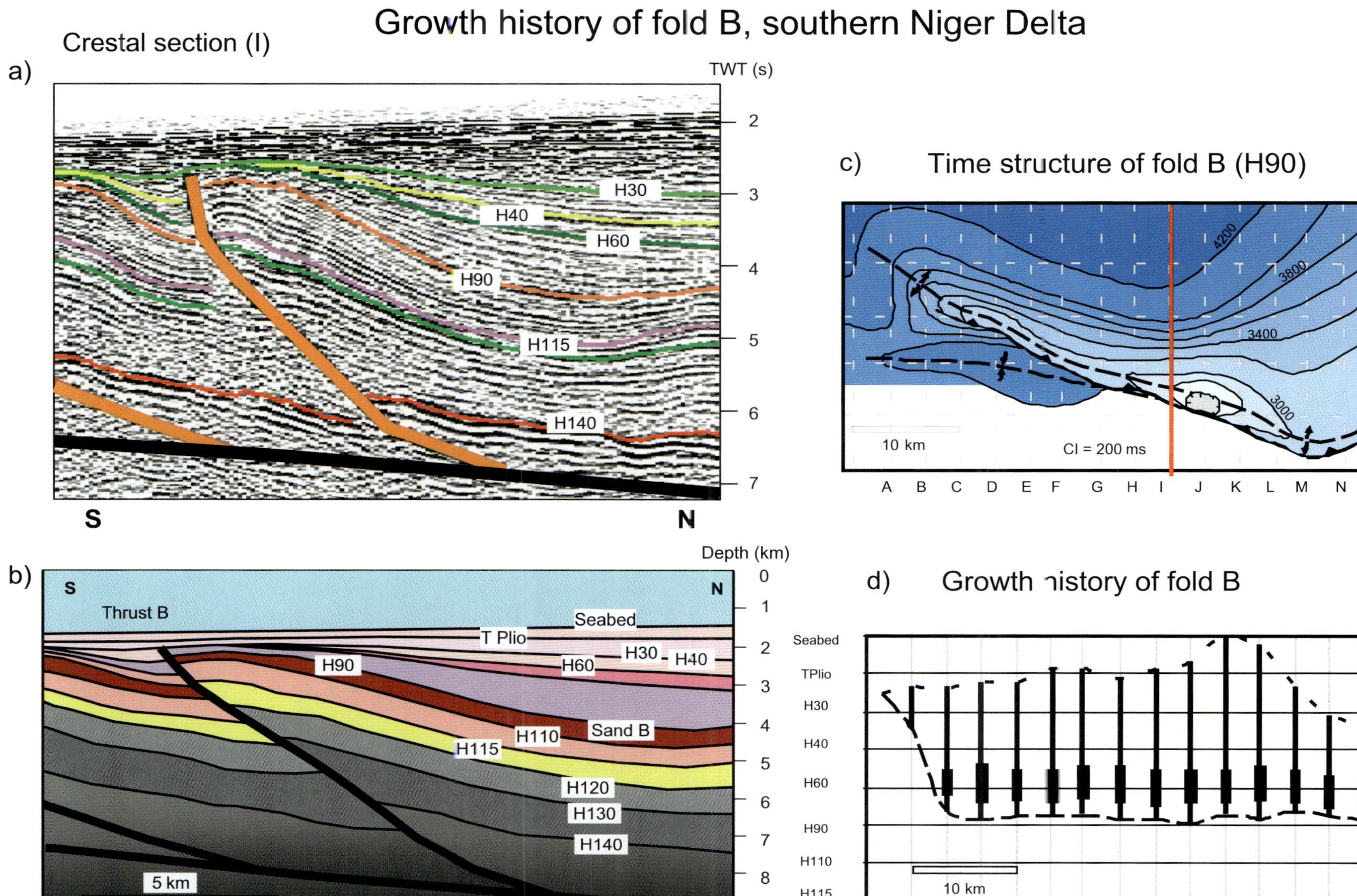

Figure 13. Growth history analysis for fold B, southern Niger Delta. (a) Two-way time (TWT) dip cross section near the crest of the fold. (b) Depth-converted section based on panel a. (c) Time structure map showing the fold geometry at the H90 horizon. (d) Growth history plot showing the relative timing of deformation along strike. The lower and upper lines represent the timing of the first and last motion, respectively, based on the analysis of growth stratigraphy. The thick bars represent times of peak activity. For this figure and Figures 12 and 14, all ages are relative based on a regular series of closely spaced picks in the back syncline of each fold correlated to the regional stratigraphy shown in Figure 4. Some interpretational errors will exist because of a lack of seismic to well ties and unavoidable jump correlation across structures using a coarse regional 2-D grid. Seismic data are courtesy of Mabon Geophysical. CI = contour interval in milliseconds; T Plio, TPlio = Top Pliocene; H = horizon.

folding before the thrust broke through. Although the fault does not link directly with any other toe thrusts, interference with an adjacent feature at one end is observed (to the right in Figure 13c) based on the deflection of the fault trace and the regional isochrones for the H115 surface. The central part of the fold and the western end (to the left in Figure 13c) appears to have grown in isolation.

The timing of onset of movement of thrust B is well constrained (Figure 13a, b). A clear isopachous pregrowth unit (up to H90 regional surface) overlain by a series of beds with clear growth on the backlimb is seen (from H90–H60). By comparing the crestal section with the more flanking sections (Figure 13d), it can be seen that the age of initiation is essentially synchronous over much of the fold. Although not shown, the equivalent sections at the extremities of the fold show a distinct younging of the age of onset near the tip region (Figure 13d).

The period of maximum activity on thrust B occurred very early in its history based on the amount of fanning in the growth strata. The bulk of the rapid growth appears to be limited to the period between H90 and H60 surfaces. More subdued growth in the younger sediments is observed, suggesting waning activity.

In contrast to the consistent age of onset, the age of last motion on thrust B is quite variable. Where the feature began to interfere with another structure (to the right in Figure 13d), it died early, just before H30, suggesting that the adjacent structure then took up the regional strain. At the free end, however, the fold continued to propagate slowly until just after the start of the Pliocene (H30). At this point, a dramatic shift in the growth history is observed. Subsequent activity records a retreat of the ongoing deformation to the central parts of the fault. As can be seen from the crestal section (Figure 13a), the very center of the fold continued to be active almost to the present day.

Thrust C (Figure 14a) is also associated with a large, forward-verging thrust. It is located within the western delta (Figure 11). The fold geometry is different from thrust A or B. The dominant thrust within the fold is blind and shows two distinct phases of growth. The earliest fault plane can be seen as an abandoned trace in the footwall of the younger trace. Similar to thrust B, fairly prominent antithetic thrusts in the hanging wall persist over most of its length. In fact, the antithetic faults appear to have propagated farther than the master thrust and are the only structures seen in the extreme parts of the fold (Figure 14b). Once again, subtle thickening of strata within the core of the feature suggests an early phase of low-relief detachment folding before the thrust developed. This fault does not link directly with any other toe thrusts but is truncated at one end by a tear fault (right end of Figure 14c). For most of its length, however, it appears to have grown in relative isolation.

The timing of onset for thrust C is again well constrained (Figure 14a). A clear isopachous prekinematic unit until just after the H90 surface exists, overlain by a series of beds with slight growth on the backlimb from H90 to H60. Much like thrusts A and B, the timing of onset appears nearly synchronous across much of the central part of the structure and was succeeded by a slower strike propagation phase, which carried it to its ultimate maximum length. The duration of this slow growth period persisted for a much longer time than in either of the other folds.

The pattern of peak activity for thrust C is similar to that seen in the other folds. In this structure, however, a distinct gap between the onset of deformation and the onset of peak motion is observed. Also, an extremely long period of peak activity in the central part of the fold persists to the present day. After an extended period of lateral propagation, activity on thrust C greatly retreated back toward the center with abandonment of the lateral tips (Figure 14d). Again, like thrust A, the onset of abandonment of the tips seemed to occur while the central part of the structure was still undergoing rapid growth. The area of active deformation has gradually collapsed back toward the crestal region, where roughly a third of the fault's strike length appears to still be active. Of the three folds, this is the only one where the crestal part still appears to be in a phase of rapid growth. As a result, the hanging wall shows dramatic sea-floor topography expressed as an emergent ridge. The height of this growing sea-floor ridge is sufficient to have triggered at least three episodes of mass failure, the most recent of which is still visible as a prominent headwall scarp on the sea floor.

Common Elements of Thrust Growth History

The three example thrust faults display a common tripartite structural growth history. The structures appear to have developed most of their strike length very early. In fact, this early phase of growth is essentially spontaneous within the uncertainties of timing inherent in the analysis. This early phase of rapid strike propagation in the upper Miocene is followed by an extensive phase of slower but steady growth in trace length. In all three cases, following the slower propagation phase, an abrupt transition from steady growth to a rapid decline in activity is observed. The activity on each of the structures died off progressively from the lateral tips toward the fault center, with the culminations being the last part of the structures to stop moving. Extensive regional mapping of the age of cessation

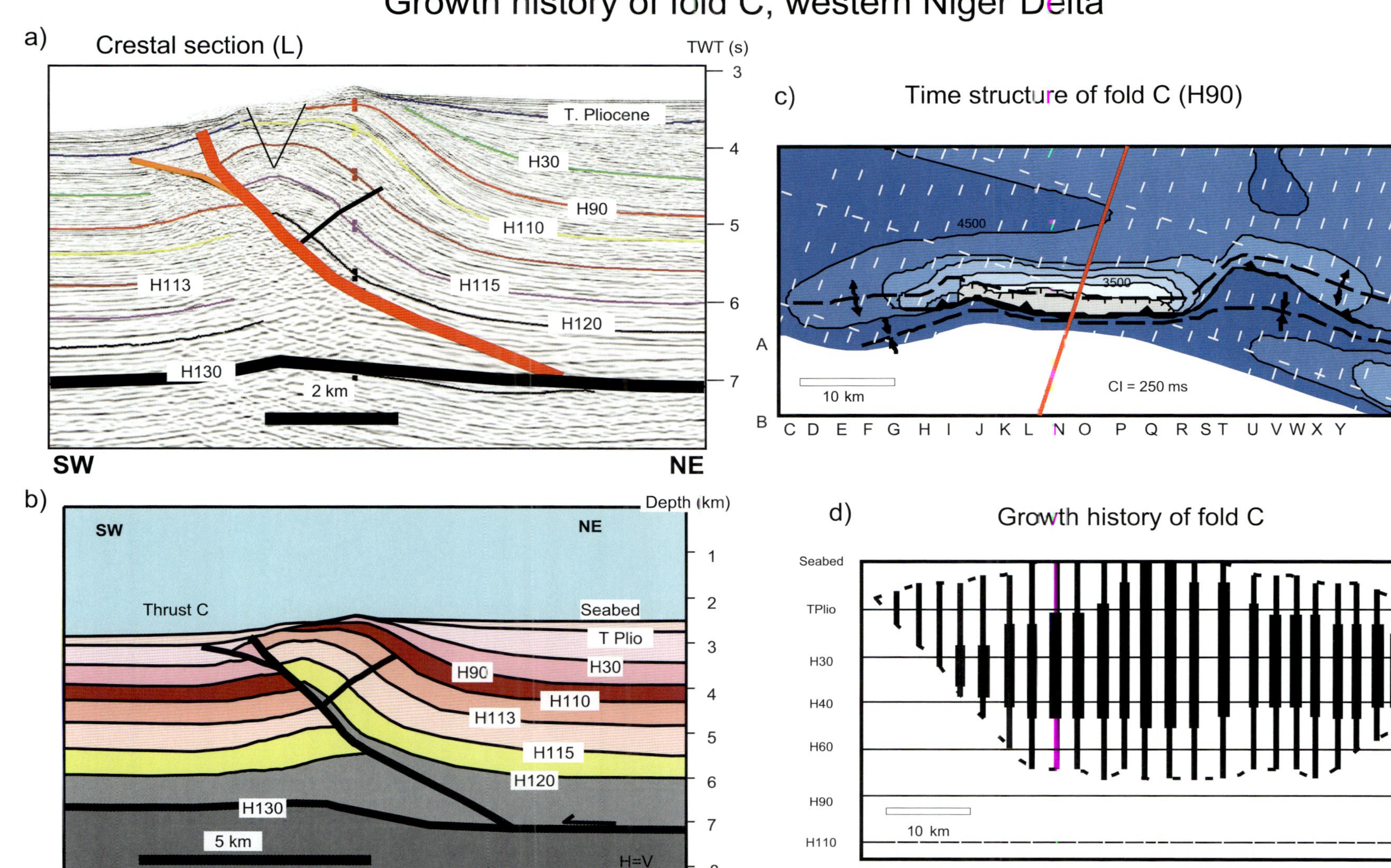

Figure 14. Growth history analysis for fold C, southern Niger Delta. (a) Two-way time (TWT) dip cross section near the crest of the fold. (b) Depth section. (c) Time structure map showing the fold geometry at the H90 horizon. (d) Growth history plot showing the relative timing of deformation along strike. The lower and upper lines represent the timing of the first and last motion, respectively, based on the analysis of growth stratigraphy. The thick bars represent times of peak activity. Seismic data are courtesy of CGGVeritas. CI = contour interval in milliseconds; T Plio, TPlio = Top Pliocene; H = horizon.

of structural growth across the entire fault network (Figures 10, 11) suggests that this late retreat toward the structural culmination is a general rule for the waning stages of toe thrusts in the deep-water Niger Delta. Although the magnitudes and duration of the individual phases vary somewhat, the fact that the same three growth stages are present (early rapid lateral propagation, slower steady growth, and final retreat toward the crest) suggests that this pattern is being driven by some underlying physical mechanisms. Some of these are discussed below.

Fluid Pressures in the Deep-water Delta

Evidence for the presence of very high fluid pressures within the sediment of the Niger Delta is expressed in several physical features. The large shale diapir belt is clearly a testament to large volumes of overpressured shale bulldozed forward by the gravitational spreading of the delta. The mobile shale has low yield strength and creeps forward and upward as large volumetric massifs driven by large lateral shear stresses created by the gravitational sliding of the delta top clastic wedge (a sort of squeeze flow). Once dewatered, mobile shale sets hard and can no longer redeform (Morley and Guerin, 1996). For continual growth, mobile shale must be fed into the base of diapir to cause continued inflation and rise of the overlying shale mass. Clearly, this needs the long-term maintenance of high fluid pressure at depth to support continued growth of the active diapirs if they are to reach the sea floor.

Mud volcanoes and sea-floor craters are also common features in the deep-water delta. Their presence reflects the escape of overpressured fluids upward toward, and commonly onto, the sea floor above growing shale diapirs and detachment folds. The presence of seabed pockmarks and craters are also prevalent features in the deep water of the Niger Delta and are commonly associated with the crestal parts of thrust structures. Their presence points to the release of overpressured fluid during structural growth episodes. This can occur only when the local fluid pressure exceeds the overburden and intrudes upward. Locally, this may be associated with slump episodes that result in unloading of overpressured zones within the fold cores.

The main driving force for generating excess fluid pressures in Tertiary deltas is disequilibrium compaction (Osborne and Swarbrick, 1997). The presence of mostly fine-grained, low-permeability strata coupled with rapid sedimentation rates combines to enable fluid overpressure development. The inability of low-permeability strata to expel their pore fluid upward transfers the vertical sediment load to the fluid within the shale pore space, preventing further compaction. This typically occurs below a certain depth in the basin, the depth at which permeability is low enough to prevent vertical fluid escape occurring at a rate that allows maintenance of hydrostatic pressures. In this chapter, the depth is referred to as the top shale overpressure (TSOP).

The TSOP depth can be evaluated from well data using shale sonic transit times (Stump et al., 1998). Figure 15a shows sonic-derived porosity for clay-rich rocks from a well in the deep-water Niger Delta. The shale porosity declines until a depth of around 2.3 km (1.4 mi) below the mud line is reached, when it can be seen to increase. This inversion points to the presence of significant fluid retention and a cessation of compaction. The pore pressure in the shale can be estimated using the concept borrowed from soil mechanics that in normally pressured sediment the porosity ϕ is related to the depth of burial by the following algorithm:

$$\phi = \phi_o \times \mathrm{EXP}(-\beta \times \sigma_{\mathrm{eff}})$$

where ϕ_o is the surface porosity (no effective stress), β is the matrix compressibility, and σ_{eff} is the vertical effective stress. A log-linear plot of porosity against effective stress for the normally pressured interval in the deep-water well is shown in Figure 15b. The linear compaction trend displayed on the plot suggests an initial (reference) porosity of 0.45 and a matrix compressibility of -0.0487. The calculated shale pore pressure vs. depth plot is shown in Figure 15c. The TSOP occurs around 2300 m (7546 ft) below mud line. Below the TSOP, the calculated shale pore pressure increases along a line subparallel to the overburden gradient. Below the TSOP, the vertical effective stress within the shale is constant and no more compaction occurs.

The measured reservoir pressures in deep-water sands from the well consistently fall above the best-fit shale pore-pressure curve. This is illustrated in Figure 15d where the pressures are plotted with reference to the estimated shale pore-pressure curve. The excess pressure within the aquifer units increases with depth. Points on this graph that fall on lines of decreasing excess pressure record the hydrostatic gradient within each of the thicker aquifer units. When pore pressures in sands exceed the ambient shale pressure at the crest of a structure, it is a strong indication that overpressure is being transferred in the permeable sands from off structure, in this case, upward along the hinterland-dipping backlimb of the fold. This process is commonly referred to as the centroid effect and, as will be discussed, may be an important process during fault-bend fold growth history.

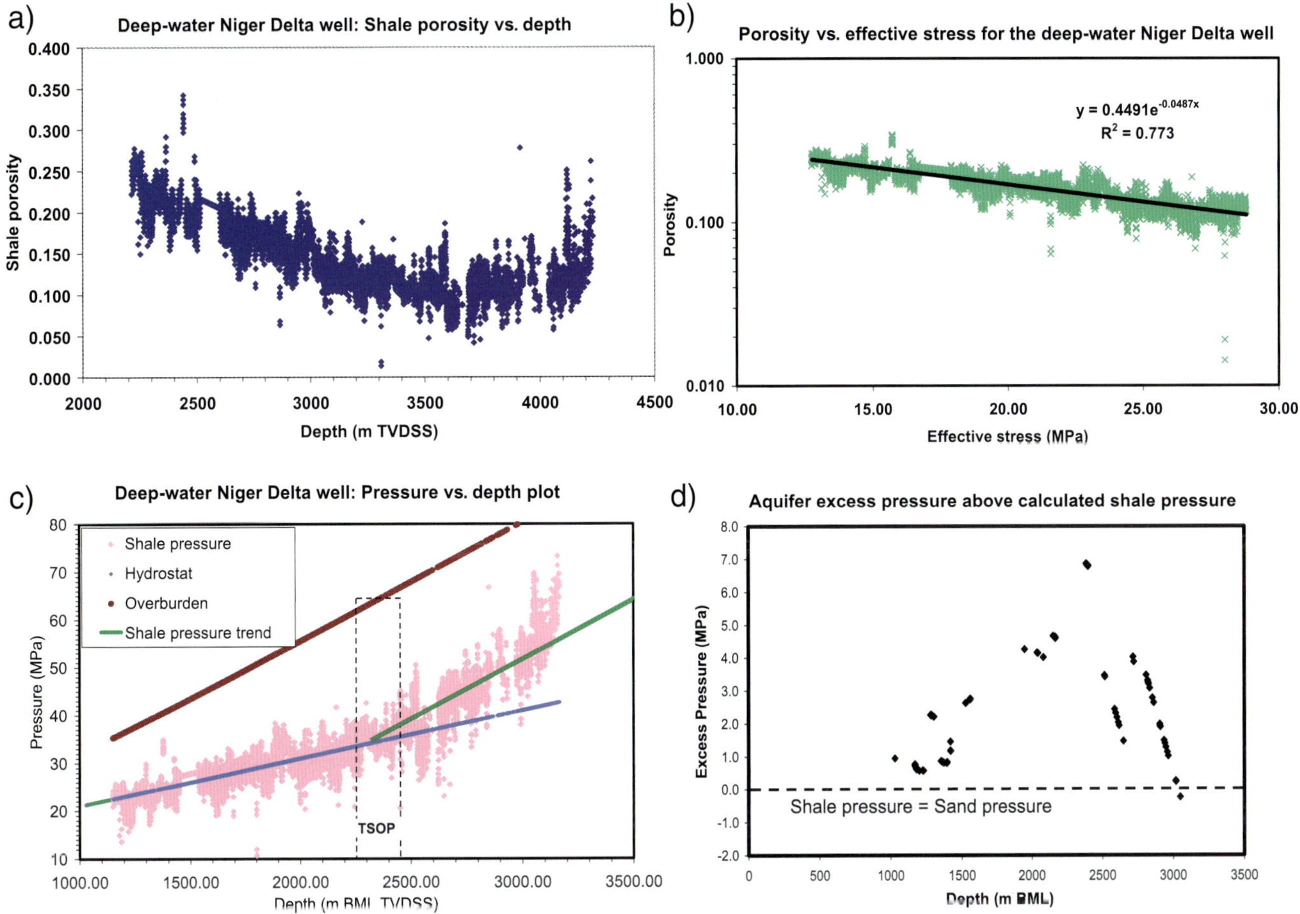

Figure 15. Shale porosity and calculated shale pressure plots from a deep-water well in the Niger Delta. (a) Shale porosity vs. depth below mud line. The inflection in porosity at around 2300-m (7546-ft) burial records the onset of overpressure and a cessation of compaction. (b) Log-linear plot of shale porosity vs. vertical effective stress in the hydrostatic zone for the well. The lithostatic stress was calculated by integrating bulk-density log data. Hydrostatic pressure was calculated from a gradient of 0.0105 MPa/m (1.5 psi/m) for seawater. The solid line represents the regression fit to the data. (c) Calculated shale pressure profile for the well. The top shale overpressure is interpreted at 2300 m (7546 ft) in the well. (d) Plot showing the difference between measured formation pressures in interbedded sands in the well and the interpreted shale pressure curve. The sands have higher pressures than the enclosing shale, reflecting a component of excess pressure interpreted to be the result of lateral pressure transfer. BML = below mud line; TVDSS = true vertical depth subsea level; TSOP = top shale overpressure.

DISCUSSION

Role of Pore Pressure

The pore-pressure evolution within a growing deep-water thrust sheet is a function of many complex processes. These processes reflect the interplay of sedimentation rate, compaction, fault displacement rate, and the mechanical behavior of the strata. Some of these processes are discussed below.

The observed compaction disequilibrium below depths of approximately 2–2.5 km (1.2–1.5 mi) in the outer Niger Delta undoubtedly is a factor in influencing the thrust belt behavior and the pore-pressure pattern within the thrust sheet. Fault movement will in theory be facilitated by the increases in pore pressure that develop as a result of rapid sediment loading and sediment deformation. In this instance, the increased pore pressure will reduce the frictional resistance to fault movement, allowing slip events to occur at lower shear stresses. Lateral pressure transfer should in theory contribute to this behavior and will be considered in more detail later.

A recent analysis by Bilotti and Shaw (2005) described and modeled the Niger Delta as a critically tapered sediment wedge. The theory of critical taper wedge mechanics states that once a wedge reaches a critical taper (a geometry defined by the surface slope and the opposing dip of the basal detachment), it grows self-similarly as material is added to (by sediment deposition or frontal thrust propagation) or removed (by erosion) from the wedge (Dahlen et al., 1984). This process can occur because the whole sediment wedge is at or near a state of macroscopic failure. It responds to externally controlled changes in its geometry by bulk deformation both at the front of the wedge (the toe) and also internally to achieve a force balance constrained by the wedge mechanical properties and the strength of the basal detachment. High fluid pressures are a key factor in this respect and can weaken the wedge and so promote lower taper angles. The analysis of 10 regional sections from the deep-water Niger Delta by Bilotti and Shaw (2005) indicated an average taper of 2.5°, comprising an average surface slope of 1° and an average basal detachment dip of 1.5°. They argue that the Niger Delta wedge is at critical taper because it has grown wider by propagation into the abyssal plain deposits of the Atlantic Ocean. A noncritical wedge would grow by internal deformation first, building taper before it propagates forward. Further analysis by Suppe (2007) suggested that the Niger Delta is floored by an extremely weak detachment, although internally, the wedge is moderately strong. The weak basal detachment is explained by invoking a very high fluid pressure within the mobile Akata shales that form the detachment horizon. The mechanical analysis of Bilotti and Shaw suggests a fluid-pressure ratio, λ, of 0.9. As they point out, this ratio is higher than can be explained by simple compactional disequilibrium alone and requires the contribution of other fluid over pressure mechanisms, such as shale dewatering or perhaps fluid expansion during hydrocarbon generation.

In the normal or poorly consolidated sediments prevalent in the delta toe, pore pressures may also increase during active deformation as a result of undrained shear. This is caused by the restructuring of the sediment matrix by microscopic grain slippage. In this process, the higher the porosity, the greater the propensity for pore collapse during deformation and the potentially higher pore-pressure response (Yassir, 1998). Zones of deforming sediment can contract (compact) during shear generating very high localized fluid pressures. These can either drain into the surrounding more permeable strata or may be pumped along permeable channels within fault zones, for example, toward lower pressure zones (Sibson, 1992). These may ultimately be responsible for the mud volcano activity associated with many growing structures.

The presence of underconsolidated zones along toe thrusts as evidenced by the fault plane reflections imaged in seismic sections points to a dilatational process being active within the Akata Formation. These soft zones presumably have higher porosities (lower densities) than the surrounding predominantly shale wall rocks and thus form negative acoustic impedance interfaces. This porosity is probably in the form of dilatant fractures. The deformation of originally high-porosity rocks by localized fault slip should in theory cause compactive shear. The fact that the fault planes (ramps) can be imaged as potentially high-porosity zones suggest that high fluid pressures are maintained within the faults. The corresponding low permeability in surrounding Akata shale may prevent fluid diffusion from resulting in a pore-pressure decrease and thus fracture closure.

Dilatancy hardening can result in a fault zone during brittle deformation if fluid diffusivity from the (impermeable) wall rocks into the newly created fracture voids is low (Rudniki, 1977). The associated pore-pressure decrease during fracture opening would result in an increased effective stress and an increased frictional resistance to continued slip. If this process were important in the toe-thrust belts, it would, by its design, be a dynamic process and is unlikely to leave a signature such as a low-impedance fault zone. The deformation processes within the fault zone and specifically the pore-pressure response, however, may modulate thrust behavior and may help explain the intermittent movement of thrusts and their resultant variable accumulated slip distribution.

Numerical modeling of thrust wedges emphasizes that fluid flow is focused into areas of active deformation (Strayer et al., 2001). Fluid movement along thrusts has been documented from accretionary prisms (Vrolijk, 1987). As a general statement, it is to be expected that thrust faults and detachment zones act as major conduits for fluid flow within the delta toe and may ultimately facilitate fluid escape (dewatering) from the deep overpressured zones within the Akata shale.

Fault movement in the outer toe-thrust belt of the Niger Delta can therefore be viewed as a response of the critically tapered clastic wedge to changes in its geometry related to continuous sedimentation. Deposition increases the taper of the sediment wedge and promotes internal deformation by gravitational downslope movement. The increased basal shear tractions transfer stress into the frontal toe-thrust belts, promoting fault slip. Sediment loading on the deep-water toe of slope may reduce slope taper and thereby stabilize the wedge. The burial of the thrust belt itself at the base of slope may also stabilize fault movement by increasing vertical load, increasing frictional work, and thus rendering the faults inactive. This has been shown to occur in 2-D kinematic models of thrust ramps and imbricate wedges (Goff and Wiltschko, 1992; Hardy et al., 1998). A complex set of deformation processes operate at the delta toe controlled ultimately by sedimentation, but strongly influenced by the pore-pressure response of the strata and the fault zones.

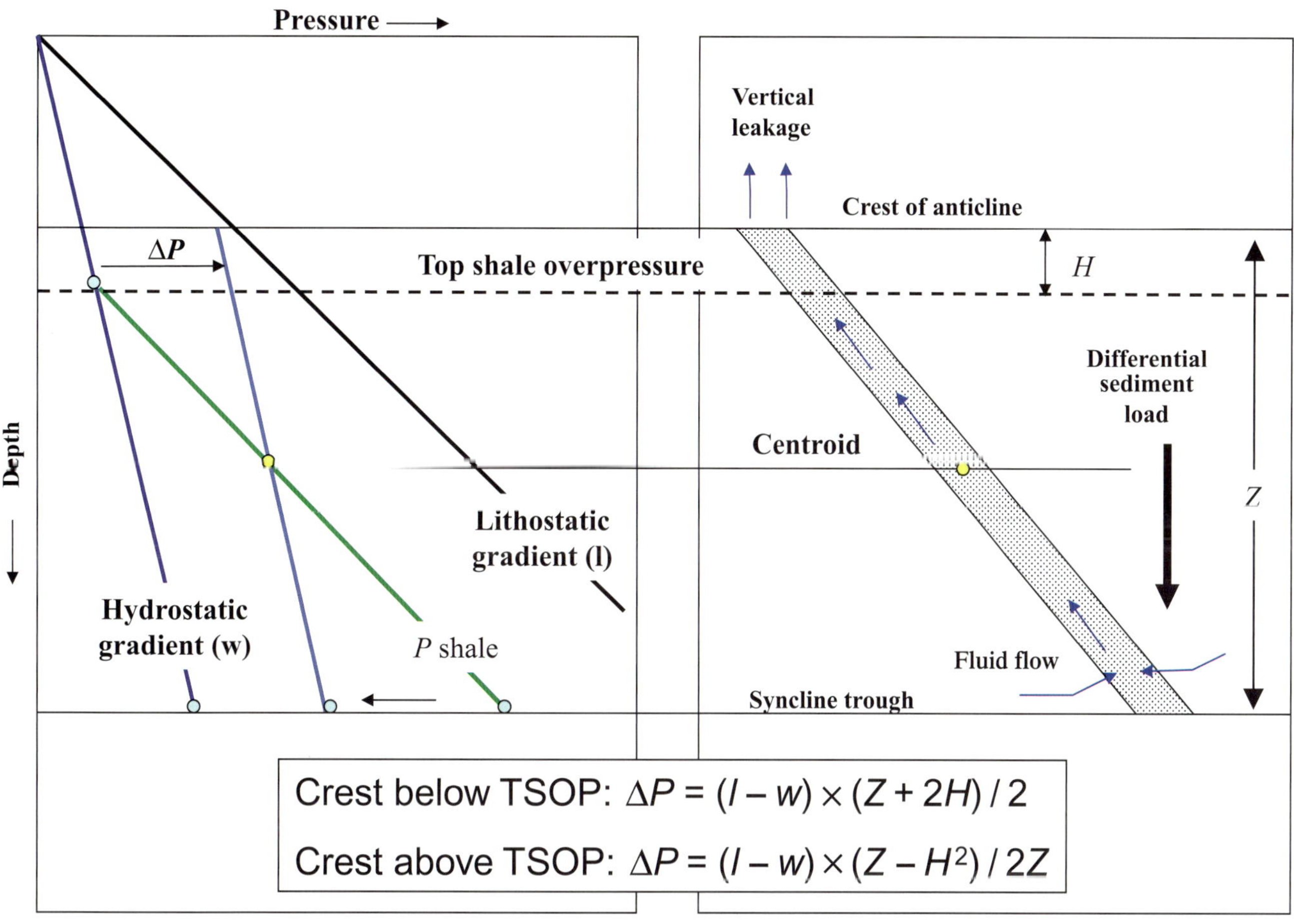

Figure 16. Simple model for pore-pressure transfer in a dipping permeable reservoir layer: the so-called 2-D centroid effect. Fluid flows along the dipping layer along a pressure gradient from a syncline trough to an anticline crest. At equilibrium, the permeable layer has a hydrostatic gradient. The centroid is the depth at which the shale and permeable layer have the same pore pressure. At pressure balance (assuming no escape of fluids from the crest of the structure), the magnitude of overpressure is a simple relation between the hydrologic column height, Z, and the depth from the crest to the top shale overpressure (TSOP), H. The two relationships, describing the cases where the anticline crest resides above and below the TSOP, are shown.

Model for Pore-pressure Evolution in a Ramping Thrust Sheet

As the above discussion emphasizes, pore-fluid pressure has an important and perhaps defining function in the evolution of the Niger Delta compressional toe structures. This chapter has primarily been concerned with the description of the deformation within these structures (particularly with regard to fault displacements and how these have evolved through time). The overall displacement patterns are complex, but a common pattern is developed in which later movement episodes concentrate toward the central parts of thrusts whereas the lateral tips cease movement earlier. Can this behavior possibly be linked to fluid pressures within the thrust belt complex? Or is the behavior controlled by the feedback between thrust-sheet geometry and erosional unloading?

To help understand and illustrate how pore pressure may vary within a growing thrust structure, a simple one-dimensional (1-D) model (Figures 16, 17) has been built using the centroid concept (Traugott and Bowers, 1996). As already mentioned, this concept models the transfer of fluid pressure along inclined aquifers that link deep basinal areas (in the case of a thrust system, the backlimb synclines) where overpressures are generated, with shallow structural culminations (fold crests) where pressures may be dissipated by fluid leakage through more permeable formations. The centroid is the point where the fluid pressures in the aquifer and enclosing overpressured shale are equal (Figure 16). The centroid point itself is not that important; it has lent its

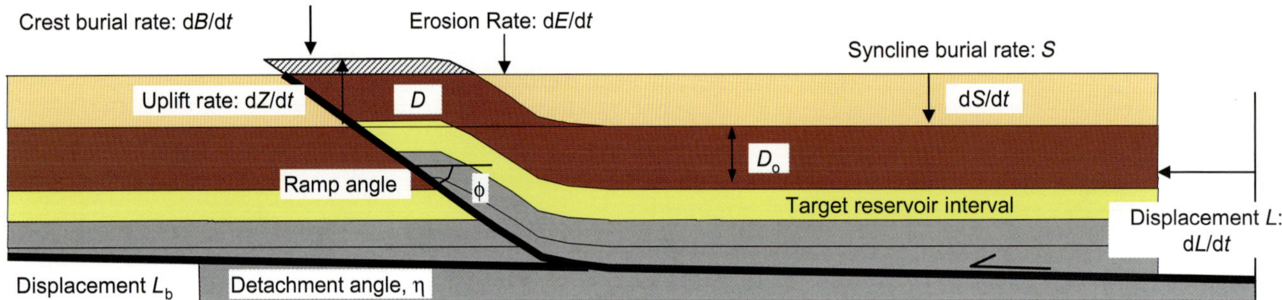

Displacement $L = \Sigma\, dL/dt \times dt$

Crest burial $B = \Sigma\, dB/dt \times dt$

Uplift $Z = L \times \sin(\phi) + L_b \times \sin(\eta)$, where ϕ is the ramp inclination, η = basal detachment inclination

L and L_b are the displacements on the ramp and on the basal detachment

Erosion: $E = a \times (Z - B)$, where $Z > B$; Coefficient a = erosion efficiency

Slump criterion: If $(Z - B)/Z > t$, then $E = a \times S \times (Z - B)$ where S = slump enhancement factor

D_o = Depth of burial of crest at start of deformation

D = current burial

Syncline burial = $Z + D - D'_o$ where D'_o is the compacted thickness of D_o

Figure 17. Formulation of the simple thrust growth model used to evaluate the progressive development of later pressure transfer as a thrust moves up an inclined ramp. The displacement rate is expressed as a step function and integrated over a time length when the fault is moving (slip episode) to generate total displacement. Multiple movement episodes are used. The burial rate at the crest of the anticline is also modeled as a simple step function. It is calibrated to the syncline, which is assumed to be a site of constant deposition. A reduction factor is then applied to the anticline crest to reflect the spatial decrease in sedimentation rate related to the uplift. Compaction and Airy isostasy are incorporated into the model. The compaction is implicitly modeled to switch off below the TSOP. If the uplift rate exceeds the burial rate, no deposition is modeled to occur at the crest of the structure. Instead, it is eroded. The amount eroded is expressed as a proportion of the difference in magnitude between uplift and burial (referred to as the erosion efficiency). An additional erosion factor (referred to as the slump enhancement factor) is switched on if a threshold is crossed in the magnitude of uplift. The slump enhancement factor used is simply a scalar quotient of the erosion efficiency and aims to mimic the rapid erosion that can be observed from the frontal parts of emergent folds on the sea floor. TSOP = top shale overpressure.

name to the concept of fluid-pressure transfer along inclined strata. Yardley and Swarbrick (2000) used Integrated Exploration Systems (IES) 2-D PetroMod™ to model the evolution of pore-pressure transfer along an inclined aquifer using the typical shale properties found in standard basin models. The results suggested that the centroid model is an oversimplification of a more complex process in which the redistribution of fluids along the inclined aquifer occurs throughout the burial and compaction process. The model predicts that no abrupt changes in pore pressure at sand shale interfaces exist. Instead, the pressures equilibrate by drainage from the low- but finite-permeability shale near the base of the model, with fluid escape into the enclosing shale at the structural crest. Several factors affect the efficiency of the mechanism, the most important being the aquifer relief and the burial rate (Yardley and Swarbrick, 2000).

The simple model outlined here uses the centroid model to balance overpressures between updip and downdip parts of an inclined aquifer. Although it is at best a rather oversimplistic approximation, it serves to illustrate the concept without the need to address more complex basin modeling and geomechanical software. In the formulation used here, the pressures in the permeable layer, when equilibrated, lie between the shale pore pressure at the crest and that in the adjacent syncline.

The magnitude of the resultant overpressure in the aquifer can be expressed as a function of three variables.

1) The depth to TSOP (S)
2) The relief of the structure (from syncline to anticline crest) referred to as the hydrologic height (Z)
3) The depth of burial to the crest of the structure below mud line (D)

From this simple geometry, two algorithms can be defined to predict the magnitude of crestal overpressure: one for the case where the crest of the structure lies above the TSOP and one when the crest lies below the TSOP (Figure 16). For a structure below the TSOP

$$\mathrm{DP} = (Z + 2H) \times (l - w)/2$$

where $H = D - S$, l is the overburden pressure gradient (related to bulk density), and w is the water gradient.

For a structure above the TSOP

$$\mathrm{DP} = (Z + H^2) \times (l - w)/2$$

The TSOP can be evaluated from shale porosity trends (as in Figure 15). It can also be calibrated by comparing the model to measured formation pressures in wells. For the Niger Delta, this calibration to deep-water basin-floor sands (not shown) yields a depth to TSOP of 2.1–2.2 km (1.3–1.4 mi). This is slightly shallower than that observed in the shale porosity trends observed in the same wells (typically 2.3–2.4 km [1.4–1.5 mi]). This discrepancy may be explained by the fact that wells are drilled near structural crests where the sands are typically losing fluid to the surrounding shale (depressing the TSOP), whereas the model is predicting the TSOP in the synclines where pressures are probably less affected by fluid drainage into aquifer units.

The centroid model is by its very nature a static model. Because it relies solely on simple geometrical measurements, it can, however, be applied to a dynamic growth model for a thrust sheet and thus serve to illustrate the potential formation pressure history within an aquifer layer through time (Figure 17). To do this, certain assumptions have to be made, the most important of which is that the TSOP occurs at a constant depth below mud line throughout the burial history. In other words, mudrocks in the delta always become overpressured when they attain a depth of burial greater than the TSOP threshold depth. For this to occur, the mudrocks ideally require constant composition and grain size and a near-constant time-averaged rate of sedimentation. In addition, to model pore-pressure redistribution through time, it is assumed that any fluid movement (pressure redistribution) occurs significantly quicker than the rate of deformation, allowing pressure equilibration to occur through all stages of structural growth.

The model has been formulated in a simple spreadsheet that tracks the growth in structural amplitude of a thrust moving up a planar ramp. Two points are tracked in the model: one at the crest of the growing structure and the other in the neighboring synclinal trough. No account is made of lateral changes in structural amplitude that no doubt affect the overall pressure redistribution. In other words, the structure modeled is assumed to be cylindrical. For elongate thrusts of 40 km (24.8 mi) in length or more, this simplification is believed acceptable. Uplift is calculated as

$$U = L \times \sin\phi + L_b \sin(\eta)$$

where L is the displacement on the thrust, ϕ is the ramp angle, L_b is the displacement on the underlying detachment, and η is the dip of the detachment. The model includes syntectonic sedimentation, erosion, compaction, and isostasy. Sedimentation in the deep-water delta is assumed to occur primarily by advection (i.e., by mass-transport processes) with deposition by sediment dropout at changes in slope. This results in sediment ponding in synclinal troughs and reduced sedimentation at the crest of growing anticlines, features that are typical of the Niger Delta deep-water thrust systems (Hooper et al., 2002). In the model, the rate of sedimentation is calibrated to the adjacent syncline and is assumed to be a time-averaged constant. Compaction is modeled using the exponential porosity decay model (Allen and Allen, 1990). An Airy isostatic correction is applied after each increment of sedimentation to model the new syncline bathymetry.

Erosion by mass wasting or by traction current activity is seen from the crests and frontal limbs of emergent fold structures in the deep-water toe-thrust belt. Example failure surfaces are seen in the forelimb of the thrust C structure (Figure 14). The presence of major slumping appears to correlate with the ramp angle, suggesting that it is related to the magnitude of uplift, with steeper ramps yielding more uplift for a given increment of displacement. Erosion is modeled to occur when the rate of uplift exceeds the time-average rate of sedimentation. If the rate exceeds a user-defined threshold, erosion is enhanced to mimic the function of slumping and mass wastage from increased seabed slopes (Figure 14).

The model has been applied to the three thrust structures described in detail above: thrust A, thrust B, and thrust C, to investigate the potential magnitude of pressure transfer and what effects slumping and significant erosion may have on fault stability. Near-crestal depth sections of these structures have been used as

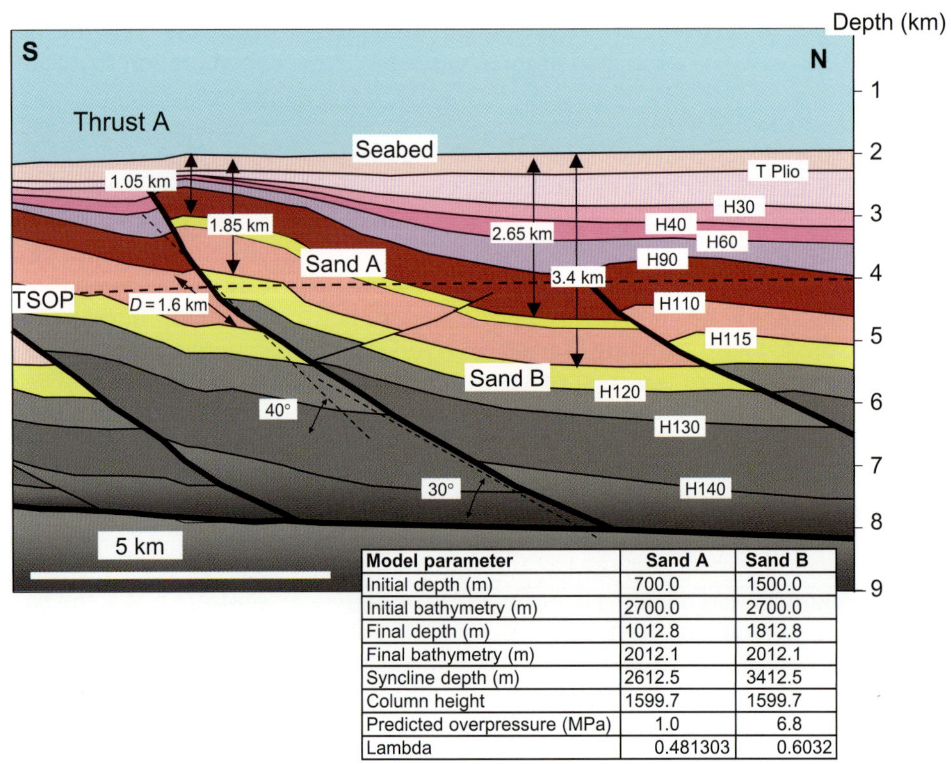

Model parameter	Sand A	Sand B
Initial depth (m)	700.0	1500.0
Initial bathymetry (m)	2700.0	2700.0
Final depth (m)	1012.8	1812.8
Final bathymetry (m)	2012.1	2012.1
Syncline depth (m)	2612.5	3412.5
Column height	1599.7	1599.7
Predicted overpressure (MPa)	1.0	6.8
Lambda	0.481303	0.6032

Figure 18. Application of the pressure transfer model to the thrust A structure in the southern Niger Delta described in Figure 11. The cross section shows a depth-converted profile of the thrust from near its structural culmination (based on Figure 12b). The depth conversion used an average velocity function derived from check-shot data from a set of deep-water wells. Geometrical parameters employed in the model are shown. The pressure transfer growth model has been applied to two notional sand units: sand A and sand B, corresponding approximately to the H110–H112 (12.0–13.5 Ma) and H115–H120 (15.8–27.8 Ma) intervals. The accompanying table shows some of the model inputs and the resultant pore-pressure predictions in the sands from the crest of the anticline. TSOP = top shale overpressure; H = horizon; T Plio = Top Pliocene.

input for the model. The main aquifer used in the analysis is assumed to occur within the early–middle Miocene H115–H120 interval (referred to as sand B). This stratigraphic section is known to contain extensive basin-floor fan systems in the outer Niger Delta. The model results are shown in detail for the A structure (Figures 18–20) and illustrated for the B and C structures (Figures 21–23).

For the thrust A, the onset of movement is dated from sediment wedging in the backlimb to be around 9.0 Ma, the H90 horizon representing the top of the prekinematic section (Figure 12). The thrust has then moved intermittently until recently, accumulating up to a maximum of nearly 1.6 km (0.99 mi) of offset. For the purposes of this exercise, this displacement is modeled as occurring in three discrete phases. Such a movement history can commonly be discerned from the backlimb growth wedge patterns using high-resolution 3-D seismic data (not shown). The episodes synthesized are H90–H60 (9.0–6.5 Ma), H40–H30 (6.0–3.5 Ma), and a final episode from 3.0 to 0.5 Ma. In the model, displacement rate is described using a sinusoidal function (see Figure 19a) that is suggestive of the idea that discrete deformation events (involving numerous slip episodes) first wax and then wane. A sinusoidal function is used because it is easy to model mathematically. Figure 18 shows a depth section of thrust A, showing the key geometrical attributes of the structure, together with a table of model parameters. The thrust can be seen to ramp at an angle of 40° through the H115 interval. The basal ramp angle in the Akata shale is around 30°. The underlying detachment has a dip of around 1.5°. Two sands are evaluated: sand A represents a sand layer at the H110 surface, whereas sand B represents the H115–H120 interval. The depth section illustrates some of the input geometrical parameters used to model the growth of the structure and its overpressure development. Figure 19 summarizes the modeled displacement history, the accumulated displacement pattern, the resultant uplift of the H115 aquifer (expressed as the hydrologic height), and the crestal burial of the aquifer through time. A prekinematic burial of 1.5 km (0.93 mi) is assumed for the H115 aquifer (sand B) and 0.7 km (0.43 mi) for the H110 aquifer units (sand A). Other model parameters used are summarized in Table 1.

The overpressure history predicted for the H115 aquifer using the centroid thrust growth model is shown in Figure 20b. The onset of fluid-pressure transfer occurs at around 6.5 Ma and builds slightly erratically through to present day. Periods of increased rate of fluid-pressure transfer are associated with quiescent intervals when the structure is buried by continued sedimentation. The magnitude of overpressures predicted by the model is 7.0 MPa (1015 psi). In this example, the modeled H115 aquifer just crosses the TSOP at the structural crest (depth of burial = 1.85 km [1.15 mi]), so

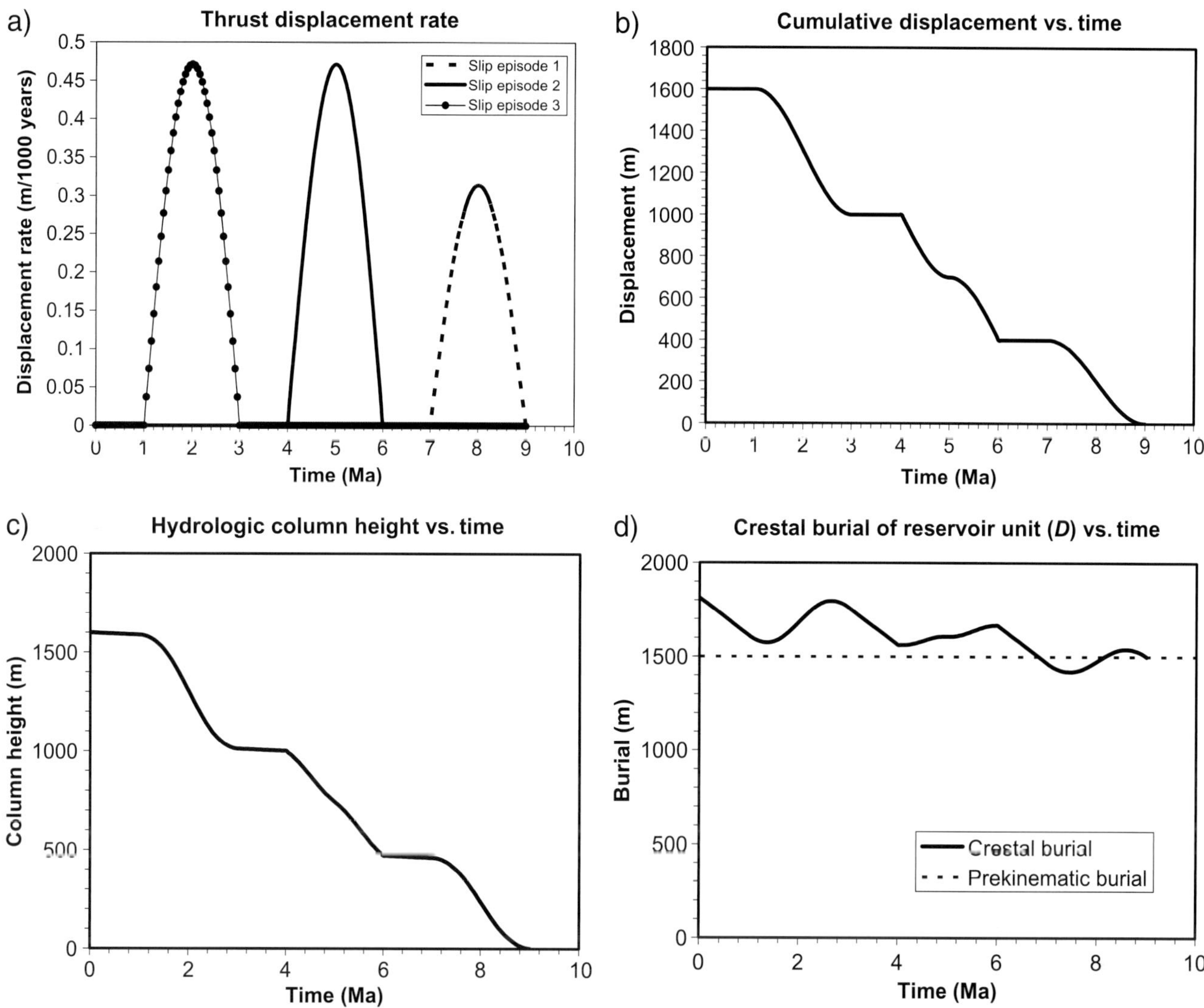

Figure 19. Graphical representation of the suite of inputs and outputs from the pressure transfer growth model based on the thrust A structure shown in Figure 18. (a) Modeled (input) displacement history. The measured cumulative displacement on thrust A is 1.65 km (1.02 mi). This has been modeled occurring in three discrete episodes, each using a sinusoidal function. (b) Integrated displacement vs. time plot based on the three discrete displacement episodes. (c) The resultant uplift (hydrologic column height) for sand B amounts to just over 1500 m (4921 ft). (d) Resultant crestal burial vs. time plot for sand B. The initial prekinematic burial is estimated to be 1500 m (4921 ft) below mud line. The effect of thrust-sheet uplift is modeled to result in little subsequent net deposition across the structure until the last 2 Ma when uplift ceased. The final burial depth is 1850 m (6069 ft). Erosive episodes (net removal of overburden) correspond to peak uplift events based on the displacement history modeled.

it is unlikely that much of the excess fluid pressure in this interval would be lost by leakage through shallower buried, more permeable shale. A shallower reservoir, for example, the H110 sand A, initially at a depth of 700 m (2296 ft) below mud line at the onset of deformation, is predicted to only experience 1.1 MPa (159.5 psi) of overpressure at its present-day depth of burial (1.0 km [0.62 mi]). These estimates are in line with the pressures measured from the deep-water well shown in Figure 15, where the pore pressures in the sands can be seen to step upward with increasing depth.

Thrust B is a much larger structure than thrust A. However, it is not associated with an increased fault displacement. We estimated that the ramp-related displacement of this thrust is 1.9 km (1.2 mi) (Figure 21). Additional uplift has occurred on the structure related to the dip of the underlying basal detachment, which is estimated to be 5°. The estimated 10 km (6.2 mi) of additional displacement on this detachment (this is required to accrue to the 12 km [7 mi] of shortening seen in Figure 6) results in an additional 872 m (2861 ft) of uplift. Sand B is buried to a depth of 3.7 km (2.3 mi) in

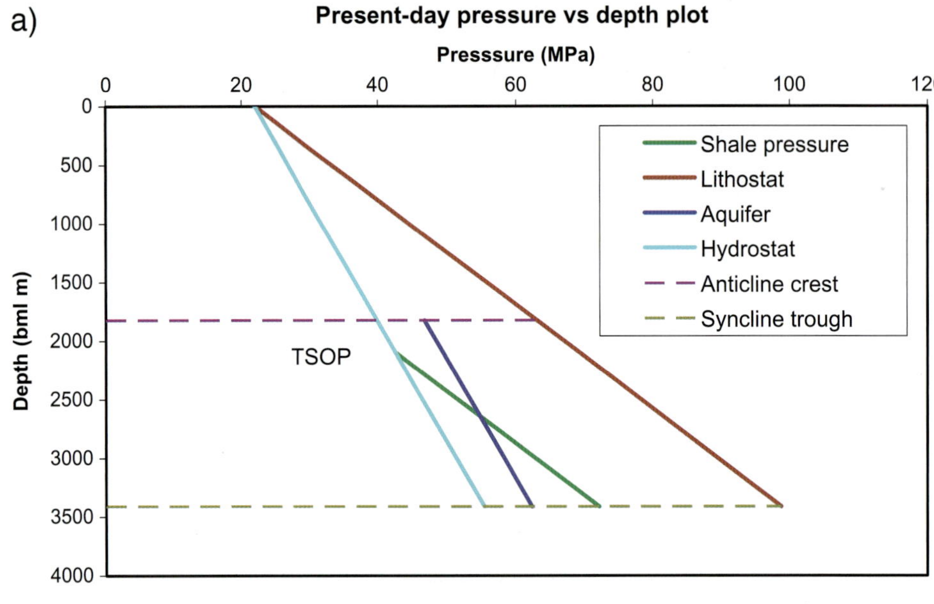

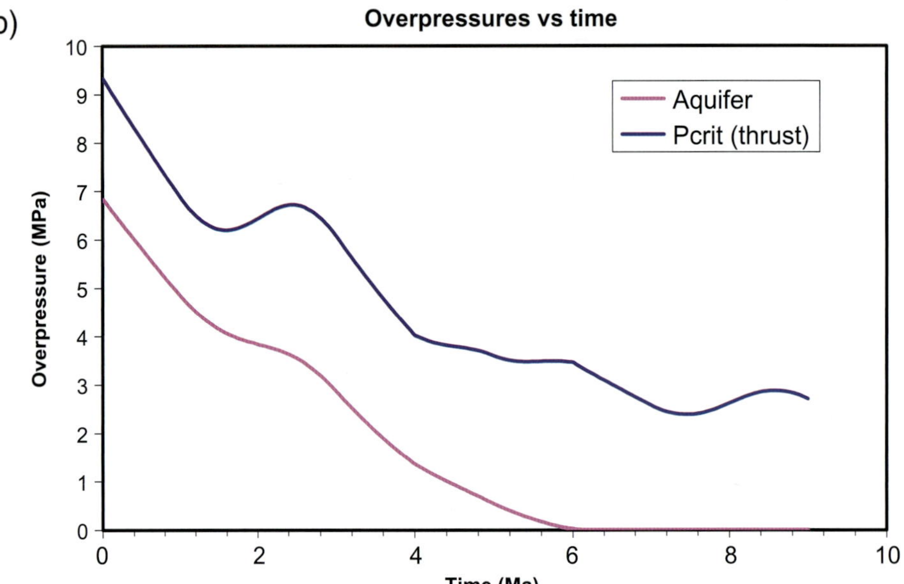

Figure 20. (a) Pressure plot for sand B from the model of Figure 18. The plot shows the resultant aquifer pressure trend relative to the hydrostatic, shale pressure, and lithostatic pressure (overburden) curves. (b) Graph showing the predicted lateral pressure transfer in sand B. The pressure builds up almost monotonically from 6 Ma. Periods of increased pore-pressure increase reflect more quiescent periods when the structure was being buried by sedimentation. Also shown is the critical pore pressure to cause fault reactivation for a frictional coefficient of 0.6. This has been calculated using the Karig (1990) algorithm to estimate maximum horizontal stress (see the text for discussion). Pcrit = the critical pore pressure to cause frictional slip (reactivation) of a fault (assuming the fault has no coherency).

the hanging-wall syncline, whereas at the crest, it is only 1.3 km (0.81 mi) below the seabed. The centroid model predicts an overpressure in the sand of 6.9 MPa (1000 psi) (assuming this does not leak off because of the relatively high permeability of the surrounding formations). Pressures are higher than thrust A because of the increased depth of burial in the syncline trough to the north, although the structure is shallower because of the additional detachment-related uplift.

A depth section across the crest of the thrust C structure is shown in Figure 22. The thrust ramps from a near-horizontal detachment surface at an angle of 45° where it cuts the H115–H120 interval (sand B). The estimated displacement is 1.9 km (1.2 mi). Sand B occurs at a depth of 2.9 km (1.8 mi) in the hanging-wall syncline, whereas it is only 900 m (2953 ft) below the seabed at the crest of the structure because of significant crestal erosion. The results of the centroid model for thrust C are shown in Figure 23. The onset of deformation is modeled at 7.0 Ma with an early slow phase of movement (from 7.0 to 5.0 Ma) succeeded by two more prolonged and rapid displacement events that account for the current significant sea-floor expression of the structure. Significant erosion from the structure, coupled with the uplift of the predicted H115 aquifer above the TSOP to a depth of only 0.9 km (0.56 mi) below mud line, has mostly prevented any buildup of overpressure. The calculated final overpressure developed

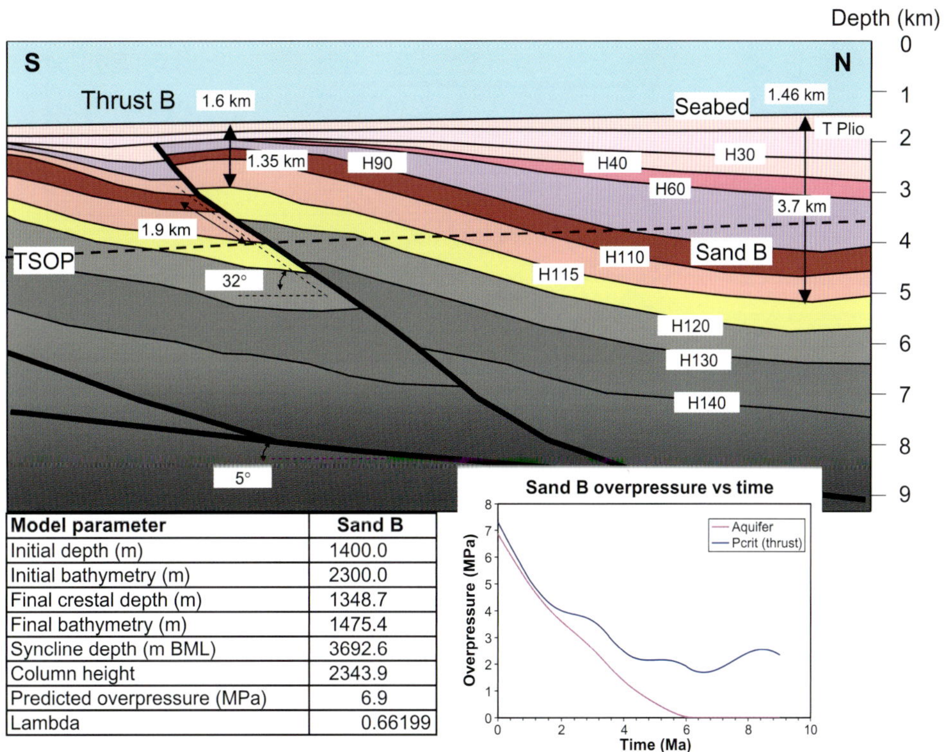

Model parameter	Sand B
Initial depth (m)	1400.0
Initial bathymetry (m)	2300.0
Final crestal depth (m)	1348.7
Final bathymetry (m)	1475.4
Syncline depth (m BML)	3692.6
Column height	2343.9
Predicted overpressure (MPa)	6.9
Lambda	0.66199

Figure 21. Application of the centroid model to the thrust B structure in the southern Niger Delta described in Figure 13. The cross section shows a depth-converted profile of the thrust from near its structural culmination. Geometrical parameters employed in the model are shown. The pressure transfer growth model has been applied to sand B, corresponding approximately to the H115–H120 (15.8–27.8 Ma) intervals, similar to that used in Figure 18. The accompanying table shows some of the model inputs and the resultant pore-pressure predictions in the sands from the crest of the anticline. The calculated aquifer overpressure evolution is shown in the graph. The growth model shows that overpressure started to be transferred to the growing fold crest at 6 Ma. The model predicts that the greater depth of burial of the syncline in this structure (3.7 km [2.3 mi]) relative to thrust A (3.4 km [2.1 mi]) results in a greater magnitude of overpressure buildup although sand B at the crest is not as deeply buried. The pore pressure is at the critical limit for fault reactivation using the Karig (1990) maximum stress estimate, suggesting that the fault could reactivate at the crest of the structure if the stresses were to reach this magnitude. TSOP = top shale overpressure; T Plio = Top Pliocene; H = horizon; m BML = meter below mud line; Pcrit = the critical pore pressure to cause frictional slip (reactivation) of a fault (assuming the fault has no coherency).

is 2.0 MPa (290 psi), and given the low depth of burial and the expected high relative permeability of the enclosing strata, it is doubtful whether this would have been preserved. Unloading of the thrust by erosion has a very large impact, however, on this structure, reducing the critical overpressure for fault slip (Figure 23d). The resultant overpressure in the aquifer gradually converges toward that required to reactivate slip for a thrust (assuming a conservative coefficient of sliding friction of 0.6). For a weaker fault, the aquifer pressure would exceed the critical pore pressure when critically stressed and mean that the fault would be likely to remain active. In this example, pressure transfer has probably had minimal impact on the fault displacement history for the H115 and shallower horizons.

The models and the parameters used in this analysis are nonunique solutions. Similar results can be achieved by varying many of the parameters that have to be incorporated into the model, particularly the displacement rates and duration of the major slip episodes, both of which are mostly conjectural. The approximately equivalent stratigraphic interval (referred to as sand B) has been modeled in each structure. Ignoring depth uncertainties, it appears that at the onset of thrust deformation, the bathymetry of the deep-water Niger Delta mimicked that seen today, with deeper water found in the west. This is consistent with the rate of sedimentation being slower in the western delta and explains why the onset of outer toe-thrust development is slightly later in this side of the delta.

The purpose of this model has been to investigate in a rudimentary fashion the magnitude of overpressures that can be expected to develop in a toe thrust related to fluid-pressure transfer during structural growth. It

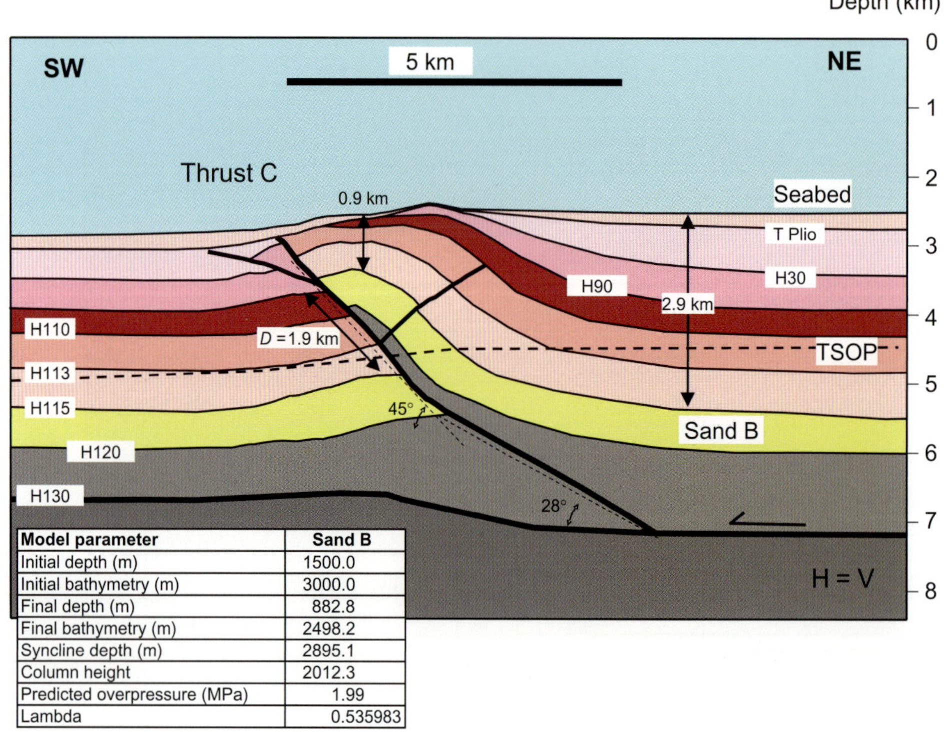

Model parameter	Sand B
Initial depth (m)	1500.0
Initial bathymetry (m)	3000.0
Final depth (m)	882.8
Final bathymetry (m)	2498.2
Syncline depth (m)	2895.1
Column height	2012.3
Predicted overpressure (MPa)	1.99
Lambda	0.535983

Figure 22. Depth cross section from the crest of the thrust C structure shown in Figure 14. The thrust is interpreted to climb at an angle of 28° through the Akata shale (gray) and then steepen to 45° through the Agbada H120–H30 interval. This change in orientation is a common feature of toe thrusts and may reflect changes in frictional strength of the sediment layers (related to overpressure and presence of stiffer deep-water sandstones in the Oligocene–Miocene deposits). The resultant fault-bend fold has a 2.0-km (1.2-mi) elevation for the sand B (H115–H120) unit. The crest is considerably shallower than the other two modeled structures and has suffered considerable erosional unloading. The table lists model inputs and predicted pore-pressure response. Sand B in this case is predicted to be only weakly overpressured (2 MPa [290 psi]) due in part to the relatively shallow depth of burial of the syncline (2.9 km [1.8 mi]). TSOP = top shale overpressure T Plio = Top Pliocene; H = horizon; H = V = horizontal = vertical scales.

suggests that significant overpressures can be exported along inclined aquifers within the thrust system but are influenced by the geometry of an individual thrust sheet together with the absolute depth of burial of the backlimb syncline trough below the TSOP. One possible effect of the increased aquifer pressures is to increase the shale pressure in the vicinity of the aquifer layers at the crest of the structure and simultaneously decrease these pressures downdip and toward the more deeply buried lateral tips of the thrust. This increased pore-fluid pressure can in theory help reduce the effective stress on the fault plane and bring the stress tensor nearer to the shear failure envelope. This effect is more pronounced for the deeper buried thrust like thrust A (Figure 19) because as modeled, the magnitude of pressure transfer is a direct function of the absolute depth of burial of the aquifer (at crest and syncline) as well as of the magnitude of the hydrologic column height. To better understand how the predicted overpressures can affect fault stability, modeling the stress in the toe-thrust belt is necessary.

State of Stress in the Niger Delta Toe-thrust Province

Ideally, to understand the effect that pressure transfer might have on faulting in the toe-thrust belt, it is necessary to have an estimate of the stress tensor and how this varies with depth. The present-day state of stress affecting the outer toe-thrust belt of the Niger Delta has not been measured. Some limits can, however, be placed on the stress based on certain mechanical criteria such as rock strength and the Mohr Coulomb failure limit. These constraints enable some first-order estimates to be made of the stress tensor that illustrate the possible function of fluid pressures in influencing the location of deformation. This deformation is ultimately the product of the gravitational push from behind orchestrated by internal adjustments within what will be assumed to be the critically stressed delta sediment prism.

Karig (1990) published a suite of algorithms that describe the state of stress within an accretionary prism thrust belt. These were based on lab-based mechanical

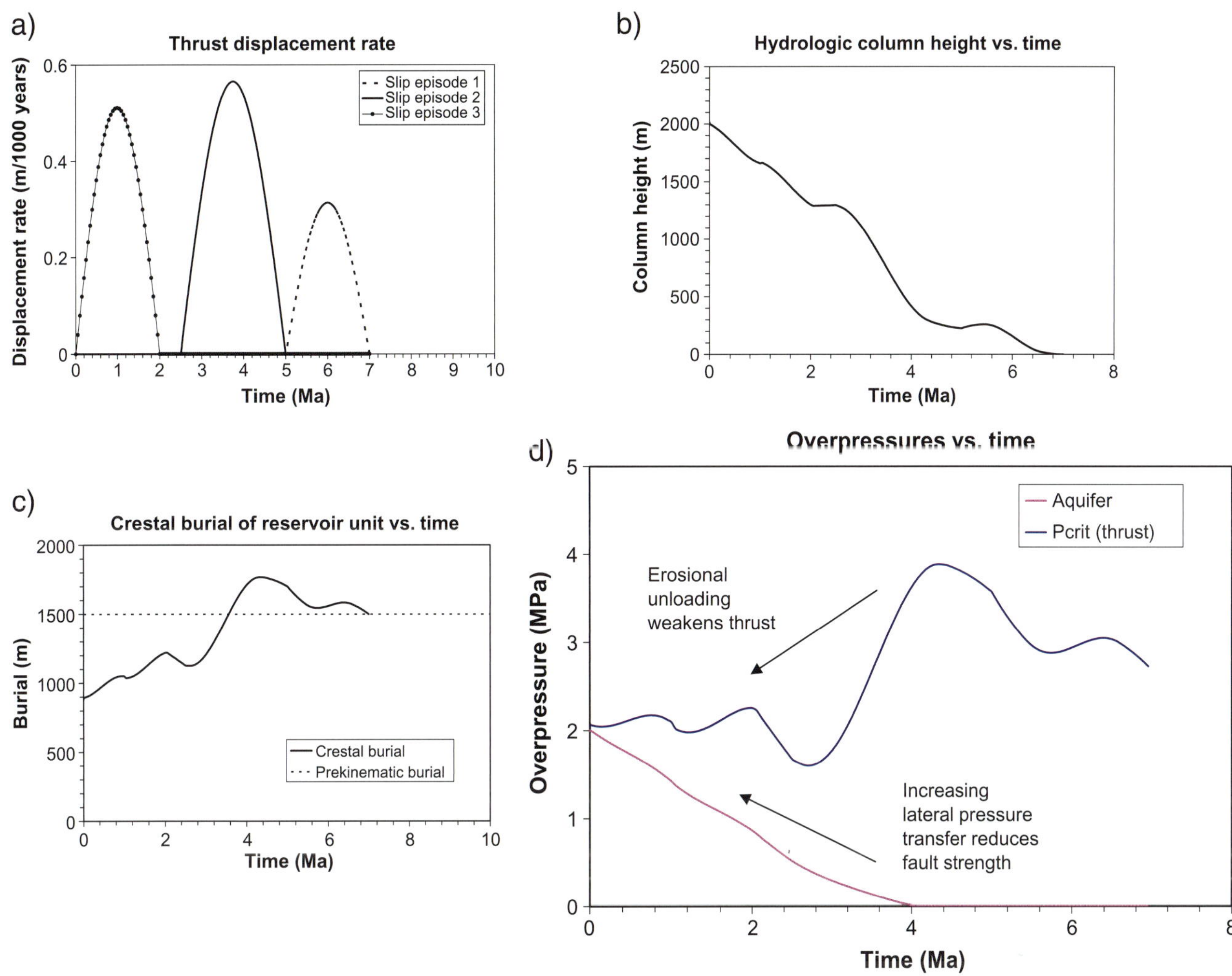

Figure 23. (a) Modeled displacement history for the thrust C structure. (b) Resultant growth in hydrologic column height. (c) Modeled crestal burial of sand B from the onset of thrust-related deformation at 7 Ma to present. The rapid phase of uplift on the thrust from about 5 Ma has resulted in significant erosion. (d) Pore-pressure history for thrust C. The modeled overpressure within sand C develops from about 4 Ma to a maximum of 2 MPa (290 psi) at present day. The effect of erosion can be seen in the reduction in critical overpressure to cause frictional slip on the fault at the equivalent depth of sand B. The convergence of transported pressure and the critical overpressure for thrust reactivation result in enhanced weakening of the thrust. Pcrit = the critical pore pressure to cause frictional slip (reactivation) of a fault (assuming the fault has no coherency).

testing of shale core samples together with principals from critical state failure theory from soil mechanics.

$$S'_{\text{max}} = 2.65 \times (1 - \lambda) \times S_{\text{v}}$$

$$S'_{\text{mean}} = 1.65 \times (1 - \lambda) \times S_{\text{v}}$$

where λ is the ratio of pore-fluid pressure to S_{v}, the vertical load (without correction for the water column). This measure of the pore-pressure ratio is different from that defined by Hubbert and Rubey (1959). The minimum stress in the toe-thrust belt is assumed to correspond to the vertical stress. This can be directly calculated from well data using integrated density logs. A lithostatic gradient of 0.225 MPa/m (32.6 psi/m)has been found to correspond closely to the integrated density data from wells and has been used in this exercise to aid evaluation of the critical state stress tensor.

The principal plane stress tensor predicted by the calculated S_{v} and the Karig algorithm is shown on a Mohr diagram for both sands and shale using the thrust examples described in the previous section (Figure 24). The stress circles for both the leading (crest) and trailing (syncline) parts of the modeled sand B aquifer unit are shown. Given that the Niger Delta sediment prism can probably be modeled as a critically tapered wedge (Bilotti and Shaw, 2005), it makes some sense to assume

Table 1. Centroid growth model parameters.

General Parameters	
Time step (years)	50,000
Erosion efficiency	0.55
Slump threshold	0.75
Slump factor	2
TSOP (m BML)	2100
Proportion sand	0.1
Proportion shale	0.9
Sand reference porosity	0.45
Shale reference porosity	0.51
Sand compaction coefficient (m-1)	0.00027
Shale compaction coefficient (m-1)	0.00066
Hydrostatic gradient (MPa/m)	0.0098
Seawater gradient (MPa/m)	0.0104
Lithostatic gradient (MPa/m)	0.0225
Sediment density (kg/m3)	2.3
Water density (kg/m3)	1.1
Mantle density (kg/m3)	3.3
Structure Specific Parameters	
Burial rates (m/1000 yrs)	
Thrust A	0.33
Thrust B	0.40
Thrust C	0.31

TSOP = top shale overpressure; m BML = meters below mud line.

that the state of stress in the toe-thrust belt is such that the toe thrusts are currently at or near to being critically stressed. In this instance, they are primed for slip as soon as any increase in basal traction increases the maximum horizontal stress. The Karig algorithms fall close to this state and in the absence of other data have been used to estimate S1′, the maximum principal effective stress. For a weak basal detachment, the principal stress should be close to horizontal; thus it is probably acceptable to assume $S'_{max} = SH'_{max} = S'_1$. At failure, the maximum horizontal stress can be constrained, assuming the fault friction angle is known. This can be expressed using the relationship

$$SH_{max} = (S_v - P_p) \times [(\mu^2 + 1)^{0.5} + \mu]^2 + P_p$$

(Finkbeiner et al., 2001) where S_v is the vertical stress, μ is the coefficient of sliding friction, and P_p is the pore pressure.

The coefficient of sliding friction is estimated from Byerlee's law to be in the range of 0.6–1.0 for typical crustal rocks. For overpressured and relatively unconsolidated shale and shale gouge within a fault zone, it might be lower, perhaps 0.4. This latter value is consistent with estimates from thrust faults within accretionary prisms based on the Coulomb failure criterion (Davis and Von Huene, 1987; Adam et al., 2004). This yields $SH_{max} = 3.12 \times (S_v - P_p) + P_p$ for intact rock and $SH_{max} = 2.18 \times (S_v - P_p) + P_p$ for shale-cored faults. The former algorithm has been used to estimate the critical pore pressure for thrust faulting shown in the centroid model overpressure plots of Figures 20, 21, and 23.

Structural Effects of Pressure Transfer

Clearly, pore pressure can be a significant factor in controlling the state of stress in the toe-thrust belt. Changes in pore pressure during deformation or related to burial compaction can affect the deformation response. In such a situation, the transfer of fluid pressure within inclined thrust-sheet backlimb aquifers may be an important factor in weakening a fault zone or focusing its pattern of slip. The analysis presented above suggests that the transfer of pressure has its greater effect (magnitude) when the permeable layers are plumbed into deeper parts of the basin because of the higher extant overpressures developed there. This implies that if the transfer were to influence where a thrust might slip, it should do so first nearer the leading part of the thrust sheet instead of the lateral tips where pressures are reduced. The effect of this pore-pressure transfer is illustrated in Figure 24. Two Mohr circle plots are shown for sand B in the thrust A structure, one at the crest of the fold and one from the syncline trough, equivalent in depth to the lateral fault tip. At the crest, the pore pressure can enhance fault slip, the sand B stress circle straddling the fault friction line ($\mu = 0.4$). At the lateral tip, the fault zone is strengthened in the vicinity of sand B as the stress circle moves away from the friction line. A thrust is therefore more likely to slip (reactivate) in the hydrologically elevated regions based on this analysis.

Conceptually, the three-stage growth history of a toe thrust documented earlier, characterized by initial rapid fault propagation, a prolonged period of stable slip accumulation, and a final regression toward the central parts of the thrust, may be directly influenced by the evolving pore-pressure history of the structure. Initially, the growing pore pressure in the prodelta strata during rapid burial leads to the weakening of a detachment horizon in the Akata shales. This ultimately triggers failure. Because the surrounding subhorizontally bedded shales are all near failure, a structure initially propagates quite rapidly. As the structure begins to gain elevation by ramp displacement, lateral pressure transfer in interbedded aquifer units leads to a stiffening of the tip region (Figure 25). Subsequent propagation is now envisaged to be more difficult, and the structure accumulates most of its slip with lesser or

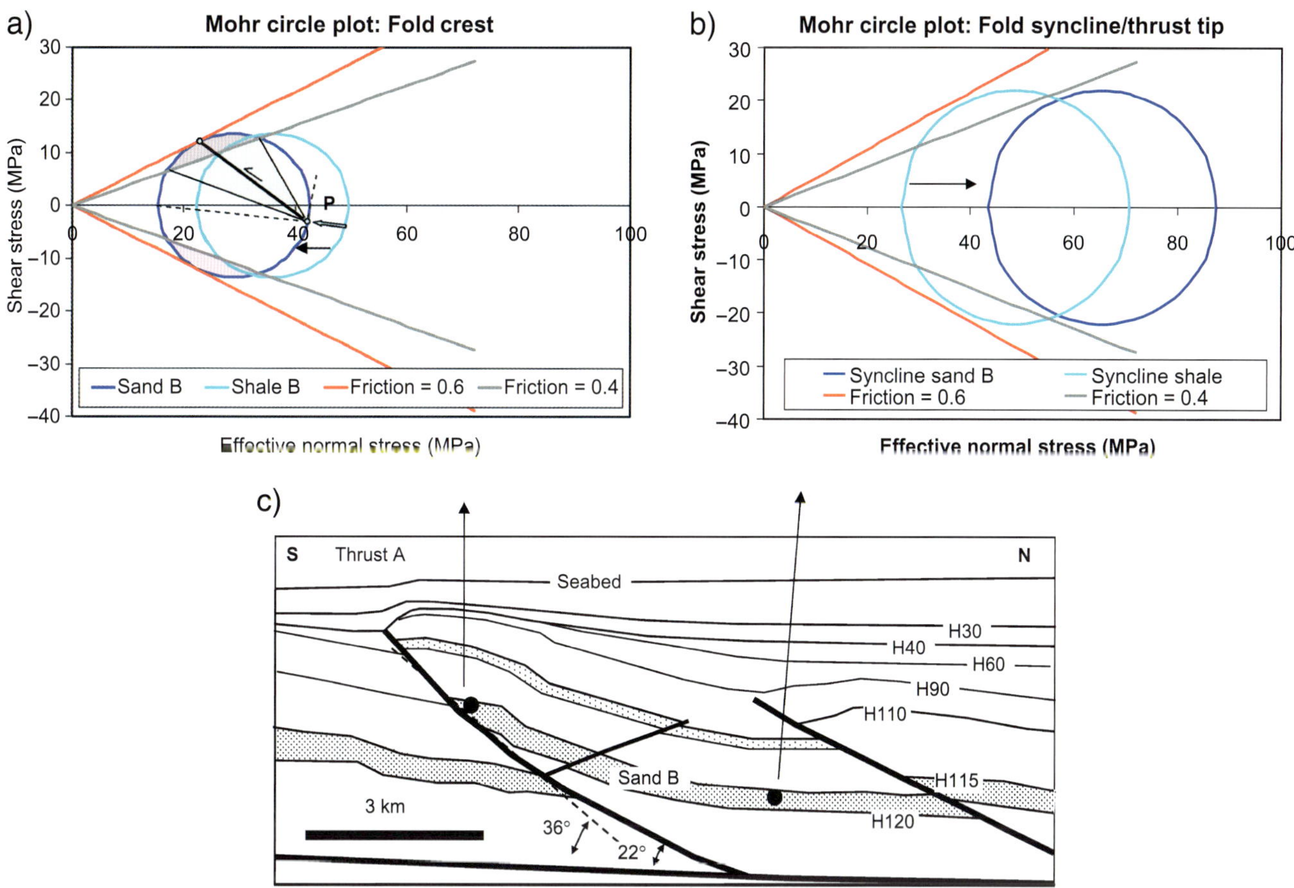

Figure 24. Mohr circle plots for sand B from two locations from the thrust A structure. (a) Mohr circle for fold the crest. Relative to the shale at the same depth, the sand B stress circle has moved toward the friction line for faulting. (b) Mohr circle for sand B from the syncline (equivalent in depth to the lateral fault tip). The stress in sand B has moved away from the frictional failure line, indicating a strengthening of the strata. This strengthening, the direct consequence of pore-pressure transfer, may be an important factor in arresting fault propagation and focusing slip toward the structural culminations along the thrust. (c) Cross section of thrust A showing location of Mohr circles. See the text for further discussion. H = horizon.

no increase in strike length. Finally, as the thrust begins to die, the stiffened outer fringes are first to lock up and the waning stages of deformation retreat to the culminations promoted by the maintenance of relatively higher pore pressure in the crestal areas than at the lateral fault tips. This pore-pressure response may be due in part to the increased aquifer relief promoting pressure transfer. Thus, in a positive feedback process, a thrust might continue to move only where the pore pressures are above a threshold that is itself influenced by the thrust-sheet displacement pattern. As the displacement accumulates, the area of slip is modified and contracts toward higher relief areas. The lower relief lateral sections of a thrust sheet see lower relative fluid pressures because these are (in the presence of a laterally connected and permeable aquifer unit) wicked toward the culminations in the thrust sheet, strengthening the more deeply buried parts of the fault zone (Figure 24).

Thrust displacement episodes are ultimately related to the gravitational movement of the whole Niger Delta wedge toward the basin, promoted by episodes of rapid shelfal deposition (Damuth, 1994). Which fault is reactivated during successive displacement episodes may be influenced by the pore-pressure distribution within the thrust belt. Where pore pressures are high, faults are more likely to be close to being critically stressed. When the push comes from the hinterland, these faults will slip first. In such an environment, a mechanism such as pressure transfer may locally promote thrust reactivation.

The deep-water Niger Delta is essentially a low net-to-gross deep-water fan system. The amount of sand in the basin is generally no more than about 10% and restricted to specific intervals (lowstand deposits) transported into the deep water as mass flows presumably sourced from slope failure events. The ability of the pressure transfer mechanism to promote fault movement

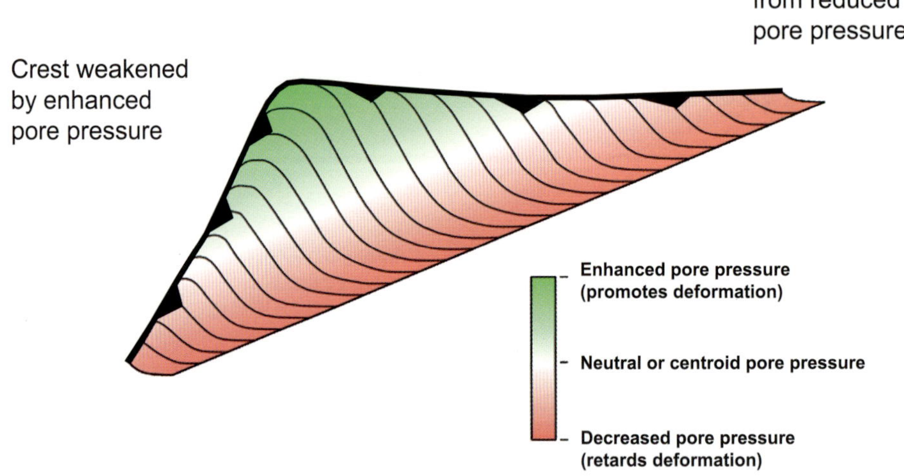

Figure 25. Conceptual view of a thrust fault plane and the effect that lateral pressure transfer can have on fault mechanics. The surface is color coded to represent the pressure differential caused by pressure transfer effects in interbedded sheet sands. The effect of this pressure transfer is to vary the strength of the strata. Increased pore pressure at the crest promotes further deformation, whereas the reduction in pressure at the downdip lateral fault tips stiffens the strata and retards or arrests both fault propagation and slip. See the text for discussion.

is tempered by just how effectively it can operate in a sediment pile that is predominantly fine grained. Presumably where laterally extensive basin-floor fan systems are present (such as in the lower Miocene H115–H120 interval of the deep-water delta), it may be very important. However, in slope environments where ribbonlike channel sand bodies are more typical, for example, above the H90 horizon, it may not have such an influence on fault behavior.

It is likely that a combination of processes ultimately controls the displacement history and pattern on individual thrust faults. The focusing of slip toward fault centers is probably a product of interactions between neighboring imbricate thrusts, the sediment load on the fault plane, the local pore-pressure conditions influenced by pore-pressure transfer, the mechanical properties of the sediments and how these change as a function of burial and absolute strain, and the fluid-pressure response of the fault zone itself. Pore-pressure transfer offers a simple and elegant mechanism to help explain the thrust movement history and should not be overlooked when investigating the mechanical behavior of critically tapered wedges.

CONCLUSIONS

Regional analysis of the timing of thrust movement in the deep-water Niger Delta toe-thrust belts has revealed a complex growth history. Movement of the outer toe-thrust belt commenced around 9 Ma. The resultant imbricate stack shows polyphase growth with complex sequencing of thrust movements. The overall pattern suggests that fault activity is distributed instead of localized within the imbricate system. Detailed analysis of the growth history of several fault-propagation folds suggests that once initiated, they propagated rapidly, gaining most of their ultimate strike length. Subsequent deformation increased fault slip and structural relief with lesser or no significant lateral propagation. With the fault tips abandoned, the loci of continued deformation can be shown to have gradually retreated to near the structural crests. We propose that this three-stage pattern of deformation is caused in part by the effects of pore-pressure distribution in the evolving fault-fold structures.

Compactional disequilibrium is an important process in the Niger Delta promoted by rapid deep-water shale deposition. Well data suggest that compaction disequilibrium starts to develop from around a depth of 2 km (1.2 mi) burial, presumably when shale permeability has been reduced to such low values that the rate of compactional dewatering cannot keep pace with the rate of sediment loading. This depth is referred to in this chapter as the TSOP. Below it, compaction ceases, and the vertical effective stress within the shale remains almost constant. Thrust propagation in this environment is facilitated by the development of high overpressures and, based on the geometrical analysis summarized here, appears to have been rapid once failure occurred.

After the initial phase of rapid lateral propagation, individual fault-bend folds began to develop structural height. Because the interbedded porous sands and silts cannot maintain the elevated pore-pressure gradients of the surrounding matrix shale, they develop a different pressure profile controlled by lateral pressure transfer. Sands near the structure crest experience a relative increased pressure, whereas those in the backlimb synclinal trough experience a relative pressure reduction (the so-called centroid effect). A simple 1-D model that tracts how the magnitude of lateral pressure transfer evolves as thrust displacement grows on a fault ramp has been developed to illustrate this effect. This model shows that even in the simple case of a cylindrical structure, the pore-pressure evolution can be complex and strongly influenced by the rate of thrust-sheet uplift relative to ongoing sedimentation, the absolute elevation of the structure above regional, the depth of burial in the backlimb syncline trough, and the depth at which compaction disequilibrium starts to influence shale pressure in the basin.

Conceptually, the resultant elevated pressure at the fold crest will tend to weaken the highest parts of the structure and promote continued deformation. The relatively decreased pressure at the base of the thrust sheet will tend to strengthen the lower parts of the structure and retard deformation. With the lateral tip line of the fault pinned, it is the lowest part of the structure that experiences the greatest stabilizing effect of lateral pressure transfer. The predicted strengthening of the matrix shale at the lateral tips of the fault will act to suppress further fault propagation. The initial phase of rapid lateral propagation with little structural relief is arrested and transitions into a more slow and steady phase of growth with slip accumulation.

Once the toe-thrust fault stops propagating laterally, whether caused by waning stresses or the initiation of neighboring structures, the tip regions become dormant. The tectonic driving stress is no longer sufficient to deform the tip region where the burial load and therefore the vertical effective stress are highest. It can continue to deform the higher parts of the structure where the lowered effective stress induces structural weakening. The culminations of the structure, where the relative pore-pressure weakening is greatest, will therefore in theory be the last to stop deforming.

We therefore envisage that pore-pressure effects are affecting rock strength at every stage of growth. From the initial weakening of the detachment to allow failure, to the strengthening of the tip line region that slows lateral propagation, to the retreat of waning deformation toward the structural culmination, the influence of pore-pressure evolution is expressed in the pattern of deformation. If this pattern proves to be representative of the growth patterns of other deltaic toe-thrust systems, then this may provide a useful analog for the likely growth patterns and physical processes behind contractional features in onshore thrust belts where active erosion prevents preservation of such detailed growth histories.

ACKNOWLEDGMENTS

We would like to thank ConocoPhillips and its partners for their support of regional studies of the Niger Delta and for the permission to publish this work. No regionally based analysis is possible without the previous efforts of countless past interpreters from the Nigeria Exploration Team, and their efforts are gratefully acknowledged. The manuscript benefited greatly from thorough reviews by John Suppe, Ken McClay, and Juergen Adam. Technical support from Ron Martinussen is greatly appreciated.

REFERENCES CITED

Adam, J., D. Klaeschen, N. Kukowski, and E. Flueh, 2004, Upward delamination of Cascadia Basin sediment infill with landward frontal accretion thrusting caused by rapid glacial age material flux: Tectonics, v. 23, p. TC3009, doi:10.1029/2002TC001475.

Allen, P. A., and J. R. Allen, 1990, Basin analysis: Principles and applications: Oxford, United Kingdom, Blackwell Science, 451 p.

Bilotti, F., and J. H. Shaw, 2005, Deep-water Niger Delta fold and thrust belt modeled as a critical-taper wedge: The influence of elevated basal fluid pressure on structural styles: AAPG Bulletin, v. 89, no. 11, p. 1475–1491.

Cohen, H. A., and K. R. McClay, 1996, Sedimentation and shale tectonics of the northwestern Niger delta front: Marine and Petroleum Geology, v. 13, p. 313–328.

Corredor, F., J. H. Shaw, and F. Bilotti, 2005, Structural styles in the deep-water fold-and-thrust belts of the Niger Delta: AAPG Bulletin, v. 89, p. 753–780.

Dahlen, F. A., J. Suppe, and D. Davis, 1984, Mechanics of fold-and-thrust belts and accretionary wedges: Cohesive Coulomb theory: Journal of Geophysical Research, v. 89, p. 10,087–10,101.

Damuth, J. E., 1994, Neogene gravity tectonics and depositional processes on the deep Niger Delta continental margin: Marine and Petroleum Geology, v. 11, p. 320–346.

Davis, D. M., and R. von Huene, 1987, Inferences on sediment strength and fault friction from structures at the Aleutian Trench: Geology, v. 15, p. 517–522.

Doust, H., 1990, Petroleum geology of the Niger Delta, *in* J. Brooks, ed., Classic petroleum provinces: Geological Society (London) Special Publication 50, p. 365.

Doust, H., and E. Omatsola, 1990, Niger Delta, *in* J. D. Edwards and P. A. Santogrossi, eds., Divergent/passive margins: AAPG Memoir 48, p. 239–248.

Finkbeiner, T., M. Zoback, P. Flemings, and B. Stump, 2001, Stress, pore pressure, and dynamically constrained hydrocarbon columns in the south Eugene Island 330 field, northern Gulf of Mexico: AAPG Bulletin, v. 85, p. 1007–1031.

Goff, D., and D. V. Wiltschko, 1992, Stresses beneath a

ramping thrust sheet: Journal of Structural Geology, v. 14, p. 437–450.
Hardy, S., C. Duncan, J. Masek, and D. Brown, 1998, Minimum work, fault activity and the growth of critical wedges in fold and thrust belts: Basin Research, v. 10, p. 365–373.
Hooper, R. J., R. J. Fitzsimmons, N. Grant, and B. C. Vendeville, 2002, The role of deformation in controlling depositional patterns in the south-central Niger Delta, west Africa: Journal of Structural Geology, v. 24, p. 847–859.
Hubbert, M. K., and W. W. Rubey, 1959, Role of fluid pressure in mechanics of overthrust faulting: Geological Society of America Bulletin, v. 70, p. 115–166.
Karig, D. E., 1990, Experimental and observational constraints on the mechanical behavior in the toes of accretionary prisms, *in* R. J. Knipe and E. H. Rutter, eds., Deformation mechanisms, rheology and tectonics: Geological Society (London) Special Publication 54, p. 383–398.
Lehner, P., and P. A. C. De Ruiter, 1977, Structural history of the Atlantic margin of Africa: AAPG Bulletin, v. 61, p. 961–981.
Morley, C. K., and G. Guerin, 1996. Comparison of gravity driven deformation styles and behavior associated with mobile shales and salt: Tectonics, v. 15, p. 1154–1170.
Osborne, M. J., and R. E. Swarbrick, 1997, Mechanisms for generating overpressure in sedimentary basins: An overview: AAPG Bulletin, v. 81, p. 1023–1041.
Rudniki, J. W., 1977, The inception of faulting in a rock mass with a weakened zone: Journal of Geophysical Research, v. 82, p. 844–854.
Sibson, R. H., 1992, Implications of fault-valve behavior for rupture nucleation and recurrence: Tectonophysics, v. 211, p. 283–293.
Storti, F., and K. McClay, 1995, Influence of syntectonic sedimentation on thrust wedges in analog models: Geology, v. 23, p. 999–1002.
Strayer, L. M., P. J. Hudleston, and L. J. Lorig, 2001, A numerical model of deformation and fluid flow in an evolving thrust wedge: Tectonophysics, v. 335, p. 121–145.
Stump, B. B., P. B. Flemings, T. Finkbeiner, and M. D. Zoback, 1998, Pressure differences between overpressured sands and bounding shales of the Eugene Island 330 field (offshore Louisiana, U.S.A.) with implications for fluid flow induced by sediment loading, *in* A. Mitchell and D. Grauls, eds., Overpressures in petroleum exploration: workshop proceedings: Elf Exploration Production Memoir 22, p. 83–92.
Suppe, J., 2007, Absolute fault and crustal strength from wedge tapers: Geology, v. 35, p. 1127–1130.
Traugott, M., and G. Bowers, 1996, Pore/fracture pressures in deep water (abs.): 1st World Oil Deepwater Technology International Symposium, Houston, Texas, December 2–5, 1997, Paper No. 3, 10 p.
Vrolijk, P., 1987, Tectonically driven fluid flow in the Kodial accretionary complex, Alaska: Geology, v. 15, p. 466–469.
Yardley, G. S., and R. E. Swarbrick, 2000, Lateral transfer— A source of additional overpressure?: Marine and Petroleum Geology, v. 17, p. 523–538.
Yassir, N., 1998, Overpressuring in compressional stress regimes— causes and detection, *in* A. Mitchell and D. Grauls, eds., Overpressures in petroleum exploration: workshop proceedings: Elf Exploration Production Memoir 22, p. 13–18.